Neuroergonomics

OXFORD SERIES IN HUMAN-TECHNOLOGY INTERACTION

SERIES EDITOR

ALEX KIRLIK

Adaptive Perspectives on Human-Technology Interaction:
Methods and Models for Cognitive Engineering and Human-Computer Interaction
Edited by Alex Kirlik

Computers, Phones, and the Internet:
Domesticating Information Technology
Edited by Robert Kraut, Malcolm Brynin, and Sara Kiesler

Neuroergonomics:
The Brain at Work
Edited by Raja Parasuraman and Matthew Rizzo

OXFORD
UNIVERSITY PRESS

Oxford University Press, Inc., publishes works that further
Oxford University's objective of excellence
in research, scholarship, and education.

Oxford New York
Auckland Cape Town Dar es Salaam Hong Kong Karachi
Kuala Lumpur Madrid Melbourne Mexico City Nairobi
New Delhi Shanghai Taipei Toronto

With offices in
Argentina Austria Brazil Chile Czech Republic France Greece
Guatemala Hungary Italy Japan Poland Portugal Singapore
South Korea Switzerland Thailand Turkey Ukraine Vietnam

Copyright © 2007 by Oxford University Press, Inc.

Published by Oxford University Press, Inc.
198 Madison Avenue, New York, New York 10016

www.oup.com

Oxford is a registered trademark of Oxford University Press

All rights reserved. No part of this publication may be reproduced,
stored in a retrieval system, or transmitted, in any form or by any means,
electronic, mechanical, photocopying, recording, or otherwise,
without the prior permission of Oxford University Press.

Library of Congress Cataloging-in-Publication Data
Neuroergonomics : the brain at work / edited by Raja Parasuraman and Matthew Rizzo.
      p. cm.
   Includes index.

   ISBN-13 978-0-19-536865-9
   1. Neuroergonomics. I. Parasuraman, R. II. Rizzo, Matthew.
   QP360.7.N48 2006
   620.8'2—dc22   2005034758

Printed in the United States of America

Neuroergonomics

The Brain at Work

EDITED BY

Raja Parasuraman and Matthew Rizzo

OXFORD
UNIVERSITY PRESS

2007

Preface

There is a growing body of research and development work in the emerging field of neuroergonomics. For the first time, this book brings together this body of knowledge in a single volume. In composing this book, we sought to show how an understanding of brain function can inform the design of work that is safe, efficient, and pleasant. *Neuroergonomics: The Brain at Work* shows how neuroergonomics builds upon modern neuroscience and human factors psychology and engineering to enhance our understanding of brain function and behavior in the complex tasks of everyday life, assessed outside the confines of the standard research laboratory, in natural and naturalistic settings.

The book begins with an overview of key issues in neuroergonomics and ends with a view toward the future of this new interdisciplinary field. Specific topics are covered in 22 intervening chapters. The subject matter is wide ranging and addresses scientific and clinical approaches to difficult questions about brain and behavior that continue to drive our investigations and the search for solutions. This composition required the input of specialists with a variety of insights on medicine, human factors engineering, physiology, psychology, neuroimaging, public health policy, and the law. Effective response to these issues requires coordinated efforts of many relevant specialists, utilizing shared knowledge and cross-fertilization of ideas. We hope this book contributes to these ends.

The breadth and depth of this volume would not have been possible without the steady influence and vision of Series Editor Alex Kirlik and the Oxford University Press. We are also extremely indebted to the authors for their creative contributions and timely responses to our extensive editorial advice. Raja Parasuraman was supported by grants from the National Institutes of Health and DARPA and Matthew Rizzo by the National Institutes of Health and the Centers for Disease Control and Prevention. Raja Parasuraman is grateful to former members of the Cognitive Science Laboratory, especially Francesco DiNocera, Yang Jiang, Bernd Lorenz, Ulla Metzger, and Sangy Panicker, for stimulating debates in the early days of neuroergonomics, many carried out online and to continuing discussions with current members including Daniel Caggiano, Shimin Fu, Pamela Greenwood, Reshma Kumar, Ericka Rovira, Peter Squire, and Marla Zinni, and to the other members of the Arch Lab at George Mason University. Matt Rizzo thanks his colleagues in neurology, engineering, public health, and the Public Policy Center for their open-minded collaboration and is especially obliged to the past and present members of the Division of Neuroergonomics (http://www.uiowa.edu/~neuroerg/) for their good humor, great ideas, and hard work. He is deeply grateful to Michael and Evelyn for nurturing his curiosity, to Annie, Ellie, and Frannie for their enduring support, and to Big Bill and Margie, now at peace. Both of us are also grateful to Constance Kadala and to Cheryl Moores for their editorial assistance.

Contents

Contributors *xi*

I Introduction

1 Introduction to Neuroergonomics
   Raja Parasuraman and Matthew Rizzo 3

II Neuroergonomics Methods

2 Electroencephalography (EEG) in Neuroergonomics
   Alan Gevins and Michael E. Smith 15

3 Event-Related Potentials (ERPs) in Neuroergonomics
   Shimin Fu and Raja Parasuraman 32

4 Functional Magnetic Resonance Imaging (fMRI): Advanced Methods and Applications to Driving
   Vince D. Calhoun 51

5 Optical Imaging of Brain Function
   Gabriele Gratton and Monica Fabiani 65

6 Transcranial Doppler Sonography
   Lloyd D. Tripp and Joel S. Warm 82

7 Eye Movements as a Window on Perception and Cognition
   Jason S. McCarley and Arthur F. Kramer 95

8 The Brain in the Wild: Tracking Human Behavior in Natural and Naturalistic Settings
   Matthew Rizzo, Scott Robinson, and Vicki Neale 113

III Perception, Cognition, and Emotion

9   Spatial Navigation
    Eleanor A. Maguire   131

10 Cerebral Hemodynamics and Vigilance
    Joel S. Warm and Raja Parasuraman   146

11 Executive Functions
    Jordan Grafman   159

12 The Neurology of Emotions and Feelings, and Their Role in Behavioral Decisions
    Antoine Bechara   178

IV Stress, Fatigue, and Physical Work

13 Stress and Neuroergonomics
    Peter A. Hancock and James L. Szalma   195

14 Sleep and Circadian Control of Neurobehavioral Functions
    Melissa M. Mallis, Siobhan Banks, and David F. Dinges   207

15 Physical Neuroergonomics
    Waldemar Karwowski, Bohdana Sherehiy, Wlodzimierz Siemionow, and Krystyna Gielo-Perczak   221

V Technology Applications

16 Adaptive Automation
    Mark W. Scerbo   239

17 Virtual Reality and Neuroergonomics
    Joseph K. Kearney, Matthew Rizzo, and Joan Severson   253

18 The Role of Emotion-Inspired Abilities in Relational Robots
    Cynthia Breazeal and Rosalind Picard   275

19 Neural Engineering
    Ferdinando A. Mussa-Ivaldi, Lee E. Miller, W. Zev Rymer, and Richard Weir   293

VI Special Populations

20 EEG-Based Brain-Computer Interface
    Gert Pfurtscheller, Reinhold Scherer, and Christa Neuper   315

21 Artificial Vision
    Dorothe A. Poggel, Lotfi B. Merabet and Joseph F. Rizzo III   329

22 Neurorehabilitation Robotics and Neuroprosthetics
    Robert Riener   346

23 Medical Safety and Neuroergonomics
    Matthew Rizzo, Sean McEvoy, and John Lee   360

VII Conclusion

24 Future Prospects for Neuroergonomics
Matthew Rizzo and Raja Parasuraman 381

Glossary 389
Author Index 397
Subject Index 419

Contributors

Siobhan Banks
University of Pennsylvania

Antoine Bechara
University of Iowa

Cynthia Breazeal
Massachusetts Institute of Technology

Vince D. Calhoun
Yale University

David F. Dinges
University of Pennsylvania

Monica Fabiani
University of Illinois at Urbana–Champaign

Shimin Fu
George Mason University

Alan Gevins
SAM Technology, Inc.

Krystyna Gielo-Perczak
Liberty Mutual Research Center

Jordan Grafman
National Institute of Neurological Disorders and Stroke

Gabriele Gratton
University of Illinois at Urbana–Champaign

Peter A. Hancock
University of Central Florida

Waldemar Karwowski
University of Louisville

Joseph K. Kearney
University of Iowa

Arthur F. Kramer
University of Illinois at Urbana–Champaign

John Lee
University of Iowa

Eleanor A. Maguire
University College, London

Melissa M. Mallis
Alertness Solutions, Inc.

Contributors

Jason S. McCarley
University of Illinois at Urbana–Champaign

Sean McEvoy
University of Iowa

Lotfi B. Merabet
Northwestern University

Lee E. Miller
Northwestern University

Ferdinando A. Mussa-Ivaldi
Northwestern University

Vicki Neale
Virginia Polytechnic and State University

Christa Neuper
Graz University of Technology

Raja Parasuraman
George Mason University

Gert Pfurtscheller
VA Medical Center, Boston

Rosalind Picard
Massachusetts Institute of Technology

Dorothe A. Poggel
VA Medical Center, Boston

Robert Riener
Graz University of Technology

Joseph F. Rizzo III
VA Medical Center, Boston

Matthew Rizzo
University of Iowa

Scott Robinson
University of Iowa

W. Zev Rymer
Northwestern University

Mark W. Scerbo
Old Dominion University

Reinhold Scherer
Graz University of Technology

Joan Severson
Digital Artifacts

Bohdana Sherehiy
University of Louisville

Wlodzimierz Siemionow
Cleveland Clinic Foundation

Michael E. Smith
Sam Technology, Inc.

James L. Szalma
University of Central Florida

Lloyd D. Tripp
Air Force Research Lab

Joel S. Warm
University of Cincinnati

Richard Weir
Northwestern University

Introduction

Raja Parasuraman and Matthew Rizzo

Introduction to Neuroergonomics

Neuroergonomics is the study of brain and behavior at work (Parasuraman, 2003). This interdisciplinary area of research and practice merges the disciplines of neuroscience and ergonomics (or human factors) in order to maximize the benefits of each. The goal is not just to study brain structure and function, which is the province of neuroscience, but also to do so in the context of human cognition and behavior at work, at home, in transportation, and in other everyday environments. Neuroergonomics focuses on investigations of the neural bases of such perceptual and cognitive functions as seeing, hearing, attending, remembering, deciding, and planning in relation to technologies and settings in the real world. Because the human brain interacts with the world via a physical body, neuroergonomics is also concerned with the neural basis of physical performance—grasping, moving, or lifting objects and one's limbs.

Whenever a new interdisciplinary venture is proposed, it is legitimate to ask whether it is necessary. To answer this query, we show how the chapters in this book, as well as other work, demonstrate that neuroergonomics provides added value, beyond that available from "traditional" neuroscience and "conventional" ergonomics, to our understanding of brain function and behavior as it occurs in the real world. The guiding principle of neuroergonomics is that examining how the brain carries out the complex tasks of everyday life—and not just the simple, artificial tasks of the research laboratory—can provide important benefits for both ergonomics research and practice. An understanding of brain function can lead to the development and refinement of theory in ergonomics, which in turn will promote new, far-reaching types of research. For example, knowledge of how the brain processes visual, auditory, and tactile information can provide important guidelines and constraints for theories of information presentation and task design. The basic premise is that the neuroergonomic approach allows the researcher to ask different questions and develop new explanatory frameworks about humans and work than an approach based solely on the measurement of the overt performance or subjective perceptions of the human operator. The added value that neuroergonomics provides is likely to be even greater for work settings such as modern semiautomated systems (Parasuraman & Riley, 1997), where measures of overt user behavior can be difficult to obtain (Kramer & Weber, 2000).

Some Examples of Neuroergonomics Research

Aviation

The following examples illustrate the value of the neuroergonomic approach. Historically, the greatest influence of human factors on technological design has been in the domain of aviation, specifically in the design of displays and controls in the aircraft cockpit (Fitts, Jones, & Milton, 1950; Wiener & Nagel, 1988). With the worldwide growth in airline travel, new proposals for air traffic management have been put forward. Implementing these proposals requires new cockpit technologies. Consider a new traffic-monitoring system that is to be installed in the cockpit to portray to the pilot other aircraft that are in the immediate vicinity, showing their speed, altitude, flight path, and so on, using color-coded symbols on a computer display. Various types of neuroergonomic research, both basic and applied, can inform the design of this system. For example, designers may wish to know what features of the symbols (e.g., shape, intensity, motion, etc.) serve to best attract the pilot's attention to a potential intruder in the immediate airspace. At the same time, there may be a concern that the presentation of traffic information, while helping the pilot monitor the immediate airspace, may increase the pilot's overall mental workload, thereby degrading the performance of the primary flight task. Although subjective or performance measures could be used to evaluate this possibility, a neuroergonomic approach can provide more sensitive evaluation of any impact on flight performance. It may also lead the researcher to ask new and potentially more profitable questions about attention allocation than before. Measures of brain function that reflect visual attention and oculomotor control can help determine the impact of the new display on the pilot's visual scanning and attentional performance (see chapter 7, this volume). Finally, neuroergonomic evaluation of the manual and physical demands involved in interacting with the information panels and controls of the new traffic-monitoring system would also be required for this system to be used effectively and safely by pilots (see chapter 15, this volume).

Driving

Neuroergonomics is also relevant to assessing interactions between the eye, the brain, and the automobile (Rizzo & Kellison, 2004). Functional magnetic resonance imaging (fMRI) permits noninvasive dynamic imaging of the human brain (see chapter 4, this volume). Analytic approaches to fMRI data, such as independent component analysis, can reveal meaningful patterns in data sets collected in subjects performing complex tasks that capture elements of automobile driving. Preliminary application of these approaches suggests that multiple neural regions, including the frontoparietal, cerebellar, and occipital areas, are differentially activated by various aspects of the driving task, such as speed control. It is also possible to relate physiological correlates of impending sleep (microsleeps) derived from electroencephalographic (EEG) activity recordings of brain activity to imminent declines in driver performance (Paul, Boyle, Rizzo, & Tippin, 2005). Finally, naturalistic studies of driver behavior provide unique evidence of long-range human interactions, strategies, and tactics of "the brain in the wild" (see chapter 8, this volume).

Neuroenginering

A third example concerns the use of brain signals as an additional communication channel for human interaction with both the natural and the human-made environment. This area of research and practice, sometimes also called *neuroengineering* or *brain-computer interface* (BCI), has had significant progress in recent years. In this approach, different types of brain signals are used to control external devices without the need for motor output, which would be advantageous for individuals who either have only limited motor control or, as in the case of "locked-in" patients with amyotrophic lateral sclerosis, virtually no motor control. The idea follows naturally from the work on "biocybernetics" in the 1980s pioneered by Donchin and others (Donchin, 1980; Gomer, 1981) but has progressed beyond the earlier achievements with technical developments in recording of brain activity in real time.

BCIs allow a user to interact with the environment without engaging in any muscular activity, for example, without the need for hand, eye, foot, or mouth movement. Instead, the user is trained to engage in a specific type of mental activity that is associated with a unique brain electrical "signature." The resulting brain potentials are recorded, processed, and classified in such a way as to provide a control signal in real time for an external device. Applications have used a variety of different measures of brain electrical activity. Invasive methods include recording of field potentials and multiunit neuronal activity from implanted electrodes; this technique has been reported to be successful in controlling robotic arms (Nicolelis, 2003). Such invasive recording techniques have superior signal-to-noise ratio but are obviously limited in use to animals or to patients with no motor functions in whom electrode implantation is clinically justified. Noninvasive BCIs have used a variety of brain signals derived from scalp EEG recordings. These include quantified EEGs from different frequency bands such as beta and mu waves (Pfurtscheller & Neuper, 2000), event-related potentials (ERPs) such as P300 (Donchin, Spence, & Wijesinghe, 2000), and contingent negative variation (Birbaumer et al., 1999). BCIs based on these signals have been used to operate voice synthesizers, control cursor movements on a computer display, and move robotic arms.

Virtual Reality

Virtual reality (VR) is particularly relevant to neuroergonomics because VR can replicate situations with greater control than is possible in the real world, allowing behavioral and neural measures of the mind and brain at work in situations that are impractical or impossible to observe in the real world. In doing so, VR can be used to study the performance of human operators engaged in hazardous tasks without putting them and others at risk for injury (see chapter 17, this volume). For example, VR can be used to study the influence of disease, drugs, fatigue, or in-vehicle technologies (such as cell phones) on aircraft piloting and automobile driving, to study how to reduce the risk of falls in the elderly, and to train students to avoid novice misjudgments and errors in performing critical medical procedures, flying aircraft, and operating heavy machinery. VR is particularly useful in workers whose jobs require spatial awareness, complex motor skills, or decisions that require evaluation of multiple possible responses amid changing contingencies, and is also proving to be useful for therapy and rehabilitation of persons with motor, cognitive, and psychiatric impairments.

Conceptual, Theoretical, and Philosophical Issues

The constituent disciplines of neuroergonomics—neuroscience and ergonomics/human factors research—are both twentieth-century, post–World War II fields. The spectacular rise of neuroscience toward the latter half of that century and the smaller but no less important growth in human factors research can both be linked to technological developments in computer science and engineering. The brain imaging technologies that have revolutionized modern neuroscience (e.g., fMRI) and the sophisticated automated systems that have stimulated much human factors work (e.g., the aircraft flight management system) were both made possible by these engineering developments. Nevertheless, the two fields have developed independently.

Traditionally, ergonomics has not paid much attention to neuroscience or to the results of studies concerning brain mechanisms underlying human perceptual, cognitive, affective, and motor processes. At the same time, neuroscience and its more recent offshoot, cognitive neuroscience, has only been recently been concerned with whether its findings bear any relation to human functioning in real (as opposed to laboratory) settings. Recent calls to move neuroscience "beyond the bench" ("Taking Neuroscience beyond the Bench," 2002) include studies of group social behavior (Cacciopo, 2002) and the development of neural prosthetics for control of robots, home automation, and other technologies for physically disabled people (see chapter 19, this volume).

The relative neglect by ergonomics of human brain function is understandable given that this discipline had its roots in a psychology of the 1940s that was firmly in the behaviorist camp. More recently, the rise of cognitive psychology in the 1960s influenced human factors, but for the most part neuroscience continued to be ignored by cognitive theorists, a state of affairs consistent with a functionalist approach to the philosophy of mind (Dennett, 1991). Such an approach implies that the characteristics of neural structure and functioning are largely irrelevant to the development of theories of mental functioning. Cognitive psychology (and cognitive science) also went through a functionalist period in the 1970s and 1980s, mainly due to the influence of researchers from artificial intelligence and computer science. However, the recent influence of cognitive neuroscience has led to a retreat from this position. Cognitive neuroscience proposes that neural structure and function constrain and in some cases determine theories of human mental processes (Gazzaniga, 2000).

If neuroscience has freed cognitive science from rigid functionalism, then ergonomics may serve to liberate it from a disembodied existence devoid of context and provide it an anchor in the real world. Even though researchers are aware of the importance of ecological validity, modern cognitive psychology (with a few exceptions) tends to study mental processes in isolation, apart from the artifacts and technologies of the world that require the use of those processes. Technology, particularly computers, can be viewed as an extension of human cognitive capability. Related to this view is the framework of cognitive engineering, in which humans and intelligent computer systems constitute "joint cognitive systems" (Hutchins, 1995; Roth, Bennett, & Woods, 1987). Furthermore, much human behavior is situated and context dependent. Context is often defined by and even driven by technological change. How humans design, interact with, and use technology—the essence of ergonomics—should therefore also be central to cognitive science.

The idea that cognition should be considered in relation to action in the world has many antecedents. Piaget's (1952) work on cognitive development in the infant and its dependence on exploration of the environment anticipated the concept of situated or embodied cognition. Clark (1997) also examined the characteristics of an embodied mind that is shaped by and helps shape action in a physical world. If cognitive science should therefore study the mind not in isolation but in interaction with the physical world, then it is a natural second step to ask how to design artifacts in the world that best facilitate that interaction. This is the domain of ergonomics or human factors. Neuroergonomics goes one critical step further. It postulates that the human brain, which implements cognition and is itself shaped by the physical environment, must also be examined in interaction with the environment in order to understand fully the interrelationships of cognition, action, and the world of artifacts.

Currently, a coherent body of concepts and empirical evidence that constitutes neuroergonomics theory does not exist. Of course, broad theories in the human sciences are also sparse, whether in ergonomics (Hancock & Chignell, 1995) or in neuroscience (Albright, Jessell, Kandel, & Posner, 2001). Sarter & Sarter (2003) proposed that neuroergonomics must follow the same reductionist approach of cognitive neuroscience in order to develop viable theories. There are small-scale theories that could be integrated into a macrotheory, but which would still pertain only to a specific domain of human functioning. For example, neural theories of attention are becoming increasingly well specified, both at the macroscopic level of large-scale neural networks (Parasuraman, 1998; Posner, 2004) and at the level of neuronal function and gene expression (Parasuraman, Greenwood, Kumar, & Fossella, 2005; Sarter, Givens, & Bruno, 2001). At the same time, psychological theories of attention have informed human factors research and design (Wickens & Hollands, 2000). Difficult though the task may be, one can envisage amalgamation of these respective theories into a neuroergonomic theory of attention. Integration across a broader range of functional domains, however, is as yet premature.

Methods

A number of methods have been developed for use in neuroergonomic research and practice. Among these are brain imaging techniques, which have been influential in the development of the field of cognitive neuroscience. Brain imaging techniques can be roughly divided into two classes. The first group of techniques is based on measurement of cerebral hemodynamics (blood flow), such as positron emission tomography (PET), fMRI, and transcranial Doppler sonography (TCD). The second group of methods involves measurement of the electromagnetic activity of the brain, including EEG, ERPs, and magnetoencephalography (MEG). For a review of brain imaging techniques for use in studies of cognition and human performance, see Cabeza and Kingstone (2001).

RESOLUTION SPACE OF BRAIN IMAGING TECHNIQUES FOR ERGONOMIC APPLICATIONS

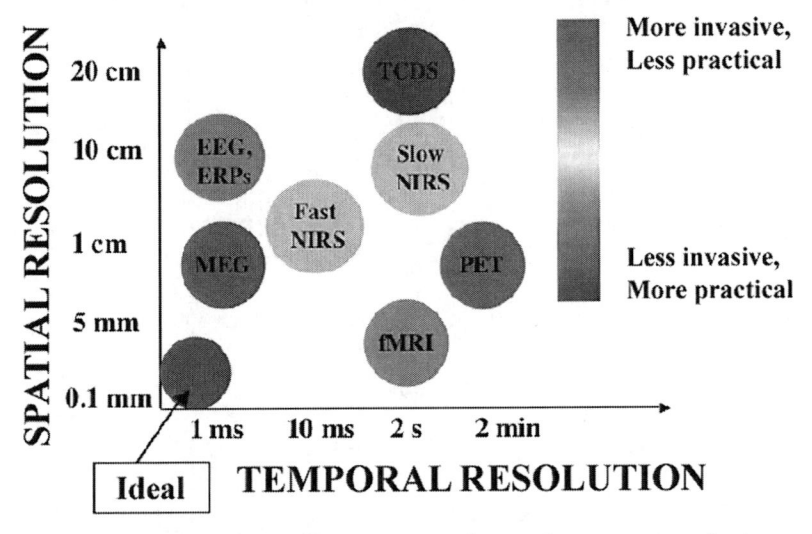

Figure 1.1. Resolution space of brain imaging techniques for ergonomic applications. Trade-offs between the criteria of the spatial resolution (y-axis) and temporal resolution (x-axis) of neuroimaging methods in measuring neuronal activity, as well as the relative noninvasiveness and ease of use of these methods in ergonomic applications (color code). EEG = electronencephalography; ERPs = event-related potentials; fMRI = functional magnetic resonance imaging; MEG = magnetoencephalography; NIRS = near-infrared spectroscopy; PET = positron emission tomography; TCDS = transcranial doppler sonography. See also color insert.

PET and fMRI currently provide the best noninvasive imaging techniques for the evaluation and localization of neural activity. However, these methods suffer from two drawbacks. First, their temporal resolution is poor compared to electrophysiological techniques such as ERPs. Second, their use is restricted to highly controlled lab environments in which participants must not move. This limits their use for examining the neural basis of performance in more complex tasks with a view to ergonomic applications, as in flight, driving simulation, or the use of virtual reality systems, although components of complex task performance are being studied (Peres, Van de Moortele, & Pierard, 2000; Calhoun et al., 2002; see also chapter 4, this volume). Optical imaging techniques such as fast near-infrared spectroscopy (NIRS) may provide both spatial and temporal resolution and the ability to be used in neuroergonomic applications (see chapter 5, this volume).

An overview of the relative merits and disadvantages of these various techniques is shown in figure 1.1. This illustration is a variant of a representation of the spatiotemporal resolution of brain imaging methods first described by Churchland and Sejnowski (1988). The ease of ergonomic application (color code) has been added to the trade-off between the criteria of spatial resolution and temporal resolution in measuring neuronal activity. Currently, there is no one technique that reaches the ideal (blue circle) of 0.1 mm spatial resolution, 1 ms temporal resolution, and ease of use in ergonomics.

In addition to brain imaging methods, oculomotor techniques can provide additional tools for neuroergonomics researchers. With the advent of low-cost, high-speed systems for measuring different types of eye movements and increasing knowledge of the underlying neural systems, oculomotor measures can provide important information not available from traditional measurement of response accuracy and speed (see chapter 7, this volume).

It should be noted that the use of brain imaging or oculomotor measures need not be a defining characteristic of neuroergonomic research. A neuroergonomic study may use behavioral measures or a computational analysis; however, in each case the performance measure or the computational model is related to a theory of brain function.

Consider the following example. Suppose that as a result of the manipulation of some factor, performance on a target discrimination task (e.g., detection of an intruder aircraft in the cockpit traffic-monitoring example discussed previously) in which location cues are provided prior to the target yields the following results: reaction time (RT) to the target when preceded by an invalid location cue is disproportionately increased, while that to a valid cue is not. This might happen, for example, if the cue is derived from the output of an automated detection system that is not perfectly reliable (Hitchcock et al., 2003). In simple laboratory tasks using such a cueing procedure, there is good evidence linking this performance pattern to a basic attentional operation and to activation of a specific distributed network of cortical and subcortical regions on the basis of previous research using noninvasive brain imaging in humans, invasive recordings in animals, and performance data from individuals who have suffered damage to these brain regions (Posner & Dehaene, 1994). One could then conduct a study using the same cueing procedure and performance measures as a behavioral assay of the activation of the neural network in relation to performance of a more complex task in which the same basic cognitive operation is used. If the characteristic performance pattern was observed—a disproportionate increase in RT following an invalid location cue, with a normal decrease in RT following a valid cue—then one could argue that the distributed cortical/subcortical network of brain regions is likely to have been involved in task performance. This would then enable the researcher to link the full body of neuroscience work on this aspect of attentional function to performance on the complex intruder-detection task. Thus, even though no physiological index was used, and although the same performance measure (RT) was used as in a traditional ergonomic analysis, the type of question asked and the explanatory framework can be quite different in the neuroergonomic approach.

Finally, a neuroergonomic study could also involve a computational analysis of brain or cognitive function underlying performance of a complex task. So long as the analysis was theoretically driven and linked to brain function, the study would qualify as neuroergonomic even though no physiological index was used. Several computational models of human performance have been developed for use in human factors (Pew & Mavor, 1998). Of these, models that can be linked, in principle, to brain function—such as neural network (connectionist) models (O'Reilly & Munakata, 2000)—would be of relevance to neuroergonomics.

Neuroergonomics and Neuropsychology

Neuropsychology and related fields (e.g., behavioral neurology, clinical and health psychology, neuropsychiatry, and neurorehabilitation) have also helped pave the way for neuroergonomics. Hebb (1949) used the term *neuropsychology* in his classic book *The Organization of Behavior: A Neuropsychological Theory*. The field broadly aims to understand how brain structure and function are related to specific psychological processes. The neuropsychological approach uses statistical techniques for standardizing psychological tests and scales to provide clinical diagnostic and assessment tools in normal and impaired individuals (de Oliveira Souza, Moll, & Eslinger, 2004).

Like neuroergonomics, neuropsychology is dedicated to a psychometric approach, holding that human behavior can be quantified with objective tests of verbal and nonverbal behavior, including neural states, and that these data reflect a person's states of mind and information processing. These processes can be divided into different domains, such as perception, attention, memory, language, executive functions (decision making and implementation), and motor abilities, and they can be assessed using a wide variety of techniques.

Both neuropsychology and neuroergonomics rely on principles of reliability (how repeatable a behavioral measure is) and validity (what a measure really shows about human brain and behavior). Neuropsychology has traditionally relied on paper-and-pencil tests, many of which are standardized and well understood (e.g., Lezak, 1995). The neuroergonomics approach is more rooted in technology, as indicated in this book. Novel techniques and tests are developing at a rapid pace, and guidelines and standards are going to be needed.

Contributions to Neuroergonomics from Other Fields: Genetics, Biotechnology, and Nanotechnology

While we have emphasized the contribution of neuroscience to neuroergonomics in this chapter, developments in other fields are also affecting the study of human brain function at work. Three such fields are molecular genetics, biotechnology, and nanotechnology, and we briefly consider their relevance here.

As discussed previously, cognitive psychology has increasingly capitalized on findings from neuroscience. More recently, the study of individual differences in cognitive function is being influenced by developments in molecular genetics and, in particular, the impressive results of the Human Genome Project. Much of what we know about the genetics of cognition has come from twin studies in which identical and fraternal twins are compared to assess the heritability of a trait. This paradigm has been widely used in behavioral genetics research for over a century. For example, the method has been used to show that general intelligence, or g, is highly heritable (Plomin & Crabbe, 2000). However, this approach cannot identify the particular genes involved in intelligence or the cognitive components of g. Recent advances in molecular genetics now allow a different, complementary approach to behavioral genetics, called *allelic association*. This method has been applied to the study of individual differences in cognition in healthy individuals, revealing evidence of modulation of cognitive task performance by specific neurotransmitter genes (Fan, Fossella, Sommer, Wu, & Posner, 2003; Greenwood, Sunderland, Friz, & Parasuraman, 2000; Parasuraman et al., 2005). This work is likely to provide the basis not only for improved understanding of the neural basis of cognition, but also for better characterization of individual differences in cognition. That, in turn, can have an impact on important human factors issues such as selection and training.

Reliable quantification of individual differences in cognitive function will have obvious implications for selection of operators for occupations that demand a high workload. While it would be premature to state that the molecular genetic approach to cognition has immediate applications to selection, further programmatic research on more complex cognitive tasks will undoubtedly lead to progress in such an endeavor. The postgenomic era has clearly demonstrated that inheritance of a particular genotype only sets a range for the phenotypic expression of that genotype, with the exact point within that range being determined by other genetic and environmental factors. Genomic analysis allows for a much more precise specification of that range for any phenotype, and for linking phenotypic variation to specific genetic polymorphisms. Selection and training have traditionally been considered together in human factors research and practice (e.g., Sanders & McCormick, 1983) but rarely in terms of a common biological framework. Examining the effects of normal genetic variation and of various training regimens on brain function may provide such a common framework.

The goal of neuroergonomics is to better understand the brain's functional structures and activities in relation to work and technology. In addition to molecular genetics, biotechnology can contribute to this effort by providing a means to study neuronal activities down to the molecular level. Biomimetic studies also allow for precise modeling of the human brain's activities. If the validity of such models can be established in the near future, then researchers could examine various manipulations of brain function that are not ethically possible with human participants.

The currently available measures of brain function are limited by sensor size and the inability to monitor brain function and influence function simultaneously. Nanotechnology provides the measurement tools that can achieve such dual-purpose needs. It can also provide new sensors for monitoring changes in neuronal function in otherwise undetectable brain structures. In addition, nanotechnology has the appropriate scale of operations necessary to deliver chemicals needed to precisely monitor and modify effects of neurotransmitters or encourage targeted neurogenesis, with the objective of improving human performance in certain work environments.

Although there are few current examples of the influence of biotechnology and nanotechnology on neuroergonomics, these fields are likely to have greater impact in the near future. De Pontbriand (2005) provided a cogent discussion of the potential benefits that biotechnology and nanotechnology can bring to neuroergonomics.

Overview of *Neuroergonomics: The Brain at Work*

This book represents a collective examination of the major theoretical, empirical, and practical issues raised by neuroergonomics. In this opening chapter, which forms part I, we have provided an overview of the field, covering theoretical and conceptual issues involved in the merging of cognitive neuroscience and human factors research. We have also briefly described neuroergonomic methods, but these are covered in more detail in part II, which consists of seven chapters describing different cognitive neuroscience methods: fMRI, EEG, ERPs, NIRS, TCD, and oculomotor measures. In addition, measures to track behavior and brain function in naturalistic environments are also described. Each chapter outlines the major features of each method, describes its principal merits and limitations, and gives illustrative examples of its use to address issues in neuroergonomics. We understand that readers will bring a variety of technical backgrounds to the examination of these methodological issues. Accordingly, key readings provided at the end of each chapter provide additional background for understanding some of the more technical details of each method, as needed.

Part III examines basic research in a number of different domains of cognition that have particular relevance for the understanding of human performance at work. We did not attempt to be comprehensive. Rather, we chose areas of cognition in which significant progress has been made in identifying the underlying neural mechanisms, thereby allowing for theory-driven application to human factors issues. The cognitive domains discussed are spatial cognition, vigilance, executive functions, and emotion and decision making. In addition, working memory, planning, and prospective memory are variously described in some of these chapters as well as in other sections of the book.

As the study of the brain at work, neuroergonomics must also examine the work environment. It is an undeniable fact that many work settings are stressful, induce fatigue, and are poorly designed in terms of workspace layout. Accordingly, part IV examines issues of stress, sleep loss, and fatigue, as well as the effects of the physical work environment.

Part V consists of four chapters that discuss several different domains of application of neuroergonomics. Again, we did not attempt to cover all of the application areas that are emerging as a result of the use of neuroergonomic research. We chose four: adaptive automation, virtual reality, robotics, and neuroengineering.

Neuroengineering applications are designed in part to help individuals with different disabilities that make it difficult for them to communicate effectively with the world. This area of work is covered in more detail in part VI. Four chapters describe neuroergonomic technologies that can be used to help the paralyzed, individuals with low or no vision, and those who require prostheses. A final chapter in this section is concerned with the evaluation of medical safety in health care settings.

Finally, in part VII, we close the volume by surveying prospects for the future of neuroergonomics.

Conclusion

Neuroergonomics represents a deliberate merger of neuroscience and ergonomics with the goal of advancing understanding of brain function underlying human performance of complex, real-world tasks. A second major goal is to use existing and emerging knowledge of human performance and brain function to design technologies and work environments for safer and more efficient operation. More progress has been made on the first goal than on the second, but both neuroergonomic research and practice should flourish in the future, as the value of the approach is appreciated. The basic enterprise of ergonomics—how humans design, interact with and use technology—can be considerably enriched if we also consider the human brain that makes such activities possible.

MAIN POINTS

1. Neuroergonomics is the study of brain and behavior at work.
2. Neuroergonomics attempts to go beyond its constituent disciplines of neuroscience and ergonomics by examining brain function and cognitive processes not in isolation but in relation to the technologies and artifacts of everyday life.
3. Some examples of neuroergonomics include research in the areas of aviation, driving,

brain-computer interfaces, and virtual reality.
4. Neuroergonomics is inconsistent with a purely functional philosophy of mind, in which brain structure and function are deemed irrelevant. In addition, neuroergonomics views brain and mind as influenced by context and technology.
5. Neuroergonomic methods include behavioral and performance studies, brain imaging, oculomotor measures, and computational techniques. These methods have different relative merits and disadvantages.

Key Readings

Cabeza, R. M., & Kingstone, A. (2001). *Handbook of functional neuroimaging of cognition.* Cambridge, MA: MIT Press.

Kramer, A. F., & Weber, T. (2000). Applications of psychophysiology to human factors. In J. T. Cacioppo, L. G. Tassinary, & G. G. Berntson (Eds.), *Handbook of psychophysiology* (2nd ed., pp. 794–814). New York: Cambridge University Press.

Mussa-Ivaldi, F. A., & Miller, L. E. (2003). Brain-machine interfaces: Computational demands and clinical needs meet basic neuroscience. *Trends in Cognitive Sciences, 26,* 329–334.

Parasuraman, R. (2003). Neuroergonomics: Research and practice. *Theoretical Issues in Ergonomics Science, 4,* 5–20.

References

Albright, T. D., Jessell, T. M., Kandel, E. R., & Posner, M. I. (2001). Progress in the neural sciences in the century after Cajal (and the mysteries that remain). *Annals of the New York Academy of Sciences, 929,* 11–40.

Birbaumer, N., Ghanayim, N., Hinterberger, T., Iversen, I., Kotchoubey, B., Kubler, A., et al. (1999). A spelling device for the paralysed. *Nature, 398,* 297–298.

Cabeza, R., & Kingstone, A. (2001). *Handbook of functional neuroimaging of cognition.* Cambridge, MA: MIT Press.

Cacioppo, J. T. (Ed.). (2002). *Foundations in social neuroscience.* Cambridge, MA: MIT Press.

Calhoun, V. D., Pekar, J. D., McGinty, V. B., Adali, T., Watson, T. D., & Pearlson, G. D. (2002). Different activation dynamics in multiple neural systems during simulated driving. *Human Brain Mapping, 16,* 158–167.

Churchland, P. S., & Sejnowski, T. J. (1988). Perspectives in cognitive neuroscience. *Science, 242,* 741–745.

Clark, A. (1997). *Being there: Putting brain, body, and the world together again.* Cambridge, MA: MIT Press.

Dennett, D. (1991). *Consciousnesss explained.* Cambridge, MA: MIT Press.

de Oliveira Souza, R., Moll, J., & Eslinger, P. J. (2004). Neuropsychological assessment. In M. Rizzo & P. J. Eslinger (Eds.), *Principles and practice of behavioral neurology and neuropsychology.* (pp. 343–367). New York: Saunders.

de Pontbriand, R. (2005). Neuro-ergonomics support from bio- and nano-technologies (p. 2512). In *Proceedings of the 11th International Conference on Human Computer Interaction.* Las Vegas, NV: HCI International.

Donchin, E. (1980). Event-related potentials: Inferring cognitive activity in operational settings. In F. E. Gomer (Ed.), *Biocybernetic applications for military systems* (pp. 35–42) (Technical Report MDC EB1911). Long Beach, CA: McDonnell Douglas.

Donchin, E., Spencer, K. M., & Wijesinghe, R. (2000). The mental prosthesis: Assessing the speed of a P300-based brain-computer interface. *IEEE Transactions on Rehabilitation Engineering, 8,* 174–179.

Fan, J., Fossella, J. A., Sommer, T., Wu, Y., & Posner, M. I. (2003). Mapping the genetic variation of attention onto brain activity. *Proceedings of the National Academy of Sciences USA, 100*(12), 7406–7411.

Fitts, P. M., Jones, R. E., & Milton, J. L. (1950). Eye movements of aircraft pilots during instrument-landing approaches. *Aeronautical Engineering Review, 9,* 24–29.

Gazzaniga, M. S. (2000). *The cognitive neurosciences.* Cambridge, MA: MIT Presss.

Gomer, F. (1981). Physiological systems and the concept of adaptive systems. In J. Moraal & K. F. Kraiss (Eds.), *Manned systems design* (pp. 257–263). New York: Plenum Press.

Greenwood, P. M., Sunderland, T., Friz, J. L., & Parasuraman, R. (2000). Genetics and visual attention: Selective deficits in healthy adult carriers of the varepsilon 4 allele of the apolipoprotein E gene. *Proceedings of the National Academy of Sciences USA, 97*(21), 11661–11666.

Hancock, P. A., & Chignell, M. H. (1995). On human factors. In J. Flach, P. Hancock, J. Caird, & K. Vicente (Eds.), *Global perspectives on the ecology of human-machine systems* (pp. 14–53). Mahwah, NJ: Erlbaum.

Hebb, D. O. (1949). *The organization of behavior: A neuropsychological theory.* New York: Wiley.

Hitchcock, E. M., Warm, J. S., Matthews, G., Dember, W. N., Shear, P. K., Tripp, L. D., et al. (2003). Automation cueing modulates cerebral blood flow and vigilance in a simulated air traffic control task. *Theoretical Issues in Ergonomics Science, 4,* 89–112.

Hutchins, E. (1995). *Cognition in the wild.* Cambridge, MA: MIT Press.

Kramer, A. F., & Weber, T. (2000). Applications of psychophysiology to human factors. In J. T. Cacioppo, L. G. Tassinary, & G. G. Berntson (Eds.), *Handbook of psychophysiology* (2nd ed., pp. 794–814). New York: Cambridge University Press.

Lezak, M. D. (1995). *Neuropsychological assessment* (3rd ed.). New York: Oxford.

Nicolelis, M. A. (2003). Brain-machine interfaces to restore motor function and probe neural circuits. *Nature Reviews Neuroscience, 4,* 417–422.

O'Reilly, R. C., & Munukata, Y. (2000). *Computational explorations in cognitive neuroscience.* Cambridge, MA: MIT Press.

Parasuraman, R. (1998). The attentive brain: Issues and prospects. In R. Parasuraman (Ed.) *The attentive brain* (pp. 3–15). Cambridge, MA: MIT Press.

Parasuraman, R. (2003). Neuroergonomics: Research and practice. *Theoretical Issues in Ergonomics Science, 4,* 5–20.

Parasuraman, R., Greenwood, P. M., Kumar, R., & Fossella, J. (2005). Beyond heritability: Neurotransmitter genes differentially modulate visuospatial attention and working memory. *Psychological Science, 16,* 200–207.

Parasuraman, R., & Riley, V. (1997). Humans and automation: Use, misuse, disuse, abuse. *Human Factors, 39,* 230–253.

Paul, A., Boyle, L., Rizzo, M., & Tippin, J. (2005). Variability of driving performance during microsleeps. In L. Boyle, J. D. Lee, D. V. McGehee, M. Raby, & M. Rizzo (Eds.) *Proceedings of Driving Assessment 2005: The Second International Driving Symposium on Human Factors in Driver Assessment, Training and Vehicle Design* (pp. 18–24). Iowa City: University of Iowa.

Peres, M., Van De Moortele, P. F., & Pierard, C. (2000). Functional magnetic resonance imaging of mental strategy in a simulated aviation performance task. *Aviation, Space and Environmental Medicine, 71,* 1218–1231.

Pew, R., & Mavor, A. (1998). *Modeling human and organizational behavior.* Washington, DC: National Academy Press.

Pfurtuscheller, G., & Neuper, C. (2001). Motor imagery and direct brain-computer communication. *Proceedings of the IEEE, 89,* 1123–1134.

Piaget, J. (1952). *The origins of intelligence in children.* New York: Longman.

Plomin, R., & Crabbe, J. (2000). DNA. *Psychological Bulletin, 126,* 806–828.

Posner, M. I. (2004). *Cognitive neuroscience of attention.* New York: Guilford.

Posner, M. I., & Deheane, S. (1994). Attentional networks. *Trends in Neuroscience, 17,* 75–79.

Rizzo, M., & Kellison, I. L. (2004). Eyes, brains, and autos. *Archives of Ophthalmology, 122,* 641–647.

Roth, E. M., Bennett, K. B., & Woods, D. D. (1987). Human interaction with an "intelligent" machine. *International Journal of Man-Machine Studies, 27,* 479–525.

Sanders, M. S., & McCormick, E. F. (1983). *Human factors in engineering and design.* New York: McGraw-Hill.

Sarter, M., Givens, B., & Bruno, J. P. (2001). The cognitive neuroscience of sustained attention: Where top-down meets bottom-up. *Brain Research Reviews, 35,* 146–160.

Sarter, N., & Sarter, M. (2003). Neuroergonomics: Opportunities and challenges of merging cognitive neuroscience with cognitive ergonomics. *Theoretical Issues in Ergonomics Science, 4,* 142–150.

Taking neuroscience beyond the bench [Editorial]. (2002). *Nature Neuroscience, 5*(Suppl.), 1019.

Wickens, C. D., & Hollands, J. G. (2000). *Engineering psychology and human performance.* New York: Longman.

Wiener, E. L., & Nagel, D. C. (1988). *Human factors in aviation.* San Diego, CA: Academic Press.

Neuroergonomics Methods

Alan Gevins and Michael E. Smith

Electroencephalography (EEG) in Neuroergonomics

This chapter considers the utility of the ongoing, scalp-recorded, human electroencephalogram (EEG) as a tool in neuroergonomics research and practice. The EEG has been extensively documented to be a sensitive index of changes in neuronal activity due to variations in the amount or type of mental activity an individual engages in, or to changes in his or her overall state of alertness and arousal. The EEG is recorded as a time-varying difference in voltage between an active electrode attached to the scalp and a reference electrode placed elsewhere on the scalp or body. In the healthy waking brain, the peak-to-peak amplitude of this scalp-recorded signal is usually well under 100 microvolts, and most of the signal power comes from rhythmic oscillations below a frequency of about 30 Hz. In many situations, the EEG is recorded simultaneously from multiple electrodes at different positions on the scalp, often placed over frontal, parietal, occipital, and temporal lobes of the brain according to a conventional placement scheme.

The scalp-recorded EEG signal reflects postsynaptic (dendritic) potentials rather than action (axonal) potentials. Since the laminar structure of the cerebral cortex facilitates a large degree of electrical summation (rather than mutual cancellation) of these postsynaptic potentials, the extracellular EEG recorded from a distance represents the passive conduction of currents produced by summating synchronous activity over large neuronal populations. Several factors determine the degree to which potentials arising in the cortex will be recordable at the scalp, including the amplitude of the signal at the cortex, the size of a region over which postsynaptic potentials are occurring in a synchronous fashion, the proportion of cells in that region that are in synchrony, the location and orientation of the activated cortical regions in relation to the scalp surface, and the amount of signal attenuation and spatial smearing produced by conduction through the highly resistive skull and other intervening tissue layers. While most of the scalp-recordable signal in the ongoing EEG presumably originates in cortical regions near the recording electrode, large signals originating at more distal cortical locations can also make a significant contribution to the activity observed at a given scalp recording site. For example, because of the orientation of the primary auditory cortices, some EEG signals generated in them project more toward the top of the head than to the geometrically closer lateral scalp surfaces.

The decomposition of an instantaneous scalp-recorded voltage measure into the constituent set of neuronal events throughout the brain that contributed to it is a mathematically ill-conditioned inverse problem that has no unique solution. Because of this indeterminacy, the EEG has significant limitations with respect to its use as a method for three-dimensional anatomical localization of neural activity in the same sense in which functional magnetic resonance imagining (fMRI) or positron emission tomography (PET) are used. However, the EEG has obvious advantages relative to other functional neuroimaging techniques as a method for continuous monitoring of brain function, either over long periods of time or in environments such as a hospital bed. Indeed, it is often the method of choice for some clinical monitoring tasks. For example, continuous EEG monitoring is an essential tool in the diagnostic evaluation of epilepsy and in the evaluation and treatment of sleep disorders. It is also coming to play an increasingly important role in neurointensive care unit monitoring and in gauging level of awareness during anesthesia.

For many years, efforts have also been under way to evaluate the extent to which the EEG might be useful as a monitoring modality in the context of human factors research. To be most useful in such settings, a monitoring method should be robust enough to be reliably measured under relatively unstructured task conditions, sensitive enough to consistently vary with some dimension of interest, unobtrusive enough to not interfere with operator performance, and inexpensive enough to eventually be deployable outside of specialized laboratory environments. It should also have reasonably good time resolution to allow tracking of changes in mental status as complex behaviors unfold. The EEG appears to meet such requirements. Furthermore, the compactness of EEG technology also means that, unlike other functional neuroimaging modalities (which typically require large expensive measuring instruments and complete immobilization of the subject), EEGs can even be collected from an ambulatory subject wearing a lightweight and nonencumbering headset.

A monitoring capability with such characteristics could provide unique value in the context of neuroergonomics research that seeks to better understand the neurobiological impact of task conditions that impose excessive cognitive workload or that result in significant mental fatigue. The need for expansion of knowledge in this area is evidenced by the extensive literature indicating that task conditions that impose cognitive overload often lead to performance errors even in alert individuals working under routine conditions. The potential for compromised performance in such circumstances can be exacerbated in individuals who are debilitated because of fatigue or sleep loss, illness or medication, or intoxication or hangover. In fact, even modest amounts of sleep loss can degrade performance on tests that require contributions from prefrontal cortical regions that control attention functions (Harrison & Horne, 1998, 1999; Harrison, Horne, & Rothwell, 2000; Linde & Bergstrom, 1992; Smith, McEvoy, & Gevins, 2002; see also chapter 14, this volume) and the magnitude of the behavioral impairment observed on such tasks can exceed that observed following a legally intoxicating dose of alcohol (Arendt, Wilde, Munt, & MacLean, 2001; Krull, Smith, Kalbfleisch, & Parsons, 1992; Williamson & Feyer, 2000).

While most often just a barrier to productivity, some critical jobs are particularly demanding in terms of the fatigue and cognitive workload they impose, and are particularly unforgiving in terms of the severe negative consequences that can be incurred when individuals performing those jobs make mistakes. For instance, in medical triage and crowded emergency room contexts the patient's life often hinges on a physician's ability to manage complex, competing demands, often after long hours on the job (Chisholm, Collison, Nelson, & Cordell, 2000). Similarly, the sleep deprivation and circadian desynchronization imposed by shift work scheduling has been noted to be a source of severe performance decrements (Scott, 1994) and has been implicated as a probable cause in a number of aviation (Price & Holley, 1990) and locomotive (Tepas, 1994) accidents. The high personal and societal costs associated with such performance failures motivate efforts to develop advanced methods for detecting states of cognitive overload or mental fatigue.

In this chapter, we review progress in developing EEG methods for such purposes. We first describe how the spectral composition of the EEG changes in response to variations in task difficulty or level of alertness during highly controlled cognitive tasks. We also consider methods for analysis of such signals that might be suitable for use in a continuous monitoring context. Finally, we review generalizations of those methods to assess complex, computer-based tasks that are more representative of real-world tasks.

EEG Signals Sensitive to Variations in Task Difficulty and Mental Effort

A significant body of literature exists concerning the EEG changes that accompany increases in cognitive workload and the allocation of mental effort. One approach to this topic has focused on EEG changes in response to varying working memory (WM) demands. WM can be construed as an outcome of the ability to control attention and sustain its focus on a particular active mental representation (or set of representations) in the face of distracting influences (Engle, Tuholski, & Kane, 1999). In many ways, this notion is nearly synonymous with what we commonly understand as effortful concentration on task performance. WM plays an important role in comprehension, reasoning, planning, and learning (Baddeley, 1992). Indeed, the effortful use of active mental representations to guide performance appears critical to behavioral flexibility (Goldman-Rakic, 1987, 1988), and measures of it tend to be positively correlated with performance on psychometric tests of cognitive ability and other indices of scholastic aptitude (Carpenter, Just, & Shell, 1990; Gevins & Smith, 2000; Kyllonen & Christal, 1990).

Many EEG studies of WM have required subjects to perform controlled n-back-style tasks (Gevins & Cutillo, 1993; Gevins et al., 1990, 1996) that demand sustained attention to a train of stimuli. In these tasks, the load imposed on WM varies while perceptual and motor demands are kept relatively constant. For example, in a spatial variant of the n-back task, stimuli are presented at different spatial positions on a computer monitor once every 4 or 5 seconds while the subject maintains a central fixation. Subjects must compare the spatial location of each stimulus with that of a previous stimulus, indicating whether a match criterion is met by making a key press response on a computer mouse or other device. In an easy, low-load version of the task, subjects compare each stimulus to the first stimulus presented in each block of trials (0-back task). In more difficult, higher-load versions, subjects compare the position of the current stimulus with that presented one, two, or even three trials previously (1-, 2-, or 3-back tasks).

These require constant updating of the information stored in WM on each trial, as well as constant attention to new stimuli and maintenance of previously presented information. To be successful in such tasks when WM demands are high, subjects typically must exert significant and continuous mental effort. Similar n-back tasks have been used to activate WM networks in a controlled fashion in the context of functional neuroimaging studies employing PET or fMRI methods (Braver et al., 1997; Cohen et al., 1994; Jansma, Ramsey, Coppola, & Kahn, 2000).

The spectral composition of the ongoing EEG displays regular patterns of load-related modulation during n-back task performance. For example, figure 2.1 displays spectral power in the 4–14 Hz range at a frontal midline (Fz) and a parietal midline (Pz) scalp location computed from the continuous

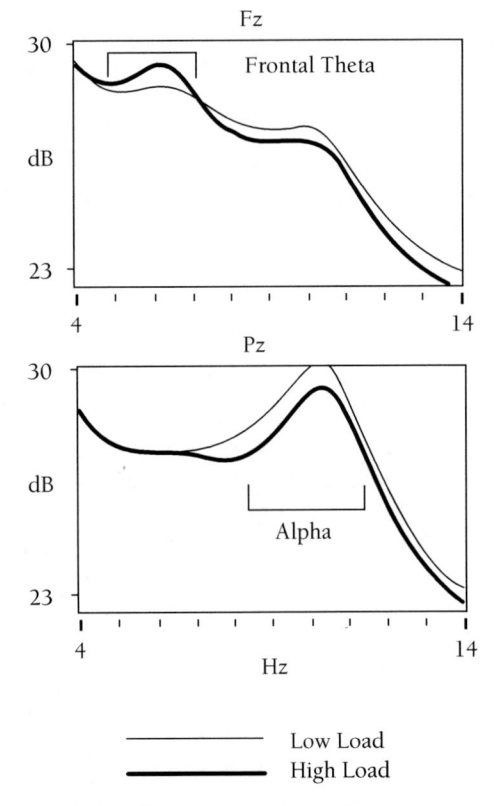

Figure 2.1. Effect of varying the difficulty of an n-back working memory task on the spectral power of EEG signals. The figure illustrates spectral power in dB of the EEG in the 4–14 Hz range at frontal (Fz) and parietal (Pz) midline electrodes, averaged over all trials of the tasks and collapsed over 80 subjects. Data from Gevins and Smith (2000).

EEG during performance of low-load (0-back) and moderately high-load (2-back) versions of a spatial *n*-back task. The data represent the average response from a group of 80 subjects in a study of individual differences in cognitive ability (Gevins & Smith, 2000) and show significant differences in spectral power as a function of task load that vary between electrode locations and frequency bands.

More specifically, at the Fz site a 5–7 Hz or theta-band spectral peak is increased in power during the high-load task relative to the low-load task. This type of frontal midline theta signal has frequently been reported to be enhanced in difficult, attention-demanding tasks, particularly those requiring a sustained focus of concentration (Gevins et al., 1979; Gevins et al., 1998; Gevins, Smith, McEvoy, & Yu, 1997; Miyata, Tanaka, & Hono, 1990; Mizuki, Tanaka, Iogaki, Nishijima, & Inanaga, 1980; Yamamoto & Matsuoka, 1990). Topographic analyses have indicated that this task loading-related theta signal tends to have a sharply defined potential field with a focus in the anterior midline region of the scalp (Gevins et al., 1997; Inouye et al., 1994); such a restricted topography is unlikely to result from distributed generators in dorsolateral cortical regions. Instead, attempts to model the generating source of the frontal theta rhythm from both EEG (Gevins et al., 1997) and magnetoencephalographic (Ishii et al., 1999) data have implicated the anterior cingulate cortex as a likely region of origin. This cortical region is thought to be part of an anterior brain network that is critical to attention control mechanisms and that is activated by the performance of complex cognitive tasks (Posner & Rothbart, 1992). Indeed, in a review of over 100 PET activation studies that examined anterior cingulate cortex activity, Paus and colleagues found that the major source of variance that affected activation in this region was associated with changes in task difficulty (Paus, Koski, Caramanos, & Westbury, 1998). The EEG results are thus consistent with these views, implying that performance of tasks that require significant mental effort places high demands on frontal brain circuits involved with attention control.

Figure 2.1 also indicates that signals in the 8–12 Hz or alpha band tend to be attenuated in the high-load task relative to the low-load task. This inverse relationship between task difficulty and alpha power has been observed in many studies in which task difficulty has been systematically manipulated. In fact, this task correlate of the alpha rhythm has been recognized for over 70 years (Berger, 1929). Because of this load-related attenuation, the magnitude of alpha activity during cognitive tasks has been hypothesized to be inversely proportional to the fraction of cortical neurons recruited into a transient functional network for purposes of task performance (Gevins & Schaffer, 1980; Mulholland, 1995; Pfurtscheller & Klimesch, 1992). This hypothesis is consistent with current understanding of the neural mechanisms underlying generation of the alpha rhythm (reviewed in Smith, Gevins, Brown, Karnik, & Du, 2001). Convergent evidence for this view is also provided by observations of a negative correlation between alpha power and regional brain activation as measured with hemodynamic measures (Goldman, Stern, Engel, & Cohen, 2002; Moosmann et al., 2003) and the frequent finding from neuroimaging studies of greater and more extensive brain activation during task performance when task difficulty increases (Bunge, Klingberg, Jacobsen, & Gabrieli, 2000; Carpenter, Just, & Reichle, 2000).

In addition to signals in the theta and alpha bands, other spectral components of the EEG have also been reported to be sensitive to changes in effortful attention. These include slow-wave activity in the delta (<3 Hz) band (McCallum, Cooper, & Pocock, 1988), high-frequency activity in the beta (15–30 Hz) and gamma (30–50 Hz) band (Sheer, 1989), and rarely studied phenomena such as the kappa rhythm that occurs around 8 Hz in a small percentage of subjects (Chapman, Armington, & Bragden, 1962).

Automated Detection of Mental Effort or Fatigue-Related Changes in the EEG

The results reviewed above indicate that spectral components of the EEG vary in a predictable fashion in response to variations in the cognitive demands of tasks. While this is a necessary condition for the development of EEG-based methods for monitoring cognitive workload, a number of other issues must also be addressed if such laboratory observations are to be transitioned into practical tools. Foremost among them is the problem of EEG artifact. That is, in addition to brain activity, signals recorded at the scalp include contaminating potentials from eye movements and blinks, muscle activity, head movements, and other physiological and instrumental sources of artifact. Such contaminants can easily mask cognition-related EEG signals (Barlow, 1986; Gevins, Doyle, Schaffer, Callaway, & Yeager, 1980; Gevins, Zeitlin, Doyle, Schaffer, & Callaway, 1979; Gevins, Zeitlin, Doyle, Yingling, et al., 1979; Gevins, Zeitlin, Yingling, et al., 1979), an essential but difficult and often subtle issue that, unfortunately, is too often given lip service but not actually dealt with. In laboratory studies, human experts review the raw data, identify artifacts and eliminate any contaminated EEG segments to ensure that data used in analyses represent actual brain activity. For large amounts of data, this is an expensive, labor-intensive process which itself is both subjective and variable. To be practical in more routine applied contexts, such decisions must be made algorithmically.

We have directed a great deal of research toward automated artifact detection. This has led to the development and testing of multicriteria spectral detectors (Gevins et al., 1975; Gevins, Yeager, Zeitlin, Ancoli, & Dedon, 1977), sharp transient waveform detectors (Gevins et al., 1976), and detectors using neural networks (Gevins & Morgan, 1986, 1988). In some cases, automated detection algorithms can perform about as well as the consensus of expert human judges. For example, in a database of about 40,000 eye movement, head/body movement, and muscle artifacts, we found that algorithmic methods successfully detected 98.3% of the artifacts with a false detection rate of 2.9%, whereas on average expert human judges found 96.5% of the artifacts with a 1.7% false detection rate. Thus, while further work on the topic is needed, it is reasonable to expect that the problem of automated artifact detection will not be an insurmountable barrier to the development of EEG-based cognitive monitoring methods.

A closely related problem is the fact that, in subjects actively performing tasks with significant perceptuomotor demands in a normal fashion, the incidence of data segments contaminated by artifacts can be high. As a result, it can be difficult to obtain enough artifact-free data segments for analysis. To minimize data loss, effective digital signal processing methods must also be developed to filter contaminants out of the EEG when possible. One powerful approach to this problem has been to implement adaptive filtering methods to decontaminate artifacts from EEG signals (Du, Leong, & Gevins, 1994). We have found such methods to be effective at recovering most of the artifact-contaminated data recorded in typical laboratory studies of subjects working on computer-based tasks. A variety of other methods have been employed by different investigators in response to this problem, including such techniques as autoregressive modeling (Van den Berg-Lensssen, Brunia, & Blom, 1989), source-modeling approaches (Berg & Scherg, 1994), and independent components analysis (Jung et al., 2000). A difficult issue with contaminant removal is that bona fide brain signals can also be removed with the artifacts. As with the problem of artifact detection, continued progress in this area suggests that, at least under some conditions and for some types of artifacts, decontamination strategies will evolve that will enable the automation of EEG processing for continuous monitoring applications.

Presuming then that automated preprocessing of the EEG can yield sufficient data for subsequent analyses, questions still remain as to whether the type of load-related changes in EEG signals can be measured in a reliable fashion in individual subjects, and whether such measurements can be accomplished with a temporal granularity suitable for tracking complex behaviors. That is, in the experiments described above, changes in the theta and alpha bands in response to variations in WM load were demonstrated by collapsing over many minutes of data recorded from a subject at each load level, and then comparing the mean differences between load levels across groups of subjects using conventional parametric statistical tests. Under normal waking conditions, such task-related EEG measures have high test-retest reliability when compared across a group of subjects measured during two sessions with a week between them (McEvoy, Smith, & Gevins, 2000). However, for the development of automated EEG analysis techniques suitable for monitoring applications, load-related changes in the EEG would ideally also be replicable when computed over short segments of data and would need to have high enough signal-to-noise ratios to be measurable within such segments.

Prior work has demonstrated that multivariate combinations of EEG variables can be used to accurately discriminate between specific cognitive states (Gevins, Zeitlin, Doyle, Schaffer, et al., 1979; Gevins, Zeitlin, Doyle, Yingling, et al., 1979; Gevins, Zeitlin, Yingling, et al., 1979; Wilson & Fisher, 1995). Furthermore, neural network-based pattern classification algorithms trained on data from individual subjects can also be used to automatically discriminate data recorded during different load levels of versions of the type of *n*-back WM task described above. For example, in one experiment (Gevins et al., 1998) eight subjects performed both spatial and verbal versions of 3-, 2-, and 1-back WM tasks on test sessions conducted on different days. For each single trial of data in each subject, spectral power estimates were computed in the theta and alpha bands for each electrode site. Pattern recognition was performed with the classic Joseph-Viglione neural network algorithm (Gevins, 1980; Gevins & Morgan, 1988; Joseph, 1961; Viglione, 1970). This algorithm iteratively generates and evaluates two-layered feedforward neural networks from the set of signal features, automatically identifying small subsets of features that produce the best classification of examples from the sample of data set aside for training. The resulting classifier networks were then cross-validated on the remaining data not included in the training sample.

Utilizing these procedures, test data segments from 3-back versus 1-back load levels could be discriminated with over 95% ($p < .001$) accuracy. Over 80% ($p < .05$) of test data segments associated with a 2-back load could also be discriminated from data segments in the 3-back or 1-back task loads. Such results provide initial evidence that, at least for these types of tasks, it is possible to develop algorithms capable of discriminating different cognitive workload levels with a high degree of accuracy. Not surprisingly, they also indicated that relatively large differences in cognitive workload are easier to detect than smaller differences, and that there is an inherent trade-off between the accuracy of classifier performance and the temporal length of the data segments being classified.

High levels of accurate classification were also achieved when applying networks trained with data from one day to data from another day and when applying networks trained with data from one task (e.g., spatial WM) to data from another task (e.g., verbal WM). It was also possible to successfully apply networks trained with data from a group of subjects to data from new subjects. Such generic networks were found on average to yield statistically significant classification results when discriminating the 1-back from the 3-back task load conditions, but their accuracy was much reduced from that achievable with subject-specific networks. On the one hand, such results indicate that there is a fair amount of commonality across days, tasks, and subjects in the particular set of EEG frequency-band measures that are sensitive to increases in cognitive workload. Such commonalities can be exploited in efforts to design efficient sensor montages and signal-processing methods. Nonetheless, they also indicate that to achieve optimal performance using EEG-based cognitive load-monitoring methods, it will likely be necessary to calibrate algorithms to accommodate individual differences. Such conclusions are also consistent with the observation that patterns of task-related EEG changes vary in conjunction with individual differences in cognitive ability and cognitive style (Gevins & Smith, 2000).

In addition to being sensitive to variations in attention and mental effort, the EEG also changes in a predictable fashion as individuals become sleepy and fatigued, or when they experience other forms of transient cognitive impairment. For example, it has long been known that the EEG of drowsy subjects has diffusely increased lower theta band activity and decreased alpha band activity (Davis, Davis, Loomis, Harvey, & Hobart, 1937; Gevins, Zeitlin, Ancoli, & Yeager, 1977). These changes are distinct from those described above characterizing increasing task load based on topography and spectral characteristics. Because such EEG changes are robust and reliable, a number of laboratories have developed and tested computerized algorithms for automated detection of drowsiness (Gevins, Zeitlin, et al., 1977; Hasan, Hirkoven, Varri, Hakkinen, & Loula, 1993). Such methods have produced highly promising results. For example, in one study we used neural network-based methods to compare task-related EEG features between alert and drowsy states in individual subjects performing the *n*-back WM tasks described above (Gevins & Smith, 1999). Utilizing EEG features in the alpha and theta bands, average test set classification accuracy was 92% (range 84–100%, average binomial $p < .001$). In another study, we explicitly compared metrics based on either behavioral response measures during an *n*-back WM task, EEG recordings during task performance and control conditions, or combinations of behavioral and EEG variables with respect to their relative sensitivity for discriminating conditions of drowsiness associated with sleep loss from alert, rested conditions (Smith et al., 2002). Analyses based on behavior alone did not yield a stable pattern of results when viewed over test intervals. In contrast, analyses that incorporated both behavioral and neurophysiological measures displayed a monotonic increase in discriminability from alert baseline with increasing amounts of sleep deprivation. Such results indicate that fairly modest amounts of sleep loss can induce neurocognitive changes detectable in individual subjects performing computer-based tasks, and that the sensitivity for detecting such states is significantly improved by the addition of EEG measures to behavioral indices.

Extension of EEG-Based Cognitive State Monitoring Methods to More Realistic Task Conditions

The results described above provide evidence for the basic feasibility of using EEG-based methods for monitoring cognitive task load, mental fatigue, and drowsiness in individuals engaged in computer-based work. However, the *n*-back WM task makes minimal demands on perceptual and motor systems, and only requires that a subject's effort be focused on a single repetitive activity. In more realistic work environments, task demands are usually less structured and mental resources often must be divided between competing activities, raising questions as to whether results obtained with the *n*-back task could generalize to such contexts.

Studies have demonstrated that more complicated forms of human-computer interaction (such as videogame play) produce mental effort-related modulation of the EEG that is similar to that observed during *n*-back tasks (Pellouchoud, Smith, McEvoy, & Gevins, 1999; Smith, McEvoy, & Gevins, 1999). This implies that it might be possible to extend EEG-based multivariate methods for monitoring task load to such circumstances. To evaluate this possibility, a subsequent study (Smith et al., 2001) was performed in which the EEG was recorded while subjects performed the Multi-Attribute Task Battery (MATB; Comstock & Arnegard, 1992). The MATB is a personal computer-based multitasking environment that simulates some of the activities a pilot might be required to perform. It has been used in several prior studies of mental workload and adaptive automation (e.g., Fournier, Wilson, & Swain, 1999; Parasuraman, Molloy, & Singh, 1993; Parasuraman, Mouloua, & Molloy, 1996). The data collected during performance of the MATB were used to test whether it is possible to derive combinations of EEG features that can be used for indexing task loading during a relatively complex form of human-computer interaction.

The MATB task included four concurrently performed subtasks in separate windows on a computer screen (for graphic depictions of the MATB visual display, see Fournier et al., 1999; Molloy & Parasuraman, 1996). These included a systems-monitoring task that required the operator to monitor and respond to simulated warning lights and gauges, a resource management task in which fuel levels in two tanks had to be maintained at a certain level, a communications task that involved receiving audio messages and making frequency adjustments on virtual radios, and a compensatory tracking task that simulated manual control of aircraft position. Manipulating the difficulty of each subtask served to vary load; such manipulations were made in a between-blocks fashion. Subjects learned to perform low-, medium-, and high-load (LL, ML, and HL) versions of the tasks. For comparison purposes they also performed a passive watching (PW) condition in which they observed the tasks unfolding without actively performing them.

Subjects engaged in extensive training on the tasks on one day, and then returned to the laboratory on a subsequent day for testing. On the test day, subjects performed multiple 5-minute blocks of each task difficulty level. Behavioral and subjective workload ratings provided evidence that on average workload did indeed increase in a monotonic fashion across the PW, LL, ML, and HL task conditions. This increase in workload was associated with systematic changes in the EEG. In particular, as in the prior study of workload changes in the *n*-back task paradigm, frontal theta band activity tended to increase with increasing task difficulty, whereas alpha band activity tended to decrease. Such results indicated that the workload manipulations were successful, and that spectral features in the theta and alpha range might be useful in attempting to automatically monitor changes in workload with EEG measures.

Separate blocks of data were thus used to derive and then independently validate subject-specific,

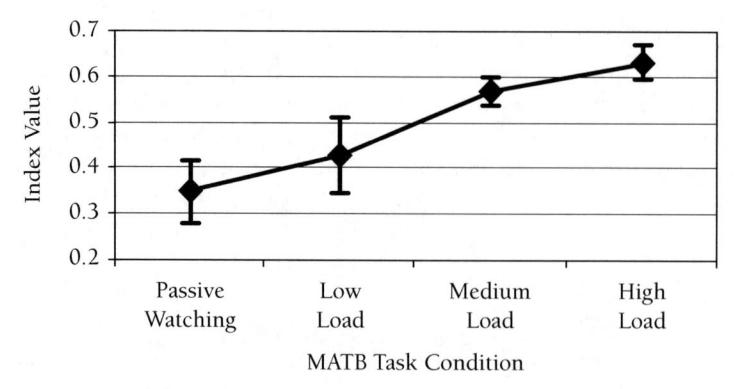

Figure 2.2. Mean and *SEM* (*N* = 16 subjects) EEG-based cognitive workload index values during performance of the MATB flight simulation task. Data are presented for each of four task versions (PW = passive watch, LL = low load, ML = moderate load, HL = high load). Average cognitive workload index scores increased monotonically with increasing task difficulty. Data from Smith., Gevins, Brown, Karnik and Du (2001).

EEG-based, multivariate cognitive workload functions. In contrast to the two-class pattern detection functions that were employed to discriminate between different task load levels in the prior study, we evaluated a different technique that results in a single subject-specific function that produces a continuous index of cognitive workload and hence could be applied to data collected at each difficulty level of the task. In this procedure, the EEG data were first decomposed into short windows and a set of spectral power estimates of activity in the theta and alpha frequency ranges was extracted from each window. A unique multivariate function was then defined for each subject that maximized the statistical divergence between a small sample of data from low and high task load conditions. To cross-validate the function, it was tested on new data segments from the same subject. Across subjects (figure 2.2), mean task load index values were found to increase systematically with increasing task difficulty and differed significantly between the different versions of the task (Smith et al., 2001). These results provide encouraging initial evidence that EEG measures can indeed provide a modality for measuring cognitive workload during more complex forms of computer interaction. Although complex, the signal processing and pattern classification algorithms employed in this study were designed for real time implementation. In fact, a prototype online system running on a circa 1997 personal computer performed the requisite calculations online and provided an updated estimate of cognitive workload at 4-second intervals while subjects were engaged in task performance.

It is worth reiterating here the critical role that effective automated artifact detection and filtering plays in such analyses. Effective artifact detection and filtering is particularly important during complex computer-based activities such as videogame play, as these types of behaviors tend to be associated with a great deal of artifact-producing head, body, and eye movement that might confound EEG-derived estimates of cognitive state. For example, figure 2.3 illustrates the average workload indices obtained from data from a single electrode (frontal central site Fz) in an individual subject during the MATB, obtained after calibrating a multivariate index function for that electrode using artifact-decontaminated examples of data from the low-load and high-load MATB conditions and then applying the resulting function to new samples of EEG data that were either decontaminated with state-of-the-art EEG artifact detection and filtering algorithms (leftmost and center columns) or without systematic artifact detection and correction (rightmost column), with *N* = 50 4-second index function scores per task condition. A linear discriminant function applied to the data was able to correctly discriminate 95% of the individual clean samples of LL MATB data as coming from that category rather than from the HL category (binomial $p < .000001$). In contrast, an equivalent linear discriminant function applied to the artifact-contaminated LL data performed at chance level.

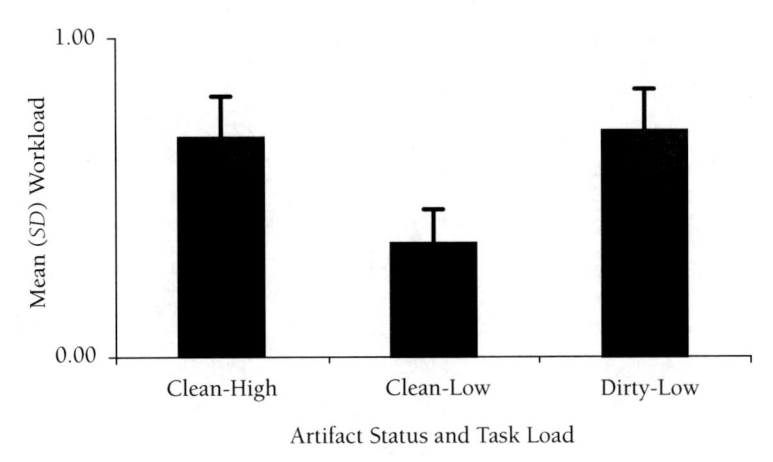

Figure 2.3. Individual subject workload index function scores from a single EEG channel (frontal central electrode Fz) can discriminate low from high load levels during MATB task performance when effective EEG artifact decontamination is employed (left and center columns), but load can be misclassified without such correction (right column).

Analogous methods have also been used in a small exploratory study that involved more naturalistic computer tasks. In that experiment (Gevins & Smith, 2003), the EEG data were recorded while subjects performed more common computer-based tasks that were performed under time pressure and that were more or less intellectually demanding. These more naturalistic activities required subjects to perform word processing, take a computer-based aptitude test, and search for information on the Web. The word processing task required subjects to correct as many misspellings and grammatical errors as they could in the time allotted, working on a lengthy text sample using a popular word processing program. The aptitude test was a practice version of the Computer-Adaptive GMAT test. Subjects were asked to solve as many data-sufficiency problems as possible in the time allotted; such problems make a high demand on logical and quantitative reasoning skills and require significant mental effort to complete in a timely fashion. The Web-searching task required subjects to use a popular Web browser and search engine to find as many answers as possible in the time allotted to a list of trivia questions provided by the experimenter. For example, subjects were required to use the browser and search engine to "convert 98.6 degrees Fahrenheit into degrees Kelvin," "find the population of the 94105 area code in the 1990 U.S. Census," and "find the monthly mortgage payment on a $349,000, 30-year mortgage with a 7.5% interest rate." Each type of task was structured such that subjects would be unlikely to be able to complete it in the time allotted. Data were also recorded from subjects as they performed easy and difficult n-back working memory tasks, and as they rested quietly, for comparison with the more naturalistic tasks.

The same basic analysis procedure described above that was applied to the EEG data recorded during MATB performance was also employed in this study to derive personalized continuous functions indexing cognitive workload. The resulting functions were then applied to new samples of that subject's data.

A summary of the results from these analyses, averaged across data segments within each task condition and compared between conditions, is presented in figure 2.4. These comparisons indicate that the cognitive load index performed in a predictable fashion. That is, the condition in which the subject was asked to passively view a blank screen produced an average EEG-based cognitive workload load around the zero point of the scale. Average index values during 0-back task performance were slightly higher than those during the resting condition, and average index values during the 3-back task were significantly higher than those recorded either during the 0-back WM task or during the resting state. All three naturalistic tasks produced workload index values slightly higher than those obtained in the 3-back task, which might be expected given that the n-back tasks had been practiced and

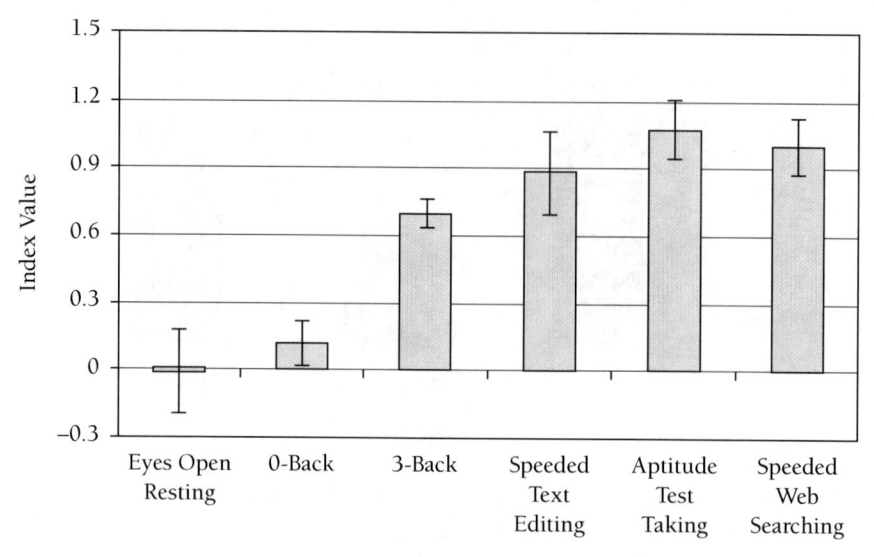

Figure 2.4. Mean and *SEM* (*N* = 7 subjects) EEG-based cognitive workload index values during resting conditions, easy and difficult versions of the *n*-back working memory tasks, and a few naturalistic types of computer-based work (see text for full description of tasks and procedure). The data represent average index values over the course of each type of task. The easy WM and resting conditions produced significantly lower values than the more difficult WM condition or the naturalistic tasks.

were repetitive in nature, whereas the other tasks were novel and required the use of strategies of information gathering, reasoning, and responding that were less stereotyped in form. Among the naturalistic tasks, the highest levels of cognitive workload were recorded during the computerized aptitude-testing task—the condition that was also subjectively experienced as the most difficult.

This pattern of results is interesting not only because it conforms with a priori expectations about how workload would vary among the different tasks, but also because it provides data relevant to the issue of how the workload measure is affected by differences in perceptuomotor demands across conditions. Since in the *n*-back tasks stimuli and motor demands are kept constant between the 0-back and 3-back load levels, the observed EEG differences in those conditions are clearly closely related to differences in the amounts of mental work demanded by the two task variants rather than other factors. However, in the study of MATB task performance described above, the source of variation in the index is somewhat less clear. On the one hand, performance and subjective measures unambiguously indicated that the mental effort required to perform the high-load version of the MATB was substantially greater than that required by the low-load (or passive watching) versions. On the other hand, the perceptuomotor requirements in the high-load version were also substantially greater than those imposed by the other version. In this latter experiment, such confounds were less of a concern. Indeed, both the text editing task and the Web searching task required more effortful visual search and more active physical responding than the aptitude test, whereas the aptitude test had little reading and less responding and instead required a great deal of thinking and mental evaluation of possibilities. Thus, the fact that the average cognitive workload values during performance of the aptitude test were higher than those observed in the other tasks provides convergent support for the notion that the subject-specific indices were more closely tracking variations in mental demands rather than variations in perceptuomotor demands in these instances. Nevertheless, the results remain ambiguous in this regard.

From Unidimensional to Multidimensional Neurophysiological Measures of Workload

Another approach to resolving the inherent ambiguity of the sort of unidimensional "whole brain" metric used to quantify mental workload in the studies described above is to generalize the metric to separate index loading of different functional brain systems. That is, the applied psychology and ergonomics literature has long posited a relative independence of the resources involved with higher-order executive processes and those involved with perceptual processing and motor activity (Gopher, Brickner, & Navon, 1982; Wickens, 1991). Furthermore, related topographic differences can be observed in regional patterns of EEG modulation. For example, it is clear that alpha band activity over posterior regions is particularly sensitive to visual stimulation and that increases in motor demands are associated with suppression of alpha and beta band activity over sensorimotor cortex (Arroyo et al., 1993; Jasper & Penfield, 1949; Mulholland, 1995). Such regional differences can also be observed during performance of complex tasks. In one study, the EEG was recorded from subjects while they either actively played a videogame or watched the screen while someone else played the game (Pellouchoud et al., 1999). Across the group of subjects, the amplitude of the frontal midline theta rhythm was larger in the active performance condition than in the resting or passive watching conditions. In contrast, a posterior alpha band signal was attenuated during both the playing and the watching conditions relative to the resting condition, suggesting that it was responding primarily to the presence of complex visual stimulation rather than active task performance. Finally, a central mu (10–13 Hz) rhythm recorded over sensorimotor cortex was attenuated during the active game-playing condition, but not during the passive watching condition, presumably reflecting activation related to the game's hand and finger motor control requirements (Pellouchoud et al., 1999). In another study where subjects were allowed to practice a videogame until they were skilled at it, the alpha rhythm recorded over frontal regions increased in amplitude with progressive amounts of practice, suggesting that smaller neuronal populations were required to regulate attention as the task became automated. In contrast, the alpha rhythm recorded over posterior regions displayed no such effect, suggesting that neural activation related to visual processing did not diminish (Smith et al., 1999).

Such considerations have led to an extension of the method described above to create multidimensional indices that provide information about the relative activation of a local neocortical region. In particular, instead of defining a single load-sensitive multivariate function for the whole head, we have worked toward extracting three independent topographically regionalized metrics from multielectrode data (Smith & Gevins, 2005) recorded in the MATB experiment described above. One metric was derived from data recorded over frontal cortical areas. Since this region of the brain is known to be involved in executive attention control and working memory processes, we refer to this metric as a measure of cortical activation related to frontal executive workload. A second metric was derived from data recorded from central and parietal regions. Since these regions are activated by motor control functions, somatosensory feedback, and the coordination of action plans with representations of extra personal space, we refer to this second metric as a measure of sensorimotor activation. A third metric was derived from electrodes over occipital regions. Since this region includes primary and secondary visual cortices, we refer to this third metric as representing variation in cortical activation due to visuoperceptual functions. While these labels are convenient for discussion, they of course are highly oversimplified with regard to describing the actual operations performed by the underlying cortical systems. They may, however, be seen as consistent with the results of fMRI studies of simulator (driving) operation (Calhoun et al., 2002; Walter et al., 2001), which have also reported activation in frontal attention networks, sensorimotor cortex, and visual cortices (see also chapter 4, this volume).

Figure 2.5 summarizes how the three regional cortical activation metrics changed as a result of task manipulations, describing the mean output of the regional metrics computed across all of the cross-validation data segments for each task difficulty level for each subject. Each regional metric was found to be significantly affected by the task difficulty manipulation, consistent with the notion that the MATB task increased workload on multiple brain systems in parallel. Furthermore, both subjective workload estimates and overt task performance were found to covary with the regional EEG-derived workload estimates, indicating the metrics were tracking changes in brain activity that were functionally significant.

In a second experiment, these regional workload metrics were tracked over the course of an

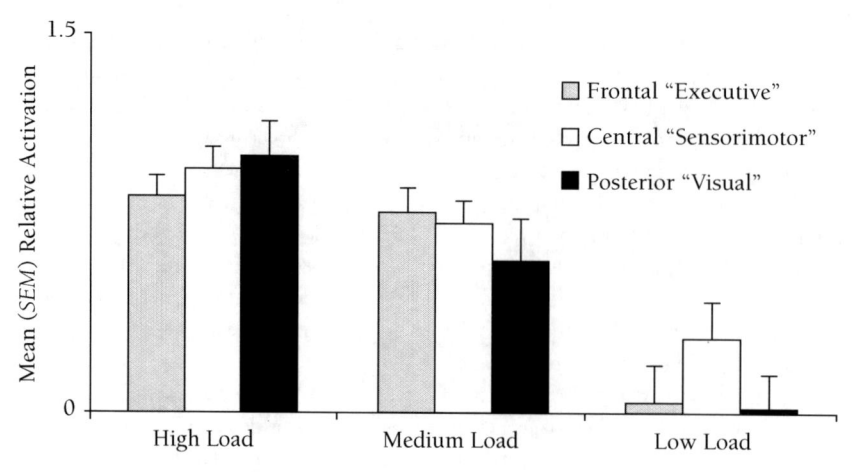

Figure 2.5. Average normalized (*SEM*) values for regional cortical activation metrics over frontal, central, and posterior regions of the scalp derived from multivariate combinations of EEG spectral features recorded from $N = 16$ participants performing 5-minute blocks of high-, medium-, and low-load versions of the MATB task.

all-night experiment during which subjects performed the HL version of the MATB and other tasks in a more or less continuous fashion without sleeping since early the prior morning (Smith & Gevins, 2005; Smith et al., 2002). During this extended wakefulness session, cortical activation as indexed by the regional EEG workload scores was observed to change with time on task despite task difficulty being held constant and despite the fact that subjects were highly practiced in the task. The changes are illustrated in figure 2.6. The daytime values are contrasted with values representing the first block of data from the overnight session, where testing on average began around 11:00 p.m. They are also contrasted with late night values from the time period within the last four test inter-

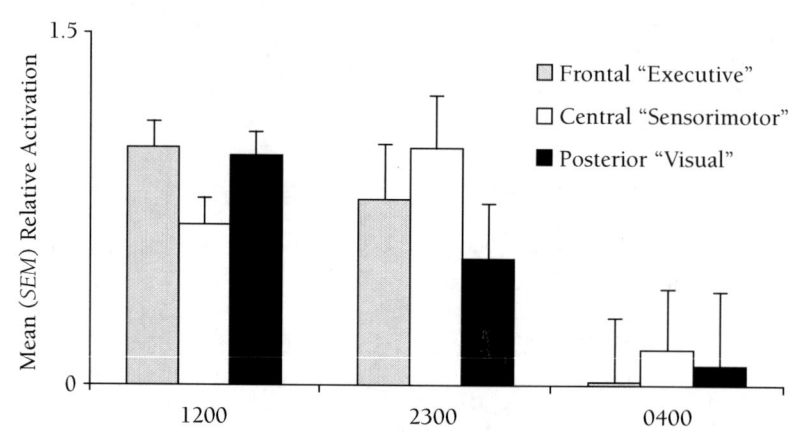

Figure 2.6. Average normalized (*SEM*) values for regional cortical activation metrics over frontal, central, and posterior regions of the scalp derived from multivariate combinations of EEG spectral features recorded from $N = 16$ participants performing 5-minute blocks of the high-load MATB task during alert daytime baseline periods (around noon or 1200 hrs), during the first test interval of an all-night recording session (2300 hrs or 11:00 p.m.), and during the test interval between 1:30 and 5:30 a.m. in which they displayed a cortical activation minima (across subjects this minimum occurred on average at 0400 hrs or 4:00 a.m.).

vals for each subject when he or she displayed a minimum in total cortical activation (for 15/16 subjects this minimum occurred between 1:30 and 5:30 a.m.). Average values for each region declined with sleep deprivation, with the largest overall declines for the frontal region.

Interestingly, subjective workload was found to be negatively correlated with the magnitude of the fatigue-related decline in the frontal region—but not the other regions—suggesting that as frontal activation decreased the subjects found it increasingly difficult to confront the demands of the high-load MATB task. The fact that perceived mental effort was observed to be positively correlated with changes in frontal cortical activity in alert individuals, yet negatively correlated with frontal cortical activation with increases in mental fatigue, might be seen as problematic for the eventual development of adaptive automation systems that aim to dynamically modulate the cognitive task demands placed on an individual in response to momentary variations in the availability of mental resources as reflected by real-time analysis of neural activity. That is, it has sometimes been suggested that it might be possible to use measures of brain activation as a basis for automated systems to off-load tasks from an individual if he or she was detected to be in a state of high cognitive workload, or allocate more tasks to an individual that appeared to have ample reserve processing capacity and was in danger of becoming bored or inattentive. The current results indicate that a decrease in cortical activation in frontal regions may reflect either a decrease in mental workload or an increase in mental fatigue and a heightened sense of mental stress. Assigning more tasks to an individual in the former case may indeed serve to increase his or her cognitive throughput. In the latter case, it may result in the sort of tragic accident that is too often reported to occur when fatigued personnel are confronted with unexpected increases in task demands (Dinges, 1995; Miller, 1996; Rosekind, Gander, & Miller, 1994). Thus, while measures of brain function during complex task performance may serve to accelerate research into the sources of performance failure under stress, it seems likely that a great deal of future research will be needed before such measures can be adapted to the problem of developing technology for adaptively augmenting the capabilities of mission-critical personnel working in demanding and stressful computerized-task environments.

Conclusion

In summary, the results reviewed above indicate that the EEG changes in a highly predictable way in response to sustained changes in task load and associated changes in the mental effort required for task performance. It also changes in a reliable fashion in response to variations in mental fatigue and level of arousal. It appears that such changes can be automatically detected and measured using algorithms that combine parameters of the EEG power spectra into multivariate functions. While such EEG metrics lack the three-dimensional spatial resolution provided by neuroimaging methods such as PET or fMRI, they can nonetheless provide useful information about changes in regional functional brain systems that may have important implications for ongoing task performance. Such methods can be effective both in gauging the variations in cognitive workload imposed by highly controlled laboratory tasks and in monitoring differences in the mental effort required to perform tasks that more closely resemble those that an individual might encounter in a real-world work environment. Because this sensitivity can in principle be obtained with technology suitable for use in real-world work environments, the EEG can be seen as a critical tool for research in neuroergonomics.

Acknowledgments. This research was supported by the U.S. Air Force Research Laboratory, the National Science Foundation, the National Aeronautics and Space Administration, the Defense Advanced Research Projects Agency, and the National Institutes of Health.

Main Points

1. The EEG recorded at the scalp is a record of instantaneous fluctuations of mass electrical activity in the brain, primarily summated postsynaptic (dendritic) potentials of large cortical neuronal populations.
2. Spectral components of the EEG signal show characteristic changes in response to variations in mental demands or state of alertness. As

with all other means of measuring brain function, EEG signals are also sensitive to the perceptual and motor activities of the subject in addition to mental activity. It is essential to separately measure these perceptual and motoric neural processes to have a strong inference that the brain function signals one would like to use as a measure of mental activity actually in fact do so.
3. The high temporal resolution of the EEG in combination with the simplicity and portability of the technology used to record and analyze it make it suitable for use in unrestrained subjects in a relatively wide range of environments, including real-world work contexts.
4. The sensitivity of EEG signals to particular task demands differs depending on the spatial positioning of scalp electrodes and, in many but not all cases, reflects functional specialization of nearby underlying cortical regions.
5. As with all brain function measurement technologies, the EEG signal is sensitive to artifactual contaminants not generated in the brain, which must be removed from the signal in order to make valid inferences about mental function. This is easier said than done.
6. There is no simple one-to-one mapping between a change in a measure of brain activation and the cognitive loading of an individual. Additional factors, such as the state of alertness, must be taken into account. Simplistic approaches to neuroadaptive automation that do not take this complexity into account will fail.

Key Readings

Gevins, A. S., & Cutillo, B. C. (1986). Signals of cognition. In: F. Lopes da Silva, W. Storm van Leeuwen, & A. Remond (Eds.), *Handbook of electroencephalography and clinical neurophysiology:. Vol. 2. Clinical applications of computer analysis of EEG and other neurophysiological signals* (pp. 335–381). Amsterdam: Elsevier.

Gevins, A. S., & Remond, A. (Eds.). (1987). *Handbook of electroencephalography and clinical neurophysiology: Vol. 1. Methods of analysis of brain electrical and magnetic signals.* Amsterdam: Elsevier.

Regan, D. (1989). *Human brain electrophysiology.* New York: Elsevier.

References

Arendt, J. T., Wilde, G. J., Munt, P. W., & MacLean, A. W. (2001). How do prolonged wakefulness and alcohol compare in the decrements they produce on a simulated driving task? *Accident Analysis and Prevention, 33,* 337–344.

Arroyo, S., Lesser, R. P., Gordon, B., Uematsu, S., Jackson, D., & Webber, R. (1993). Functional significance of the mu rhythm of human cortex: An electrophysiological study with subdural electrodes. *Electroencephalography and Clinical Neurophysiology, 87,* 76–87.

Baddeley, A. (1992). Working memory. *Science, 255,* 556–559.

Barlow, J. S. (1986). Artifact processing rejection and reduction in EEG data processing. In F. H. Lopes da Silva, W. Storm van Leeuwen, & A. Remond (Eds.), *Handbook of electroencephalography and clinical neurophysiology* (Vol. 2, pp. 15–65). Amsterdam: Elsevier.

Berg, P., & Scherg, M. (1994). A multiple source approach to the correction of eye artifacts. *Electroencephalography and Clinical Neurophysiology, 90,* 229–241.

Berger, H. (1929). Uber das Elektroenzephalogramm des Menschen. *Archives of Psychiatry, 87*(Nervenk), 527–570.

Braver, T. S., Cohen, J. D., Nystrom, L. E., Jonides, J., Smith, E. E., & Noll, D. C. (1997). A parametric study of prefrontal cortex involvement in human working memory. *NeuroImage, 5,* 49–62.

Bunge, S. A., Klingberg, T., Jacobsen, R. B., & Gabrieli, J. D. (2000). A resource model of the neural basis of executive working memory. *Proceedings of the National Academy of Sciences USA, 97,* 3573–3578.

Calhoun, V. D., Pekar, J. J., McGinty, V. B., Adali, T., Watson, T. D., & Pearlson, G. D. (2002). Different activation dynamics in multiple neural systems during simulated driving. *Human Brain Mapping, 16*(3), 158–167.

Carpenter, P. A., Just, M. A., & Reichle, E. D. (2000). Working memory and executive function: Evidence from neuroimaging. *Current Opinion in Neurobiology, 10,* 195–199.

Carpenter, P. A., Just, M. A., & Shell, P. (1990). What one intelligence test measures: A theoretical account of the processing in the Raven Progressive Matrices Test. *Psychological Review, 97,* 404–431.

Chapman, R. M., Armington, J. C., & Bragden, H. R. (1962). A quantitative survey of kappa and alpha

EEG activity. *Electroencephalography and Clinical Neurophysiology, 14*, 858–868.

Chisholm, C. D., Collison, E. K., Nelson, D. R., & Cordell, W. H. (2000). Emergency department workplace interruptions: Are emergency physicians "interrupt-driven" and "multitasking"? *Academy of Emergency Medicine, 8*, 686–688.

Cohen, J. D., Forman, S. D., Braver, T. S., Casey, B. J., Servan-Schreiber, D., & Noll, D. C. (1994). Activation of prefrontal cortex in a non-spatial working memory task with functional MRI. *Human Brain Mapping, 1*, 293–304.

Comstock, J. R., & Arnegard, R. J. (1992). *The multiattribute task battery for human operator workload and strategic behavior research* (No. 104174): NASA Technical Memorandum.

Davis, H., Davis, P. A., Loomis, A. L., Harvey, E. N., & Hobart, G. (1937). Human brain potentials during the onset of sleep. *Journal of Neurophysiology, 1*, 24–37.

Dinges, D. F. (1995). An overview of sleepiness and accidents. *Journal of Sleep Research, 4*(Suppl. 2), 4–14.

Du, W., Leong, H. M., & Gevins, A. S. (1994). Ocular artifact minimization by adaptive filtering. In *Proceedings of the Seventh IEEE SP Workshop on Statistical Signal and Array Processing* (pp. 433–436). Quebec City, Canada: IEEE Press.

Engle, R. W., Tuholski, S., & Kane, M. (1999). Individual differences in working memory capacity and what they tell us about controlled attention, general fluid intelligence and functions of the prefrontal cortex. In A. Miyake & P. Shah (Eds.), *Models of working memory* (pp. 102–134). Cambridge, UK: Cambridge University Press.

Fournier, L. R., Wilson, G. F., & Swain, C. R. (1999). Electrophysiological, behavioral, and subjective indexes of workload when performing multiple tasks: Manipulations of task difficulty and training. *International Journal of Psychophysiology, 31*, 129–145.

Gevins, A. S. (1980). Pattern recognition of brain electrical potentials. *IEEE Transactions on Pattern Analysis and Machine Intelligence, PAMI-2*, 383–404.

Gevins, A. S., Bressler, S. L., Cutillo, B. A., Illes, J., Miller, J. C., Stern, J., et al. (1990). Effects of prolonged mental work on functional brain topography. *Electroencephalography and Clinical Neurophysiology, 76*, 339–350.

Gevins, A. S., & Cutillo, B. (1993). Spatiotemporal dynamics of component processes in human working memory. *Electroencephalography and Clinical Neurophysiology, 87*, 128–143.

Gevins, A. S., Doyle, J. C., Schaffer, R. E., Callaway, E., & Yeager, C. (1980). Lateralized cognitive processes and the electroencephalogram. *Science, 207*, 1005–1008.

Gevins, A. S., & Morgan, N. (1986). Classifier-directed signal processing in brain research. *IEEE Transactions on Biomedical Engineering, BME-33*(12), 1058–1064.

Gevins, A. S., & Morgan, N. H. (1988). Applications of neural-network (NN) signal processing in brain research. *IEEE Transactions on Acoustics, Speech, and Signal Processing, 36*(7), 1152–1161.

Gevins, A. S., & Schaffer, R. E. (1980). A critical review of electroencephalographic EEG correlates of higher cortical functions. *CRC Critical Reviews in Bioengineering, 4*, 113–164.

Gevins, A. S., & Smith, M. E. (1999). Detecting transient cognitive impairment with EEG pattern recognition methods. *Aviation Space and Environmental Medicine, 70*, 1018–1024.

Gevins, A. S., & Smith, M. E. (2000). Neurophysiological measures of working memory and individual differences in cognitive ability and cognitive style. *Cerebral Cortex, 10*(9), 829–839.

Gevins, A. S., & Smith, M. E. (2003). Neurophysiological measures of cognitive workload during human-computer interaction. *Theoretical Issues in Ergonomics Science, 4*, 113–131.

Gevins, A. S., Smith, M. E., Le, J., Leong, H., Bennett, J., Martin, N., et al. (1996). High-resolution evoked potential imaging of the cortical dynamics of human working memory. *Electroencephalography and Clinical Neurophysiology, 98*, 327–348.

Gevins, A. S., Smith, M. E., Leong, H., McEvoy, L., Whitfield, S., Du, R., et al. (1998). Monitoring working memory load during computer-based tasks with EEG pattern recognition methods. *Human Factors, 40*, 79–91.

Gevins, A. S., Smith, M. E., McEvoy, L., & Yu, D. (1997). High-resolution EEG mapping of cortical activation related to working memory: Effects of task difficulty, type of processing, and practice. *Cerebral Cortex, 7*, 374–385.

Gevins, A. S., Yeager, C. L., Diamond, S. L., Spire, J. P., Zeitlin, G. M., & Gevins, A. H. (1975). Automated analysis of the electrical activity of the human brain (EEG): A progress report. *Proceedings of the Institute of Electrical and Electronics Engineers, Inc.*, 1382–1399.

Gevins, A. S., Yeager, C. L., Diamond, S. L., Spire, J. P., Zeitlin, G. M., & Gevins, A. H. (1976). Sharp-transient analysis and thresholded linear coherence spectra of paroxysmal EEGs. In P. Kellaway & I. Petersen (Eds.), *Quantitative analytic studies in epilepsy* (pp. 463–481). New York: Raven.

Gevins, A. S., Yeager, C. L., Zeitlin, G. M., Ancoli, S., & Dedon, M. (1977). On-line computer rejection of

EEG artifact. *Electroencephalography and Clinical Neurophysiology, 42,* 267–274.

Gevins, A. S., Zeitlin, G. M., Ancoli, S., & Yeager, C. L. (1977). Computer rejection of EEG artifact. II: Contamination by drowsiness. *Electroencephalography and Clinical Neurophysiology, 42,* 31–42.

Gevins, A. S., Zeitlin, G. M., Doyle, J. C., Schaffer, R. E., & Callaway, E. (1979). EEG patterns during "cognitive" tasks. II. Analysis of controlled tasks. *Electroencephalography and Clinical Neurophysiology, 47,* 704–710.

Gevins, A. S., Zeitlin, G. M., Doyle, J. C., Yingling, C. D., Schaffer, R. E., Callaway, E., et al. (1979). Electroencephalogram correlates of higher cortical functions. *Science, 203,* 665–668.

Gevins, A. S., Zeitlin, G. M., Yingling, C. D., Doyle, J. C., Dedon, M. F., Schaffer, R. E., et al. (1979). EEG patterns during "cognitive" tasks. I. Methodology and analysis of complex behaviors. *Electroencephalography and Clinical Neurophysiology, 47,* 693–703.

Goldman, R. I., Stern, J. M., Engel, J. J., & Cohen, M. S. (2002). Simultaneous EEG and fMRI of the alpha rhythm. *Neuroreport, 13,* 2487–2492.

Goldman-Rakic, P. (1987). Circuitry of primate prefrontal cortex and regulation of behavior by representational memory. In F. Plum & V. Mountcastle (Eds.), *Handbook of physiology: The nervous system: higher functions of the brain* (Vol. 5, pp. 373–417). Bethesda, MD: American Physiological Society.

Goldman-Rakic, P. (1988). Topography of cognition: Parallel distributed networks in primate association cortex. *Annual Review of Neuroscience, 11,* 137–156.

Gopher, D., Brickner, M., & Navon, D. (1982). Different difficulty manipulations interact differently with task emphasis: Evidence for multiple resources. *Journal of Experimental Psychology: Human Perception and Performance, 8,* 146–157.

Harrison, Y., & Horne, J. A. (1998). Sleep loss impairs short and novel language tasks having a prefrontal focus. *Journal of Sleep Research, 7,* 95–100.

Harrison, Y., & Horne, J. A. (1999). One night of sleep loss impairs innovative thinking and flexible decision making. *Organizational Behavior and Human Decision Processes, 78,* 128–145.

Harrison, Y., Horne, J. A., & Rothwell, A. (2000). Prefrontal neuropsychological effects of sleep deprivation in young adults—a model for healthy aging? *Sleep, 23,* 1067–1073.

Hasan, J., Hirkoven, K., Varri, A., Hakkinen, V., & Loula, P. (1993). Validation of computer analysed polygraphic patterns during drowsiness and sleep onset. *Electroencephalography and Clinical Neurophysiology, 87,* 117–127.

Inouye, T., Shinosaki, K., Iyama, A., Matsumoto, Y., Toi, S., & Ishihara, T. (1994). Potential flow of frontal midline theta activity during a mental task in the human electroencephalogram. *Neuroscience Letters, 169,* 145–148.

Ishii, R., Shinosaki, K., Ukai, S., Inouye, T., Ishihara, T., Yoshimine, T., et al. (1999). Medial prefrontal cortex generates frontal midline theta rhythm. *Neuroreport, 10,* 675–679.

Jansma, J. M., Ramsey, N. F., Coppola, R., & Kahn, R. S. (2000). Specific versus nonspecific brain activity in a parametric n-back task. *NeuroImage, 12,* 688–697.

Jasper, H. H., & Penfield, W. (1949). Electrocorticograms in man: Effect of the voluntary movement upon the electrical activity of the precentral gyrus. *Archives of Psychiatry, Z. Neurology, 183,* 163–174.

Joseph, R. D. (1961). *Contributions to perceptron theory.* Ithaca, NY: Cornell University Press.

Jung, T. P., Makeig, S., Humphries, C., Lee, T. W., McKeown, M. J., Iragui, V., et al. (2000). Removing electroencephalographic artifacts by blind source separation. *Psychophysiology, 37,* 163–178.

Krull, K. R., Smith, L. T., Kalbfleisch, L. D., & Parsons, O. A. (1992). The influence of alcohol and sleep deprivation on stimulus evaluation. *Alcohol, 9,* 445–450.

Kyllonen, P. C., & Christal, R. E. (1990). Reasoning ability is little more than working memory capacity?! *Intelligence, 14,* 389–433.

Linde, L., & Bergstrom, M. (1992). The effect of one night without sleep on problem-solving and immediate recall. *Psychological Research, 54,* 127–136.

McCallum, W. C., Cooper, R., & Pocock, P. V. (1988). Brain slow potential and ERP changes associated with operator load in a visual tracking task. *Electroencephalography and Clinical Neurophysiology, 69,* 453–468.

McEvoy, L. K., Smith, M. E., & Gevins, A. (2000). Test-retest reliability of cognitive EEG. *Clinical Neurophysiology, 111*(3), 457–463.

Miller, J. C. (1996, April). Fit for duty? *Ergonomics in Design,* 11–17.

Miyata, Y., Tanaka, Y., & Hono, T. (1990). Long term observation on Fm-theta during mental effort. *Neuroscience, 16,* 145–148.

Mizuki, Y., Tanaka, M., Iogaki, H., Nishijima, H., & Inanaga, K. (1980). Periodic appearances of theta rhythm in the frontal midline area during performance of a mental task. *Electroencephalography and Clinical Neurophysiology, 49,* 345–351.

Molloy, R., & Parasuraman, R. (1996). Monitoring an automated system for a single failure: Vigilance and task complexity effects. *Human Factors, 38,* 311–322.

Moosmann, M., Ritter, P., Krastel, I., Brink, A., Thees, S., Blankenburg, F., et al. (2003). Correlates of alpha rhythm in functional magnetic resonance imaging and near infrared spectroscopy. *Neuroimage, 20*(1), 145–158.

Mulholland, T. (1995). Human EEG, behavioral stillness and biofeedback. *International Journal of Psychology, 19,* 263–279.

Parasuraman, R., Molloy, R., & Singh, I. L. (1993). Performance consequences of automation-induced "complacency." *International Journal of Aviation Psychology, 3*(1), 1–23.

Parasuraman, R., Mouloua, M., & Molloy, R. (1996). Effects of adaptive task allocation on monitoring of automated systems. *Human Factors, 38,* 665–679.

Paus, T., Koski, L., Caramanos, Z., & Westbury, C. (1998). Regional differences in the effects of task difficulty and motor output on blood flow response in the human anterior cingulate cortex: A review of 107 PET activation studies. *Neuroreport, 9,* R37–R47.

Pellouchoud, E., Smith, M. E., McEvoy, L., & Gevins, A. (1999). Mental effort-related EEG modulation during video-game play: Comparison between juvenile subjects with epilepsy and normal control subjects. *Epilepsia, 40*(Suppl. 4), 38–43.

Pfurtscheller, G., & Klimesch, W. (1992). Functional topography during a visuoverbal judgment task studied with event-related desynchronization mapping. *Journal of Clinical Neurophysiology, 9,* 120–131.

Posner, M. I., & Rothbart, M. K. (1992). Attentional mechanisms and conscious experience. In A. D. Milner & M. D. Rugg (Eds.), *The neuropsychology of consciousness* (pp. 91–111). San Diego: Academic Press.

Price, W. J., & Holley, D. C. (1990). Shiftwork and safety in aviation. *Occupational Medicine, 5,* 343–377.

Rosekind, M. R., Gander, P. H., & Miller, D. L. (1994). Fatigue in operational settings: Examples from aviation environment. *Human Factors, 36,* 327–338.

Scott, A. J. (1994). Chronobiological considerations in shiftworker sleep and performance and shiftwork scheduling. *Human Performance, 7,* 207–233.

Sheer, D. E. (1989). Sensory and cognitive 40 Hz event-related potentials. In E. Basar & T. H. Bullock (Eds.), *Brain dynamics 2* (pp. 339–374). Berlin: Springer.

Smith, M. E., & Gevins, A. (2005). *Neurophysiologic monitoring of cognitive brain function for tracking mental workload and fatigue during operation of a PC-based flight simulator.* Paper presented at SPIE International Symposium on Defense and Security: Symposium on Biomonitoring for Physiological and Cognitive Performance During Military Operations, Orlando, FL.

Smith, M. E., Gevins, A., Brown, H., Karnik, A., & Du, R. (2001). Monitoring task load with multivariate EEG measures during complex forms of human computer interaction. *Human Factors, 43*(3), 366–380.

Smith, M. E., McEvoy, L. K., & Gevins, A. (1999). Neurophysiological indices of strategy development and skill acquisition. *Cognitive Brain Research, 7,* 389–404.

Smith, M. E., McEvoy, L. K., & Gevins, A. (2002). The impact of moderate sleep loss on neurophysiologic signals during working memory task performance. *Sleep, 25,* 784–794.

Tepas, D. I. (1994). Technological innovation and the management of alertness and fatigue in the workplace. *Human Performance, 7,* 165–180.

Van den Berg-Lensssen, M. M., Brunia, C. H., & Blom, J. A. (1989). Correction of ocular artifacts in EEGs using an autoregressive model to describe the EEG; a pilot study. *Electroencephalography and Clinical Neurophysiology, 73,* 72–83.

Viglione, S. S. (1970). Applications of pattern recognition technology. In J. M. Mendel & K. S. Fu (Eds.), *Adaptive learning and pattern recognition systems* (pp. 115–161). New York: Academic Press.

Walter, H., Vetter, S. C., Grothe, J., Wunderlich, A. P., Hahn, S., & Spitzer, M. (2001). The neural correlates of driving. *Neuroreport, 12*(8), 1763–1767.

Wickens, C. D. (1991). Processing resources and attention. In D. L. Damos (Ed.), *Multiple-task performance* (pp. 1–34). London: Taylor and Francis.

Williamson, A. M., & Feyer, A. M. (2000). Moderate sleep deprivation produces impairments in cognitive and motor performance equivalent to legally prescribed levels of alcohol intoxication. *Occupational and Environmental Medicine, 57,* 649–655.

Wilson, G. F., & Fisher, F. (1995). Cognitive task classification based upon topographic EEG data. *Biological Psychology, 40,* 239–250.

Yamamoto, S., & Matsuoka, S. (1990). Topographic EEG study of visual display terminal VDT performance with special reference to frontal midline theta waves. *Brain Topography, 2,* 257–267.

Shimin Fu and Raja Parasuraman

Event-Related Potentials (ERPs) in Neuroergonomics

Event-related potentials (ERPs) represent the brain's neural response to specific sensory, motor, and cognitive events. ERPs are computed by recording the electroencephalogram (EEG) from the scalp of a human participant and by averaging EEG epochs time-locked to a particular event. The use of ERPs to examine various aspects of human cognitive processes has a long history. Pioneering work on ERP correlates of cognitive processes such as attention (Hillyard, Hink, Schwent, & Picton, 1973), working memory (Donchin, 1981), and language (Kutas & Hillyard, 1984) were carried out in the 1970s and 1980s. These studies were important because they established the use of ERPs as a tool for *mental chronometry* (Posner, 1978), or the examination of the timing of the neural events associated with different components of information processing. However, these landmark studies did not greatly influence theory or empirical research in cognitive psychology in the era in which they were carried out. Moreover, because of their poor spatial resolution in localizing sources of neuronal activity underlying scalp electrical potentials, ERPs were not well regarded by neuroscientists accustomed to the spatial precision of single-cell recording in animals. The mid-1980s were a period when the cognitive neuroscience revolution was in its early phases (Gazzaniga, 1995). Consequently, ERP research did not enjoy much currency in the mainstream of either cognitive psychology or neuroscience.

The situation changed a few years later. The development of other neuroimaging techniques such as positron emission tomography (PET) and functional magnetic resonance imaging (fMRI) led to their growing use to examine the neural basis of human cognitive processes, beginning with the seminal work of Posner, Petersen, Fox, and Raichle (1988). Neuroimaging allowed for the rediscovery of ERPs in cognitive psychology and cognitive neuroscience. As a result, ERPs made a comeback in relation to both psychology and neuroscience and today enjoy an acknowledged status in both fields. The importance of ERPs as a tool in cognitive neuroscience was further enhanced with the realization that PET, fMRI, and related neuroimaging techniques had serious limitations in their temporal resolution of assessing neural processing, despite their great advantage over ERPs with respect to spatial localization of neuronal activity.

At the present time, therefore, ERPs hold a unique position in the toolshed of cognitive neuroscientists. Because of the inherent sluggishness (several seconds) of neuroimaging techniques (PET and fMRI) based on cerebral hemodynamics compared to the time scale of neuronal activity (milliseconds), ERPs are being increasingly used whenever there is a need to examine the relative timing of neural mechanisms underlying cognitive processes. In a number of cases, the timing information provided by ERPs has been critical in resolving a major theoretical issue in cognitive science. Probably the best-known example involves the use of ERPs to address the early versus late selection debate in selective attention research (Hillyard et al., 1973; Luck & Girelli, 1998; Mangun, 1995).

ERP research has also been applied to practical issues in a number of domains, most notably in neurology and psychiatry. In general, most applications of ERPs have involved the investigation of abnormal behavior, as in ERP studies of neuropsychiatric conditions such as schizophrenia and Alzheimer's disease. ERP studies of normal populations have also been carried out, but primarily in relation to issues of child development and normal aging. In contrast, applications (as opposed to basic research) of ERPs to problems of human performance in normal everyday situations have been relatively infrequent. Nevertheless, there is a small but noteworthy literature on ERP applications to problems of human performance in work situations—the province of human factors or ergonomics.

Applications of ERP to issues relevant to human factors and ergonomics involve a number of topics. These include the assessment of mental workload, evaluation of mechanisms of vigilance decrement, and monitoring of operator fatigue in human-machine systems. Other areas include the use of ERPs to assess the influence of stressors, automation, and online adaptive aiding on human operator performance. These studies have been reviewed elsewhere (see Byrne & Parasuraman, 1996; Kramer & Belopolsky, 2003; Kramer & Weber, 2002; Parasuraman, 1990; Wickens, 1990) and so are not revisited here. Rather, the purpose of this chapter is to provide a methodological overview of the use of ERPs and their potential application to the examination of issues in human factors and ergonomics.

In this chapter, we describe how ERPs can be recorded and analyzed, and discuss research on a number of ERP components, focusing on those that are particularly relevant to human factors issues: the early-latency, attention-related P1 and N1; the long-latency P3 or P300; the mismatch negativity (MMN); the lateralized readiness potential (LRP); and the error-related negativity (ERN). We discuss the use of these ERP components to address four neuroergonomic issues: (1) assessment of mental workload; (2) understanding the neural basis of error detection and performance monitoring; (3) response readiness; and (4) studies of automatic processing. We begin, however, by outlining the specific advantages offered by ERPs in neuroergonomics research.

ERPs in Relation to Other Neuroimaging Techniques

A core feature of neuroergonomics is an interest in brain mechanisms in relation to human performance at work (Parasuraman, 2003). To this end, researchers may make use of physiological measures that reflect, more or less directly, aspects of brain function. Alternatively, neuroergonomic research may not directly involve such measures but rely on the results of studies of brain function conducted by others to guide hypotheses regarding the design of human-machine interfaces or the training of operators of such systems. When physiological measures are used, the most direct are those derived from the brain itself: the EEG, which represents the summated activity of dendritic (postsynaptic) populations of neuronal cells as recorded on the scalp; magnetoencephalography (MEG), which consists of the associated magnetic flux that is recorded at the surface of the head; and ERPs and event-related magnetic fields, which constitute the brain's specific response to sensory, cognitive, and motor events. In addition to these electromagnetic measurements, measures of the brain's metabolic and vascular response (PET and fMRI), which can be linked to neuronal activity, also provide a noninvasive window on human brain function.

Currently, electromagnetic measures such as ERPs provide the best temporal resolution (1 ms or better) for evaluating neural activity in the human brain, and metabolic measures such as fMRI have the best spatial resolution (1 cm or better). No single technique combines both high temporal and spatial resolution. Furthermore, some of these techniques (e.g., fMRI) are expensive and restrict participant movement, which makes them difficult to use for neuroergonomic studies. However, new imaging technologies are being developed, such as functional near-infrared spectroscopy (fNIRS) and other forms of optical imaging, that promise to provide high temporal and spatial resolution (Gratton & Fabiani, 2001). These techniques have the additional advantage of being more portable and less expensive than fMRI, which will therefore add them to the catalog of available methods that are appropriate for neuroergonomic research. (For a review of brain imaging techniques and their application to psychology, see Cabeza & Kingstone, 2001; see also chapters 4 and 5, this volume).

Fundamentals of ERPs

ERPs are recorded using sensors (tin or silver electrodes) placed on the scalp of human participants and by extracting and signal averaging samples of the EEG. The ERP represents the average of the EEG samples that are time-locked to a particular event. The time-locking event can be an external stimulus (e.g., sounds, words, faces, etc.), the behavioral response elicited by the participant (e.g., a button press, speech, or other motor movement), or an internal cognitive operation such as (silent) identification or making a decision choice. The signal averaging can be done either forward or backward in time with respect to the time-locking event; typically backward averaging is done for response-related ERPs. For time-locking to an internal, unobservable cognitive event, some external temporal marker indicating when that event is likely to have occurred is necessary in order to obtain an ERP.

The signal averaging technique assumes that any EEG activity that is not time-locked to the event sums to zero if sufficient numbers of samples are taken, and that the resulting ERP waveform reflects the brain's specific response to the eliciting event. This assumption has been questioned over the years, as there is evidence that EEG components (such as alpha desynchronization) do not cancel out with repeated averaging. There is also evidence that such event-related EEG responses may carry important information regarding perceptual or cognitive processing (Makeig et al., 2002). However, a review of this research is beyond the scope of this chapter, in which we focus on the use of specific ERP components in relation to neuroergonomic issues.

The amplitude of the ERP is usually on the order of about several microvolts. This is smaller than that of the background EEG that is recorded during resting conditions, which tends to be in the range of tens of microvolts. Because the ERP is time-locked to events that are embedded into the noisy background of EEG, signal averaging over several samples is necessary. Depending on the ERP component of interest, typically tens or hundreds of samples need to be included in the average.

Signal averaging of the EEG must be accompanied by recording of eye and body movements, because the electrical signals from these sources can contaminate the EEG. For a better estimation of the true neural responses to the time-locked stimulus, participant motor action, or cognitive event, EEG epochs with eye blinks or body movements need to be rejected before averaging. The difficulties, particularly in field settings, of obtaining artifact-free samples of EEG are frequently underestimated. Consequently, reliable correction of EEG for artifacts is a prerequisite for use of ERPs in neuroergonomics. The magnitude of the electrooculogram (EOG) is typically used for artifact correction. The EOG is recorded from electrodes about 1 cm outside the canthi (near the eyes) and from electrodes above and below the left or right eye to monitor horizontal eye movements and blinks. If muscle movements occur, these can generally be seen as large, high-frequency activity and can be relatively easily recognized in the raw EEG waveform. An alternative to rejection of a particular EEG epoch that is contaminated with eye movement is to use a subtraction technique prior to signal averaging. In this method, a model-estimated contribution of the EOG to the EEG epoch is subtracted from the EEG sample, and that sample is retained in the average (Gratton, Coles, & Donchin, 1983). However, this method requires an extra assumption of the reliability of the EOG correction model, and so a more conservative method is to simply reject all trials with excessive EOG and to increase the overall number of samples to improve the signal-to-noise ratio (SNR) of the ERP.

Obtaining an ERP waveform with a good SNR is a key methodological issue in ERP research. The best recordings occur in an electrically shielded and soundproofed room. However, this is not essential, and ERPs have been recorded in nonshielded environments and even in the field. An important requirement is to maintain the impedance of the

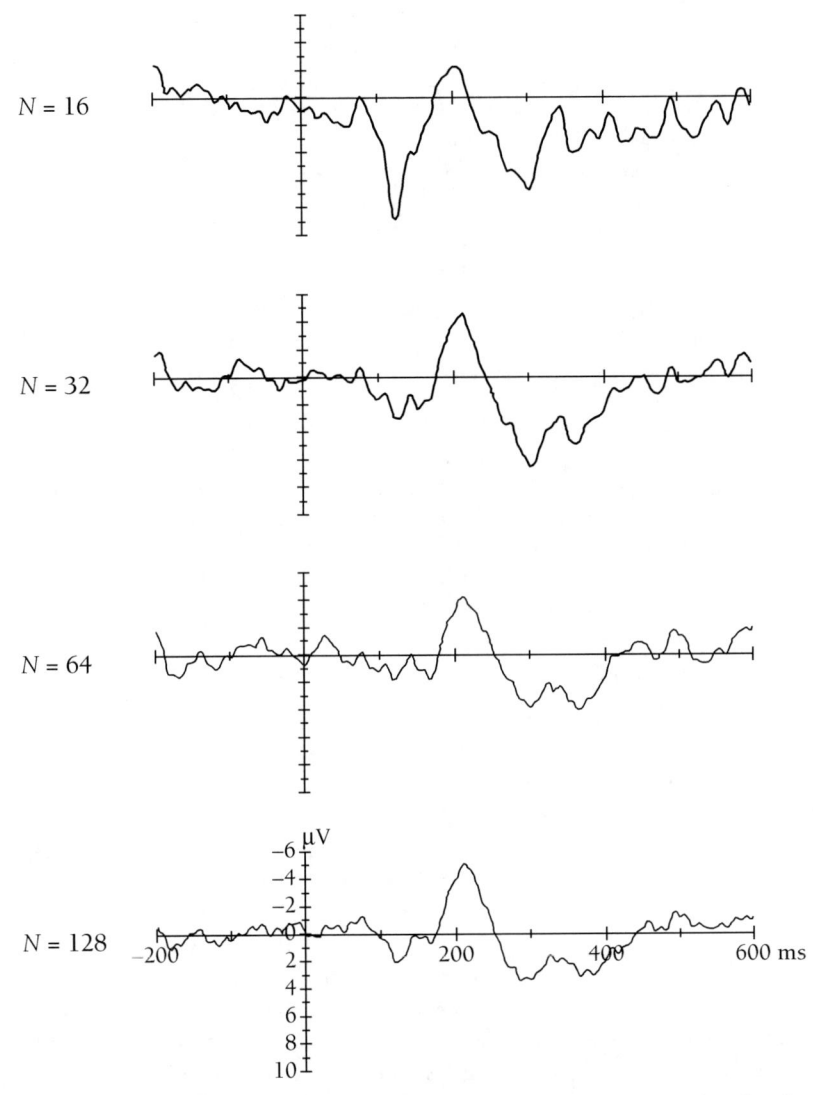

Figure 3.1. Illustration of the relationship between the signal-to-noise ratio (SNR) and the number of trials used for averaging an ERP. The SNR is progressively improved, or the ERP waveforms are made less noisy, by increasing the number of trials for averaging from 16 to 32, 64, and 128. The single-trial data were averaged across EEGs elicited by a visual stimulus (four dots with each on the corner of a virtual rectangle) presented in the left visual field. The recording electrode, the right occipital site, overlies the occipitotemporal area contralateral to the stimulus visual field. Data from one participant from Fu, Caggiano, Greenwood, and Parasuraman (2005).

scalp electrodes below 5,000 ohms and to use low-noise, high-impedance amplifiers (Picton et al., 2000). In theory, the SNR of an ERP is directly proportional to the square root of the number N of trials in the average, or $\sqrt{N}$. What this function indicates is that doubling the SNR requires that N should be quadrupled. For example, obtaining twice the SNR for an ERP with 16 trials would require 64 trials. Figure 3.1 shows how the SNR of an averaged ERP waveform can be improved by increasing the number of trials N.

In practice, stable ERPs with adequate SNR can be obtained without an excessively large number of trials, assuming that artifact-free trials are averaged, but the precise number of trials depends on the ERP component of interest. Roughly speaking,

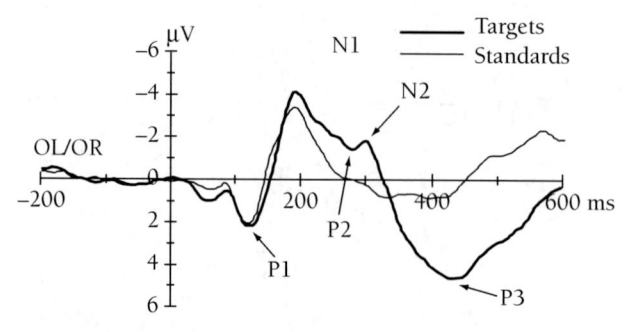

Figure 3.2. Illustration of the visual ERP components P1, N1, P2, N2, and P3 in response to the standard (thin line) and target (thick line) stimuli. Stimuli were arrows pointing to the northeast or northwest that were flashed randomly to the left and right visual field. Participants were asked to respond to one type of arrow (targets, 10%) on both sides and to make no response to the other (standards, 90%). The size of stimuli could be large or small. The cues, which were presented 100 to 300 ms before the stimuli, could also be large or small. Data were averaged across 14 participants and across stimulus size and cue validity. OR = right occipital site; OL = left occipital site.

10–30 trials might be sufficient for an average of a large component like P3 in a laboratory recording situation, whereas 100 or even more trials might be needed for an average of relatively early and small components such as visual P1 or C1 (see below for definitions of these ERP components). However, it may not be practical to administer several hundred trials to a participant, because of the probability of participant fatigue and the resultant increased chance of muscle artifacts. Thus, a practical issue in designing an ERP study is to find a good balance between obtaining good SNR for the ERP components of interest and keeping participants alert and fatigue free.

Once a stable, artifact-free ERP waveform with good SNR has been obtained, the peak amplitudes and latencies of specific ERP components—both positive and negative—can be measured. Typically, peak amplitude is measured relative to the averaged amplitude of a pre-event baseline (generally, 100 or 200 ms before the onset of the event). Alternatively, the peak-to-peak amplitude difference between two consecutive positive and negative components, or the mean amplitude of a component, are sometimes reported. Figure 3.2 illustrates the ERPs elicited by a visual nontarget stimulus (thin line) and a target (thick line) stimulus in a selective attention paradigm. The ERP components of P1, N1, P2, N2, and P3 are shown.

The convention for naming ERP components has generally been based on the polarity (positive or negative) and the sequence or order of a specific component in the ERP waveform. For example, P1 is the first visual cortical response with a positive polarity and N1 is the first negative potential. Unfortunately, there are exceptions to this convention, which can occasionally cause some confusion. For example, in visual ERPs, there is a component that precedes P1, known as C1 (for component 1, or the first visual ERP component). This component is generated in the striate cortex, prior to P1 (thought to be generated in extrastriate cortex), and because the polarity of C1 can be either positive or negative, depending upon the retinal location that is stimulated (Clark, Fan, & Hillyard, 1995; Clark & Hillyard, 1996). Another convention used to name ERP components is based on polarity and peak latency. For example, P70 represents a positive electrical response with a peak latency of 70 ms after stimulus onset, and P115 represents another positive component with a peak latency of 115 ms. This convention has the advantage of clarifying that the same component can have different latencies, because P70 and P115 might both characterize the P1 component. However, the convention is also potentially misleading, because P180 could represent either a P1 component for a slowly developing visual process (e.g., object shape that is extracted from visual motion) or a faster P2 component elicited by a flashing visual figure. Finally, some ERP components are derived by computing the difference in ERP waveforms between two conditions. In this case, a descriptive terminology is typically used. For example, the auditory MMN is the difference wave obtained by subtracting ERPs of standard (frequent) stimuli from ERPs of deviant (occasional) stimuli (Naatanen, 1990). Other ERP components using this descriptive nomenclature are the LRP and the ERN.

ERPs can provide important information about the time course of mental processes, the brain regions that mediate these processes, and even communication between different brain areas. ERP patterns can be observed at different levels. At the single-electrode level, the amplitude, latency, and duration of different ERP components may vary across conditions. These differences may be compared across electrodes over different brain areas to get a clue as to the brain areas involved. At the two-dimensional scalp level, the scalp ERP voltage maps may vary in a specific time range when a certain component of mental processing occurs. At a three-dimensional level, the location, strength, and orientation of the intracortical neural generator(s) corresponding to a specific mental processing stage may vary across conditions.

ERPs are powerful for tracking the time course of mental processes, but not accurate in inferring the anatomy of the underlying neural activity, as compared with brain imaging techniques such as PET and fMRI. It can be misleading to infer from the observation of ERP activation at a particular scalp electrode site that the neural generator of that activity is located directly underneath the electrode. The reason is that neural activity generated in the brain is volume conducted and can be recorded at quite a distance from the source. For example, the so-called brain stem ERPs, which refer to a series of very short latency (<10 ms) ERP components following high-frequency auditory stimulation, and which are known from animal and lesion studies to be generated in the brain stem, are best recorded at a very distant scalp site, Cz, which is located at the top of the head.

Source localization is a better solution to determining the functional anatomy of neural activity as reflected in the scalp ERP. A number of source localization methods have been proposed. Most of these methods try to solve the inverse problem—how to estimate the source of a volume-conducted signal from the distant pattern. One of the best known is called brain electrical source analysis (BESA; Scherg, 1990). In this method, successive forward solutions are projected in an attempt to solve the inverse problem: An initial location and orientation for a dipole (neural generator) are assumed, then the projected scalp pattern of that dipole is compared to the actual ERP pattern, and the error is used to correct the location and/or orientation of the dipole, with the process being repeated until the error is minimized. Unfortunately, this technique faces the difficulty that no unique solution can be obtained while inferring a source from surface recordings. This is because any specific scalp voltage distribution can be caused by numerous configurations derived from different combinations of the number, orientation, and strength of neural generators in the brain. On the other hand, the distributed-solution-based LORETA algorithm (see figure 3.3) reveals the authentic but "blurred" three-dimensional point source with certain dispersions at the activation location (Pascual-Marqui, 1999), and thus, no accurate anatomic localization can be obtained.

The temporal information that ERPs provide concerning the neural activity associated with cognitive processes is sufficient for many human factors applications. However, as discussed previously, neuroergonomic research can also benefit from studies in which neural activity can be localized to specific brain locations, given that there may be other knowledge about the function of those brain regions to guide hypotheses relevant to issues in human factors problems. Hence, if ERPs can be combined with other imaging techniques with high spatial resolution, both temporal and localizing information might be gained. Accordingly, several studies have combined ERP with PET or fMRI techniques that provide accurate anatomical information (Heinze et al., 1994; Mangun, Hopfinger, Kussmaul, Fletcher, & Heinze, 1997; Mangun et al., 2001). The localization information provided by these imaging techniques can be used as seeds or start points for initial dipole placement, instead of an arbitrarily chosen location. This constrained or seeded solution of ERP dipole modeling allows one to test the difference between the derived data from the seeded generator and the observed scalp voltage data with high

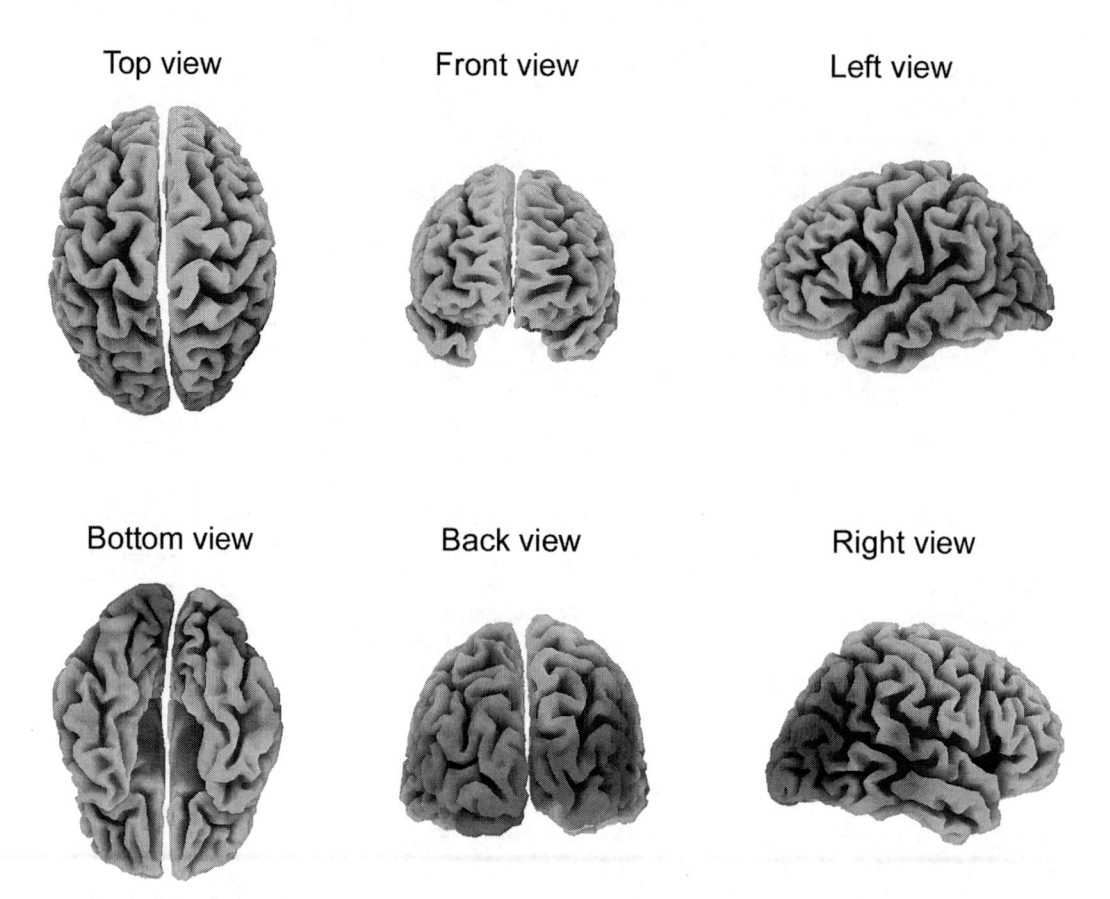

Figure 3.3. Illustration of results obtained from LORETA (Pascual-Marqui, 1999), a distributed localization method. Data were for a left visual field target array under valid and invalid cue conditions, averaged across large and small cue conditions. The attentional effects on the P1 and N1 range were obtained by subtracting ERPs of invalid trials from ERPs of valid trials. The brain activations in response to these attentional effects are displayed at 154 ms, when the global field power is maximal. It is clear that these attentional effects are distributed on the posterior brain regions contralateral to the stimulus visual field. That is, for the left visual field (LVF) stimulus, attentional effects are more pronounced in right posterior brain regions. See also color insert.

temporal resolution. However, most studies using this combined approach have recorded neuroimaging and ERP data at separate times, although on the same group of participants and the same task. Therefore, an important issue for future research is to be able to simultaneously combine high-temporal-resolution ERP recording with high spatial-resolution imaging such as fMRI in the same subjects at the same time.

There are a number of other methodological and experimental design issues related to the use of ERPs to examine the neural basis of cognitive processes. For space reasons, we have discussed only the most basic issues. For recent volumes that cover all the major methodological issues in ERP research, see Handy (2005) and Luck (2005).

ERPs and Neuroergonomics

ERPs can provide neural measures of the operator's mental state with millisecond temporal resolution and a certain degree of spatial resolution. Such information can be gleaned even when the operator does not make an overt behavioral response. This can be important under conditions when the operator plays a supervisory role over automated systems. This type of information obtained from ERP recordings is therefore specific and not available from other sources, such as expert judgment, biomechanical measures, or motor responses. Of course, an important issue in applying ERPs to human factors issues is that the ERP assessment must be reliable, efficient, and valid (Kramer & Belopolsky,

2003). In the following, we discuss a number of different human factors issues where ERP studies have been conducted.

Assessment of Mental Workload

By far the largest number of ERP studies in human factors have addressed the issue of the assessment of mental workload. The topic continues to be of importance in human factors research because the design of an efficient human-machine system is not just a matter of assessing performance, but also of evaluating how well operators can meet the workload demands imposed on them by the system. A major question that must be addressed is whether human operators are overloaded and can meet additional unexpected demands (Moray, 1979; Wickens, 2002). Behavioral measures such as accuracy and speed of response to probe events have been widely used to assess mental workload, but measures of brain function as revealed by ERPs offer some unique advantages that can be exploited in particular applications (Kramer & Weber, 2000; Parasuraman, 2003). Such measures can also be linked to emerging cognitive neuroscience knowledge on attention (Parasuraman & Caggiano, 2002; Posner, 2004), thereby allowing for the development of neuroergonomic theories of mental workload.

Of the ERP studies of mental workload, a large number have examined the P3 or P300 component, which was discovered by Sutton, Braren, Zubin, and John (1965). P300 is characterized by a slow positive wave with a mean latency following stimulus onset of about 300 ms, depending on stimulus complexity and other factors (Polich, 1987). The P300 typically has a centroparietal scalp distribution, with a maximum over parietal cortex. P300 amplitude is relatively large and therefore easily measured, sometimes even on single trials. The amplitude of the P300 is very sensitive to the probability of presentation of the eliciting stimulus. In the typical oddball paradigm used to elicit the P300, two stimuli drawn from different perceptual or conceptual categories (e.g., red vs. green colors, low vs. high tones, or nouns vs. verbs) are presented in a random sequence. The stimulus that is presented with a low probability (e.g., 20%) elicits a larger P300 than the high-probability (e.g., 80%) stimulus. Since P300 amplitude is sensitive to the probability of a task-defined category, it is thought to reflect postperceptual or postcategorical processes. In support of this view, increasing the difficulty of identifying the target stimulus (e.g, by masking) increases the latency of the P300 wave (Kutas, McCarthy, & Donchin, 1977), whereas increases in the difficulty of response selection do not affect P300 latency (Magliero, Bashore, Coles, & Donchin, 1984). This has led to the view that the latency of the P300 provides a relatively pure measure of perceptual processing and categorization time, independent of response selection and execution stages (Kutas et al., 1977; McCarthy & Donchin, 1981). Consistent with this view, P300 is sensitive to uncertainty both in detecting and identifying a masked target in noise (Parasuraman & Beatty, 1980).

It has also been proposed that the amplitude of P300 is proportional to the amount of attentional resources allocated to the task (Johnson, 1986). Thus any diversion of processing resources away from target discrimination in a dual-task situation will lead to a reduction in P300 amplitude. For example, Wickens, Isreal, and Donchin (1977) showed that the amplitude of P300 decreased when a primary task, tone counting, was combined with a secondary task, visual tracking. However, increases in the difficulty of the tracking task (by increasing the bandwidth of the forcing function) did not lead to a further reduction in P300 amplitude (Isreal, Chesney, Wickens, & Donchin, 1980). This pattern of findings was taken to support the view that the P300 reflects processing resources associated with perceptual processing and stimulus categorization, but not response-related processes, which were manipulated in the tracking task. In another study in which the tone-counting task was paired with a visual monitoring task whose perceptual difficulty was manipulated (monitoring four vs. eight aircraft in a simulated air traffic control task), P300 amplitude was reduced with increases in difficulty.

These and other related studies (Kramer, Wickens, & Donchin, 1983, 1985) have clearly established that P300 amplitude provides a reliable index of resource allocation related to perceptual and cognitive processing. Similar results have been obtained when P300 has been used in conjunction with performance in high-fidelity fixed-wing (Kramer, Sirevaag, & Braune, 1987) and rotary-wing aircraft (Sirevaag et al., 1993). P300 therefore provides a reliable and valid index of mental workload, to the extent that perceptual and cognitive aspects of information processing are major contributors to workload in a given performance setting. Response-related contributions to workload, however, are not reflected in P300 amplitude.

Two recent examples of human factors studies that exploited these characteristics of P300 to assess cognitive workload are briefly described here. Schultheis and Jameson (2004) conducted a study of the difficulty of text presented in hypermedia systems, with a view to investigating whether the text difficulty could be adaptively varied dependent on the cognitive load imposed on the user. They paired an auditory oddball task with easy and difficult versions of text and measured pupil diameter and the P300 to the oddball task. They found that P300 amplitude, but not pupil diameter, was significantly reduced for the difficult hypermedia condition. The authors concluded that P300 amplitude and other measures, such as reading speed, may be combined to evaluate the relative ease of use of different hypermedia systems.

Baldwin, Freeman, and Coyne (2004) used a visual color-discrimination oddball task to elicit P300 during simulated driving under conditions of normal and reduced visibility (fog). P300 amplitude, but not discrimination accuracy or reaction time (RT), was reduced when participants drove in fog compared to driving with normal visibility. In contrast, P300 was not sensitive to changes in traffic density, while the behavioral measures were sensitive to this manipulation of driving difficulty. The authors concluded that neither neural nor behavioral measures alone are sufficient for assessing cognitive workload during different driving scenarios, but that multiple measures may be needed. They speculated that such combinatorial algorithms could be used to trigger adaptive automation during high workload by engaging driver-aiding systems and disengaging secondary systems such as cellular phones or entertainment devices.

Resource allocation theories propose that the allocation of processing resources to one task in a dual-task pairing leads to a withdrawal of resources to the second task (Navon & Gopher, 1979; Wickens, 1984). The P300 studies showed that if the second task was an oddball task, then P300 amplitude was reduced. At the same time, resource allocation theories would predict an increase in P300 amplitude with the allocation of resources to the primary task. This was demonstrated by Wickens, Kramer, Vanesse, and Donchin (1983), who showed that while P300 to a secondary task decreased with increase in task difficulty, P300 to an event embedded within the primary task increased. Thus there is a reciprocal relation between P300 amplitude and resource allocation between primary and secondary tasks in a dual-task situation. Further evidence for reciprocity of primary and secondary task resources was provided in a subsequent study by Sirevaag, Kramer, Coles, and Donchin (1989). This pattern of P300 reciprocity is consistent with resource trade-offs predicted by the multiple resource theory of Wickens (1984).

A unique feature of the P300 is that it is simultaneously sensitive to the allocation of attentional resources (P300 amplitude) and to the timing of stimulus identification and categorization processes (P300 latency). These features allow not only for assessing the workload associated with dual- or multitask performance, but also allow for identifying the sources that contribute to workload and dual-task interference. These features were elegantly exploited in a dual-task study by Luck (1998) using the psychological refractory period (PRP) task (Pashler, 1994). In the PRP paradigm, two targets are presented in succession and the participant must respond to both. The typical finding is that when the interval between the two targets is short, RT to the second target is substantially delayed. Luck (1998) identified the source of this interference by recording the P300 to the second target stimulus, which was either the letter X or the letter O, with one of the letters being presented only 25% of the time (the oddball). Luck (1998) found that whereas RT was significantly increased when the interval between the two stimuli was short, P300 latency was only slightly increased. This suggested that the primary source of dual-task interference was the response selection stage, which affected RT but not P300 latency.

Despite the large body of evidence confirming the sensitivity of P300 to perceptual and cognitive workload, and the smaller body of work on the sensitivity of N1 (or Nd), the issue of whether P300 can be used to track dynamic variation in workload, in real time, or in near-real time, has not been fully resolved. On the one hand, P300 is a relatively large ERP component, so that its measurement on single (or a few) trials is easier than for other ERP components. However, the question remains whether P300 amplitude computed on the basis of just a few trials can reliably discriminate between different levels of cognitive workload. Humphrey and Kramer (1994) provided important information relevant to this issue. They had participants perform two complex, continuous tasks, gauge-monitoring and mental arithmetic, and recorded ERPs to discrete events from each task. The difficulty of each was manipulated to create low- and high-workload conditions. The amplitude of the ERP (averaged over many samples) was sensitive to increased processing demands in each task. Humphrey and Kramer then used a stepwise discriminant analysis to ascertain how the number of ERP samples underlying P300 affected accuracy of classification of the low- and high-workload conditions. They found that classification accuracy increased monotonically with the number of ERP samples, but that high accuracy (approximately 90%) could be achieved with relatively few samples (5–10). Such a small number of samples can be collected relatively quickly, say in about 25–50 seconds, assuming a 20% oddball target frequency and a 1-second interstimulus interval. These results are encouraging with respect to the use of single-trial P300 for dynamic assessment of cognitive workload.

The studies conducted to date indicate that the sensitivity of ERPs to temporal aspects of neural activation has been put to good use in dissecting sources of dual-task interference and in assessing mental workload. Furthermore, as illustrated by the research of Humphrey and Kramer (1994), there has also been some success in using ERP components as near real-time measures of mental workload. Furthermore, flight and driving simulation studies have shown the added value that ERPs can provide in assessment of workload in complex tasks. However, additional work is needed with single-trial ERPs to further validate the use of ERPs for real-time workload assessment.

Further progress in this area may come sooner rather than later because of developments in the related field of brain-computer interface (BCI). This refers to the use of brain signals to control external devices without the need for motor output. Such brain-based control would be advantageous for individuals who either have only limited or no motor control, such as "locked-in" patients with amyotrophic lateral sclerosis. The idea of BCIs follows naturally from the work on biocybernetics in the 1980s pioneered by Donchin and others (Donchin, 1980; Gomer, 1981). The goal of biocybernetics was to use EEG and ERPs as an additional communication channel in human-machine interaction. Similarly, BCI researchers hope to provide those with limited communication abilities additional means of communicating and interacting with the world. BCIs require that brain signals for control be analyzed in real time within a short period of time, so as to give the patient adequate control of an external device. Hence, many BCI researchers are examining various EEG and ERP components that can be detected in single or very few trials. These include EEG mu rhythms (Pfurtscheller & Neuper, 2001), P300 (Donchin, Spence, & Wijesinghe, 2000), steady-state evoked potentials (McFarland, Sarnacki, & Wolpaw, 2003), and contingent negative variation and other slow potentials (Birnbaumer et al., 1999). The outcome of this program of research and development is likely to have a great influence on the development of real-time ERP and EEG-based systems for aiding nondisabled persons (see also chapter 20, this volume).

Attentional Resources and the P1 and N1 Components

Although most ERP research on mental workload has focused on the P300 component, other ERP components have also been examined in a few studies. Attentional resource allocation can also be manifested in the amplitude of ERP components earlier than P300, such as N1. N1 is an early component of ERPs with a peak latency of about 100 ms in audition and about 160 ms in vision (Naatanen, 1992). It has been proposed that the N1 component provides an index of resource allocation under high information load conditions (Hink, Van Voorhis, Hillyard, & Smith, 1977; Parasuraman, 1978, 1985).

Parasuraman (1985) had participants perform a visual and an auditory discrimination task concurrently, and systematically varied the priority to be placed on one task relative to the other, from 0% to 100%. The effect of information load on attention allocation was also investigated by manipulating the presentation rates (slow vs. fast). Performance operating characteristics showed that the priority instructions were successful, in that performance on one task varied directly with task priority to that task, whereas performance on the other task varied inversely with priority. The amplitudes of the visual N160 and P250 and the auditory N100 components also varied directly and in a graded manner with task priority. However, these graded changes in amplitude for visual or auditory ERP components occurred only when the stimulus presentation rate was high, suggesting that a certain amount of workload is necessary for resource allocation between the two channels. These results complement those described earlier for P300 and show that these early latency components also exhibit resource reciprocity.

The auditory N100 component is known to consist of both an exogenous N100 component and an endogenous Nd component that is modulated by attention (Naatanen, 1992). It is therefore of interest to note that the Nd component is also resource sensitive. Singhal, Doerfling, and Fowler (2002) had participants perform a dichotic listening task while engaged in a simulated flight task of varying levels of difficulty. The amplitudes of both Nd and P300 were reduced at the highest level of difficulty.

Visual selective attention can also modulate the amplitude of the early P1 component (70–110 ms; Fu, Caggiano, Greenwood, & Parasuraman, 2005; Fu, Fan, Chen, & Zhuo, 2001; Hillyard & Munte, 1984; Mangun, 1995). This early P1/N1 attentional modulation contributes to the debate on early versus late selection in cognitive psychology by providing evidence that attention selects visual information at the early processing stage, rather than at later a selection stage such as decision making or response. For example, in a sustained attention study by Hillyard and Munte (1984), participants were asked to selectively attend to the left or right visual field and ignore the other visual field during separate blocks while maintaining their gaze on the central fixation point. It was found that the attended stimulus elicited larger P1 and N1 relative to the same stimulus at the same visual field when it was ignored or unattended (i.e., participants attended to the stimulus in the opposite visual field). This P1/N1 enhancement by visual selective attention is considered to reflect an early "sensory gating" mechanism of visual-spatial attention (Hillyard & Anllo-Vento, 1998), and has been replicated across a variety of tasks, such as sustained attention and trial-by-trial cueing tasks (Mangun, 1995) and visual search tasks (Luck, Fan, & Hillyard, 1993).

Some ERP evidence suggests that the earliest attentional modulation occurs in the extrastriate cortex (the source of the P1 attentional effect), but not in the striate cortex (the source for the C1 component; Clark et al., 1995; Clark & Hillyard, 1996). A feedback mechanism has been proposed to account for the role of striate cortex in the visual processing, which suggests an anatomically early but temporally later striate cortex activity after the reentrant process from higher visual cortex onto V1 (Martinez et al., 1999, 2001; Mehta, Ulbert, & Schroeder, 2000a, 2000b). By such feedback or reentrant processes, neural refractoriness may be reduced and the perceptual salience of attended stimuli may be enhanced (Mehta et al., 2000a, 2000b). Alternatively, the figure-ground contrast and the salience of attended stimuli may be enhanced (Lamme & Spekreijse, 2000; Super, Spekreijse, & Lamm, 2001). This is consistent with a recent finding of striate cortex activation by attention in the brain imaging literature (Gandhi, Heeger, & Boynton, 1999; Somers, Dale, Seiffert, & Tootell, 1999; Tootell et al., 1998). However, whether the initial processing in the striate cortex (indexed by the C1 component) is modulated by attention is still controversial. It is possible that the previous studies might have not adopted the optimal experimental conditions for investigating the role of striate cortex in early visual processing (Fu, Caggiano, Greenwood, & Parasuraman, 2005; Fu, Huang, Luo, Greenwood, & Parasuraman, 2004).

While the results of these studies on the locus of attentional selection in the brain are primarily relevant to basic theoretical issues in cognitive neuroscience, there are implications for neuroergonomics as well. First, the results indicate that several more ERP candidate components, such as C1 and N1, in addition to the later P300, might be sensitive to the effects of workload or stimulus load on the operator and may contribute to successful evaluation of operator mental state. Second, the P1 component is a very sensitive index of the allocation of visuospatial attention. Consequently, it would be very informative to monitor the operator's attentional allocation in tasks involving spatial selection, such as when a driver is driving (see figure 3.2 for attentional modulation of the P1 component). Finally, the earlier ERP components such as C1, P1, and N1 usually have a relatively well-understood psychological meaning and involve less complex brain mechanisms than the later ERP components such as P300; in terms of source localization, these earlier components can be also localized more easily and with less error. Thus these earlier ERP components can provide an assessment of neural mechanisms in a more accurate way under conditions when a detailed evaluation of the operator's mental status is needed. However, to what extent these ERP components can be reliably used in complex tasks remains to be seen. One potential limitation is the requirement of a higher number of trials for EEG averaging as compared to the later and larger components such as P300. Thus the use of early latency ERP components may require longer recording times and may be less efficient in real-time monitoring of a human operator's mental state as compared with the P300.

Error Detection and Performance Monitoring

Another important area for neuroergonomic research using ERPs is the analysis and possible prediction of human error. Cognitive scientists and human factors analysts have proposed many different approaches to the classification, description, and explanation of human error in complex human-machine systems (Norman & Shallice, 1986; Reason, 1990; Senders & Moray, 1991). Analysis of brain activity associated with errors can help refine these taxonomies, particularly in leading to testable hypotheses concerning the elicitation of error. The neural signature of an error may provide important information in this regard.

One such neural sign is a specific ERP component associated with error, error-related negativity or ERN. The ERN, which has a frontocentral distribution over the scalp, reaches a peak about 100–150 ms after the onset of the erroneous response (as revealed by measures of electromyographic activity) and is smaller or absent following a correct response (Gehring, Goss, Coles, Meyer, & Donchin, 1993). The ERN amplitude is related to perceived accuracy, or the extent to which participants are aware of their errors (Scheffers & Coles, 2000). Importantly, the ERN seems to reflect central mechanisms and is relatively independent of output modality. Holroyd, Dien, and Coles (1998) found that errors made in a choice reaction time task in which either the hand or the foot was used to respond led to nearly identical ERN.

The ERN was first identified in studies in which ERPs were selectively averaged to correct and incorrect responses in discrimination tasks (Gehring et al., 1993). In choice RT tasks, a negative-going potential is observed at anterior recording sites on trials in which participants make errors. The amplitude of this potential was found to be larger when the task instruction emphasized response accuracy over response speed—hence the label error-related negativity (Gehring et al., 1993). The ERN appears to be a manifestation of this error signal, with its amplitude reflecting the degree of mismatch between the two representations of the error response and the correct response, or the degree of error detected by the participants. That is, the greater the difference between the error response and the correct response, the larger the ERN amplitude; or the more the participants realize their errors, the larger the ERP amplitude.

The ERN may also be considered to be a neural response related to performance monitoring activities, or a comparison between representations of the appropriate response and the response actually made (Bernstein, Scheffers, & Coles, 1995; Scheffers & Coles, 2000). The amplitude of the ERN therefore could provide a useful index of the perceived inaccuracy of the operator's performance, given that Scheffers and Coles showed that larger ERNs are associated with errors due to premature responding (known errors), whereas smaller ERNs are associated with errors due to data limitations (uncertain errors). Scheffers and Coles used the Eriksen flanker task, in which a central letter in a five-letter array was designated as the target for detection (H or S). The flanker stimuli could be compatible (HHHHH and SSSSS) or incompatible (HHSHH, SSHSS) with the central target letter. Following each target detection response, participants were asked to rate their confidence using a 5-point scale of *sure correct, unsure correct, don't know, unsure incorrect,* or *sure incorrect*. Response-locked ERPs were separately averaged for trials with correct and incorrect responses. The typical ERN was observed, with its amplitude being larger for incorrect than for correct trials at frontal scalp sites. Furthermore, on incorrect trials, the greater the confidence participants had that their responses were wrong, the larger the ERN amplitude. For correct trials, the less that participants felt confident about their response (i.e., rating their correct response from *sure correct* to *sure incorrect*), the larger the ERN amplitude. Therefore, there was a systematic relationship between the participants' subjective perception of their response accuracy and ERN amplitude, with the smallest ERN associated with perceived-correct responses and increasingly larger ERN associated with perceived-incorrect responses. These results were consistent with the idea that the ERN is associated with the detection of errors during task performance.

Scheffers and Coles (2000) also investigated the relationship between type of errors and ERN amplitude. When participants judged their incorrect response as *sure incorrect*, they must have had enough information to be aware of their wrong response, indicating that this type of error was due to a premature response. On the other hand, responses judged as *don't know* were probably due to data limitations associated with insufficient stimulus quality or by the presence of incompatible flanking stimuli. Assuming that ERN is elicited by the comparison between the appropriate and the actual response, premature responses should be associated with this comparison and therefore should elicit larger ERN. On the other hand, data limitations should be associated with a compromised representation of the appropriate response, so that only a partial mismatch is involved, and therefore should elicit smaller ERN. Their results confirmed that errors due to premature responses elicited larger ERN, whereas those due to data limitations were associated with ERN of intermediate amplitude, supporting the view that ERN is a manifestation of a process that monitors the accuracy of performance.

The ERN may be generated in the anterior cingulate cortex (ACC) in the frontal lobe (Dehaene, Posner, & Tucker, 1994; Miltner, Braun, & Coles, 1997). As such, the ERN might reflect monitoring processes at a lower level of control rather than being associated with the detection and evaluation of errors at the higher level of supervisory executive control that is mediated by the prefrontal cortex. Moreover, ERN, and the ACC activation it represents, might be a manifestation of a monitoring process that specifically detects errors rather than conflict in general (Scheffers & Coles, 2000).

Rollnik et al. (2004) combined ERP and repetitive transcranial magnetic stimulation (rTMS) techniques to investigate the role of medial frontal cortex (including ACC) and dorsolateral prefrontal cortex (DLPFC) in performance monitoring. They found that ERN errors in an Eriksen flanker task elicited the ERN. Moreover, rTMS of the medial frontal cortex attenuated ERN amplitude and increased the subsequent error positivity (Pe) relative to a no-stimulation (control) condition, whereas no such effect was observed after lateral frontal stimulation, suggesting that the medial frontal cortex is important for error detection and correction.

The relevance of ERN to neuroergonomic research and applications is straightforward. The ERN allows identification, prediction, and perhaps prevention of operator errors in real time. For example, the ERN could be used to identify the human operator tendency to either commit, recognize, or correct an error. This could potentially be detected covertly by online measurement of ERN prior to the actual occurrence of the error, given that the ERN could be reliably measured on a single trial. Theoretically a system could be activated by an ERN detector in order to either take control of the situation (e.g., in those cases where time to act is an issue) or to notify the operator about the error he or she committed, even providing an adaptive interface that selectively presents the critical subsystems or function. Such a system would have the advantage of keeping the operator still in control of the entire system, while providing an anchor for troubleshooting when the error actually occurs (and having the possibility, for the system, to correct it by itself if needed).

Response Readiness

It has long been known that scalp ERPs can be recorded prior to the onset of a motor action. Such potentials are known as readiness potentials (Bereitschaftpotential; Kornhuber & Deecke, 1965). The lateralized readiness potential (LRP) is derived from the readiness potentials that occur several hundred milliseconds before a hand movement (Gratton, Coles, Sirevaag, Eriksen, & Donchin, 1988). The LRP is normally more pronounced on the contralateral scalp site of the responding hand, relative to the ipsilateral site. Typically, scalp electrodes at the C3 and C4 sites overlying motor cortex are used to record the LRP. These characteristics suggest that the LRP might be a good index of the covert process of movement preparation (Kutas & Donchin, 1980). The asymmetry between contra- and ipsilateral readiness potential specific to hand movement is defined by LRP, which is the average of the two difference waves obtained by subtracting the readiness potential from the ipsilateral to that of the contralateral hemisphere for left- and right-hand responses, respectively (Coles, 1989; Gratton et al., 1988). The averaging procedure is applied to remove all the non-movement-related asymmetric effects because they remain constant when the side of movement changes. The resulting LRP is thought to reflect the time course of the activation of the left and right motor responses; that is, the LRP can indicate whether and when a motor response is selected. It has been further proposed that the interval between the stimulus onset and LRP onset indicates the processes prior to the response hand selection, whereas the interval between the onset of LRP and the completion of the motor response reflects response organization and execution processes (Osman & Moore, 1993).

The LRP has been combined with other ERP components such as ERN to obtain converging evidence of response selection and execution processes in relation to error. For example, Van Schie, Mars, Coles, and Bekkering (2004) analyzed both the ERN and the LRP to investigate the neural mechanisms underlying error processing and showed that activity in both the medial frontal cortex (as measured by ERN) and the motor cortices (as measured by LRP) was modulated by the correctness of both self-executed and observed action.

Even though modern human-machine systems are becoming increasingly automated (Parasuraman & Mouloua, 1996), so that human operator actions are limited, many tasks of everyday life continue to require significant amounts of motor output. To the extent that such motor actions contribute to system performance and safety, motor-related brain potentials such as the LRP can be used to assess the speed efficiency of such actions and their underlying neural basis. Thus, the LRP is another useful tool for neuroergonomics in systems such as driving or keyboard entry, where motor actions are a key feature.

Assessment of Automatic Processing

One other potential ERP index that can be of use to neuroergonomics is the MMN. The MMN, which was discovered by Naatanen and colleagues, is considered to reflect automatic processing of auditory information (for reviews, see Naatanen, 1990; Naatanen & Winkler, 1999). The typical paradigm used to elicit the MMN is the oddball paradigm, in which an infrequent deviant auditory stimulus (tone, click, etc.) is embedded in a sequence of repeated standard stimuli. The difference wave obtained by subtracting the ERPs to deviant stimuli from the standard stimuli produces the MMN. The MMN has an onset latency of 100 ms and has maximal distribution over the frontal area. Recent neuroimaging studies have confirmed that the source of the MMN can be localized to the bilateral supratemporal cortex and the right frontal cortex (Naatanen, 2001).

The MMN might reflect, at least partially, automatic auditory change detection between the occasional deviant stimuli and the memory trace developed by the repetition of the standard stimuli, because it shows a similar amplitude when subjects attend to stimuli and when they ignore them, and it can be obtained when participants' attention is directed to another modality, such as vision. But the MMN might not completely reflect automatic processing, because its amplitude under certain conditions can vary with the load in the attended modality (Woldorff, Hackley, & Hillyard, 1991).

Several characteristics of the MMN might be useful for neuroergonomics. First, there is a correlation between MMN latency and RT, because both diminish when the physical difference between the deviant and standard stimuli is increased. Therefore, the MMN latency might be used as an index of response speed when no overt responses are available from the operator. The other is that the latency of MMN is inversely related to, and the amplitude of MMN is positively related to, the magnitude of the difference between the deviants and the standards. Furthermore, the discrimination process generating MMN might be important to involuntary orienting or attentional switching to a change in the acoustic environment.

Whether MMN is specific to the auditory modality is controversial (for a recent review, see Pazo-Alvarez, Cadaveira, & Amenedo, 2003). Some recent studies indicate that there might be a visual MMN that is an analog of the auditory MMN (Czigler & Csibra, 1990, 1992; Fu, Fan, & Chen, 2003). However, the studies for visual MMN should include the necessary experimental controls to affirm that there is a reliable homologue of the auditory MMN, by illustrating its underlying neural sources, the optimal parameters and property deviations to elicit it, and its characteristics as compared with auditory MMN. Confirmation of a visual MMN would be of considerable interest, because of the greater range of potential applications to visual processing issues in human factors research.

Conclusion

We have discussed the contribution of ERPs to human factors research and practice by describing the specific methodological advantages that ERPs provide over other measures of human brain function. ERPs have provided some specific insights not available from other measures in such areas as mental workload assessment, attentional resource allocation, error detection, and brain-computer interfaces. In a previous review, Parasuraman (1990) concluded that while the role of ERPs in human factors is modest, that role in a small number of cases is highly significant. Years later, with the prosperity of cognitive neuroscience and its penetration to fields such as cognitive psychology, the high-temporal-resolution and moderate-spatial-resolution ERP technique may be of greater importance now with the development of neuroergonomics. At the same time, it must be acknowledged that neuroergonomic applications of ERPs are still small in number. With further technical developments in miniaturization and portability of ERP systems, such practical applications may prosper. Given the continued development of automated systems in which human operators monitor rather than actively control system functions, there will be numerous additional opportunities for the use of ERP-based neuroergonomic measures and theories to provide insights into the role of brain function at work.

Main Points

1. ERPs represent the brain's neural response to specific sensory, motor, and cognitive events and are computed by averaging EEG epochs time-locked to a particular event.
2. Compared to other neuroimaging techniques, ERPs provide the best temporal resolution (1 ms or better) for evaluating neural activity in the human brain.
3. The spatial resolution of neuronal activity provided by ERPs is poor compared to fMRI, but can be improved through the use of various source localization techniques.
4. ERP components such as P300, N1, P1, ERN, and LRP can provide information on the neural basis of functions critical to a number of human factors issues. These include the assessment of mental workload, attention resource allocation, dual-task performance, error detection and prediction, and motor control.

Key Readings

Handy, T. (2005). *Event-related potentials: A methods handbook*. Cambridge, MA: MIT Press.
Luck, S. (2005). *An introduction to the event-related potential technique*. Cambridge, MA: MIT Press.
Parasuraman, R. (2003). Neuroergonomics: Research and practice. *Theoretical Issues in Ergonomics Science, 4*, 5–20.
Zani, A., & Proverbio, A. M. (2002). *The cognitive electrophysiology of mind and brain*. New York: Academic Press.

References

Baldwin, C. L., Freeman, F. G., & Coyne, J. T. (2004). Mental workload as a function of road type and visibility: Comparison of neurophysiological, behavioral, and subjective measures. In *Proceedings of the Human Factors and Ergonomics Society 48th annual meeting* (pp. 2309–2313). Santa Monica, CA: Human Factors and Ergonomics Society.
Bernstein, P. S., Scheffers, M. K., & Coles, M. G. (1995). "Where did I go wrong?" A psychophysiological analysis of error detection. *Journal of Experimental Psychology: Human Perception and Performance, 21*, 1312–1322.
Birnbaumer, N., Ghanayim, N., Hinterberger, T., Iversen, I., Kotchoubery, B., Kubler, A., et al. (1999). A spelling device for the paralysed. *Nature, 398*, 297–298.
Byrne, E. A., & Parasuraman, R. (1996). Psychophysiology and adaptive automation. *Biological Psychology, 42*, 249–268.
Cabeza, R., & Kingstone, A. (2001). *Handbook of functional neuroimaging of cognition*. Cambridge, MA: MIT Press.
Clark, V. P., Fan, S., & Hillyard, S. A. (1995). Identification of early visual evoked potential generators by retinotopic and topographic analyses. *Human Brain Mapping, 2*, 170–187.
Clark, V. P., & Hillyard, S. A. (1996). Spatial selective attention affects early extrastriate but not striate components of the visual evoked potential. *Journal of Cognitive Neuroscience, 8*, 387–402.
Coles, M. G. (1989). Modern mind-brain reading: Psychophysiology, physiology, and cognition. *Psychophysiology, 26*, 251–269.

Czigler, I., & Csibra, G. (1990). Event-related potentials in a visual discrimination task: Negative waves related to detection and attention. *Psychophysiology, 27,* 669–676.

Czigler, I., & Csibra, G. (1992). Event-related potentials and the identification of deviant visual stimuli. *Psychophysiology, 29,* 471–485.

Dehaene, S., Posner, M. I., & Tucker, D. M. (1994). Localization of a neural system for error detection and compensation. *Psychological Science, 5,* 303–305.

Donchin, E. (1980). Event-related potentials: Inferring cognitive activity in operational settings. In F. E. Gomer (Ed.), *Biocybernetic applications for military systems* (Technical Report MDC EB1911, pp. 35–42). Long Beach, CA: McDonnell Douglas.

Donchin, E. (1981). Surprise! . . . Surprise? *Psychophysiology, 18,* 493–513.

Donchin, E., Spencer, K. M., & Wijesinghe, R. (2000). The mental prosthesis: Assessing the speed of a P300-based brain-computer interface. *IEEE Transactions on Neural Systems and Rehabilitation Engineering, 8,* 174–179.

Fu, S., Caggiano, D., Greenwood, P. M., & Parasuraman, R. (2005). Event-related potentials reveal dissociable mechanisms for orienting and focusing visuospatial attention. *Cognitive Brain Research, 23,* 341–353.

Fu, S., Fan, S., & Chen, L. (2003). Event-related potentials reveal involuntary processing to orientation change in the visual modality. *Psychophysiology, 40,* 770–775.

Fu, S., Fan, S., Chen, L., & Zhuo, Y. (2001). The attentional effects of peripheral cueing as revealed by two event-related potential studies. *Clinical Neurophysiology, 112,* 172–185.

Fu, S., Greenwood, P. M., & Parasuraman, R. (2005). Brain mechanisms of involuntary visuospatial attention: An event-related potential study. *Human Brain Mapping, 25,* 378–390.

Fu, S., Huang, Y., Luo, Y., Greenwood, P. M., & Parasuraman, R. (2004). The role of perceptual difficulty in visuospatial attention: An event-related potential study. *Annual meeting of the Cognitive Neuroscience Society,* San Francisco, S87.

Gandhi, S. P., Heeger, D. J., & Boynton, G. M. (1999). Spatial attention affects brain activity in human primary visual cortex. *Proceedings of the National Academy of Sciences USA, 96,* 3314–3319.

Gazzaniga, M. S. (1995). *The cognitive neurosciences.* Cambridge, MA: MIT Press.

Gehring, W. J., Goss, B., Coles, M. G. H., Meyer, D. E., & Donchin, E. (1993). A neural system for error detection and compensation. *Psychological Science, 4,* 385–390.

Gomer, F. (1981). Physiological systems and the concept of adaptive systems. In J. Moraal & K. F. Kraiss (Eds.), *Manned systems design* (pp. 257–263). New York: Plenum Press.

Gratton, G., Coles, M. G., & Donchin, E. (1983). A new method for off-line removal of ocular artifact. *Electroencephalography and Clinical Neurophysiology, 55,* 468–484.

Gratton, G., Coles, M. G., Sirevaag, E. J., Eriksen, C. W., & Donchin, E. (1988). Pre- and poststimulus activation of response channels: A psychophysiological analysis. *Journal of Experimental Psychology: Human Perception and Performance, 14,* 331–344.

Gratton, G., & Fabiani, M. (2001). Shedding light on brain function: The event-related optical signal. *Trends in Cognitive Science, 5,* 357–363.

Heinze, H. J., Mangun, G. R., Burchert, W., Hinrichs, J., Scholz, M., Munte, T. F., et al. (1994). Combined spatial and temporal imaging of brain activity during visual selective attention in humans. *Nature, 372,* 543–546.

Hillyard, S. A., & Anllo-Vento, L. (1998). Event-related brain potentials in the study of visual selective attention. *Proceedings of the National Academy of Sciences U.S.A., 95,* 781–787.

Hillyard, S. A., Hink, R. F., Schwent, V. L., & Picton, T. W. (1973). Electrical signs of selective attention in the human brain. *Science, 182,* 177–180.

Hillyard, S. A., & Munte, T. F. (1984). Selective attention to color and location: An analysis with event-related brain potentials. *Perception and Psychophysics, 36,* 185–198.

Hink, R. F., Van Voorhis, S. T., Hillyard, S. A., & Smith, T. S. (1977). The division of attention and the human auditory evoked potential. *Neuropsychologia, 15,* 497–505.

Holroyd, C. B., Dien, J., & Coles, M. G. H. (1998). Error-related scalp potentials elicited by hand and foot movements: Evidence for an output-independent error-processing system in humans. *Neuroscience Letters, 242,* 65–68.

Humphrey, D., & Kramer, A. F. (1994). Towards a psychophysiological assessment of dynamic changes in mental workload. *Human Factors, 36,* 3–26.

Isreal, J. B., Wickens, C. D., Chesney, G. L., & Donchin, E. (1980). The event-related brain potential as an index of display-monitoring workload. *Human Factors, 22,* 211–224.

Johnson, R., Jr. (1986). A triarchic model of P300 amplitude. *Psychophysiology, 23,* 367–384.

Kornhuber, H. H., & Deecke, L. (1965). Hirnpotentialanderungen bei WillKuerbewegungen und passiven Bewegungen des Menschen: Bereitschaftspotential und reafferente Potentiale. *Pfluegers Archives Ges. Physiologie, 8,* 529–566.

Kramer, A. F., & Belopolsky, A. (2003). Assessing brain function and metal chronometry with event-related brain potentials. In K. Brookhuis (Ed.), *Handbook of human factors and ergonomics methods*. (pp. 365–374). New York: Taylor and Francis.

Kramer, A. F., Sirevaag, E. J., & Braune, R. (1987). A psychophysiological assessment of operator workload during simulated flight missions. *Human Factors, 29*, 145–160.

Kramer, A. F., & Weber, T. (2000). Applications of psychophysiology to human factors. In J. T. Cacioppo, L. G. Tassinary, & G. G. Berntson (Eds.), *Handbook of psychophysiology* (2nd ed.). Cambridge, UK: Cambridge University Press.

Kramer, A. F., Wickens, C. D., & Donchin, E. (1983). An analysis of the processing requirements of a complex perceptual-motor task. *Human Factors, 25*, 597–621.

Kramer, A. F., Wickens, C. D., & Donchin, E. (1985). Processing of stimulus properties: Evidence for dual-task integrality. *Journal of Experimental Psychology: Human Perception and Performance, 11*, 393–408.

Kutas, M., & Donchin, E. (1980). Preparation to respond as manifested by movement-related brain potentials. *Brain Research, 202*, 95–115.

Kutas, M., & Hillyard, S. A. (1984). Brain potentials during reading reflect word expectancy and semantic association. *Nature, 307*, 161–163.

Kutas, M., McCarthy, G., & Donchin, E. (1977). Augmenting mental chronometry: The P300 as a measure of stimulus evaluation time. *Science, 197*, 792–795.

Lamme, V. A., & Spekreijse, H. (2000). Contextual modulation in primary visual cortex and scene perception. In M. S. Gazzaniga (Ed.), *The new cognitive neurosciences* (pp. 279–290). Cambridge, MA: MIT Press.

Luck, S. (1998). Sources of dual-task interference: Evidence from human electrophysiology. *Psychological Science, 9*, 223–227.

Luck, S. (2005). *An introduction to the event-related potential technique*. Cambridge, MA: MIT Press.

Luck, S., Fan, S., & Hillyard, S. A. (1993). Attention-related modulation of sensory-evoked brain activity in a visual search task. *Journal of Cognitive Neuroscience, 5*, 188–195.

Luck, S. J., & Girelli, M. (1998). Electrophysiological approaches to the study of selective attention in the human brain. In R. Parasuraman (Ed.), *The attentive brain* (pp. 71–94). Cambridge, MA: MIT Press.

Magliero, A., Bashore, T. R., Coles, M. G., & Donchin, E. (1984). On the dependence of P300 latency on stimulus evaluation processes. *Psychophysiology, 21*, 171–186.

Makeig, S., Westerfield, M., Jung, T.-P., Enghoff, S., Townsend, J., Courchesne, E., et al. (2002). Dynamic brain sources of visual evoked responses. *Science, 295*, 690–694.

Mangun, G. R. (1995). Neural mechanisms of visual selective attention. *Psychophysiology, 32*, 4–18.

Mangun, G. R., Hopfinger, J. B., Kussmaul, C. L., Fletcher, E., & Heinze, H. J. (1997). Covariations in ERP and PET measures of spatial selective attention in human extrastriate visual cortex. *Human Brain Mapping, 5*, 273–279.

Mangun, G. R., Hinrichs, H., Scholz, M., Mueller-Gaertner, H. W., Herzog, H., Krause, B. J., et al. (2001). Integrating electrophysiology and neuroimaging of spatial selective attention to simple isolated visual stimuli. *Vision Research, 41*, 1423–1435.

Martinez, A., Anllo-Vento, L., Sereno, M. I., Frank, L. R., Buxton, R. B., Dubowitz, D. J., et al. (1999). Involvement of striate and extrastriate visual cortical areas in spatial attention. *Nature Neuroscience, 2*, 364–369.

Martinez, A., DiRusso, F., Anllo-Vento, L., Sereno, M. I., Buxton, R. B., & Hillyard, S. A. (2001). Putting spatial attention on the map: Timing and localization of stimulus selection processes in striate and extrastriate visual areas. *Vision Research, 41*, 1437–1457.

McCarthy, G., & Donchin, E. (1981). A metric for thought: A comparison of P300 latency and reaction time. *Science, 211*, 77–80.

McFarland, D. J., Sarnacki, W. A., & Wolpaw, J. R. (2003). Brain computer interface (BCI) operation: Optimizing information transfer rates. *Biological Psychology, 63*, 237–251.

Mehta, A. D., Ulbert, I., & Schroeder, C. E. (2000a). Intermodal selective attention in monkeys: I. Distribution and timing of effects across visual areas. *Cerebral Cortex, 10*, 343–358.

Mehta, A. D., Ulbert, I., & Schroeder, C. E. (2000b). Intermodal selective attention in monkeys: II. Physiological mechanisms of modulation. *Cerebral Cortex, 10*, 359–370.

Miltner, W. H. R., Braun, C. H., & Coles, M. G. (1997). Event-related brain potentials following incorrect feedback in a time-production task: Evidence for a "generic" neural system for error detection. *Journal of Cognitive Neuroscience, 9*, 787–797.

Moray, N. (1979). *Mental workload*. New York: Plenum.

Naatanen, R. (1990). The role of attention in auditory information processing as revealed by event-related brain potentials. *Behavioral and Brain Sciences*, 199–290.

Naatanen, R. (1992). *Attention and brain function*. Hillsdale, NJ: Erlbaum.

Naatanen, R. (2001). The perception of speech sounds by the human brain as reflected by the mismatch negativity (MMN) and its magnetic equivalent (MMNm). *Psychophysiology, 38,* 1–21.

Naatanen, R., & Winkler, I. (1999). The concept of auditory stimulus representation in cognitive neuroscience. *Psychological Bulletin, 125,* 826–859.

Navon, D., & Gopher, D. (1979). On the economy of the human information processing system. *Psychological Review, 86,* 214–255.

Norman, D. A., & Shallice, T. (1986). Attention to action: Willed and automatic control of behavior. In R. Davidson, G. Schwartz, & D. Shapiro (Eds.), *Consciousness and self-regulation: Advances in theory and research* (Vol. 4, pp. 1–18). New York: Plenum.

Osman, A., & Moore, C. M. (1993). The locus of dual-task interference: Psychological refractory effects on movement-related brain potentials. *Journal of Experimental Psychology: Human Perception and Performance, 19,* 1292–1312.

Parasuraman, R. (1978). Auditory evoked potentials and divided attention. *Psychophysiology, 15,* 460–465.

Parasuraman, R. (1985). Event-related brain potentials and intermodal divided attention. *Proceedings of the Human Factors Society, 29,* 971–975.

Parasuraman, R. (1990). Event-related brain potentials and human factors research. In J. W. Rohrbaugh, R. Parasuraman, & R. Johnson, Jr. (Eds.), *Event-related brain potentials: Basic issues and applications* (pp. 279–300). New York: Oxford University Press.

Parasuraman, R. (2003). Neuroergonomics: Research and practice. *Theoretical Issues in Ergonomics Science, 4,* 5–20.

Parasuraman, R., & Beatty, J. (1980). Brain events underlying detection and recognition of weak sensory signals. *Science, 210,* 80–83.

Parasuraman, R., & Caggiano, D. (2002). Mental workload. In V. S. Ramachandran (Ed.), *Encyclopedia of the human brain* (Vol. 3, pp. 17–27). San Diego: Academic Press.

Parasuraman, R., & Mouloua, M. (1996). *Automation and human performance: Theory and applications.* Hillsdale, NJ: Erlbaum.

Pascual-Marqui, R. D. (1999). Review of methods for solving the EEG inverse problem. *International Journal of Bioelectromagnetism, 1,* 75–86.

Pashler, H. (1994). Dual-task interference in simple tasks: Data and theory. *Psychological Bulletin, 116,* 220–244.

Pazo-Alvarez, P., Cadaveira, F., & Amenedo, E. (2003). MMN in the visual modality: A review. *Biological Psychology, 63,* 199–236.

Pfurtscheller, G., & Neuper, C. (2001). Motor imagery and direct brain-computer communication. *Proceedings of the IEEE, 89,* 1123–1134.

Picton, T. W., Bentin, S., Berg, P., Donchin, E., Hillyard, S. A., Johnson, R., Jr., et al. (2000). Guidelines for using human event-related potentials to study cognition: Recording standards and publication criteria. *Psychophysiology, 37,* 127–152.

Polich, J. (1987). Task difficulty, probability, and interstimulus interval as determinants of P300 from auditory stimuli. *Electroencephalography and Clinical Neurophysiology, 68,* 311–320.

Posner, M. I. (1978). *Chronometric explorations of mind.* Hillsdale, NJ: Erlbaum.

Posner, M. I. (2004). *Cognitive neuroscience of attention.* New York: Guilford.

Posner, M. I., Petersen, S. E., Fox, P. T., & Raichle, M. E. (1988). Localization of cognitive functions in the human brain. *Science, 240,* 1627–1631.

Reason, J. T. (1990). *Human error.* New York: Cambridge University Press.

Rollnik, J. D., Bogdanova, D., Krampfl, K., Khabirov, F. A., Kossev, A., Dengler, R., et al. (2004). Functional lesions and human action monitoring: Combining repetitive transcranial magnetic stimulation and event-related brain potentials. *Clinical Neurophysiology, 115,* 356–360.

Scheffers, M. K., & Coles, M. G. (2000). Performance monitoring in a confusing world: Error-related brain activity, judgments of response accuracy, and types of errors. *Journal of Experimental Psychology: Human Perception and Performance, 26,* 141–151.

Scherg, M. (1990). Fundamentals of dipole source potential analysis. In F. Grandori, M. Hoke, & G. L. Romani (Eds.), *Auditory evoked magnetic fields and electric potentials. Advances in audiology* (Vol. 6, pp. 40–69). Basel: Karger.

Schultheis, H., & Jameson, A. (2004). Assessing cognitive load in adaptive hypermedia systems: Physiological and behavioral methods. In P. De Bra & W. Nejdl (Eds.), *Adaptive hypermedia and adaptive web-based systems.* (pp. 18–24). Eindhoven, Netherlands: Springer.

Senders, J. W., & Moray, N. P. (1991). *Human error: Cause, prediction and reduction.* Hillsdale, NJ: Erlbaum.

Singhal, A., Doerfling, P., & Fowler, B. (2002). Effects of a dual task on the N100–P200 complex and the early and late Nd attention waveforms. *Psychophysiology, 39,* 236–245.

Sirevaag, E. J., Kramer, A. F., Coles, M. G. H., & Donchin, E. (1989). Resource reciprocity: An event-related brain potentials analysis. *Acta Psychologica, 70,* 77–97.

Sirevaag, E. J., Kramer, A. F., Wickens, C. D., Reisweber, M., Strayer, D., & Grenell, J. (1993). Assessment of pilot performance and mental workload in rotary wing aircraft. *Ergonomics, 36,* 1121–1140.

Somers, D. C., Dale, A. M., Seiffert, A. E., & Tootell, R. B. (1999). Functional MRI reveals spatially specific attentional modulation in human primary visual cortex. *Proceedings of the National Academy of Sciences USA, 96,* 1663–1668.

Super, H., Spekreijse, H., & Lamme, V. A. (2001). Two distinct modes of sensory processing observed in monkey primary visual cortex (V1) [see comment]. *Nature Neuroscience, 4,* 304–310.

Sutton, S., Braren, M., Zubin, J., & John, E. R. (1965). Evoked-potential correlates of stimulus uncertainty. *Science, 150,* 1187–1188.

Tootell, R. B., Hadjikhani, N., Hall, E. K., Marrett, S., Vanduffel, W., Vaughan, J. T., et al. (1998). The retinotopy of visual spatial attention. *Neuron, 21,* 1409–1422.

Van Schie, H. T., Mars, R. B., Coles, M. G., & Bekkering, H. (2004). Modulation of activity in medial frontal and motor cortices during error observation. *Nature Neuroscience, 7,* 549–554.

Wickens, C. D. (1984). Processing resources in attention. In R. Parasuraman & D. R. Davies (Eds.), *Varieties of attention* (pp. 63–102). San Diego: Academic Press.

Wickens, C. D. (1990). Applications of event-related potential research to problems in human factors. In J. W. Rohrbaugh, R. Parasuraman, & R. Johnson (Eds.), *Event-related brain potentials: Basic and applied issues* (pp. 301–309). New York: Oxford University Press.

Wickens, C. D. (2002). Multiple resources and performance prediction. *Theoretical Issues in Ergonomics Science, 3,* 159–177.

Wickens, C. D., Isreal, J. B., & Donchin, E. (1977). The event-related cortical potential as an index of task workload. *Proceedings of the Human Factors Society, 21,* 282–287.

Wickens, C. D., Kramer, A. F., Vanesse, L., & Donchin, E. (1983). The performance of concurrent tasks: A psychophysiological analysis of the reciprocity of information processing. *Science, 221,* 1080–1082.

Woldorff, M. G., Hackley, S. A., & Hillyard, S. A. (1991). The effects of channel-selective attention on the mismatch negativity wave elicited by deviant tones. *Psychophysiology, 28,* 30–42.

Vince D. Calhoun

Functional Magnetic Resonance Imaging (fMRI)
Advanced Methods and Applications to Driving

Functional magnetic resonance imaging (fMRI) is now able to provide an unprecedented view inside the living human brain. Early studies focused upon carefully designed, but often artificial, paradigms for probing specific neural systems and their modulation (Cabeza & Nyberg, 2000). Such approaches, though important, do not speak directly to how the brain is performing a complex real-world task. Relatively real-world or naturalistic tasks often involve multiple cognitive domains and may well result in emergent properties not present in tasks that probe separate cognitive domains. Performing a naturalistic task is obviously quite complex, and the difficulty of analyzing and interpreting the data produced by these paradigms has hindered the performance of brain imaging studies in this area. However, there is growing interest in studying complex tasks using fMRI, and new analytic techniques are enabling this exciting possibility. In this chapter, I focus upon the collection and analysis of fMRI data during the performance of naturalistic tasks. Specifically, I use the exemplar of simulated driving to demonstrate the development of a brain activation model for simulated driving.

Essentials of fMRI

fMRI is a brain imaging technique in which relative changes in cerebral oxygenation can be assessed noninvasively while a person is engaged in performing cognitive tasks. Participants lie supine in the MRI scanner while exposed to a static magnetic field, typically 1.5 or 3 Teslas (1.5 or 3 T). A radio frequency coil placed near the head is then briefly pulsed to temporarily disturb the aligned magnetic field passing through the brain, and the resulting magnetic resonance signal that is emitted as the field settles back into its static state is detected. fMRI exploits the fact that oxygenated blood has different magnetic properties than deoxygenated blood or the surrounding tissues, as a result of which the strength of the magnetic resonance signal varies with the level of blood oxygenation. As a given brain region increases its neural activity, there is initially a small depletion in the local oxygenated blood pool. A short while later, however, the cerebrovasculature responds to this oxygen depletion by increasing the flow of oxygenated blood into the region for a short time, before returning oxygenated blood levels back to normal. For reasons that are not completely understood, the supply of oxygenated blood exceeds the neural demand, so that the ratio of oxygenated to deoxygenated blood is altered. It is this disparity—the blood oxygen level-dependent (BOLD) signal—that is measured and tracked throughout the brain. Researchers can then use the BOLD signal and associate it with performance of a cognitive task, in comparison to a baseline or resting state, or to another cognitive task that differs in requiring an additional cognitive operation. Subtraction is used to isolate the activation patterns associated with the cognitive operation. This is the basic block design of many fMRI studies and allows the researcher to identify cortical regions that are involved with perceptual, cognitive, emotional, and motor operations (see Cabeza & Nyberg, 2000; Frackowiak et al., 2003, for reviews). This chapter focuses on more sophisticated fMRI analytical techniques to assess changes in brain activity in persons engaged in a complex activity of daily living, driving.

Why Study Simulated Driving With fMRI?

Driving is a complex behavior involving interrelated cognitive elements including selected and divided attention, visuospatial interpretation, visuomotor integration, and decision making. Several cognitive models have been proposed for driving (Ranney, 1994), especially for visual processing aspects and driver attributes (Ballard, Hayhoe, Salgian, & Shinoda, 2000; Groeger, 2000). However, such models are complicated and hard to translate into imaging studies. Many of the cognitive elements expected to be involved in driving have been studied separately using imaging paradigms designed to probe discrete brain systems (Cabeza & Nyberg, 2000). Imaging studies have used subtractive methods to study the neural correlates of driving (Walter et al., 2001) but have not attempted to study the complex temporal dynamics of driving.

Challenges in fMRI Data Analysis

The BOLD technique yields signals that are sensitized to hemodynamic sequelae of brain activity. These time-dependent data are typically reduced, using knowledge of the timing of events in the task paradigm, to graphical depictions of where in the brain the MRI time course resembled the paradigm time courses (e.g., the task-baseline cycle in the block design is convolved with the hemodynamic impulse response). The data are then analyzed using a variety of statistical techniques, from simple t tests and analysis of variance (ANOVA) to the general linear model (GLM; Worsley & Friston, 1995). The GLM method is a massively univariate inferential approach in which the same hypothesis (the brain time course resembles the paradigm time course) is tested repeatedly at every location (voxel) in the brain. Many fMRI studies use the block design, in which cognitive tasks typically employ subtraction between two types of tasks modified in slight increments (Cabeza & Nyberg, 2000) and provide visualization of brain regions that differ.

While much basic knowledge has been gained from the block design method, it suffers from the limitation that there is no attempt to study the temporal dynamics of brain activation in response to a cognitive challenge. Event-related fMRI, in which MRI signals are correlated with the timing of individual stimulus and response events, can be used to examine temporal dynamics (Friston et al., 1998). An advantage of using an event-related design to examine fMRI data collected during simulated driving would be the ability to assess the times of unique discrete events such as car crashes. As most commonly implemented, however, event-related fMRI typically requires fairly stringent assumptions about the hemodynamic response resulting from such an event (Friston et al., 1998). More flexible modeling can be done by estimating the hemodynamic impulse response through deconvolution approaches (Glover, 1999). However, all of these approaches require a well-defined set of stimuli. For a simulated driving task, the stimuli (e.g., many types of visual stimuli, motor responses, crashes, driving off the road, etc.) are overlapping and continuous, and it is difficult to generate a well-defined set of stimuli.

Data-Driven Approach

In addition to univariate model-based approaches, a variety of multivariate data-driven approaches have been used to reveal patterns in fMRI data. Among the data-driven approaches applied to fMRI data, we and others (Biswal & Ulmer, 1999; Calhoun, Adali, Pearlson, & Pekar, 2001c; Calhoun,

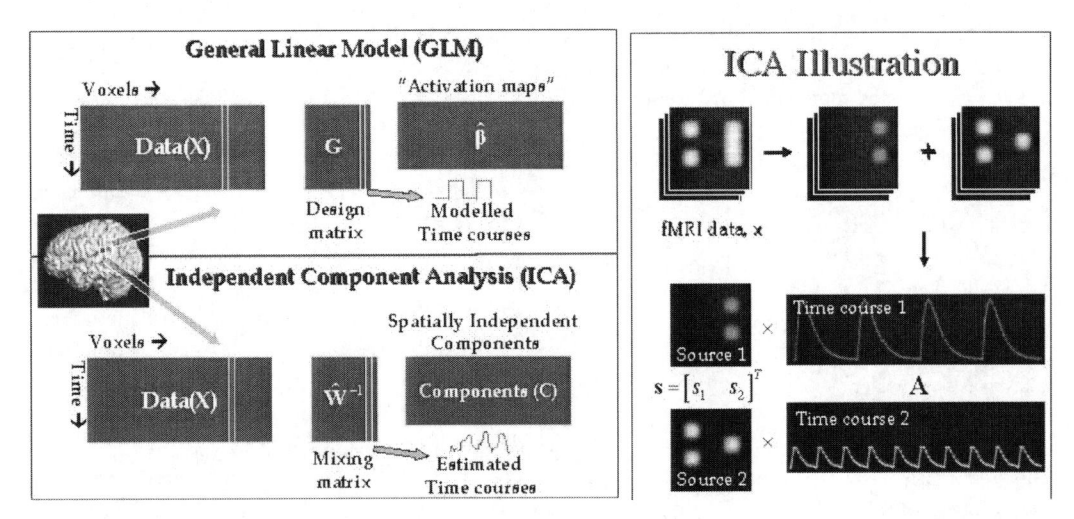

Figure 4.1. Comparison of general linear model (GLM) and independent component analysis (ICA; left) and ICA illustration (right). The GLM (top left) is by far the most common approach to analyzing fMRI data, and to use this approach, one needs a model for the fMRI time course, whereas in spatial ICA (bottom left), there is no explicit temporal model for the fMRI time course. This is estimated along with the hemodynamic source locations (right). The ICA model assumes the fMRI data, x, is a linear mixture of statistically independent sources s, and the goal of ICA is to separate the sources given the mixed data and thus determine the s and A matrices. See also color insert.

Adali, Hansen, Larsen, & Pekar, 2003; McKeown, Hansen, & Sejnowski, 2003; McKeown et al., 1998) have shown that spatial independent component analysis (ICA) is especially useful at revealing brain activity not anticipated by the investigator, and at separating brain activity into networks of brain regions with synchronous BOLD fMRI signals. Independent component analysis was originally developed to solve problems similar to the "cocktail party" problem (Bell & Sejnowski, 1995). The ICA algorithm, assuming independence in time (independence of the voices), can separate mixed signals into individual sources (voices). In our application, we assume independence of the hemodynamic source locations from the fMRI data (independence in space), resulting in maps for each of these regions, as well as the time course representing the fMRI hemodynamics.

The principle of ICA of fMRI is illustrated in figure 4.1. Two sources are presented along with their representative hemodynamic mixing functions. In the fMRI data, these sources are mixed by their mixing functions and added to one another. The goal of ICA is then to unmix the sources using some measure of statistical independence. The ICA approach does not assume a form for the temporal behavior of the components and thus is called a blind source separation method. Semiblind ICA methods, which have been applied to fMRI data, can be used to provide some of the advantages of both the general linear model and ICA, although they have not yet been applied to fMRI of simulated driving studies (Calhoun, Adali, Stevens, Kiehl, & Pekar, 2005).

The ICA approach has also been extended to allow for the analysis of multiple subjects (Calhoun, Adali, Pearlson, & Pekar, 2001a; Woods, 1996). We applied this method to our driving activation data, analyzing the data from all subjects in a single ICA estimation. This provides a way to extract behavioral correlates without having an a priori hemodynamic model. It also enables us to estimate the significance of the voxels in each component by computing the variance across individuals.

The Driving Paradigm

The driving scenario consisted of three repeating conditions, resembling a standard block design. While these three conditions provided a way to compare behavior, we did not rely upon simple comparison of the images between different conditions, as in most fMRI block designs, but rather examined the source locations and the modulation of the temporal dynamics. This approach thus provided a useful way of analyzing complex behaviors not possible using traditional (between-epoch) fMRI comparisons. Importantly, the activity represented by several of these components (e.g., anterior cingulate and frontoparietal) was not found when we analyzed our data using the GLM, or in a similar fMRI of driving study using only GLM analysis (Walter et al., 2001). This suggests that ICA may be especially useful for analysis of fMRI data acquired during such rich naturalistic behavior, as opposed to the experimental paradigms typical of the fMRI literature where GLM analysis is so prevalent.

We also investigated whether driving speed would modulate the associated neural correlates differently. Driving at a faster speed was not expected to modulate primary visual regions much since the visual objects seen are no more complicated than for the slower speed. Likewise, a speed change was expected to make little difference in the rate of finger movement or motor activation. However, increasing speed may increase activation in regions subserving eye movements or visual attention (Hopfinger, Buonocore, & Mangun, 2000) or anxiety. We thus scanned two groups of subjects driving at different rates of speed.

Experiments and Methods

Participants

Subjects (2 female, 10 male; mean age 22.5 years) were approved by the institutional review board and were compensated for their participation. Subjects were given additional compensation if they stayed within a prespecified speed range. Subjects were screened with a complete physical and neurological examination, and urine toxicological testing, as well as the SCAN interview (Janca, Ustun, & Sartorius, 1994), to eliminate participants with Axis I psychiatric disorders. Subjects were divided into two groups (three subjects were in both groups), one group driving faster ($N = 8$) and the other group driving slower ($N = 7$). Three subjects were included in both groups and thus scanned during both speeds. The slower and faster driving groups were delineated by changing the speedometer display units from kilometers per hour (kph) to miles per hour (mph), resulting in an actual speed range of 100–140 or 160–224 kph, respectively.

Experimental Design

We obtained fMRI scans of subjects as they twice performed a 10-minute task consisting of 1-minute epochs of (a) an asterisk fixation task, (b) active simulated driving, and (c) watching a simulated driving scene (while randomly moving fingers over the controller). Epochs (b) and (c) were switched in the second run, and the order was counterbalanced across subjects. During the driving epoch, participants were performing simulated driving using a modified game pad controller with buttons for left, right, acceleration, and braking. The paradigm is illustrated in figure 4.2.

The simulator used was a commercially available driving game, Need for Speed II (Electronic Arts, 1998). Simulators of this type are both inexpensive and relatively realistic. We have previously demonstrated the validity of a similar simulated driving environment compared directly to real on-road driving (McGinty, Shih, Garrett, Calhoun, & Pearlson, 2001). The controller was shielded in copper foil and connected to a computer outside the scanner room though a waveguide in the wall. All ferromagnetic screws were removed and replaced by plastic components. An LCD projector outside the scanner room and behind the scanner projected through another waveguide to a translucent screen, which the subjects saw via a mirror, attached to the head coil of the fMRI scanner. The screen subtended approximately 25 degrees of visual field. The watching epoch was the same for all subjects (a playback of a previously recorded driving session). For the driving epoch, subjects started at the same point on the track with identical conditions (e.g., car type, track, traffic conditions). They were instructed to stay in the right lane except in order to pass, to avoid collisions, to stay within a speed range of 100–140 (the units were not specified) and to drive normally.

Image Acquisition

Data were acquired at the FM Kirby Research Center for Functional Brain Imaging at Kennedy Krieger Institute on a Philips NT 1.5 Tesla scanner. A sagittal localizer scan was performed first, followed by a T1-weighted anatomical scan (TR = 500 ms, TE = 30 ms, field of view = 24 cm, matrix = 256 × 256, slice thickness = 5 mm, gap =

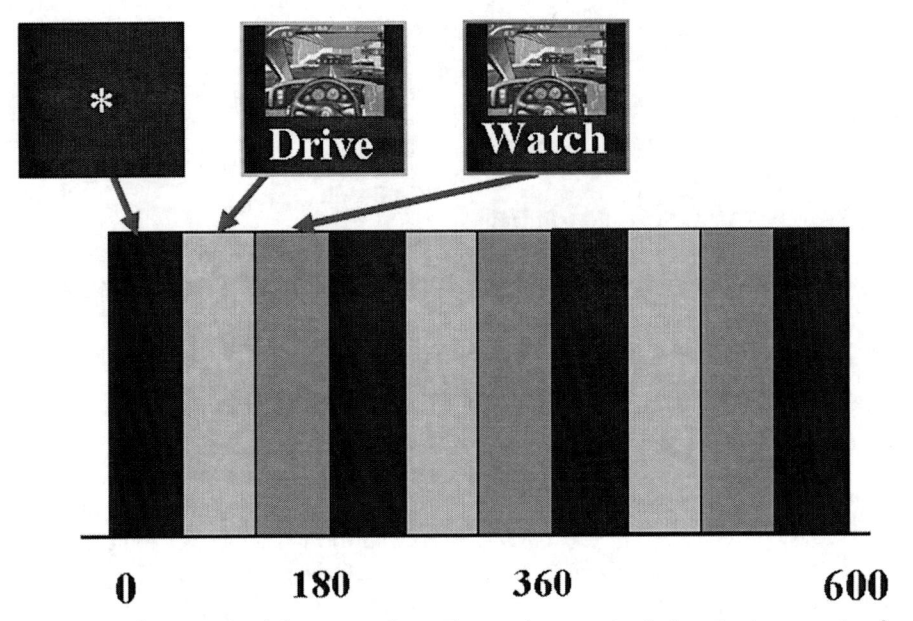

Figure 4.2. fMRI simulated driving paradigm. The paradigm consisted of ten 1-minute epochs of (a) a fixation target, (b) driving the simulator, and (c) watching a simulation while randomly moving fingers over the controller. The paradigm was presented twice, changing the order of the (b) and (c) epochs and counterbalancing the first order across subjects. See also color insert.

0.5 mm) consisting of 18 slices through the entire brain, including most of the cerebellum. Next, we acquired the functional scans consisting of an echo-planar scan (TR = 1 s, TE = 39 ms, field of view = 24 cm, matrix = 64 × 64, slice thickness = 5 mm, gap = 0.5 mm) obtained consistently over a 10-minute period per run for a total of 600 scans. Ten dummy scans were performed at the beginning to allow for longitudinal equilibrium, after which the simulated driving paradigm was begun.

fMRI Data Analysis

The images were first corrected for timing differences between the slices using windowed Fourier interpolation to minimize the dependence upon which reference slice is used (Calhoun, Adali, Kraut, & Pearlson, 2000; Van de Moortele et al., 1997). Next the data were imported into the statistical parametric mapping software package SPM99 (Worsley & Friston, 1995). Data were motion corrected, spatially smoothed with an 8 × 8 × 11 mm Gaussian kernel, and spatially normalized into the standard space of Talairach and Tournoux (1998). The data were slightly subsampled to 3 × 3 × 4 mm,

resulting in 53 × 63 × 28 voxels. For display, slices 2–26 were presented.

Independent Component Analysis

Data from each subject were reduced from 600 to 30 dimensions using principal component analysis (PCA; representing greater than 99% of the variance in the data). This prereduction (i.e., the incorporation of an initial PCA stage for each subject) is a pragmatic step, reduces the amount of memory required to perform the ICA estimation, and does not have a significant effect on the results, provided the number chosen is not too small (Calhoun, Adali, Pearlson, & Pekar, 2001a). To enable group averaging, data from all subjects were then concatenated and this aggregate data set reduced to 25 time points using PCA, followed by group independent component estimation using a neural network algorithm that attempted to minimize the mutual information of the network outputs (Bell & Sejnowski, 1995). Time courses and spatial maps were then reconstructed for each subject and the random effects spatial maps thresholded at $p < 0.00025$ ($t = 4.5$, $df = 14$;). The methods we used have been implemented in a user-friendly software package available for download (http://icatb.sourceforge.net).

Results

For the fMRI motion correction, the translation corrections were less than half a voxel for any participant. Pitch, roll, and yaw rotations were generally within 2.0 degrees or less. There were no visually apparent differences between the average motion parameters for the driving runs. Changes were also not significant as assessed by a t test on the sum of the squares of the parameters over time.

GLM Results

We performed a multiple regression analysis using the general linear model approach in which the driving and watch conditions were modeled with separate regressors for each participant and aggregated across subjects using a one-sample t test of the single subject images for each condition (Holmes & Friston, 1998) Results are presented in figure 4.3. Relative increases are in red and decreases in blue. Networks correlated with the drive regressor are depicted on the left and networks correlated with the watch regressor are on the right. We found, as reported by others, that motor and cerebellar networks were significantly different in the contrast between drive and watch (Walter et al., 2001).

ICA Results

ICA results are summarized in figure 4.4, with different colors coding for each component. The resulting 25 time courses were sorted according to their correlation with the driving paradigm and then visually inspected for task-related or transiently task-related activity. Of the 25 components, only six demonstrated such relationships. The fMRI data comprise a linear mixture of each of the six depicted components. That is, if a given voxel had a high value for a given component image, the temporal pattern of the data resembled the temporal pattern depicted for that component. Additionally, some areas consisted of more than one temporal pattern. For example, the pink and blue components overlapped heavily in the anterior cingulate and medial frontal regions. Selected Talairach coordinates of the volume and maxima of each anatomical region within the maps are presented in table 4.1. The ICA results were generally consistent with the GLM results; however, the

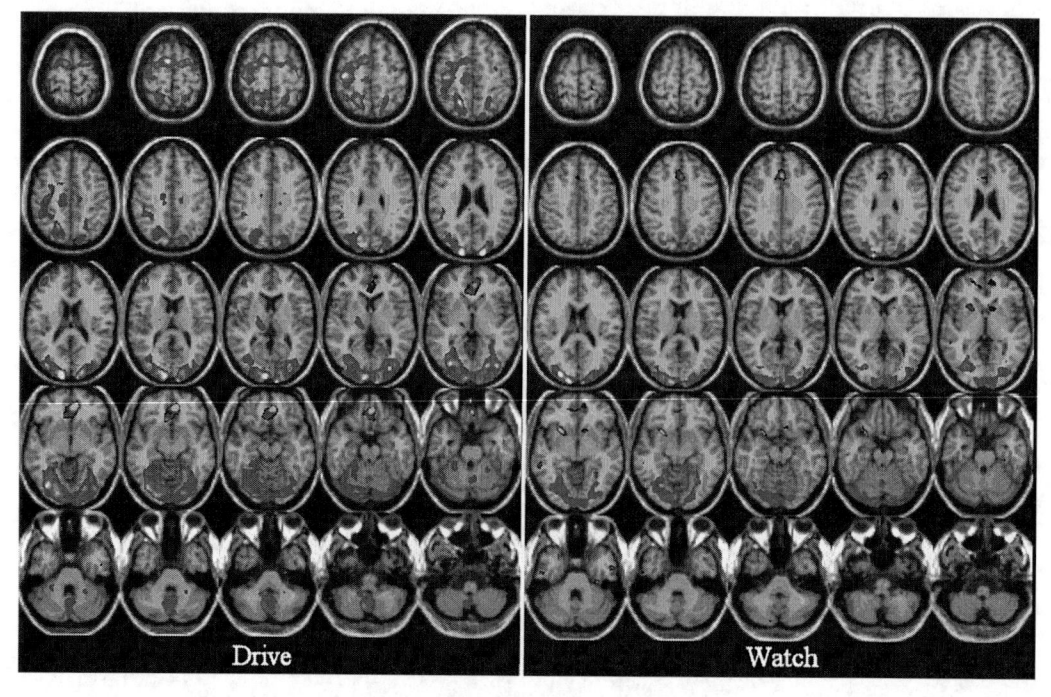

Figure 4.3. Results from the GLM-based fMRI analysis. See also color insert.

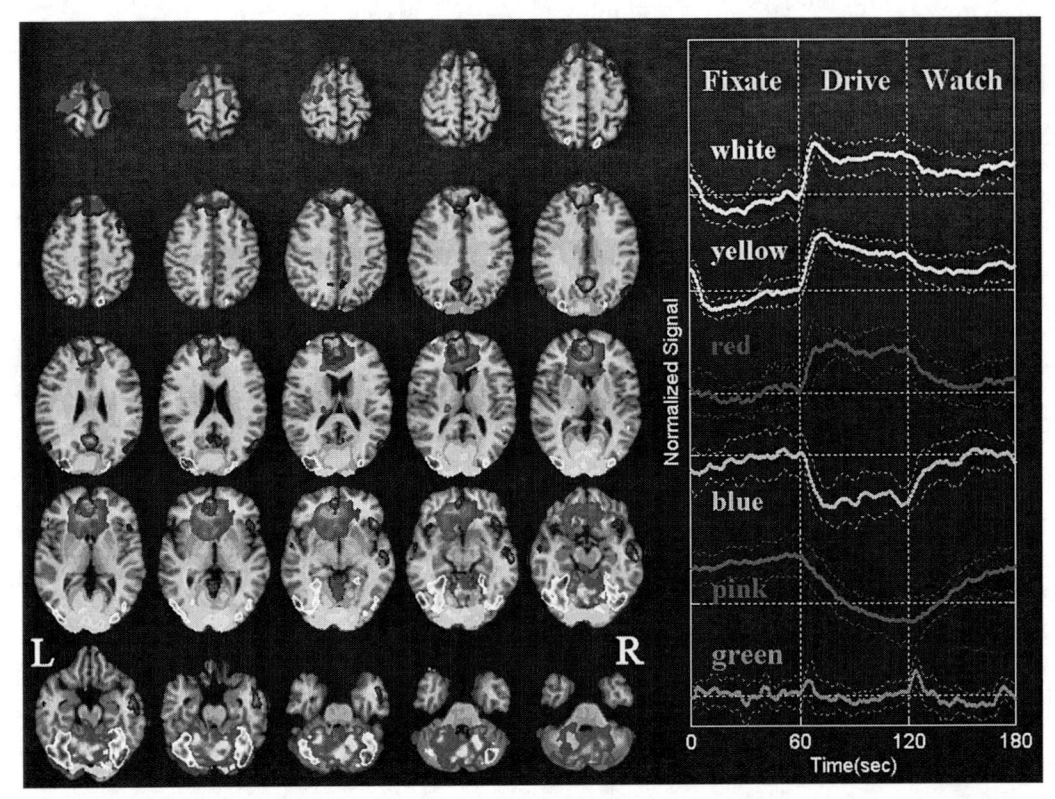

Figure 4.4. Independent component analysis results. Random effects group fMRI maps are thresholded at $p < 0.00025$ ($t = 4.5$, $df = 14$). A total of six components are presented. A green component extends on both sides of the parieto-occipital sulcus including portions of cuneus, precuneus, and the lingual gyrus. A yellow component contains mostly occipital areas. A white component contains bilateral visual association and parietal areas, and a component consisting of cerebellar and motor areas is depicted in red. Orbitofrontal and anterior cingulate areas identified are depicted in pink. Finally, a component including medial frontal, parietal, and posterior cingulate regions is depicted in blue. Group averaged time courses (right) for the fixate-drive-watch order are also depicted with similar colors. Standard deviation across the group of 15 scans is indicated for each time course with dotted lines. The epochs are averaged and presented as fixation, drive, and watch. See also color insert.

GLM approach failed to reveal the complexity of the temporal dynamics present within the data set.

The group-averaged time course for the fixation-drive-watch paradigm (with the standard deviation across the 15 subjects) for each component is presented on the right side of figure 4.4, with color use as in the spatial component maps. The three epoch cycles are averaged together and are presented as fixation, drive, and watch. Though it is intriguing to try to separate finer temporal correlates within the ICA time courses (such as the time of a crash), we did not have exact timing for individual events in the simulator used and thus limited our exploration to epoch-based averages. The ordering of the epochs did not significantly change the results. Each of the time courses depicted was modulated by the driving paradigm. Four main patterns are apparent: (1) primary visual (yellow) and higher-order visual/cerebellar (white) areas were most active during driving and less active during watching (as compared to fixation) (W,Y); (2) cerebellar/motor (red) and frontoparietal (blue) areas were only increased or decreased during driving; (3) anterior cingulate, medial frontal, and other frontal (pink) areas demonstrated exponential decrements during driving and rebounded during fixation; and (4) visual (green) areas transiently activated when the driving or watching paradigms were changed.

We also examined relationships between driving speed and activation. We measured driving speed to verify that the speed was within the range specified (100–140 or 220–308 kph). These data

Table 4.1. Selected Talairach Coordinates of the Volume and Maxima of Each Anatomic Region

Component	Area	Brodmann	Volume (cc)	Max t (x,y,z)
Pink	L/R anterior cingulate	10,24,25,32,42	12.6/11.1	18.94(−9,−35,−6)/15.38(3,43,−6)
	L/R medial frontal	9,10,11,25,32	8.2/11.3	16.97(−12,35,6)/14.45(6,40,−6)
	L/R inferior frontal	11,25(L),47	5.4/2.2	13.20(−9,40,−15)/9.02(12,−40,−15)
	L/R middle frontal	10,11,47	2.7/2.1	15.33(−18,37,−10)/9.34(12,43,−15)
	L/R superior frontal	9,10,11	1.6/3.8	11.60(−18,43,−11)/9.70(18,50,16)
Blue	L/R sup., med. frontal	6,8,9,10	33.4/29.6	14.68(−9,62,15)/ 15.83(3,50,2)
	L/R anterior cingulate	10,24,32,42	5.2/2.5	11.44(−3,49,−2)/14.22(3,47,7)
	L/R precuneus (parietal)	7,23,31	1.6/3.1	8.72(−3,−54,30)/11.38(6,−54,30)
	L/R inf. parietal, sup. marg., angular	39,40	1.4/2.0	7.91(−53,−59,35)/7.29(53,−68,31)
	L/R inferior frontal	11(R),45(R),47	0.8/3.4	6.23(−39,23,−14)/11.89(48,26,−10)
	L/R middle frontal	6,8	0.3/1.2	6.94(−18,34,44)/7.10(27,40,49)
	L/R cingulate/post. cing.	23,30,31,32	1.2/1.7	8.59(−3,−54,26)/11.88(6,−57,30)
Pink/blue (overlap)	L/R anterior cingulate	10,24,32,42	4.5/2.4	17.52(−9,38,3)/15.38(3,43,−6)
	L/R medial frontal	9,10,11,32	4.1/3.8	15.79(−6,38,−6)/15.83(3,50,2)
	L/R superior frontal	9,10	.2/.2	7.60(−9,52,−3)/8.43(12,51,20)
Red	L/R precentral	4(L),6	3.7/0.5	5.24(−33,−13,60)/3.45(24,−11,65)
	L/R thalamus	—	0.5/0.1	7.30(−15,−20,−10)/4.78(9,−14,5)
	L/R cerebellum	—	31.7/34.1	11.18(−6,−59,−10)/13.86(6,−71,−22)
Yellow	L/R lingual, cuneus	17,18,19(L)	18.4/16.6	15.1(−9,−96,−9)/18.21(18,−93,5)
	L/R fusiform, mid. occip.	18,19	7.2/5.9	13.07(−9,−95,14)/11.38(18,−98,14)
	L/R inferior occipital	17,18	1.6/1.3	8.91(−15,−91,−8)/10.48(21,−88,−8)
Green	L/R cuneus	17,18,19,23,30	9.7/7.2	22.88(−9,−81,9)/23.19(6,−78,9)
	L/R lingual	17,18,19	4.6/2.7	21.98(−6,−78,4)/16.12(15,−58,−1)
	L/R posterior cingulate	30,31	0.8/0.6	13.06(−12,−64,8)/12.33(9,−67,8)
White	L/R fusiform, inf. occip.	18,19,20,36(R),37	15.6/12.3	9.48(−48,−70,−9)/10.56(42,−74,−17)
	L/R cerebellum	—	13.0/13.5	9.36(−33,−50,−19)/12.08(42,−74,−22)
	L/R middle occipital	18,29,37	6.9/4.0	10.74(−33,−87,18)/7.11(33,−87,18)
	L/R lingual, cuneus	7,18,19	2.7/1.5	8.57(−27,−83,27)/7.22(24,−92,23)
	L/R parahippocampal	19,36,37	1.7/0.8	8.15(−24,−47,−10)/7.16(27,−44,−10)
	L/R superior occipital	19	1.1/0.1	12.04(−33,−86,27)/6.09(33,−86,23)
	L/R precuneus (parietal)	7,19(L)	0.9/7.7	6.91(−27,−77,36)/7.90(18,−70,50)
	L/R inferior temporal	19,37	0.7/0.4	6.26(−48,−73,−1)/6.04(48,−71,−1)
	L/R superior parietal	7	0.4/0.5	6.38(−18,−67,54)/6.77(18,−70,54)

Note. For each component, all the voxels above the threshold were entered into a database to provide anatomical and functional labels for the left (L) and right (R) hemispheres. The volume of activated voxels in each area is provided in cubic centimeters (cc). Within each area, the maximum t value and its coordinate is provided.

are summarized for each subject in table 4.2. Note that the average speed for the faster group was slower than the range specified. This was because there were more collisions at the faster speed, thus reducing the average (although it was still significantly higher than the slower group). The increase in collisions indicates that subjects are making more errors, presumably because of a desire to comply with the specified speed range.

The frontoparietal (blue) component was decreased during the drive epoch only. Calculating the change in activation for the two speed groups and

Table 4.2. Average and Maximum Driving Speeds for the 15 Subjects (Kilometers per Hour)

	Slow		Fast	
Subject	Average	Maximum	Average	Maximum
1	106	142	140	248
2	106	126	125	190
3	117	138	124	220
4	113	131	122	206
5	114	140	151	196
6	120	140	137	192
7	98	141	155	246
8			159	213
Avg.	111	137	139	214

Note. All subjects performed the task as instructed and the slow drivers all had average speeds lower than those of the fast drivers.

comparing them with a t test revealed significantly ($p < 0.02$) greater driving-related changes when subjects were driving faster. This is depicted in figure 4.5a and is consistent with an overall increase in vigilance while driving at the faster speed. Previous imaging studies have implicated similar frontal and parietal regions in visual awareness (Rees, 2001).

Examination of the orbitofrontal/anterior cingulate (pink) time courses during the driving epoch revealed an exponentially decaying curve, displayed in figure 4.4. We extracted the signal decreases during the 180 (3 × 60) time points of driving $s(n)$ and normalized so that the average of the asterisk epoch was 1 and the minimum of the driving epoch was 0 (results were not sensitive to this normalization). Next we transformed the orbitofrontal time course to a linear function through the equation $an = \ln[s(n) + 1]$ and fit a line to it such that $x(n) = ay(n) + b$. The resultant fit was very good (average $R = 0.9$).

Comparing the subjects driving slower with those driving faster using a t test revealed a significant difference ($p < .02$) in the rate parameter $\hat{a}$. That is, the *rate* of decrease in anterior cingulate pink signal is faster in the subjects driving at a faster rate (a more difficult task). This speed-related rate change was also consistent with the involvement of the orbitofrontal and anterior cingulate involvement in disinhibition (i.e., "taking off the brake"; Blumer & Benson, 1975; Rauch et al., 1994). A comparison of the rates is depicted in figure 4.5b. No significant correlations were found for any of the other revealed components.

The components identified by our analysis lend themselves naturally to interpretation in terms of well-known neurophysiological networks. This interpretation is depicted graphically in figure 4.6 using colors corresponding to those in the imaging results. Results are divided into six domains containing one or more networks: (1) vigilance, (2) error monitoring and inhibition, (3) motor, (4) visual, (5) higher order visual/motor, and (6) visual monitoring. Components are grouped according to their modulation by driving with the speed-modulated components indicated as well. In this figure, we also show additional refinements that have been made in a subsequent analysis involving alcohol intoxication (Calhoun, Pekar, & Pearlson, 2004).

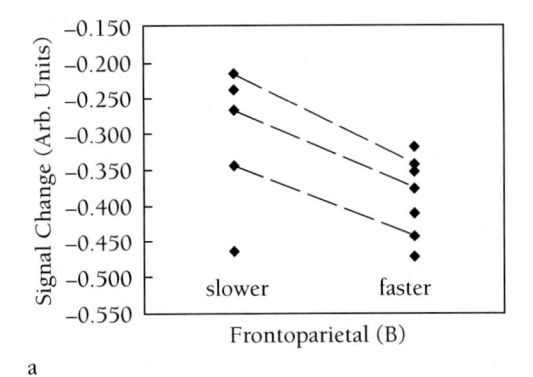

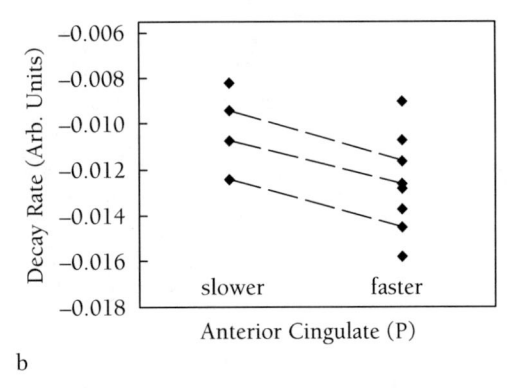

Figure 4.5. Components demonstrating significant speed-related modulation. The frontoparietal component (a) demonstrated decreases during the driving epoch and greater decreases for the faster drivers. The anterior cingulate/orbitofrontal component (b) decreased more rapidly for the faster drivers. The lines connect the three subjects who were scanned during both faster and slower driving.

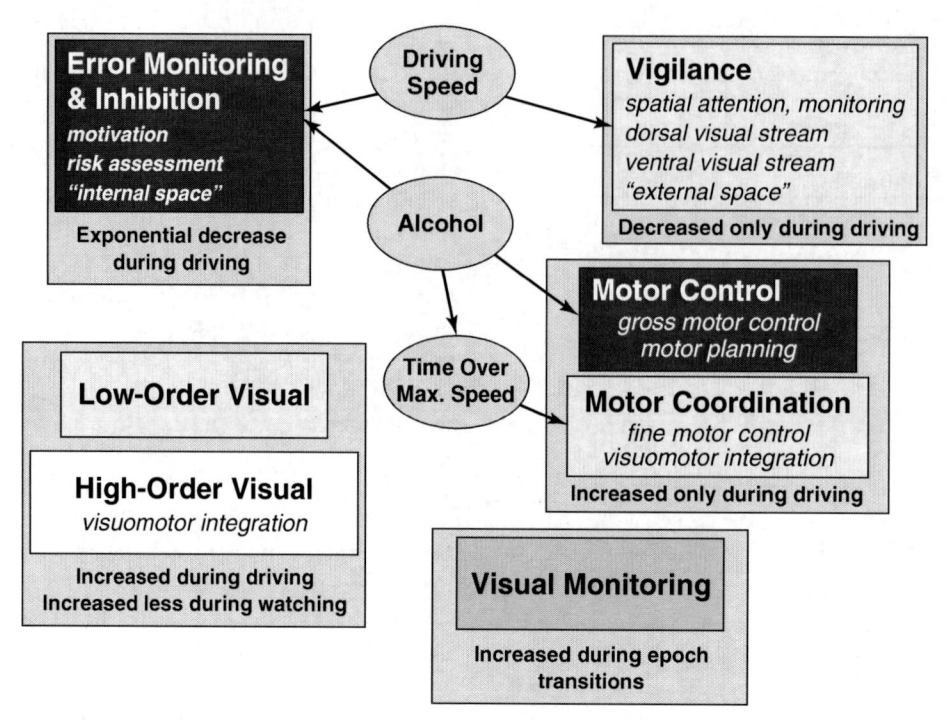

Figure 4.6. Interpretation of imaging results involved in driving. The colors correspond to those used in figure 2.4. Components are grouped according to the averaged pattern demonstrated by their time courses. The speed-modulated components are indicated with arrows. See also color insert.

Discussion

A previous fMRI study involving simulated aviation found speed-related changes in frontal areas similar to those that we have observed (Peres et al., 2000). Attentional modulation may explain the speed-related changes as these two components (blue, pink) include areas implicated in both the anterior and posterior divisions of the attention system (Posner, DiGirolamo, & Fernandez-Duque, 1997; Schneider, Pimm-Smith, & Worden, 1994). Orbitofrontal cortex has been demonstrated to exhibit fMRI signal change during breaches of expectation (i.e., error detection) in a visual task (Nobre, Coull, Frith, & Mesulam, 1999). Our finding of anterior cingulate cortex in the pink component and both anterior and posterior cingulate cortex in the blue component is consistent with recent studies demonstrating functionally distinct anterior and posterior cingulate regions in spatial attention (Mesulam, Nobre, Kim, Parrish, & Gitelman, 2001). The frontal and parietal regions identified in the blue component have also been implicated in attentional tasks involving selected and divided attention (Corbetta, Miezin, Dobmeyer, Shulman, & Petersen, 1991; Corbetta et al., 1998). The angular gyrus, superior parietal gyrus, and posterior cingulate gyrus were also identified in the simulated aviation task. In our study, these areas were contained within the same component and are thus functionally connected to one another; that is, they demonstrate similar fMRI signal changes, and in that sense are distinct from areas involved in other components.

It is also informative to consider the components identified in the context of their interactions. The overlapping areas of the blue and pink components, consisting mainly of portions of the anterior cingulate and the medial frontal gyrus, are indicated in figure 4.4 and table 4.1. The anterior cingulate has been divided into rostral affect and caudal cognition regions (Devinsky, Morrell, & Vogt, 1995), consistent with the division between the pink and blue components. Note that activity in the blue regions decreases rapidly during the driving epoch, whereas the pink regions decrease slowly during the driving condition. One interpretation of these results is that a narrowing of attention (vigilance) is initiated once the driving condition begins. Error correction and disinhibition are revealed as a gradual decline of this component at a rate determined in part by the vigilance network. During the fast driving condition, the vigilance component changes more; thus the error correction and disinhibition component decreases at a faster rate. An EEG study utilizing the Need for Speed simulation software revealed greater alpha power in the frontal lobes during replay than during driving and was interpreted as being consistent with a reduction of attention during the replay task (Schier, 2000). Our results are consistent with this interpretation, as neither error monitoring nor focused vigilance is presumably prominent during replay (watching).

The inverse correlation between speed and frontal/cingulate activity suggests an alternative interpretation. Driving fast should engage reflex or overlearned responses more than critical reasoning resources. As the cingulate gyrus has been claimed to engage in the monitoring of such resources, such a view implies that the cingulate gyrus should be less activated as well. Such interpretations, while consistent with our results, require further investigation. For example, it would be interesting to study the effect of speed at more than two levels, in a parametric manner. In addition, physiological measures such as galvanic skin conductance could be used to test for correlations with stress levels.

Other activated components we observed were consistent with prior reports. For example, the visual association/cerebellar component (white) demonstrates activation in regions previously found to be involved in orientation (Allen, Buxton, Wong, & Courchesne, 1997) and complex scene interpretation or memory processing (Menon, White, Eliez, Glover, & Reiss, 2000). This component also appears to contain areas involved in the modulation of connectivity between primary visual (V2) and motion-sensitive visual regions (V5/area MT), such as parietal cortex (Brodmann area 7), along with visual association areas (Friston & Buchel, 2000). These areas, along with the primary visual areas (yellow), have been implicated in sensory acquisition (Bower, 1997) and attention or anticipation (Akshoomoff, Courchesne, & Townsend, 1997). Activation in both the white and yellow components was increased above fixation during watching and further increased during driving. This is in contrast to Walter et al. (2001), but is consistent with sensory acquisition (present in both driving and watching) combined with the attentional and motor elements of driving (present in the driving epoch). That is, the further increase in these areas during driving appears to be an attentional modulated increase (Gandhi, Heeger, & Boynton, 1999).

The transient visual areas (green) demonstrate an increase at the transitions between epochs. We identified similar areas, also transiently changing between epochs, in a simple visual task (Calhoun, Adali, Pearlson, & Pekar, 2001b). Similar areas have been detected in a meta-analysis of transient activation during block transitions (Konishi, Donaldson, & Buckner, 2001) and may be involved in switching tasks in general. The red component was mostly in the cerebellum in areas implicated in motor preparation (Thach, Goodkin, & Keating, 1992). Primary motor contributions were low in amplitude, presumably due to the small amount of motor movement involved in controlling the driving task (0.1/0.9 Hz rate on average for left or right hand, no significant difference with speed). This would also explain why there was little activation during the watching epoch as during this time motor preparation and visuomotor integration are presumably minimal.

The goal of the example fMRI study described in this chapter was to capture the temporal nature of the neural correlates of the complex behavior of driving. We decomposed the activation due to a complex behavior into interpretable pieces using a novel, generally applicable approach, based upon ICA (see figure 4.6). Several components were identified, each modulated differently by our imaging paradigm. Regions that increased or decreased consistently, increased transiently, or which exhibited gradual signal decay during driving were identified. Additionally, two of the components in regions implicated in vigilance and error monitoring or inhibition processes were significantly associated with driving speed. Imaging results are grouped into cognitive domains based upon the areas recruited and their modulation with our paradigm.

In summary, the combination of driving simulation software and an advanced analytic technique enabled us to study simulated driving and develop a model for simulated driving and brain activation that did not previously exist. It is clear that driving is a complex task. The ability to study, with fMRI, a complex behavior such as driving, in conjunction with paradigms studying more specific aspects of cognition, may enhance our overall understanding of the neural correlates of complex behaviors.

Acknowledgments. Supported by the National Institutes of Health under Grant 1 R01 EB 000840-02, by an Outpatient Clinical Research Centers Grant M01-RR00052, and by P41 RR 15241-02.

Main Points

1. Functional MRI provides an unprecedented view of the living human brain.
2. The use of fMRI paradigms involving naturalistic behaviors has proven challenging due to difficulties in software design and analytic approaches.
3. Current simulation environments can be imported into the fMRI environment and used as a paradigm.
4. Simulated driving is one such paradigm, with important implications for both brain research and the understanding of how the impaired brain functions in a real-world environment.
5. Advanced analytic approaches, including independent component analysis, are enabling the analysis and interpretation of the complex data sets generated during naturalistic behaviors.

Key Readings

Calhoun, V. D., Adali, T., Pearlson, G. D., & Pekar, J. J. (2001). A method for making group inferences from functional MRI data using independent component analysis. *Human Brain Mapping, 14,* 140–151.

Calhoun, V. D., Pekar, J. J., & Pearlson, G. D. (2004). Alcohol intoxication effects on simulated driving: Exploring alcohol-dose effects on brain activation using functional MRI. *Neuropsychopharmacology, 29,* 2097–2107.

McKeown, M. J., Makeig, S., Brown, G. G., Jung, S. S., Kindermann, S., Bell, A. J., et al. (1998). Analysis of fMRI data by blind separation into independent spatial components. *Human Brain Mapping, 6,* 160–188.

Turner, R., Howseman, A., Rees, G. E., Josephs, O., & Friston, K. (1998). Functional magnetic resonance imaging of the human brain: Data acquisition and analysis. *Experimental Brain Research, 123,* 5–12.

References

Akshoomoff, N. A., Courchesne, E., & Townsend, J. (1997). Attention coordination and anticipatory control. *International Review of Neurobiology, 41,* 575–598.

Allen, G., Buxton, R. B., Wong, E. C., & Courchesne, E. (1997). Attentional activation of the cerebellum independent of motor involvement. *Science, 275,* 1940–1943.

Ballard, D. H., Hayhoe, M. M., Salgian, G., & Shinoda, H. (2000). Spatio-temporal organization of behavior. *Spatial Vision, 13,* 321–333.

Bell, A. J., & Sejnowski, T. J. (1995). An information maximization approach to blind separation and blind deconvolution. *Neural Computing, 7,* 1129–1159.

Biswal, B. B., & Ulmer, J. L. (1999). Blind source separation of multiple signal sources of fMRI data sets using independent components analysis. *Journal of Computer Assisted Tomography, 23,* 265–271.

Blumer, D., & Benson, D. F. (1975). Psychiatric aspects of neurologic diseases. In D. F. Benson & D. Blumer (Eds.), *Personality changes with frontal and temporal lobe lesions* (pp. 151–170). New York: Grune and Stratton.

Bower, J. M. (1997). Control of sensory data acquisition. *International Review of Neurobiology, 41,* 489–513.

Cabeza, R., & Nyberg, L. (2000). Imaging cognition II: An empirical review of 275 PET and fMRI studies. *Journal of Cognitive Neuroscience, 12,* 1–47.

Calhoun, V. D., Adali, T., Hansen, J. C., Larsen, J., & Pekar, J. J. (2003). ICA of fMRI: An overview. In *Proceedings of the International Conference on ICA and BSS,* Nara, Japan.

Calhoun, V. D., Adali, T., Kraut, M., & Pearlson, G. D. (2000). A weighted-least squares algorithm for estimation and visualization of relative latencies in event-related functional MRI. *Magnetic Resonance in Medicine, 44,* 947–954.

Calhoun, V. D., Pekar, J. J., & Pearlson, G. D. (2004). Alcohol intoxication effects on simulated driving: Exploring alcohol-dose effects on brain activation using functional MRI. *Neuropsychopharmacology, 29,* 2097–2107.

Calhoun, V. D., Adali, T., Pearlson, G. D., & Pekar, J. J. (2001a). A method for making group inferences from functional MRI data using independent component analysis. *Human Brain Mapping, 14,* 140–151.

Calhoun, V. D., Adali, T., Pearlson, G. D., & Pekar, J. J. (2001c). Spatial and temporal independent component analysis of functional MRI data containing a pair of task-related waveforms. *Human Brain Mapping, 13,* 43–53.

Calhoun, V. D., Adali, T., Stevens, M., Kiehl, K. A., & Pekar, J. J. (2005). Semi-blind ICA of fMRI: A method for utilizing hypothesis-derived time courses in a spatial ICA analysis. *Neuroimage, 25,* 527–538.

Corbetta, M., Akbudak, E., Conturo, T. E., Snyder, A. Z., Ollinger, J. M., Drury, H. A., et al. (1998). A common network of functional areas for attention and eye movements. *Neuron, 21,* 761–773.

Corbetta, M., Miezin, F. M., Dobmeyer, S., Shulman, G. L., & Petersen, S. E. (1991). Selective and divided attention during visual discriminations of shape, color and speed: Functional anatomy by positron emission tomography. *Journal of Neuroscience, 11,* 2383–2402.

Devinsky, O., Morrell, M. J., & Vogt, B. A. (1995). Contributions of anterior cingulated cortex to behaviour. *Brain, 118*(Pt. 1), 279–306.

Duann, J. R., Jung, T. P., Kuo, W. J., Yeh, T. C., Makeig, S., Hsieh, J. C., et al. (2002). Single-trial variability in event-related BOLD signals. *Neuroimage, 15,* 823–835.

Electronic Arts. (1998). *Need for Speed II* [computer driving game]. Redwood City, CA: Author.

Frackowiak, R. S. J., Friston, K. J., Frith, C., Dolan, R., Price, P., Zeki, S., et al. (2003). *Human brain function.* New York: Academic Press.

Friston, K. J., & Buchel, C. (2000). Attentional modulation of effective connectivity from V2 to V5/MT in humans. *Proceedings of the National Academy of Sciences USA, 97,* 7591–7596.

Friston, K. J., Fletcher, P., Josephs, O., Holmes, A., Rugg, M. D., & Turner, R. (1998). Event-related FMRI: Characterizing differential responses. *Neuroimage, 7,* 30–40.

Gandhi, S. P., Heeger, D. J., & Boynton, G. M. (1999). Spatial attention affects brain activity in human primary visual cortex. *Proceedings of the National Academy of Sciences USA, 96,* 3314–3319.

Glover, G. H. (1999). Deconvolution of impulse response in event-related BOLD fMRI. *Neuroimage, 9,* 416–429.

Groeger, J. (2000). *Understanding driving: Applying cognitive psychology to a complex everyday task.* New York: Psychology Press.

Holmes, A. P., & Friston, K. J.. (1998). Generalizability, random effects, and population inference. *NeuroImage, 7,* S754.

Hopfinger, J. B., Buonocore, M. H., & Mangun, G. R. (2000). The neural mechanisms of top-down attentional control. *Nature Neuroscience, 3,* 284–291.

Janca, A., Ustun, T. B., & Sartorius, N. (1994). New versions of World Health Organization instruments for the assessment of mental disorders. *Acta Psychiatrica Scandinavica, 90,* 73–83.

Konishi, S., Donaldson, D. I., & Buckner, R. L. (2001). Transient activation during block transition, *Neuroimage, 13,* 364–374.

McGinty, V. B., Shih, R. A., Garrett, E. S., Calhoun, V. D., & Pearlson, G. D. (2001). Assessment of intoxicated driving with a simulator: A validation study with on road driving. In *Proceedings of the Human Centered Transportation and Simulation Conference,* November 4–7, 2001; Iowa City, IA.

McKeown, M. J., Hansen, L. K., & Sejnowski, T. J. (2003). Independent component analysis of functional MRI: What is signal and what is noise? *Current Opinion in Neurobiology, 13,* 620–629.

McKeown, M. J., Makeig, S., Brown, G. G., Jung, T. P., Kindermann, S. S., Bell, A. J., et al. (1998). Analysis of fMRI data by blind separation into independent spatial components. *Human Brain Mapping, 6,* 160–188.

Menon, V., White, C. D., Eliez, S., Glover, G. H., & Reiss, A. L. (2000). Analysis of a distributed neural system involved in spatial information, novelty, and memory processing. *Human Brain Mapping, 11,* 117–129.

Mesulam, M. M., Nobre, A. C., Kim, Y. H., Parrish, T. B., & Gitelman, D. R. (2001). Heterogeneity of cingulated contributions to spatial attention. *Neuroimage, 13,* 1065–1072.

Nobre, A. C., Coull, J. T., Frith, C. D., & Mesulam, M. M. (1999). Orbitofrontal cortex is activated during breaches of expectation in tasks of visual attention. *Nature Neuroscience, 2,* 11–12.

Peres, M., Van de Moortele, P. F., Pierard, C., Lehericy, S., LeBihan, D., & Guezennez, C. Y. (2000). Functional magnetic resonance imaging of mental strategy in a simulated aviation performance task. *Aviation, Space and Environmental Medicine, 71,* 1218–1231.

Posner, M. I., DiGirolamo, G. J., & Fernandez-Duque, D. (1997). Brain mechanisms of cognitive skills. *Consciousness and Cognition, 6,* 267–290.

Ranney, T. A. (1994). Models of driving behavior: A review of their evolution. *Accident Analysis and Prevention, 26,* 733–350.

Rauch, S. L., Jenike, M. A., Alpert, N. M., Baer, L., Brieter, H. C., Savage, C. R., et al. (1994). Regional cerebral blood flow measured during symptom provocation in obsessive-compulsive disorder using oxygen 15-labeled carbon dioxide and positron emission tomography. *Archives of General Psychiatry, 51,* 62–70.

Rees, G. (2001). Neuroimaging of visual awareness in patients and normal subjects. *Current Opinion in Neurobiology, 11,* 150–156.

Schier, M. A. (2000). Changes in EEG alpha power during simulated driving: A demonstration. *International Journal of Psychophysiology, 37,* 155–162.

Schneider, W., Pimm-Smith, M., & Worden, M. (1994). Neurobiology of attention and automaticity. *Current Opinion in Neurobiology, 4,* 177–182.

Talairach, J., & Tournoux, P. (1988). *A co-planar stereotaxic atlas of a human brain.* Stuttgart: Thieme.

Thach, W. T., Goodkin, H. P., & Keating, J. G. (1992). The cerebellum and the adaptive coordination of movement. *Annual Review of Neuroscience, 15,* 403–442.

Van de Moortele, P. F., Cerf, B., Lobel, E., Paradis, A. L., Faurion, A., & Le Bihan, D. (1997). Latencies in FMRI time-series: Effect of slice acquisition order and perception. *NMR Biomedicine, 10,* 230–236.

Walter, H., Vetter, S. C., Grothe, J., Wunderlich, A. P., Hahn, S., & Spitzer, M. (2001). The neural correlates of driving. *Neuroreport, 12,* 1763–1767.

Woods, R. P. (1996). Modeling for intergroup comparisons of imaging data. *Neuroimage, 4,* S84–S94.

Worsley, K. J., & Friston, K. J. (1995). Analysis of fMRI time-series revisited—again. *Neuroimage, 7,* 30–40.

Gabriele Gratton and Monica Fabiani

Optical Imaging of Brain Function

In this chapter, we review the use of noninvasive optical imaging methods in studying human brain function, with a view toward their possible applications to neuroergonomics. Fast optical imaging methods make it possible to image brain activity with subcentimeter spatial resolution and millisecond-level temporal resolution. In addition, these methods are also relatively inexpensive and adaptable to different experimental situations. These properties make optical imaging a promising new approach for the study of brain function in experimental and possibly applied situations. The main limitations of optical imaging methods are their penetration (a few centimeters from the surface of the head) and, at least at present, a relatively low signal-to-noise ratio.

Optical imaging methods are a class of techniques that investigate the way light interacts with tissue. If the tissue is very superficial, light of many wavelengths can be used for the measurements. However, if the tissue is deep, like the brain, only a narrow wavelength range is usable for measurement—a window between 680 and 1,000 nm (commonly called the near-infrared or NIR range). This is because at other wavelengths water and hemoglobin absorb much of the light, whereas in the NIR range these substances absorb relatively little. Near-infrared photons can penetrate deeply into tissue (up to 5 cm). The main obstacle to the movement of NIR photons through the head is represented by the fact that most head tissues (such as the skull and gray and white matter) are all highly scattering. Thus, the movement of photons through the head can be represented as a diffusion process, and noninvasive optical imaging is sometimes called diffusive optical imaging. Besides limiting the penetration of the photons, the scattering process also limits the spatial resolution of the technique to a few millimeters.

Work conducted in our laboratory and others during the last few years has shown that optical imaging can be used to study the time course of activity in specific cortical areas. We were the first to show that noninvasive optical imaging is sensitive to neuronal activity (Gratton, Corballis, Cho, Fabiani, & Hood, 1995). This result has been replicated many times in our laboratory (for a review, see Gratton & Fabiani, 2001), and in at least four other laboratories (Franceschini, & Boas, 2004; Steinbrink et al., 2000; Tse, Tien, & Penney, 2004; Wolf, Wolf, Choi, Paunescu, 2003; Wolf, Wolf, Choi, Toronov, et al., 2003). In a number of different studies, we have found that this phenomenon can be observed in all cortical areas investigated (visual cortex, auditory cortex, somatosensory cortex, motor areas, prefrontal and parietal cortex). Further, we have developed a technology for full-head imaging (using 1,024 channels) that has been successfully applied to cognitive paradigms. We have also developed a full array of recording and analysis tools that make noninvasive cortical imaging a practical method for research and potentially for some applied settings. In the remainder of this chapter, we review the main principles of optical imaging and some of the types of optical signals that can be used to study brain function, as well as aspects of recording and analysis and examples of applications to cognitive neuroscience issues.

Principles of Noninvasive Optical Imaging

Noninvasive optical imaging is based on two major observations:

1. Near-infrared light penetrates several centimeters into tissue. Because of the scattering properties of most head tissues and the size of the adult human head, in the case of noninvasive brain imaging this process approximates a diffusion process.
2. The parameters of this diffusion process are influenced by physiological events in the brain. These parameters are related to changes in the scattering and absorption properties of the tissue itself.

Noninvasive optical imaging studies are based on applying small, localized sources of light to points on the surface of the head and measuring the amount or delay of the light reaching detectors located also on the surface at some distance (typically a few centimeters). Note that, because of the diffusion process, this procedure is sensitive not only to superficial events but also to events occurring at some depth in the tissue. To understand this phenomenon, consider that the photons emitted by the sources will propagate in a random fashion through the tissue. Because the movement of the photons is random, there is no preferred direction, and if the medium was infinite and homogenous, the movement of the photons could describe a sphere. However, in reality the diffusive tissue (e.g., the head) is surrounded by a nondiffusive medium (e.g., air). In this situation, photons that are very close to the surface are likely to move outside the diffusive medium. Once they enter the nondiffusive medium (air), their movement becomes rectilinear. Therefore, they are not likely to enter the head again and will not reach the detector. This means that only photons traveling at some depth inside the tissue are likely to reach the detector. For this reason, diffusive optical imaging methods (also called photon migration methods) become sensitive to changes in optical properties that occur at some depth inside the tissue (see Gratton, Maier, Fabiani, Mantulin, & Gratton, 1994). The depth investigated by a particular source-detector configuration depends on many parameters, one of the most important of which is the source-detector distance (see figure 5.1; see also Gratton, Sarno, Maclin, Corballis, & Fabiani, 2000).

Currently, two main families of methods for noninvasive optical imaging are in use: (a) simpler methods based on continuous light sources (continuous wave or CW methods); and (b) more elaborate techniques based on light sources varying in intensity over time (time-resolved or TR methods). Although more complex and expensive, TR methods provide important additional information about the movement of photons through tissue, and specifically the ability to measure the photons' time-of-flight (i.e., the time required by photons to move between a source and a detector). This parameter is very important for the accurate measurement of optical properties (scattering and absorption) of the tissue and, in our observations, has been particularly useful for studying fast changes in optical parameters presumably related to neuronal activity.

Optical Signals

A large body of data supports the idea that at least two categories of signals can be measured with noninvasive optical methods: (a) fast signals, occurring simultaneously with neuronal activity; and (b) slow signals, mostly related to the changes in the hemodynamic and metabolic properties of the tissue that lag neuronal activity of several seconds.

Fast signals are related to changes in the scattering of neural tissue that occur simultaneously with electrical events related to neuronal activity. The first demonstration of these scattering changes was obtained by Hill and Keynes (1949). Subsequent studies have confirmed these original findings,

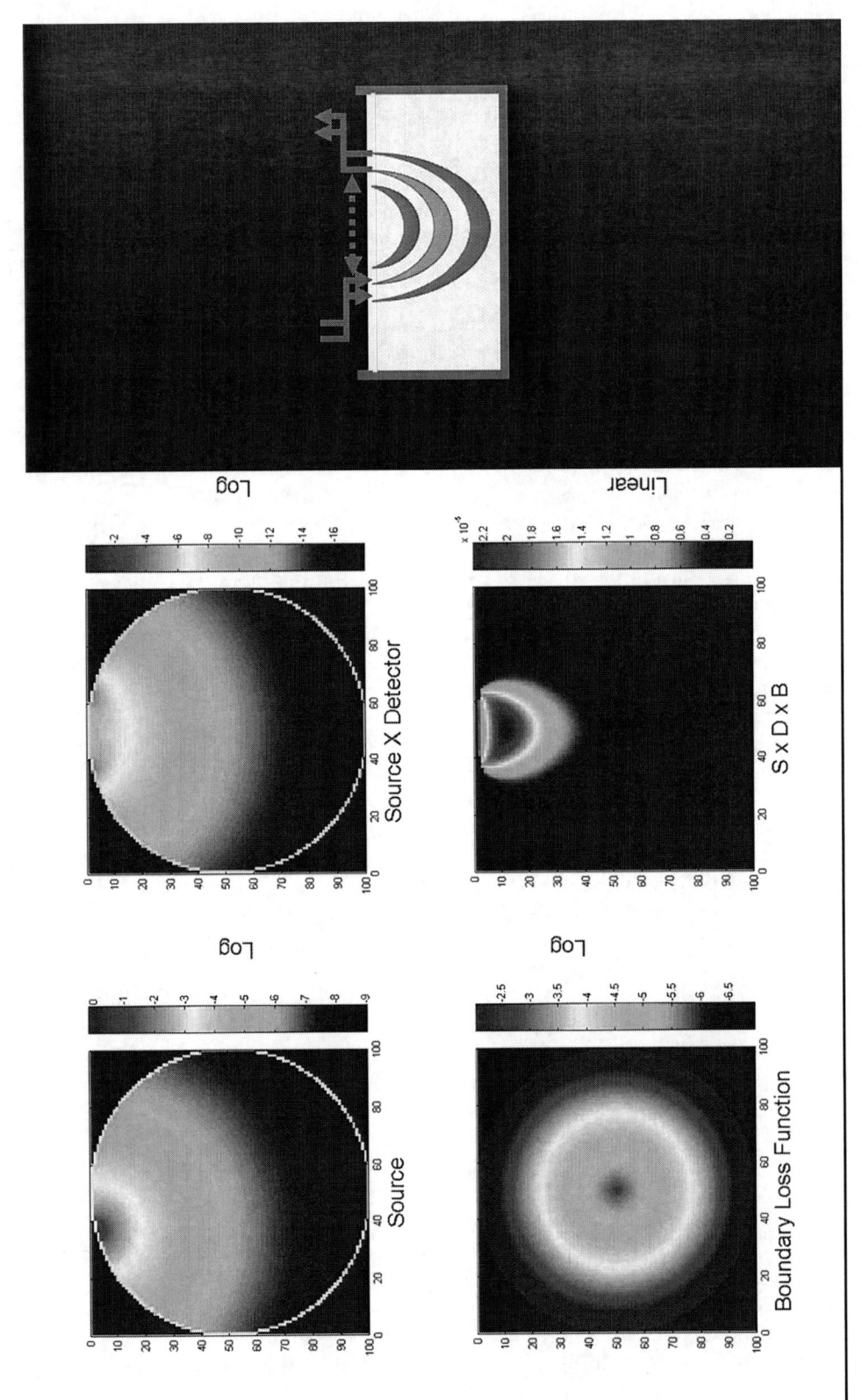

Figure 5.1. Mathematical model of the propagation of light into a scattering medium. Top left, diffusion of light in the scattering medium from a surface source; top middle, area of diffusion for photons produced by a surface source and reaching a source detector; top right, schematic representation of variations in the depth of the diffusion spindle as a function of variations in source-detector distance. See also color insert.

first in isolated neurons (Cohen, Hille, Keynes, Landowne, & Rojas, 1971), then in vitro in hippocampal slices (Frostig, Lieke, Ts'o, & Grinvald, 1990), and finally in the central nervous system of living invertebrates (Stepnoski et al., 1991) and mammals (Rector, Poe, Kristensen, & Harper, 1997). The evidence that this phenomenon can also be observed noninvasively in humans is reviewed in a later section. These scattering changes are assumed to be directly associated with postsynaptic depolarization and hyperpolarization (MacVicar & Hochman, 1991). Two phenomena probably contribute to these scattering changes: (a) cell swelling and shrinking, due to the movement of water in and out of the cell (associated with the movement of ions; MacVicar & Hochman, 1991); and (b) membrane deformation, associated with reorientation of membrane proteins and phospholipids (Stepnoski et al., 1991). Which of these two phenomena is most critical is still unclear. The evidence suggests that, at least at the macro level relevant for noninvasive imaging, the fast signal can be considered as a scalar, isotropic change in the scattering properties of a piece of nervous tissue, with decrease in scattering related to excitation of a particular area, and increase in scattering related to inhibition of the same area (Rector et al., 1997).

The results obtained by Rector et al. (1997) with fibers implanted in the rat hippocampus led to specific predictions about the type of phenomenon that should be observed using sources and detectors on the same side of the relevant tissue, as is the case with noninvasive cortical measures in the adult human brain. In fact, cortical excitation should be associated with an increase in the average photon time of flight (because, on average, the photons penetrate deeper into the tissue), and cortical inhibition should be associated with a reduction in the average time of flight (because, on average, the photons penetrate less into the tissue). As shown later, this is exactly the result we typically obtain in our studies.

It should be noted that these predictions only hold for situations in which the activated area is at least 5–10 mm deep, because at shallower depths light propagation may not follow a diffusive regime. It should further be noted that these predictions are related to TR measurements. In fact, fast optical effects can also be recorded with CW methods (as demonstrated by Steinbrink et al., 2000; and by Franceschini & Boas, 2004), typically as increases in the transparency of active cortical areas. However, intensity estimates (the only ones available with CW methods) are more sensitive to superficial (nonbrain) effects and to slow hemodynamic effects (see below) than time-of-flight measures (available only with TR methods). Therefore, in our experience, TR methods appear to be particularly suited for the study of fast optical signals. In any case, fast effects are relatively small and typically require extensive averaging for their measurement.

Intrinsic changes in the coloration of the brain due to variations in blood flow in active brain areas have been shown in an extensive series of studies on optical imaging of exposed animal cortex conducted in the last 20 years (see Frostig et al., 1990; Grinvald, Lieke, Frostig, Gilbert, & Wiesel, 1986). The intrinsic optical signal is typically observed over several seconds after the stimulation of a particular cortical area. This signal capitalizes in part on differences in the absorption spectra of oxy- and deoxyhemoglobin. Using a spectroscopic approach, Malonek and Grinvald (1996) demonstrated three major types of optical changes that develop over several seconds after stimulation of a particular cortical area: (a) a relatively rapid blood deoxygenation occurring during the first couple of seconds (this signal has been associated with the initial dip in the functional magnetic resonance imaging (fMRI) signal reported by Menon et al., 1995); (b) a large increase in oxygenation occurring several seconds after stimulation (the drop in deoxyhemoglobin concentration associated with this signal corresponds to the blood oxygenated level-dependent [BOLD] effect observed with fMRI); and (c) changes in scattering beginning immediately after the cortex is excited and lasting until the end of stimulation, whose nature is not completely understood (these changes in scattering may include the fast effects reported above). Of these three signals, the oxygenation effect has been first reported in noninvasive imaging from the human cortex by Hoshi and Tamura (1993) and subsequently replicated in a large number of studies (for an early review of this work, see Villringer & Chance, 1997). The other slow optical signals are more difficult to observe noninvasively, probably because they are dwarfed by the large oxygenation effect. Slow effects are easy to observe sometimes even on single trials. They can be detected using both CW and TR methods, although TR methods provide a more precise quantification of oxy- and deoxyhemoglobin concentration.

In summary, two types of signals can be detected using noninvasive optical imaging methods: (a) fast scattering effects related to neuronal activity, and (b) slow absorption effects related to hemodynamic changes. These signals are reviewed in more details in the following sections.

Fast Optical Signals

The Event-Related Optical Signal

During the last decades, our laboratory has carried out a large number of studies demonstrating the possibility of detecting the fast optical signal with noninvasive methods. Our initial observations were in response to visual stimulation (Gratton, Corballis, et al., 1995; for a replication, see Gratton & Fabiani, 2003). This work, using a frequency-domain TR method, revealed the occurrence of a fast optical signal marked by an increase in the photons' time-of-flight parameter (also called phase delay, because it is obtained by measuring the phase delay of the photon density wave of NIR light modulated at high—MHz range—frequencies). This study was based on the stimulation (obtained through pattern reversals) of each of the four quadrants of the visual field in different blocks. Recordings were obtained from an array of locations over the occipital cortex. The prediction was made that each stimulation condition should generate a response in a different set of locations, following the contralateral inverted representation of the visual field in the occipital cortex. In this way, for each location of the occipital cortex, we could compare the optical response obtained when that region was expected to be stimulated with the optical response obtained when other regions were expected to be stimulated. This paradigm controlled for a number of possible alternative explanations for the results (i.e., nonspecific brain changes, movements, or luminance artifacts). In addition, it provided an initial estimate of the spatial resolution of the technique, since it implied that the optical response was limited to the stimulated cortical area. The original article showed a clear differentiation between the four responses (Gratton, Corballis, et al., 1995), but was based on a small number of subjects, due to the fact that the recording equipment available at the time was based on one channel, and 12 sessions were required to derive a map for one subject. For these reasons, we later replicated this study using multichannel equipment and a larger number of subjects (Gratton & Fabiani, 2003). In this second study, we also increased the temporal sampling (from 20 to 50 Hz). Thus, we could determine that the peak latency of the response in visual (possibly V1) cortex (which occurred between 50 and 100 ms in the original study) is in fact close to 80 ms—corresponding to the peak latency of the C1 component of the visual evoked potential (see figure 5.2). Further, when fMRI and EROS data from the same subjects are compared, it is apparent that there

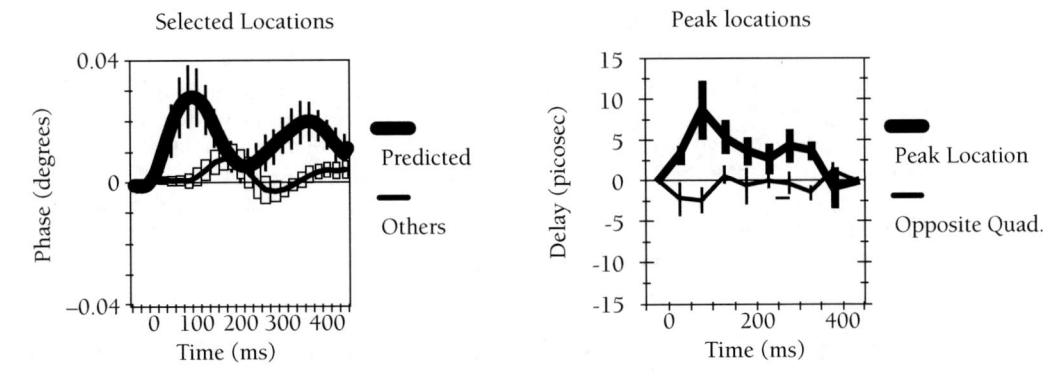

Figure 5.2. Left panel: Time course of the fast optical response (EROS) observed at the predicted (thick line) and control (thin line) locations in Gratton and Fabiani (2003). The error bars indicate the standard error of estimate computed across subjects. Right panel: Time course of the fast optical response (EROS) for locations with maximum response (thick line) and, from the same locations, when the opposite visual field quadrant was stimulated (thin line). From Gratton, G., Corballis, P. M., Cho, E., Fabiani, M., & Hood, D.C. (1995). Shades of gray matter: Noninvasive optical images of human brain responses during visual stimulation. *Psychophysiology, 32,* 505–509. Reprinted with permission from Blackwell Publishing.

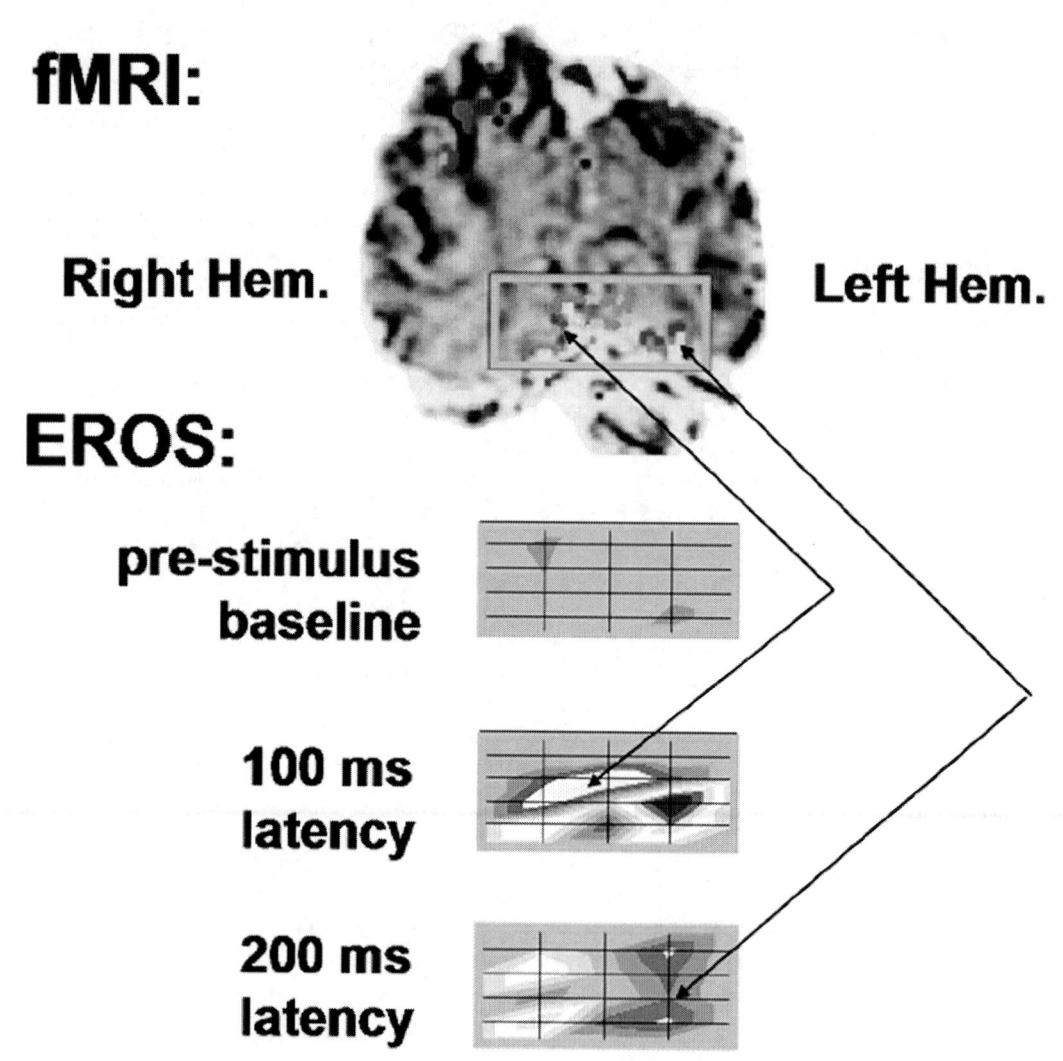

Figure 5.3. Comparison of fMRI and EROS responses in a visual stimulation experiment. The data refer to stimulation of the upper left quadrant of the visual field in one subject. The upper panel reports the BOLD-fMRI slice with maximum activity in this condition. Voxels in yellow are those exhibiting the most significant BOLD response. The green rectangle corresponds to the area explored with EROS. Maps of the EROS response before stimulation and 100 ms and 200 ms after grid reversal are shown at the bottom of the figure, with areas in yellow indicating locations of maximum activity. Note that the location of maximum EROS response varies with latency: The earlier response (100 ms) corresponds in location to the more medial fMRI response, whereas the later response (200 ms) corresponds in location to the more lateral fMRI response. The data are presented following the radiological convention (i.e., the brain is represented as if it were seen from the front). See also color insert.

is a good spatial correspondence between the two maps, with EROS further allowing for temporal discrimination for the order of events (see figure 5.3).

In the last few years, we have replicated this basic finding (i.e., localized increase in phase delay indicating the activation of a particular cortical area) several times in the visual modality (Fabiani, Ho, Stinard, & Gratton, 2003; Gratton, 1997; Gratton, Fabiani, Goodman-Wood, & DeSoto, 1998; Gratton, Goodman-Wood, & Fabiani, 2001) using stimuli varying along a number of dimensions, including frequency, size, shape, onset, and so on. Further, we showed that similar effects are observed in preparation for motor responses (De Soto, Fabiani, Geary, & Gratton, 2001; Gratton, Fabiani, et al., 1995), as well as in response to auditory (Fabiani, Low, Wee,

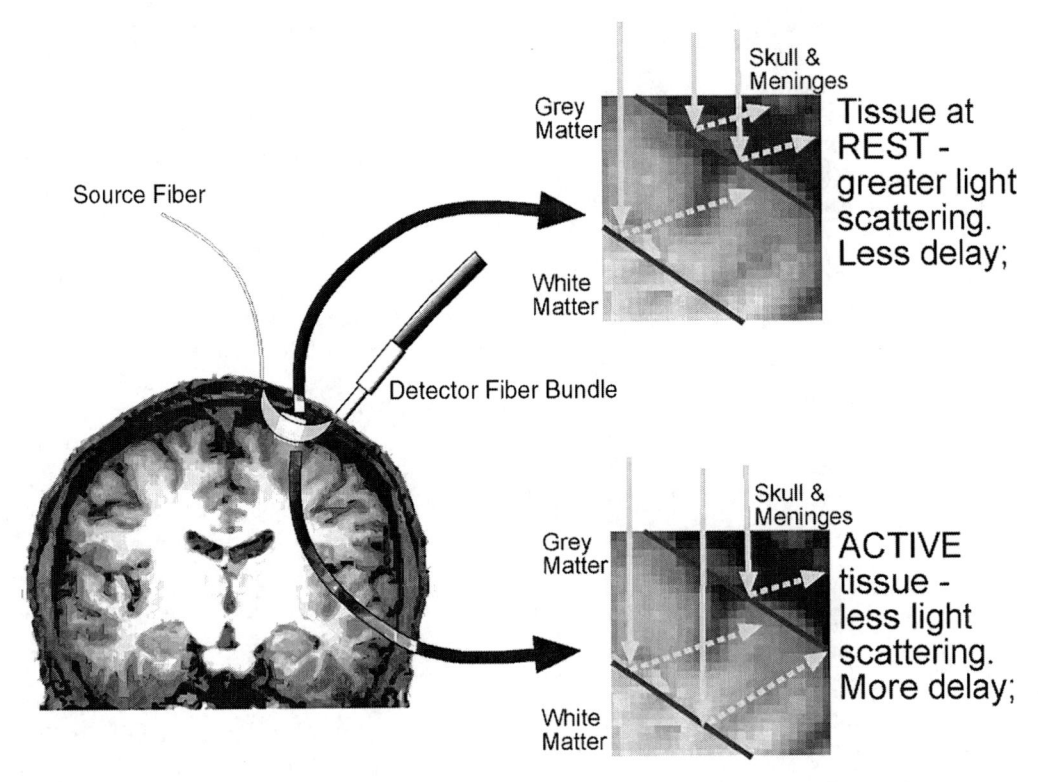

Figure 5.4. Schematic representation of the backscattering of photons under conditions of rest and activity in the cortex. Reprinted with permission from Blackwell Publishing.

Sable, & Gratton, 2006; Rinne et al., 1999; Sable et al., 2006) and somatosensory (Maclin, Low, Sable, Fabiani, & Gratton, 2004) stimuli. The feasibility of recording fast optical signals noninvasively has also been reported by four other laboratories: Franceschini and Boas (2004) using motor and somatosensory modalities, Steinbrink et al. (2000) using somatosensory stimulation, Wolf and colleagues using both visual (Wolf, Wolf, Choi, Paunescu, 2003; Wolf, Wolf, Choi, Toronov, et al., 2003) and motor modalities (Morren et al., 2004), and Tse et al. (2004) using auditory stimulation. In contrast to this positive evidence, only one report of difficulties in obtaining a fast optical signal has been reported (Syre et al., 2003).

In summary, the data overwhelmingly support the claim that fast optical signals can be recorded noninvasively in all modalities that have been attempted. In general, the signal corresponds to an increase in phase delay (available with TR methods) and a reduction in light (available with both TR and CW methods; see figure 5.4 for a schematic illustration). Monte Carlo simulations (presented in Gratton, Fabiani, et al., 1995) suggest that the increase in phase delay is consistent with a reduction in scattering in deep (cortical) areas. This reduction in scattering determines an increase in tissue transparency, so that photons on average penetrate deeper before reaching the detector. The reduction in intensity is to be attributed to the relative lack of a reflective surface conducting the photons to the detector (see figure 5.4). In this sense, the functional signal can be conceptualized as a variation of the echo from the cortex of the light produced by the source. This conceptualization also helps explain why TR methods may be more suitable for studying the fast optical signal (although in some cases CW methods can also detect fast signals; see Franceschini & Boas, 2004; Steinbrink et al., 2000).

Because the fast optical signal obtained with TR methods is recorded in response to internal or external events (and visible in time-locked averages, similarly to the event-related brain potential, ERP) we have labeled it the event-related optical signal (EROS). Our work indicates that EROS can be considered as a measure of neuronal activity in a localized cortical area. We have conducted several studies to further characterize this signal. These studies have showed the following:

1. EROS activity temporally corresponds with ERP activity obtained simultaneously or in similar paradigms (e.g., De Soto et al., 2001; Gratton et al., 1997, 2001; Rinne et al., 1999).
2. EROS activity is spatially colocalized with the BOLD fMRI response (Gratton et al., 1997, 2000) and with the slow optical response (see below; Gratton et al., 2001).
3. The amplitude of EROS varies with experimental manipulations such as stimulation frequency in a manner very similar to ERP and hemodynamic responses (Gratton et al., 2001).
4. The spatial resolution of EROS is approximately 0.5–1 cm (Gratton & Fabiani, 2003).
5. The temporal resolution of EROS is similar to that of ERPs, allowing for the identification of peaks of activity with a millisecond scale (Gratton et al., 2001). For example, the peak of the auditory N1 is the same as observed in concurrently recorded ERP data (Fabiani et al., 2006). The limits of this resolution are determined by the characteristics of the neurophysiological phenomena under study (i.e., aggregate cell activity rather than action potentials) and by the temporal sampling used, rather than by the inherent properties of the signal.
6. By using multiple source-detector distances, it is possible to estimate the depth of EROS responses and to show that these estimates correspond to the depth of activity as measured with fMRI (Gratton et al., 2000). These measures are possible up to a depth of 2.5 cm from the surface of the head.

These findings, taken together, suggest that EROS can be used to study the time course of activity in localized cortical areas. Our recent use of a full-head recording system promises to make the measure a general method for studying brain function. Currently, two main limitations should be noted:

1. Like all noninvasive optical methods, EROS cannot measure deep structures.
2. With current technology, the signal-to-noise ratio of EROS is relatively low, and averaging across a large number of trials is required. However, application of new methodologies (such as the "pi detector," Wolf et al., 2000; a high modulation frequency, Toronov et al., 2004; and appropriate frequency filters, Maclin, Gratton, and Fabiani, 2003) may at least partially address this problem.

Applications of EROS in Cognitive Neuroscience

The combination of spatial and temporal information provided by EROS can be of great utility in cognitive neuroimaging. In this area of research, it is of great importance to establish not only which areas of the brain are active, but the order of their activation, as well as the general architecture of the information processing system. To illustrate these points, we will discuss here the results from two cognitive neuroimaging studies we have conducted with EROS, both related to concepts of early and late levels of selective attention.

Psychologists have focused on selective attention for a number of years and have presented arguments in favor of or against early and late selective attention models (Johnston & Dark, 1986). Here we focus on two specific questions: (a) At which point within the information processing flow is it possible to separate the processing of attended and unattended items? (b) How long within the information processing flow is irrelevant information processed before being discarded? Within a cognitive neuroscience framework, the first issue can be restated as the identification of the earliest response in sensory cortical regions showing modulation by attention. The second issue can be cast in terms of demonstrating whether simultaneous activation of conflicting motor responses is possible. We describe two studies showing how EROS can be useful to address these questions.

Locus of Early Selective Attention. ERPs have been used for a number of years to address the issue of the onset of early selective attention effects (for a review, see Hackley, 1993). On the basis of ERP data, for instance, Martinez et al. (1999) have shown that the earliest visually evoked brain activity, such as the C1 response, is not modulated by attention, but that attention modulation effects are evident in subsequent ERP components, such as the P1 (see also chapter 3, this volume). These data contrast with reports from fMRI studies indicating attention effects occurring in Brodmann Area 17

(BA 17; primary visual cortex), typically considered to be the first cortical station of the visual pathway (Gandhi, Heeger, & Boynton, 1999). Martinez et al. (1999) proposed that this apparent contrast is determined by the fact that BA 17 is initially unable to separate attended and unattended information, and that the attention effects visible in this region are due to reafferent activity. They argue that fMRI, given its low temporal resolution, would not be able to distinguish between the earliest response and that associated with the reafferent response. In their study, however, their argument was based on source modeling of ERP activity, an approach that has yet to gain universal acceptance in the field. A technique combining spatial and temporal resolution would be very useful here, because it would allow us to be able to ascertain whether the early response in BA 17 is in fact similar for attended and unattended stimuli, and that attention modulation effects become evident only later in processing. With this idea in mind, we (Gratton, 1997) recorded EROS in a selective attention task very similar to that used by Martinez and colleagues. Two streams of stimuli appeared to the left and right of fixation, and subjects were instructed to monitor one of them for the rare (20%) occurrence of a deviant stimulus (a rectangle instead of a square). EROS activity was monitored over occipital areas. We found two types of EROS responses, with latencies between 50 and 150 ms from stimulation—a more medial response, presumably from BA 17, similar for attended and not attended stimuli, and a more lateral response, presumably from BA 18 and BA 19, which was significantly larger for attended than for unattended stimuli. These data support the claim that the initial response in BA 17 is unaffected by attention, and that early visual selective attention effects begin in the extrastriate cortex.

Parallel Versus Serial Processing. We rephrased the question of how far within the processing system irrelevant information is processed in terms of evidence for the simultaneous (parallel) activation of conflicting motor responses. This question is relevant to the basic discussion present in cognitive psychology about whether the information processing system is organized serially or in parallel. Several pieces of evidence (in particular from conflict tasks, such as the Stroop paradigm) have been used to argue in favor of the parallel organization of the system, but serial accounts for these findings have also been presented (Townsend, 1990). Recasting this issue in terms of brain activity may make it more easily treatable. Specifically, the ultimate test of the parallelism of the information processing system would come from the demonstration that two output systems, such as the left and right motor cortices, can be activated simultaneously, one by relevant and the other by irrelevant information. This would demonstrate that irrelevant information is processed (at least in part) all the way up to the response system. Unfortunately, the most commonly used brain imaging methods have had problems providing this demonstration. In fact, this issue requires concurrent spatial resolution (so that activity in the two motor cortices can be measured independently) and temporal resolution (so that the simultaneous—or parallel—timing of the two activations can be ascertained). Therefore, neither ERPs nor fMRI can be used to provide this evidence. We (De Soto et al., 2001) used EROS for this purpose. Specifically, we recorded EROS from the left and right motor cortex while subjects were performing a spatial Stroop task. In this task, subjects are presented with either the word *above* or the word *below*, presented above or below a fixation cross. In different trials, subjects were asked to respond (with one or the other hand) on the basis of either the meaning or the position of the word. On half of the trials, the two dimensions conflicted. If parallel processing were possible, on these correct conflict trials we expected subjects to activate simultaneously both the motor cortex associated with the correct response and that associated with the incorrect response (hence the delay in RT). Only unilateral activation was expected on no-conflict trials. This is in fact the pattern of results we obtained. These results provide support for parallel models of information processing and show that irrelevant information is processed all the way up to the response system, concurrently with relevant information. Similarly to the selective attention study, this study shows the usefulness of a technique combining spatial and temporal information in investigating the architecture of cognition.

Slow Optical Signals

Slow optical signals are due to the hemodynamic changes occurring in active neural areas (see figure 5.5 for an example). These changes are well

Hemoglobin Changes

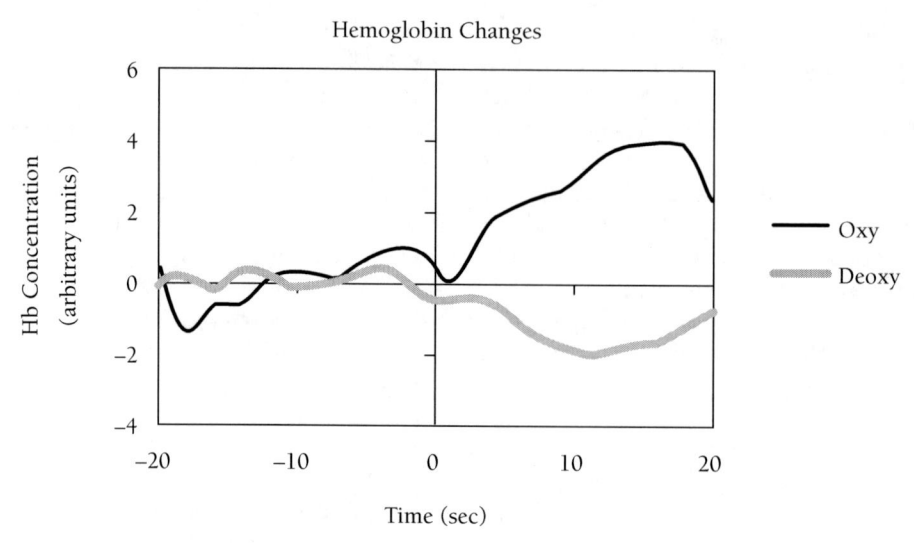

Figure 5.5. Changes in the concentration of oxyhemoglobin (solid black line) and deoxyhemoglobin (gray line) in occipital areas during visual stimulation (beginning of stimulation marked by the vertical line at time 0) measured with near-infrared spectroscopy.

demonstrated and are the basis for the BOLD fMRI signal and for O_{15} positron emission tomography (PET) as well as for invasive optical imaging in animals. In humans, slow optical signals are recorded in a number of laboratories and are often referred to as near-infrared spectroscopy (NIRS). This is because slow signal recordings typically employ multiple light wavelengths in the NIR range and decompose the signal into changes in oxy- and deoxyhemoglobin using a spectroscopic approach. Many of the considerations that can be made about BOLD fMRI studies are also valid about the NIRS signals. Slow optical signals occur with a latency of approximately 4–5 seconds compared to the eliciting neural activity. The quantitative relationship between neuronal and hemodynamic signals (neurovascular coupling) has been the subject of investigation. For instance, we used a visual stimulation paradigm in which the frequency of visual stimulation was varied between 1 and 10 Hz (Gratton et al., 2001). We measured both fast and slow optical signals concurrently and found that the amplitude of the slow signal could be predicted on the basis of the amplitude of the fast signal integrated over time. The relationship between the two signals was approximately linear.

Most studies using NIRS report a relatively low spatial resolution for the slow optical signal (2 cm or so) and typically attribute this low resolution to the diffusive properties of the method. However, the recording methodology may play a significant role in the low resolution reported. In fact, these studies are typically based on a very sparse recording montage, so that channels rarely or never overlap with each other. Wolf et al. (2000) have shown that a much better spatial resolution (and signal-to-noise ratio) can be obtained by using a number of partially overlapping channels (pi detectors). Further, very few of the NIRS studies published to date take into account individual differences in brain anatomy or coregister the optical data with anatomical MR images. It is therefore likely that the spatial resolution of slow optical measures can be greatly improved by the use of appropriate methodologies. In summary, slow optical measures (NIRS) can provide data that are quite comparable to other hemodynamic imaging methods with the following main differences:

1. Data from deep structures are not available.
2. The NIRS technology is more flexible and portable and can be applied to many normal and even bedside settings.

When compared to the fast signal (EROS), the following differences should be noted:

1. NIRS is based on slow hemodynamic changes and therefore has an intrinsically limited temporal resolution.

2. NIRS has a better signal-to-noise ratio than EROS, as changes in oxy- and deoxyhemoglobin can be sometimes seen in individual trials.

It is important to note that fast and slow optical signals can be easily recorded simultaneously, and therefore there is no need to choose one over the other, although experimental conditions can be created that optimize the recording conditions for one or the other of these methods.

Methodology

In this section, we describe some of the methods used for recording noninvasive optical imaging data. Methods vary greatly across laboratories, so for simplicity we focus on the methodology used in our lab.

Recording

Our optical recording apparatus (Imagent, built by ISS Inc., Champaign, IL) uses a frequency-domain (FD) TR method and is flexible both in the number of sources and detectors that can be used, and in a number of other recoding parameters. This equipment uses laser diodes as light sources, emitting light at 690 and 830 nm (note that other types of sources, including larger lasers, lamps, or LEDs are used by other optical recording devices). Our particular Imagent device has 128 possible sources, although in practice only about half of them are used in a particular study because of the need to avoid cross talk of different sources activated at the same time. The detectors (16 for the Imagent housed in our lab) are photomultiplier tubes (PMTs; note that other types of detectors are possible and are used in other devices, including light-sensitive diodes and CCD cameras). Sources and detectors are mounted on the head by means of a modified bike helmet (see figure 5.6). Care is taken to move the hair from under the detector fibers (which are 3-mm fiber bundles). Source fibers are small (400 microns in diameter) and can easily be placed in between hair.

Note that the number of sources greatly exceeds the number of detectors. Thus, each detector picks

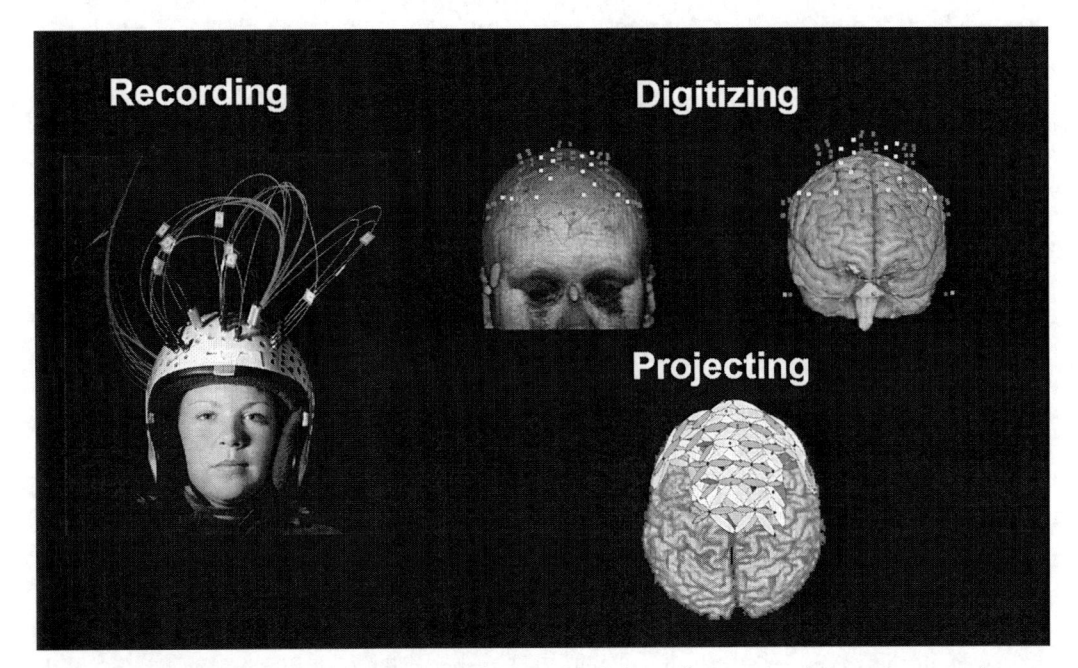

Figure 5.6. Photograph of modified helmet used for recording (left panel); examples of digitized montage (top right) and surface channel projections (bottom right). From Griffin, R. (1999). Illuminating the brain: The love and science of EROS. *Illumination*, 2(2), 20–23. Used with permission from photographer Rob Hill and the University of Missouri-Columbia. See also color insert.

up light emitted by different sources. To distinguish which source is providing the light received by the detector, sources are time-multiplexed. Although the use of a large number of sources yields a large number of recording channels—which is very useful to increase the spatial resolution and area investigated by the optical recordings—it has the disadvantage of reducing the amount of time each light source is used (because of the time-multiplexing procedure), thus reducing the signal-to-noise ratio of the measurement. In the future, development of machines with a larger number of detectors may address this problem. The peak power of each diode is between 5 and 10 mW. However, since the diodes are time-multiplexed, on average each source provides only between .3 and .6 mW.

In this approach, a particular source-detector combination is referred to as a channel. Currently, in order to provide enough recording time for each channel and maintain a sufficient sampling rate (40–100 Hz), we can record data from 256 channels during an optical recording session. However, to obtain sufficient spatial sampling and concurrently cover most of the scalp, we require something on the order of 500–1,000 channels. For this reason, we combine data from different (typically four) recording sessions (for a total of 1,024 channels). In the future, the increase in the number of detectors may also help address this issue.

To carry the light from the source to the head and from the head to the detectors, we use optic fibers. Those for the sources can be quite small (0.4 mm diameter); however, it is convenient to have large fibers for the detectors (3 mm diameter) to collect more light from the head. The fibers are held on the head using appropriately modified motorcycle helmets of different sizes, depending on the subject's head size. If the interest is in comparing different wavelengths, source fibers connected to different diodes are paired together. Prior to inserting the detector fibers into the helmet, hair is combed away from the relevant locations using smooth instruments; the sources are so thin that this process is not necessary.

Because the Imagent is an FD device, the sources are modulated at radio frequencies (typically 100–300 MHz). A heterodyning frequency (typically 5–10 kHz) is used to derive signals at a much lower frequency than that at which the sources are modulated. This is obtained by feeding the PMTs with a slightly different frequency than the source (for example, 100.005 MHz instead of 100 MHz). The use of a heterodyning frequency greatly reduces the cost of the machine.

Frequency domain methods provide three parameters for each data point and channel: (a) DC amplitude (average amount of light reaching the detector); (b) AC amplitude (oscillation at the cross-correlation frequency); and (c) phase delay (an estimate of photons' time of flight). AC or DC measures can be used to estimate the slow hemodynamic effects (spectroscopic approach), whereas phase delay measures can be used to estimate the fast neuronal effects (although intensity measures can also be used for this purpose). Using AC oscillations makes the measurements quite insensitive to environmental noise sources. Note that for devices based on CW methods, only one parameter is obtained (light intensity, equivalent to the DC amplitude).

Given the constraints imposed by our current recording apparatus, an important consideration with respect to the measurement of optical imaging data is to decide where on the head to place the source and detectors fibers. We refer to a particular set of sources and detector fibers as a *montage*. The design of an appropriate montage needs to take several considerations into account. First, it is crucial to avoid cross talk of different sources that can be picked up by the same detector. This implies that the light arriving at a particular detector at a particular time needs to come from only one source. This is facilitated by the fact that the properties of the head make it very unlikely for photons to reach detectors that are located more than 7–8 cm distance from a particular source. Another aspect of the design is that it is useful to maximize the number of channels in which the distance between the source and detector is between 2 and 5 cm. This range is determined by the fact that (a) when the source-detector distance is less than 2 cm, most of the photons traveling between the source and the detector will not penetrate deep enough to reach the cortex; and (b) when the source-detector distance is greater than 5 cm, too few photons will reach the detector for stable measurements. Figure 5.6 shows an example of a montage covering the frontal cortex, as well as a map of the projections of each channel color coded based on the source-detector distance.

Digitization and Coregistration

Coregistration is the procedure used to align the optical data with the anatomy of the brain. This step is necessary to relate the optical measurements to specific brain structures. Note that, as mentioned earlier, EROS has a spatial resolution on the order of 5–10 mm. This means that using scalp landmarks alone (such as those exploited by the 10/20 system used for EEG recordings) cannot be sufficient for appropriate representation of the data. This is because the relationship between scalp landmarks and underlying brain structures is very approximate (typically varying by several centimeters between one person and another). Thus, in order to exploit the spatial resolution afforded by the EROS, we need a technology capable of achieving a subcentimeter level of coregistration with individual brains' anatomy. This is particularly important when optical imaging is used with older adults, as individual differences in anatomy will increase with age.

We have developed a procedure that reduces the coregistration error to 3–8 mm (Whalen, Fabiani, & Gratton, 2006). This procedure is based on the digitization of the source and detector locations (using a magnetic-field 3-D digitizer), as well as of three fiducials and a number of other points on the surface of the head. Using an iterative least square fitting procedure (Marquardt-Levenberg), the locations of each individual subject are then fitted to the T1 MR images for the same subjects (in which the fiducial locations are marked with Beekley spots; see figure 5.6). The first step of the procedure involves fitting the fiducials, with the following steps fitting the rest of the points on the head surface. The locations for each source and detector are then transformed into Talairach coordinates, separately for each subject.

Analysis

In our laboratory, the analysis of optical data is conducted in two steps. The first step involves preprocessing of the optical data and is conducted separately for each channel. This involves: (a) correction of phase wrapping around the 360-degree value; (b) polynomial detrending of slow drifts (less than .01 Hz); (c) correction of pulse artifacts (see Gratton & Corballis, 1995); (d) band-pass frequency filtering (if needed; Maclin et al., 2003); (e) segmentation into epochs; (f) averaging; and (g) baseline correction. The second step involves (a) 2-D or 3-D reconstruction of the data from each channel, (b) combining together data from different channels according to their locations on the surface of the head, and (c) statistical analyses across subjects using predetermined contrasts (following the same model used for analysis of fMRI and PET data—statistical parametric mapping [SPM]; Friston et al., 1995). These analyses can be carried out independently for each data point in space or time. In many cases, a greater power is obtained by restricting the analysis (in both time and space) using regions and intervals of interest (ROIs and IOIs). The ROIs and IOIs can be derived from the literature or from the fMRI and electrophysiological data obtained with the same paradigms and subjects.

Conclusion and Considerations for Neuroergonomics

Noninvasive optical imaging is a set of new tools for studying human brain function. Two main types of phenomena can be studied: (a) neuronal activity, occurring immediately after the presentation of the stimuli or in preparation for responses; and (b) hemodynamic phenomena occurring with a few seconds' delay. These two phenomena can be measured together. In addition, optical imaging is compatible with other techniques, such as ERPs and fMRI. Since the recording systems are potentially portable, optical imaging methods can in principle be used in a number of applied settings to study brain function. Their main limitation is the penetration of the signal (a few centimeters from the surface of the head), which precludes the measurement of deep structures (at least in adults).

Fast signals appear particularly attractive for their ability to study the time course of brain activity coupled with spatial localization. In principle, this can be used to construct dynamic models of the brain in action, which would make it feasible to identify processing bottlenecks—possibly at the individual subject level. For these research purposes, fast optical signals can be applied immediately in experimental settings. However, more work is needed before they can be employed in applied settings. First, the current recording systems are relatively cumbersome, requiring multiple sessions (and a laborious setup) to obtain the desired spatial sampling. This problem may be addressed through machine design (e.g., by increasing the number of detectors and making the setup easier). Also, selection of target cortical regions may make the setup procedure much simpler. Second, the signal-to-noise ratio of the measures is relatively low. This problem may also be addressed (at least in part) through appropriate system design (including the use of more detectors, pi detectors, and higher-modulation frequencies) as well as analytical methods (including appropriate filtering methods).

Slow signals can be practically useful because of their ease of recording and good signal-to-noise ratio. Some laboratories have investigated the possibility of building machines combining good recording capabilities with very low weight, so that they can be mounted directly on the head of a subject and be operated remotely through telemetry. Even with current technology, it is possible to record useful data with very compact machines mounted on a chair or that the subject can carry in a backpack. This would make it possible to perform hemodynamic imaging in a number of applied settings—something of clear importance for neuroergonomics. If it were possible to use similar devices also for the recording of fast optical signals, than the possibility of obtaining fast and localized optical measures of brain activity in applied settings would open new and exciting possibilities in neuroergonomics.

In a series of studies, we have investigated the use of EROS to study brain preparatory processes (Agran, Low, Leaver, Fabiani, & Gratton, 2005; Leaver, Low, Jackson, Fabiani, & Gratton, 2005). The basic rationale of this work is that information about how the brain prepares for upcoming stimuli may be useful for detecting appropriate preparatory states during real-life operations, and that this information may be used for designing appropriate human-machine interfaces that would adapt to changing human cognitive requirements. Specifically, we have focused on the question of whether general preparation for information processing can be separated from specific preparation for particular types of stimuli and responses. Preparatory states were investigated using cueing paradigms in which cues, presented some time in advance of an imperative stimulus, predicted which particular stimulus or response dimension was to be used on an upcoming trial. Preliminary data indicate that EROS activity occurring between the cue and the imperative stimulus signaled the occurrence of both general preparatory states and specific preparatory states for particular stimulus or response dimensions. If this type of measurement could be applied to real-life situations online, it could provide useful information for a neurocognitive human-machine interface.

Main Points

1. Noninvasive optical imaging possesses several key features that make it a very useful tool for studying brain function, including:
   a. Concurrent sensitivity to both neuronal and hemodynamic phenomena
   b. Good combination of spatial and temporal resolution
   c. Adaptability to a number of different environments and experimental conditions
   d. Relatively low cost (when compared to MEG, PET, and fMRI)
2. Limitations of this approach include:
   a. Limited tissue penetration (up to 5 cm or so from the surface of the adult head)
   b. Relatively low signal-to-noise ratio (for the fast neuronal signal—the slow hemodynamic signal is large and can be potentially observed on single trials)
3. The characteristics of optical methods, and of fast signals in particular, make them promising methods in neuroimaging, especially for mapping the time course of brain activity over the head and its relationship with hemodynamic responses.

Acknowledgments Preparation of this chapter was supported in part by NIBIB grant #R01 EB002011-08 to Gabriele Gratton, NIA grant #AG21887 to Monica Fabiani, and DARPA grant (via NSF EIA 00-79800 AFK) to Gabriele Gratton and Monica Fabiani.

Key Readings

Bonhoeffer, T., & Grinvald, A. (1996). Optical imaging based on intrinsic signals: The methodology. In A. W. Toga & J. C. Mazziotta (Eds.), *Brain mapping: The methods* (pp. 75–97). San Diego, CA: Academic Press.

Frostig, R. (Ed.). (2001). *In vivo optical imaging of brain function*. New York: CRC Press.

Gratton, G., Fabiani M., Elbert, T., & Rockstroh, B. (Eds.). (2003). Optical imaging. *Psychophysiology, 40*, 487–571.

References

Agran, J., Low, K. A., Leaver, E., Fabiani, M., & Gratton, G. (2005). Switching between input modalities: An event-related optical signal (EROS) study. *Journal of Cognitive Neuroscience,* (Suppl.), *17*, 89.

Cohen, L. B., Hille, B., Keynes, R. D., Landowne, D., & Rojas, E. (1971). Analysis of the potential-dependent changes in optical retardation in the squid giant axon. *Journal of Physiology, 218*(1), 205–237.

DeSoto, M. C., Fabiani, M., Geary, D. L., & Gratton, G. (2001). When in doubt, do it both ways: Brain evidence of the simultaneous activation of conflicting responses in a spatial Stroop task. *Journal of Cognitive Neuroscience, 13*, 523–536.

Fabiani, M., Ho, J., Stinard, A., & Gratton, G. (2003). Multiple visual memory phenomena in a memory search task. *Psychophysiology, 40*, 472–485.

Fabiani, M., Low, K. A., Wee, E., Sable, J. J., & Gratton, G. (2006). Lack of suppression of the auditory N1 response in aging. *Journal of Cognitive Neuroscience, 18*(4), 637–650.

Franceschini, M. A., & Boas, D. A. (2004). Noninvasive measurement of neuronal activity with near-infrared optical imaging. *NeuroImage, 21*(1), 372–386.

Friston, K. J., Holmes, A. P., Worsley, K. J., Poline, J.-P., Frith, C. R., & Frackowiack, R. S. J. (1995). Statistical parametric maps in function neuroimaging: A general linear approach. *Human Brain Mapping, 2*, 189–210.

Frostig, R. D., Lieke, E. E., Ts'o, D. Y., & Grinvald, A. (1990). Cortical functional architecture and local coupling between neuronal activity and the microcirculation revealed by in vivo high-resolution optical imaging of intrinsic signals. *Proceedings of the National Academy of Sciences USA, 87*, 6082–6086.

Gandhi, S. P., Heeger, D. J., & Boynton, G. M. (1999). Spatial attention affects brain activity in human primary visual cortex. *Proceedings of the National Academy of Sciences USA, 96*, 3314–3319.

Gratton, G. (1997). Attention and probability effects in the human occipital cortex: An optical imaging study. *NeuroReport, 8*, 1749–1753.

Gratton, G., & Corballis, P. M. (1995). Removing the heart from the brain: Compensation for the pulse artifact in the photon migration signal. *Psychophysiology, 32*, 292–299.

Gratton, G., Corballis, P. M., Cho, E., Fabiani, M., & Hood, D. (1995). Shades of gray matter: Noninvasive optical images of human brain responses during visual stimulation. *Psychophysiology, 32*, 505–509.

Gratton, G., & Fabiani, M. (2001). Shedding light on brain function: The event-related optical signal. *Trends in Cognitive Sciences, 5*, 357–363.

Gratton, G., & Fabiani, M. (2003). The event related optical signal (EROS) in visual cortex: Replicability, consistency, localization and resolution. *Psychophysiology, 40*, 561–571.

Gratton, G., Fabiani, M., Corballis, P. M., Hood, D. C., Goodman-Wood, M. R., Hirsch, J., et al. (1997). Fast and localized event-related optical signals (EROS) in the human occipital cortex: Comparison with the visual evoked potential and fMRI. *NeuroImage, 6*, 168–180.

Gratton, G., Fabiani, M., Friedman, D., Franceschini, M. A., Fantini, S., Corballis, P. M., et al. (1995). Rapid changes of optical parameters in the human brain during a tapping task. *Journal of Cognitive Neuroscience, 7*, 446–456.

Gratton, G., Fabiani, M., Goodman-Wood, M. R., & DeSoto, M. C. (1998). Memory-driven processing in human medial occipital cortex: An event-related optical signal (EROS) study. *Psychophysiology, 38*, 348–351.

Gratton, G., Goodman-Wood, M. R., & Fabiani, M. (2001). Comparison of neuronal and hemodynamic measure of the brain response to visual stimulation: An optical imaging study. *Human Brain Mapping, 13*, 13–25.

Gratton, G., Maier, J. S., Fabiani, M., Mantulin, W., & Gratton, E. (1994). Feasibility of intracranial near-infrared optical scanning. *Psychophysiology, 31*, 211–215.

Gratton, G., Sarno, A. J., Maclin, E., Corballis, P. M., & Fabiani, M. (2000). Toward non-invasive 3-D imaging of the time course of cortical activity: Investigation of the depth of the event-related optical signal (EROS). *Neuroimage, 11*, 491–504.

Grinvald, A., Lieke, E., Frostig, R. D., Gilbert, C. D., & Wiesel, T. N. (1986). Functional architecture of

cortex revealed by optical imaging of intrinsic signals. *Nature, 324,* 361–364.
Hackley, S. A. (1993). An evaluation of the automaticity of sensory processing using event-related potentials and brain-stem reflexes. *Psychophysiology, 30,* 415–428.
Hill, D. K., & Keynes, R. D. (1949). Opacity changes in stimulated nerve. *Journal of Physiology, 108,* 278–281.
Hoshi, Y., & Tamura, M. (1993). Dynamic multichannel near-infrared optical imaging of human brain activity. *Journal of Applied Physiology, 75,* 1842–1846.
Johnston, W. A., & Dark, V. J. (1986). Selective attention. *Annual Review of Psychology, 37,* 43–75.
Leaver, E., Low, K., Jackson, C., Fabiani, M., & Gratton, G. (2005). Exploring the spatiotemporal dynamics of global versus local visual attention processing with optical imaging. *Journal of Cognitive Neuroscience,* (Suppl.), *17,* 55.
Maclin, E., Gratton, G., & Fabiani, M. (2003). Optimum filtering for EROS measurements. *Psychophysiology, 40,* 542–547.
Maclin, E. L., Low, K. A., Sable, J. J., Fabiani, M., & Gratton, G. (2004). The event related optical signal (EROS) to electrical stimulation of the median nerve. *Neuroimage, 21,* 1798–1804.
MacVicar, B. A., & Hochman, D. (1991). Imaging of synaptically evoked intrinsic optical signals in hippocampal slices. *Journal of Neuroscience, 11,* 1458–1469.
Malonek, D., & Grinvald, A. (1996). Interactions between electrical activity and cortical microcirculation revealed by imaging spectroscopy: Implications for functional brain mapping. *Science, 272,* 551–554.
Martinez, A., Anllo-Vento, L., Sereno, M. I., Frank, L. R., Buxton, R. B., Dubowitz, D.J., et al. (1999). Involvement of striate and extrastriate visual cortical areas in spatial attention. *Nature Neuroscience, 2,* 364–369.
Menon, R. S., Ogawa, S., Hu, X., Strupp, J. P., Anderson, P., & Ugurbil, K. (1995). BOLD-based functional MRI at 4 tesla includes a capillary bed contribution: Echo-planar imaging correlates with previous optical imaging using intrinsic signals. *Magnetic Resonance Medicine, 33,* 453–459.
Morren, G., Wolf, U., Lemmerling, P., Wolf, M., Choi, J. H., Gratton, E., et al. (2004). Detection of fast neuronal signals in the motor cortex from functional near infrared spectroscopy measurements using independent component analysis. *Medical and Biological Engineering and Computing, 42,* 92–99.

Rector, D. M., Poe, G. R., Kristensen, M. P., & Harper, R. M. (1997). Light scattering changes follow evoked potentials from hippocampal Schaeffer collateral stimulation. *Journal of Neurophysiology, 78,* 1707–1713.
Rinne, T., Gratton, G., Fabiani, M., Cowan, N., Maclin, E., Stinard, A., et al. (1999). Scalp-recorded optical signals make sound processing from the auditory cortex visible. *Neuroimage, 10,* 620–624.
Sable, J. J., Low, K. A., Whalen, C. J., Maclin, E. L., Fabiani, M., & Gratton, G. (2006). *Optical imaging of perceptual grouping in human auditory cortex.* Manuscript submitted for publication.
Steinbrink, J., Kohl, M., Obrig, H., Curio, G., Syre, F., Thomas, F., et al. (2000). Somatosensory evoked fast optical intensity changes detected noninvasively in the adult human head. *Neuroscience Letters, 291,* 105–108.
Stepnoski, R. A., LaPorta, A., Raccuia-Behling, F., Blonder, G. E., Slusher, R. E., & Kleinfeld, D. (1991). Noninvasive detection of changes in membrane potential in cultured neurons by light scattering. *Proceedings of the National Academy of Sciences USA, 88,* 9382–9386.
Syre, F., Obrig, H., Steinbrink, J., Kohl, M., Wenzel, R., & Villringer, A. (2003). Are VEP correlated fast optical changes detectable in the adult by non invasive near infrared spectroscopy (NIRS)? *Advances in Experimental Medicine and Biology, 530,* 421–431.
Toronov, V. Y., D'Amico, E., Hueber, D. M., Gratton, E., Barbieri, B., & Webb, A. G. (2004). Optimization of the phase and modulation depth signal-to-noise ratio for near-infrared spectroscopy of the biological tissue. *Proceedings of SPIE, 5474,* 281–284.
Townsend, J. T. (1990). Serial and parallel processing: Sometimes they look like Tweedledum and Tweedledee but they can (and should) be distinguished. *Psychological Science, 1,* 46–54.
Tse, C.-Y., Tien, K.-R., & Penney, T. B. (2004). Optical imaging of cortical activity elicited by unattended temporal deviants. *Psychophysiology, 41,* S72.
Villringer, A., & Chance, B. (1997). Non-invasive optical spectroscopy and imaging of human brain function. *Trends in Neuroscience, 20,* 435–442.
Whalen, C., Fabiani, M., & Gratton, G. (2006). *3D coregistration of the event-related optical signal (EROS) with anatomical MRI.* Manuscript in preparation.
Wolf, M., Wolf, U., Choi, J. H., Paunescu, L. A., Safonova, L. P., Michalos, A., & Gratton, E. (2003). Detection of the fast neuronal signal on the motor cortex using functional frequency domain near infrared spectroscopy. *Advances in Experimental Medical Biology, 510,* 225–230.

Wolf, M., Wolf, U., Choi, J. H., Toronov, V., Paunescu, L. A., Michalos, A., & Gratton, E. (2003). Fast cerebral functional signal in the 100ms range detected in the visual cortex by frequency-domain near-infrared spectrophotometry. *Psychophysiology, 40*, 542–547.

Wolf, U., Wolf, M., Toronov, V., Michalos, A., Paunescu, L. A., & Gratton, E. (2000). Detecting cerebral functional slow and fast signals by frequency-domain near-infrared spectroscopy using two different sensors. *OSA Biomedical Topical Meeting, Technical Digest*, 427–429.

Lloyd D. Tripp and Joel S. Warm

Transcranial Doppler Sonography

Doppler ultrasound has been used in medicine for many years. The primary long-standing applications include monitoring of the fetal heart rate during labor and delivery and evaluating blood flow in the carotid arteries. Applications developed largely in the last two decades have extended its use to virtually all medical specialties (Aaslid, Markwalder, & Nornes, 1982; Babikian & Wechsler, 1999; Harders, 1986; Risberg, 1986; Santalucia & Feldmann, 1999). More recently, Doppler ultrasound has been employed in the field of psychology to measure the dynamic changes in cerebral blood perfusion that occur during the performance of a wide variety of mental tasks. That application is the focus of this chapter.

The close coupling of cerebral blood flow with cerebral metabolism and neural activation (Deppe, Ringelstein, & Knecht, 2004; Fox & Raichle, 1986; Fox, Raichle, Mintun, & Dence, 1988) provides investigators with a mechanism by which to study brain systems and cognition via regional cerebral blood flow. Local distribution patterns of blood perfusion within the brain can be evaluated with high spatial resolution using invasive neuroimaging techniques like positron emission tomography (PET), functional magnetic resonance imaging (fMRI), and single photon emission computed tomography (SPECT; see Fox & Raichle, 1986; Kandel, Schwartz, & Jessel, 2000; Raichle, 1998). However, the high cost of these techniques coupled with a need for the injection of radioactive material into the bloodstream and the requirement that observers remain relatively still during testing limits their application in measuring the intracranial blood flow parameters that accompany neuronal activation during the performance of mental tasks, especially over long periods of time (Duschek & Schandry, 2003). One brain imaging supplement or surrogate for the field of neuroergonomics is transcranial Doppler sonography (TCD). Although its ability to provide detailed information about specific brain loci is limited in comparison to the other imaging procedures, TCD offers good temporal resolution and, compared to the other procedures, it can track rapid changes in blood flow dynamics that can be related to functional changes in brain activity in near-real time under less restrictive and invasive conditions (Aaslid, 1986). This chapter summarizes the technical and anatomical elements of ultrasound measurement, the important methodological issues to be considered in its use, and the general findings regarding TCD-measured hemovelocity changes during the performance of mental tasks.

Basic Principles of Ultrasound Doppler Sonography

Doppler Fundamentals

The transcranial Doppler method, first described by Aaslid et al. (1982), enables the continuous noninvasive measurement of blood flow velocities within the cerebral arteries. The cornerstone of the transcranial Doppler technique can be traced back to the Austrian physicist Christian Doppler, who developed the principle known as the Doppler effect in 1843. The essence of this effect is that the frequency of light and sound waves is altered if the source and the receiver are in motion relative to one another. Transcranial Doppler employs ultrasound from its source or transducer, which is directed toward an artery within the brain. The shift in frequency occurs when ultrasound waves or signals are reflected by erythrocytes (red blood cells) moving through a blood vessel. As described by Harders (1986), the Doppler shift is expressed by the formula:

$$F = 2(F_0 \times V \times \cos \alpha)/C \qquad (1)$$

where F_0 = the frequency of the transmitted ultrasound, V = the real blood flow velocity, α = the angle between the transmitted sound beam and the direction of blood flow, and C = the velocity of the sound in the tissue (1,550 m/s in the soft tissue). The magnitude of the frequency shift is directly proportional to the velocity of the blood flow (Duschek & Schandry, 2003).

Blood vessels within the brain that are examined routinely with TCD include the middle cerebral artery (MCA), the anterior cerebral artery (ACA), and the posterior cerebral artery (PCA). The MCA carries 80% of the blood flow within each cerebral hemisphere (Toole, 1984). Hemovelocity information from this artery provides a global index within each hemisphere of blood flow changes that accompany neuroactivation. More region-specific information regarding hemovelocity and neuronal activity can be obtained from the ACA, which feeds blood to the frontal lobes and to medial regions of the brain, and from the PCA, which supplies blood to the visual cortex. The locations of these arteries within the brain are illustrated in figure 6.1. In general, blood flow velocities, measured in centimeters per second, are fastest in the MCA followed in turn by the ACA and the PCA (Sortberg, Langmoen, Lindegaard, & Nornes, 1990).

Dynamic adaptive blood delivery to neuronal processes, "neurovascular coupling," is controlled by the contraction and dilation of small cerebral vessels that result from the changing metabolic demands of the neurons. When an area of the brain becomes metabolically active, as in the performance of mental tasks, by-products of this activity, such as carbon dioxide, increase. This results in elevation of blood flow to the region to remove the waste product (Aaslid, 1986; Risberg, 1986). Therefore, TCD offers the possibility of measuring changes in metabolic resources during task performance (Stroobant & Vingerhoets, 2000). In addition to CO_2 changes, the cerebral vascular regulating mechanisms also

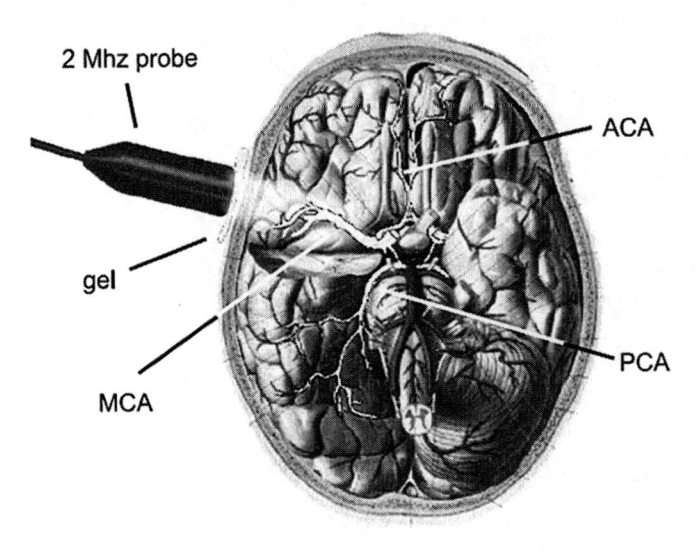

Figure 6.1. Measurement of cerebral blood flow in the middle cerebral artery (MCA). The anterior and posterior cerebral arteries (ACA and PCA, respectively) can be insonated as well. Reprinted from *Neuroimage, 21*, Deppe, M., Ringelstein, E. B., & Knecht, S., The investigation of functional brain lateralization by transcranial Doppler sonography, 1124–1146, copyright 2004, with permission from Elsevier.

involve biochemical mediators that include potassium (+K), hydrogen (+H), lactate, and adenosine. More detailed descriptions of the biochemical mediators can be found in Hamman and del Zoppo (1994) and Watanabe, Nobumasa, and Tadafumi (2002). One might assume that the increase in blood flow also serves to deliver needed glucose, but this possibility is open to question (Raichle, 1989, 1998). It is important to note, as Duschek and Schandry (2003) have done, that the diameters of the larger arteries, the MCA, ACA, and PAC, remain largely unchanged under varying task demands, indicating that hemovelocity changes in the large arteries do not result from their own vascular activity but instead from changes in the blood demanded by their perfusion territories and thus, changes in local neuronal activity.

TCD Instrumentation

Two types of examination procedures are available with most modern transcranial Doppler devices—the continuous wave (CW) and pulsed wave (PW) procedures. The former utilizes ultrasound from a single crystal transducer that is transmitted continuously at a variety of possible frequencies to a targeted blood vessel while the returning signal (backscatter) is received by a second crystal. Measurement inaccuracies can exist with this procedure because it does not screen out signals from vessels other than the one to be specifically examined or insonated. The pulsed-wave transcranial Doppler procedure employs a single probe or transducer that is used both to transmit ultrasound waves at a frequency of 2 MHz and to receive the backscatter from that signal. Dopplers that employ pulsed ultrasound allow the user to increase or decrease the depth of the ultrasound beam to isolate a single vessel and thus increase the accuracy of the data. Consequently, the pulsed Doppler is the one employed most often in behavioral studies.

The TCD device uses a mathematical expression known as a fast Fourier transform to provide a pictorial representation of blood flow velocities in real time on the TCD display as illustrated in figure 6.2 (Deppe, Knecht, Lohmann, & Ringelstein, 2004). This information can then be stored in computer memory for later analysis.

Modern Transcranial Doppler devices have the capacity to measure blood flow velocities simultaneously in both cerebral hemispheres. When performing an ultrasound examination, the transducer is positioned above one of three ultrasonic windows. Illustrated in figure 6.3 are the anterior

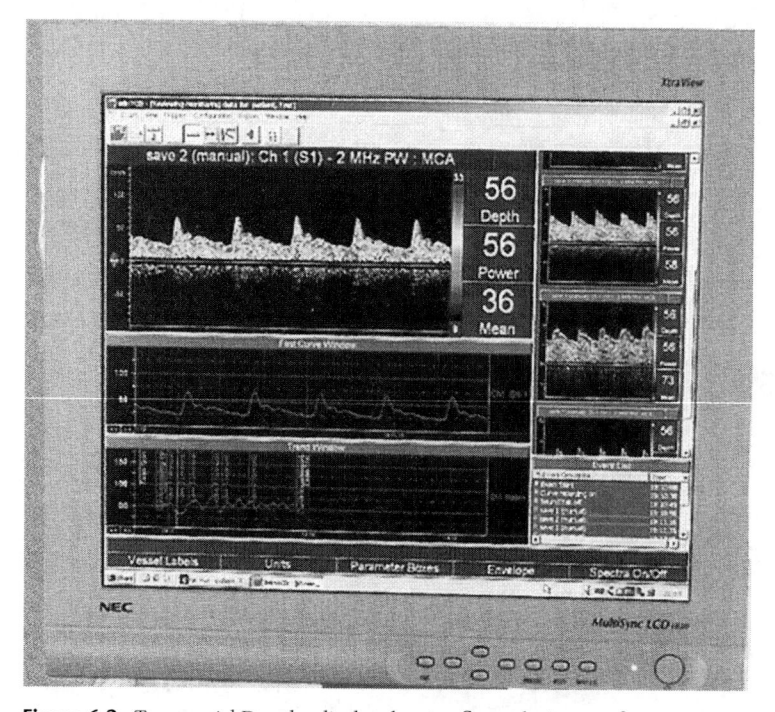

Figure 6.2. Transcranial Doppler display showing flow velocity waveforms, mean cerebral blood flow velocity, and depth of the ultrasound signal. See also color insert.

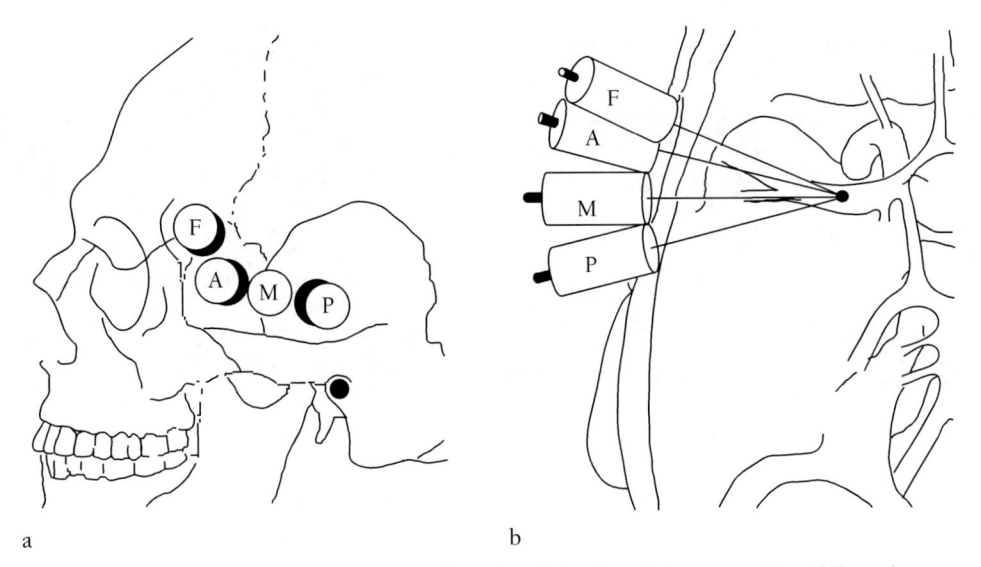

Figure 6.3. Subareas of the transcranial temporal window. (A): F, frontal; A, anterior; M, middle; and P, posterior. (B): Transducer angulations vary according to the transtemporal window utilized. Reprinted with permission from Aaslid (1986).

temporal, middle temporal, and posterior temporal windows, located just above the zygomatic arch along the temporal bone (Aaslid, 1986). These areas of the skull allow the ultrasound beam to penetrate the cranium and enter into the brain, thereby allowing for the identification of blood flow activity in the MCA, PCA, and ACA (Aaslid, 1986). The cerebral arteries can be identified by certain characteristics that are specific to each artery, including depth of insonation, the direction of blood flow, and blood flow velocity, as described in table 6.1 (after Aaslid, 1986).

In clinical uses, the patient is supine on the examination table. Participants are seated in an upright position during behavioral testing. To begin the TCD examination, the power setting on the Doppler is increased to 100%, and the depth is set to 50 mm. The transducer is prepared by applying ultrasonic gel to its surface. The gel serves as a coupling medium that allows the ultrasound to be transmitted from the transducer to the targeted blood vessel in the cranium. The transducer is then placed in front of the opening of the ear above the zygomatic arch and slowly moved toward the anterior window until the MCA is found. Once the MCA is located, the examination of other arteries can be accomplished by adjusting both the angle of the transducer and depth of the ultrasound beam. Due to the salience of the transducer, participants can accurately remember its position on the skull. Accordingly, that information can be reported to the investigator and utilized in a subsequent step in which the investigator mounts the transducer in a headband to be worn by the participant during testing.

Headband devices can vary in design from an elastic strap and mounting bracket that holds a single transducer, as shown in figure 6.4 (left), to a more elaborate design that is fashioned after a welder's helmet headband, illustrated in figure 6.4 (right). Additionally, the headband may be designed with transducer mounting brackets on the left and right sides accommodating two transducers that are used in simultaneous bilateral blood flow measurement. This configuration is also shown in figure 6.4. (right). The low weight and small size of the transducer and the ability to insert it conveniently into a headband

Table 6.1. Criteria for Artery Identification

Blood Vessel	Depth (mm)	Direction of Flow in Relation to Transducer	Mean Velocity (cm/s)
Middle cerebral artery	30–67	Toward	62 +/–12
Anterior cerebral artery	60–80	Away	50 +/–11
Posterior cerebral artery	55–70	Toward	39 +/–10

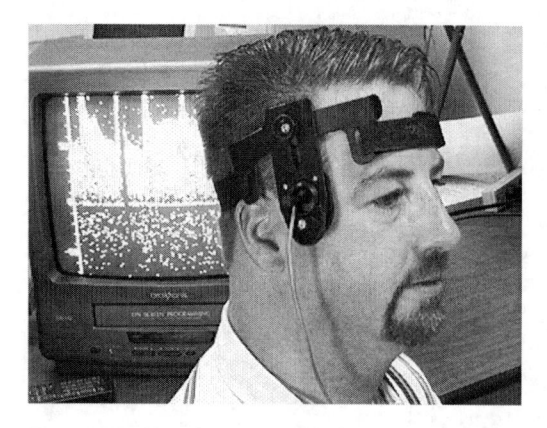

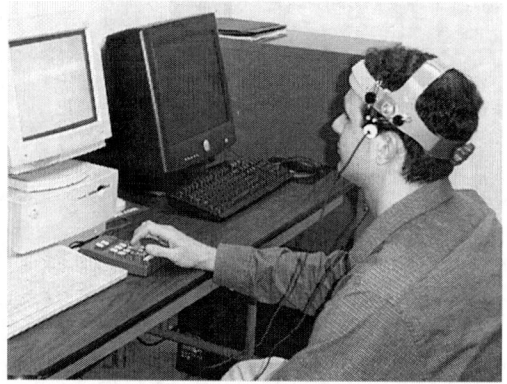

Figure 6.4. Transducer mounting designs: (left) elastic headband mounting bracket configured with a single transducer for unilateral recording; (right) welder's helmet mounting bracket affording bilateral recording capability.

mounting bracket permit real-time measurement of cerebral blood flow velocity while not limiting mobility or risking vulnerability to body motion.

Applying the headband to the participant is the most critical step in the instrumentation process. The mounting brackets shown in figure 6.4 must be aligned with the ultrasonic window where the original signal was found; care must also be taken to secure the headband to the participant's skull in a comfortable manner that is neither too tight nor too loose. Once the headband is in place, the transducer is again prepared by applying ultrasonic gel and then inserted into the mounting bracket until it makes contact with the skull. At this point, positioning adjustments are made to the transducer to reacquire the Doppler signal. As previously described, the MCA is typically used as the target artery; the depth and angle of the transducer can then be altered to acquire the ACA and PCA, if desired (see figure 6.4). Upon reacquiring the signal, the transducer is secured in the mounting bracket using a tightening device that accompanies the mounting bracket system. After the transducer has been tightened, data collection can begin.

Most studies measure task-related variations in blood flow as a percentage of a resting baseline value. Consequently, the acquisition of baseline blood flow velocity data is a key element in TCD research. Several strategies are possible for determining baseline values. The simplest is to have participants maintain a period of relaxed wakefulness for an interval of time, e.g., 5 minutes, and use hemovelocity during the final 60 seconds of the baseline period as the index against which to compare blood flow during or following task performance

(Aaslid, 1986; Hitchcock et al., 2003). Such a strategy is most useful when only a single mental task is involved. Strategies become more complex when multiple tasks are employed. As described by Stroobant and Vingerhoets (2000), one approach in the multiple-task situation is to use an alternating rest-activation strategy in which a rest condition precedes each task. Each rest phase can then be used as the baseline measurement for its subsequent mental task, or the average of all rest conditions can be employed to determine blood flow changes in each task. An advantage of the first option is the likelihood that artifactual physiological changes between measurements will be minimized because the resting and testing phases occur in close temporal proximity. The second option offers the advantage of increased reliability in the baseline index, since it is based upon a larger sample of data. Stroobant and Vingerhoets (2000) noted that an alternate strategy under multiple-task conditions is to make use of successive cycles of rest and activation within tasks. They pointed out that while the temporal proximity between rest and activation is high with this technique, care must be taken to ensure that task-induced activation has subsided before the beginning of a rest phase is defined.

Additional issues to be considered in developing baseline measures are (1) whether participants should have their eyes open or closed during the baseline phase (Berman & Weinberger, 1990) and (2) whether they should sit calmly, simply looking at a blank screen (Bulla-Hellwig, Vollmer, Gotzen, Skreczek, & Hartje, 1996), or watching the display of the criterion task unfold on the screen without a work imperative (Hitchcock et al., 2003). At present, there is no universal standard for baseline measurement in TCD research. As Stroobant and Vingerhoets (2000) suggested, future studies are needed to compare the different methodological approaches to this crucial measurement issue.

Additional Methodological Concerns

As with any physiological measure, certain exclusion criteria apply in the use of the TCD procedure. In a careful review of this issue, Stroobant and Vingerhoets (2000) pointed out that participants in TCD investigations are frequently required to abstain from caffeine, nicotine, and medication. The duration of abstention has varied widely across studies, with times ranging from 1 to 24 hours prior to taking part in the investigation. Stroobant and Vingerhoets (2000) recommended that the abstention period be at least 12 hours prior to the study.

Due to concerns about handedness and lateralized functions that are prevalent in the neuroscience literature (Gazzaniga, Ivry, & Mangun, 2002; Gur et al., 1982; Markus & Boland, 1992; Purves et al., 2001), investigators employing the TCD procedure tend to restrict their participant samples to right-handed individuals. Moreover, since age is known to influence cerebral blood flow, with lower levels of blood flow at rest and during task performance in older people (Droste, Harders, & Rastogi, 1989b; Orlandi & Murri, 1996), and since there is some evidence pointing to higher blood flow velocities in females than males of similar age (Arnolds & Von Reutern, 1986; Rihs et al., 1995; Vingerhoets & Stroobandt, 1999; Walter, Roberts, & Brownlow, 2000), it is important to control for age and gender effects in TCD studies. Evidence is also available to show that blood flow is sensitive to emotional processes (Gur, Gur, & Skolnick, 1988; Troisi et al., 1999; Stoll, Hamann, Mangold, Huf, & Winterhoff-Spurk, 1999), leading Stroobant and Vingerhoets (2000) to suggest that investigators should try to reduce participants' levels of anxiety in TCD studies. Finally, since variations in cerebral hemodynamics during task performance have been found to differ between controls and patients suffering from migraine headache, cerebral ischemia, and cardiovascular disease (Backer et al., 1999; Stroobant, Van Nooten, & Vingerhoets, 2004), participants' general health is also of concern in TCD investigations.

Reliability and Validity

As with any measurement procedure, the potential usefulness of the TCD technique will rest upon its reliability and validity. Since the baseline values comprise the standard from which task-determined effects are derived, one would expect the Doppler technique to yield baseline data that are reproducible over time and independent of spontaneous autoregulatory fluctuations in the brain. Several investigations have reported a reasonably satisfactory range of baseline reliability coefficients between .71 and .98 (Baumgartner, Mathis, Sturzenegger, & Mattle, 1994; Bay-Hansen, Raven, & Knudsen, 1997; Maeda et al., 1990; Matthews, Warm, & Washburn, 2004; Totaro, Marini, Cannarsa, & Prencipe, 1992). In addition, Matthews et al. (2004) and Schmidt et al. (2003) have also reported significant interhemispheric correlations for resting baseline values. Results such as these indicate that individual differences in baseline values are highly consistent. In addition to baseline reliability, other studies have demonstrated strong test-retest reliabilities in task-induced blood flow changes. Knecht et al. (1998) reported a reliability coefficient of .90 for blood flow responses obtained from two consecutive examinations on a word-fluency task; Matthews et al. (2004) reported significant intertask correlations ranging between .42 and .66 between blood flow responses induced by a battery of high-workload tasks including line discrimination, working memory, and tracking, and Stroobant and Vingerhoets (2001) reported test-retest reliability values ranging from .61 to .83 for eight different verbal and visuospatial tasks.

Partial convergent validity for the TCD procedure comes from studies showing that factors known to decrease cerebral blood volume, such as when observers are passively exposed to increased gravitational force, also reduce TCD measured blood flow (Tripp & Chelette, 1991). More definitive convergent validity has been determined for the TCD technique by comparing task-induced blood flow changes obtained with this procedure to those secured by PET, fMRI, and the WADA test (Wada & Rasmussen, 1960), a procedure often used in identifying language lateralization. Similar results have been obtained with these techniques (Duschek & Schandry, 2003; Jansen et al., 2004; Schmidt et al., 1999), and substantial correlations, some as high as .80, have been reported (Deppe et al., 2000; Knake et al., 2003; Knecht et al., 1998;

Rihs, Sturzenegger, Gutbrod, Schroth, & Mattle, 1999). Further validation of the TCD procedure comes from studies showing that changes in hemovelocity during the performance of complex mental tasks that are noted with this procedure are not evident when observers simply gaze at stimulus displays without a work imperative (Hitchcock et al., 2003) and that such changes are not related to more peripheral reactions involving changes in breathing rate, heart rate, blood pressure, or end-tidal carbon dioxide (Kelley et al., 1992; Klingelhofer et al., 1997; Schnittger, Johannes, & Munte, 1996; Stoll et al., 1999).

TCD and Psychological Research

Roots in the Past

The possibility that examining changes in brain blood flow during mental activities could lead to an understanding of the functional organization of the brain was recognized in the nineteenth century by Sir Charles Sherrington (Roy & Sherrington, 1890) and discussed by William James (1890) in his classic text *Principles of Psychology*. The first empirical observation linking brain activity and cerebral blood flow was made in a clinical study by John Fulton (1928), who reported changes in blood flow during reading in a patient with a bony defect over the primary visual cortex. Since those early years, research with the PET and fMRI procedures has provided considerable evidence for a close tie between cerebral blood flow and neural activity during the performance of mental tasks (Raichle, 1998; Risberg, 1986).

Work with the fTDC procedure has added substantially to this evidence by showing that changes in blood flow velocity occur in a wide variety of tasks varying from simple signal detection to complex information processing. The literature is extensive and space limitations permit us to only describe some of these findings. More extensive coverage can be found in reviews by Duschek and Schandry (2003), Klingelhofer, Sander, and Wittich (1999), and Stroobant and Vingerhoets (2001).

Basic Perceptual Effects

The presentation of even simple visual stimuli can cause variations in flow velocity in the PCA. Studies have shown that visual signals are accompanied by increases in blood flow in this artery (Aaslid, 1987; Conrad & Klingelhofer, 1989, 1990; Kessler, Bohning, Spelsperg, & Kompf, 1993), that these hemovelocity changes will occur to stimuli as brief as 50 ms in duration, and that the magnitude of the hemovelocity increase is directly related to the intensity of the stimuli and to the size of the visual field employed (Sturzenegger, Newell, & Aaslid, 1996; Wittich, Klingelhofer, Matzander, & Conrad, 1992). Hemovelocity increments tend to be greater to intermittent than to continuous light (Conrad & Klingelhofer, 1989, 1990; Gomez, Gomez, & Hall, 1990), and there is evidence that flow velocity increments might be lateralized to the right hemisphere in response to blue, yellow, and red light (Njemanze, Gomez, & Horenstein, 1992). It is important to note that the flow velocities in the PCA accompanying visual stimulation are greater when observers are required to search the visual field for prespecified targets or to recognize symbols than when they simply look casually at the displays (Schnittger et al., 1996; Wittich et al., 1992). The fact that the dynamics of the blood flow response are affected by how intently observers view the visual display indicates that more than just the physical character of the stimulus determines the nature of the blood flow response. As Duschek and Shandry (2003) pointed out, the Doppler recordings of blood flow changes in the PCA clearly reflect visually evoked activation processes.

Changes in cerebral blood flow are not limited to visual stimuli, however; they occur to acoustic stimuli as well. Klingelhofer and his associates (1997) have shown a bilateral increase in the MCA to a white noise signal and an even larger hemovelocity increase when observers listened to samples of speech. In the latter case, the blood flow changes were lateralized to the left MCA, as might be anticipated from left-hemispheric dominance in the processing of auditory language (Hellige, 1993). In other studies, observers listened to musical passages in either a passive mode or under conditions in which they had to identify or recognize the melodies that they heard. Passive listening led to bilateral increases in blood flow in the MCA, while in the active mode, blood flow increments were lateralized to the right hemisphere (Matteis, Silvestrini, Troisi, Cupini, & Caltagirone, 1997; Vollmer-Haase, Finke, Hartje, & Bulla-Hellwig, 1998). The latter result coincides with PET studies indicating right hemisphere dominance in the perception of melody (Halpern & Zatorre, 1999; Zatorre, Evans, & Meyer, 1994).

In still another experiment involving acoustic stimulation, Vingerhoets and Luppens (2001) examined blood flow dynamics in conjunction with a number of dichotic listening tasks that varied in difficulty. Blood flow responses in the MCAs of both hemispheres varied directly with task difficulty, though with a greater increase in the right hemisphere. However, unlike prior results with the PET technique in which directing attention to stimuli in one ear led to more pronounced blood flow in the contralateral hemisphere during dichotic listening (O'Leary et al., 1996), blood flow asymmetries were not noted in this study. The authors suggested that hemodynamic changes caused by attentional strategies in the dichotic listening task may be too subtle to be detected as lateralized changes in blood flow velocity. Nevertheless, the more pronounced right hemispheric response to the attention tasks employed in this study is consistent with the dominant role of the right hemisphere in attention described in PET studies (Pardo, Fox, & Raichle, 1991) and in other TCD experiments involving short-term attention (Droste, Harders, & Rastogi, 1989a; Knecht, Deppe, Backer, Ringelstein, & Henningsen, 1997) and vigilance or long-term attention tasks (Hitchcock et al., 2003). Transcranial Doppler research with vigilance tasks is described more fully in chapter 10.

Hemovelocity changes occur not only to the presentation of stimuli but also to their anticipation. Along this line, Backer, Deppe, Zunker, Henningsen, and Knecht (1994) performed an experiment in which observers were led to expect threshold-level tactual stimulation on the tip of either their right or left index finger 5 seconds after receiving an acoustic cue. Blood flow in the MCA contralateral to the side of the body in which stimulation was expected increased about 3 seconds prior to the stimulus, indicating an attention-determined increase in cortical activity. These findings were supported in a similar study by Knecht et al. (1997) in which blood flow in the contralateral MCA also increased following the acoustic cue, but only when subsequent stimulation was anticipated. No changes were observed in conditions in which observers were presented with tactual stimulation alone, indicating higher cortical activation in connection with an anticipation of stimulation than for activation associated with the tactual stimulus itself. In still another experiment, this group (Backer et al., 1999) demonstrated that the expectancy effect was tied to stimulus intensity. Using near-threshold tactual stimuli, they again found more pronounced blood flow in the MCA contralateral to the side where the stimulus was expected. In contrast, anticipation for stimuli well above threshold showed right-dominant blood flow increases that were independent of the side of stimulation. The finding of anticipatory blood flow with the TCD procedure is similar to that in PET studies showing increased blood flow in the postcentral cortex during the anticipation of tactual stimuli (Roland, 1981).

Complex Information Processing

Studies requiring the relatively simple detection and recognition of stimuli have been accompanied by experiments that focused on more complex tasks. As a case in point, we might consider a series of studies by Droste, Harders, and Rastogi (1989a, 1989b) in which recordings were made of blood flow velocities in the left and right MCAs during the performance of six different tasks: reading aloud, dot distance estimation, noun finding, spatial perception, multiplication, and face recognition. All of the tasks were associated with increases in blood flow velocity, with the largest increase occurring when observers were required to read abstract four-syllable nouns aloud. In a related study, this group of investigators measured hemovelocity changes bilaterally while observers performed tasks involving spatial imaging and face recognition. Increments in blood flow velocity were also noted in this study (Harders, Laborde, Droste, & Rastogi, 1989a).

The findings described above regarding the ability of linguistic tasks to increase blood flow in the MCA have been reported in several other investigations, which essentially revealed left hemispheric dominance in right-handed observers and higher variability in left-handers (Bulla-Hellwig et al., 1996; Markus & Boland, 1992; Njemanze, 1991; Rihs et al., 1995; Varnadore, Roberts, & McKinney, 1997; Vingerhoets & Stroobant, 1999). Moreover, Doppler-measured blood flow increments while solving arithmetic problems have also been described by Kelley et al. (1992) and Thomas and Harer (1993). Similarly, hemovelocity increments have also been described when observers are engaged in a variety of spatial processing tasks. Vingerhoets and Stroobant (1999) have reported right hemispheric specialization in a task involving the mental rotation of symbols, while in a study involving the mental rotation of cubes, Serrati et al. (2000) reported that blood flow was directly related to the difficulty of the rotation task. Other spatial tasks in which blood flow increments were obtained are compensatory tracking (Zinni & Parasuraman, 2004) and the Multi-Attribute Task Battery (Comstock & Arnegard, 1992), a mah-jongg tile sorting task Kelly et al. (1993), and a simulated flying task (Wilson, Finomore, & Estepp, 2003). In the latter case, blood flow was found to vary directly with task difficulty. Along with blood flow changes occurring during the execution of complex performance tasks, studies by Schuepbach and his associates have reported increments in blood flow velocity in the MCA and the ACA that were linked to problem difficulty on the Tower of Hanoi and Stockings of Cambridge Tests, which measure executive functioning and planning ability (Frauenfelder, Schuepbach, Baumgartner, & Hell, 2004; Schuepbach et al., 2002). Clearly, in addition to detecting rapid blood flow changes in simple target detection or recognition tasks, the TCD procedure is also capable of detecting such changes during complex cognitive processing, and in some cases increments in cerebral blood flow have been shown to vary directly with task difficulty (Frauenfelder et al., 2004; Schuepbach et al., 2002; Serrati et al., 2000; Wilson et al., 2003). These studies involving variations in difficulty are critical because they demonstrate that the linkage between blood flow and complex cognitive processing is more than superficial. It is worth noting, however, that the studies that have established the blood flow–difficulty linkage have done so using only two levels of difficulty. To establish the limits of the blood flow–difficulty linkage, studies are needed that manipulate task difficulty in a parametric manner.

Conclusions and Neuroergonomic Implications

The research reviewed in this chapter indicates that fTDC is an imaging tool that allows for fast and mobile assessment of task-related brain activation. Accordingly, TCD could be used to assess task difficulty and task engagement, to evaluate hemispheric asymmetries in perceptual representation, and perhaps to determine when operators are in need of rest or replacement. It may also be useful in determining when operators might benefit from adaptive automation, an area of emerging interest in the human factors field in which the allocation of function between human operators and computer systems is flexible and adaptive assistance is provided when participants are in need of it due to increased workload and fatigue (Parasuraman, Mouloua, & Hilburn, 1999; Scerbo, 1996; Wickens & Hollands, 2000; Wickens, Mavor, Parasuraman, & McGee, 1998; chapter 16, this volume). In these ways, the TCD procedure could lead to a greater understanding of the interrelations between neuroscience, cognition, and action that are basic elements in the neuroergonomic approach to the design of technologies and work environments for safer and more efficient operation (Parasuraman, 2003; Parasuraman & Hancock, 2004).

Main Points

1. Ultrasound can be used to measure brain blood flow velocity during the performance of mental tasks via the transcranial Doppler procedure.
2. The transcranial Doppler procedure is a flexible, mobile, relatively inexpensive, and noninvasive technique to measure brain blood flow during the performance of mental tasks.
3. Transcranial Doppler sonography offers the possibility of measuring changes in metabolic resources during task performance.
4. Transcranial Doppler measurement has good temporal resolution, substantial reliability, and convergent validity with other brain imaging procedures such as PET and fMRI.
5. Transcranial Doppler measurement has been successful in linking brain blood flow to performance in a wide variety of tasks ranging from simple signal detection to complex information processing and in identifying cerebral laterality effects.
6. From a neuroergonomic perspective, transcranial Doppler measurement may be

useful in determining when operators are task engaged and when they are in need of rest or replacement.

Key Readings

Aaslid, R. (Ed.). (1986). *Transcranial Doppler sonography*. New York: Springer-Verlag.

Babikian, V. L., & Wechsler, L. R. (Eds.). (1999). *Transcranial Doppler ultrasonography* (2nd ed.). Boston: Butterworth Heinemann.

Duschek, S., & Schandry, R. (2003). Functional transcranial Doppler sonography as a tool in psychophysiological research. *Psychophysiology, 40*, 436–454.

Klingelhofer, J., Sander, D., & Wittich, I. (1999). Functional ultrasonographic imaging. In V. L. Babikian & L. R. Wechsler (Eds.), *Transcranial Doppler ultrasonography* (2nd ed., pp. 49–66). Boston: Butterworth Heinemann.

Stroobant, N., & Vingerhoets, G. (2000). Transcranial Doppler ultrasonography monitoring of cerebral hemodynamics during performance of cognitive tasks: A review. *Neuropsychology Review, 10*, 213–231.

References

Aaslid, R. (Ed.). (1986). *Transcranial Doppler sonography*. New York: Springer-Verlag.

Aaslid, R. (1987). Visually evoked dynamic blood flow response on the human cerebral circulation. *Stroke, 18*, 771–775.

Aaslid, R., Markwalder, T. M., & Nornes, H. (1982). Noninvasive transcranial Doppler ultrasound recording of flow velocity in basal cerebral arteries. *Journal of Neurosurgery, 57*, 769–774.

Arnolds, B. J., & Von Reutern, G. M. (1986). Transcranial Doppler sonography: Examination technique and normal reference values. *Ultrasound in Medicine and Biology, 12*, 115–123.

Babikian, V. L., & Wechsler, L. R. (Eds.). (1999). *Transcranial Doppler ultrasonography* (2nd ed.). Boston: Butterworth Heinemann.

Backer, M., Deppe, M., Zunker, P., Henningsen, H., & Knecht, S. (1994). Tuning to somatosensory stimuli during focal attention. *Cerebrovascular Disease, 4*(Suppl. 3), 3.

Backer, M., Knecht, S., Deppe, M., Lohmann, H., Ringelstein, E. B., & Henningsen, H. (1999). Cortical tuning: A function of anticipated stimulus intensity. *Neuroreport, 10*, 293–296.

Baumgartner, R. W., Mathis, J., Sturzenegger, M., & Mattle, H. P. (1994). A validation study on the intraobserver reproducibility of transcranial color-coded duplex sonography velocity measurements. *Ultrasound in Medicine and Biology, 20*, 233–237.

Bay-Hansen, J., Raven, T., & Knudsen, G. M. (1997). Application of interhemispheric index for transcranial Doppler sonography velocity measurements and evaluation of recording time. *Stroke, 28*, 1009–1014.

Berman, K. F., & Weinberger, D. R. (1990). Lateralization of cortical function during cognitive tasks: Regional cerebral blood flow studies of normal individuals and patients with schizophrenia. *Journal of Neurology, Neurosurgery, and Psychiatry, 53*, 150–160.

Bulla-Hellwig, M., Vollmer, J., Gotzen, A., Skreczek, W., & Hartje, W. (1996). Hemispheric asymmetry of arterial blood flow velocity changes during verbal and visuospatial tasks. *Neuropsychologia, 34*, 987–991.

Comstock, J. R., & Arnegard, R. J. (1992). *The multiattribute task battery for human operator workload and strategic behavior* (NASA Technical Memorandum 104174). Hampton, VA: NASA Langley Research Center.

Conrad, B., & Klingelhofer, J. (1989). Dynamics of regional cerebral blood flow for various visual stimuli. *Experimental Brain Research, 77*, 437–441.

Conrad, B., & Klingelhofer, J. (1990). Influence of complex visual stimuli on the regional cerebral blood flow. In L. Deecke, J. C. Eccles, & V. B. Mountcastel (Eds.), *From neuron to action* (pp. 227–281). Berlin: Springer.

Deppe, M., Knecht, S., Lohmann, H., & Ringelstein, E. B. (2004). A method for the automated assessment of temporal characteristics of functional hemispheric lateralization by transcranial Doppler sonography. *Journal of Neuroimaging, 14*, 226–230.

Deppe, M., Knecht, S., Papke, K., Lohmann, H., Fleischer, H., Heindel, W., et al. (2000). Assessment of hemispheric language lateralization: A comparison between fMRI and fTCD. *Journal of Cerebral Blood Flow and Metabolism, 20*, 263–268.

Deppe, M., Ringelstein, E. B., & Knecht, S. (2004). The investigation of functional brain lateralization by transcranial Doppler sonography. *Neuroimage, 21*, 1124–1146.

Droste, D. W., Harders, A. G., & Rastogi, E. (1989a). A transcranial Doppler study of blood flow velocity in the middle cerebral arteries performed at rest and during mental activities. *Stroke, 20*, 1005–1011.

Droste, D. W., Harders, A. G., & Rastogi, E. (1989b). Two transcranial Doppler studies on blood flow velocity in both middle cerebral arteries during rest and the performance of cognitive tasks. *Neuropsychologia, 27,* 1221–1230.

Duschek, S., & Schandry, R. (2003). Functional transcranial Doppler sonography as a tool in psychophysiological research. *Psychophysiology, 40,* 436–454.

Fox, P. T., & Raichle, M. E. (1986). Focal physiological uncoupling of cerebral blood flow and oxidative metabolism during somatosensory stimulation in human subjects. *Proceedings of the National Academy of Science USA, 83,* 1140–1144.

Fox, P. T., Raichle, M. E., Mintun, M. A., & Dence, C. (1988). Nonoxidative glucose consumption during focal physiologic neural activity. *Science, 241,* 462–464.

Frauenfelder, B. A., Schuepback, D., Baumgartner, R. W., & Hell, D. (2004). Specific alterations of cerebral hemodynamics during a planning task: A transcranial Doppler sonography study. *Neuroimage, 22,* 1223–1230.

Fulton, J. F. (1928). Observations upon the vascularity of the human occipital lobe during visual activity. *Brain, 51,* 310–320.

Gazzaniga, M. S., Ivry, R. B., & Mangun, G. R. (2002). *Cognitive neuroscience: The biology of the mind* (2nd Ed.). New York: Norton.

Gomez, S. M., Gomez, C. R., & Hall, I. S. (1990). Transcranial Doppler ultrasonographic assessment of intermittent light stimulation at different frequencies. *Stroke, 21,* 1746–1748.

Gur, R. C., Gur, R. E., Obrist, W. D., Hungerbuhler, J. P., Younkin, D., Rosen, A. D., et al. (1982). Sex and handedness differences in cerebral blood flow during rest and cognitive activity. *Science, 217,* 659–661.

Gur, R. C., Gur, R. E., & Skolnick, B. E. (1988). Effects of task difficulty on regional cerebral blood flow: Relationships with anxiety and performance. *Psychophysiology, 25,* 392–399.

Halpern, A. R., & Zatorre, R. J. (1999). When that tune runs through your head: A PET investigation of auditory imagery for familiar melodies. *Cerebral Cortex, 9,* 697–704.

Hamann, G. F., & del Zoppo, G. J. (1994). Leukocyte involvement in vasomotor reactivity of the cerebral vasculature. *Stroke, 25,* 2117–2119.

Harders, A. (1986). *Neurosurgical applications of transcranial Doppler sonography.* New York: Springer-Verlag.

Harders, A. G., Laborde, G., Droste, D. W., & Rastogi, E. (1989). Brain activity and blood flow velocity changes during verbal and visuospatial cognitive tasks. *International Journal of Neuroscience, 47,* 91–102.

Hellige, J. B. (1993). *Hemispheric asymmetry: What's right and what's left.* Cambridge, MA: Harvard University Press.

Hitchcock, E. M., Warm, J. S., Matthews, G., Dember, W. N., Shear, P. K., Tripp, L. D., et al. (2003). Automation cueing modulates cerebral blood flow and vigilance in a simulated air traffic control task. *Theoretical Issues in Ergonomics Science, 4,* 89–112.

James, W. (1890). *Principles of psychology.* New York: Henry Holt.

Jansen, A., Floel, A., Deppe, M., van Randenborgh, J., Drager, B., Kanowski, M., et al. (2004). Determining the hemispheric dominance of spatial attention: A comparison between fTCD and fMRI. *Human Brain Mapping, 23,* 168–180.

Kandel, E. R., Schwartz, J. H., & Jessel, T. M. (2000). *Principles of neural science* (4th ed.). New York: McGraw Hill.

Kelley, R. E., Chang, J. Y., Scheinman, N. J., Levin, B. E., Duncan, R. C., & Lee, S. C. (1992). Transcranial Doppler assessment of cerebral flow velocity during cognitive tasks. *Stroke, 23,* 9–14.

Kelley, R. E., Chang, J. Y., Suzuki, S., Levin, B. E., & Reyes-Iglesias, Y. (1993). Selective increase in right hemisphere transcranial Doppler velocity during a spatial task. *Cortex, 29,* 45–52.

Kessler, C., Bohning, A., Spelsberg, B., & Kompf, D. (1993). Visually induced reactivity in the posterior cerebral artery. *Stroke, 24,* 506.

Klingelhofer, J., Matzander, G., Sander, D., Schwarze, J., Boecke, H., Bischoff, C. (1997). Assessment of functional hemispheric asymmetry by bilateral simultaneous cerebral blood flow velocity monitoring. *Journal Cerebral Blood Flow Metabolism, 17,* 577–85.

Klingelhofer, J., Sander, D., & Wittich, I. (1999). Functional ultrasonographic imagining. In V. L. Babikian & L. R. Wechsler (Eds.), *Transcranial Doppler ultrasonography* (2nd ed., pp 49–66). Boston: Butterworth Heinemann.

Knake, S., Haag, A., Hamer, H. M., Dittmer, C., Bien, S., Oertel, W. H., et al. (2003). Language lateralization in patients with temporal lobe epilepsy: A comparison of functional transcranial Doppler sonography and the Wada test. *Neuroimaging, 19,* 1228–1232.

Knecht, S., Deppe, M., Backer, M., Ringelstein, E. B., & Henningsen, H. (1997). Regional cerebral blood flow increases during preparation for and processing of sensory stimuli. *Experimental Brain Research, 116,* 309–314.

Knecht, S., Deppe, M., Ebner, A., Henningsen, H., Huber, T., Jokeit, H., et al. (1998). Noninvasive determination of language lateralization by functional transcranial Doppler sonography: A comparison with the Wada test. *Stroke, 29,* 82–86.

Maeda, H., Etani, H., Handa, N., Tagaya, M., Oku, N., Kim, B. H., et al. (1990). A validation study on the reproducibility of transcranial Doppler velocimetry. *Ultrasound in Medicine and Biology, 16,* 9–14.

Markus, H. S., & Boland, M. (1992). "Cognitive activity" monitored by non-invasive measurement of cerebral blood flow velocity and its application to the investigation of cerebral dominance. *Cortex, 28,* 575–581.

Matteis, M., Silvestrini, M., Toroisi, E., Cupini, L. M., & Caltagirone, C. (1997). Transcranial Doppler assessment of cerebral flow velocity during perception and recognition of melodies. *Journal of Neurobiological Science, 149,* 57–61.

Matthews, G., Warm, J. S., & Washburn, D. (2004). *Diagnostic methods for predicting performance impairment associated with combat stress* (Report No. 1). Fort Detrick, MD: U.S. Army Medical Research and Material Command.

Njemanze, P. C. (1991). Cerebral lateralization in linguistic and nonlinguistic perception: Analysis of cognitive styles in the auditory modality. *Brain and Language, 41,* 367–380.

Njemanze, P. C., Gomez, C. R., & Horenstein, S. (1992). Cerebral lateralization and color perception: A transcranial Doppler study. *Cortex, 28,* 69–75.

O'Leary, D. S., Andreasen, N. C., Hurtig, R. R., Hichwa, R. D., Watkins, G. L., Boles-Ponto, L. L., et al. (1996). A positron emission tomography study of binaurally and dichotically presented stimuli: Effects of level of language and directed attention. *Brain and Language, 53,* 20–39.

Orlandi, G., & Murri, L. (1996). Transcranial Doppler assessment of cerebral flow velocity at rest and during voluntary movements in young and elderly healthy subjects. *International Journal of Neuroscience, 84,* 45–53.

Parasuraman, R. (2003). Neuroergonomics: Research and practice. *Theoretical Issues in Ergonomic Science, 4,* 5–20.

Parasuraman, R., & Hancock, P. (2004). Neuroergonomics: Harnessing the power of brain science for HF/E. *Bulletin of the Human Factors and Ergonomics Society, 47,* 1, 4–5.

Parasuraman, R., Mouloua, M., & Hilburn, B. (1999). Adaptive aiding and adaptive task allocation enhance human-machine interaction. In M. W. Scerbo & M. Mouloua (Eds.), *Automation technology and human performance: Current research and trends* (pp. 119–123). Mahwah, NJ: Erlbaum.

Pardo, J. V., Fox, P. T., & Raichle, M. E. (1991). Localization of a human system for sustained attention by positron emission tomography. *Nature, 349,* 61–64.

Purves, D., Augustine, G. J., Fitzpatrick, D., Katz, L. C., LaMantia, A. S., McNamara, J. O., & Williams, S. M. (2001). *Neuroscience* (2nd ed.). Sunderland, MA: Sinauer.

Raichle, M. E. (1987). Circulatory and metabolic correlates of brain function in normal humans. In F. Plum (Ed.), *Handbook of physiology: The nervous system, Vol. V. Higher functions of the brain* (pp. 643–674). Bethesda, MD: American Physiological Society.

Raichle, M. E. (1998). Behind the scenes of functional brain imaging: A historical and physiological perspective. *Proceedings of the National Academy of Science USA, 95,* 765–772.

Rihs, F., Gutbrod, K., Gutbrod, B., Steiger, H. J., Sturzenegger, M., & Mattle, H. P. (1995). Determination of cognitive hemispheric dominance by "stereo" transcranial Doppler sonography. *Stroke, 26,* 70–73

Rihs, F., Sturzenegger, M., Gutbrod, K., Schroth, G., & Mattle, H. P. (1999). Determination of language dominance: Wada test confirms functional transcranial Doppler sonography. *Neurology, 52,* 1591–1596.

Risberg, J. (1986). Regional cerebral blood flow in neuropsychology. *Neuropsychologica, 34,* 135–140.

Roland, P. E. (1981). Somatotopical tuning of postcentral gyrus during focal attention in man: A regional cerebral blood flow study. *Journal of Neurophysiology, 46,* 744–754.

Roy, C. S., & Sherrington, C. S. (1890). On the regulation of the blood supply of the brain. *Journal of Physiology (London), 11,* 85–108.

Santalucia, P., & Feldman, E. (1999). The basic transcranial Doppler examination: Technique and anatomy. In V. L. Babikian & L. R. Wechsler (Eds.), *Transcranial Doppler ultrasonography* (pp. 13–33). Woburn, MA: Butterworth-Heinemann.

Scerbo, M. W. (1996). Theoretical perspectives on adaptive automation. In R. Parasuraman & M. Mouloua (Eds.), *Automation and human performance* (pp. 37–63). Mahwah, NJ: Erlbaum.

Schmidt, E. A., Piechnik, S. K., Smielewski, P., Raabe, A., Matta, B. F., & Czosnyka, M. (2003). Symmetry of cerebral hemodynamic indices derived from bilateral transcranial Doppler. *Journal of Neuroimaging,13,* 248–254.

Schmidt, P., Krings, T., Willmes, K., Roessler, F., Reul, J., & Thron, A. (1999). Determination of cognitive hemispheric lateralization by "functional" transcranial Doppler cross-validated by functional MRI. *Stroke, 30,* 939–945.

Schnittger, C., Johannes, S., & Munte, T. F. (1996). Transcranial Doppler assessment of cerebral blood flow velocity during visual spatial selective attention in humans. *Neuroscience Letters, 214,* 41–44.

Schuepbach, D., Merlo, M. C. G., Goenner, F., Staikov, I., Mattle, H. P., Dierks, T., et al. (2002). Cerebral hemodynamic response induced by the Tower of Hanoi puzzle and the Wisconsin Card Sorting test. *Neuropsychologia, 40,* 39–53.

Serrati, C., Finocchi, C., Calautti, C., Bruzzone, G. L., Colucci, M., Gandolfo, C., et al. (2000). Absence of hemispheric dominance for mental rotation ability: A transcranial Doppler study. *Cortex, 36,* 415–425.

Sorteberg, W., Langmoen, I. A., Lindegaard, K. F., & Nornes, H. (1990). Side-to-side differences and day-to-day variations of transcranial Doppler parameters in normal subjects. *Journal of Ultrasound Medicine, 9,* 403–409.

Stoll, M., Hamann, G. F., Mangold, R., Huf, O., & Winterhoff-Spurk, P. (1999). Emotionally evoked changes in cerebral hemodynamics measured by transcranial Doppler sonography. *Journal of Neurology, 246,* 127–133.

Stroobant, N., Van Nooten, G., & Vingerhoets, G. (2004). Effect of cardiovascular disease on hemodynamic response to cognitive activation: A functional transcranial Doppler study. *European Journal of Neurology, 11,* 749–754.

Stroobant, N., & Vingerhoets, G. (2000). Transcranial Doppler ultrasonography monitoring of cerebral hemodynamics during performance of cognitive tasks: A review. *Neuropsychology Review, 10,* 213–231.

Stroobant, N., & Vingerhoets, G. (2001). Test-retest reliability of functional transcranial Doppler ultrasonography. *Ultrasound in Medicine and Biology, 27,* 509–514.

Sturzenegger, M., Newell, D. W., & Aaslid, R. (1996). Visually evoked blood flow response assessed by simultaneous two-channel transcranial Doppler using flow velocity averaging. *Stroke, 27,* 2256–2261.

Thomas, C., & Harer, C. (1993). Simultaneous bihemispherical assessment of cerebral blood flow velocity changes during a mental arithmetic task. *Stroke, 24,* 614–615.

Toole, J. F. (1984). *Cerebrovascular disease.* New York: Raven Press

Totaro, R., Marini, C., Cannarsa, C., & Prencipe, M. (1992). Reproducibility of transcranial Doppler sonography: A validation study. *Ultrasound in Medicine and Biology, 18,* 173–177.

Tripp, L. D., & Chelette, T. L. (1991). Cerebral blood flow during +Gz acceleration as measured by transcranial Doppler. *Journal of Clinical Pharmacology, 31,* 911–914.

Troisi, E., Silvestrini, M., Matteis, M., Monaldo, B. C., Vernieri, F., & Caltagirone, C. (1999). Emotion-related cerebral asymmetry: Hemodynamics measured by functional ultrasound. *Journal of Neurology, 246,* 1172–1176.

Varnadore, A. E., Roberts, A. E., & McKinney, W. M. (1997). Modulations in cerebral hemodynamics under three response requirements while solving language-based problems: A transcranial Doppler study. *Neuropsychologia, 35,* 1209–1214.

Vingerhoets, G., & Luppens, E. (2001). Cerebral blood flow velocity changes during dichotic listening with directed or divided attention: A transcranial Doppler ultrasonography study. *Neuropsychologia, 39,* 1105–1111.

Vingerhoets, G., & Stroobant, N. (1999). Lateralization of cerebral blood flow velocity changes during cognitive tasks: A simultaneous bilateral transcranial Doppler study. *Stroke, 30,* 2152–2158.

Vollmer-Haase, J., Finke, K., Hartje, W., & Bulla-Hellwig, M. (1998). Hemispheric dominance in the processing of J. S. Bach fugues: A transcranial Doppler sonography (TCD) study with musicians. *Neuropsychologia, 36,* 857–867.

Wada, W., & Rasmussen, T. (1960). Intracarotid injection of sodium amytal for the lateralization of cerebral speech dominance. *Journal of Neurosurgery, 17,* 266–282.

Walter, K. D., Roberts, A. E., & Brownlow, S. (2000). Spatial perception and mental rotation produce gender differences in cerebral hemovelocity: A TCD study. *Journal of Psychophysiology, 14,* 37–45.

Watanabe, A., Nobumasa, K., & Tadafumi, K. (2002). Effects of creatine on mental fatigue and cerebral hemoglobin oxygenation. *Neuroscience Research, 42,* 279–285.

Wickens, C. D., & Hollands, J. G. (2000). *Engineering psychology and human performance* (3rd ed.). Upper Saddle River, NJ: Prentice Hall.

Wickens, C. D., Mavor, A. S., Parasuraman, R., & McGee, J. P. (1998). *The future of air traffic control: Human operators and automation.* Washington, DC: National Academy Press.

Wilson, G., Finomore, Jr., V., & Estepp, J. (2003). Transcranial Doppler oximetry as a potential measure of cognitive demand. In *Proceedings of the 12th International Symposium on Aviation Psychology* (pp. 1246–1249). Dayton, OH:.

Wittich, I., Klingelhofer, G., Matzander, B., & Conrad, B. (1992). Influence of visual stimuli on the dynamics of reactive perfusion changes in the posterior cerebral artery territory. *Journal of Neurology, 239,* 9.

Zatorre, R. J., Evans, A. C., & Meyer, E. (1994). Neural mechanisms underlying melodic perception and memory for pitch. *Journal of Neuroscience, 14,* 1908–1919.

Zinni, M., & Parasuraman, R. (2004). The effects of task load on performance and cerebral blood flow velocity in a working memory and a visuomotor task. In *Proceedings of the 48th Annual Meeting of the Human Factors and Ergonomics Society.* (pp. 1890–1894). Santa Monica, CA: Human Factors and Ergonomics Society.

Jason S. McCarley and Arthur F. Kramer

Eye Movements as a Window on Perception and Cognition

Eye tracking has been a tool in human factors since the first days of the field, when Fitts, Jones, and Milton (1950) analyzed the eye movements of pilots flying instrument landing approaches. Lacking computerized eye tracking equipment, the researchers recorded data by filming their subjects with a movie camera mounted in the cockpit. Once collected, the data, thousands of frames, were coded by hand. The results gave insight into the pilots' scanning habits and revealed shortcomings in cockpit display design that might have gone undiscovered through any other methodology. The frequency with which pilots looked at a given instrument, for example, was determined by the importance of that instrument to the maneuver being flown; the duration of each look was determined by the difficulty of interpreting the instrument; patterns of scanning varied across individuals; and instruments were arranged in a pattern that encouraged a suboptimal pattern of transitions between channels. Such information, painstakingly acquired, provided an important foundation for the understanding of expert and novice pilot performance and pointed directly toward potential improvements in cockpit design.

Happily, sophisticated eye tracking systems are now abundant, inexpensive, easy to use, and often even portable. Oculomotor data are thus far easier to collect and analyze than they were at the time of Fitts's work (see Jacob & Karn, 2003, for discussion of developments in eye tracking technology and the history of eye tracking in engineering psychology). The cognitive and neural mechanisms that control eye movements, moreover, have become better understood with the advent of cognitive neuroscience, putting researchers in a stronger position to predict, control, and interpret oculomotor behavior. This is compatible with the neuroergonomics view (Parasuraman, 2003) that an understanding of the brain mechanisms underlying eye movements can enhance applications of this measure to investigate human factors issues. Accordingly, eye movement data are being used by ergonomics researchers in an ever-growing variety of domains—including radiological diagnosis (e.g., Carmody, Nodine, & Kundel, 1980; Kundel & LaFollette, 1972), driving (e.g., Mourant & Rockwell, 1972), reading (e.g., McConkie & Rayner, 1975; Rayner, 1998), airport baggage screening (McCarley, Kramer, Wickens, Vidoni, & Boot, 2004), and athletic performance (Abernethy, 1988)—and for a range of purposes. They are often studied for understanding of the perceptual-cognitive processes and strategies mediating performance in complex tasks. This may include identifying nonoptimal aspects of a human-machine interface (e.g., Fitts et al., 1950), delineating expert-novice differences in the performance of a given task (e.g., Kundel & LaFollette, 1972), or revealing the nature of the lapses that produce performance errors (e.g., McCarley, Kramer, et al., 2004). Such knowledge can inform future human factors interventions such as the redesign of visual displays or the development of training programs. Oculomotor data can also be used to draw inferences about an operator's cognitive state or mental workload level (May, Kennedy, Williams, Dunlap, & Brannan, 1990; Wickens & Hollands, 2000) and may even be diagnostic of the nature of mental workload imposed by a nonvisual secondary task (Recarte & Nunes, 2000). Research on interface design, finally, has suggested that eye movements may be useful as a form of control input in human-computer interaction, providing a means for users to perform onscreen point-and-click operations (Surakka, Illi, & Isokoski, 2003), for example, or to control the display of visual information (Reingold, Loschky, McConkie, & Stampe, 2003).

In this chapter, we briefly review findings from the basic study of eye movements in cognitive psychology and perception and consider implications of these findings for human factors researchers and practitioners. Because of constraints on space, discussion focuses largely on the aspects of eye movements that pertain to human performance in visual search and supervisory monitoring tasks or human interaction with machine systems. For a more comprehensive review and discussion of the eye movement literature, including consideration of the role of eye movements in reading, see Findlay and Gilchrist (2003), Hyönä, Radach, and Deubel (2003), and Rayner (1998).

Saccades and Fixations

The monocular visual field in a human observer extends over 150° in both the horizontal and vertical directions. The perception of visual detail, however, is restricted to a small rod-free region of central vision known as the *fovea*, and in particular, to the rod- and capillary-free *foveola*, a single degree of visual angle in diameter (Wandell, 1995). Away from the fovea, perception of high spatial frequencies declines rapidly, due in part to increases in the spacing between the retinal cones and in part to changes in the neural connectivity between retina, geniculate, and cortex (Wilson, Levi, Maffei, Rovamo, & DeValois, 1990). Movements of the eyes are used to shift the small region of detailed vision from one area of interest to another as necessary for visual exploration and monitoring.

Researchers have identified several distinct forms of eye movements (see Findlay & Gilchrist, 2003, for a detailed review). Three that directly subserve visual information sampling are *vergence shifts*, *pursuit movements*, and *saccades*. Vergence shifts are movements in which the left and right eyes rotate in opposite directions, moving the point of regard in depth relative to the observer. Pursuit movements are those in which the eyes travel smoothly and in conjunction so as to maintain fixation on a moving object. Of most interest to human factors specialists, finally, are saccades, ballistic movements that rapidly shift the observer's gaze from one point of interest to another. Saccadic eye movements are typically 30 to 50 ms in duration, and can reach velocities of 500° per second (Rayner, 1998). They tend to occur 3–4 times per second in normal scene viewing and are separated by *fixations* that are generally 200–300 ms in duration. Saccades can be broadly classified as *reflexive*, *voluntary*, or *memory guided*. Reflexive saccades, though they can be suppressed or modified by intentional processes, are visually guided movements programmed automatically in response to a transient signal at the saccade target location. Voluntary saccades are programmed endogenously to a location not marked by a visual transient. Memory-guided saccades are made to a cued location, but only after delay (Pierrot-Deseilligny, Milea, & Müri, 2004).

Notably, useful visual information is acquired only during fixations. During the execution of a saccade, thresholds for visual detection are highly elevated and little new information is acquired, a phenomenon known as *saccadic suppression* (Matin, 1974). This effect seems to occur in part because pre- and postsaccadic input masks the signals produced during the movement itself (Campbell & Wurtz, 1979), in part because the retinal stimulus patterns that obtain while the eye is in flight fall largely outside the visual system's spatiotemporal sensitivity range (Castet & Masson, 2000), and in part because retinal signals generated during the saccade are inhibited at a site between the retina and the first visual cortical area (Burr, Morrone, & Ross, 1994; Riggs, Merton, & Morton, 1974; Thilo, Santoro, Walsh, & Blakemore, 2004). As such, visual sampling during saccadic scanning is essentially discrete, with periods of information acquisition interrupted by transient periods of (near) blindness. Interestingly, Gilchrist, Brown, and Findlay (1997) have described a patient who is unable to move her eyes yet shows a pattern of visual scanning behavior—discrete dwells interspersed with rapid head movements—similar to that seen in normal subjects. Gilchrist et al. take this as evidence that saccadic behavior may be an optimal method of sampling information from a visual scene.

Describing Saccadic Behavior

Given the complexity of our visual behavior, it is not surprising that a myriad of dependent measures for understanding oculomotor scanning have been used (for more comprehensive discussion of such measures, see Inhoff & Radach, 1998; Jacob & Karn, 2003; Rayner, 1998). Typically, saccades are described by their amplitude and direction, though in some cases saccade duration and maximum velocity may also be of interest. Analysis of visual samples that occur between saccades is more complex. A sample is characterized in part by its location, obviously, and for some purposes it may be necessary to specify location with great precision. In human factors research, though, it is often sufficient merely to determine whether the point of regard falls within a designated area of interest (AOI). For example, an aviation psychologist may wish to know which cockpit instrument the pilot is inspecting at a given moment, without needing to know the precise location of the pilot's gaze within the instrument. Thus, multiple consecutive fixations within the same AOI might be considered part of a single look at the instrument. Any series of one or more consecutive fixations on the same AOI is termed a *dwell* or *gaze*. A sequence of multiple consecutive fixations within a single AOI may occur either because the observer reorients the eyes within the AOI (if the region is sufficiently large) or because the observer executes a corrective *microsaccade* (Rayner, 1998) to return the point of regard to its original location following an unintended drift. It is also common for an observer scanning a scene or display to return more than once to the same AOI. In such cases, the initial gaze within the AOI may be termed the *first-pass gaze*.

Given these considerations, there are multiple measures for describing the temporal and spatiotemporal properties of oculomotor sampling. At the finest level is *fixation duration*, the duration of a single pause between saccades. As noted above, fixation durations are sensitive to manipulations that affect the difficulty of extracting foveal visual information, including noise masking (van Diepen & d'Ydewalle, 2003), contrast reduction (van Diepen, 2001), and, in a visual search task, a decrease in target-distractor discriminability (Hooge & Erkelens, 1996). In cases where there are multiple consecutive fixations on the same AOI, fixation duration may be less informative than *gaze duration*. In understanding a pilot's cockpit scanning behavior, for instance, it is probably more useful to know the total amount of time spent on a gaze at a given instrument than to know the durations of individual fixations comprising the gaze. It also common, finally, to report *dwell frequencies* for various AOIs. Dwell frequency data are of particular interest in supervisory monitoring tasks, where the operator will scan a limited set of AOIs over an extended period of time. A number of studies have shown that gaze duration is sensitive to the information value of the item being fixated, with items that are unexpected or semantically inconsistent with their context receiving longer gazes than items that are expected within the context (e.g., Friedman, 1979; Loftus & Mackworth, 1978). In the context of aviation, as noted above, gaze durations in the cockpit are modulated by the difficulty of extracting information from the fixated instrument, with instruments that are more difficult to read or interpret receiving longer dwells (Fitts et al., 1950). Dwell frequency, in contrast, is determined by the relative importance of each AOI, such that a cockpit instrument is visited more often when it provides information critical to the flight maneuver being performed (Bellenkes, Wickens, & Kramer, 1997; Fitts et al., 1950).

Overt and Covert Attention Shifts

Interpretation of eye movement data is typically guided by the assumption that an observer is attending where the eyes are pointing, a premise formalized by Just and Carpenter (1980) as the *eye-mind assumption*. It is also well-known, however, that attention can at times be allocated to one location within the visual field while the eyes are focused on another. Indeed, it is common for cognitive psychologists to study attention using tachistoscopic displays expressly designed to disallow eye movements so as to avoid confounding the effects of visual acuity with those of central attentional processes. Attention researchers thus distinguish overt shifts of attention, those involving a movement of the head, eyes, or body, from covert shifts, accomplished without reorienting the sensory surface (Posner, 1980).

Of interest is the interaction between overt and covert orienting. Posner (1980) described four forms that the physiological and functional relationship between overt and covert processes might take. At one extreme was the logical possibility that eye movements and covert attention shifts might be produced by the identical neural mechanisms. The result would be full functional dependence between the two forms of orienting, a prediction disproved by the observation that attention can be oriented covertly without triggering an accompanying eye movement. At the opposite end of the theoretical spectrum was the possibility that eye movements and covert shifts are functionally independent, a hypothesis at odds with the eye-mind assumption and with the introspective sense that attention and the eyes are closely linked. In between the possibilities of full dependence and full independence, Posner identified the theories of an efference-driven relationship and of a nonobligatory functional relationship. The latter account posits that overt and covert attention are not structurally coupled but tend to travel in concert simply because they are attracted by similar objects and events. The efference theory, also known as the premotor (Rizzolatti, Riggio, Dascola, & Umiltà, 1987) or oculomotor readiness theory (Klein, 1980), holds that a covert attention shift is equivalent to an unexecuted overt shift. The neural mechanisms and processes involved in covert and overt orienting, in other words, are identical up until the point at which an overt shift is carried out.

Which of these accounts is correct? Although early findings appeared to support the theory of a nonstructural functional relationship between overt and covert orienting (Klein, 1980), more recent findings have demonstrated that covert attention is obligatorily shifted to the location of a saccade target prior to execution of the overt movement, consistent with the claims of the premotor theory. For example, Hoffman and Subramaniam (1995) asked subjects to saccade to one of four target locations (above, below, right, or left of fixation) while attempting to discriminate a visual probe letter presented prior to saccade onset presented at one of the same four locations. Probe discrimination was best when the probe appeared at the saccade target location. Likewise, Deubel and Schneider (1996) asked subjects to discriminate a target character embedded in a row of three items while simultaneously preparing a saccade to one of the three locations. Data again showed that psychophysical performance was highest when the discrimination target and saccade target were coincident (see also Sheperd, Findlay, & Hockey, 1986). Kowler, Anderson, Dosher, and Blaser (1995) had subjects report the identity of a letter that appeared at one of eight locations arranged on an imaginary circle while also programming a saccade. Performance on both tasks was best when the locations for the saccade and letter identification were the same. Similar effects have been found across stimulus modalities, interestingly, with discrimination accuracy enhanced for auditory probes that are presented near rather than far from a saccade target location (Rorden & Driver, 1999). Evidence for a presaccadic shift of covert attention has also come from findings of a *preview benefit*, a tendency for information that is available at the saccade target location prior to movement execution to facilitate postsaccadic processing (Henderson, Pollatsek, & Rayner, 1987). As discussed below, finally, positron emission tomography (PET) and functional magnetic resonance imaging (fMRI) data support the conclusion that overt and covert attentional control are rooted in overlapping neural systems. It should be noted that covert attentional processes apparently do not serve the purpose of scouting out or selecting potential saccade targets (Findlay, 1997; Peterson, Kramer, Wang, Irwin, & McCarley, 2001). Rather, the processes responsible for selecting saccade targets appear to operate in parallel across the visual field (as captured in the models of saccade generation discussed below), with covert attention accruing at the target location in the lead-up to saccade execution.

In total, data indicate that a reallocation of covert attention is a necessary but not sufficient precondition for saccade execution; attention can move without the eyes following, but the eyes cannot move unless attention has shifted first. Thus, a fixation at a given location is strong evidence that attention has been there, but the failure to fixate a location does not guarantee that the location has not been attended. For the human factors specialist, however, these conclusions come with two caveats. First, it is not clear that covert attentional shifts independent of eye movements play a significant role in many real-world tasks (Findlay & Gilchrist, 1998; Moray & Rotenberg, 1989). Indeed, subjects often rely on oculomotor scanning even in cases where a stronger reliance on covert processing might improve performance. Brown, Huey, and Findlay (1997), for example, found a tendency for subjects to commence oculomotor scanning rapidly following stimulus onset in a visual search task, despite the fact that they could improve performance by delaying the initial saccade and allowing greater time for covert processes to operate. Similarly, Shapiro and Raymond (1989) discovered that although action video game players could be trained to improve their scores by executing fewer saccades and relying instead on covert processing, they did not tend to adopt this strategy spontaneously. It may thus be that in many naturalistic tasks, covert attentional shifts are by themselves of little practical interest. Moray and Rotenberg (1989, p. 1320) speculated that "the more a task is like a real industrial task, where the operators can move their head, eyes and bodies, can slouch or sit up, and can, in general, relate physically to the task in any way they please, the more 'attention' is limited by 'coarse' mechanisms such as eye and hand movements." The second caveat is that even if an operator has fixated on an item of interest, and therefore has attended to it, there is no guarantee that the item has been processed to the point of recognition or access to working memory. For instance, "looked but failed to see" errors, in which a driver gazes in the direction of a road hazard but does not notice it, are a relatively common cause of traffic accidents (Langham, Hole, Edwards, & O'Neill, 2002). Such lapses in visual encoding can be engendered by nonvisual cognitive load, like that imposed by an auditory or verbal loading task. In one demonstration of this, Strayer, Drews, and Johnston (2003) measured subjects' incidental memory for road signs encountered in a simulated driving task. Data showed that recognition was degraded if stimulus encoding occurred while subjects were engaged in a hands-free cell phone conversation. This effect obtained even when controlling for the amount time that the road signs were fixated at the time of encoding, confirming that the encoding of information within a gaze was impaired independent of possible changes in gaze duration.

Attentional Breadth

In addition to asking questions about movements of covert attention within the course of a visual fixation, researchers have also measured the breadth of covert processing. The terms *stationary field* (Sanders, 1963), *functional field of view* (FFOV), *conspicuity area* (Engel, 1971), *visual lobe* (Courtney & Chan, 1986), *visual span* (Jacobs, 1986), and *perceptual span* (McConkie & Rayner, 1975, 1976) have all been used to describe the area surrounding the point of regard from which information is extracted in the course of a fixation. (Though researchers sometimes draw distinctions between their precise definitions, we use these various terms interchangeably.) In some cases, the perceptual span is delineated psychophysically by measuring the distance from fixation at which a target can be detected or localized within a briefly flashed display (e.g., Sekuler & Ball, 1986; Scialfa, Kline, & Lyman, 1987). In other instances, span is assessed using *gaze-contingent paradigms* in which eye movements are tracked and visual displays are modified as a function of where the observer is looking (McConkie & Rayner, 1975, 1976; Rayner, 1998). The most common of these is the *moving window* technique. Here, task-critical information is masked or degraded across the display except within a small area surrounding the point of regard. An example of such a display is shown in figure 7.1. To measure the perceptual span, researchers can manipulate the size of the moving window and determine the point at which performance reaches normal levels. In reading, the perceptual span is asymmetrical, with the direction of the asymmetry depending on the direction in which readers scan. In English, which is printed left to right, the perceptual span extends 3–4 characters to the left of the reader's fixation and 14–15 characters to the right (McConkie & Rayner, 1975, 1976). In Hebrew, printed right to left, the direction of asymmetry is reversed (Pollatsek, Bolozky, Well, & Rayner, 1981). In both cases, in

Figure 7.1. Illustration of a moving window display. Display resolution is normal within a region centered on the observer's point of regard and is degraded outside that region. Display is gaze-contingent, such that the high-resolution window follows the point of regard as the eyes move. From Reingold et al. (2003). From Reingold, E. M., Loschky, L. C., McConkie, G. W., & Stampe, D. M. (2003). Gaze-contingent multresolutional displays: An integrative review. Human Factors, 45, 307–328, with permission of Blackwell Publishing.

other words, the perceptual span extends farther in the direction that the reader is scanning than in the opposite direction. In visual search, the span of effective vision depends on the similarity of the target and distractor items, being narrower when target and distractors are more similar (Rayner & Fisher, 1987a). Span size can also be reduced by increases in nonvisual workload (Pomplun, Reingold, & Shen, 2001). Jacobs (1986) found that the size of the visual span in a search task, as manipulated through changes in target-distractor discriminability, accounts for a large proportion of variance in saccade amplitudes but a relatively small proportion of variance in fixation durations.

Distinct functional regions can also be delineated outside the central perceptual span. An experiment by Sanders (1963) required subjects to make same-different judgments of paired stimuli, one presented at the subject's initial fixation point, the other at a distance of between 10° and 100° of visual angle to the right. Data suggested that the visual field could be divided into three concentric zones. The stationary field, as noted above, was the region surrounding fixation within which both stimuli could be processed without an eye movement. The *eye field* was the region within which an eye movement was required from the left item to the right one in order to compare them. The *head field*, finally, was the region within which a combined eye and head movement was necessary to compare the two targets. Data showed that moving the right stimulus outward from the stationary field to the eye field increased response times for same-different judgments, and that moving it again from the eye field to the head field increased them further still. Importantly, the response times calculated by Sanders did not include the time necessary for execution of the eye and head movements themselves. The transition from stationary field to eye field, and then from eye field to head field, therefore seems not only to require additional motor responses of the observer, but also to impose a cost of integrating visual information across multiple gazes (cf. Irwin, 1996). The processing advantage of the eye field relative to the head field arose, Sanders speculated, because coarse preliminary processing of the information within the eye field primed recognition of stimuli that were subsequently brought within the stationary field (see also Rayner & Fisher, 1987b).

Attentional breadth appears to be an important mediator of visual performance in complex tasks and real-world domains. Pringle, Irwin, Kramer, and Atchley (2001) found a negative correlation between the size of the FFOV, as measured with a visual search task using a briefly flashed display, and the time necessary for subjects to notice visual events within free-viewed traffic scenes in a change detection task. Their results suggest that a psychophysical measure of visual span can indeed provide an index of how "far and wide" observers tend to spread covert attention while scanning a complex naturalistic stimulus. Consistent with this conclusion, Owsley and colleagues (1998) have found negative correlations between the size of FFOV and accident rates in elderly drivers.

Bottom-Up and Top-Down Control

In developing most any form of visual information display—aircraft cockpit, automobile dashboard, computer software interface, traffic sign, or warning label—one of the designer's primary goals is to predict and control the display user's scanning, ensuring that task-relevant stimuli are noticed, fixated, and processed. For this, it is necessary to understand the information that guides the selection of saccade targets in oculomotor scanning. Like the control of covert attention (Yantis, 1998), guidance of the eyes is accomplished by the interplay of bottom-up/stimulus-driven and top-down/goal-driven mechanisms. Attention shifts, overt or covert, are considered bottom-up to the extent that they are driven by stimulus properties independently of the observer's behavioral goals or expectations, and are deemed top-down to the extent that they are modulated by intentions or expectations. Among scientists studying covert attention, theorists variously argue that attentional shifts are never purely stimulus driven but are always regulated by top-down control settings (Folk, Remington, & Johnston, 1992); that attentional control is primarily stimulus driven, such that salient objects tend to capture covert processing regardless of the observer's goals or set (Theeuwes, 1994); and that some stimulus properties—specifically, the abrupt appearance of a new object within the visual scene—are uniquely capable of capturing covert attention, but that salience per se is not sufficient to trigger a reflexive attention shift against the observer's intentions (Jonides & Yantis, 1988; Yantis & Jonides, 1990). Despite much research, debate among the proponents of these competing theories has not been resolved.

In oculomotor research, bottom-up control has been studied using the *oculomotor capture* paradigm introduced by Theeuwes, Kramer, Hahn, and Irwin (1998). Stimuli and procedure from a typical experiment are illustrated in figure 7.2. Subjects in the study by Theeuwes et al. began each trial by gazing at a central fixation marker surrounded by a ring of six gray circles, each containing a small gray figure-8 premask. After a brief delay, five of the cir-

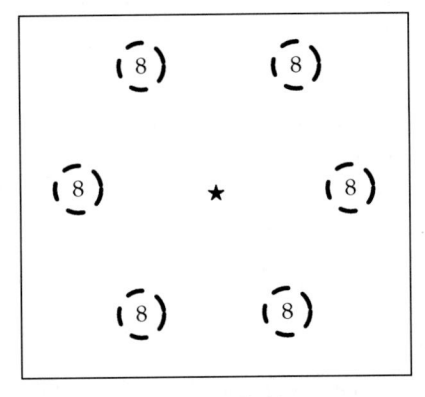

Fixation (1000ms)　　　　　　　　Target/Onset Distractor

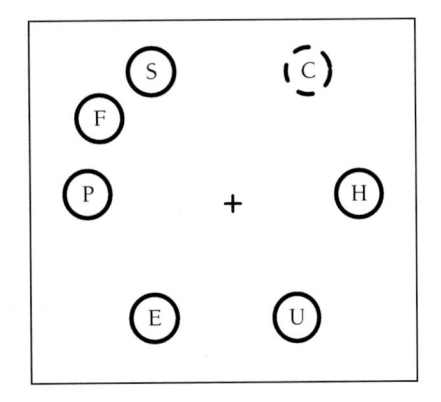

Figure 7.2. Schematic illustration of the stimuli employed in the Theeuwes et al. (1998) study of oculomotor capture. Gray circles are indicated by dashed lines; red circles are indicated by solid lines. See text for description of procedure. Reprinted with permission of Blackwell Publishing from Theeuwes et al. (1998).

a   Control (no new object)

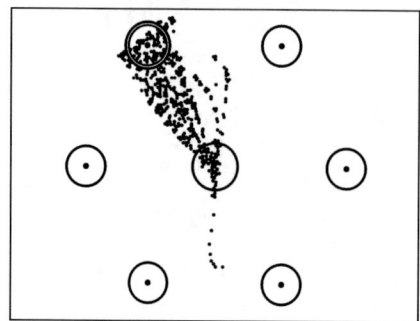

b   New object appearing at 30° from the target

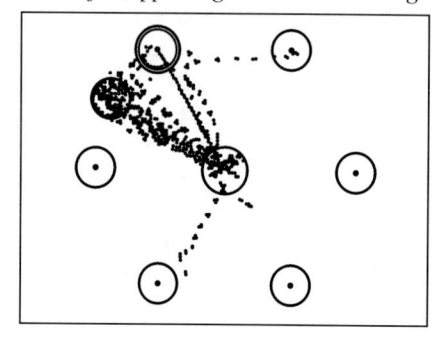

c   New object appearing at 90° from the target

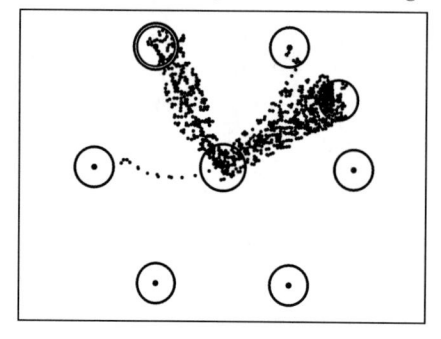

d   New object appearing at 150° from the target

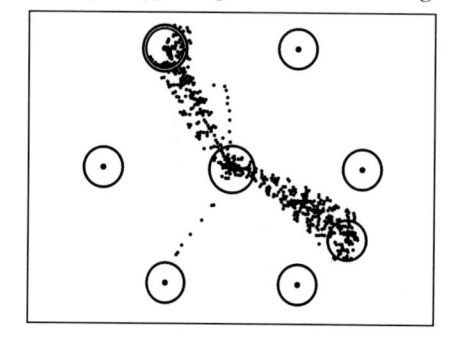

cles changed to red and segments were removed from premasks to reveal letters. Simultaneously, an additional red circle containing a letter appeared abruptly at a location that had previously been unoccupied. The subjects' task was to saccade as quickly as possible to the remaining gray circle and identify the letter within it. The onset circle was never itself the target. Despite this, remarkably, on roughly a quarter of all trials subjects made a reflexive saccade to the onset circle before moving to the gray target stimulus. Just as the sudden appearance of a new visual object seems to strongly and perhaps obligatorily attract covert attention (Yantis & Jonides, 1990), the appearance of a new distractor object captured the subjects' gaze, overriding top-down settings that specified the gray item as the desired saccade target. Figure 7.3 presents sample data. Control experiments confirmed that the additional distractor captured attention only when it appeared as an abrupt onset, not when it was present from the beginning of the trial. On those trials where capture occurred, interestingly, the eyes dwelt on the onset stimulus for less than 100 ms before shifting to the gray target, a period too brief to allow full programming of a saccade. The authors concluded that programming of two saccades—one a stimulus-driven movement toward the onset object and the other a goal-driven movement toward the gray target—had occurred in parallel (Theeuwes et al., 1998). Subsequent experiments have suggested that abrupt onsets are uniquely effective in producing oculomotor capture. Luminance increases and color singletons (i.e., uniquely colored items among homogeneously colored background stimuli), for example, produce far weaker capture of the eyes, even when matched in salience to an abrupt onset (Irwin, Colcombe, Kramer, & Hahn, 2000).

Another simple and popular method for studying bottom-up and top-down control of eye movements is the antisaccade task developed by Hallett (1978). The observer begins a trial by gazing at a central fixation mark, and a visual go signal is then

Figure 7.3. Sample data from Theeuwes et al. (1998). Data points indicate gaze position samples collected at 250 Hz during the first saccade of each trial following target onset. The target stimulus appears at the 11 o'clock position in all frames. Reprinted with permission of Blackwell Publishing from Theeuwes et al. (1998).

flashed in either the left or right periphery. The observer's task is to move the eyes in the direction opposite the go signal. To perform the task correctly, therefore, the observer must suppress the tendency to make a reflexive saccade toward the go signal and instead program a voluntary saccade in the other direction. Performance on antisaccade trials can be compared to that on prosaccade trials, where the observer is asked to saccade toward the transient signal. Data show that antisaccades are initiated more slowly than prosaccades, and that directional errors are common in the antisaccade conditions but rare in the prosaccade task (Everling & Fischer, 1998; Hallett, 1978). As discussed further below, Guitton, Buchtel, and Douglas (1985) found a selective deficit in antisaccade performance among patients with frontal lobe lesions, suggesting a role for the frontal regions in inhibiting reflexive saccades and programming voluntary movements. Consistent with this conclusion, data from healthy subjects indicate that antisaccade performance is mediated by executive working memory processes typically ascribed to the frontal lobes (Kane, Bleckley, Conway, & Engle, 2001; Roberts, Hager, & Heron, 1994).

Search is another form of visual behavior incorporating bottom-up and top-down processes. Modern theories of search (Itti & Koch, 2000; Treisman & Sato, 1990; Wolfe, 1994) generally assume that attentional scanning is guided by activation within a salience map (Koch & Ullman, 1985) representing points of interest within the visual scene. In the first stages of search, a scene or display is encoded within an array of low-level representations, known as feature maps, tuned to stimulus properties such as color, orientation, motion, and spatial frequency. Contrast between feature values (e.g., contrast between red and green regions on either side of a chromatic border) produces bottom-up signals that are fed forward to the salience map and serve to direct attention within the visual scene. In cases where physical properties of the target are specified, goal-driven control of search can be effected by top-down modulation of activation within feature maps, amplifying signals within the maps that encode known target properties or attenuating activation within the maps that encode properties that do not belong to the target (Treisman & Sato, 1990; Wolfe, 1994). The influence of feature-guided search is evident in the phenomenon of *saccade selectivity* during visual search, the tendency for observers to preferentially fixate stimuli that share features with the target (e.g., Findlay, 1997; Scialfa & Joffe, 1998; D. E. Williams & Reingold, 2001; L. G. Williams, 1967). A number of studies have found that saccade guidance based on color is more effective than that based on shape or orientation (D. E. Williams & Reingold, 2001; L. G. Williams, 1967), presumably because of the poor spatial resolution in the peripheral visual field. Under circumstances where target uncertainty or high levels of camouflage make top-down feature guidance ineffective, models based on bottom-up guidance alone may predict scanning performance well (Itti & Koch, 2000). Another form of top-down control in search, independent of feature-based guidance, is *contextual cuing*, whereby attention is biased toward potential target-rich locations within a familiar scene (Chun & Jiang, 1998; Peterson & Kramer, 2001). Interestingly, this effect occurs even within abstract stimulus displays (e.g., random arrangements of letters) and without conscious recognition that a given scene has been previously encountered. The phenomenon thus appears to be driven by implicit memory for specific scene exemplars, rather than by explicit semantic knowledge of likely target locations.

The influence of both goal-driven and stimulus-driven processes are evident in naturalistic and complex tasks, though to varying degrees across domains. A famous demonstration of top-down eye movement guidance came from Yarbus (1967), who showed that observers will scan a picture differently depending on the judgment they are asked to make of it (e.g., judging whether the family depicted in a scene is wealthy versus judging the family members' ages). In scanning images of traffic scenes, observers are more likely to fixate on locations or objects that are highly task relevant (e.g., traffic lights, pedestrians) than items of little task relevance, but at the same time are more likely to fixate on items that are highly salient than on items that are less conspicuous (McCarley, Vais, et al., 2004; Pringle, 2001; Theuwes, 1996). In supervisory monitoring, operators adapt their visual sampling to the bandwidth and information value of various channels, apparently forming a mental model of system behavior to guide their scanning top-down (Carbonell, Ward, & Senders, 1968; Senders, 1964). In performing stereotyped everyday tasks in non-dynamic environments, similarly, operators may rely almost exclusively on goal-driven sampling. Land and Hayhoe (2001), for example, recorded eye movements of subjects making sandwiches or preparing tea. Data showed a close coupling of eye movements to manipulative actions. Subjects carried out each task by performing a series of object-related actions (ORAs), each a simple manual act that brought the subject a step closer to the final goal. ORAs were typically delineated by a shift of the eyes from the object currently being manipulated (e.g., slice of bread) to the object next to be manipulated (e.g., knife). A fixation on each object generally began a fraction of a second before manipulation of the object was initiated, and persisted until gaze shifted to a new object to begin the next ORA. In both the sandwich-making and tea-making tasks, the proportion of fixations on task-irrelevant objects was less than 5%, leading the authors to conclude that eye movements were strongly goal driven.

Effort

Although visual scanning is guided largely by information content and information demand, as reflected in the stimulus-driven and goal-driven control of the eyes, it may also be constrained by *information access cost*, the effort needed to reach and sample a given channel. Gray and Fu (2004) noted that even small differences in the time or effort needed to complete a given action can prompt changes in an operator's behavioral strategy, producing large changes in overall performance levels. In complex visuomotor tasks, operators appear to trade the costs of sampling movements against the cognitive demands of maintaining information in working memory. A series of experiments by Ballard, Hayhoe, and Pelz (1995) illustrate this well. The subjects' task was to arrange colored tiles to create a copy of a mosaic pattern provided by the experimenters. Displays were divided into three areas: the *model*, the pattern to be copied; the *resource*, the space from which the colored tiles to be placed were retrieved; and the *workspace*, the region in which the copy was to be constructed. To perform the task, subjects were required to examine and remember the model, and then to retrieve colored tiles from the resource and place them within the workspace as appropriate. As noted by the experimenters, the task thus tapped a variety of sensory, cognitive, and motor skills, but was simple enough that moment-to-moment subgoals (e.g., pick up tile, drop tile) could be identified. Most important, the task allowed participants to adopt their own performance strategies. The experimenters could thus examine how variations in the demands of the task altered subjects' strategy selection.

In a first experiment (Ballard et al., 1995), the model, resource pool, and workspace were presented on a single computer monitor, and subjects performed the task with their heads fixed in a chin-rest. All visual information needed to complete each trial could be obtained with eye movements alone. Under these circumstances, subjects appeared to trade the costs of frequent eye movements for the benefits of a low working memory load. Indeed, data suggested that subjects held no more than the minimum possible amount of information in visual working memory. Rather than attempting to remember the model as a whole, or to simultaneously remember both the color and location of the next block to be placed, subjects tended to fixate the model once before selecting each block from the resource pool, to gaze at the resource pool in order to pick up a block, then to look at the model again before placing the selected block. A similar pattern of effects obtained in later experiments in which subjects performed the task by manipulating 3-D blocks, rather than computer icons, as long as the model, workspace, and resource pool were close enough (within 20° of visual angle) together for subjects to scan between them using eye movements and short head movements. Results changed, however, when the model and workspace were separated by a greater distance (70° of visual angle), such that large head movements were necessary to scan between them. In these cases, subjects relied more heavily on visual working memory, gathering and remembering a greater amount of information with each look at the model and therefore reducing the need for long-distance, high-effort gaze shifts between the model and the workspace. The implication of such findings for designers of human-machine systems are direct: To minimize demands on scanning and working memory, display channels that provide information to be compared or integrated should be arrayed near one another, and controls should be placed near the displays that they affect. These guidelines, long recognized in human factors research, have been codified as the *proximity display principle* (Wickens & Carswell, 1995) and *colocation principle* (Wickens & Hollands, 2000) of interface design.

Neurophysiology of Saccadic Behavior

Although the use of eye movements in the study of behavior and information processing has enhanced our understanding of human performance and cognition in real-world tasks, this research has to a large extent progressed in parallel with the study of the neuronal underpinnings of eye movements and their relationship to different cognitive constructs. Given the view of neuroergonomics (Parasuraman, 2003) that each domain of research may benefit from an understanding of the other, it is useful to review research that has addressed the relationship between eye movements and neuronal activity (see reviews by Gaymard, Ploner, Rivaud, Vermersch, & Pierrot-Deseilligny, 1998; Pierrot-Deseilligny et al., 2004, for additional details).

Researchers have utilized a variety of techniques to study the neural control of saccades in humans. Early research on humans attempted to localize specific saccade control functions by examining patients with circumscribed lesions. The research of Guitton et al. (1985), mentioned above, revealed deficits in the antisaccade but not the prosaccade task among patients with frontal lesions. These data and others from patient studies have been interpreted as evidence for the role of the frontal regions, and more specifically the dorsolateral prefrontal cortex (DLPFC) and frontal eye fields, in the inhibition of reflexive saccades and the programming and execution of voluntary saccades.

More recent studies have employed neuroimaging techniques such as PET and fMRI to examine the neuronal circuits that generate saccades, exploring the network of cortical and subcortical regions that contribute in common and uniquely to the planning and execution of prosaccades, antisaccades, memory-based saccades, and covert attention shifts. These studies have, by and large, found evidence for the activation of extensive networks of frontal, parietal, and midbrain regions, similar to areas revealed in studies of single-unit activity with nonhuman primates (Gottlieb, Kusunoki, & Goldberg, 1998) in different eye movement and attention-shifting tasks. Corbetta (1998) was the first to show, using fMRI, that the brain regions associated with covert orienting and overt eye movements overlap in frontal and parietal cortices. Kimmig et al. (2001) found, in a PET study, that similar brain regions were activated in pro- and antisaccade tasks but that activation levels were higher in the antisaccade task for areas including the frontal eye fields, supplementary eye fields, parietal eye fields, putamen, and thalamus. Sweeney et al. (1996) reported increased activation in the DLPFC, a brain region associated with working memory and interference control, in both a memory-guided saccade and antisaccade task.

Another technique useful in the study of the neuronal underpinnings of saccades is transcranial magnetic stimulation (TMS). TMS involves the application of a brief magnetic pulse to the scalp. This pulse induces localized electrical fields that alter the electrical field in the brain below the stimulator, in effect producing a virtual lesion that is both reversible and transient. Terao et al. (1998) administered a TMS pulse, over either the frontal or parietal cortex, at various times after the presentation of a peripheral stimulus that cued an antisaccade. Increased saccadic latencies and erroneous prosaccades (i.e., saccades toward rather than away from the eliciting stimulus) were induced by stimulation of either cortical region, though changes in saccade parameters occurred earlier for parietal than for frontal stimulation. Ro, Henik, Machado, and Rafal (1997) employed TMS to examine the neuronal organization of stimulus-driven and voluntary saccades. Subjects were asked to make saccades to either the left or right in response to either a centrally located arrow (endogenous go signal) or a peripheral (exogenous go signal) marker. TMS pulses were presented at varying intervals with respect to the saccade go signal. TMS delivered over the superior prefrontal cortex increased latencies for saccades made in response to the endogenous go signal. No effects were observed for saccade latency when the go signal was exogenous (see also Muri, Henik, Machado, & Rafal, 1996). These data suggest that TMS can be employed to study both the locus of different saccade control and implementation functions and the timing of these control processes.

The studies described above, along with a much more extensive body of research that has examined eye movements in humans as well as other animals, have begun to map out the neural circuits that are responsible for oculomotor control. Results indicate that a large number of interconnected frontal, parietal, and midbrain regions contribute to oculomotor behavior, with different subsets of these areas being responsible for the performance of prosaccade, antisaccade, and memory-guided saccade tasks. For example, visually guided saccades (e.g., prosaccades) involve a pathway from the visual cortex through the parietal cortex, to the frontal and supplementary eye fields, to the superior colliculus, and finally to the brain stem, where motor commands are generated for eye movements (Pierrot-Deseilligny et al., 2004). More direct pathways are also available from the parietal regions and separately from the frontal regions to the superior colliculus and from the frontal eye fields directly to the brain stem (Hanes & Wurtz, 2001). Although little is presently known about the functional differences among these pathways, it does appear that the direct pathway from the parietal regions is concerned with eye movements to salient stimuli of relevance to the organism and that the direct pathway from DLPFC to the superior colliculus may play an important role in the inhibition of reflexive saccades that compete with voluntary saccades.

An interesting question is whether this knowledge of the neural circuits that underlie different varieties of saccades (i.e., prosaccades, antisaccades, memory-based saccades) will be of use in modeling and predicting multitask performance decrements in situations of interest to human factors researchers and practitioners. A specific characteristic of neuroergonomics research as described by Parasuraman (2003) is the use of knowledge of brain mechanisms underlying perception and cognition to formulate hypotheses or inform models in human factors research. For example, models of multitask performance (e.g., Just, Carpenter, & Miyake, 2003; Kinsbourne & Hicks, 1978; Polson & Freidman, 1988; Wickens, 1992) have successfully utilized knowledge of brain function and structure in predicting processing bottlenecks in complex tasks. Whether the level of specificity of neural circuits controlling eye movements will enable the further refinement of multitask models as well as other psychological constructs relevant to human factors (e.g., visual search, linguistic processing, decision making, etc.) is an interesting and important topic for future research.

Computational Models of Saccadic Behavior

A partial understanding of the pathways that contribute to oculomotor behavior has led to the recent development of a number of models of eye movement control that attempt to account for a variety of well-established phenomena. These models can be distinguished on at least three dimensions: whether they are local versus global models of the oculomotor system (i.e., the extent to which the model focuses on specific components of the oculomotor circuit versus the entire system of neural components); the extent to which they are constrained by current knowledge about the physiology and anatomy of oculomotor circuits; and whether they are qualitative or quantitative in nature. For example, Trappenberg, Dorris, Munoz, and Klein (2001) described a model of the interaction of multiple signals in the intermediate levels of the superior colliculus leading to the initiation of saccades. The model is built on the integration and competition of exogenous and endogenous inputs via short-distance excitatory and long-distance inhibitory connections between the receptive fields of neurons in the superior colliculus. Exogenous signals refer to minimally processed visual inputs, such as motion and luminance transients, while endogenous signals refer to inputs based on instructions and expectancies. The importance of both of the model's components, input type and nature of connections in the colliculus, are well established in the literature. The model has been used to successfully account for both neuron firing rates in the superior colliculus of nonhuman animals and human performance data in a variety of different tasks and phenomena, including the gap effect (i.e., faster saccadic reaction times when a fixation point is turned off a couple hundred milliseconds before a prosaccade target is presented as compared to a situation in which the fixation stimulus is maintained), distractor effects on saccadic reaction times, the influence of target expectancies on saccadic reaction times, and the difference in latencies between pro- and antisaccades. A model by Godjin and Theeuwes (2002) extended the Trappenberg model by adding an additional top-down inhibitory mechanism that can diminish the activation level of some stimuli and locations based on experience and expectancies concerning task-relevant stimuli. This model has been successful in accounting for the saccade behavior associated with the oculomotor capture paradigm.

Itti and Koch (2000) designed a computational model in which shifts of attention (i.e., eye movements) are based solely on a bottom-up salience-based analysis of the visual world. Local differences in orientation, intensity, and color are combined into a master salience map that determines the order in which locations are inspected. Although efforts to incorporate goal-driven search will likely improve the model's applicability and performance (Navalpakkam & Itti, 2005), the model in its original form is interesting and unique in that top-down information such as expectancies and experience plays no role in the guidance of the eyes. Another relatively unique aspect of this effort is that the authors have tested their model with both simple laboratory-based visual displays and high-resolution photographs of natural scenes. Results have been promising.

A recently proposed model of oculomotor behavior takes a different tack in modeling saccade control by describing the role of both cortical and subcortical control circuits for a variety of different phenomena. Findlay and Walker (1999) proposed a model, like that of Trappenberg et al. (2001), that entails competitive interaction among different signals in the initiation and control of eye movements. However, in their model the competition and integration takes place between inputs concerned with both the where and the when of eye movement initiation throughout the eye movement circuit. Predictions of the model are discussed in terms of phenomena similar to those examined by Trappenberg and colleagues. The Findlay and Walker model is notable in its attempt to describe how different cortical and subcortical components of the oculomotor circuit interact to produce saccades. Unlike the Trappenberg et al. model, Findlay and Walker's model does not yet provide quantitative predictions that can be used to predict patterns of neuronal activity.

In summary, although there are clearly many unanswered questions concerning the neuronal circuits that underlie oculomotor control under different circumstances as well as the relation between oculomotor control and other forms of exploration (e.g., limb movements), we do currently have a sufficiently detailed understanding of the important neuronal circuits and functions to enable us to model and predict oculomotor behavior. Indeed, in principle there is no reason why the models described above should not be applied to oculomotor behavior in more complex tasks such as looking for defects in manufactured products or scanning an image for a camouflaged military vehicle (see Itti & Koch, 2000, for an initial attempt to do this). Applications to complex tasks such as these should provide useful information on the scalability of these models to situations outside the laboratory.

Conclusions and Future Directions

Our understanding of the role of eye movements in information extraction from displays and real-world scenes has increased substantially in recent years. Techniques such as gaze-contingent control procedures have been refined so as to enable researchers to infer operators' strategies and capabilities as they inspect the visual world, either in the service of an intended action or for the purpose of extracting information that will be encoded into memory for later use.

The near future is likely to bring theoretical and technical developments that will further enhance the utility of eye movement technology for the understanding of human perception, cognition, and action in complex simulated and real-world environments. There is an increasing trend for the development of multimodal assessment techniques (e.g., event-related potentials[ERPs], optical imaging, heart rate, respiration, etc.) that will capitalize on the relative advantages of different measures of cognition—in and out of the laboratory. Methods such as fMRI, electroencephalogram, ERPs, near-infrared spectroscopy, ultrasonography, and others in different chapters of this volume will continue to be useful. High-speed measurement of eye movements provides a significant addition to the toolbox of neuroergonomics. Given the usefulness of eye movement measurement techniques in inferring cognitive state, the collection and analysis of oculomotor data are likely to be an important component in such multimodal effects. The computational models described above are also likely to be further tested and developed such that they can be used in a predictive fashion to enhance the design of display devices for systems such as aircraft, automobile, and industrial systems. Finally, given the increasing need for real-time assessment of perceptual and cognitive state along with the capabilities of eye movement measurement procedures to tap system-relevant cognitive processes, we anticipate that the measurement of oculomotor parameters will be integrated into real-time workload and performance assessment algorithms.

Main Points

1. Eye movements pervade visual behavior. They are important to understand as an element of human performance in their own right, and also as a window on the perceptual and cognitive processes underlying behavioral performance.
2. A visual scene is typically inspected with a series of discrete fixations separated by rapid saccadic eye movements. Information is collected only during the fixations; visual input is suppressed during the movements themselves.
3. It is possible to shift visual attention without making an eye movement. In everyday visual behavior, however, attention and the eyes are closely coupled. Performance may be limited by the breadth with which the operator spreads attention during a fixation.
4. Eye movements are guided by both bottom-up/stimulus-driven and top-down/goal-driven processes. Oculomotor behavior may also be influenced by information access costs, the effort required to scan and sample information from the environment.
5. Eye movements are controlled by a broad network of cortical and subcortical brain regions, including areas of the parietal cortex, the prefrontal cortex, and the superior colliculus.
6. Computational models of eye movement control may be useful for guiding the design of visual displays and predicting the efficacy of visual scanning in applied contexts.

Key Readings

Corbetta, M. (1998). Frontoparietal cortical networks for directing attention and the eye to visual locations: Identical, independent, or overlapping neural systems? *Proceedings of the National Academy of Sciences USA, 95*, 831–838.

Findlay, J. M., & Gilchrist, I. D. (2003). *Active vision.* Oxford, UK: Oxford University Press.

Itti, L., & Koch, C. (2000). A saliency-based search mechanism for overt and covert shifts of visual attention. *Vision Research, 40*, 1489–1506.

Pierrot-Deseilligny, C., Milea, D., & Müri, R. (2004). Eye movement control by the cerebral cortex. *Current Opinions in Neurology, 17*, 17–25.

Rayner, K. (1998). Eye movements in reading and information processing: 20 years of research. *Psychological Bulletin, 124*, 372–422.

References

Abernethy, B. (1988). Visual search in sport and ergonomics: Its relationship to selective attention and performance expertise. *Human Performance, 1*, 205–235.

Ballard, D. H., Hayhoe, M. M., & Pelz, J. B. (1995). Memory representations in natural tasks. *Journal of Cognitive Neuroscience, 7*, 66–80.

Bellenkes, A. H., Wickens, C. D., & Kramer, A. F. (1997). Visual scanning and pilot expertise: The role of attentional flexibility and mental model development. *Aviation, Space, and Environmental Medicine, 68*, 569–579.

Brown, V., Huey, D., & Findlay, J. M. (1997). Face detection in peripheral vision: Do faces pop-out? *Perception, 26*, 1555–1570.

Burr, D. C. (1994). Selective suppression of the magnocellular visual pathway during saccadic eye movements. *Nature, 371*, 511–513.

Burr, D. C., Morrone, M. C., & Ross, J. (1994). Selective suppression of the magnocellular visual pathway during saccadic eye movements. *Nature, 371*, 511–513.

Campbell, F. W., & Wurtz, R. H. (1978). Saccadic omission: Why we do not see a grey-out during a saccadic eye movement. *Vision Research, 18*, 1297–1303.

Carbonell, J. R., Ward, J. L., & Senders, J. W. (1968). A queueing model of visual sampling: Experimental validation. *IEEE Transactions on Man-Machine Systems, MMS-9*, 82–87.

Carmody, D. P., Nodine, C. F., & Kundel, H. L. (1980). An analysis of perceptual and cognitive factors in radiographic interpretation. *Perception, 9*, 339–344.

Castet, E., & Masson, G. S. (2000). Motion perception during saccadic eye movements. *Nature Neuroscience, 3*, 177–183.

Chun, M. M., & Jiang, Y. (1998). Contextual cuing: Implicit learning and memory of visual context guides spatial attention. *Cognitive Psychology, 36*, 28–71.

Corbetta, M. (1998). Frontoparietal cortical networks for directing attention and the eye to visual locations: Identical, independent, or overlapping neural systems? *Proceedings of the National Academy of Sciences, 95*, 831–838.

Courtney, A. J., & Chan, H. S. (1986). Visual lobe dimensions and search performance for targets on a competing homogeneous background. *Perception and Psychophysics, 40,* 39–44.

Deubel, H., & Schneider, W. X. (1996). Saccade target selection and object recognition: Evidence for a common attentional mechanism. *Vision Research, 36,* 1827–1837.

Engel, G. R. (1971). Visual conspicuity, directed attention and retinal locus. *Vision Research, 11,* 563–576.

Everling, S., & Fischer, B. (1998). The antisaccade: A review of basic research and clinical studies. *Neuropsychologia, 36,* 885–899.

Findlay, J. M. (1997). Saccade target selection during visual search. *Vision Research, 37,* 617–631.

Findlay, J. M., & Gilchrist, I. D. (1998). Eye guidance and visual search. In G. Underwood (Ed.), *Eye guidance in reading and scene perception* (pp. 295–312). Amsterdam: Elsevier.

Findlay, J. M., & Gilchrist, I. D. (2003). *Active vision.* Oxford, UK: Oxford University Press.

Findlay, J. M., & Walker, R. (1999). A model of saccadic eye movement generation based on parallel processing and competitive inhibition. *Behavioral and Brain Sciences, 22,* 661–721.

Fitts, P. M., Jones, R. E., & Milton, J. L. (1950). Eye movements of aircraft pilots during instrument-landing approaches. *Aeronautical Engineering Review, 9,* 24–29.

Folk, C. L., Remington, R. W., & Johnston, J. C. (1992). Involuntary covert orienting is contingent on attentional control settings. *Journal of Experimental Psychology: Human Perception and Performance, 18,* 1030–1044.

Friedman, A. (1979). Framing pictures: The role of knowledge in automatized encoding and memory for gist. *Journal of Experimental Psychology: General, 108,* 316–355.

Gaymard, B., Ploner, C. J., Rivaud, S., Vermersch, A. I., & Pierrot-Deseilligny, C. (1998). Cortical control of saccades. *Experimental Brain Research, 123,* 159–163.

Gilchrist, I. D., Brown, V., & Findlay, J. M. (1997). Saccades without eye movements. *Nature, 390,* 130–131.

Godjin, R., & Theeuwes, J. (2002). Programming of endogenous and exogenous saccades: Evidence for a competitive interaction model. *Journal of Experimental Psychology: Human Perception and Performance, 28,* 1039–1054.

Gottlieb, J. P., Kusunoki, M., & Goldberg, M. E. (1998). The representation of salience in monkey parietal cortex. *Nature, 391,* 481–484.

Gray, W. D., & Fu, W. (2004). Soft constraints in interactive behavior: The case of ignoring perfect knowledge in-the-world for imperfect knowledge in-the-head. *Cognitive Science, 28,* 359–382.

Guitton, D., Buchtel, H. A., & Douglas, R. M. (1985). Frontal lobe lesions in man cause difficulties in suppressing reflexive glances and in generating goal-directed saccades. *Experimental Brain Research, 58,* 455–472.

Hallet, P. E. (1978). Primary and secondary saccades to goals defined by instructions. *Vision Research, 18,* 1279–1296.

Hanes, D. P., & Wurtz, R. H. (2001). Interaction of the frontal eye field and superior colliculus for saccade generation. *Journal of Neurophysiology, 85,* 804–815.

Henderson, J. M., Pollatsek, A., & Rayner, K. (1987). Effects of foveal priming and extrafoveal preview on object identification. *Journal of Experimental Psychology: Human Perception and Performance, 13,* 449–463.

Hoffman, J. E., & Subramaniam, B. (1995). The role of visual attention in saccadic eye movements. *Perception and Psychophysics, 57,* 787–795.

Hooge, I. T. C., & Erkelens, C. J. (Eds.). (1996). Control of fixation duration during a simple search task. *Perception and Psychophysics, 58,* 969–976.

Hyönä, J., Radach, R., & Deubel, H. (2003). *The mind's eye: Cognitive and applied aspects of eye movement research.* Amsterdam: North-Holland.

Inhoff, A. W., & Radach, R. (1998). Definition and computation of oculomotor measures in the study of cognitive processes. In G. Underwood (Ed.), *Eye guidance in reading and scene perception* (pp. 29–53). Amsterdam: Elsevier.

Irwin, D. E. (1996). Integrating information across saccadic eye movements. *Current Directions in Psychological Science, 5,* 94–100.

Irwin, D. E., Colcombe, A. M., Kramer, A. F., & Hahn, S. (2000). Attentional and oculomotor capture by onset, luminance, and color singletons. *Vision Research, 40,* 1443–1458.

Itti, L., & Koch, C. (2000). A saliency-based search mechanism for overt and covert shifts of visual attention. *Vision Research, 40,* 1489–1506.

Jacob, J. K., & Karn, K. S. (2003). Eye-tracking in human-computer interaction and usability research: Ready to deliver the promises. In J. Hyönä, R. Radach, & H. Deubel (Eds.), *The mind's eye: Cognitive and applied aspects of eye movement research* (pp. 574–605). Amsterdam: North-Holland.

Jacobs, A. M. (1986). Eye-movement control in visual search: How direct is visual span control? *Perception and Psychophysics, 39,* 47–58.

Jonides, J., & Yantis, S. (1988). Uniqueness of abrupt visual onset as an attention-capturing property. *Perception and Psychophysics, 43,* 346–354.

Just, M., & Carpenter, P. A. (1980). A theory of reading: From eye fixations to comprehension. *Psychological Review, 87*, 329–354.

Just, M., Carpenter, P. A., & Miyake, A. (2003). Neuroindices of cognitive workload: Neuroimaging, pupillometric and event-related brain potential studies of brain work. *Theoretical Issues in Ergonomic Science, 4*, 56–88.

Kane, M. J., Bleckley, M. K., Conway, A. R. A., & Engle, R. W. (2001). A controlled-attention view of WM capacity. *Journal of Experimental Psychology: General, 130*, 169–183.

Kimmig, H., Greenlee, M. W., Gondan, M., Schira, M., Kassubeck, J., & Mergner, T. (2001). Relationship between saccadic eye movements and cortical activity as measured by fMRI: Quantitative and qualitative aspects. *Experimental Brain Research, 141*, 184–194.

Kinsbourne, M., & Hicks, R. E. (1978). Functional cerebral space: A model for overflow, transfer, and interference effects in human performance. In J. Requin (Ed.), *Attention and performance VII* (pp. 345–362). Hillsdale, NJ: Erlbaum.

Klein, R. M. (1980). Does oculomotor readiness mediate cognitive control of visual attention? In R. S. Nickerson (Ed.), *Attention and performance VIII* (pp. 259–276). Hillsdale, NJ: Erlbaum.

Koch, C., & Ullman, S. (1985). Shifts in visual attention: Towards the underlying circuitry. *Human Neurobiology, 4*, 219–222.

Kowler, E., Anderson, E., Dosher, B., & Blaser, E. (1995). The role of attention in the programming of saccades. *Vision Research, 35*, 1897–1916.

Kundel, H. L., & LaFollette, P. S. (1972). Visual search patterns and experience with radiological images. *Radiology, 103*, 523–528.

Land, M. F., & Hayhoe, M. (2001). In what ways do eye movements contribute to everyday activities? *Vision Research, 41*, 3559–3565.

Langham, M., Hole, G., Edwards, J., & O'Neil, C. (2002). An analysis of "looked but failed to see" accidents involving parked police cars. *Ergonomics, 45*, 167–185.

Loftus, G. R., & Mackworth, N. H. (1978). Cognitive determinants of fixation location during picture viewing. *Journal of Experimental Psychology: Human Perception and Performance, 4*, 565–572.

Matin, E. (1974). Saccadic suppression: A review and an analysis. *Psychological Bulletin, 81*, 899–917.

May, J. G., Kennedy, R. S., Williams, M. C., Dunlap, W. P., & Brannan, J. R. (1990). Eye movement indices of mental workload. *Acta Psychologica, 75*, 75–89.

McCarley, J. S., Kramer, A. F., Wickens, C. D., Vidoni, E. D., & Boot, W. R. (2004). Visual skills in airport security inspection. *Psychological Science, 15*, 302–306.

McCarley, J. S., Vais, M. J., Pringle, H. L., Kramer, A. F., Irwin, D. E., & Strayer, D. L. (2004). Conversation disrupts visual scanning and change detection in complex traffic scenes. *Human Factors, 46*, 424–436.

McConkie, G., & Rayner, K. (1975). The span of the effective stimulus during a fixation in reading. *Perception and Psychophysics, 17*, 578–586.

McConkie, G., & Rayner, K. (1976). Asymmetry of the perceptual span in reading. *Bulletin of the Psychonomic Society, 8*, 365–368.

Moray, N., & Rotenberg, I. (1989). Fault management in process control: Eye movements and action. *Ergonomics, 32*, 1319–1342.

Mourant, R. R., & Rockwell, T. H. (1972). Strategies of visual search by novice and experienced drivers. *Human Factors, 14*, 325–335.

Muri, R. M., Vermersch, A. I., Rivaud, S., Gaymard, B., & Pierrot-Deseilligny, C. (1996). Effects of single-pulse transcranial magnetic stimulation over the prefrontal and posterior parietal cortices during memory-guided saccades in humans. *Journal of Neurophysiology, 76*, 2102–2106.

Navalpakkam, V., & Itti, L. (2005). Modeling the influence of task on attention. *Vision Research, 45*, 205–231.

Owsley, C., Ball, K., McGwin, G., Sloane, M. E., Roenker, D. L., White, M. F., et al. (1998). Visual processing impairment and risk of motor vehicle crash among older adults. *Journal of the American Medical Association, 279*, 1083–1088.

Parasuraman, R. (2003). Neuroergonomics: Research and practice. *Theoretical Issues in Ergonomics Science, 4*, 5–20.

Peterson, M. S., & Kramer, A. F. (2001). Attentional guidance of the eyes by contextual information and abrupt onsets. *Perception and Psychophysics, 63*, 1239–1249.

Peterson, M. S., Kramer, A. F., Wang, R. F., Irwin, D. E., & McCarley, J. S. (2001). Visual search has memory. *Psychological Science, 12*, 287–292.

Pierrot-Deseilligny, C., Milea, D., & Müri, R. (2004). Eye movement control by the cerebral cortex. *Current Opinions in Neurology, 17*, 17–25.

Pollatsek, A., Bolozky, S., Well, A. D., & Rayner, K. (1981). Asymmetries in the perceptual span for Israeli readers. *Brain and Language, 14*, 174–180.

Polson, M., & Friedman, A. (1988). Task sharing within and between hemispheres: A multiple resource approach. *Human Factors, 30*, 633–643.

Pomplun, M., Reingold, E. M., & Shen, J. (2001). Investigating the visual span in comparative search: The effects of task difficulty and divided attention. *Cognition, 81*, B57–B67.

Posner, M. I. (1980). Orienting of attention. *Quarterly Journal of Experimental Psychology, 32*, 3–25.

Pringle, H. L. (2001). *The role of attention and working memory in detection of changes in complex scenes.* Unpublished doctoral dissertation, University of Illinois, Urbana-Champaign.

Pringle, H. L., Irwin, D. E., Kramer, A. F., & Atchley, P. (2001). The role of attentional breadth in perceptual change detection. *Psychonomic Bulletin and Review, 8,* 89–95.

Rayner, K. (1998). Eye movements in reading and information processing: 20 years of research. *Psychological Bulletin, 124,* 372–422.

Rayner, K., & Fisher, D. L. (1987a). Eye movements and the perceptual span during visual search. In J. K. O'Regan & A. Lévy-Schoen (Eds.), *Eye movements: From physiology to cognition* (pp. 293–302). Amsterdam: North Holland.

Rayner, K., & Fisher, D. L. (1987b). Letter processing during eye fixations in visual search. *Perception and Psychophysics, 42,* 87–100.

Recarte, M. A., & Nunes, L. M. (2000). Effects of verbal and spatial-imagery tasks on eye fixations while driving. *Journal of Experimental Psychology: Applied, 6,* 31–43.

Reingold, E. M., Loschky, L. C., McConkie, G. W., & Stampe, D. M. (2003). Gaze-contingent multresolutional displays: An integrative review. *Human Factors, 45,* 307–328.

Riggs, L. A., Merton, P. A., & Morton, H. B. (1974). Suppression of visual phosphenes during saccadic eye movements. *Vision Research, 14,* 997–1010.

Rizzolatti, G., Riggio, L., Dascola, I., & Umiltà, C. (1987). Reorienting attention acros the horizontal and vertical meridians—Evidence in favor of a premotor theory of attention. *Neuropsychologia, 25,* 31–40.

Ro, T., Henik, A., Machado, L., & Rafal, R. D. (1997). Transcranial magnetic stimulation of the prefrontal cortex delays contralateral endogenous saccades. *Journal of Cognitive Neuroscience, 9,* 433–440.

Roberts, R. J., Hager, L. D., & Heron, C. (1994). Prefrontal cognitive processes: Working memory and inhibition in the antisaccade task. *Journal of Experimental Psychology: General, 123,* 347–393.

Rorden, C., & Driver, J. (1999). Does auditory attention shift in the direction of an upcoming saccade? *Neuropsychologia, 37,* 357–377.

Sanders, A. F. (1963). *The selective process in the functional visual field.* Assen, Netherlands: Van Gorcum.

Scialfa, C. T., & Joffe, K. (1998). Response times and eye movements in feature and conjunction search as a function of target eccentricity. *Perception and Psychophysics, 60,* 1067–1082.

Scialfa, C. T., Kline, D. W., & Lyman, B. J. (1987). Age differences in target identification as a function of retinal location and noise level: Examination of the useful field of view. *Psychology and Aging, 2,* 14–19.

Sekuler, R., & Ball, K. (1986). Visual localization: Age and practice. *Journal of the Optical Society of America A, 3,* 864–867.

Senders, J. (1964). The human operator as a monitor and controller of multidegree of freedom systems. *IEEE Transactions on Human Factors in in Electronics, HFE-5,* 2–6.

Shapiro, D., & Raymond, J. E. (1989). Training efficient oculomotor strategies enhances skill acquisition. *Acta Psychologica, 71,* 217–242.

Shepherd, M., Findlay, J. M., & Hockey, R. J. (1986). The relationship between eye movements and spatial attention. *Quarterly Journal of Experimental Psychology, 38A,* 475–491.

Strayer, D. L., Drews, F. A., & Johnston, W. A. (2003). Cell phone-induced failures of visual attention during simulated driving. *Journal of Experimental Psychology: Applied, 9,* 23–32.

Surakka, V., Illi, M., & Isokoski, P. (2003). Voluntary eye movements in human-computer interaction. In J. Hyönä, R. Radach, & H. Deubel (Eds.), *The mind's eye: Cognitive and applied aspects of eye movement research* (pp. 473–491). Amsterdam: Elsevier.

Sweeney, J. A., Mintun, M. A., Kwee, S., Wiseman, M. B., Brown, D. L., Rosenberg, D. R., et al. (1996). Positron emission tomography study of voluntary saccadic eye movements and spatial working memory. *Journal of Neurophysiology, 75,* 454–468.

Terao, Y., Fukuda, H., Ugawa, Y., Hikosaka, O., Hanajima, R., Furubayashi, T., et al. (1998). Visualization of the information flow through human oculomotor cortical regions by transcranial magnetic stimulation. *Journal of Neurophysiology, 80,* 936–946.

Theeuwes, J. (1994). Stimulus-driven capture and attentional set: Selective search for color and visual abrupt onsets. *Journal of Experimental Psychology: Human Perception and Performance, 20,* 799–806.

Theeuwes, J. (1996). Visual search at intersections: An eye-movement analysis. In A. G. Gale, I. D. Brown, C. M. Haslegrave, & S. P. Taylor (Eds.), *Vision in vehicles* (Vol. 5, pp. 125–234). Amsterdam: North-Holland.

Theeuwes, J., Kramer, A. F., Hahn, S., & Irwin, D. E. (1998). Our eyes do not always go where we want them to go: Capture of the eyes by new objects. *Psychological Science, 9,* 379–385.

Thilo, K. V., Santoro, L., Walsh, V., & Blakemore, C. (2004). The site of saccadic suppression. *Nature Neuroscience, 7,* 13–14.

Trappenberg, T. P., Dorris, M. C., Munoz, D. P., & Klein, R. M. (2001). A model of saccade initiation based on the competitive integration of exogenous and endogenous signals in the superior colliculus. *Journal of Cognitive Neuroscience, 13,* 256–271.

Treisman, A., & Sato, S. (1990). Conjunction search revisited. *Journal of Experimental Psychology: Human Perception and Performance, 16*, 459–478.

van Diepen, P. M. J. (2001). Foveal stimulus degradation during scene perception. In F. Columbus (Ed.), *Advances in psychology research* (Vol. 2, pp. 89–115). Huntington, NY: Nova Science.

van Diepen, P. M. J., & d'Ydewalle, G. (2003). Early peripheral and foveal processing in fixations during scene perception. *Visual Cognition, 10*, 79–100.

Wandell, B. A. (1995). *Foundations of vision*. Sunderland, MA: Sinauer.

Wickens, C. D. (1992). *Engineering psychology and human performance*. New York: HarperCollins.

Wickens, C. D., & Carswell, C. M. (1995). The proximity compatibility principle: Its psychological foundations and relevance to display design. *Human Factors, 37*, 473–494.

Wickens, C. D., & Hollands, J. G. (2000). *Engineering psychology and human performance* (3rd ed.). Upper Saddle River, NJ: Prentice Hall.

Williams, D. E., & Reingold, E. M. (1997). Preattentive guidance of eye movements during triple conjunction search tasks: The effects of feature discriminability and saccadic amplitude. *Psychonomic Bulletin and Review, 8*, 476–488.

Williams, L. G. (1967). The effect of target specification on objects fixated during visual search. *Perception and Psychophysics, 1*, 315–318.

Wilson, H. R., Levi, D., Maffei, L., Rovamo, J., & Devalois, R. (1990). The perception of form: Retina to striate cortex. In L. Spillman & J. S. Werner (Eds.), *Visual perception: The neurophysiological foundations* (pp. 231–272). San Diego: Academic Press.

Wolfe, J. M. (1994). Guided search 2.0: A revised model of visual search. *Psychonomic Bulletin and Review, 1*, 202–238.

Yantis, S. (1998). Attentional control. In H. Pashler (Ed.), *Attention* (pp. 223–256). East Sussex, NJ: Psychology Press.

Yantis, S., & Jonides, J. (1990). Abrupt visual onsets and selective attention: Voluntary versus automatic allocation. *Journal of Experimental Psychology: Human Perception and Performance, 16*, 121–134.

Yarbus, A. L. (1967). *Eye movements and vision*. New York: Plenum.

8

Matthew Rizzo, Scott Robinson, and Vicki Neale

The Brain in the Wild
Tracking Human Behavior in Natural and Naturalistic Settings

A problem for students of human behavior is that people often act differently in controlled laboratory and clinical settings than they do in real life. This is because the goals, rewards, dangers, benefits, and time frames of sampled behavior can differ markedly between the laboratory and clinic and "the wild." A laboratory test may seem artificial and frustrating, and may not be taken seriously, resulting in a misleadingly poor performance. On the other hand, subjects may be on their best behavior and perform optimally when they know they are being graded in a laboratory or clinic, yet they may behave ineffectively in real life and fail to meet their apparent performance potentials at work, home, school, or in a host of instrumental activities of daily living, such as automobile driving. Solutions to these pitfalls in the study of brain-behavior relationships, as we shall see, can be derived through rigorous observations of people at work and play in naturalistic settings, drawing from principles already being applied in studies of animal behavior and making use of great advances in sensor technology for simultaneously recording the movements of individuals, their surroundings, and their internal body and brain states.

Measuring Movement in the Real World

In the absence of field observations, most of what we know about human behavior in the wild comes from human testimony (from structured and unstructured interviews and questionnaires) and epidemiology—a partial and sometimes inaccurate account of "yesterday's history."

Questionnaire tools may be painstakingly developed to assess all manner of behavioral issues and quality of life, and these generally consist of self-reports of subjects or reports by their family members, friends, supervisors, or caregivers. Incident reports of unsafe outcomes at work, in hospitals, or on roads completed by trained observers (e.g., medical professionals, human factors experts, the police) are another source of similar information. However, these reports may be inaccurate because of poor observational skills or bias by untrained or trained observers. Subject reports may be affected by misperception, misunderstanding, deceit, and a variety of memory and cognitive impairments, including lack of self-awareness of acquired impairments caused by fatigue, drugs, aging, neurological or psychiatric disease, or systemic medical disorders. These information sources provide few data on human performance and physiology and often lack key details of what real people do in the real world.

Under these circumstances, it is important to consider more direct sources of evidence of human behavior in the real world. It would be advantageous to combine sensors that capture the movement, physiology, and even social interactions of people who are seeing, feeling, attending, deciding, erring, and self-correcting during the activities of daily living. Such a "people tracker" could provide detailed information on the behavior, physiology, and pathophysiology of individuals in key everyday situations, in settings, systems, and organizations where things may go wrong.

A people tracker system might be compared to a black box recorder in an aircraft or automobile, recording behavior sequences leading to an outcome or event of interest, such as an error in a task. Subjects will be less likely to behave out of character if the system is unobtrusive. For example, individuals driving unobtrusive instrumented vehicles (see below) may display personal grooming and hygiene habits, such as rhinotillexomania (nose picking). In this vein, privacy issues are an important consideration. No one should experience the merciless voyeurism and intrusions of privacy inflicted on the hapless character played by Jim Carrey in *The Truman Show*. Instead, the intention of such devices would be to prevent injury and improve human performance and health by developing alerting and warning systems, training programs, and rehabilitation interventions.

The Mismatch between Clinical Tests, Self-Report, and Real-Life Behavior

Performance measures obtained in a laboratory or clinic may inaccurately reflect real-world performance. People often act differently in the real world than they or their relatives indicate. That is, they do not do what they say they do, or would do.

Consider the challenge of assessing the real-world potential of individuals with decision-making impairments caused by brain damage, drugs, fatigue, or developmental disorders. Decision making requires the evaluation of immediate and long-term consequences of planned actions, and it is often included with impulse control, insight, judgment, and planning under the rubric of executive functions (e.g., Benton, 1991; Damasio, 1996, 1999; Rolls, 1999, 2000). Impairments of these functions affect tactical and strategic decisions and actions in the real world. Some of these impaired individuals have high IQs and perform remarkably well on cognitive tests, including specific tests of decision making, yet fail repeatedly in real life. One reason for this mismatch is that laboratory tests of cognition (measured by standardized neuropsychological tests) are imperfect and may not measure what we think they do. Another factor is that people perform differently when they are being observed directly. Furthermore, some individuals who appear to demonstrate the ability to generate good plans may not enact these plans in the real world because they lack discipline, motivation, or social support, or choose alternative strategies with short-term benefits that are disadvantageous in the long term. A well-described example is subject EVR, who had bilateral frontal lobe amputations associated with surgery to remove a meningioma (a non-malignant brain tumor). The brain damage in EVR resembles that in the famous case of railroad foreman Phineas Gage, whose frontal lobes were severely injured in 1848 by a large iron rod that blew through his head as he was using it to tamponade dynamite (Damasio, Grabowski, Frank, Galaburda, & Damasio, 1994), transforming him from a trustworthy, hard working, dependable worker to a capricious, irresponsible, ne'er-do-well.

Behavior clearly depends very much on the environment in which it is observed. A recent study of cardiac rehabilitation measures after a myocardial infarction (e.g., amount a patient walked in 6 minutes) did not correlate well with patient activity on discharge (Jarvis & Janz, 2005). Some of the individuals with the best scores in the hospital were among the least active at home and vice versa, raising concerns that patients with worse hearts were too active too soon, and that those with better hearts were returning to bad habits at home. Along similar lines, U.S. soldiers addicted to heroin under stress of war and cheap, pure drugs in Vietnam often abstained without treatment back home, among family, work, and costly drugs.

It is becoming increasingly evident that data collection in a naturalistic setting is a unique source for obtaining critical human factors data relevant to the brain at work in the wild. As we shall see, there are various issues of device development, sensor choice, and placement. There are needs for taxonomies for classifying devices and for classifying likely behavior from sensor outputs. To infer higher-level behaviors from sensor and video data in humans in naturalistic settings, it is also possible to apply ethological techniques that have been used to analyze behavior sequences in animals. These possibilities open large new areas of research to benefit human productivity, health, and organizational systems, and can build on research that has addressed animal behavior (Prete, 2004).

Ethology and Remote Tracking

The output of people trackers may be assessed using principles and methods now used to study animal behavior. Automated behavior monitors are widely used in laboratory neuroscience research for measuring general motor activity, tracking animals in a testing chamber (such as a maze), and distinguishing basic patterns of behavior (such as walking versus rearing in an open-field environment). Video and sensor-based tracking also plays a vital role in movement sciences, such as kinesiology and motor control, for providing detailed, quantitative records of movement of different body and limb segments during coordinated action. Indeed, most modern neuroscience laboratories involve some form of automated recording of behaviorally relevant data from external devices or physiological sensors placed on or implanted within the animal.

As technologies for monitoring animals in laboratory environments improve, we can expect that applications in real-world environments also will expand. Basic methods of ethology—the study of animal behavior in natural settings, pioneered by the Nobel laureates Konrad Lorenz, Niko Tinbergen, and Karl von Frisch—involve direct observation of behavior, including descriptive and quantitative methods for coding and recording behavioral events. Paper-and-pencil methods of the early ethologists have largely been replaced by laptop computer and PDA-based event-recording systems. Several software packages now are available that are optimized for coding behavior in real time or from video recordings, using focal subject, one-zero, instantaneous, or other structured sampling techniques (Lehner, 1996). Commercial software such as The Observer (Noldus Information Technology) offers semiautomated assistance to facilitate coding and analysis of behavior; public domain alternatives such as JWatcher (http://galliform.psy.mq.edu.au/jwatcher) also are available. Event-recording software can be interfaced with video or automated sensor data to provide synchronized records of behavior and physiology that are essential for linking overt actions to underlying mechanisms.

More recently, observational methods for describing behavior have been augmented by a variety of recording and remote monitoring technologies, including film and video; telemetry systems for monitoring basic physiological parameters such as heart rate, temperature, respiratory rate, or muscle activity; and radio transmitters for tracking large-scale movements within a home range or during migration.

The power and limitations of measuring behavior from remote sensor data may be well illustrated by studies of behavior of animal fetuses, which are not amenable to direct observation except through invasive procedures. In large animal species, such as sheep, arrays of recording instruments may be surgically placed within the uterine environment to measure key variables of behavioral relevance, including electroencephalogram (EEG) or electrocorticogram (ECoG), eye movements (electrooculogram, EOG), heart rate, blood pressure, intratracheal pressure (to detect fetal breathing), and electromyogram (EMG) of select muscles to measure swallowing, oral activity, limb activity, or changes in tone of postural muscles (useful for distinguishing quiet and active sleep states). Additional recording instruments provide data about key environmental conditions in utero, such as uterine contractions, partial pressure of oxygen (pO_2), maternal heart rate, and blood pressure. After placement of the instruments during a surgical preparation, the fetus and uterus are returned to the maternal abdomen, and the output of the recording instruments is routed to an automated data acquisition system, which rectifies, averages, and saves measurements within convenient time units (Robinson, Wong, Robertson, Nathanielsz, & Smotherman, 1995; Towell, Figueroa, Markowitz, Elias, & Nathanielsz, 1987).

The result of this automated monitoring approach is a continuous, real-time record of behavioral and physiological data, collected 24 hours a day, 7 days a week, over the last 3 to 4 weeks of gestation (normal gestation in the sheep is about

21 weeks). The advantage for researchers interested in questions about prenatal development is that such data collection can preserve a record of developmental change in real time. But the concomitant disadvantage is that the tremendous volume of data forces researchers to adopt explicit strategies for sampling, summarizing, and simplifying the data to discover and extract useful information from it (e.g., Anderson et al., 1998; Robertson et al., 1996). And the most severe limitation is that, despite the wealth of information provided from a large array of behaviorally relevant sensors, it is still impossible to draw clear correspondences between patterns of change in measured variables and actual behavior of the fetus. For instance, seeing a pattern of tongue, mouth, and esophageal activity when the fetus is experimentally exposed to a taste-odor cue in the amniotic fluid is sufficient to document a behavioral response to a test stimulus, but the pattern of measurements must be calibrated to the actual behavior of an observable subject (such as a newborn lamb) to conclude that the response involved an ingestive or aversion reaction to the test stimulus (Robinson et al., 1995).

Automated tracking of human behavior in the field, particularly when arrays of multiple sensors are used to record data on a fine time scale, will pose similar problems for analysis and interpretation. Researchers should plan in advance not only how to record sensor data, but how to identify practical strategies for summarizing, simplifying, or sampling from the collected data set. This task is analogous to learning to drink from a fire hose (Waldrop, 1990). To do this correctly, it is essential to relate patterns of data obtained from recording instruments to observable behavior recorded on video in controlled environments, as well as to use automated tracking to draw inferences about real human behavior.

Data Analysis Strategies

Automated recording from multiple remote sensors offers many possibilities for data analysis that may be useful for characterizing both normal and abnormal human behavior. Particularly if sensor data are recorded with high temporal resolution, data sets will provide more than general summaries of activity and can be used to create detailed, quantitative characterizations of behavior. Time-series data (involving any data set that preserves the continuous time sequence in which data are recorded) can be analyzed using a variety of well-established analytic methods to uncover temporal patterns in the data. Time-series analyses are particularly useful for detecting cyclical fluctuations in behavioral variables, such as daily (circadian) and higher frequency (ultradian) rhythmic patterns, and more complex patterns that are indicative of dynamic or chaotic organization in the data (e.g., Anderson et al., 1998; Kelso, 1995; Robertson & Bacher, 1995; see figure 8.1).

Another approach that may simplify large data sets is to establish criteria for identifying discrete behavioral events from continuous sensor recordings. Remote sensors are likely to produce some variable output as noise in the absence of overt behavior, but calibrating these recordings to synchronized video may help establish thresholds or other operational criteria for delineating discrete behavioral events. For example, continuous EMG data from different limb muscles can be simplified by characterizing signature patterns associated with single extension-flexion cycles of limb movement. Thus, a noisy time series that is difficult to compare between subjects may be collapsed into a simplified time series that can be meaningfully compared with conventional statistical methods. This essentially was the strategy adopted to condense EMG data from oral and limb muscles into movement frequencies to assess behavioral responses to experimental stimulation in fetal sheep (Robinson et al., 1995).

Derived time series of categorical events also are amenable to other methods for detecting and characterizing behavioral organization. One of the more powerful of these analytic approaches is the Markov analysis of sequential structure from a behavioral data set. A Markov process is defined as a sequential structure in which knowledge of a preceding category completely specifies the next category of behavior to occur. In data obtained from real animals or humans, sequential relationships between behavioral categories, or between two or more interactants, are rarely this deterministic. Rather, sequential structure emerges as a quantitative set of relations characterized by nonrandom probabilities linking successive pairs of events. Like time-series approaches, Markov sequential analyses can reveal orderliness in the structure of ongoing behavior expressed by an individual (e.g.,

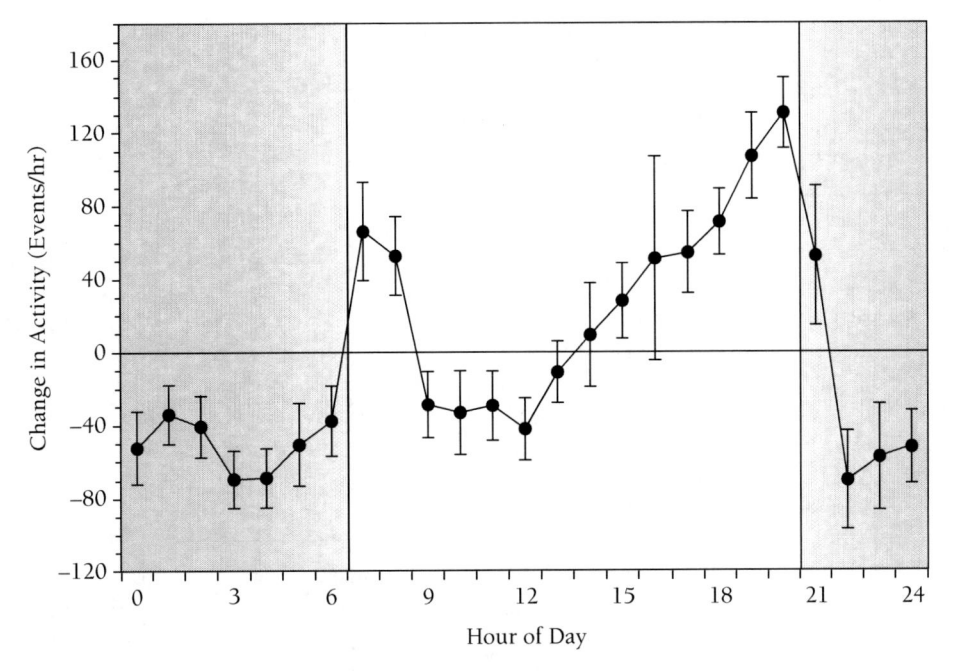

Figure 8.1. Example of 24-hour rhythmic variation in motor activity derived from continuous time-series recordings of EMG from fetal sheep between 120 and 145 days of gestation. Individual movement events of a tongue protractor muscle (geniohyoid) were derived and counted in 15-second intervals from the EMG data. The time series then was decomposed, extracting a seasonal (24 hr) trend (over 25 days) and residual components of the time series. The circadian plot identifies the period of dark (shaded) and light during the day, and reveals two peaks in oral activity just after lights on and just before lights off. Data points depict the mean change in activity per hour relative to overall average activity (horizontal line) observed in 18 fetal subjects; vertical bars depict the standard error of the mean (*SEM*).

Hailman & Sustare, 1973; Robinson & Smotherman, 1992) and in the pattern of communicative interaction between a parent and child or two individuals engaged in dialogue (e.g., Bakeman & Gottman, 1997).

Taxonomies of Tracking Systems

Mulder (1994) reviewed different configurations for recording and classifying human movements that are germane to people tracking. Briefly, these systems can be classified as inside-inside, inside-outside, and outside-inside. Sophisticated systems to record the signatures of complex behaviors could be built from different combinations of these.

Inside-inside systems employ sensors and sources located on a person's body, such as silver chloride electrodes to record surface electromyographic activity, eye movements, gastric activity (electrogastrogram), galvanic skin response (GSR; also referred to as electrodermal response), or a glove with piezoelectric transducers to sense deformation caused by changing configurations of the hand. These systems may capture body activity while a person roams over a large territory, but may be obtrusive, subject to movement artifact, and do not provide external verification of behavior and correct rejection of artifacts from unanticipated external sources of noise.

Inside-outside systems employ sensors on the body that detect external sources, either artificial or natural. Examples include a scleral coil moving in an externally generated electromagnetic field to record eye movements, or accelerometers attached to the trunk or limbs of a person moving in the earth's gravitational field. These systems provide some world-based information. For instance, it is possible to track distance traveled and energy expended from accelerometers attached to a person (as in pedometers). However, inside-outside systems share some of the same problems as inside-inside systems. Workspace and accuracy are generally limited, and there is no external validation of behavior unless the system is combined with additional body-mounted sensors, such as pinhole cameras or radar that show where a person is going.

Outside-inside systems use external sensors that process various sources of information, including images and markers or emitters. Examples include optoelectronic systems that track reflective markers attached to the body, video tracking of reflections from the eye, and global positioning system (GPS) data from GPS sensors attached to a person. The systems lose data due to occlusion when something passes between the person and the detector, such as with bifocal spectacles in front of some eye trackers and loss of GPS data when persons move indoors and walls occlude connection to the orbiting satellite. These systems are also subject to artifact, depending on the signal source. For example, infrared detecting systems are susceptible to sunlight. Artificial vision systems that track data from optical images in real time have huge information processing demands and rely on complex algorithms that are highly prone to error and artifact.

Strengths and weaknesses of different tracking systems, interfaces, and synchronization between different data streams—and the pros and cons of different sensors (e.g., optical, audio, electromagnetic, mechanical), sensor calibration and drift, and data acquisition and analysis software—are important topics for discussion and further research.

Potential Applications

A people tracker could be used for locating and tracking human activity in a variety of populations and circumstances. Such a device might use GPS, differential GPS, or arrays of accelerometers on the trunk or limb. Different accelerometer implementations could include laser/fiber optic, piezoelectric, and membrane/film. Data from these devices might be obtained in synch with other physiological indices (such as heart rate).

In general, smaller and less obtrusive sensor devices are better. Power supply to the devices and data storage and downloading are other technical issues. Different versions could have a memory for collecting data for hours to days to weeks, downloaded remotely (e.g., via cell phone, Internet, infrared transmission to a nearby collector, or by swapping out a cartridge). A taxonomy would be needed to define and set levels for trigger events that define activities such as walking, running, sleeping, and falling. The signals would need to be displayed in an understandable way.

As these problems are solved, potential neuroergonomic applications can expand, which are highly relevant to multiple research areas in healthy and impaired individuals. General applications in healthy populations could include assessing the activity of athletes, hikers, mountaineers, soldiers, and medical personnel (nurses, residents, physicians) and other individuals to assess workforce and human ergonomics in a variety of settings, including hospitals, factories, war zones, and transportation.

Potential medical applications include assessing magnitude, progression, or improvement of disease. Examples include baseline and longitudinal assessments of activity and social interactions, and of response to interventions in the following:

1. Advanced age
2. Depression
3. Stroke
4. Attention-deficit/hyperactivity disorder
5. Multiple sclerosis with remission, relapse, or progression
6. Alzheimer's disease and response to cholinesterase inhibitors and other medications
7. Parkinson's disease and response to dopaminergic agents or deep brain stimulation
8. Arthritis or fractures affecting the leg, hip, or spine, pre- and posttherapy
9. Alcohol or drug use
10. Malingering
11. Vestibular disease and other balance disorders
12. Pain syndromes (e.g., back pain, migraine, fibromyalgia)
13. Prescription drug use, for example, in cardiac disease, pulmonary disease, allergies, insomnia, and hypertension, and so on

Acute online outputs from these devices might be used identify an elderly person who has not shifted position for too long due to an illness, or in whom an abrupt change in signal suggests a fall to the floor. Longitudinal comparisons of chronic data output from these devices could be used to track disease improvement or progression (e.g., in stroke,

Alzheimer's disease, Parkinson's disease), and objectively assess outcomes of interventions (drugs, rehabilitation) in clinical trials.

Tracking Human Movement and Energy Expenditure

Early efforts on tracking human activity, outside of simple photography and video analysis, focused on the activity of children and adults and were based mostly on accelerometer ouputs. There is a body of work in exercise physiology on tracking energy expenditure in behaving humans (Welk, 2002). Researchers may calibrate instruments to energy expenditure and use Ainsworth's Physical Activity Compendium for organizing data (http://prevention.sph.sc.edu/Tools/compendium.htm). In some situations, output is calibrated to observation using observational systems such as the Children's Activity Rating Scale (e.g., Puhl, Greaves, Hoyt, & Baranowski, 1990) or the System for Observing Play and Leisure Activity in Youth of McKenzie, Marshall, Sallis, and Conway (2000).

Researchers assess the integrity of accelerometers in advance of use by testing their output before and after subjecting them to vigorous mechanical shaking. Procedures for calibrating accelerometers in pedometers strapped around the ankles involve the subject walking on a treadmill with a certain stride length and speed. Accelerometers can be placed in different locations and combinations on the body. Axial accelerometers can be placed on the hip as a reflection of movement of the main body mass. Accelerometers can be placed on the limbs to assess arm and leg movement, for example, to compare the swing of paretic and unimpaired sides during walking. For greater detail on limb coordination, it may be necessary to place at least two sensors on each limb.

Accelerometry can be combined with other physical measures such as skin temperature, sweating, heart rate, and so on, to give a more complete profile of what a person is doing. Wearable devices can be put into clothing, for example, accelerometers in pants pockets and heart sensors in bras. Contemporaneous estimates of energy expenditure using radiolabeled water can help provide an independent index of metabolic activity (Levine et al., 2005). Estimates of activity based on self-report are often inaccurate.

Levine et al. (2005) used a Tracmor triaxial accelerometer system (Maastricht, Netherlands), a validated system for detecting body movement in free-living subjects, to assess changes in posture and movement associated with the routines of daily life known as nonexercise activity thermogenesis. Results of this pilot study showed that activities that include getting up to stretch, walking to the refrigerator, and just plain fidgeting are greater in lean people than in sedentary people, account for hundreds of calories of energy expenditure per day, and may make the difference between being lean and obese.

An application of movement sensors is to gauge the activity of children with attention-deficit/hyperactivity disorder. There can be many other applications in normal and in cognitively impaired individuals (Jovanov, Milenkovic, Otto, & deGroen, 2005; Sieminski, Cowell, Montgomery, Pillai, & Gardner, 1997). Macko et al. (2002, 2005) monitored step activity of hemiparetic stroke patients moving freely. Technical issues included the algorithm for identifying slow gait (<0.5 mph), integrating physiological monitoring of multiple variables, battery life, ease of setup, and obtrusiveness and comfort of the instrumentation package to be worn. The Cyma Corp StepWatch showed acceptable accuracy, reliability, and clinical utility for step monitoring in stroke patients. Validated step activity might be expanded to coregister time-marked heart rate, GPS position, and other measures. Multiple devices might be used to assess both social and physiological interactions in circumstances such as crowds, parties, pedestrian traffic, and so on, with lessons from ethology.

Benbadis, Siegrist, Tatum, Heriaud, and Anthony (2004) found that short-term outpatient EEG video monitoring could provide evidence to discriminate between psychogenic (nonepileptic) seizures and true epileptic seizures in patients with atypical spells. In this case, the results from a simple combination of sensors eliminated the need for expensive inpatient EEG video monitoring. The results helped avoid improper drug administration in patients with alternative diagnoses that require a different approach, such as with cardiac or vestibular symptoms, anxiety attacks, somatization disorder, or malingering.

Accelerometers combined with light sensors may help tell if a person has gone outside or stayed inside during an activity. Combination with video data can help identify a specific environment and activity a person is performing. Some accelerometers can be triggered with a bar magnet to mark when a certain kind of behavior occurs. A synchronously triggered voice recorder can document a person's narrative of what may have just occurred. In combination with several of the neuroergonomics techniques described in part II of this volume, the setups have the potential to answer many questions in human factors research and medicine. What factors trigger a stroke patient to move a neglected or paretic arm, or a person with Parkinson's disease to start walking? What behaviors or internal states precede an anxiety attack, falling asleep while piloting a plane or driving a car, an industrial accident, or a medical error by health care personnel?

Tracking over Long Distances

Human behavior can be studied as people travel afoot. It can also be studied as people travel over larger distances (see chapter 9, this volume). This can be done using instrumented vehicles (IVs). Internal networks of modern vehicles make it possible to obtain detailed information from the driver's own car (Rizzo, Jermeland, & Severson, 2002). Modern vehicles report certain variables relevant to speed, emissions controls, and vehicle performance, and even seatbelt and headlight use, climate and traction control, wheel speed, and antilock braking system (ABS) activation. In addition, lane-tracking video can be processed with computer algorithms to assess lane-keeping behavior. Radar systems installed in the vehicle can gather information on the proximity, following distance, and lane-merging behavior of the driver and other neighboring vehicles on the road. GPS systems can show where and when a driver drives, takes risks, and commits safety errors. Wireless systems can check the instrumentation and send performance data to remote locations. Together, these developments can provide direct, real-time information on driver strategy, vehicle usage, upkeep, drive lengths, route choices, and decisions to drive during inclement weather and high traffic that cannot be observed any other way. Continuous data obtained in these IVs provide key information for traffic safety research and interventions and are highly relevant to the larger issues of studying human strategies, tactics and cognitive errors in the real world, in natural situations, in which people behave like they do in real settings.

Multiple studies have used IVs in traffic safety research (e.g., Dingus et al., 2002; Fancher et al., 1998; Hanowski, Wierwille, Garness, & Dingus, 2000). In most cases an experimenter is present, and drivers who are aware of being observed are liable to drive in an overly cautious and unnatural manner. Because total data collection times are often less than an hour and crashes and serious safety errors are relatively uncommon, no study until recently has captured precrash or crash data for a police-reported crash, and no information is available on general vehicle usage. Instead, insights on vehicle usage by at-risk drivers have relied on questionnaires completed by individuals who may have defective memory and cognition.

The main data that transportation researchers have on actual collisions and contributing factors are collected post hoc (in forensic or epidemiological research). These data are highly dependent upon eyewitness testimony, driver memory, and police reports, all of which have serious limitations. The best information we have had regarding near collisions in at-risk drivers comprises anecdotal reports by driving evaluators and instructors (usually testing novice drivers) and police reports of moving violations. Most of these potential crash precursors, if they are ever even recognized at all, are known only to the involved parties and are never available for further study and subsequent dissemination of safety lessons.

A driver driving his or her own IV is exposed to the usual risk of the real-world road environment that he or she is normally exposed to, without the psychological pressure that may be present when a driving evaluator is in the car. Road test conditions can vary depending on the weather, daylight, traffic, and driving course. However, this is an advantage in naturalistic testing, where repeated observations in varying real-life settings can provide a wealth of information regarding driver risk acceptance, safety countermeasures, and adaptive behaviors, and unique insights on the ranging relationships between low-frequency, high-severity driving errors and high-frequency, low-severity driver errors. These types of brain-in-the-wild relationships were explored in detail in a Virginia Tech Transportation Institute/National Highway Transportation Safety Administration study of driving performance and safety errors in 100 neurologically normal individuals, driving 100 total driver years (Dingus et al., 2006; Neale, Dingus, Klauer, Sudweeks, & Goodman, 2005).

Hidden sensors detected vehicle longitudinal acceleration and rate of yaw (lateral acceleration). Infrared sensors detected cell phone use. Readings off the internal network of each IV provided information on speed, use of controls, seatbelt and headlight use, traction control, wheel speed, and ABS activation, and air bag deployment. GPS output showed where and when a driver drove. Five miniature charge coupled devices (CCD) cameras mounted in the vehicle provided video information of the (1) driver's face plus driver's side of vehicle, (2) forward view, (3) instrument panel, (4) passenger side of vehicle, and (5) rear of vehicle.

A data acquisition system onboard each IV continuously stored measured variable data and video streams from five cameras during driving sessions. The internal hard drive had the capacity to store at least 4 weeks of 6-hour-long driving days. Each IV had a cell phone account to enable the investigators to download video and vehicle data "snippets" to verify proper systems operation, query the data system for GPS location information, and delete archived data from the vehicle hard drive. A chase vehicle allowed investigators to troubleshoot the experimental IVs and to download raw data from the IVs (at least once every 2 weeks).

Raw data obtained from each IV were filtered using specific criteria to flag where a critical incident may have occurred in the IV data stream. For example, longitudinal and lateral accelerometers measured g-forces as drivers applied the brakes or swerve to miss an obstacle, and these were used to flag critical driving situations in the data stream. Specific values (indicated by X in table 8.1) for each of these variables were used to determine when

Table 8.1. Trigger Criteria to Flag Critical Incidents

Trigger Type	Description
1. Lateral acceleration	Lateral motion equal or greater than 0.X g.* Will indicate when a driver has swerved to miss an obstacle.
2. Longitudinal acceleration	Acceleration or deceleration equal to or greater than 0.X g. Will indicate when a driver has either braked hard or accelerated hard to avoid an obstacle.
3. Lane deviation	Activated if the driver crosses the solid line border (Boolean occurrence). May indicate when a driver is either inattentive or losing control of the vehicle.
4. Normalized lane position	Activated if the driver's path deviates by X.X% of centerline. May indicate if a driver is inattentive or losing control of the vehicle.
5. Forward time to collision	Activated if the driver followed the preceding vehicle at X range/range-rate. May indicate if a driver is following another vehicle too closely and/or demonstrating unsafe or aggressive driving.
6. Rear time to collision	Activated if the driver following the IV is closing in on the IV at a rate of X range/range-rate. May indicate if a driver is being followed too closely.
7. Yaw rate	Activated if the lateral motion of the vehicle is 0.X g. Will be an indication if a driver has swerved or is rapidly turning the steering wheel.
8. ABS brake status	Activated if the ABS brakes are active. Note: only applicable to those vehicles that have ABS brakes. Will provide another indication of when a driver is braking hard.
9. Traction control	Activated if the traction control system comes on. Note: only applicable to those vehicles that have traction control and can be monitored via the in-vehicle network. May indicate when a driver may potentially lose control of the vehicle.
10. Airbag status	Activated if the airbag is deployed. Will indicate a collision has occurred.
11. RF sensor	Activated if the driver is using a cell phone or a PDA when the vehicle is on. May indicate when the driver is distracted.
12. Seat belt	Activated when car is in motion and seat belt is not fastened.

*Specific values (X) for each of these variables are used to determine possible critical incidents.

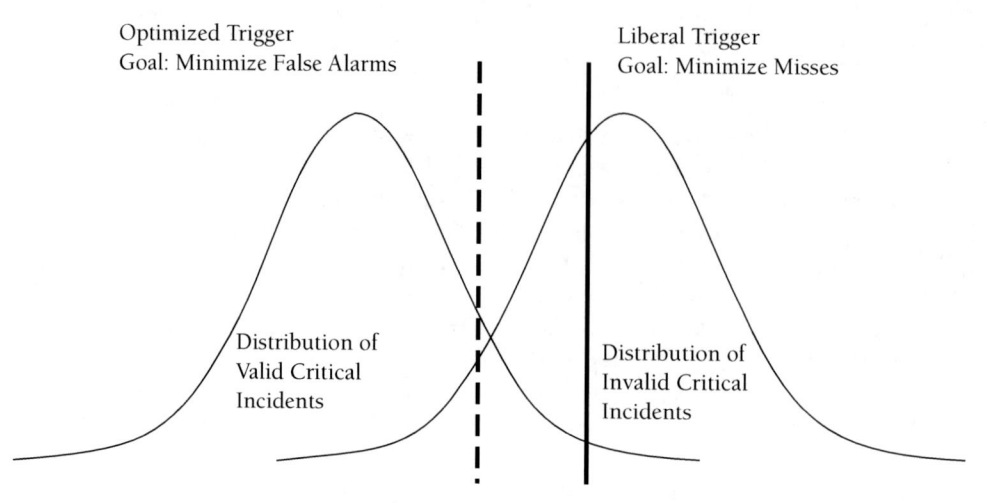

Figure 8.2. Graphical depiction of where the trigger criteria will be set to minimize misses and false alarms in classification of critical incidents.

a possible critical incident occurred. A sensitivity analysis was performed by setting the trigger criteria to a liberal level (figure 8.2, right side) to reduce the chance of a missed valid incident while allowing a high number of invalid incidents (false alarms).

As mentioned above, critical incident triggers (e.g., swerving, rapid breaking) cause the automatic flagging of that segment of the IV data stream. These flags initiate standardized review of the relevant, time-linked IV data collected immediately preceding and following the critical incident. The reviews are based on objective decision trees, which culminate in a taxonomy of driver errors and decisions. These classification schemata were developed to classify the myriad events that occur with large-scale traffic observations (e.g., Wierwille et al., 2002).

This framework allows objective and detailed analysis of preincident maneuvers, precipitating factors, and contributing factors, far more than is possible with epidemiological (postincident) studies. The flagged segments of the IV data stream are analyzed to determine the sequence of events surrounding each critical incident, allowing characterization of the following:

- Driver actions just prior to onset of the incident
- Type of incident that occurred
- Precipitating factor (initial action that started the incident sequence)
- Causal factors or any action or behavior that contributed to the outcome of the incident
- Type of evasive maneuvers performed (if any)

By applying the data reduction schemata listed above, critical incidents are classified into either (a) appropriate responses to an unsafe event, or (b) the three basic categories of inappropriate responses that serve as dependent measures in this study:

- Driver errors: Driver safety errors in the absence of a nearby object.
- Near crashes (close calls or near misses): Events that require a rapid evasive maneuver but no physical contact is made with an object (person, vehicle, guardrail, etc.).
- Crashes: Incidents in which physical contact is made.

Examples of the steps in the application of these classification tools to flagged portions of the data stream from the IV are shown in figures 8.3 and 8.4. Application of the tree in figure 8.3 could result in many possible outcomes, one of which is "Conflict with a lead vehicle." Figure 8.4 shows further classification of "Conflict with lead vehicle" in the case of a crash.

Data collection in the 100-Car Naturalistic Driving Study was completed in 2004, and the final results of the study were recently compiled. All enrolled drivers allowed installation of an instrumentation package into their own vehicle (78 cars) or to use a new model-year IV provided free of charge for

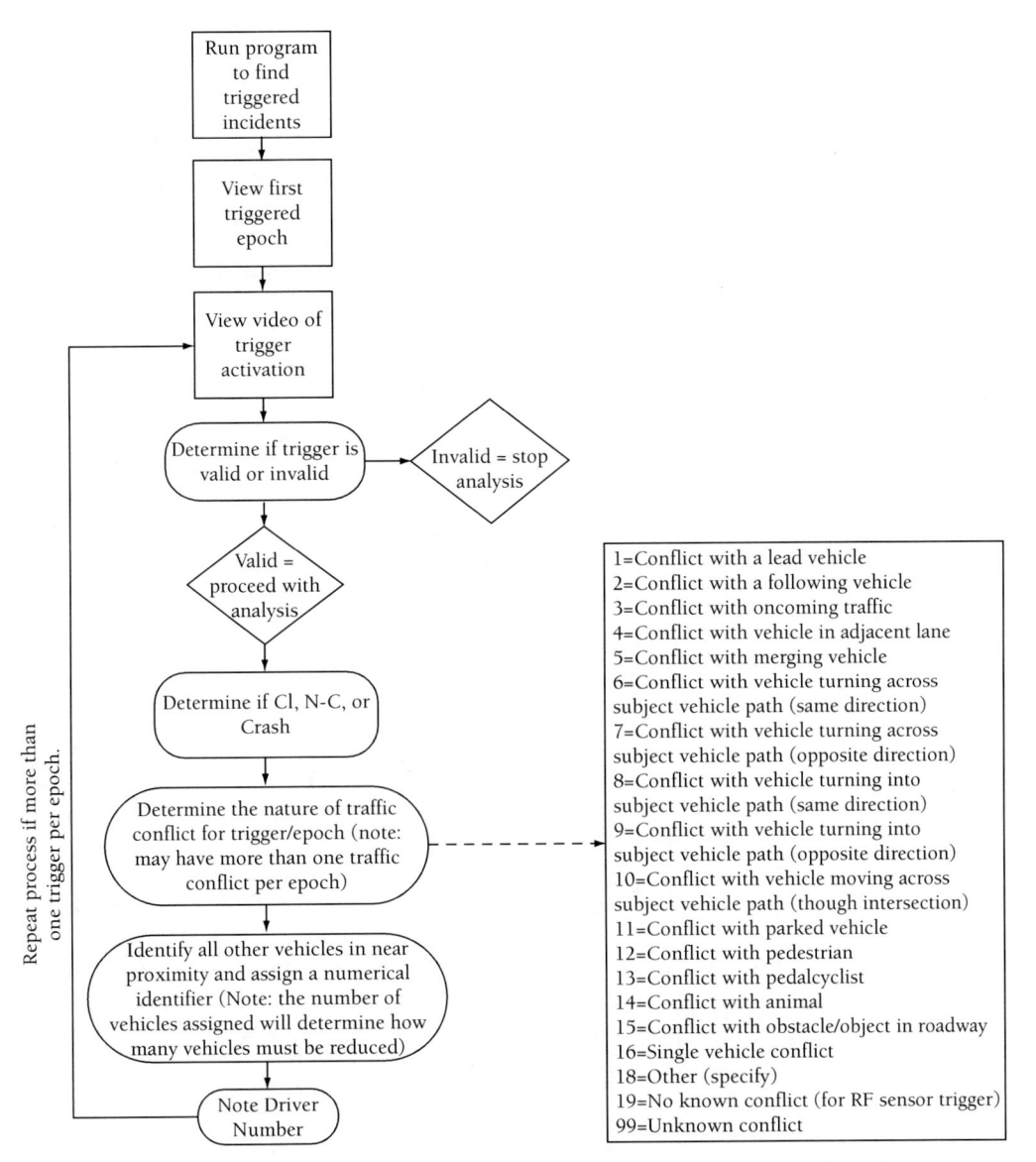

Figure 8.3. Depiction of sequence of steps used to locate and classify triggered incidents in the IV data stream. From these incidents, multiple types of traffic conflicts are identified for further analyses, as in figure 8.4. CL = conflicts, NC = no conflicts.

their own use (22 cars). The average age of the primary drivers in the study was 36, with 61% being male. Data collection for the 100-Car Study represented monitoring periods of 12–13 months per vehicle, resulting in almost 43,000 hours of actual driving data, and approximately 2,000,000 vehicle miles. Overall, there were a total of 69 crashes (see figure 8.5), 761 near crashes, and 7,479 other relevant incidents (including 5,568 driver errors) for which data could be completely reduced (see figure 8.4). The severity of crashes varied, with 75% being mild impacts, such as when tires strike curbs or other obstacles. Using taxonomy tools to classify all relevant incidents (see figures), the majority of incidents could be described as "lead vehicle"; however, several other types of conflicts (adjacent vehicle, following vehicle, single vehicle, object/obstacle) also occurred at least 100 times each.

The next phase of this work will examine how cognitive impairments are related to driver errors and crashes. Analyses of the IV data stream can show specific driver behaviors that led to critical

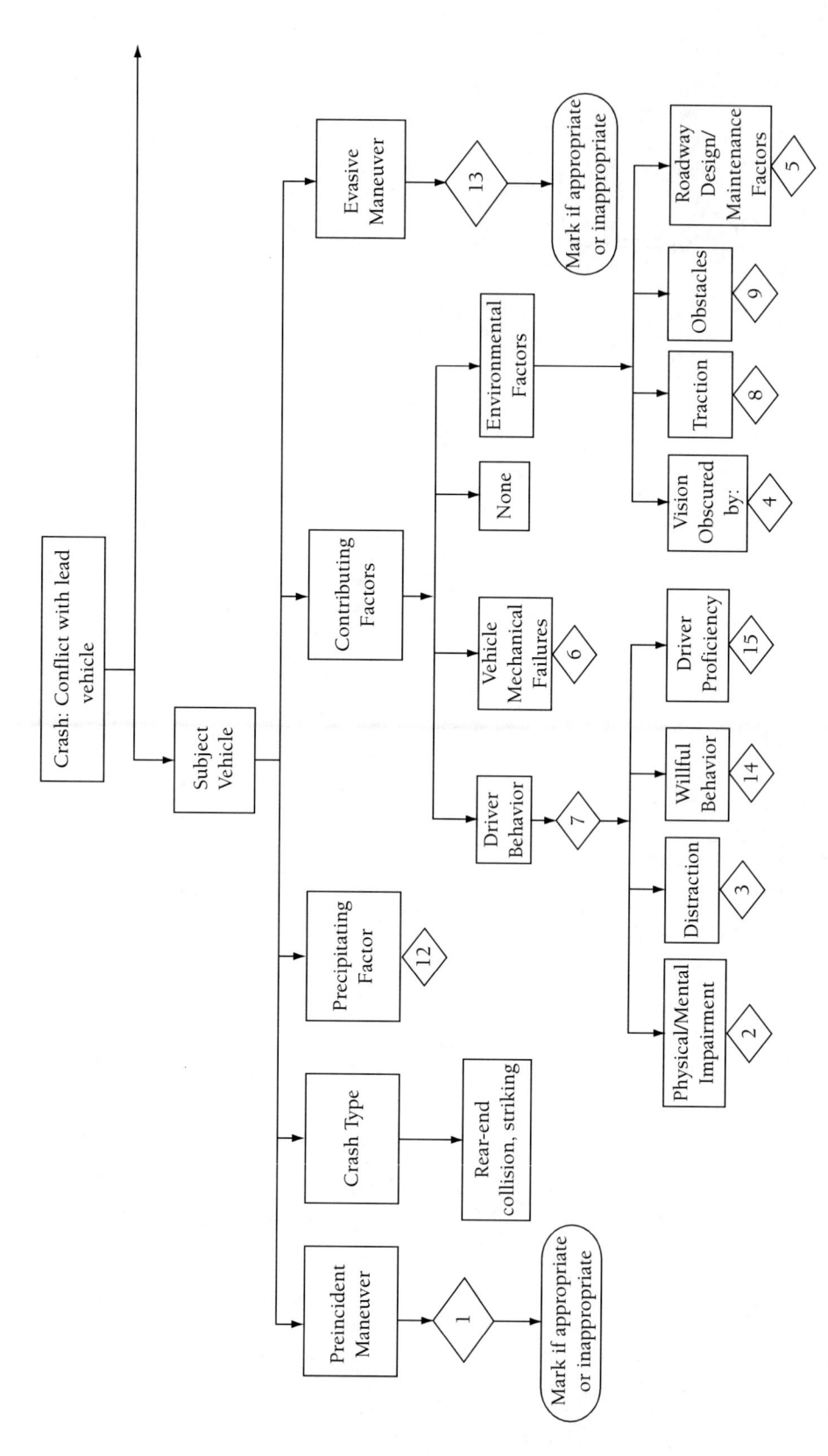

Figure 8.4. Decision tree (partial) for assessing a critical incident: conflict with lead vehicle (traffic conflict type #1).

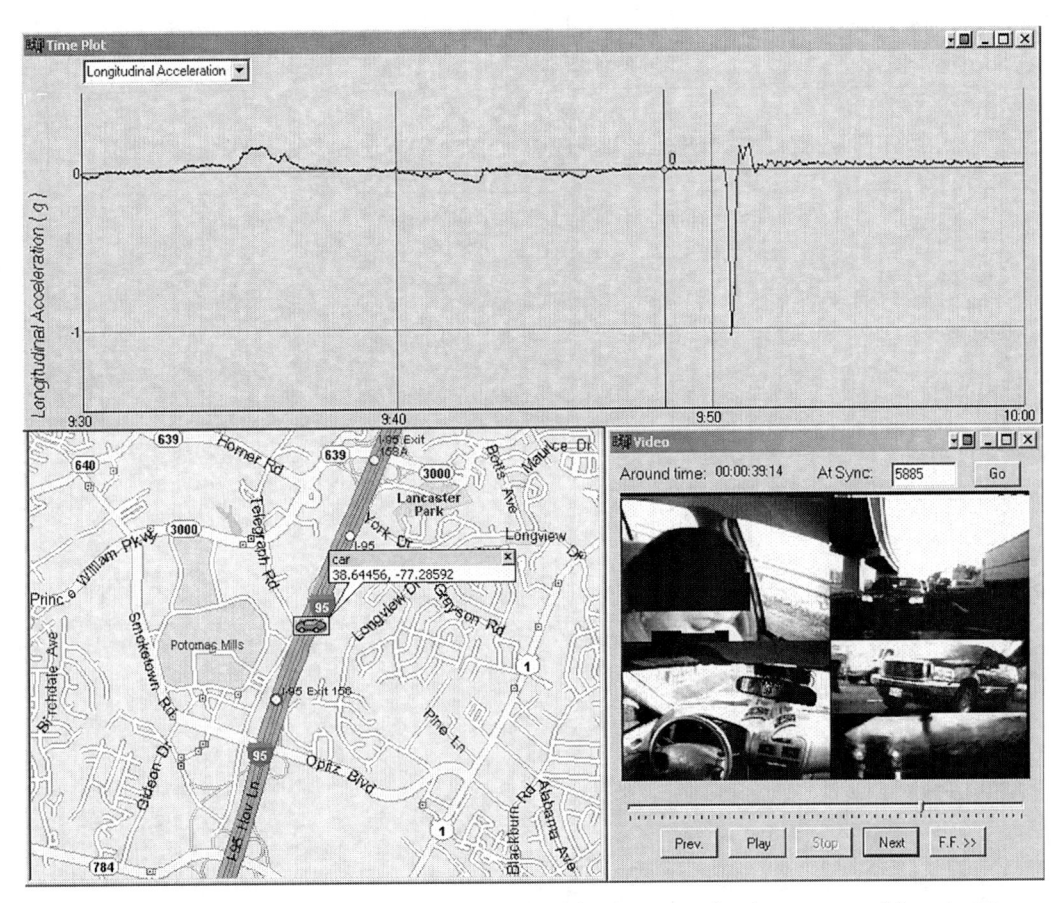

Figure 8.5. A real-life crash produces an abrupt longitudinal deceleration in the electronic record from the IV. GPS data show the exact location of the crash on the map. Corresponding video data (the quad view) show the driver was looking down (to eat a sandwich) and reacted too late to a braking lead vehicle.

incidents in impaired drivers. For example, longitudinal accelerations combined with braking behavior or lateral acceleration may indicate a driver reacted abruptly to avoid an obstacle, for example, because of an unsafe go/no-go decision to pass through an intersection or merge between lanes. Lane deviations and lateral accelerations may indicate executive misallocation of attention, as in drivers whose "eyes off-road" time increases (Dingus, 1995; Zwahlen & Balasubramanian, 1974). Critical incidents identified in the IV data stream could also be interpreted with respect to differing ambient conditions. Drivers traveling on different roads and at different times of the year face different safety contingencies, depending on traffic, atmospheric conditions, and time of day. However, a critical conceptual issue with respect to decision making is how drivers respond to safety contingencies, regardless of when or where they arise. GPS and video data for the IV of each decision-making-impaired driver can be linked to contemporaneous data from National Weather Service and Department of Transportation sites and archives. This allows monitoring of driver-specific road and weather conditions that coincide with the IV performance data and number of trips, trip lengths, total miles driven, and speed in unfavorable weather and lighting. Failure to adjust to altered contingencies is a hallmark of decision-making impairment due to brain lesions and is the focus of new naturalistic studies of driving. Table 8.2 lists examples of potential relationships between cognitive tests, the executive functions and constructs measured by these tests, and unsafe driver behaviors associated with these functions, and suggests possible answers to the question, "What if Phineas Gage could drive?"

Table 8.2. Off-Road Tests and the Measure Functions/Constructs That Predict Unsafe Driver Behaviors That May Lead to a Near Crash or Crash

Test	Function/Construct Measured	Examples of Unsafe Driver Behaviors Associated with Impaired Functions
WCST	Response to changing contingencies	Failure to adjust speed or following distance to changing road conditions
Trails B Trails A	Response alternation	Failure to alternate eye gaze appropriately between road, mirrors, and gauges
Tower of Hanoi	Planning and execution of multistep tasks	Sudden brake application; swerving across lane; running car near empty
Stroop	Susceptibility to interference	Glances of >2 s off road; e.g., with passenger present or while eating
Gambling	Decision making	Traffic violation, e.g., speeding and other traffic violations; engaging in behavior extraneous to driving
Go/no-go	Impulse control	Running red light; engaging in behavior extraneous to driving
Digit span	Working memory	Cutting off vehicles because of forgetting their location; disregard for following vehicle; driving slowly in left lane

Note. Occurrence of the unsafe behavior is assessed from flagged segments of the IV data stream. Trails = Trail-Making Test, Parts A and B; WCST = Wisconsin Card-Sorting Task.

Conclusion

Modern technology allows development of various "people trackers" using combinations of accelerometers, GPS, video, and other sensors (e.g., to measure cerebral activity, eye movement, heart rate, skin temperature) to make naturalistic observations of human movement and behavior. These devices can advance the goal of examining human performance, strategies, tactics, interactions, and errors in humans engaged in real-world tasks. Besides various issues of device development and sensor choice and placement, we need to develop taxonomies for classifying likely behavior from sensor output. We also need to be able to analyze behavior sequences using new applications of classic ethological techniques. Different implementations can provide unique data on how the brain interacts with diverse environments and systems, at work and at play, and in health, fatigue, or disease states.

Main Points

1. People often act differently in the real world than they or their relatives indicate. Consequently, it is important to consider more direct sources of evidence of human behavior in natural and naturalistic settings.
2. Modern technology allows development of various "people trackers" using combinations of accelerometers, GPS, video, and other sensors (e.g., to measure cerebral activity, eye movement, heart rate, skin temperature) to make naturalistic observations of human movement and behavior.
3. These devices can assess physiology and behavior from different vantages (such as outside looking inside and inside looking outside) to evaluate the behavior of people who are seeing, feeling, attending, deciding, erring, and self-correcting during activities of daily living.
4. Analysis of behavior sequences measured with these devices can draw from classic ethological techniques.
5. Current challenges for device development concern sensor choice and placement and development of taxonomies for classifying behavior from sensor output.
6. Different device implementations can provide unique data on how the brain interacts with diverse environments and systems, at work and at play, and in healthy and impaired states.

Key Readings

Dingus, T. A., Klauer, S. G., Neale, V. L., Petersen, A., Lee, S. E., Sudweeks, J., et al. (2006). *The 100-Car Naturalistic Driving Study: Phase II—Results of the 100-car field experiment* (Project Report for DTNH22-00-C-07007, Task Order 6; Report No. TBD). Washington, DC: National Highway Traffic Safety Administration.

Lehner, P. N. (1996). *Handbook of ethological methods* (2nd ed.). Cambridge, UK: Cambridge University Press.

Levine, J. A., Lanningham-Foste, L. M., McCrady, S. K., Krizan, A. C., Olson, L. R., Kane, P. H., et al. (2005). Interindividual variation in posture allocation: Possible role in human obesity science. *Science, 307,* 584–586.

Robinson, S. R., Wong, C. H., Robertson, S. S., Nathanielsz, P. W., & Smotherman, W. P. (1995). Behavioral responses of a chronically-instrumented sheep fetus to chemosensory stimuli presented in utero. *Behavioral Neuroscience, 109,* 551–562.

References

Anderson, C. M., Mandell, A. J., Selz, K. A., Terry, L. M., Wong, C. H., Robinson, S. R., et al. (1998). The development of nuchal atonia associated with active (REM) sleep in fetal sheep: Presence of recurrent fractal organization. *Brain Research, 787,* 351–357.

Bakeman, R., & Gottman, J. M. (1997). *Observing interaction: An introduction to sequential analysis* (2nd ed.). Cambridge: Cambridge University Press.

Benbadis, S. R., Siegrist, K., Tatum, W. O., Heriaud, L., & Anthony, K. (2004). Short-term outpatient EEG video with induction in the diagnosis of psychogenic seizures. *Neurology, 63*(9), 1728–1730.

Benton, A. L. (1991). The prefrontal region: Its early history. In H. S. Levin, H. M. Eisenberg, & A. L. Benton (Eds.), *Frontal lobe function and dysfunction* (pp. 3–12). New York: Oxford University Press.

Damasio, A. R. (1996). The somatic marker hypothesis and the possible functions of the prefrontal cortex. *Philosophical Transactions of the Royal Society of London (Biology), 351,* 1413–1420.

Damasio, A. R. (1999). *The feeling of what happens: Body and emotion in the making of consciousness.* New York: Harcourt Brace.

Damasio, H., Grabowski, T., Frank, R., Galaburda, A. M., & Damasio, A. R. (1994). The return of Phineas Gage: Clues about the brain from the skull of a famous patient. *Science, 264,* 1102–1105.

Dingus, T. A. (1995). Moving from measures of performance (MOPS) to measures of effectiveness (MOEs) in the safety evaluation of ITS products or demonstrations. In D. Nelson (Ed.), *Proceedings of the ITS Safety Evaluation Workshop.* Washington, DC: ITS America.

Dingus, T. A., Klauer, S. G., Neale, V. L., Petersen, A., Lee, S. E., Sudweeks, J., et al. (2006). *The 100-Car Naturalistic Driving Study: Phase II—Results of the 100-car field experiment* (Project Report for DTNH22-00-C-07007, Task Order 6; Report No. TBD). Washington, DC: National Highway Traffic Safety Administration.

Dingus, T. A., Neale, V. L., Garness, S. A., Hanowski, R., Keisler, A., Lee, S., et al. (2002). *Impact of sleeper berth usage on driver fatigue.* Final Project Report. (Report No. 61-96-00068). Washington, DC: U.S. Department of Transportation, Federal Motor Carriers Safety Administration.

Fancher, P., Ervin, R., Sayer, J., et al. (1998). *Intelligent cruise control field operational test: Final report* (Report No. DOT-HS-808-849). Washington, DC: U.S. Department of Transportation, National Highway Traffic Safety Administration.

Hailman, J. P., & Sustare, B. D. (1973). What a stuffed toy tells a stuffed shirt. *Bioscience, 23,* 644–651.

Hanowski, R. J., Wierwille, W. W., Garness, S. A., & Dingus, T. A. (2000). *Impact of local/short haul operations on driver fatigue: Final report* (Report No. DOT-MC-00-203). Washington, DC: U.S. Department of Transportation, Federal Motor Carriers Safety Administration.

Jarvis, R., & Janz, K. (2005). An assessment of daily physical activity in individuals with chronic heart failure. *Medicine and Science in Sports and Exercise, 37*(5, Suppl.), S323–S324.

Jovanov, E., Milenkovic, A., Otto, C., & deGroen, P. C. (2005). A wireless body area network of intelligent motion sensors for computer assisted physical rehabilitation. *Journal of NeuroEngineering and Rehabilitation, 2,* 6.

Kelso, J. A. S. (1995). *Dynamic patterns: The self-organization of brain and behavior.* Cambridge, MA: MIT Press.

Lehner, P. N. (1996). *Handbook of ethological methods* (2nd ed.). Cambridge, UK: Cambridge University Press.

Levine, J. A., Lanningham-Foste, L. M., McCrady, S. K., Krizan, A. C., Olson, L. R., Kane, P. H., et al. (2005). Interindividual variation in posture allocation: Possible role in human obesity science. *Science, 307,* 584–586.

Macko, R. F., Haeuber, E., Shaughnessy, M., Coleman, K. L., Boone, D. A., Smith, G. V., et al. (2002). Microprocessor-based ambulatory activity monitoring in stroke patients. *Medicine and Science in Sports and Exercise, 34,* 394–399.

Macko, R. F., Ivey, F. M., Forrester, L. W., Hanley, D., Sorkin, J. D., Katzel, L. I., et al. (2005). Treadmill exercise rehabilitation improved ambulatory function and cardiovascular fitness in patients with chronic stroke. A randomized, controlled trial. *Stroke, 36,* 2206–2211.

McKenzie, T. L., Marshall, S. J., Sallis, J. F., & Conway, T. L. (2000). Leisure-time physical activity in school environments: An observational study using SOPLAY. *Preventive Medicine, 30,* 70–77.

Mulder, S. (1994, July). Human movement tracking technology. Hand-Centered Studies of Human Movement Project, Simon Fraser University. Technical Report 94-1.

Neale, V. L., Dingus, T. A., Klauer, S. G., Sudweeks, J., & Goodman, M. J. (2005). An overview of the 100-car naturalistic study and findings. *International Technical Conference on the Enhanced Safety of Vehicles* (CD-ROM). Washington, DC: National Highway Traffic Safety Administration.

Prete, F. R. (Ed.). (2004). *Complex worlds from simpler nervous systems.* Cambridge, MA: Bradford MIT Press.

Puhl, J., Greaves, K., Hoyt, M., & Baranowski, T. (1990). Children's Activity Rating Scale (CARS): Description and calibration. *Research Quarterly for Exercise and Sport, 61*(1), 26–36.

Rizzo, M., Jermeland, J., & Severson, J. (2002). Instrumented vehicles and driving simulators. *Gerontechnology, 1,* 291–296.

Robertson, S. S., & Bacher, L. F. (1995). Oscillation and chaos in fetal motor activity. In J. P. Lecanuet, N. A. Krasnegor, W. P. Fifer, & W. P. Smotherman (Eds.), *Fetal development: A psychobiological perspective* (pp. 169–189). Hillsdale, NJ: Erlbaum.

Robertson, S. S., Johnson, S. L., Bacher, L. F., Wood, J. R., Wong, C. H., Robinson, S. R., et al. (1996). Contractile activity of the uterus prior to labor alters the temporal organization of spontaneous motor activity in the fetal sheep. *Developmental Psychobiology, 29,* 667–683.

Robinson, S. R., & Smotherman, W. P. (1992). The emergence of behavioral regulation during fetal development. *Annals of the New York Academy of Sciences, 662,* 53–83.

Robinson, S. R., Wong, C. H., Robertson, S. S., Nathanielsz, P. W., & Smotherman, W. P. (1995). Behavioral responses of a chronically-instrumented sheep fetus to chemosensory stimuli presented in utero. *Behavioral Neuroscience, 109,* 551–562.

Rolls, E. T. (1999). *The brain and emotion.* Oxford, UK: Oxford University Press.

Rolls, E. T. (2000). The orbitofrontal cortex and reward. *Cerebral Cortex, 10,* 284–294.

Sieminski, D. J., Cowell, L. L., Montgomery, P. S., Pillai, S. B., & Gardner, A. W. (1997). Physical activity monitoring in patients with peripheral arterial occlusive disease. *Journal of Cardiopulmonary Rehabilitation, 17*(1), 43–47.

Towell, M. E., Figueroa, J., Markowitz, S., Elias, B., & Nathanielsz, P. (1987). The effect of mild hypoxemia maintained for twenty-four hours on maternal and fetal glucose, lactate, cortisol, and arginine vasopressin in pregnant sheep at 122 to 139 days' gestation. *American Journal of Obstetrics and Gynecology, 157,* 1550–1557.

Waldrop, M. M. (1990). Learning to drink from a fire hose. *Science, 248,* 674–675.

Welk, G. (Ed.). (2002). *Physical activity assessment for health-related research.* Champaign, IL: Human Kinetics.

Wierwille, W. W., Hanowski, R. J., Hankey, J. M., Kieliszewski, C. A., Lee, S. E., Medina, A., et al. (2002). *Identification and evaluation of driver errors: Overview and recommendations* (Final Report for Federal Highway Administration contract DTFH 61-97-C-00051). Washington, DC: Federal Highway Administration.

Zwahlen, H. T., & Balasubramanian, K. N. (1974). A theoretical and experimental investigation of automobile path deviation when driver steers with no visual input. *Transportation Research Record, 520,* 25–37.

Perception, Cognition, and Emotion

Eleanor A. Maguire

Spatial Navigation

Space: Ubiquitous Yet Elusive

We are always somewhere. "Our body occupies space, it moves through space, it interacts with things in space, and it can mentally rotate and manipulate representations of space. Other objects also occupy space and maintain relations in space with one another and with us" (Kolb & Whishaw, 1990, p. 643). As humans, our ability to operate in large-scale space has been crucial to our adaptation and survival. Even now, a sizeable chunk of our day is spent trying to get from place to place, whether it is work, home, school, or the store. Many of us have experienced navigation-related arguments on a road trip, the annoyance of taking the wrong route, punctuating our cell phone conversations with inquiries about the other party's location, and the city dwellers among us are bombarded with information about traffic flow and what routes to avoid. We are a species constantly on the move.

The significance of our spatial abilities is most poignantly revealed when they become impaired. Patients who, because of brain injury or disease, are unable find their way in large-scale space can experience a devastating loss of independence and social isolation (Aguirre & D'Esposito, 1999; Barrash, 1998; Maguire, 2001). An even more fundamental role for space has also been proposed. While spatial navigation is a cross-species behavior, it has been suggested that in humans this ability has evolved into the basic scaffolding for episodic memory (Burgess, Maguire, & O'Keefe, 2002; O'Keefe & Nadel, 1978; but see Eichenbaum, 2004). The life events that comprise our individual personal history always have a spatial context, and when patients suffer dense amnesia, it often co-occurs with navigation deficits (Spiers, Burgess, Hartley, Vargha-Khadem, & O'Keefe, 2001). Given that large-scale space is the backdrop for all behavior we direct toward the external world, and that it may be fundamental to our internal memory representations, how does the human brain support our complex yet apparently seamless navigation behavior?

The study of spatial memory has a long and productive history. A common experimental approach in cognitive psychology and neuropsychology has been to extrapolate from simplified spatial stimuli studied in a laboratory setting. In this way, confounding variables are minimized and experimental control maximized, making it possible to examine relatively pure cognitive processes. However, performance on laboratory tasks has been found to dissociate from actual wayfinding ability in real environments (e.g., Nadolne & Stringer, 2001). Thus, there are clearly factors less amenable to examination using standard laboratory tasks requiring a complementary approach utilizing real-world or ecologically valid paradigms (Bartlett, 1932). That said, using naturalistic tasks presents significant challenges in terms of experimental control and data interpretation. However, in other disciplines, real large-scale environments have been studied. Geographers and urban planners have long examined different components of large-scale space such as landmarks, paths, and districts (Lynch, 1960). Environmental psychologists have studied the factors affecting wayfinding in complex spaces (Downs & Stea, 1973; Evans, 1980). Developmental psychologists have theorized about how spatial knowledge is acquired and its key determinants (Hermer & Spelke, 1994; Siegal & White, 1975). Meanwhile, animal physiologists have identified neurons (place cells) in a part of the rat brain called the hippocampus that exhibited location-specific firing (O'Keefe & Dostrovsky, 1971; O'Keefe & Nadel, 1978). In neuropsychology, as well, patients with a variety of etiologies have been reported with wayfinding deficits in the real world (Aguirre & D'Esposito, 1999; Barrash, 1998; Maguire, 2001; Uc, Rizzo, Anderson, Shi, & Dawson, 2004).

Until about 10 years ago, these different strands of spatial research were largely singular in pursuing their own particular interests, and this was not surprising. It was acknowledged that attempts should be made to consider how the navigating brain interacted with its natural context, namely the real world (Nadel, 1991; O'Keefe & Nadel, 1978). However, the means to achieve this was lacking, with no way to map functions onto specific human brain regions, or to test navigation in truly naturalistic environments while maintaining some degree of experimental control. Two breakthroughs in the last 10 years have begun to facilitate an interdisciplinary approach to spatial navigation. The first of these was the development of brain imaging technologies, particularly magnetic resonance imaging (MRI). Not only does MRI produce high-resolution structural brain images, but use of techniques such as echo planar imaging (EPI) permits inferences about changes in neural activity, making human cognition accessible in vivo. The use of functional brain imaging to study navigation is not without its problems, however.

Most obvious is how to get participants to navigate while their heads are fixed in a physically restricted and noisy brain scanner. Methods used to circumvent this have included the use of static photographs of landmarks and scenes (Epstein & Kanwisher, 1998), having subjects mentally navigate during scanning (Ghaem et al., 1997; Maguire, Frackowiak, & Frith, 1997), or showing films of navigation through environments (Maguire, Frackowiak, & Frith, 1996). While insights into the neural basis of aspects of spatial memory have certainly been gained from such studies, clearly the optimal navigation task during scanning would be dynamic and interactive, with concomitant performance measures.

The second major advance that has facilitated the study of spatial navigation is an explosion in the development of computer simulation technology. Commercially available video games are dynamic and interactive, with a first-person ground-level perspective, and have increasingly complex and naturalistic large-scale environments as their backdrops. These virtual reality (VR) games are often accompanied by editors, allowing researchers to manipulate aspects of the game to produce environments and scenarios suitable to address experimental questions (see chapter 17, this volume). The extent of immersion or presence felt in a VR environment, that is, the degree to which the user treats it as he or she does the real world and behaves in a similar manner, is obviously an important concern. With some limitations, several studies have indicated good correspondence between the spatial knowledge of an environment acquired in the real world and a model of that environment in VR (Arthur, Hancock, & Chrysler, 1997; Regian & Yadrick, 1994; Ruddle, Payne, & Jones, 1997; Witmer et al., 1996), and VR has been used to aid rehabilitation in patients with memory deficits (Brooks & Rose, 2003) and to teach disabled children (Wilson, Foreman, & Tlauka, 1996). Care must be taken in designing realistic VR environments as, for example, realistic landmarks improve navigation while abstract coloured patterns do not (Ruddle et al., 1997), and performance tends to correlate with the extent of presence felt by the subject (Witmer & Singer, 1994). The use of VR in both neuropsychological and neuroimaging contexts has opened up new avenues to explore how the brain navigates in the real world (Burgess et al., 2002; Maguire, Burgess, & O'Keefe, 1999; Spiers &

Maguire, 2006). In this chapter, I describe some of the most recent studies that exploit this approach. I believe that examining brain-environment interactions in this way is already contributing to key theoretical debates in the field of memory, and may in the future directly inform treatment and rehabilitative interventions in memory-impaired patients. Furthermore, this emerging knowledge may one day influence environmental design itself, closing the loop, so to speak, between brain and space.

The Good, the Bad, and the Lost

One of the most enduring questions about navigation concerns why some people are better at finding their way than others. Why do people get lost? In one early neuroimaging study, the brain regions involved in active navigation were directly investigated by requiring subjects to find their way between locations within a complex, texture-rich VR town while in a positron emission tomography (PET) scanner (Maguire, Burgess, Donnett, Frackowiak, et al., 1998). This town was created to include many different possible routes between any two locations. The right parahippocampus and hippocampus were activated by successful navigation between locations based on the subjects' knowledge of the layout of the town compared to following a route of arrows through the town. Most interesting of all, subjects' accuracy of navigation was found to correlate significantly with activation in the right hippocampus (see figure 9.1). Activation of the left hippocampus was associated with successful navigation but did not correlate with accuracy of navigation. By contrast, medial and right inferior parietal activation was associated with all conditions involving active movement through the town. Speed of virtual movement through the town correlated with activation in the caudate nucleus, whereas performance of novel detours was associated with additional activations in left prefrontal cortex. This study highlighted the distributed network underpinning spatial navigation in (virtual) large-scale space (see Spiers & Maguire, 2006, for more on this). Furthermore, it showed that it was possible to identify specific functions of particular brain areas, allowing us to theorize about how the navigation system as a whole might work (see also Burgess et al., 2002, for a discussion on how these

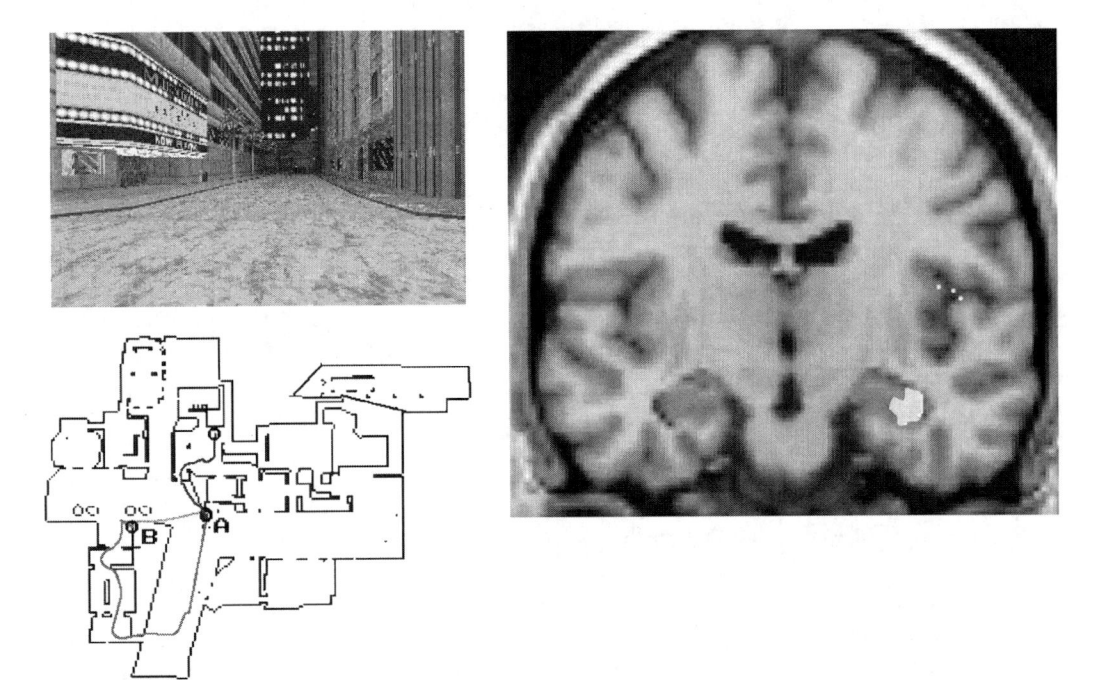

Figure 9.1. Activity in the hippocampus correlates with accuracy of path taken in a virtual reality town (see Maguire, Burgess, Donnett, Frackowiak, et al., 1998). The more direct the path (in yellow on the aerial view), the more active the hippocampus. See also color insert.

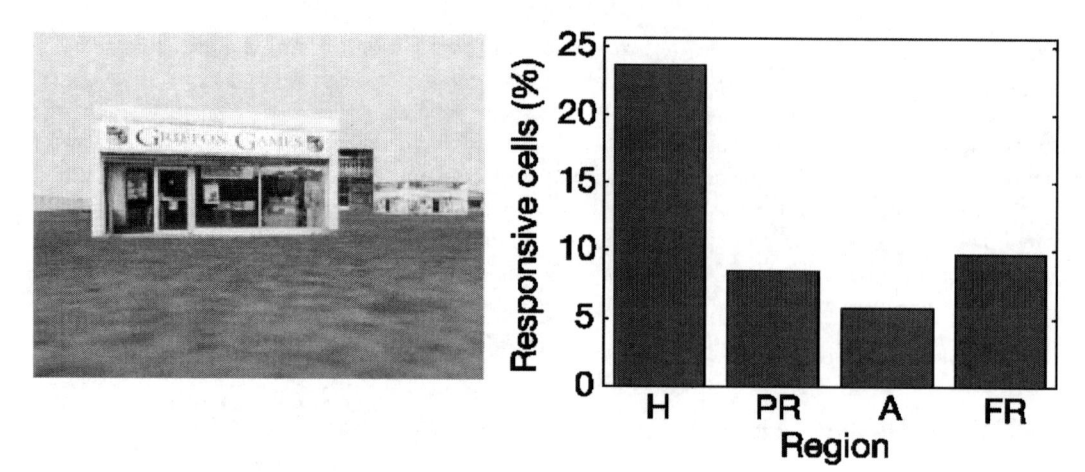

Figure 9.2. Place-responsive cells (see Ekstrom et al., 2003) from intracranial electrodes in the human brain while patients navigated in a VR townlike environment. These neurons were clustered in the hippocampus (H) compared with the amygdala (A), parahippocampal region (PR), and frontal lobes (FR). From Ekstrom et al. (2003). Reprinted by permission from Macmillan Publishers Ltd.: *Nature,* Arne D. Ekstrom et al., Cellular networks underlying human spatial navigation, volume 425, issue 6954, pages 184–188, 2003. See also color insert.

data fit with the wider navigation literature). Accordingly, we suggested that these results were consistent with right hippocampal involvement in supporting a representation of locations within the town allowing accurate navigation, left hippocampal involvement in more general mnemonic processes, posterior parietal involvement in guiding egocentric movement through space, orienting the body relative to doorways, avoiding obstacles, etc., and involvement of the caudate in movement-related aspects of navigation.

Essentially the same task was used to investigate navigation following either focal bilateral hippocampal damage (Spiers, Burgess, Hartley, et al., 2001) or unilateral anterior temporal lobectomy (Spiers, Burgess, Maguire, et al., 2001). Participants were tested on their ability to navigate accurately to 10 locations in the VR town. The right temporal lobectomy patients were impaired compared to controls, taking longer routes (Spiers, Burgess, Maguire, et al., 2001). A patient with focal bilateral hippocampal pathology, Jon (Vargha-Khadem et al., 1997) was also tested and was impaired on the navigation task (Spiers, Burgess, Hartley, et al., 2001). Interestingly, this VR town was also used as a backdrop for an episodic memory task (Spiers, Burgess, Hartley, et al., 2001; Spiers, Burgess, Maguire, et al., 2001; Burgess, Maguire, Spiers, & O'Keefe, 2001). Subjects followed a prescribed route through the VR town and along the way repeatedly met two characters who gave them different objects in two different places. In contrast to spatial navigation, the overall performance of the left temporal lobectomy patients on remembering who gave them the objects and when was significantly worse than controls. As well as being impaired on the navigation task, hippocampal patient Jon was also impaired on all the episodic memory tests. These data seem to confirm a role for the right hippocampus in supporting spatial navigation, and possibly the left hippocampus in more general aspects of context-specific episodic memory.

Ekstrom et al. (2003) provided striking confirmatory evidence for this navigation view at the single-neuron level (see figure 9.2). Patients with pharmacologically intractable seizures had intracranial electrodes implanted in the hippocampus, parahippocampal, and frontal regions. Responses from single neurons were recorded while patients navigated around a small VR town. Cells were identified that fired when a patient was in a particular location in the town, irrespective of his or her orientation. These neurons were mostly found in the hippocampus and may be similar to the place cells identified in the rat hippocampus (e.g., O'Keefe & Dostrovsky, 1971). By contrast, cells that responded to a particular view, that is, a particular shop front, were mostly located in the parahippocampal cortex. Finally, cells that responded to the goal of the patient, such as picking up a particular passenger, were distributed through frontal and temporal cortices. Previous functional neuroimaging studies that found hippocampal activation during navigation

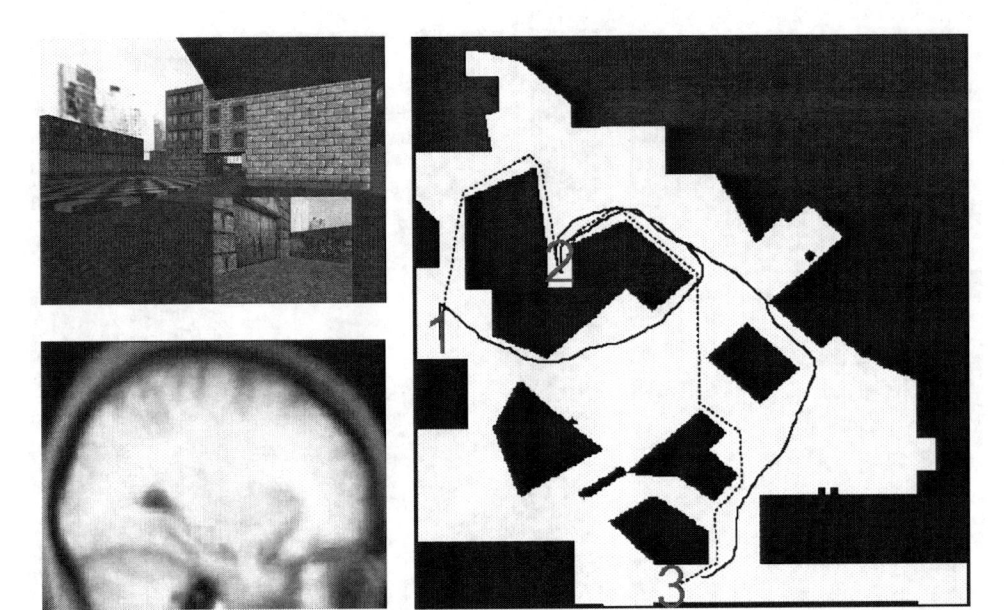

Figure 9.3. Town 1 from Hartley et al. (2003). In the wayfinding task, the current target location is indicated in the lower right corner of the VR display. The map shows an example path followed by a subject (solid line) between the first three target locations. The corresponding ideal path is shown as a dotted line. Accuracy of performance was correlated with activity in the hippocampus. See also color insert.

(e.g., Maguire, Burgess, Donnett, Frackowiak, 1998) are consistent with the high density of place cells found in the patients' hippocampi. Not only does the Ekstrom study confirm previous animal and human neuroimaging work, but in one way it also extends it. The demonstration of goal-related neurons highlights an important aspect of navigation that has received some attention in environmental psychology (e.g., Magliano, Cohen, Allen, & Rodrigue, 1995), but little in human neuroscientific navigation studies (see Spiers & Maguire, 2006, for a recent example). Every journey we make has a purpose, and different goals may influence the cognitive processes engaged, and the brain systems activated.

While measuring the firing of single neurons is highly desirable, such studies are very rare, extremely difficult to execute, and by their nature are highly constrained in terms of the brain areas that can be sampled. Functional neuroimaging therefore offers the next best means to study the navigating brain. In the last several years there have been further improvements in the realism of VR environments coupled with the better spatial and temporal resolution of functional MRI (fMRI; see also chapter 4, this volume). Capitalizing on these developments, a study has provided additional insights into the differences between good and less good navigators. Hartley et al. (2003) were interested in contrasting two different kinds of navigation we all experience in our everyday lives. On the one hand, we often navigate along very familiar routes, for instance taking the same route from work to home (route following). By contrast, we sometimes have to use novel routes locating new places in a familiar environment (wayfinding). Prior to being scanned, subjects learned two distinct but similar VR towns. In one, subjects repeatedly followed the same route, while in the other they were allowed to explore freely. During fMRI scanning, subjects found their way to specified destinations in the freely explored town or followed the well-worn route in the other town. Behavioral performance was measured by comparing the path taken during each task with the ideal path, with the distance error calculated as the additional distance traveled by the subject. The hippocampus was more active in good navigators and less active in poorer navigators during wayfinding (see figure 9.3). Good navigators activated the head of the caudate while navigating along a well-learned route (see figure 9.4). Hartley et al. (2003) suggested that good navigators select the appropriate representation for the task at hand, the hippocampal representation for wayfinding, and the caudate representation for route following. Consistent with

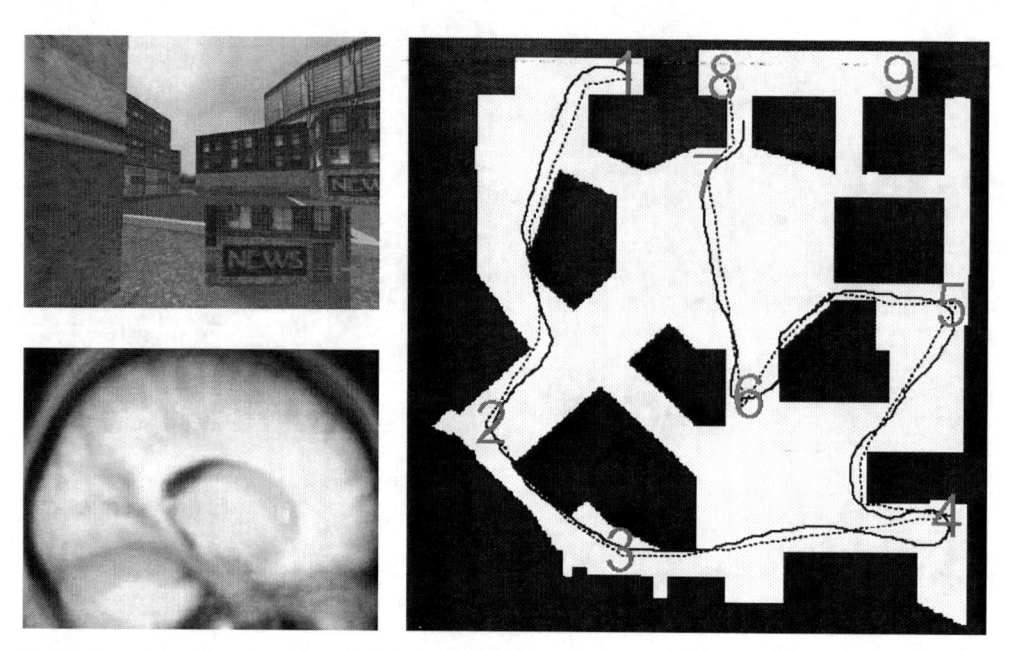

Figure 9.4. Town 2 from Hartley et al. (2003). In the route-following task, the current target location is indicated in the lower right corner of the VR display. The map shows the well-worn route followed by a subject (solid line). Accuracy of performance was correlated with activity in the caudate nucleus. See also color insert.

this, the authors also noted that in the poorest navigators, activation in the head of the caudate was greatest during wayfinding, suggesting the use of an inappropriate (route-following) representation.

In another fMRI study, subjects navigated in a virtual radial eight-arm mazelike environment (Iaria, Petrides, Dagher, Pike, & Bohbot, 2003). It was found that subjects spontaneously adopted one of two strategies during navigation. Some subjects used the relationship between landmarks to guide navigation. Other subjects used a nonspatial strategy whereby they counted the arms of the maze clockwise or counterclockwise from the start position or a single landmark. The authors noted that this suggests a natural variability in the strategies adopted by humans faced with a navigation task. They went on to report that increased activity was apparent in the right hippocampus only in those subjects using the spatial landmark strategy. By contrast, the group that adopted the nonspatial strategy instead showed sustained activity in the caudate nucleus. This and the Hartley et al. (2003) results suggest that the hippocampus and caudate systems both offer a means to support navigation in humans. The engagement of the most appropriate system for the task at hand may be a fundamental contributor to the success or otherwise of the navigation. Of course, this begs several additional questions: Are the systems competing or complementary, and what factors influence their engagement in the first place?

Voermans et al. (2004) provided some further insights combining VR and fMRI with a lesion approach. Patients with preclinical Huntington's disease (HD) are a useful model of relatively selective caudate dysfunction. Subjects, both healthy controls and preclinical HD patients, had to memorize and recognize well-defined routes through VR homes in a navigational memory task. A noncompetitive interaction was found such that the hippocampus compensated for gradual caudate nucleus dysfunction with a gradual increase in activity, maintaining normal behavior. Although characterized here as a complementary relationship, others have described it as competitive (Poldrak & Packard, 2003). It is possible that parallel systems might have a cooperative capacity when one system is damaged. However, in healthy individuals, the two may still compete and possibly impede one another. Given that the hippocampal and caudate systems have very different operating mechanisms (Packard & McGaugh, 1996; White & McDonald, 2002), it may be that the flexible hippocampus can compensate for a compromised caudate system, but the more functionally constrained caudate cannot fulfill a flexible navigation role.

In the absence of pathology, what might drive a subject to adopt a particular navigation strategy, engaging one brain system rather than another? One factor that is often cited anecdotally in relation to navigation is sex differences. Do men and women differ in how they navigate? In recent years VR, among other methods, has been recruited to provide more controlled means of investigating this question. It would seem that women might be more reliant on landmarks within the environment when navigating, with men tending to focus on the Euclidean properties of the environment (Sandstrom, Kaufman, & Huettel, 1998). Grön, Wunderlich, Spitzer, Tomczak, and Riepe (2000) suggested there might be a brain correlate of this difference. They compared a group of men with a group of women who navigated a VR environment during fMRI. They reported that the men activated the hippocampus more than the women, while the women activated right parietal and prefrontal cortices. It may be that the hippocampal activation reflects the male use of Euclidean information, and the cortical activations in the women a landmark-based strategy. However, in this study, women also performed significantly less well than the males on the VR task. Given that previous fMRI studies of exclusively male subjects (e.g., Hartley et al., 2003) found that hippocampal activity is correlated with performance, the sex differences seen by Grön et al. might be explicable in terms of a more general tendency for individual variations in performance to be correlated with hippocampal activation. It is certainly the case that women and men can perform comparably on VR navigation tasks, particularly when one controls for factors such as familiarity with video game playing. However, that women consistently do less well than men in navigation tests (Astur, Ortiz, & Sutherland, 1998; Moffat, Hampson, & Hatzipantelis, 1998; Sakthivel, Patterson, & Cruz-Neira, 1999) leaves open the question of why the female group performed less well in Grön et al.'s study. The possibility remains that the performance-related effects observed have a physiological basis that affects men and women differently (perhaps for evolutionary reasons; Ecuyer-Dab & Robert, 2004; Saucier et al., 2002). Indeed, there are structural differences between the medial temporal lobes of men and women that might index such a difference (Good et al., 2001).

Understanding both sex and individual differences in navigation is important, not least in order to know if one's navigation capability is preordained or whether it can change. Recent findings from studying navigation experts suggest that the human brain's navigation system is more plastic than hitherto thought.

The Experts

Licensed London taxi drivers are unique. They engage in years of training (on average 2–4 years) in order to pass the very stringent examinations that enable them to qualify for a license. Some 320 routes linking over 25,000 streets in greater London have to be memorized and, in addition, thousands of famous landmarks, buildings, and places of interest have to be learned. Acquiring The Knowledge, as it is known, is a truly amazing accomplishment for the 23,000 licensed taxi drivers currently operating in London. Licensed London taxi drivers represent an ideal opportunity to examine complex and successful navigation, and a previous PET study found that they activated the hippocampus when accurately recalling complex routes around the city (Maguire et al., 1997). By studying London taxi drivers, insights might also be gleaned into the effects of training on the adult human brain. Interestingly, neurogenesis in the hippocampus has now been associated with spatial memory and learning in birds and small mammals (Lavenex, Steele, & Jacobs, 2000; Patel, Clayton, & Krebs, 1997; Shors et al., 2001), and has been found in adult primates (Gould, Reeves, Graziano, & Gross, 1999). In general, hippocampal volume has been shown to be related to spatial ability in several species of birds and small mammals (Lee, Miyasato, & Clayton, 1998; Sherry, Jacobs, & Gaulin, 1992) in terms of their ability to keep track of large numbers of stored food items or large home ranges. Furthermore, variations in hippocampal volume in birds and small mammals have been found to track seasonal changes in the need for spatial memory (Lavenex et al., 2000; Smulders, Sasson, & DeVoogd, 1995). Would hippocampal volume changes be apparent in humans who had undergone intensive navigation training?

The structural MRI brain scans of male licensed London taxi drivers have been compared with that of age-matched non–taxi drivers (Maguire et al.,

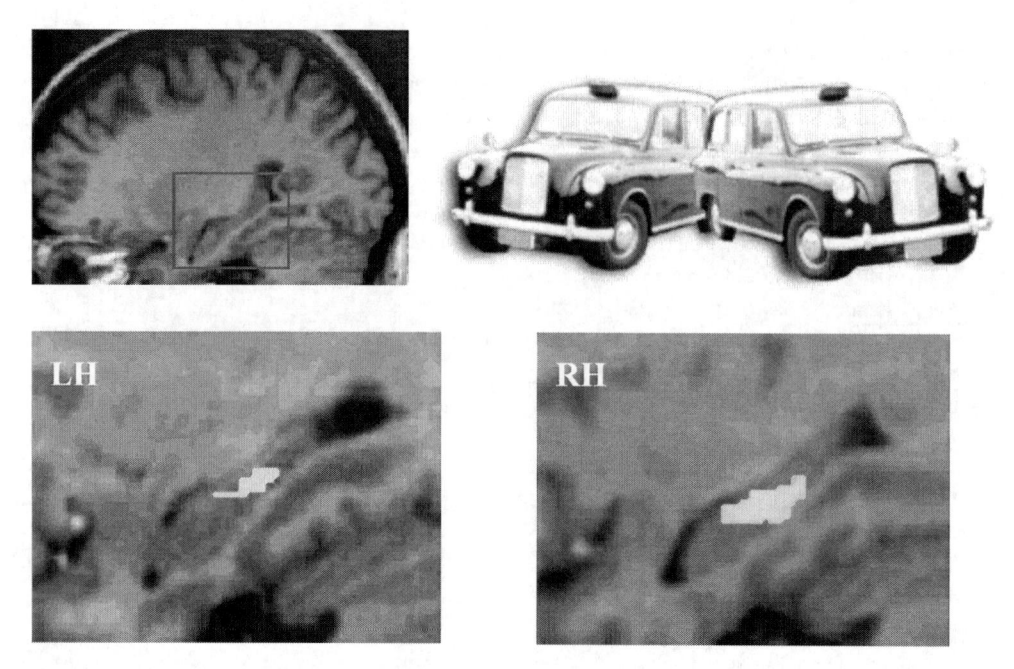

Figure 9.5. MRI sagittal brain sections. Yellow areas indicate where there was increased gray matter density in the left (LH) and right (RH) hippocampi of licensed London taxi drivers compared with non–taxi driver control subjects (see Maguire et al., 2000). See also color insert.

2000). Significant differences in gray matter volume between the two groups were found in the hippocampus, with the posterior hippocampus being larger on both sides in taxi drivers (see figure 9.5). Interestingly, the anterior hippocampus was smaller in the taxi drivers. Moreover, the increase in right posterior hippocampus correlated positively with the time spent in the job, while the anterior hippocampus decreased in volume the longer the time taxi driving. This study provides an intriguing hint of experience-dependent structural plasticity in the human brain and further suggests an intimate link with navigation and the hippocampus in humans as well as other animals. It also suggests that while training may have positive effects by increasing gray matter volume in one area, there may be a price to pay for this with a gray matter decrease elsewhere. The possible neuropsychological sequelae (if any) of the anterior hippocampal gray matter decrease in taxi drivers are currently under investigation.

Is job training really the key factor that is driving the gray matter changes in taxi drivers? We hypothesised that the correlation finding suggests that increased posterior hippocampal gray matter volume is acquired in response to increased taxi driving experience, perhaps reflecting their detailed spatial representation of the city. However, an alternate hypothesis is that the difference in hippocampal volume is instead associated with innate navigational expertise, leading to an increased likelihood of becoming a taxi driver. To investigate such a hypothesis requires the examination of gray matter in non–taxi driver navigation experts. If increased hippocampal gray matter volume were found to be taxi driver–specific, this would be further evidence that hippocampal structure can be changed by interaction with large-scale space. To investigate this possibility, we examined the structural MRI brain scans of subjects who were not taxi drivers (Maguire, Spiers, et al., 2003). We assembled a group of subjects who were well matched on a range of pertinent variables, but who showed wide variation across the group in terms of their ability to learn to find their way in a VR town. Despite this wide range of navigational expertise, there was no association between expertise and posterior hippocampal gray matter volume (or, indeed, gray matter volume throughout the brain). This failure to find an association between hippocampal volume and navigational expertise thus suggests that structural differences in the human hippocampus are acquired in response to intensive environmental stimulation.

What about other experts, particularly those who also have the ability to remember vast amounts of information? We examined individuals renowned for outstanding memory feats in forums such as the World Memory Championships (Maguire, Valentine, et al., 2003). These individuals practice for hours every day to hone the strategies that allow them to achieve superior memory. Using neuropsychological, structural, and functional brain imaging measures, we found that superior memory is not driven by exceptional intellectual ability or structural brain differences. Rather, we found that superior memorizers used a spatial learning strategy (the method of loci; Yates, 1966) while preferentially engaging brain regions critical for memory and for spatial memory in particular, including the hippocampus. It is interesting to note that, although very proficient in the use of this route-based spatial mnemonic, no structural brain changes were detected in the right posterior hippocampus such as were found in the London taxi drivers. This may be because taxi drivers develop and have a need to store a large and complex spatial representation of London, while the memorizers use and reuse a much more constrained set of routes. Structural brain differences have been reported for other professionals compared with nonexpert control subjects, such as orchestra musicians (Sluming et al., 2002), pianists (Schlaug, Jancke, Huang, & Steinmetz, 1995), and bilinguals (Mechelli et al., 2004). However, the cause and effect of these differences is unclear. What is required is a study that tracks brain changes and neuropsychological performance in the same individuals over time while they acquire a skill. A study of this kind is currently underway with individuals training to be licensed London taxi drivers.

Clearly, much remains to be understood about how training affects the brain, and many factors need to be considered, not just the type of representation being acquired but also the nature of the job itself. For example, in a study examining the temporal lobes of an airline cabin crew, Cho (2001) found that chronic jet lag resulted in deficits in spatial memory and decreased volume in the temporal lobe. Elevated basal levels of the stress hormone cortisol were also found in the saliva of the flight crews and correlated with temporal lobe volume reduction. Cho reasoned that the stress associated with disruption of circadian rhythms and sleep patterns produced the elevated cortisol, which in turn affected temporal lobe volume. Thus in terms of how the brain interacts with job training and the occupational environment in general, the limited number of studies so far suggest that at the very least both cognitive and emotional factors, with potential positive and negative effects, need to be considered.

The Structure of the Environment

Successful navigation in virtual or real environments may therefore depend on the kind of brain doing the navigating, the type of strategy it adopts, and the amount of navigation exposure and training it has undergone. In addition to these features of the navigator, as we might class them, several other factors are clearly relevant, not least of which is the environment itself. Environmental psychologists have spent decades examining how the physical structure of our buildings, towns, and cities influences how we navigate within them. How regular or complex the layout (Evans, 1980), how integrated the environment (Peponis, Zimring, & Choi, 1990), and the presence of salient divisions (such as rivers and parks; Lynch, 1960), are just some of the many aspects of the physical environment they have studied. The relative youth of neuroscientific investigations of real-world environments means that so far there is a dearth of information about how the environment's physical structure influences the brain. The single most robust finding is that the parahippocampal gyrus seems to be particularly responsive to features of the environment such as landmarks and buildings (e.g., Epstein & Kanwisher, 1998; Maguire, Burgess, Donnett, O'Keefe, & Frith, 1998), in contrast to the hippocampus, which is more concerned with representing the overall spatial layout (Burgess et al., 2002).

An fMRI study gives some further insights into the factors affecting the parahippocampal responsivity to landmarks. Janzen and van Turennout (2004) had subjects watch footage of a route through a VR museum that they were instructed to learn (see figure 9.6). The museum contained landmarks consisting of objects on tables, positioned at two types of junctions, either points where a navigational decision had to be made or at simple turns where no decision was required. Volunteers were then scanned using fMRI during an old-new recognition test in which the landmarks from the

Toy at decision point **Toy at non-decision point**

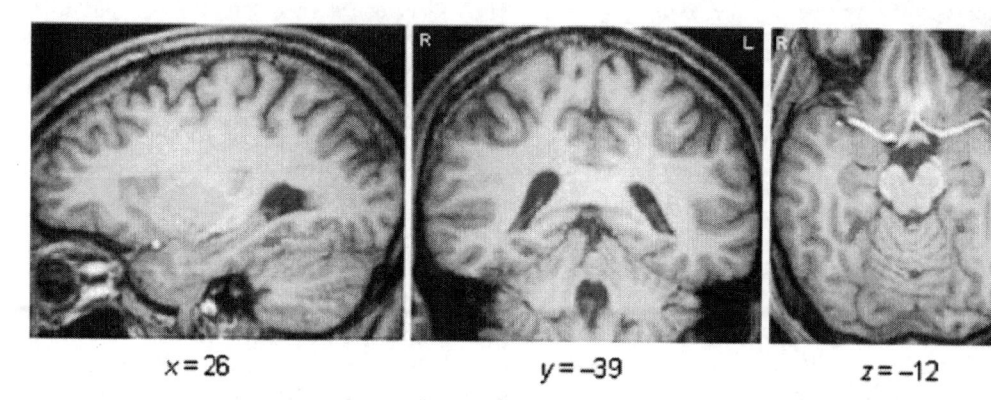

$x = 26$ $y = -39$ $z = -12$

Figure 9.6. Views from the virtual museum from Janzen and van Turennout (2004). The left panel shows a toy placed at a decision point, and the right scene a toy at a nondecision point. A parahippocampal region was more active for toys at decision points, that is, that had navigational relevance. From Janzen and van Turennout (2004). Reprinted by permission from Macmillan Publishers Ltd.: *Nature Neuroscience,* Gabriele Janzen and Miranda van Turennout, Selective neural representation of objects relevant for navigation, volume 7, issue 6, pages 673–677, 2004. See also color insert.

museum and new landmarks were shown from a canonical perspective on a white background. The authors found that the parahippocampal gyrus was more active for landmarks that had been seen at decision points than those that had been seen at simple turns. The parahippocampal signal was apparent even when the navigationally relevant landmarks were lost to conscious awareness. This study shows that the brain identifies landmarks at key decision points, and it does so automatically, requiring just one exposure. This mechanism, where the association with navigational relevance can be made despite a change in perspective, could be an important basis for successful and flexible wayfinding (Spiers & Maguire, 2004).

Another feature of the environment that has been found to elicit a specific brain response is when suddenly a previously used route becomes blocked and one is required to replan and seek an alterative route. When this occurred in a VR town during PET scanning, the left prefrontal cortex was more active when subjects successfully took a detour and reached their destination by this alternative route (Maguire, Burgess, Donnett, Frackowiak, 1998; Spiers & Maguire, 2006). This ability to take a detour is fundamental to flexible navigation, and much more remains to be understood about how frontal executive processes interact with the mnemonic role of the medial temporal lobe to achieve this.

The Learning Context

While only tentative steps have as yet been made into probing the brain–physical environment relationship, there has been some interest in examining the effect of learning context on brain responses. How we acquire spatial knowledge might influence how well we subsequently navigate. Learning to find our way in a new area is normally accomplished by exploration at the ground level. Alternatively, we often use an aerial or survey perspective in the form of maps to aid navigation. Shelton and Gabrieli (2002; see also Mellet et al., 2000) scanned subjects using fMRI while they learned two different VR environments, one from the ground level and the other from a survey perspective. Behaviorally, in these rather simple environments, subsequently tested performance accuracy was similar for the two types of perspective. While there were commonalities in some of the brain areas active during the two kinds of learning, there were also differences. Learning from the ground perspective activated additional areas including the hippocampus and the right inferior parietal and posterior cingulate cortices. By contrast, learning from an aerial perspective activated fusiform and inferior temporal areas, superior parietal cortex, and the insula. The authors suggested that the different patterns of brain activation might reflect the psychological differences between ground and aerial encoding. Subjects reported that at the ground level they had a sense of immersion in the environment, whereas survey learners had no such feeling. Ground-level learning also requires much more updating of information as the local environment changes as one moves through it (see also Wolbers, Weiller, & Buchel, 2004), while the aerial perspective allows more direct access to the global environmental structure. The greater hippocampal activation in the ground-level learning may therefore reflect the spatial updating and map-building properties of this brain region (O'Keefe & Nadel, 1978). While the accuracy of performance did not differ between the two learning perspectives in this instance, in more complex and realistic environments differences might emerge. This study highlights that how an environment is initially experienced may influence the type of spatial representation acquired and the kinds of purposes for which it might be suited.

Virtual reality provides an opportunity to investigate another feature of the learning context in human spatial memory, namely viewpoint dependence. While we often learn a landmark or route from a particular point of view or direction, truly flexible navigation requires the ability to find our way from any direction. King, Burgess, Hartley, Vargha-Khadem, and O'Keefe (2002) provided subjects with a view from the rooftops surrounding a richly textured VR courtyard. During presentation, objects appeared in different locations around the courtyard. During testing, several copies of each object were presented in different locations, with the subject asked to indicate which was in the same location as at presentation. Between presentation and testing, the subject's viewpoint might remain the same or be changed to another location overlooking the courtyard. A patient with focal bilateral hippocampal pathology, Jon (Vargha-Khadem et al., 1997) was mildly impaired on the same-view condition. In contrast, he was massively impaired in the shifted viewpoint condition, suggesting a crucial role for the human hippocampus in representing a world-centered or allocentric view of large-scale space. A recent fMRI study has also documented brain activation differences between viewer-centered, object-centered and landmark-centered frames of reference in a VR environment (Committeri et al., 2004).

Conclusion

Even 15 years ago, being able to make meaningful inferences from observing brain activity while people navigate around complex environments seemed unthinkable. This highly selective review illustrates that today we are able to do just that. Real-world settings are no longer out of bounds for the experimentalist. Rather, I would argue they are essential to fully appreciate the true operation of the human brain. Technical advances in brain imaging hardware, increasing sophistication in fMRI experimental design and data analyses, as well as ever-more realistic VR towns and cities have all played their part in providing this exciting opportunity. It is still early and the work so far has mainly focused on addressing basic questions and assessing convergence of evidence with other established fields such as animal physiology and human neuropsychology. However, we are now beginning to move on from this, appreciating the plasticity and dynamics within the brain's navigation systems. It is not a stretch to hope that in the next few years, a much more fruitful exchange will be possible whereby technological and environmental improvements might be driven by an informed understanding of how the brain finds its way in the real world.

Main Points

1. It is very difficult to investigate experimentally the neural basis of realistic navigation in humans.
2. In the last decade or so, the development of virtual reality and brain scanning techniques such as fMRI has opened up new opportunities.
3. We are starting to understand the distributed brain networks that underpin our ability to navigate in large-scale space, including the specific contributions of regions such as the hippocampus, parahippocampal cortex, and caudate nucleus.
4. The success or otherwise of navigation may be influenced by the extent to which the hippocampus is engaged.
5. The type of navigation strategy employed, as well as factors such as gender, amount of navigation training, the learning context, and the structure of the physical environment, are also key factors influencing how the brain navigates.

Acknowledgments. The author is supported by the Wellcome Trust.

Key Readings

Burgess, N., Maguire, E. A., & O'Keefe, J. (2002). The human hippocampus and spatial and episodic memory. *Neuron, 35,* 625–641.
Eichenbaum, H. (2004). Hippocampus: Cognitive processes and neural representations that underlie declarative memory. *Neuron, 44,* 109–120.
O'Keefe, J., & Nadel, L. (1978). *The hippocampus as a cognitive map.* New York: Oxford University Press.
Spiers, H. J., & Maguire, E. A. (2006). Thoughts, behavior, and brain dynamics during navigation in the real world. *NeuroImage* (in press; early view).

References

Aguirre, G. K., & D'Esposito, M. (1999). Topographical disorientation: A synthesis and taxonomy. *Brain, 122,* 1613–1628.
Arthur, E. J., Hancock, P. A., & Chrysler, S. T. (1997). The perception of spatial layout in real and virtual worlds. *Ergonomics, 40,* 69–77.
Astur, R. S., Ortiz, M. L., & Sutherland, R. J. (1998). A characterization of performance by men and women in a virtual Morris water task: A large and reliable sex difference. *Behavioral Brain Research, 93,* 185–190.
Barrash, J. (1998). A historical review of topographical disorientation and its neuroanatomical correlates. *Journal of Clinical and Experimental Neuropsychology, 20,* 807–827.
Bartlett, F. C. (1932). *Remembering: A study in experimental and social psychology.* Cambridge, UK: Cambridge University Press.
Brooks, B. M., & Rose, F. D. (2003). The use of virtual reality in memory rehabilitation: Current findings and future directions. *Neurorehabilitation, 18,* 147–157.
Burgess, N., Maguire, E. A., & O'Keefe, J. (2002). The human hippocampus and spatial and episodic memory. *Neuron, 35,* 625–641.
Burgess, N., Maguire, E. A., Spiers, H., & O'Keefe, J. (2001). A temporoparietal and prefrontal network for retrieving the spatial context of life-like events. *NeuroImage, 14,* 439–453.
Cho, K. (2001). Chronic "jet lag" produces temporal lobe atrophy and spatial cognitive deficits. *Nature Neuroscience, 4,* 567–568.
Committeri, G., Galati, G., Paradis, A.-L., Pizzamiglio, L., Berthoz, A., & LeBihan, D. (2004). Reference frames for spatial cognition: Different brain areas are involved in viewer-, object- and landmark-centered judgments about object location. *Journal of Cognitive Neuroscience, 16,* 1517–1535.
Downs, R. M., & Stea, D. (1973). *Image and environments.* Chicago: Aldine.
Ecuyer-Dab, I., & Robert, M. (2004). Have sex differences in spatial ability evolved from male competition for mating and female concern for survival? *Cognition, 91,* 221–257.
Eichenbaum, H. (2004). Hippocampus: Cognitive processes and neural representations that underlie declarative memory. *Neuron, 44,* 109–120.
Ekstrom, A., Kahana, M. J., Caplan, J. B., Fields, T. A., Isham, E. A., Newman, E. L., et al. (2003). Cellular networks underlying human spatial navigation. *Nature, 425,* 184–187.

Epstein, R., & Kanwisher, N. (1998). A cortical representation of the local visual environment. *Nature, 392,* 598–601.

Evans, G. W. (1980). Environmental cognition. *Psychological Bulletin, 88,* 259–287.

Ghaem, O., Mellet, E., Crivello, F., Tzourio, N., Mazoyer, B., Berthoz, A., et al. (1997). Mental navigation along memorized routes activates the hippocampus, precuneus, and insula. *Neuroreport, 8,* 739–744.

Good, C. D., Johnsrude, I., Ashburner, J., Henson, R. N., Friston, K., & Frackowiak, R. S. J. (2001). Cerebral asymmetry and the effects of sex and handedness on brain structure: A voxel based morphometric analysis of 465 normal adult human brains. *Neuroimage, 14,* 685–700.

Gould, E., Reeves, A. J., Graziano, M. S., & Gross, C. G. (1999). Neurogenesis in the neocortex of adult primates. *Science, 286,* 548–552.

Grön, G., Wunderlich, A. P., Spitzer, M., Tomczak, R., & Riepe, M. W. (2000). Brain activation during human navigation: Gender-different neural networks as substrate of performance. *Nature Neuroscience, 3,* 404–408.

Hartley, T., Maguire, E. A., Spiers, H. J., & Burgess, N. (2003). The well-worn route and the path less traveled: Distinct neural bases of route following and wayfinding in humans. *Neuron, 37,* 877–888.

Hermer, L., & Spelke, E. S. (1994). A geometric process for spatial reorientation in young children. *Nature, 370,* 57–59.

Iaria, G., Petrides, M., Dagher, A., Pike, B., & Bohbot, V. D. (2003). Cognitive strategies dependent on the hippocampus and caudate nucleus in human navigation: Variability and change with practice. *Journal of Neuroscience, 23,* 5945–5952.

Janzen, G., & van Turennout, M. (2004). Selective neural representation of objects relevant for navigation. *Nature Neuroscience, 7,* 673–677.

King, J. A., Burgess, N., Hartley, T., Vargha-Khadem, F., & O'Keefe, J. (2002). The human hippocampus and viewpoint dependence in spatial memory. *Hippocampus, 12,* 811–820.

Kolb, B., & Whishaw, I. Q. (1990). *Fundamentals of human neuropsychology.* New York: Freeman.

Lavenex, P., Steele, M. A., & Jacobs, L. F. (2000). The seasonal pattern of cell proliferation and neuron number in the dentate gyrus of wild adult eastern grey squirrels. *European Journal of Neuroscience, 12,* 643–648.

Lee, D. W., Miyasato, L. E., & Clayton, N. S. (1998). Neurobiological bases of spatial learning in the natural environment: Neurogenesis and growth in the avian and mammalian hippocampus. *Neuroreport, 9,* R15–R27.

Lynch, K. (1960). *The image of the city.* Cambridge, MA: MIT Press.

Magliano, J. P., Cohen, R., Allen, G. L., & Rodrigue, J. R. (1995). The impact of wayfinder's goal on learning a new environment: Different types of spatial knowledge as goals. *Journal of Environmental Psychology, 15,* 65–75.

Maguire, E. A. (2001). The retrosplenial contribution to human navigation: A review of lesion and neuroimaging findings. *Scandinavian Journal of Psychology, 42,* 225–238.

Maguire, E. A., Burgess, N., Donnett, J. G., Frackowiak, R. S., Frith, C. D., & O'Keefe, J. (1998). Knowing where and getting there: A human navigation network. *Science, 280,* 921–924.

Maguire, E. A., Burgess, N., Donnett, J. G., O'Keefe, J., & Frith, C. D. (1998). Knowing where things are: Parahippocampal involvement in encoding object locations in virtual large-scale space. *Journal of Cognitive Neuroscience, 10,* 61–76.

Maguire, E. A., Burgess, N., & O'Keefe, J. (1999). Human spatial navigation: Cognitive maps, sexual dimorphism, and neural substrates. *Current Opinion in Neurobiology, 9,* 171–177.

Maguire, E. A., Frackowiak, R. S., & Frith, C. D. (1996). Learning to find your way: A role for the human hippocampal formation. *Proceedings of the Royal Society of London, B, Biological Sciences, 263,* 1745–1750.

Maguire, E. A., Frackowiak, R. S. J., & Frith, C. D. (1997). Recalling routes around London: Activation of the right hippocampus in taxi drivers. *Journal of Neuroscience, 17,* 7103–7110.

Maguire, E. A., Gadian, D. G., Johnsrude, I. S., Good, C. D., Ashburner, J., Frackowiak, R. S., et al. (2000). Navigation-related structural change in the hippocampi of taxi drivers. *Proceedings of the National Academy of Sciences USA, 97,* 4398–4403.

Maguire, E. A., Spiers, H. J., Good, C. D., Hartley, T., Frackowiak, R. S. J., & Burgess, N. (2003). Navigation expertise and the human hippocampus: A structural brain imaging analysis. *Hippocampus, 13,* 208–217.

Maguire, E. A., Valentine, E. R., Wilding, J. M., & Kapur, N. (2003). Routes to remembering: The brains behind superior memory. *Nature Neuroscience, 6,* 90–95.

Mechelli, A., Crinion, J. T., Noppeney, U., O'Doherty, J., Ashburner, J., Frackowiak, R. S., et al. (2004). Neurolinguistics: Structural plasticity in the bilingual brain. *Nature, 431,* 757.

Mellet, E., Briscogne, S., Tzourio-Mazoyer, N., Ghaem, O., Petit, L., Zago, L., et al. (2000). Neural correlates of topographic mental exploration: The impact of route versus survey perspective learning. *Neuroimage, 12,* 588–600.

Moffatt, S. D., Hampson, E., & Hatzipantelis, M. (1998). Navigation in a "virtual" maze: Sex differences and correlation with psychometric measures of spatial ability in humans. *Evolution and Human Behavior, 19,* 73–87.

Nadel, L. (1991). The hippocampus and space revisited. *Hippocampus, 1,* 221–229.

Nadolne, M. J., & Stringer, A. Y. (2001). Ecological validity in neuropsychological assessment: Prediction of wayfinding. *Journal of the International Neuropsychological Society, 7,* 675–682.

O'Keefe, J., & Dostrovsky, J. (1971). The hippocampus as a spatial map: Preliminary evidence from unit activity in the freely-moving rat. *Brain Research, 34,* 171–175.

O'Keefe, J., & Nadel, L. (1978). *The hippocampus as a cognitive map.* New York: Oxford University Press.

Packard, M. G., & McGaugh, J. L. (1996). Inactivation of hippocampus or caudate nucleus with lidocaine differentially affects expression of place and response learning. *Neurobiology of Learning and Memory, 65,* 65–72.

Patel, S. N., Clayton, N. S., & Krebs, J. R. (1997). Spatial learning induces neurogenesis in the avian brain. *Behavioral Brain Research, 89,* 115–128.

Peponis, J., Zimring, C., & Choi, Y. K. (1990). Finding the building in wayfinding. *Environment and Behavior, 22,* 555–589.

Poldrack, R. A., & Packard, M. G. (2003). Competition among multiple memory systems: Converging evidence from animal and human brain studies. *Neuropsychologia, 41,* 245–251.

Regian, J. W., & Yadrick, R. M. (1994). Assessment of configurational knowledge of naturally and artificially acquired large-scale space. *Journal of Environmental Psychology, 14,* 211–223.

Ruddle, R. A., Payne, S. J., & Jones, D. M. (1997). Navigating buildings in "desk-top" virtual environments: Experimental investigations using extended navigational experience. *Journal of Experimental Psychology: Applied, 3,* 143–159.

Sakthivel, M., Patterson, P. E., & Cruz-Neira, C. (1999). Gender differences in navigating virtual worlds. *Biomedical Science and Instrumentation, 35,* 353–359.

Sandstrom, N. J., Kaufman, J., & Huettel, S. A. (1998). Males and females use different distal cues in a virtual environment navigation task. *Cognitive Brain Research, 6,* 351–360.

Saucier, D. M., Green, S. M., Leason, J., MacFadden, A., Bell, S., & Elias, L. J. (2002). Are sex differences in navigation caused by sexually dimorphic strategies or by differences in the ability to use the strategies? *Behavioral Neuroscience, 116,* 403–410.

Schlaug, G., Jancke, L., Huang, Y., & Steinmetz, H. (1995). In vivo evidence of structural brain asymmetry in musicians. *Science, 267,* 699–701.

Shelton, A. L., & Gabrieli, J. D. E. (2002). Neural correlates of encoding space from route and survey perspectives. *Journal of Neuroscience, 22,* 2711–2717.

Sherry, D. F., Jacobs, L. F., & Gaulin, S. J. (1992). Spatial memory and adaptive specialization of the hippocampus. *Trends in Neuroscience, 15,* 298–303.

Shors, T. J., Miesagaes, G., Beylin, A., Zhao, M., Rydel, T., & Gould, E. (2001). Neurogenesis in the adult is involved in the formation of trace memories. *Nature, 410,* 372–376.

Siegel, A. W., & White, S. H. (1975). The development of spatial representation in of large-scale environments. In H. W. Reese (Ed.), *Advances in child development and behavior* (pp. 9–55). New York: Academic Press.

Sluming, V., Barrick, T., Howard, M., Cezayirli, E., Mayes, A., & Roberts, N. (2002). Voxel-based morphometry reveals increased gray matter density in Broca's area in male symphony orchestra musicians. *Neuroimage, 17,* 1613–1622.

Smulders, T. V., Sasson, A. D., & DeVoogd, T. J. (1995). Seasonal variation in hippocampal volume in a food-storing bird, the black-capped chickadee. *Journal of Neurobiology, 27,* 15–25.

Spiers, H. J., Burgess, N., Hartley, T., Vargha-Khadem, F., & O'Keefe, J. (2001). Bilateral hippocampal pathology impairs topographical and episodic but not recognition memory. *Hippocampus, 11,* 715–725.

Spiers, H. J., Burgess, N., Maguire, E. A., Baxendale, S. A., Hartley, T., Thompson, P., et al. (2001). Unilateral temporal lobectomy patients show lateralised topographical and episodic memory deficits in a virtual town. *Brain, 124,* 2476–2489.

Spiers, H. J., & Maguire, E. A. (2004). A "landmark" study in understanding the neural basis of navigation. *Nature Neuroscience, 7,* 572–574.

Spiers, H.J., & Maguire, E.A. (2006). Thoughts, behavior, and brain dynamics during navigation in the real world. *NeuroImage* (in press; early view).

Uc, E. Y., Rizzo, M., Anderson, S. W., Shi, Q., & Dawson, J. D. (2004). Driver route-following and safety errors in early Alzheimer disease. *Neurology, 63,* 832–837.

Vargha-Khadem, F., Gadian, D. G., Watkins, K. E., Connelly, A., Van Paesschen, W., & Mishkin, M. (1997). Differential effects of early hippocampal pathology on episodic and semantic memory. *Science, 277,* 376–380.

Voermans, N. C., Petersson, K. M., Daudey, L., Weber, B., van Spaendonck, K. P., Kremer, H. P. H., et al. (2004). Interaction between the human hippocampus and the caudate nucleus during route recognition. *Neuron, 43*, 427–435.

White, N. M., & McDonald, R. J. (2002). Multiple parallel memory systems in the brain of the rat. *Neurobiology of Learning and Memory, 77*, 125–184.

Wilson, P. N., Foreman, N., & Tlauka, M. (1996). Transfer of spatial information from a virtual to a real environment in physically disabled children. *Disability Rehabilitation, 18*, 633–637.

Witmer, B. G., Bailey, J. H., Knerr, B. W., & Parsons, K. C. (1996). Virtual spaces and real world places: Transfer of route knowledge. *International Journal of Human Computer Studies, 45*, 413–428.

Witmer, B.G. & Singer, M.J. (1994). Measuring presence in virtual environments. *ARI technical report 1014*. US Army Research Institute for the Behavioral and Social Sciences, Alexandria, VA, USA.

Wolbers, T., Weiller, C., & Buchel, C. (2004). Neural foundations of emerging route knowledge in complex spatial environments. *Cognitive Brain Research, 21*, 401–411.

Yates, F. A. (1966). *The art of memory*. London: Pimlico.

Joel S. Warm and Raja Parasuraman

Cerebral Hemodynamics and Vigilance

The efficiency and safety of many complex human-machine systems can be critically dependent on the mental workload and vigilance of the operators of such systems. As pointed out by Wickens and Hollands (2000), it has long been recognized that the design of a high-quality human-machine system is not just a matter of assessing performance but also of evaluating how well operators can meet the workload demands imposed on them by the system. Major questions that must be addressed are whether human operators can meet additional unexpected demands when they are otherwise overloaded (Moray, 1979; Wickens, 2002) and whether they are able to maintain vigilance and respond effectively to critical events that occur at unpredictable intervals (Warm & Dember, 1998).

These considerations point to the need for sensitive and reliable measurement of human mental workload and vigilance. Behavioral measures, such as accuracy and speed of response to probe events, have been widely used to assess these psychological functions. However, as discussed by Kramer and Weber (2000), Parasuraman (2003), and Wickens (1990), measures of brain function offer some unique advantages that can be exploited in particular applications. Among these is the ability to extract covert physiological measures continuously in complex system operations in which overt behavioral measures may be relatively sparse. Perhaps a more compelling rationale is that measures of brain function can be linked to emerging cognitive neuroscience knowledge on attention (Parasuraman & Caggiano, 2002; Posner, 2004), thereby allowing for the development of neuroergonomic theories that in turn can advance practical applications of research on mental workload and vigilance.

In this chapter, we describe a series of recent neuroergonomic studies from our research group on vigilance, focusing on the use of noninvasive measurement of cerebral blood flow velocity. We use a theoretical framework of attentional resources (Kahneman, 1973; Moray, 1967; Navon & Gopher, 1979; Norman & Bobrow, 1975; Posner & Tudela, 1997; Wickens, 1984). Resource theory is the dominant theoretical approach to the assessment of human mental workload (Wickens, 2002) and also provides a major conceptual framework for understanding human vigilance performance (Parasuraman, 1979; Warm & Dember, 1998). Consistent with the view first proposed by Sir Charles Sherrington (Roy & Sherrington, 1890), a considerable amount of research on brain imaging indicates that there is a close tie between cerebral blood flow and neural activity in the performance of mental tasks

(Raichle, 1998; Risberg, 1986). Consequently, changes in blood flow velocity and oxygenation in our studies are considered to reflect the availability and utilization of information processing assets needed to cope with the vigilance task.

The Hemodynamics of Vigilance

Brain Systems and Vigilance

Vigilance involves the ability of observers to detect transient and infrequent signals over prolonged periods of time. That aspect of human performance is of considerable concern to human factors and ergonomic specialists because of the critical role that vigilance occupies in many complex human-machine systems, including military surveillance, air-traffic control, cockpit monitoring and airport baggage inspection, industrial process and quality control, and medical functions such as cytological screening and vital sign monitoring during surgery (Hancock & Hart, 2002; Parasuraman, 1986; Warm & Dember, 1998). Thus, it is important to understand the neurophysiological factors that control vigilance performance.

In recent years, brain imaging studies using positron emission tomography (PET) and functional magnetic resonance imaging (fMRI) techniques have been successful in demonstrating that changes in cerebral blood flow and glucose metabolism are involved in the performance of vigilance tasks (see review by Parasuraman, Warm, & See, 1998). These studies have also identified several brain regions that are active in such tasks, including the right frontal cortex and the cingulate gyrus, as well as subcortical nuclei such as the locus coeruleus. Although these studies have identified brain regions involved in vigilance, Parasuraman et al. (1998) have pointed out some major limitations of this research. With the exception of PET studies by Paus et al. (1997) and by Coull, Frackowiak, and Frith (1998), the brain imaging studies have neglected to link the systems they have identified to performance efficiency, perhaps because of the high cost associated with using PET and fMRI during the prolonged running times characteristic of vigilance research. Thus, the functional role of the brain systems identified in the imaging studies remains largely unknown. Gazzaniga, Ivry, and Mangun (2002) have also emphasized the necessity of linking neuroimaging results to human performance for enhanced understanding of research in cognitive neuroscience.

Other problems with the PET and fMRI procedures are that they feature restrictive environments in which observers need to remain almost motionless throughout the scanning procedure so as not to compromise the quality of the brain images, and fMRI acquisition is accompanied by loud noise. Observers in vigilance experiments rarely remain motionless, however. Instead, research has shown that they tend to fidget during the performance of a vigilance task, with the amount of motor activity increasing with time on task (Galinsky, Rosa, Warm, & Dember, 1993). Moreover, the noise of fMRI is one of several environmental variables that can degrade vigilance performance. For example, Becker, Warm, Dember, and Hancock (1995) showed that noise lowered perceptual sensitivity in a vigilance task, interfered with the ability of observers to profit from knowledge of results, and elevated perceived mental workload. Accordingly, the conditions required for the effective use of the PET and fMRI techniques may not provide a suitable environment for linking changes in brain physiology with vigilance performance over a prolonged period of time. To meet this need, we turned to two other imaging procedures, transcranial Doppler sonography (TCD) and transcranial cerebral oximetry.

Transcranial Doppler Sonography

TCD is a noninvasive neuroimaging technique that employs ultrasound signals to monitor cerebral blood flow velocity or hemovelocity in the mainstem intracranial arteries—the middle, anterior, and posterior arteries. These arteries are readily isolated through a cranial "transtemporal window" and exhibit discernible measurement characteristics that facilitate their identification (Aaslid, 1986). The TCD technique uses a small 2 MHz pulsed Doppler transducer to gauge arterial blood flow. The transducer is placed just above the zygomatic arch along the temporal bone, a part of the skull that is functionally transparent to ultrasound. The depth of the pulse is adjusted until the desired intracranial artery (e.g., the middle cerebral artery, MCA) is isonated. TCD measures the difference in frequency between the outgoing and reflected energy as it strikes moving erythrocytes.

The low weight and small size of the transducer and the ability to embed it conveniently in a headband permit real-time measurement of cerebral blood flow while not limiting, or being hampered by, body motion. Therefore, TCD enables inexpensive, continuous, and prolonged monitoring of cerebral blood flow velocity concurrent with task performance. Blood flow velocities, measured in centimeters per second, are typically highest in the MCA, and the MCA carries about 80% of the blood flow within each cerebral hemisphere (Toole, 1984). Consequently, our TCD studies of mental workload and vigilance assess blood flow velocity in the MCA, but other TCD studies, particularly those examining perceptual processes, also measure blood flow in the posterior cerebral artery (PCA). For further methodological details of the TCD technique, see chapter 6.

When a particular area of the brain becomes metabolically active, as in the performance of mental tasks, by-products of this activity such as carbon dioxide (CO_2) increase. This increase in CO_2 leads to a dilation of blood vessels serving that area, which in turn results in blood flow to that region (Aaslid, 1986). Consequently, TCD offers the possibility of measuring changes in metabolic activity during task performance. The use of TCD in brain imaging performance applications is limited, in part, by its low spatial resolution: TCD can supply gross hemispheric data, but it does not provide information about changes in specific brain loci, as is the case with PET and fMRI. Nevertheless, TCD offers good temporal resolution (Aaslid, 1986) and, compared to PET and fMRI, it can track rapid changes in blood flow dynamics that can be followed in real time under less restrictive and invasive conditions. The use of TCD to index blood flow changes in a wide variety of cognitive, perceptual, and motor tasks has been reviewed elsewhere (Duschek & Schandry, 2003; Klingelhofer, Sander, & Wittich, 1999; Stroobant & Vingerhoets, 2000; see also, chapter 6).

Transcranial Cerebral Oximetry

The TCD technique provides a very economical way to assess cerebral blood flow in relatively unrestricted environments. However, TCD does not directly provide information on oxygen utilization in the brain, which would be useful to assess as another indicator of the activation of neuronal populations recruited in the service of cognitive processes. Optical imaging, in particular near-infrared spectroscopy (NIRS), can be used in the assessment of cerebral oximetry. There are several types of NIRS technology, including the recent development of so-called fast NIRS, as discussed in chapter 5. The standard NIRS technique has several advantages over TCD, including the ability to assess activation in several brain regions, and not just in the left and right hemispheres as with TCD. Previous research using NIRS has shown that tissue oxygenation increases with the information processing demands of the task being performed by an observer (Punwani, Ordidge, Cooper, Amess, & Clemence, 1998; Toronov et al., 2001). Hence, one might expect that along with cerebral blood flow, cerebral oxygenation would also be related to both mental workload and to vigilance.

TCD Studies of Vigilance

Working Memory and Vigilance

The initial study from our research group to examine TCD in relation to vigilance was conducted by Mayleben (1998). That study was guided by the finding that working memory demand can be a potent influence on vigilance performance (Davies & Parasuraman, 1982; Parasuraman, 1979). Parasuraman (1979) had first showed that successive discrimination tasks, in which the detection of critical targets requires comparison of information in working memory, are more susceptible to performance decrement over time than simultaneous discrimination tasks, which have no such memory imperative. The role of memory representation in the vigilance decrement was confirmed in a study by Caggiano and Parasuraman (2004). Moreover, Warm and Dember (1998) conducted a series of studies showing that other psychophysical and task factors that reduce attentional resources (e.g., low signal salience, dual-task demands) have a greater detrimental effect on successive than on simultaneous vigilance tasks. These findings can be interpreted in terms of the resource model described earlier in this chapter. According to that model, a limited-capacity information processing system allocates resources to cope with situations that confront it. The vigilance

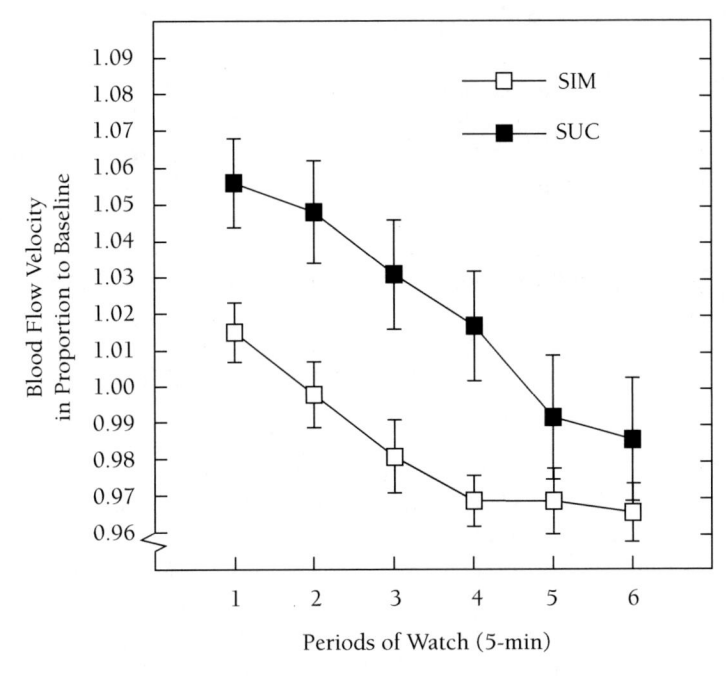

Figure 10.1. Mean cerebral blood flow velocity as a function of periods of watch for simultaneous-type (SIM) and successive-type (SUC) vigilance tasks. Error bars are standard errors. After Mayleben (1998).

decrement, the decline in signal detections over time that characterizes vigilance performance (Davies & Parasuraman, 1982; Warm & Jerison, 1984), reflects the depletion of information processing resources or reservoirs of energy that cannot be replenished in the time available. Given that changes in blood flow might reflect the availability and utilization of the information processing assets needed to cope with a vigilance task, Mayleben (1998) hypothesized that the vigilance decrement should be accompanied by a decline in cerebral hemovelocity and that the overall level of blood flow should be greater for a memory-demanding successive task than for a memory-free simultaneous task.

Participants in this study were asked to perform either a successive or a simultaneous vigilance task during a 30-minute vigil. Critical signals for detection in the simultaneous task were cases in which one of two lines on a visual display was slightly taller than the other. In the successive task, critical signals were cases in which both lines were slightly taller than usual. Pilot work ensured that the tasks were equated for difficulty under alerted conditions. In this and in all of the subsequent studies from our laboratory described in this chapter, blood flow or hemovelocity is expressed as a percentage of the last 60 seconds of a 5-minute resting baseline, as recommended by Aaslid (1986). As illustrated in figure 10.1, Mayleben (1998) found that the vigilance decrement in detection rate over time was accompanied by a parallel decline in cerebral hemovelocity. Also consistent with expectations from a resource model, the overall level of blood flow velocity was significantly higher for observers who performed the successive task than for those who performed the simultaneous task.

An important additional finding of the Mayleben (1998) study was that the blood flow effects were lateralized—hemovelocity was greater in the right than in the left hemisphere, principally in the performance of the memory-based successive task. A result of this sort is consistent with earlier PET and psychophysical studies showing right-brain superiority in vigilance (Parasuraman et al., 1998) and with studies by Tulving, Kapur, Craik, Moscovitch, and Houle (1994) indicating that memory retrieval is primarily a right-brain function. Schnittger, Johannes, Arnavaz, and Munte (1997) also reported the performance–blood flow relation over time described in the Mayleben (1998) study.

However, a clear coupling of blood flow and performance could not be determined in their investigation because of the absence of a control for the possibility of spontaneous declines in blood flow over time, such as may result from systemic declines in arousal. Following a suggestion by Parasuraman (1984), Mayleben (1998) employed such a control by exposing a group of observers to the dual-line display for 30 minutes in the absence of a work imperative. Blood flow remained stable over the testing period under such conditions. Thus, the decline in cerebral blood flow was closely linked to the need to maintain attention to the visual display and not merely to the passage of time.

A potential challenge to an interpretation of these results along resource theory lines comes from the findings that blood flow velocity is sensitive to changes in blood pressure and cardiac output (Caplan et al., 1990) and that changes in heart rate variability are correlated with vigilance performance (Parasuraman, 1984). Accordingly, one could argue that the performance and hemovelocity findings in this study do not reflect information processing per se but rather a gross change in systemic vascular activity that covaried with blood flow. The lateralization of the performance and hemovelocity findings challenges such a view, since gross changes in vascular activity are not likely to be hemisphere dependent.

Controlling the Vigilance Decrement with Signal Cueing

Signal detection in vigilance can be improved by providing observers with consistent and reliable cues to the imminent arrival of critical signals. As previous experiments have shown, the principal consequence of such forewarning is the elimination of the vigilance decrement (Annett, 1996; Wiener & Attwood, 1968). The cueing effect can be linked to resource theory as follows. Observers need to monitor a display only after having been prompted about the arrival of a signal, and therefore can husband their information processing resources over time. In contrast, when no cues are provided, observers are never certain of when a critical signal might appear, and consequently have to process information on their displays continuously across the watch, thereby consuming more of their resources over time than cued observers. Thus, one can predict that in the presence of perfectly reliable cueing, the temporal decline in cerebral blood flow would be attenuated in comparison to a noncued condition and also in comparison to conditions in which cueing was less than perfectly reliable, since observers in such conditions would not be relieved of the need to attend continuously to the vigilance display.

This prediction was tested by Hitchcock et al. (2003) using a simulated air-traffic control (ATC) display. Critical signals for detection were pairs of aircraft traveling on a potential collision course. Observers monitored the simulated ATC display for 40 minutes. To manipulate perceptual difficulty, signal salience, varied by changing the Michaelson contrast ratio of the aircraft to their background, was high (98%, dark black aircraft on a light background) or low (2%, light gray aircraft on a light background). Signal salience was combined factorially with four levels of cue reliability—100% reliable, 80% reliable, 40% reliable, and a no-cue control. Observers in the cueing groups were instructed that a critical signal would occur within one of the five display updates immediately following the verbal prompt *look* provided through a digitized male voice. Observers in each of the cue groups were advised about the reliability of the cues they would receive. To control for accessory auditory stimulation, observers in the no-cue group received acknowledgment after each response in the form of the word *logged* spoken in the same male voice.

As can be seen in figure 10.2, the detection scores for the several cueing conditions in this study were similar to each other during the early portion of the vigil and diverged by the end of the session. More specifically, performance efficiency remained stable in the 100% reliable cueing condition but declined over time in the remaining conditions, so that by the end of the vigil, performance efficiency was clearly best in the 100% group followed in order by the 80%, 40%, and no-cue groups.

The hemovelocity scores from the left hemisphere showed a significant decline over time but no effect for cueing with either high-salience or low-salience signals. A similar result was found for high-salience signals in the right hemisphere. Cueing effects emerged, however, in the hemovelocity scores for the right hemisphere with low-salience signals. As was the case with detection probability, the hemovelocity scores for the several cueing conditions were similar to each other during the early portions of the vigil, but showed differential rates

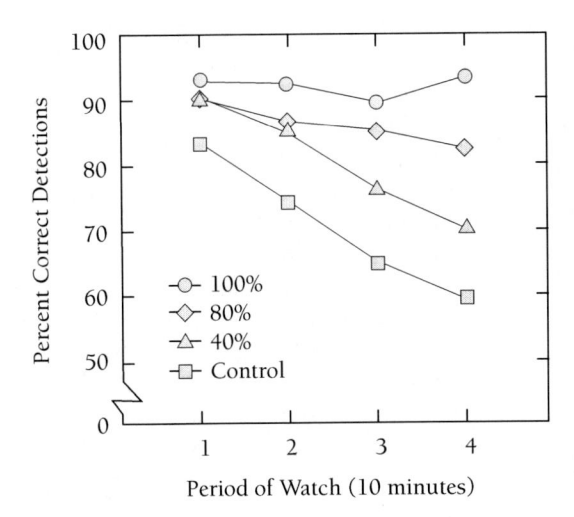

Figure 10.2. Percentages of correct detections as a function of periods of watch for four cue-reliability conditions. After Hitchcock et al. (2003).

of decline over time, so that by the end of the vigil, blood flow was clearly highest in the 100% group followed in order by the 80%, 40%, and no cue groups. This result is illustrated in figure 10.3.

In summary, the hemovelocity scores taken from the right hemisphere under low salience almost exactly mirrored the effects of cuing on performance efficiency. The finding that this result was limited to low-salience signals is consistent with a study by Korol and Gold (1998) indicating that brain systems involving glucose metabolism need to be sufficiently challenged in order for measurable physiological changes to emerge in cognitive and attentional processing tasks. Restriction of the cue-time-salience hemovelocity findings to the right hemisphere is consistent with expectations about right hemisphere control of vigilance. As in

the initial study, blood flow remained stable over time in both hemispheres throughout the watch when observers were exposed to the simulated air-traffic display without a work imperative.

Visual Search

The ATC task used in the Hitchcock et al. (2003) study was such that critical signals could be detected without any substantial need for searching the display. In contrast, many real-world environments, both in ATC and elsewhere, require that operators conduct a visual search of displays in order to detect critical signals. A well-established finding from laboratory studies of visual search is the *search asymmetry* phenomenon (Treisman & Gormican, 1988). This effect refers to more rapid detections

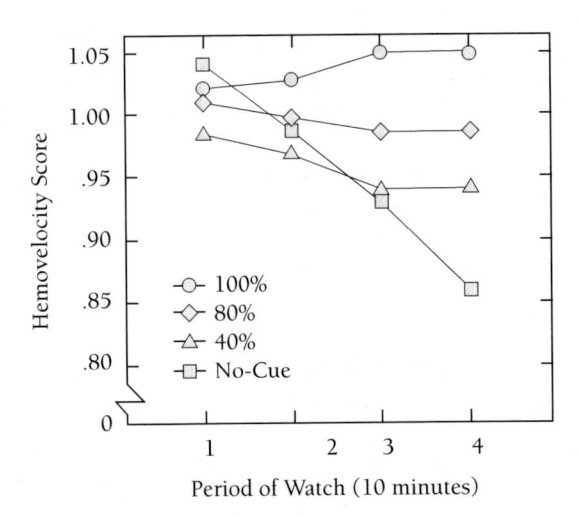

Figure 10.3. Mean hemovelocity scores as a function of periods of watch for four cue-reliability conditions. Data are from the right hemisphere/low-signal salience condition. After Hitchcock et al. (2003).

when searching for the presence of a distinguishing feature in an array of stimuli as opposed to its absence. Indeed, when searching for presence, the distinguishing feature appears to be so salient that it seems to pop out of the display. The phenomenon of search asymmetry has been accounted for by the feature integration model (Treisman & Gormican, 1988), which suggests that searching for the presence of a feature is guided by preattentive, parallel processing, while more deliberate serial processing is required for determining its absence.

Studies by Schoenfeld and Scerbo (1997, 1999) have extended the presence-absence distinction to the accuracy of signal detections in long-duration sustained attention or vigilance tasks. Performance efficiency in vigilance tasks varies inversely with the information processing demand imposed by the task, as indexed by the number of stimulus elements that must be scanned in search of critical signals (Grubb, Warm, Dember, & Berch, 1995; Parasuraman, 1986). The view that detecting the absence of a feature is more capacity demanding than detecting its presence led Schoenfeld and Scerbo (1997, 1999) to predict that increments in the number of array elements to be scanned in separating signals from noise in vigilance would have a more negative effect upon signal detection in the feature absence than in the feature presence case. Consistent with that prediction, they found that when observers were required to detect the absence of a figure, signal detectability declined as the size of the stimulus array to be scanned was increased from two to five elements. Increasing array size, however, had no effect on performance when observers were required to monitor for the presence of that feature. In addition, observers rated the workload of their assignment to be greater when monitoring for feature absence than presence on the NASA Task Load Index (TLX) scale, a standard subjective report measure of the perceived mental workload imposed by a task (Hart & Staveland, 1988; Warm, Dember, & Hancock, 1996; Wickens & Hollands, 2000).

In most vigilance tasks, critical signals for detection are embedded in a background of repetitive nonsignal or neutral events. Several studies have demonstrated that signal detections vary inversely with the background event rate and that this effect is more prominent in tasks requiring high information processing demand than low (Lanzetta, Dember, Warm, & Berch, 1987; Parasuraman, 1979; Warm & Jerison, 1984). Given that detecting the absence of a feature is more capacity demanding than detecting its presence, Hollander and his associates (2004) hypothesized that the degrading effects of increments in background event rate would be more pronounced when observers monitored for the absence than for the presence of a feature. As in the studies by Schoenfeld and Scerbo (1997, 1999), perceived mental workload was also anticipated to be greater when observers monitored for feature absence than for feature presence. With regard to blood flow, Hollander et al. (2004) predicted that blood flow would be higher when observers were required to detect feature absence than presence and would show a greater decline over time in the absence than in the presence condition. The two types of tasks, presence and absence, were combined factorially with three levels of event rate, 6, 12, and 24 events per minute, to provide six experimental conditions. In all conditions, observers participated in a 40-minute vigil divided into four 10-minute periods of watch during which they monitored an array of five circles positioned around the center of a video display terminal at the 3, 5, 7, 9, and 12 o'clock locations. The critical signal for detection in the presence condition was the appearance of a vertical 4 mm line intersecting the 6 o'clock position within one of the circles in the array. In the absence condition, the vertical line was present in all circles but one. Ten critical signals were presented in each watchkeeping period in all experimental conditions.

As anticipated, the event rate effect was indeed more pronounced in the absence than in the presence condition. Detection probability remained stable across event rates in the presence condition but declined with increments in event rate in the absence condition. Another important aspect of the performance data was the finding that signal detections declined significantly over time in both the feature presence and absence conditions.

The finding that the event rate effect was more pronounced when observers had to monitor for stimulus absence than presence is reminiscent of the findings in the earlier reports by Schoenfeld and Scerbo (1997, 1999) that the effect of another information processing factor in vigilance, the size of the element set that must be scanned in search of critical signals, is also more notable in the feature absence than the feature presence condition. Also consistent with Schoenfeld and Scerbo (1997, 1999), perceived mental workload was greater

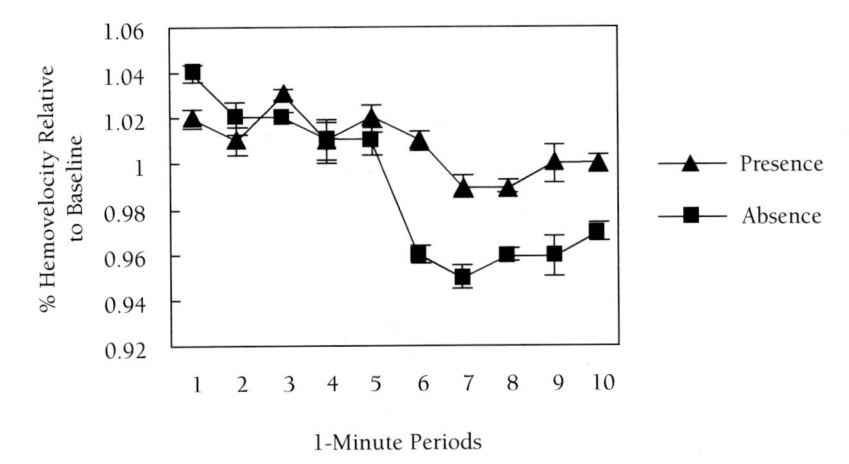

Figure 10.4. Mean hemovelocity scores in feature presence or absence conditions for successive 1-minute intervals during the initial period of watch. Data are for the right hemisphere. Error bars are standard errors. After Hollander et al. (2004).

when participants monitored for feature absence than presence. Thus, the results support the notion of differential capacity demand in detecting feature absence than presence. However, the finding of a vigilance decrement in the feature presence condition suggests that counter to the early claim in feature-integration theory that feature detection is preattentive, some information processing cost must be associated with detecting feature presence. This interpretation is supported by the fact that even though the mental workload of the presence condition was less than that of the absence condition, it still fell in the upper level of the NASA TLX scale. Spatial cueing studies in which the size of a precue is varied prior to the presentation of the search display have also found cue-size effects on target identification time for both feature and conjunction search (Greenwood & Parasuraman, 1999, 2004). Results such as these are consistent with the emerging view in the search literature that the alignment of feature detection with purely preattentive processing may no longer be tenable (Pashler, 1998; Quinlan, 2003).

When measured over the 10-minute intervals of watch, blood flow in the Hollander et al. (2004) study was found to be greater in the presence than in the absence condition. That result was counter to expectations based on the view that detecting feature absence is more capacity demanding than detecting feature presence. It is conceivable, however, that this apparent reversal of the expected effect reflected the fact than the demands of feature absence were great enough to tax information processing resources very early in the vigil and that those resources were not replenished over time. An account along that line would be supported if it could be shown that while differences in blood flow are greater in the feature absence than in the feature presence case at the very outset of the vigil, the reverse effect emerged as the vigil continued. Toward that end, a fine-grained minute-by-minute analysis was performed on the blood flow scores of the presence and absence conditions during the initial watchkeeping period in the left and right hemispheres. No task differences were noted for the left hemisphere. However, as shown in figure 10.4, blood flow velocity in the right hemisphere was greater in the feature absence than in the feature presence condition at the outset of the vigil, and the reverse effect emerged after observers had performed the task for 6 minutes. Statistical tests indicated that there were no significant differences between the two tasks in the first 5 minutes of watch but that there were statistically significant differences between the conditions in the 6th through the 10th minutes of watch. Similar fine-grained examination of the data for the remaining periods of watch revealed that the reduced level of blood flow in the absence condition that emerged halfway through the initial watchkeeping period remained consistent throughout each minute of all of the remaining periods of watch.

Evidently, Hollander et al.'s (2004) initial expectation of greater blood flow in the absence condition

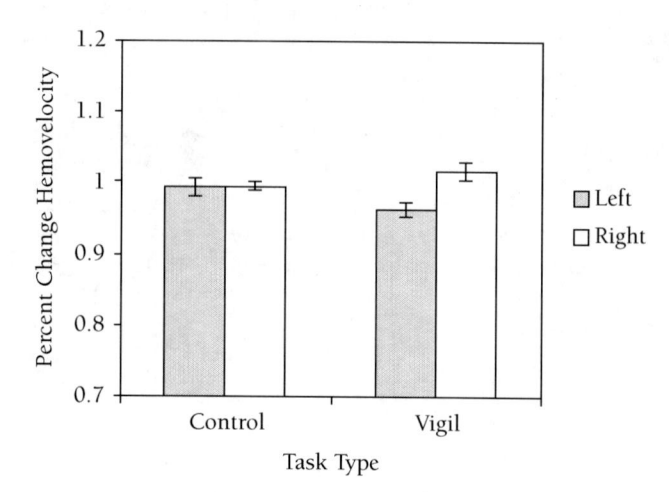

Figure 10.5. Percentage change in hemovelocity relative to resting baseline in the left and right cerebral hemispheres for the control and active vigilance conditions. After Helton et al. (in press).

underestimated the degree to which that condition taxed information processing resources in the vigilance task. Rather than being reflected in an overall elevation in blood flow, the greater information processing demand exerted by the absence condition was evident in an early-appearing drain on resources. Although initially unanticipated, this finding from the fine-grained analysis of the data is consistent with the expectation of a steeper decline in blood flow associated with the absence condition. As in the Hitchcock et al. (2003) and Mayleben (1998) studies, hemovelocity in the Hollander et al. (2004) study remained stable over the course of the watch in both hemispheres among control observers who viewed the displays without a work imperative, indicating once again that the blood flow effects were indeed task dependent.

The Abbreviated Vigil

Thus far we have discussed TCD findings in studies that made use of traditional vigilance tasks lasting 30 minutes or more. Because of their long duration, investigators have found it inconvenient to incorporate such tasks in test batteries or, as discussed previously, to link vigilance performance with brain imaging metrics such as PET and fMRI. Accordingly, it is of interest to examine whether the TCD-vigilance findings can be replicated in shorter-duration vigilance tasks (Nuechterlein, Parasuraman, & Jiang, 1983; Posner, 1978; Temple et al., 2000). Toward that end, Helton et al. (in press) used a 12-minute vigilance task developed by Temple et al. (2000) in which participants were asked to inspect the repetitive presentation on a VDT of light gray capital letters consisting of an O, a D, and a backward D. The letters were presented for only 40 ms at a rate of 57.5 events per minute and exposed against a visual mask consisting of unfilled circles on a white background. Critical signals for detection were the appearances of the letter O (signal probability = 0.20/period of watch). In addition to TCD measurement of blood flow, Helton et al. (in press) also employed the NIRS procedure to measure transcranial cerebral oximetry via a Somanetics INVOS 4100 Cerebral Oximeter.

As can be seen in figures 10.5 and 10.6, both blood flow velocity (figure 10.5) and oxygenation (figure 10.6) were found to be significantly higher in the right than in the left cerebral hemisphere among observers who performed the vigilance task while there were no hemispheric differences in the blood flow and oxygenation measures among control observers who viewed the vigilance display without a work imperative. In this study, the oxygenation measure was based on a percentage of a 3-minute resting baseline.

Clearly, the results of this study indicated that performance in the abbreviated vigil was right lateralized, a finding that coincides with the outcome of earlier blood flow studies featuring more traditional long-duration vigils and with PET and fMRI investigations (Parasuraman et al., 1998). This parallel has several important implications. It provides strong support for the argument that the abbreviated vigil is a valid analog of long-duration vigilance tasks. The fact that the NIRS procedure yielded laterality effects similar to those of the TDC

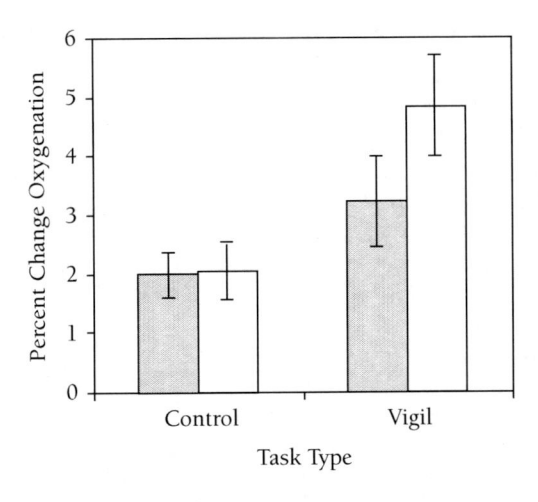

Figure 10.6. Percentage change frontal lobe oxygenation relative to resting baseline in the left and right cerebral hemispheres for the control and active vigilance conditions. After Helton et al. (in press).

procedure further implies that laterality in vigilance is a generalized effect that appears in terms of both hemovelocity and blood oxygenation. It also implies that the NIRS procedure may be a useful supplement to the TCD approach in providing a noninvasive imaging measure of brain activity in the performance of a vigilance task.

It is important to note that while Helton et al.'s (in press) results regarding laterality of function with the abbreviated vigil were consistent with those found with its long-duration analogs, their findings regarding the decrement function were not. Both the TCD and the NIRS indices remained stable over the course of the watch while performance efficiency declined over time. It is possible that cerebral hemodynamics are structured so that overall hemispheric dominance emerges early in the time course of task performance but that temporally based declines in cerebral blood flow and blood oxygen levels require a considerable amount of time to become observable. Thus, the abbreviated 12-minute vigil, which is only about 30% as long as the vigils employed in earlier vigilance studies, may not be long enough to permit time-based declines in blood flow or blood oxygen levels.

Conclusion

One of the goals of neuroergonomics is to enhance understanding of aspects of human performance in complex systems with respect to the underlying brain mechanisms and to provide measurement tools to study these mechanisms (Parasuraman, 2003). From this perspective, the use of TCD-based measures of cerebral blood flow to assess human mental workload and vigilance can be considered a success. The vigilance studies have revealed a close coupling between vigilance performance and blood flow and they provide empirical support for the notion that blood flow may represent a metabolic index of information processing resource utilization during sustained attention. The demonstration of systematic modulation of blood flow in the right cerebral hemisphere with time on task, memory load, signal salience and cueing, the detection of feature absence or presence, and target detection in Temple et al.'s (2000) abbreviated vigil provides evidence for a right hemispheric brain system that is involved in the functional control of vigilance performance over time.

Another goal of neuroergonomics research is to use knowledge of brain function to enhance human-system performance. In additional to the theoretical and empirical contributions of TCD research, there are also some potentially important ergonomic ramifications. TCD may offer a noninvasive and inexpensive tool to "monitor the monitor" and to help decide when operator vigilance has reached a point where task aiding is necessary or operators need to be rested or removed. NIRS-based measurement of blood oxygenation may provide similar information.

Main Points

1. Transcranial Doppler sonography and near-infrared spectroscopy can be used to measure cerebral blood flow velocity and cerebral

oxygenation, respectively, during the performance of a vigilance task.
2. The temporal decline in signal detections that characterizes vigilance performance is accompanied by a similar decline in brain blood flow over time.
3. Brain blood flow is greater in the performance of a memory-demanding successive-type vigilance task than for a memory-free simultaneous-type task.
4. Changes in signal detection in vigilance brought about by variations in cue reliability are paralleled by changes in brain blood flow.
5. The temporal decline in blood flow accompanying vigilance performance is greater when critical signals for detection are defined by the absence than by the presence of a target element.
6. Both the TCD and NIRS measures point to a right hemispheric system in the control of vigilance performance.

Key Readings

Aaslid, R. (1986). Transcranial Doppler examination techniques. In R. Aaslid (Ed.), *Transcranial Doppler sonography* (pp. 39–59). New York: Springer-Verlag.

Duschek, S., & Schandry, R. (2003). Functional transcranial Doppler sonography as a tool in psychophysiological research. *Psychophysiology, 40,* 436–454.

Hitchcock, E. M., Warm, J. S., Matthews, G., Dember, W. N., Shear, P. K., Tripp, L. D., et al. (2003). Automation cueing modulates cerebral blood flow and vigilance in a simulated air traffic control task. *Theoretical Issues in Ergonomics Science, 4,* 89–112.

Parasuraman, R., Warm, J. S., & See, J. W. (1998). Brain systems of vigilance. In R. Parasuraman (Ed.), *The attentive brain* (pp. 221–256). Cambridge, MA: MIT Press.

References

Aaslid, R. (1986). Transcranial Doppler examination techniques. In R. Aaslid (Ed.), *Transcranial Doppler sonography* (pp. 39–59). New York: Springer-Verlag.

Annett, J. (1996). Training for perceptual skills. *Ergonomics, 9,* 459–468.

Becker, A. B., Warm, J. S., Dember, W. N., & Hancock, P. A. (1995). Effects of jet engine noise and performance feedback on perceived workload in a monitoring task. *International Journal of Aviation Psychology, 5,* 49–62.

Caggiano, D. M., & Parasuraman, R. (2004). The role of memory representation in the vigilance decrement. *Psychonomic Bulletin and Review, 11,* 932–937.

Caplan, L. R., Brass, L. M., DeWitt, L. D., Adams, R. J., Gomex, C., Otis, S., et al. (1990). Transcranial Doppler ultrasound: Present status. *Neurology, 40,* 496–700.

Coull, J. T., Frackowiak, R. J., & Frith, C. D. (1998). Monitoring for target objects: Activation of right frontal and parietal cortices with increasing time on task. *Neuropsychologia, 36,* 1325–1334.

Davies, D. R., & Parasuraman, R. (1982). *The psychology of vigilance.* London: Academic Press.

Duschek, S., & Schandry, R. (2003). Functional transcranial Doppler sonography as a tool in psychophysiological research. *Psychophysiology, 40,* 436–454.

Galinsky, T. L., Rosa, R. R., Warm, J. S., & Dember, W. N. (1993). Psychophysical determinants of stress in sustained attention. *Human Factors, 35,* 603–614.

Gazzaniga, M. S., Ivry, R., & Mangun, G. R. (2002). *Cognitive neuroscience: The biology of the mind* (2nd ed.). New York: Norton.

Greenwood, P. M., & Parasuraman, R. (1999). Scale of attentional focus in visual search. *Perception & Psychophysics, 61,* 837–859.

Greenwood, P. M., & Parasuraman, R. (2004). The scaling of spatial attention in visual search and its modification in healthy aging. *Perception & Psychophysics, 66,* 3–22.

Grubb, P. L., Warm, J. S., Dember, W. N., & Berch, D. B. (1995). Effects of multiple signal discrimination on vigilance performance and perceived workload. *Proceedings of the Human Factors and Ergonomics Society, 39th annual meeting,* 1360–1364.

Hancock, P. A., & Hart, G. (2002). Defeating terrorism: What can human factors/ergonomics offer? *Ergonomics and Design, 10,* 6–16.

Hart, S. G., & Staveland, L. E. (1988). Development of the NASA-TLX (Task Load Index): Results of empirical and theoretical research. In P. A. Hancock & N. Meshkati (Eds.), *Human mental workload* (pp. 139–183). Amsterdam: North Holland.

Helton, W. S., Hollander, T. D., Warm, J. S., Tripp, L. D., Parsons, K., Matthews, G., et al. (in press). The abbreviated vigilance task and cerebral hemodynamics. *Journal of Clinical and Experimental Neuropsychology.*

Hitchcock, E. M., Warm, J. S., Matthews, G., Dember, W. N., Shear, P. K., Tripp, L. D., et al. (2003). Automation cueing modulates cerebral blood flow and vigilance in a simulated air traffic control task. *Theoretical Issues in Ergonomics Science, 4,* 89–112.

Hollander, T. D., Warm, J. S., Matthews, G., Shockley, K., Dember, W. N., Weiler, E. M., et al. (2004). Feature presence/absence modifies the event rate effect and cerebral hemovelocity in vigilance. *Proceedings of the Human Factors and Ergonomics Society, 48th annual meeting,* 1943–1947.

Kahneman, D. (1973). *Attention and effort.* Englewood Cliffs, NJ: Prentice Hall.

Klingelhofer, J., Sander, D., & Wittich, I. (1999). Functional ultrasonographicic imaging. In V. L. Babikian & L. R. Wechsler (Eds.), *Transcranial Doppler ultrasonography* (2nd ed., pp. 49–66). Boston: Butterworth Heinemann.

Korol, D. L., & Gold, P. E. (1998). Glucose, memory, and aging. *American Journal of Clinical Nutrition, 67,* 764–771.

Kramer, A. F., & Weber, T. (2000). Applications of psychophysiology to human factors. In J. T. Cacioppo, L. G. Tassinary, & G. G. Berntson (Eds.), *Handbook of psychophysiology* (2nd ed., pp. 794–814). New York: Cambridge University Press.

Lanzetta, T. M., Dember, W. N., Warm, J. S., & Berch, D. B. (1987). Effects of task type and stimulus homogeneity on the event rate function in sustained attention. *Human Factors, 29,* 625–633.

Mayleben, D. W. (1998). *Cerebral blood flow velocity during sustained attention.* Unpublished doctoral dissertation, University of Cincinnati, OH.

Moray, N. (1967). Where is capacity limited? A survey and a model. *Acta Psychologica, 27,* 84–92.

Moray, N. (1979). *Mental workload.* New York: Plenum.

Navon, D., & Gopher, D. (1979). On the economy of human processing systems. *Psychological Review, 86,* 214–255.

Norman, D. A., & Bobrow, D. G. (1975). On data-limited and resource-limited processes. *Cognitive Psychology, 7,* 44–64.

Nuechterlein, K., Parasuraman, R., & Jiang, Q. (1983). Visual sustained attention: Image degradation produces rapid sensitivity decrement over time. *Science, 220,* 327–329.

Parasuraman, R. (1979). Memory load and event rate control sensitivity decrements in sustained attention. *Science, 205,* 924–927.

Parasuraman, R. (1984). The psychobiology of sustained attention. In J. S. Warm (Ed.), *Sustained attention in human performance* (pp. 61–101). London: Wiley.

Parasuraman, R. (1986). Vigilance, monitoring, and search. In K. Boff, L. Kaufman, & J. Thomas (Eds.), *Handbook of perception: Vol. 2. Cognitive processes and performance* (pp. 43.1–43.39). New York: Wiley.

Parasuraman, R. (2003). Neuroergonomics: Research and practice. *Theoretical Issues in Ergonomics Science, 4,* 5–20.

Parasuraman, R., & Caggiano, D. (2002). Mental workload. In V. S. Ramachandran (Ed.), *Encyclopedia of the human brain* (Vol. 3, pp. 17–27). San Diego: Academic Press.

Parasuraman, R., Warm, J. S., & See, J. W. (1998). Brain systems of vigilance. In R. Parasuraman (Ed.), *The attentive brain* (pp. 221–256). Cambridge, MA: MIT Press.

Pashler, H. (1998). *The psychology of attention.* Cambridge, MA: MIT Press.

Paus, T., Zatorre, R. J., Hofle, N., Caramanos, Z., Gotman, J., Petrides, M., et al. (1997). Time-related changes in neural systems underlying attention and arousal during the performance of an auditory vigilance task. *Journal of Cognitive Neuroscience, 9,* 392–408.

Posner, M. I. (1978). *Chronometric explorations of mind.* Hillsdale, NJ: Erlbaum.

Posner, M. I. (2004). *Cognitive neuroscience of attention.* New York: Guilford.

Posner, M. I., & Tudela, P. (1997). Imaging resources. *Biological Psychology, 45,* 95–107.

Punwani, S., Ordidge, R. J., Cooper, C. E., Amess, P., & Clemence, M. (1998). MRI measurements of cerebral deoxyhaemoglobin concentration (dhB)—correlation with near infrared spectroscopy (NIRS). *NMR in Biomedicine, 11,* 281–289.

Quinlan, P. T. (2003). Visual feature integration theory: Past, present, and future. *Psychological Bulletin, 129,* 643–673.

Raichle, M. E. (1998). Behind the scenes of functional brain imaging: A historical and physiological perspective. *Proceedings of the National Academy of Sciences USA, 95,* 765–772.

Risberg, J. (1986). Regional cerebral blood flow in neuropsychology. *Neuropsychologica, 34,* 135–140.

Roy, C. S., & Sherrington, C. S. (1890). On the regulation of the blood supply of the brain. *Journal of Physiology (London), 11,* 85–108.

Schnittger, C., Johannes, S., Arnavaz, A., & Munte, T. F. (1997). Relation of cerebral blood flow velocity and level of vigilance in humans. *NeuroReport, 8,* 1637–1639.

Schoenfeld, V. S., & Scerbo, M. W. (1997). Search differences for the presence and absence of features in sustained attention. *Proceedings of the Human*

Factors and Ergonomics Society, 41st annual meeting, 1288–1292.

Schoenfeld, V. S., & Scerbo, M. W. (1999). The effects of search differences for the presence and absence of features on vigilance performance and mental workload. In M. W. Scerbo & M. Mouloua (Eds.), *Automation technology and human performance: Current research and trends* (pp. 177–182). Mahwah, NJ: Erlbaum.

Stroobant, N., & Vingerhoets, G. (2000). Transcranial Doppler ultrasonography monitoring of cerebral hemodynamics during performance of cognitive tasks: A review. *Neuropsychology Review, 10,* 213–231.

Temple, J. G., Warm, J. S., Dember, W. N., Jones, K. S., LaGrange, C. M., & Matthews, G. (2000). The effects of signal salience and caffeine on performance, workload, and stress in an abbreviated vigilance task. *Human Factors, 42,* 183–194.

Toole, J. F. (1984). *Cerebralvascular disorders* (3rd ed.). New York: Raven.

Toronov, V., Webb, A., Choi, J. H., Wolf, M., Michalos, A., Gratton, E., et al. (2001). Investigation of human brain hemodynamics by simultaneous near-infrared spectroscopy and functional magnetic resonance imaging. *Medical Physics, 28,* 521–527.

Treisman, A. M., & Gormican, S. (1988). Feature analysis in early vision: Evidence from search asymmetries. *Psychological Bulletin, 95,* 15–48.

Tulving, E., Kapur, S., Craik, F. I., Moscovitch, M., & Houle, S. (1994). Hemispheric encoding/retrieval asymmetry in episodic memory: Positron emission topography findings. *Proceedings of the National Academy of Sciences USA, 91,* 2016–2020.

Warm, J. S., & Dember, W. N. (1998). Tests of a vigilance taxonomy. In R. R. Hoffman, M. F. Sherick, & J. S. Warm (Eds.), *Viewing psychology as a whole: The integrative science of William N. Dember* (pp. 87–112). Washington, DC: American Psychological Association.

Warm, J. S., Dember, W. N., & Hancock, P. A. (1996). Vigilance and workload in automated systems. In R. Parasuraman & M. Mouloua (Eds.), *Automation and human performance: Theory and applications* (pp. 183–200). Mahwah, NJ: Erlbaum.

Warm, J. S., & Jerison, H. J. (1984). The psychophysics of vigilance. In J. S. Warm (Ed.), *Sustained attention in human performance* (pp. 15–59). Chichester, UK: Wiley.

Wickens, C. D. (1984). Processing resources in attention. In R. Parasuraman & D. R. Davies (Eds.), *Varieties of attention* (pp. 63–102). New York: Academic Press.

Wickens, C. D. (1990). Applications of event-related potential research to problems in human factors. In J. W. Rohrbaugh, R. Parasuraman, & R. Johnson (Eds.), *Event-related brain potentials: Basic and applied issues* (pp. 301–309). New York: Oxford University Press.

Wickens, C. D. (2002). Multiple resources and performance prediction. *Theoretical Issues in Ergonomics Science, 3,* 159–177.

Wickens, C. D., & Hollands, J. G. (2000). *Engineering psychology and human performance* (3rd ed.). Upper Saddle River, NJ: Prentice-Hall.

Wiener, E. L., & Attwood, D. A. (1968). Training for vigilance: Combined cueing and knowledge of results. *Journal of Applied Psychology, 52,* 474–479.

Jordan Grafman

Executive Functions

There is no region of the human cerebral cortex whose functional assignments are as puzzling to us as the human prefrontal cortex (HPFC). Over 100 years of observation and experimentation has led to several general conclusions about its overall functions. The prefrontal cortex is important for modulating higher cognitive processes such as social behavior, reasoning, planning, working memory, thought, concept formation, inhibition, attention, and abstraction. Each of these processes is very important for many aspects of human ergonomic study. Yet, unlike the research conducted in other cognitive domains such as object recognition or word storage, there has been little effort to propose, and investigate in detail, the underlying cognitive architecture that would capture the essential features and computational properties of the higher cognitive processes presumably modulated by the HPFC. Since the processes that are attributed to the HPFC appear to constitute the most complex and abstract of human cognitive functions, many of which are responsible for the internal guidance of behavior, a critical step in understanding the functions of the human brain requires an adequate description of the cognitive topography of the HPFC.

In this chapter, I argue for the validity of a representational research framework to understand HPFC functioning in humans. My colleagues and I have labeled the set of HPFC representational units as a structured event complex (SEC). I briefly summarize the key elements of the biology and structure of the HPFC, the evidence of its importance in higher-level cognition based on convergent evidence from lesion and neuroimaging studies, and some key models postulating the functions of the HPFC, and finally offer some suggestions about how HPFC functions are relevant for ergonomic study.

Anatomical Organization of the Human Prefrontal Cortex

What we know about the anatomy and physiology of the HPFC is inferred almost entirely from work in the primate and lower species. It is likely that the connectivity already described in other species also exists in the HPFC (Petrides & Pandya, 1994). The HPFC is composed of Brodmann's areas 8–14 and 24–47. Grossly, it can be subdivided into lateral, medial, and orbital regions with Brodmann's areas providing morphological subdivisions within (and occasionally across) each of the gross regions (Barbas, 2000). Some regions of the prefrontal cortex have a total of six layers; other regions are agranular,

159

meaning that the granule cell layer is absent. The HPFC has a columnar design like other cortical regions. All regions of the HPFC are interconnected. The HPFC is also richly interconnected with other areas of brain and has at least five distinct regions that are independently involved in separate corticostriatal loops (Alexander, Crutcher, & DeLong, 1990). The functional role of each relatively segregated circuit has been described (Masterman & Cummings, 1997). The HPFC also has strong limbic system connections via its medial and orbital efferent connections that terminate in the amygdala, thalamus, and parahippocampal regions (Groenewegen & Uylings, 2000; Price, 1999). Finally, the HPFC has long pathway connections to association cortex in the temporal, parietal, and occipital lobes. Almost all of these pathways are reciprocal.

When compared with the prefrontal cortex of other species, most investigators have claimed that the HPFC is proportionally (compared to the remainder of the cerebral cortex) much larger (Rilling & Insel, 1999; Semendeferi, Armstrong, Schleicher, Zilles, & Van Hoeseall, 2001). Other recent research indicates that the size of the HPFC is not proportionally larger than that of other primates, but that its internal neural architecture must be more sophisticated, or at least differentially organized in order to support superior human functions (Chiavaras, LeGoualher, Evans, & Petrides, 2001; Petrides & Pandya, 1999). The functional argument is that in order to subserve such higher-order cognitive functions as extended reactive planning and complex reasoning that are not obviously apparent in other primates or lower species, the HPFC must have a uniquely evolved neural architecture (Elston, 2000).

The HPFC is not considered fully developed until the mid to late 20s. This is later than almost all other cortical association areas. The fact that the HPFC does not fully mature until young adulthood suggests that those higher cognitive processes mediated by the prefrontal cortex are still developing until that time (Diamond, 2000).

The HPFC is innervated by a number of different neurotransmitter and peptide systems—most prominent among them being the dopaminergic, serotonergic, and cholinergic transmitters and their varied receptor subtypes (Robbins, 2000). The functional role of each of these neurotransmitters in the HPFC is not entirely clear. Mood disorders that involve alterations in serotonergic functions lead to reduced blood flow in HPFC. Several degenerative neurological disorders are at least partially due to disruption in the production and transfer of dopamine from basal ganglia structures to the HPFC. This loss of dopamine may cause deficits in cognitive flexibility. Serotonergic receptors are distributed throughout the HPFC and have a role in motivation and intention. Finally, the basal forebrain in ventral and posterior HPFC is part of the cholinergic system, whose loss can cause impaired memory and attention. These modulating chemical anatomical systems may be important for adjusting the "gain" within and across representational networks in order to facilitate or inhibit activated cognitive processes.

A unique and key property of neurons in the prefrontal cortex of monkeys (and presumably humans) is their ability to fire during an interval between a stimulus and a delayed probe (Levy & Goldman-Rakic, 2000). Neurons in other brain areas are either directly linked to the presentation of a single stimulus or the probe itself, and if they demonstrate continuous firing, it is probable that they are driven by neurons in the prefrontal cortex or by continuous environmental input. If the firing of neurons in the prefrontal cortex is linked to activity that moves the subject toward a goal rather than reacting to the appearance of a single stimulus, then potentially those neurons could continuously fire across many stimuli or events until the goal was achieved or the behavior of the subject disrupted. This observation of sustained firing of prefrontal cortex neurons across time and events has led many investigators to suggest that the HPFC must be involved in the maintenance of a stimulus across time, that is, working memory (Fuster, Bodner, & Kroger, 2000).

Besides the property of sustained firing, Elston (2000) has demonstrated a unique structural feature of neurons in the prefrontal cortex. Elston (2000) found that pyramidal cells in the prefrontal cortex of macaque monkeys are significantly more spinous than pyramidal cells in other cortical areas, suggesting that they are capable of handling a larger amount of excitatory input than pyramidal cells elsewhere. This could be one of several structural explanations for the HPFC's ability to integrate input from many sources in order to implement more abstract behaviors.

Thus, the HPFC is a proportionally large cortical region that is extensively and reciprocally interconnected with other associative, limbic, and basal ganglia brain structures. It matures somewhat later than other cortex, is richly innervated with modulatory chemical systems, and may have some unique structural features not found in other cortical networks. Finally, neurons in the prefrontal cortex appear to be particularly able to fire over extended periods of time until a goal is achieved. These features of the HPFC map nicely onto some of the cognitive attributes of the HPFC identified in neuropsychological and neuroimaging studies.

Functional Studies of Human Prefrontal Cortex

The traditional approach to understanding the functions of the HPFC is to perform cognitive studies testing the ability of normal and impaired humans on tasks designed to induce the activation of processes or representational knowledge presumably stored in the HPFC (Grafman, 1999). Both animals and humans with brain lesions can be studied to determine the effects of a prefrontal cortex lesion on task performance. Lesions in humans, of course, are due to an act of nature, whereas lesions in animals are precisely and purposefully made. Likewise, intact animals can be studied using precise electrophysiological recordings of single neurons or neural assemblies. In humans, powerful new neuroimaging techniques such as functional magnetic resonance imaging (fMRI) have been used to demonstrate frontal lobe activation during the performance of a range of tasks in normal subjects and patients (see also chapter 4). A potential advantage in studying humans (instead of animals) comes from the presumption that since the HPFC represents the kind of higher-order cognitive processes that distinguish humans from other primates, an understanding of its underlying cognitive and neural architecture can only come from the study of humans.

Patients with frontal lobe lesions are generally able to understand conversation and commands, recognize and use objects, express themselves adequately to navigate through some social situations in the world, learn and remember routes, and even make decisions. On the other hand, they have documented deficits in sustaining their attention and anticipating what will happen next, in dividing their resources, inhibiting prepotent behavior, adjusting to some situations requiring social cognition, processing the theme or moral of a story, forming concepts, abstracting, reasoning, and planning (Arnett et al., 1994; Carlin et al., 2000; Dimitrov, Grafman, Soares, & Clark, 1999; Dimitrov, Granetz, et al., 1999; Goel & Grafman, 1995; Goel et al., 1997; Grafman, 1999; Jurado, Junque, Vendrell, Treserras, & Grafman, 1998; Vendrell et al., 1995; Zahn, Grafman, & Tranel, 1999). These deficits have been observed and confirmed by investigators over the last 50 years of clinical and experimental research.

Neuroimaging investigators have published studies that show prefrontal cortex activation during encoding, retrieval, decision making and response conflict, task switching, reasoning, planning, forming concepts, understanding the moral or theme of a story, inferring the motives or intentions of others, and similar high-level cognitive processing (Goel, Grafman, Sadato, & Hallett, 1995; Koechlin, Basso, Pietrini, Panzer, & Grafman, 1999; Koechlin, Corrado, Pietrini, & Grafman, 2000; Nichelli et al., 1994; Nichelli, Grafman, et al., 1995; Wharton et al., 2000). The major advantage, so far, of these functional neuroimaging studies is that they have generally provided convergent evidence for the involvement of the HPFC in controlling endogenous and exogenous-sensitive cognitive processes, especially those that are engaged by the abstract characteristics of a task.

Neuropsychological Frameworks to Account for HPFC Functions

Working Memory

Working memory has been described as the cognitive process that allows for the temporary activation of information in memory for rapid retrieval or manipulation (Ruchkin et al., 1997). It was first proposed some 30 years ago to account for a variety of human memory data that were not addressed by contemporary models of short-term memory (Baddeley, 1998b). Of note is that subsequent researchers have been unusually successful in describing the circumstances under which the so-called slave systems employed by working memory would be used. These slave systems allowed for the maintenance of the stimuli in a number of different forms that could be manipulated by the central executive component of the working memory system (Baddeley, 1998a). Neuroscience support for their model followed quickly. Joaquin Fuster was among the first neuroscientists to recognize that neurons in the prefrontal cortex appeared to have a special capacity to discharge over time intervals when the stimulus was not being shown prior to a memory-driven response by the animal (Fuster et al., 2000). He interpreted this neuronal activity as being concerned with the cross-temporal linkage of information processed at different points in an ongoing temporal sequence. Goldman-Rakic and her colleagues later elaborated on this notion and suggested that these same PFC neurons were fulfilling the neuronal responsibility for working memory (Levy & Goldman-Rakic, 2000). In her view, PFC neurons temporarily hold in active memory modality-specific information until a response is made. This implies a restriction on the kind of memory that may be stored in prefrontal cortex. That is, this point of view suggests that there are no long-term representations in the prefrontal cortex until an explicit intention to act is required, and then a temporary representation is created. Miller has challenged some of Goldman-Rakic's views about the role of neurons in the prefrontal cortex and argued that many neurons in the monkey prefrontal cortex are modality nonspecific and may serve a broader integrative function rather than a simple maintenance function (Miller, 2000). Fuster, Goldman-Rakic, and Baddeley's programs of research have had a major influence on the functional neuroimaging research programs of Courtney (Courtney, Petit, Haxby, & Ungerleider, 1998), Smith and Jonides (1999), and Cohen (Nystrom et al., 2000)—all of whom have studied normal subjects in order to remap the HPFC in the context of working memory theory.

Executive Function and Attentional/Control Processes

Although rather poorly described in the cognitive science literature, it is premature to simply dismiss the general notion of a central executive (Baddeley, 1998a; Grafman & Litvan, 1999b). Several investigators have described the prefrontal cortex as the seat of attentional and inhibitory processes that govern the focus of our behaviors and therefore, why not ascribe the notion of a central executive operating within the confines of the HPFC?

Norman and Shallice (1986) proposed a dichotomous function of the central executive in HPFC. They argued that the HPFC was primarily specialized for the supervision of attention toward unexpected occurrences. Besides this supervisory attention system, they also hypothesized the existence of a contention scheduling system that was specialized for the initiation and efficient running of automatized behaviors such as repetitive routines, procedures, and skills. Shallice, Burgess, Stuss, and others have attempted to expand this idea of the prefrontal cortex as a voluntary control device and have further fractionated the supervisory attention system into a set of parallel attention processes that work together to manage complex multitask behaviors (Burgess, 2000; Burgess, Veitch, de Lacy Costello, & Shallice, 2000; Shallice & Burgess, 1996; Stuss et al., 1999).

Social Cognition and Somatic Marking

The role of the HPFC in working memory and executive processes has been extensively examined, but there is also substantial evidence that the prefrontal cortex is involved in controlling certain aspects of social and emotional behavior (Dimitrov, Grafman, & Hollnagel, 1996; Dimitrov, Phipps, Zahn, & Grafman, 1999). Although the classic story of the 19th-century patient Phineas Gage, who suffered a penetrating prefrontal cortex lesion, has been used to exemplify the problems that patients with ventromedial prefrontal cortex lesions have in obeying social rules, recognizing social cues, and making appropriate social decisions, the details of this social cognitive impairment have occasionally been inferred or even embellished to suit the enthusiasm of the storyteller—at least regarding Gage (Macmillan, 2000). On the other hand, Damasio and his colleagues have consistently confirmed the association of ventromedial prefrontal cortex lesions and social behavior and decision-making abnormalities (Anderson, Bechara, Damasio, Tranel, & Damasio, 1999; Bechara, Damasio, & Damasio, 2000; Bechara, Damasio, Damasio, & Lee, 1999; Damasio, 1996; Eslinger, 1998; Kawasaki et al., 2001). The exact functional assignment of that area of HPFC is still subject to dispute, but convincing evidence has been presented that indicates it serves to associate somatic markers (autonomic nervous system modulators that bias activation and decision making) with social knowledge, enabling rapid social decision making—particularly for overlearned associative knowledge. The somatic markers themselves are distributed across a large system of brain regions, including limbic system structures such as the amygdala (Damasio, 1996).

Action Models

The HPFC is sometimes thought of as a cognitive extension of the functional specialization of the motor areas of the frontal lobes (Gomez Beldarrain, Grafman, Pascual-Leone, & Garcia-Monco, 1999) leading to the idea that it must play an essential cognitive role in determining action sequences in the real world. In keeping with that view, a number of investigators have focused their investigations on concrete action series that have proved difficult for patients with HPFC lesions to adequately perform. By analyzing the pattern of errors committed by these patients, it is possible to construct cognitive models of action execution and the role of the HPFC in such performance. In some patients, while the total number of errors they commit is greater than that seen in controls, the pattern of errors committed by patients is similar to that seen in controls (Schwartz et al., 1999). Reduced arousal or effort can also contribute to a breakdown in action production in patients (Schwartz et al., 1999). However, other studies indicate that action production impairment can be due to a breakdown in access to a semantic network that represents aspects of action schema and prepotent responses (Forde & Humphreys, 2000). Action production must rely upon an association between the target object or abstract goal and specific motoric actions (Humphreys & Riddoch, 2000). In addition, the magnitude of inhibition of inappropriate actions appears related to the strength in associative memory of object-goal associations (Humphreys & Riddoch, 2000). Retrieving or recognizing appropriate actions may even help subjects subsequently detect a target (Humphreys & Riddoch, 2001). It should be noted that action disorganization syndromes in patients are usually elicited with tasks that have been traditionally part of the examination of ideomotor or ideational praxis, such as brushing your teeth, and it is not clear whether findings in patients performing such tasks apply to a breakdown in action organization at a higher level such as planning a vacation.

Computational Frameworks

A number of computational models of potential HPFC processes as well as of the general architecture of the HPFC have been developed in recent years. Some models have offered a single explanation for performance on a wide range of tasks. For example, Kimberg and Farah (1993) showed that the weakening of associations within a working memory component of their model led to impaired simulated performance on a range of tasks such as the Wisconsin Card Sorting Test and the Stroop Test that patients with HPFC lesions are known to perform poorly on. In contrast, other investigators have argued for a hierarchical approach to modeling HPFC functions that incorporates a number of layers, with the lowest levels regulated by the environment and the highest levels regulated by internalized rules and plans (Changeux & Dehaene, 1998). In addition to the cognitive levels of their model, Changeux and Dehaene, relying on simulations, suggested that control for transient "prerepresentations" that are modulated by reward and punishment signals improved their model's ability to predict patient performance data on the Tower of London test. Norman and Shallice (1986) first ascribed two major control systems to the HPFC. As noted earlier in this chapter, one system was concerned with rigid, procedurally based, and overlearned behaviors, whereas the other system was concerned with supervisory control over novel situations. Both systems could be simultaneously active, although one system's activation usually dominated performance. The Norman and Shallice model has been incorporated into a hybrid computational model that blends their control system idea with a detailed description of selected action sequences and their errors (Cooper & Shallice, 2000). The Cooper and Shallice model can account for sequences of response, unlike some recurrent network models, and like the Changeux and Dehaene model is hierarchical in nature and based on interactive activation principles. It also was uncanny in predicting the kinds of errors of action disorganization described by Schwartz and Humphreys in their patients. Other authors have implemented interactive control models that use production rules with scheduling strategies for activation and execution to simulate executive control (Meyer & Kieras, 1997). Tackling the issue of how the HPFC mediates schema processing, Botvinick and Plaut (2000) have argued that schemas are emergent system properties rather than explicit representations. They developed a multilayered recurrent connectionist network model to simulate action sequences that is somewhat similar to the Cooper and Shallice model described above. In their simulation, action errors occurred when noise in the system caused an internal representation for one scenario to resemble a pattern usually associated with another scenario. Their model also indicated that noise introduced in the middle of a sequence of actions was more disabling than noise presented closer to the end of the task.

The biological plausibility of all these models has not been formally compared yet but it is just as important to determine whether these models can simulate the behaviors and deficits of interest. The fact that models such as the ones described above are now being implemented is a major advance in the study of the functions of the HPFC.

Commonalities and Weaknesses of the Frameworks Used to Describe HPFC Functions

The cognitive and computational models briefly described above have commonalities that point to the general role of the prefrontal cortex in maintaining information across time intervals and intervening tasks, in modulating social behavior, in the integration of information across time, and in the control of behavior via temporary memory representations and thought rather than allowing behavior to depend upon environmental contingencies alone. None of the major models have articulated in detail the domains and features of a representational knowledge base that would support such HPFC functions, making these models difficult to reject using error or response time analysis of patient data or functional neuroimaging.

Say I was to describe cognitive processing in the cerebral cortex in the following way. The role of the cortex is to rapidly process information and encode its features, and to bind these features together. This role is rather dependent on bottom-up environmental input but represents the elements of this processed information in memory. Perhaps this is not too controversial a way to describe the role of the occipital, parietal, or temporal cortex in processing objects or words. For the cognitive neuropsychologist, however, it would be critical to define the features of the word or object, the characteristics of the memory representation that lead to easier encoding or retrieval of the object or word, and the psychological structure of the representational neighborhood (how different words or objects are related to each other in psychological and potentially neural space). Although there are important philosophical, psychological, and biological arguments about the best way to describe a stored unit of memory (be it an orthographic representation of a word, a visual scene, or a conceptualization), there is general agreement that memories are representations. There is less agreement as to the difference between a representation and a cognitive process. It could be argued that processes are simply the sustained temporary activation of one or more representations.

My view is that the descriptions of the functional roles of the HPFC summarized in most of the models and frameworks already described in this chapter are inadequate to obtain a clear understanding of its role in behavior. To obtain a clear understanding of the HPFC, I believe that a theory or model must describe the cognitive nature of the representational networks that are stored in the prefrontal cortex, the principles by which the representations are stored, the levels and forms of the representations, hemispheric differences in the representational component stored based on the underlying computational constraints imposed by the right and left prefrontal cortex, and it must lead to predictions about the ease of retrieving representations stored in the prefrontal cortex under normal conditions, when normal subjects divide their cognitive resources or shift between tasks, and after various forms of brain injury. None of the models noted above were intended to provide answers to any of these questions except in the most general manner.

Process Versus Representation—How to Think About Memory in the HPFC

My framework for understanding the nature of the knowledge stored in the HPFC depends upon the idea that unique forms of knowledge are stored in the HPFC as representations. In this sense, a representation is an element of knowledge that, when activated, corresponds to a unique brain state signified by the strength and pattern of neural activity in a local brain sector. This representational element is a "permanent" unit of memory that can be strengthened by repeated exposure to the same or a similar knowledge element and is a member of a local psychological and neural network composed of multiple similar representations. Defining the specific forms of the representations in HPFC so that a cognitive framework can be tested is crucial since an inappropriate representational depiction can compromise a model or theory as a description of a targeted phenomenon. It is likely that these HPFC representations are parsed at multiple grain sizes (that are shaped by behavioral, environmental, and neural constraints).

What should a representational theory claim? It should claim that a process is a representation (or set of representations) in action, essentially a representation that, when activated, stays activated over a limited or extended time domain. In order to be activated, a representation has to be primed by input from a representation located outside its region or by associated representations within its region. This can occur via bottom-up or top-down information transfer. A representation, when activated, may or may not fit within the typical time window described as working memory. When it does, we are conscious of the representation. When it does not, we can still process that representation, but we may not have direct conscious access to all of its contents.

The idea that representations are embedded in computations performed by local neural networks and are permanently stored within those networks so that they can be easily resurrected in a similar form whenever that network is stimulated by the external world's example of that representation or via associated knowledge is not novel nor free of controversy. But similar ideas of representation have dominated the scientific understanding of face, word, and object recognition and have been recognized as an acceptable way to describe how the surface and lexical features of information could be encoded and stored in the human brain. Despite the adoption of this notion of representation to the development of cognitive architectures for various stimuli based on "lower-level" stimulus features, the application of similar representational theory to better understand the functions of the HPFC has moved much more slowly and in a more limited way.

Evolution of Cognitive Abilities

There is both folk wisdom about, and research support for, the idea that certain cognitive abilities are uniquely captured in the human brain, with little evidence for these same sophisticated cognitive abilities found in other primates. Some examples of these cognitive processes include complex language abilities, social inferential abilities, or reasoning. It is not that these and other complex abilities are not present in other species but probably that they exist only in a more rudimentary form.

The HPFC, as generally viewed, is most developed in humans. Therefore, it is likely that it has supported the transition of certain cognitive abilities from a rudimentary level to a more sophisticated one. I have already touched upon what kinds of abilities are governed by the HPFC. It is likely, however, that such abilities depend upon a set of fundamental computational processes unique to humans that support distinctive representational forms in the prefrontal cortex (Grafman, 1995). My goal in the remainder of my chapter is to suggest the principles by which such unique representations would be distinctively stored in the HPFC.

The Structured Event Complex

The Archetype SEC

There must be a few fundamental principles governing evolutionary cognitive advances from other primates to humans. A key principle must be the ability of neurons to sustain their firing and code the temporal and sequential properties of ongoing events in the environment or the mind over longer and longer periods of time. This sustained firing has enabled the human brain to code, store, and retrieve the more abstract features of behaviors whose goal or end stage would not occur until well after the period of time that exceeds the limits of consciousness in the present. Gradually in evolution, this period of time must have extended itself to encompass and encode all sorts of complex behaviors (Nichelli, Clark, Hollnagel, & Grafman, 1995; Rueckert & Grafman, 1996, 1998). Many aspects of such complex behaviors must be translated into compressed (and multiple modes of) representations (such as a verbal listing of a series of things to do and the same set of actions in visual memory) while others may have real-time representational unpacking (unpacking means the amount of time and resources required to activate an entire representation and sustain it for behavioral purposes over the length of time it would take to actually perform the activity—for example, an activity composed of several linked events that take 10 minutes to perform would activate some component representations of that activity that would be active for the entire 10 minutes).

The Event Sequence

Neurons and assemblies firing over extended periods of time in the HPFC process sets of input that can be defined as events. Along with the extended firing of neurons that allows the processing of behaviors across time, there must have also developed special neural parsers that enabled the editing of these behaviors into linked sequential but individual events (much the way speech can be parsed into phonological units or sentences into grammatical constituents) (Sirigu et al., 1996, 1998). The event sequences, in order to be goal oriented and cohere, must obey a logical sequential structure within the constraints of the physical world, the culture that the individual belongs to, and the individual's personal preferences. These event sequences, as a whole, can be conceptualized as units of memory within domains of knowledge (e.g., a social attitude, a script that describes cooking a dinner, or a story that has a logical plot). We purposely labeled the archetype event sequence the SEC in order to emphasize that we believed it to be the general form of representation within the HPFC and to avoid being too closely tied to a particular description of higher-level cognitive processes contained in story, narrative processing, script, or schema frameworks.

Goal Orientation

Structured event complexes are not random chains of behavior performed by normally functioning adults. They tend to have boundaries that signal their onset and offset. These boundaries can be determined by temporal cues, cognitive cues, or environmental or perceptual cues. Each SEC, however, has some kind of goal whose achievement precedes the offset of the SEC. The nature of the goal can be as different as putting a bookshelf together or choosing a present to impress your child on her birthday. Some events must be more central or important to an SEC than others. Subjects can have some agreement on which ones they are when explicitly asked. Some SECs are well structured, with all the cognitive and behavioral rules available for the sequence of events to occur, and there is a clear definable goal. Other SECs are ill-structured, requiring the subject to adapt to unpredictable events using analogical reasoning or similarity judgment to determine the sequence of actions online (by retrieving a similar SEC from memory) as well as developing a quickly fashioned goal. Not only are SEC goals central to its execution, but the process of reaching the goal can be rewarding. Goal achievement itself is probably routinely accompanied by a reward that is mediated by the brain's neurochemical systems. Depending on the salience of this reward cue, it can become essential to the subject's subsequent competent execution of that same or similar SEC. Goal attainment is usually obvious, and subjects can consciously move onto another SEC in its aftermath.

Representational Format of the SEC

I hypothesize that SECs are composed of a set of differentiated representational forms that would be stored in different regions of the HPFC but are activated in parallel to reproduce all the SEC elements of a typical episode. These distinctive memories would represent thematic knowledge, morals, abstractions, concepts, social rules, features of specific events, and grammars for the variety of SECs embodied in actions, stories and narratives, scripts, and schemas.

Memory Characteristics

As just described, SECs are essentially distributed memory units with different components of the SEC stored in various regions within the prefrontal cortex. The easiest assumption to make, then, is that they obey the same principles as other memory units in the brain. These principles revolve around frequency of activation based on use or exposure, association with other memory units, category specificity of the memory unit, plasticity of the representation, priming mechanisms, and binding of the memory unit and its neighborhood memory units to memory units in more distant representational networks both in, and remote from, the territory of the prefrontal cortex.

Frequency of Use and Exposure

As a characteristic that predicts a subject's ability to retrieve a memory, frequency is a powerful variable. For the SEC, the higher the frequency of the memory units composing the SEC components, the more resilient they should be in the face of prefrontal cortex damage. That is, it is predicted that patients with frontal lobe damage would be most preserved performing or recognizing those SECs that they usually do as a daily routine and most impaired when asked to produce or recognize novel or rarely executed SECs. This retrieval deficit would be affected by the frequency of the specific kind of SEC component memory units stored in the damaged prefrontal cortex region.

Associative Properties Within an HPFC Functional Region

In order to hypothesize the associative properties of an SEC, it is necessary to adapt some general information processing constraints imposed by each of the hemispheres (Beeman, 1998; Nichelli, Grafman, et al., 1995; Partiot, Grafman, Sadato, Flitman, & Wild, 1996). A number of theorists have suggested that hemispheric asymmetry of information coding revolves around two distinct notions. The left hemisphere is specialized for finely tuned rapid encoding that is best at processing within-event information and coding for the boundaries between events. For example, the left prefrontal cortex might be able to best process the primary meaning of an event. The right hemisphere is thought to be specialized for coarse slower coding, allowing for the processing of information that is more distantly related (to the information currently being processed) and could be adept at integrating or synthesizing information across events in time. For example, the right prefrontal cortex might be best able to process and integrate information across events in order to obtain the theme or moral of a story that is being processed for the first time. When left hemisphere fine-coding mechanisms are relied upon, a local memory element would be rapidly activated along with a few related neighbors with a relatively rapid deactivation. When right hemisphere coarse coding mechanisms are relied upon, there should be weaker activation of local memory elements but a greater spread of activation across a larger neighborhood of representations and for a sustained period of time—even corresponding to the true duration of the SEC currently being processed. This dual form of coding probably occurs in parallel with subjects shifting between the two depending on task and strategic demands. Furthermore, the organization of a population of SEC components within a functionally defined region, regardless of coding mechanisms, should be based on the same principles argued for other forms of associative representation with both inhibition of unrelated memory units and facilitation of neighboring (and presumably related) memory units following activation.

Order of Events

The HPFC is specialized for the processing of events over time. One aspect of the SEC that is key to its representation is event order. Order is coded by the sequence of events. The stream of action must be parsed as each event begins and ends in order to explicitly recognize the nature, duration, and number of events that compose the event sequence (Hanson & Hanson, 1996; Zacks & Tversky, 2001). I hypothesize that in childhood, because of the neural constraints of an immature HPFC, individual events are initially represented as independent memory units and only later in development are they linked together to form an SEC. Thus, in adults, there should be some redundancy of representation of the independent event (formed in childhood) and the membership of that same event within the SEC. Adult patients with HPFC lesions would be expected to commit errors of order in developing or executing SECs but could wind up defaulting to retrieving the independently stored events in an attempt to slavishly carry out fragments of an activity. Subjects are aware of the sequence of events that make up an SEC and can even judge their relative importance or centrality to the overall SEC theme or goal. Each event has a typical duration and an expected onset and offset time within the time frame of the entire SEC that is coded. The order of the independent events that make up a particular SEC must be routinely adhered to by the performing subject in order to develop a more deeply stored SEC representation and to improve the subject's ability to predict the sequence of events. The repeated performance of an SEC leads to the systematic and rigidly ordered execution of events—an observation compatible with the AI notion of total order planning. In contrast, new SECs are constantly being encoded, given the variable and occasionally unpredictable nature of strategic thought or environmental demands. This kind of adaptive planning in AI is known as partial-order planning, since event sequences are composed online, with the new SEC consisting of previously experienced events now interdigitating with novel events. Since there must be multiple SECs that are activated in a typical day, it is likely that they too (like the events within an SEC) can be activated in sequence, or additionally in a cascading or parallel manner (to manage two or more tasks at the same time).

Category Specificity

There is compelling evidence that the HPFC can be divided into regions that have predominant connectivity with specific cortical and subcortical brain sectors. This has led to the hypothesis that SECs may be stored in the HPFC on a category-specific basis. For example, it appears that patients with ventral or medial prefrontal cortex lesions are especially impaired in performing social and reward-related behaviors, whereas patients with lesions to the dorsolateral prefrontal cortex appear most impaired on mechanistic planning tasks (Dimitrov, Phipps, et al., 1999; Grafman et al., 1996; Partiot, Grafman, Sadato, Wachs, & Hallett, 1995; Pietrini, Guazzelli, Basso, Jaffe, & Grafman, 2000; Zalla et al., 2000). Further delineation of category specificity within the HPFC awaits more precise testing using various SEC categories as stimuli (Crozier et al., 1999; Sirigu et al., 1998).

Neuroplasticity of HPFC

We know relatively little about the neurobiological rules governing plasticity of the HPFC. It is probable that the same plasticity mechanisms that accompany learning and recovery of function in other cortical areas operate in the frontal lobes too (Grafman & Litvan, 1999a; see also chapter 22, this volume, for related discussion of neuroplasticity). For example, a change in prefrontal cortex regional functional map size with learning has been noted. Shrinkage of map size is usually associated with learning of a specific element of many within a category of representation, whereas an increase in map size over time may reflect the general category of representational form being activated (but not a specific element of memory within the category). After left brain damage, right homologous HPFC assumption of at least some of the functions previously associated with Broca's area can occur. How the unique characteristics of prefrontal cortex neurons (e.g., sustained reentrant firing patterns or idiosyncratic neural architectures) interact with the general principles of cortical plasticity has been little explored to date. In terms of the flexibility of representations in the prefrontal cortex, it appears that this area of cortex can rapidly reorganize itself to respond to new environmental contingencies or rules. Thus, although the general underlying principles of how information is represented may be similar within and across species, individual experience manifested by species or individuals within a species will be influential in what is stored in prefrontal cortex and important to control for when interpreting the results of experiments trying to infer HPFC functional organization.

Priming

At least two kinds of priming (Schacter & Buckner, 1998) should occur when an SEC is activated. First of all, within an SEC, there should be priming of forthcoming adjacent and distant events by previously occurring events. Thus, in the case of the event that indicates you are going into a restaurant, subsequent events such as paying the bill or ordering from the menu may be primed at that moment. This priming would activate those event representations even though they had not occurred yet. The activation might be too far below threshold for conscious recognition that the event has been activated, but there is a probably a relationship between the intensity of the primed activation of a subsequent event and the temporal and cognitive distance the current event is from the primed event. The closer the primed event is in sequence and time to the priming event, the more activated it should be. The second kind of priming induced by SEC activation would involve

SECs in the immediate neighborhood of the one currently activated. Closely related SECs (or components of SECs) in the immediate neighborhood should be activated to a lesser degree than the targeted SEC regardless of hemisphere. More distantly related SECs (or components of SECs) would be inhibited in the dominant hemisphere. More distantly related SECs (or components of SECs) would be weakly activated, rather than inhibited, in the nondominant hemisphere.

Binding

Another form of priming, based on the principle of binding (Engel & Singer, 2001) of distinct representational forms across cortical regions, should occur with the activation of an SEC. The sort of representational forms I hypothesize are stored in the HPFC, such as thematic knowledge, should be linked to more primitive representational forms such as objects, faces, words, stereotyped phrases, scenes, and emotions. This linkage or binding enables humans to form a distributed episodic memory for later retrieval. The binding also enables priming across representational forms to occur. For example, by activating an event within an SEC that is concerned with working in the office, activation thresholds should be decreased for recognizing and thinking about objects normally found in an office, such as a telephone. In addition, the priming of forthcoming events within an SEC referred to above would also result in the priming of the objects associated with the subsequent event. Each additional representational form linked to the SEC should improve the salience of the bound configuration of representations. Absence of highly SEC-salient environmental stimuli or thought processes would tend to diminish the overall activation of the SEC-bound configuration of representations and bias which specific subset of prefrontal cortex representational components are activated.

Hierarchical Representation of SECs

I have previously argued for a hierarchy of SEC representation (Grafman, 1995). That is, I predicted that SECs, within a domain, would range from specific episodes to generalized events. For example, you could have an SEC representing the actions and themes of a single evening at a specific restaurant, an SEC representing the actions and themes of how to behave at restaurants in general, and an SEC representing actions and themes related to eating that are context independent. In this view, SEC episodes are formed first during development of the HPFC, followed by more general SECs, and then the context-free and abstract SECs. As the HPFC matures, it is the more general, context-free, and abstract SECs that allow for adaptive and flexible planning. Since these SECs do not represent specific episodes, they can be retrieved and applied to novel situations for which a specific SEC does not exist.

Relationship to Other Forms of Representation

Basal Ganglia Functions

The basal ganglia receive direct connections from different regions of the HPFC, and some of these connections may carry cognitive commands. The basal ganglia, in turn, send back to the prefrontal cortex, via the thalamus, signals that reflect their own processing. Even if the basal ganglia work in concert with the prefrontal cortex, their exact role in cognitive processing is still debatable. They appear to play a role in the storage of visuomotor sequences (Pascual-Leone et al., 1993; Pascual-Leone, Grafman, & Hallett, 1995), in reward-related behavior (Zalla et al., 2000), and in automatic cognitive processing such as overlearned word retrieval. It is likely that the SECs in the prefrontal cortex bind with the visuomotor representations stored in the basal ganglia to produce an integrated set of cognitive and visuomotor actions (Koechlin et al., 2000; Koechlin et al., 2002; Pascual-Leone, Wassermann, Grafman, & Hallett, 1996) relevant to particular situations.

Hippocampus and Amygdala Functions

Both the amygdala and the hippocampus have reciprocal connections with the prefrontal cortex. The amygdala, in particular, has extensive connections with the ventromedial prefrontal cortex (Price, 1999; Zalla et al., 2000). The amygdala's signals may provide a somatic marker or cue to the stored representational ensemble in the ventromedial prefrontal cortex representing social attitudes, rules, and knowledge. The more salient the input provided by the somatic cue, the more important the somatic marker becomes for biasing the activation of social knowledge and actions.

The connections between the prefrontal cortex and the hippocampus serve to enlist the SEC as a contextual cue that forms part of an episodic ensemble of information (Thierry, Gioanni, Degenetais, & Glowinski, 2000). The more salient the context, the more important it becomes for enhancing the retrieval or recognition of episodic memories. Thus, the hippocampus also serves to help bind the activation of objects, words, faces, scenes, procedures, and other information stored in posterior cortices and basal structures to SEC-based contextual information such as themes or plans. Furthermore, the hippocampus may be involved in the linkage of sequentially occurring events. The ability to explicitly predict a subsequent event requires conscious recollection of forthcoming events, which should require the participation of a normally functioning hippocampus. Since the hippocampus is not needed for certain aspects of lexical or object priming, for example, it is likely that components of the SEC that can also be primed (see above) do not require the participation of the hippocampus. Thus subjects with amnesia might gain confidence and comfort in interactions in a context if they were reexposed to the same context (SEC) that they had experienced before. In that case, the representation of that SEC would be strengthened even without later conscious recollection of experiencing it. Thus, SEC representational priming in amnesia should be governed by the same restraints that affect word or object priming in amnesia.

Temporal-Parietal Cortex Functions

The computational processes representing the major components of what we recognize as a word, object, face, or scene are stored in the posterior cortex. These representations are crucial components of a context and can provide the key cue to initiate the activation of an SEC event or its inhibition. Thus, the linkage between anterior and posterior cortices is very important for providing evidence that contributes to identifying the temporal and physical boundaries delimiting the independent events that make up an SEC.

Evidence For and Against the SEC Framework

The advantage of the SEC formulation of the representations stored in the HPFC is that it resembles other cognitive architecture models that are constructed so as to provide testable hypotheses regarding their validity. When hypotheses are supported, they lend confidence to the structure of the model as predicated by its architects. When hypotheses are rejected, they occasionally lead to the rejection of the entire model but may also lead to a revised view of a component of the model.

The other major driving forces in conceptualizing the role of the prefrontal cortex have, in general, avoided the level of detail required of a cognitive or computational model and instead have opted for functional attributions that can hardly be disproved. This is not entirely the fault of the investigator as the forms of knowledge or processes stored in the prefrontal cortex have perplexed and eluded investigators for more than a century. What I have tried to do by formulating the SEC framework is to take the trends in cognitive capabilities observed across evolution and development, including greater temporal and sequential processing and more capacity for abstraction, and assume what representational states those trends would lead to.

The current evidence for an SEC-type representational network is supportive but still rather sparse. SECs appear to be selectively processed by anterior prefrontal cortex regions (Koechlin et al., 1999, 2000). Errors in event sequencing can occur with preservation of aspects of event knowledge (Sirigu, Zalla, Pillon, Grafman, Agid, et al., 1995). Thematic knowledge can be impaired even though event knowledge is preserved (Zalla et al., 2002). Frequency of the SEC can affect the ease of retrieval of SEC knowledge (Sirigu, Zalla, Pillon, Grafman, Agid, et al., 1995; Sirigu, Zalla, Pillon, Grafman, Dubois, et al., 1995). There is evidence for category specificity in that the ventromedial prefrontal cortex appears to be specialized for social knowledge processing (Dimitrov, Phipps, et al., 1999). The HPFC is a member of many extended brain circuits. There is evidence that the hippocampus and the HPFC cooperate when the sequence of events have to be anticipated (Dreher et al., 2006). The amygdala and the HPFC cooperate when SECs are goal and reward oriented or emotionally relevant (Zalla et al., 2000). The basal ganglia, cerebellum, and HPFC cooperate as well (Grafman et al., 1992; Hallett & Grafman, 1997; Pascual-Leone et al., 1993) in the transfer of performance responsibilities between cognitive and visuomotor representations. When the SEC is novel or multitasking is involved, the anterior frontopolar prefrontal cortex is recruited, but when SECs are overlearned, slightly more posterior frontomedial prefrontal cortex is recruited (Koechlin et al., 2000). When subjects rely upon the visuomotor components of a task, the basal ganglia and cerebellum are more involved but when subjects have to rely upon the cognitive aspects of the task, the HPFC is more involved in performance (Koechlin et al., 2002). Thus, there is positive evidence for the representation of several different SEC components within the HPFC. There has been little in the way of negative studies of this framework, but many predictions of the SEC framework in the areas of goal orientation, neuroplasticity, priming, associative properties, and binding have not been fully explored to date and could eventually be falsified. For the purposes of understanding the role of the prefrontal cortex in ergonomic understanding and decision making and learning, researchers should focus on the functions and representations of the HPFC as detailed above.

Future Directions for the SEC Model

The representational model of the structured event complex described above lends itself to the generation of testable predictions or hypotheses. To reiterate, like the majority of representational formats hypothesized for object, face, action, and word stores, the SEC subcomponents can each be characterized by the following features: frequency of exposure or activation, imaginableness, association to other items or exemplars in that particular representational store, centrality of the feature to the SEC (i.e., what proportional relevance the feature has to recognizing or executing the SEC), length of the SEC in terms of number of events and duration of each event and the SEC as a whole, implicit or explicit activation, and association to other representational forms that are stored in other areas of the HPFC or in more posterior cortex or subcortical regions.

All these features can be characterized psychometrically by quantitative values based on normative studies using experimental methods that have obtained similar values for words, photographs, objects, and faces. Unfortunately, there have been only a few attempts to collect some of this data for SECs such as scripts, plans, and similar stimuli. If these values for all of the features of interest of an SEC were obtained, one could then make predictions about changes in SEC performance after HPFC lesions. For example, one hypothesis from the SEC representational model described above is that the frequency of activation of a particular representation will determine its accessibility following HPFC lesions. Patients with HPFC lesions will have had many different experiences eating dinner, including eating food with their hands as a child, later eating more properly at the dining room table, eating at fast-food restaurants, eating at favorite regular restaurants, and eventually eating occasionally at special restaurants or a brand-new restaurant. After an HPFC lesion of moderate size, a patient should be limited in retrieving various subcomponents of the SEC stored in the lesioned sector of the HPFC. Thus, such a patient would be expected to behave more predictably and reliably when eating dinner at home than when eating in a familiar restaurant, and worst of all when eating in a new restaurant with an unusual seating or dining procedure for the first time. The kinds of errors that would characterize the inappropriate behavior would depend on the particular subcomponents of the SEC (and thus regions within or across hemispheres) that were damaged. For example, if the lesion were in the right dorsolateral prefrontal cortex, the patient might have difficulty integrating knowledge across dining events so that he or she would be impaired in determining the (unstated) theme of the dinner or restaurant, particularly if the restaurant procedures were unfamiliar enough that the patient could not retrieve an analogous SEC. Only one study has attempted to directly test this general idea of frequency sensitivity with modest success (Sirigu, Zalla, Pillon, Grafman, Agid, et al., 1995a). This is just one example of many predictions that emerge from the SEC model with components that have representational features. The claim that SEC representational knowledge is stored in the HPFC in various cognitive subcomponents is compatible with claims made for models for other forms of representational knowledge stored in other areas of the brain and leads to the same kind of general predictions regarding SEC component accessibility made for these other forms of knowledge following brain damage. Future studies need to test these predictions.

Representation Versus Process Revisited

The kind of representational model I have proposed for the SEC balances the overreliance upon so-called process models such as working memory that dominate the field today. Process models rely upon a description of performance (holding or manipulating information) without necessarily being concerned about the details of the form of representation (i.e., memory) activated that is responsible for the performance.

Promoting a strong claim that the prefrontal cortex is concerned with processes rather than permanent representations is a fundamental shift of thinking away from how we have previously tried to understand the format in which information is stored in memory. It suggests that the prefrontal cortex has little neural commitment to long-term storage of knowledge, in contrast to the posterior cortex. Such a fundamental shift in brain functions devoted to memory requires a much stronger philosophical, neuropsychological, and neuroantomical defense for the process approach than has been previously offered by its proponents. The representational point of view that I offer regarding HPFC knowledge stores is more consistent with previous cognitive neuroscience approaches to understanding how other forms of knowledge such as words or objects are represented in the brain. It also allows for many hypotheses to be derived for further study and therefore can motivate more competing representational models of HPFC functions.

Neuroergonomic Applications

There is no doubt that the impairments caused by lesions to the HPFC can be very detrimental to people's ability to maintain their previous level of work, responsibility to their family, and social commitments (Grafman & Litvan, 1999b). These are all key ergonomic and social issues. In turn, the general role of the prefrontal cortex in maintaining information across time intervals and intervening tasks, in modulating social behavior, in the integration of information across time, and in the control of behavior via temporary memory representations and thought rather than allowing behavior to depend upon environmental contingencies alone appears critical for high-level ergonomic functioning. We know that deficits in executive functions can have a more profound effect on daily activities and routines than sensory deficits, aphasia, or agnosia (Schwab et al., 1993). Rehabilitation specialists are aware of the seriousness of deficits in executive impairments, but there are precious few group studies detailing specific or general improvements in executive functions that are maintained in the real world and that lead to a positive functional outcome (Levine et al., 2000; Stablum et al., 2000). Likewise, performance in various work situations that rely upon HPFC cognitive processes have been rarely studied by cognitive neuroscientists. This is one reason why the development of neuroergonomics is an encouraging sign of further interaction between cognitive neuroscience and human factors (Parasuraman, 2003). Hopefully, a deeper understanding of HPFC functions will lead to even more applications. I will briefly describe three examples of how executive functions, associated with the HPFC, are used in day-to-day life.

The first example involves driving. Driving is often conceived of as requiring skills that engage perceptual, tactical, and strategic processes. Both tactical and strategic processes would require the development and execution of stored plans that favor long-term success (e.g., no tickets, no accidents) over short-term gain (e.g., expression of anger at other drivers by cutting them off, driving fast to make an appointment, looking at a map while driving instead of pulling off to the side of the road). Complicating matters these days is the use of cell phones that require divided attention. Dividing attention almost always results in a decrement in performance in one of the tasks being performed, and this no doubt takes place with driving skills when a driver is carrying on a cell phone conversation while driving. So even though we can use a skill associated with the frontal lobes like multitasking that results in an increase in the quantity of tasks simultaneously performed, it does not mean the quality of your task performance will improve—in fact, it is likely to decline. Overlearned strategies, rapid tactical decision making, and dividing attention are all abilities primarily governed by the HPFC and are likely to be needed in situations ranging from driving to managerial decisions, warfare, and air-traffic control towers.

A second example of how executive functions are used in daily life involves making judgments of others' behaviors. The ventromedial prefrontal cortex might be concerned with storing attitudes and stereotypes about others (e.g., this would include rapid judgments about an unknown person's abilities, depending on their sex, age, ethnicity, and racial identity). Your ability to determine the intention of others might depend on your ability to access your stored knowledge of plans and other behavioral sequences I referred to in this chapter as SECs, with the most frequent and typical SECs being stored in the medial prefrontal cortex. Your acquired knowledge about the moral behaviors of known individuals would also be stored in the anterior HPFC. Thus, many regions of our HPFC that give us a sense of inner control, wisdom, insight, sensitivity, and hunches are involved in judging a person under circumstances ranging from meeting a stranger to choosing a political party or candidate to vote for, solidifying a work relationship, or getting married.

The third example involves stock investment. Making the assumption that the investor is either a professional or an informed day trader, many abilities associated with the human prefrontal cortex are involved in the needed judgments and performance. These abilities include modulating tendencies for short-term investments governed by great fluctuations in worth in favor of more secure long-term investments, having a rational overall plan for investments given an individual's needs, and planning how much time to devote to researching different kinds of investments. All of these abilities would depend heavily upon the HPFC.

It is also apparent that the HPFC functions optimally between the late 20s and mid-50s, and so expert decision making with optimal high-level skills should occur primarily between those ages. Both younger and older individuals are more at risk for performing at a disadvantage as high-level cognitive skills are demanded. In addition, there are wide individual differences in the ability to utilize high-level cognitive skills governed by the HPFC even within the optimal age range. Although academic ability can be assessed through traditional tests and measures of achievement like graduating with a higher degree or proven performance skills, the higher-level cognitive abilities associated with the HPFC are rarely directly assessed. It might be that a new discipline of neuroergonomics would supplement traditional human factors research for specific tasks (such as driving, economic decision making, and social dynamics) if the addition of cognitive neuroscience techniques would improve both assessment and implementation of skills. An obvious model for this application would be the introduction of neuroergonomics to the training practices of government agencies concerned with improving the skills and abilities of soldiers.

Conclusion

In this chapter, I have shown that key higher-level cognitive functions known as the executive functions are strongly associated with the HPFC. I have argued that an important way to understand the functions of the HPFC is to adapt the representational model that has been the predominant approach to understanding the neuropsychological aspects of, for example, language processing and object recognition. The representational approach I developed is based on the structured event complex framework. This framework claims that there are multiple subcomponents of higher-level knowledge that are stored throughout the HPFC as distinctive domains of memory. I also have argued that there are topographical distinctions in where these different aspects of knowledge are stored in the HPFC. Each memory domain component of the SEC can be characterized by psychological features such as frequency of exposure, category specificity, associative properties, sequential dependencies, and goal orientation, which governs the ease of retrieving an SEC. In addition, when these memory representations become activated via environmental stimuli or by automatic or reflective thought, they are activated for longer periods of time than knowledge stored in other areas of the brain, giving rise to the impression that performance dependent upon SEC activation is based on a specific form of memory called working memory. Adapting a representational framework such as the SEC framework should lead to a richer corpus of predictions about subject performance that can be rejected or validated via experimental studies. Furthermore, the SEC framework lends itself quite easily to rehabilitation practice. Regarding issues of importance in ergonomics, the HPFC is important for managing aspects of decision making, social cognition, planning, foresight, goal achievement, and risk evaluation—all the kinds of cognitive processes that contribute to work-related decision making and judgment. Whether the application of cognitive neuroscience techniques substantially increase the success of training and evaluation methods currently adapted by the human factors community remains to be seen.

MAIN POINTS

1. The human prefrontal cortex is the last brain area to develop in humans and is one of the few areas of the brain most evolved in humans.
2. Executive functions including reasoning, planning, and social cognition are primarily mediated by the prefrontal cortex.
3. There is a scientific debate regarding whether the prefrontal cortex primarily is a processor of knowledge stored in other brain areas or has its own unique knowledge stores.
4. Executive functions (and therefore the prefrontal cortex) mediate many of the skills and abilities that are considered essential for high-level work performance.

Acknowledgments. Portions of this chapter were adapted from J. Grafman (2002), The human prefrontal cortex has evolved to represent components of structured event complexes. In F. Boller and J. Grafman (Eds.), *Handbook of Neuropsychology*, 2nd ed., Vol. 7 *The Frontal Lobes*, pp. 157–174). Amsterdam: Elsevier Science. B.V.

Key Readings

Cacioppo, J. T. (Ed.). (2002). *Foundations in social neuroscience*. Cambridge: MIT Press.
Stuss, D. T., & Knight, R. T. (Eds.). (2002). *Principles of frontal lobe function*. New York: Oxford University Press.

Wood, J. N., & Grafman, J. (2003). Human prefrontal cortex: Processing and representative perspectives. *Nature Reviews Neuroscience, 4*(2), 139–147.

References

Alexander, G. E., Crutcher, M. D., & DeLong, M. R. (1990). Basal ganglia-thalamocortical circuits: Parallel substrates for motor, oculomotor, "prefrontal" and "limbic" functions. *Progress in Brain Research, 85,* 119–146.
Anderson, S. W., Bechara, A., Damasio, H., Tranel, D., & Damasio, A. R. (1999). Impairment of social and moral behavior related to early damage in human prefrontal cortex. *Nature Neuroscience, 2,* 1032–1037.
Arnett, P. A., Rao, S. M., Bernardin, L., Grafman, J., Yetkin, F. Z., & Lobeck, L. (1994). Relationship between frontal lobe lesions and Wisconsin Card Sorting Test performance in patients with multiple sclerosis. *Neurology, 44,* 420–425.
Baddeley, A. (1998a). The central executive: A concept and some misconceptions. *Journal of the International Neuropsychological Society, 4,* 523–526.
Baddeley, A. (1998b). Recent developments in working memory. *Current Opinion in Neurobiology, 8,* 234–238.
Barbas, H. (2000). Complementary roles of prefrontal cortical regions in cognition, memory, and emotion in primates. *Advances in Neurology, 84,* 87–110.
Bechara, A., Damasio, H., & Damasio, A. R.. (2000). Emotion, decision making and the orbitofrontal cortex. *Cerebral Cortex, 10,* 295–307.
Bechara, A., Damasio, H., Damasio, A. R., & Lee, G. P. (1999). Different contributions of the human amygdala and ventromedial prefrontal cortex to decision-making. *Journal of Neuroscience, 19,* 5473–5481.
Beeman, M. (1998). Coarse semantic coding and discourse comprehension. In M. Beeman & C. Chiarello (Eds.), *Right hemisphere language comprehension* (pp. 255–284.) Mahwah, NJ: Erlbaum.
Botvinick, M., & Plaut, D. C. (2000, April). *Doing without schema hierarchies: A recurrent connectionist approach to routine sequential action and its pathologies.* Paper presented at the annual meeting of the Cognitive Neuroscience Society, San Francisco, CA.
Burgess, P. W. (2000). Strategy application disorder: The role of the frontal lobes in human multitasking. *Psychological Research, 63,* 279–288.
Burgess, P. W., Veitch, E., de Lacy Costello, A., & Shallice, T. (2000). The cognitive and neuroanatomical correlates of multitasking. *Neuropsychologia, 38,* 848–863.

Carlin, D., Bonerba, J., Phipps, M., Alexander, G., Shapiro, M., & Grafman, J. (2000). Planning impairments in frontal lobe dementia and frontal lobe lesion patients. *Neuropsychologia, 38*, 655–665.

Changeux, J. P., & Dehaene, S. (1998). Hierarchical neuronal modeling of cognitive functions: From synaptic transmission to the Tower of London. *Comptes Rendus de l'Académie des Sciences III, 321*, 241–247.

Chiavaras, M. M., LeGoualher, G., Evans, A., & Petrides, M. (2001). Three-dimensional probabilistic atlas of the human orbitofrontal sulci in standardized stereotaxic space. *Neuroimage, 13*, 479–496.

Cooper, R., & Shallice, T. (2000). Contention scheduling and the control of routine activities. *Cognitive Neuropsychology, 7*, 297–338.

Courtney, S. M., Petit, L., Haxby, J. V., & Ungerleider, L. G. (1998). The role of prefrontal cortex in working memory: Examining the contents of consciousness. *Philosophical Tranactions of the Royal Society of London, B, Biological Sciences, 353*, 1819–1828.

Crozier, S., Sirigu, A., Lehericy, S., van de Moortele, P. F., Pillon, B., Grafman, J., et al. (1999). Distinct prefrontal activations in processing sequence at the sentence and script level: An fMRI study. *Neuropsychologia, 37*, 1469–1476.

Damasio, A. R. (1996). The somatic marker hypothesis and the possible functions of the prefrontal cortex. *Philosophical Tranactions of the Royal Society of London, B, Biological Sciences, 351*, 1413–1420.

Diamond, A. (2000). Close interrelation of motor development and cognitive development and of the cerebellum and prefrontal cortex. *Child Development, 71*, 44–56.

Dimitrov, M., Grafman, J., & Hollnagel, C. (1996). The effects of frontal lobe damage on everyday problem solving. *Cortex 32*, 357–366.

Dimitrov, M., Grafman, J., Soares, A. H., & Clark, K. (1999). Concept formation and concept shifting in frontal lesion and Parkinson's disease patients assessed with the California Card Sorting Test. *Neuropsychology, 13*, 135–143.

Dimitrov, M., Granetz, J., Peterson, M., Hollnagel, C., Alexander, G., & Grafman, J. (1999). Associative learning impairments in patients with frontal lobe damage. *Brain and Cognition, 41*, 213–230.

Dimitrov, M., Phipps, M., Zahn, T. P., & Grafman, J. (1999). A thoroughly modern Gage. *Neurocase, 5*, 345–354.

Dreher, J. C., Koechlin, E., Ali, O., & Grafman, J. (2006). *Dissociation of task timing expectancy and task order anticipation during task switching.* Manuscript submitted for publication.

Elston, G. N. (2000). Pyramidal cells of the frontal lobe: All the more spinous to think with. *Journal of Neuroscience, 20*(RC95), 1–4.

Engel, A. K., & Singer, W. (2001). Temporal binding and the neural correlates of sensory awareness. *Trends in Cognitive Science, 5*, 16–25.

Eslinger, P. J. (1998). Neurological and neuropsychological bases of empathy. *European Neurology, 39*, 193–199.

Forde, E. M. E., & Humphreys, G. W. (2000). The role of semantic knowledge and working memory in everyday tasks. *Brain and Cognition, 44*, 214–252.

Fuster, J. M., Bodner, M., & Kroger, J. K. (2000). Cross-modal and cross-temporal association in neurons of frontal cortex. *Nature, 405*, 347–351.

Goel, V., & Grafman, J. (1995). Are the frontal lobes implicated in "planning" functions? Interpreting data from the Tower of Hanoi. *Neuropsychologia, 33*, 623–642.

Goel, V., Grafman, J., Sadato, N., & Hallett, M. (1995). Modeling other minds. *Neuroreport, 6*, 1741–1746.

Goel, V., Grafman, J., Tajik, J., Gana, S., & Danto, D. (1997). A study of the performance of patients with frontal lobe lesions in a financial planning task. *Brain, 120*, 1805–1822.

Gomez-Beldarrain, M., Grafman, J., Pascual-Leone, A., & Garcia-Monco, J. C. (1999). Procedural learning is impaired in patients with prefrontal lesions. *Neurology, 52*, 1853–1860.

Grafman, J. (1995). Similarities and distinctions among current models of prefrontal cortical functions. *Annals of the New York Academy of Sciences, 769*, 337–368.

Grafman, J. (1999). Experimental assessment of adult frontal lobe function. In B. L. Miller & J. Cummings (Eds.), *The human frontal lobes: Function and disorder* (pp. 321–344). New York: Guilford.

Grafman, J., & Litvan, I. (1999a). Evidence for four forms of neuroplasticity. In J. Grafman & Y. Christen (Eds.), *Neuronal plasticity: Building a bridge from the laboratory to the clinic* (pp. 131–140). Berlin: Springer.

Grafman, J., & Litvan, I. (1999b). Importance of deficits in executive functions. *Lancet, 354*, 1921–1923.

Grafman, J., Litvan, I., Massaquoi, S., Stewart, M., Sirigu, A., & Hallett, M. (1992). Cognitive planning deficit in patients with cerebellar atrophy. *Neurology, 42*, 1493–1496.

Grafman, J., Schwab, K., Warden, D., Pridgen, A., Brown, H. R., & Salazar, A. M. (1996). Frontal lobe injuries, violence, and aggression: A report of the Vietnam Head Injury Study. *Neurology, 46*, 1231–1238.

Groenewegen, H. J., & Uylings, H. B. (2000). The prefrontal cortex and the integration of sensory, limbic and autonomic information. *Progress in Brain Research, 126*, 3–28.

Hallett, M., & Grafman, J. (1997). Executive function and motor skill learning. *International Review of Neurobiology, 41*, 297–323.

Hanson, C., & Hanson, S. E. (1996). Development of schemata during event parsing: Neisser's perceptual cycle as a recurrent connectionist network. *Journal of Cognitive Neuroscience, 8*, 119–134.

Humphreys, G. W., & Riddoch, M. J. (2000). One more cup of coffee for the road: Object-action assemblies, response blocking and response capture after frontal lobe damage. *Experimental Brain Research, 133*, 81–93.

Humphreys, G. W., & Riddoch, M. J. (2001). Detection by action: Neuropsychological evidence for action-defined templates in search. *Nature Neuroscience, 4*, 84–88.

Jurado, M. A., Junque, C., Vendrell, P., Treserras, P., & Grafman, J. (1998). Overestimation and unreliability in "feeling-of-doing" judgments about temporal ordering performance: Impaired self-awareness following frontal lobe damage. *Journal of Clinical and Experimental Neuropsychology, 20*, 353–364.

Kawasaki, H., Kaufman, O., Damasio, H., Damasio, A. R., Granner, M., Bakken, H., et al. (2001). Single-neuron responses to emotional visual stimuli recorded in human ventral prefrontal cortex. *Nature Neuroscience, 4*, 15–6.

Kimberg, D. Y., & Farah, M. J. (1993). A unified account of cognitive impairments following frontal lobe damage: The role of working memory in complex, organized behavior. *Journal of Experimental Psychology: General, 122*, 411–428.

Koechlin, E., Basso, G., Pietrini, P., Panzer, S., & Grafman, J. (1999). The role of the anterior prefrontal cortex in human cognition. *Nature, 399*, 148–151.

Koechlin, E., Corrado, G., Pietrini, P., & Grafman, J. (2000). Dissociating the role of the medial and lateral anterior prefrontal cortex in human planning. *Proceedings of the National Academy of Sciences USA, 97*, 7651–7656.

Koechlin, E., Danek, A., Burnod, Y., & Grafman, J. (2002). Medial prefrontal and subcortical mechanisms underlying the acquisition of behavioral and cognitive sequences. *Neuron, 35*(2), 371–381.

Levine, B., Robertson, I. H., Clare, L., Carter, G., Hong, J., Wilson, B. A., et al. (2000). Rehabilitation of executive functioning: An experimental-clinical validation of goal management training. *Journal of the International Neuropsychological Society, 6*, 299–312.

Levy, R., & Goldman-Rakic, P. S. (2000). Segregation of working memory functions within the dorsolateral prefrontal cortex. *Experimental Brain Research, 133*, 23–32.

Macmillan, M. (2000). *An odd kind of fame: Stories of Phineas Gage.* Cambridge, MA: MIT Press.

Masterman, D. L., & Cummings, J. L. (1997). Frontal-subcortical circuits: The anatomic basis of executive, social and motivated behaviors. *Journal of Psychopharmacology, 11*, 107–114.

Meyer, D. E., & Kieras, D. E. (1997). A computational theory of executive cognitive processes and multiple-task performance: Part 1. Basic mechanisms. *Psychological Review, 104*, 3–65.

Miller, E. K. (2000). The prefrontal cortex and cognitive control. *Nature Reviews Neuroscience, 1*, 59–65.

Nichelli, P., Clark, K., Hollnagel, C., & Grafman, J. (1995). Duration processing after frontal lobe lesions. *Annals of the New York Academy of Sciences, 769*, 183–190.

Nichelli, P., Grafman, J., Pietrini, P., Always, D., Carton, J. C., & Miletich, R. (1994). Brain activity in chess playing. *Nature, 369*, 191.

Nichelli, P., Grafman, J., Pietrini, P., Clark, K., Lee, K. Y., & Miletich, R. (1995). Where the brain appreciates the moral of a story. *Neuroreport, 6*, 2309–2313.

Norman, D. A., & Shallice, T. (1986). Attention to action: Willed and automatic control of behavior. In R. J. Davidson, G. E. Schwartz, & D. Shapiro (Eds.), *Consciousness and self-regulation* (Vol 4., pp. 1–18). New York: Plenum.

Nystrom, L. E., Braver, T. S., Sabb, F. W., Delgado, M. R., Noll, D. C., & Cohen, J. D. (2000). Working memory for letters, shapes, and locations: fMRI evidence against stimulus-based regional organization in human prefrontal cortex. *Neuroimage, 11*, 424–446.

Parasuraman, R. (2003). Neuroergonomics: Research and practice. *Theoretical Issues in Ergonomics Science, 4*, 5–20.

Partiot, A., Grafman, J., Sadato, N., Flitman, S., & Wild, K. (1996). Brain activation during script event processing. *Neuroreport, 7*, 761–766.

Partiot, A., Grafman, J., Sadato, N., Wachs, J., & Hallett, M. (1995). Brain activation during the generation of non-emotional and emotional plans. *Neuroreport, 6*, 1397–1400.

Pascual-Leone, A., Grafman, J., Clark, K., Stewart, M., Massaquoi, S., Lou, J. S., et al. (1993). Procedural learning in Parkinson's disease and cerebellar degeneration. *Annals of Neurology, 34*, 594–602.

Pascual-Leone, A., Grafman, J., & Hallett, M. (1995). Procedural learning and prefrontal cortex. *Annals of the New York Academy of Sciences, 769*, 61–70.

Pascual-Leone, A., Wassermann, E. M., Grafman, J., & Hallett, M. (1996). The role of the dorsolateral prefrontal cortex in implicit procedural learning. *Experimental Brain Research, 107*, 479–485.

Petrides, M., & Pandya, D. N. (1994). Comparative architectonic analysis of the human and macaque frontal cortex. In F. Boller & J. Grafman (Eds.),

Handbook of neuropsychology (Vol 9., pp. 17–58). Amsterdam: Elsevier.

Petrides, M., & Pandya, D. N. (1999). Dorsolateral prefrontal cortex: Comparative cytoarchitectonic analysis in the human and the macaque brain and corticocortical connection patterns. *European Journal of Neuroscience, 11*, 1011–1036.

Pietrini, P., Guazzelli, M., Basso, G., Jaffe, K., & Grafman, J. (2000). Neural correlates of imaginal aggressive behavior assessed by positron emission tomography in healthy subjects. *American Journal of Psychiatry, 157*, 1772–1781.

Price, J. L. (1999). Prefrontal cortical networks related to visceral function and mood. *Annals of the New York Academy of Sciences, 877*, 383–396.

Rilling, J. K., & Insel, T. R. (1999). The primate neocortex in comparative perspective using magnetic resonance imaging. *Journal of Human Evolution, 37*, 191–223.

Robbins, T. W. (2000). Chemical neuromodulation of frontal-executive functions in humans and other animals. *Experimental Brain Research, 133*, 130–138.

Ruchkin, D. S, Berndt, R. S., Johnson, R., Ritter, W., Grafman, J., & Canoune, H. L. (1997). Modality-specific processing streams in verbal working memory: Evidence from spatio-temporal patterns of brain activity. *Cognitive Brain Research, 6*, 95–113.

Rueckert, L., & Grafman, J. (1996). Sustained attention deficits in patients with right frontal lesions. *Neuropsychologia, 34*, 953–963.

Rueckert, L., & Grafman, J. (1998). Sustained attention deficits in patients with lesions of posterior cortex. *Neuropsychologia, 36*, 653–660.

Schacter, D. L., & Buckner, R. L. (1998). Priming and the brain. *Neuron, 20*, 185–195.

Schwab, K., Grafman, J., Salazar, A. M., & Kraft, J. (1993). Residual impairments and work status 15 years after penetrating head injury: Report from the Vietnam Head Injury Study. *Neurology, 43*, 95–103.

Schwartz, M. F., Buxbaum, L. J., Montgomery, M. W., Fitzpatrick-DeSalme, E., Hart, T., Ferraro, M., et al. (1999). Naturalistic action production following right hemisphere stroke. *Neuropsychologia, 37*, 51–66.

Semendeferi, K., Armstrong, E., Schleicher, A., Zilles, K., & Van Hoesen, G. W. (2001). Prefrontal cortex in humans and apes: A comparative study of Area 10. *American Journal of Physical Anthropology, 114*, 224–241.

Shallice, T., & Burgess, P. W. (1996). The domain of supervisory processes and temporal organization of behavior. *Philosophical Transactions of the Royal Society of London, B, 351*, 1405–1412.

Sirigu, A., Cohen, L., Zalla, T., Pradat-Diehl, P., Van Eeckhout, P., Grafman, J., et al. (1998). Distinct frontal regions for processing sentence syntax and story grammar. *Cortex, 34*, 771–778.

Sirigu, A., Zalla, T., Pillon, B., Grafman, J., Agid, Y., & Dubois, B. (1995). Selective impairments in managerial knowledge following pre-frontal cortex damage. *Cortex, 31*, 301–316.

Sirigu, A., Zalla, T., Pillon, B., Grafman, J., Agid, Y., & Dubois, B. (1996). Encoding of sequence and boundaries of scripts following prefrontal lesions. *Cortex, 32*, 297–310.

Sirigu, A., Zalla, T., Pillon, B., Grafman, J., Dubois, B., & Agid, Y. (1995). Planning and script analysis following prefrontal lobe lesions. *Annals of the New York Academy of Sciences, 769*, 277–288.

Smith, E. E., & Jonides, J. (1999). Storage and executive processes in the frontal lobes. *Science, 283*, 1657–1661.

Stablum, F., Umilta, C., Mogentale, C., Carlan, M., & Guerrini, C. (2000). Rehabilitation of executive deficits in closed head injury and anterior communicating artery aneurysm patients. *Psychological Research, 63*, 265–278.

Stuss, D. T., Toth, J. P., Franchi, D., Alexander, M. P., Tipper, S., & Craik, F. I. (1999). Dissociation of attentional processes in patients with focal frontal and posterior lesions. *Neuropsychologia, 37*, 1005–1027.

Thierry, A. M., Gioanni, Y., Degenetais, E., & Glowinski, J. (2000). Hippocampo-prefrontal cortex pathway: Anatomical and electrophysiological characteristics. *Hippocampus, 10*, 411–419.

Vendrell, P., Junque, C., Pujol, J., Jurado, M. A., Molet, J., & Grafman, J. (1995). The role of prefrontal regions in the Stroop task. *Neuropsychologia, 33*, 341–352.

Wharton, C. M., Grafman, J., Flitman, S. S., Hansen, E. K., Brauner, J., Marks, A., et al. (2000). Toward neuroanatomical models of analogy: A positron emission tomography study of analogical mapping. *Cognitive Psychology, 40*, 173–197.

Zacks, J. M., & Tversky, B. (2001). Event structure in perception and conception. *Psychological Bulletin, 127*, 3–21.

Zahn, T. P., Grafman, J., & Tranel, D. (1999). Frontal lobe lesions and electrodermal activity: Effects of significance. *Neuropsychologia, 37*, 1227–1241.

Zalla, T., Koechlin, E., Pietrini, P., Basso, G., Aquino, P., Sirigu A., et al. (2000). Differential amygdala responses to winning and losing: A functional magnetic resonance imaging study in humans. *European Journal of Neuroscience, 12*, 1764–1770.

Zalla, T., Phipps, M., & Grafman, J. (2002). Story processing in patients with damage to the prefrontal cortex. *Cortex, 38*(2), 215–231.

12

Antoine Bechara

The Neurology of Emotions and Feelings, and Their Role in Behavioral Decisions

The 1970s and 1980s were replete with studies by decision-making researchers identifying phenomena that systematically violated normative principles of economic behavior (Kahneman & Tversky, 1979). Decision-making research in the 1990s began to see a shift in emphasis from not merely demonstrating violations of normative principles to attempting to shed light on the underlying psychological mechanisms responsible for the various effects. Today, several researchers agree that the next phase of exciting research in this area is likely to emerge from building on recent advances in the field of neuroscience.

Modern economic theory assumes that human decision making involves rational Bayesian maximization of expected utility, as if humans were equipped with unlimited knowledge, time, and information-processing power. The influence of emotions on decision making has been ignored for the most part. However, the development of what became known as the *expected utility theory* was really based on the idea that people established their values for wealth on the basis of the pain and pleasure that it would give them. So *utility* was conceived as a balance between pleasure and pain. These notions about emotions in human decisions were eliminated from notions of utility in subsequent economic models. To the extent that current economic models of expected utility exclude emotion from their vocabulary, it is really inconsistent with their foundations.

Thus the prevalent assumption, which is perhaps erroneous, is that a direct link exists between knowledge and the implementation of behavioral decisions, that is, one does what one actually knows. This is problematic in view of the fact that normal people often deviate from rational choice, despite having the relevant knowledge. This deviation is even more pronounced in patients with certain neurological or psychiatric disorders, who often engage in behaviors that could harm them, despite clear knowledge of the consequences. Therefore, the neuroscience of decision making in general, and understanding the processes by which emotions exert an influence on decision making in particular, provides a neural road map for the intervening physiological processes between knowledge and behavior, and the potential interruptions that lead to a disconnection between what one knows and what one does not. Many of these intervening steps involve hidden physiological processes, many of which are emotional in nature, and neuroscience can enrich our understanding of a variety of decision-making phenomena. Given the importance of emotion in the understanding of human suffering, its value in the management of disease, its role in social interactions, and its relevance to fundamental neuroscience and cognitive science, a comprehensive understanding of human cognition requires far greater knowledge of the neurobiology of emotion. The aim of this chapter is to provide a neuroscientific perspective in support of the view that the process of decision making is influenced in many important ways by neural substrates that regulate homeostasis, emotion, and feeling. Implications are also drawn for decision making in everyday and work situations.

The Neurology of Emotions and Feelings

Suppose you see a person you love bringing you a red rose. The encounter may cause your heart to race, your skin to flush, and facial muscles to contract with a happy facial expression. The encounter may also be accompanied by some bodily sensations, such as hearing your heartbeat, feeling butterflies in your stomach. However, there is also another kind of sensation, the emotional feelings of love, ecstasy, and elation directed toward your loved one. Neuroscientists and philosophers have debated whether these two sensations are fundamentally the same. The psychological view of James-Lange (James, 1884) implied that the two were the same. However, philosophers argued that emotions are not just bodily sensations; the two have different objects. Bodily sensations are about awareness of the internal state of the body. Emotional feelings are directed toward objects in the external world. Neuroscientific evidence based on functional magnetic resonance imaging (fMRI) tends to provide important validation of the theoretical view of James-Lange that neural systems supporting the perception of bodily states provide a fundamental ingredient for the subjective experience of emotions. This is consistent with contemporary neuroscientific views (e.g., see Craig, 2002), which suggest that the anterior insular cortex, especially on the right side of the brain, plays an important role in the mapping of bodily states and their translation into emotional feelings. The view of Damasio (1999, 2003) is consistent with this notion, but it suggests further that emotional feelings are not just about the body, but they are also about things in the world as well. In other words, sensing changes in the body would require neural systems, of which the anterior insular cortex is a critical substrate. However, the feelings that accompany emotions require additional brain regions. In Damasio's view, feelings arise in conscious awareness through the representation of bodily changes in relation to the object or event that incited the bodily changes. This second-order mapping of the relationship between organism and object occurs in brain regions that can integrate information about the body with information about the world. Such regions include the anterior cingulate cortex (figure 12.1), especially its dorsal part.

According to Damasio (1994, 1999, 2003), there is an important distinction between *emotions* and *feelings*. Emotions are a collection of changes in bodily and brain states triggered by a dedicated brain system that responds to the content of one's perceptions of a particular entity or event. The responses toward the body proper enacted in a bodily state involve physiological modifications that range from changes that are hidden from an external observer (e.g., changes in heart rate, smooth muscle contraction, endocrine release, etc.) to changes that are perceptible to an external observer (e.g., skin color, body posture, facial expression, etc.). The signals generated by these changes toward the brain itself produce changes that are mostly perceptible to the individual in whom they were enacted, which then provide the essential ingredients for what is ultimately perceived as a feeling. Thus *emotions* are what an outside observer can see, or at least can measure through neuroscientific tools. *Feelings* are what the individual senses or subjectively experiences.

An emotion begins with appraisal of an emotionally competent object. An emotionally competent object is basically the object of one's emotion, such as the person you are in love with. In neural terms, images related to the emotional object are represented in one or more of the brain's sensory processing systems. Regardless of how short this presentation is, signals related to the presence of that object are made available to a number of emotion-triggering sites elsewhere in the brain. Some of these emotion-triggering sites are the amygdala and the orbitofrontal cortex (see figure 12.1). Evidence suggests that there may be some difference in the way the amygdala and the orbitofrontal cortex process emotional information: The amygdala is more engaged in the triggering of emotions when the emotional object is present in the environment; the orbitofrontal cortex is more important when the emotional object is recalled from memory (Bechara, Damasio, & Damasio, 2003; see also chapter 11, this volume).

In order to create an emotional state, the activity in triggering sites must be propagated to execution sites by means of neural connections. The emotion execution sites are visceral motor structures that include the hypothalamus, the basal forebrain, and some nuclei in the brain stem tegmentum (figure 12.1).

Feelings result from neural patterns that represent changes in the body's response to an emotional object. Signals from body states are relayed back to the brain, and representations of these body states are formed at the level of visceral sensory nuclei in the brain stem. Representations of these body signals also form at the level of the insular cortex and lateral somatosensory cortex (figure 12.1). It is

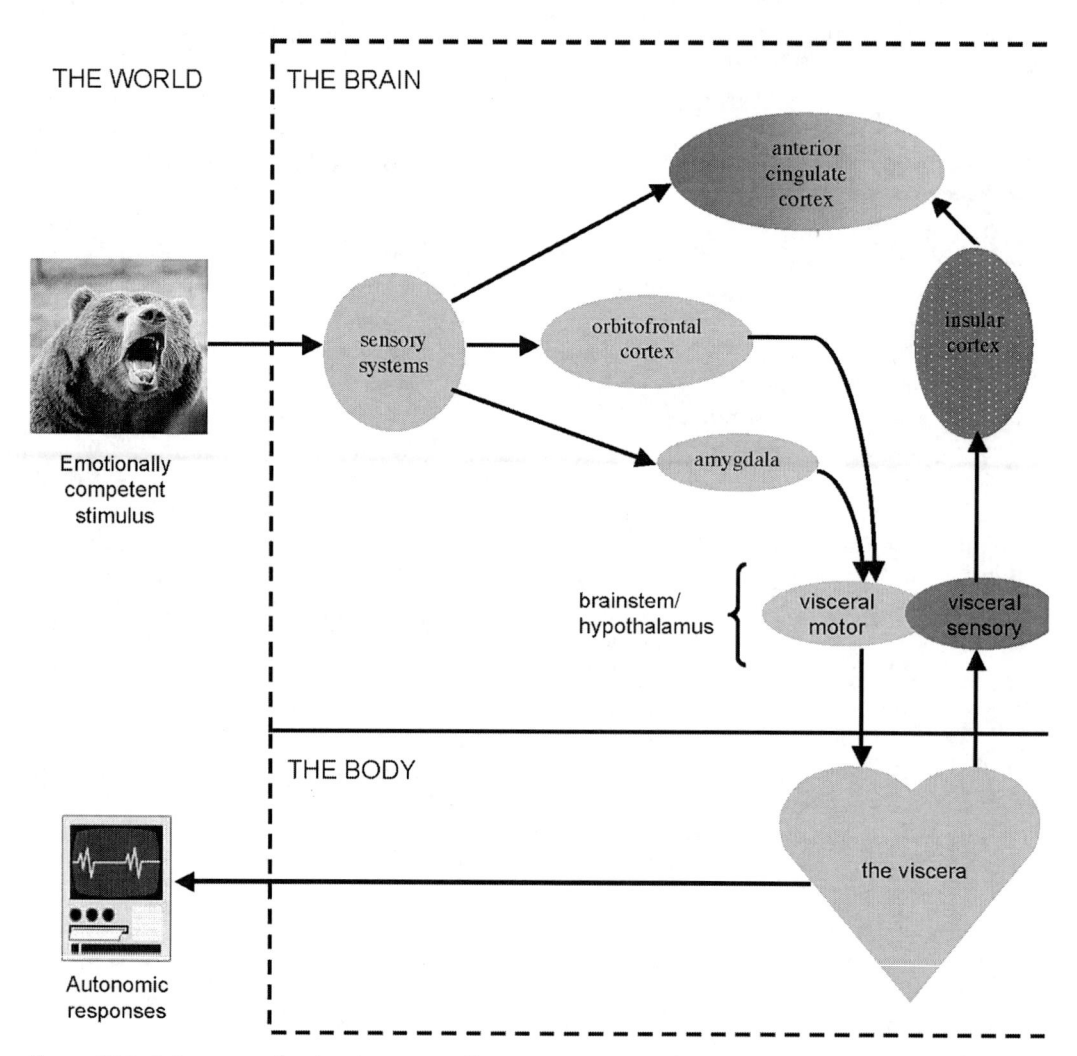

Figure 12.1. Information related to the emotionally competent object is represented in one or more of the brain's sensory processing systems. This information, which can be derived from the environment or recalled from memory, is made available to the amygdala and the orbitofrontal cortex, which are trigger sites for emotion. The emotion execution sites include the hypothalamus, the basal forebrain, and nuclei in the brain stem tegmentum. Only the visceral response is represented, although emotion comprises endocrine and somatomotor responses as well. Visceral sensations reach the anterior insular cortex by passing through the brain stem. Feelings result from the re-representation of changes in the viscera in relation to the object or event that incited them. The anterior cingulate cortex is a site where this second-order map is realized.

most likely that the reception of bodily signals at the level of the brain stem does not give rise to conscious feelings as we know them, but the reception of these signals at the level of the cortex does so. The anterior insular cortex plays a special role in mapping visceral states and in bringing interoceptive signals to conscious perception. It is less clear whether the anterior insular cortex also plays a special role in translating the visceral states into subjective feeling and self-awareness. In Damasio's view, feelings arise in conscious awareness through the representation of bodily changes in relation to the emotional object (present or recalled) that incited the bodily changes. A first-order mapping of self is supported by structures in the brain stem, insular cortex, and somatosensory cortex. However, additional regions, such as the anterior cingulate cortex, are required for a second-order mapping of the relationship between organism and emotional object, and the integration of information about the body with information about the world.

Disturbances of Emotional Experience after Focal Brain Damage

There are many instances of disturbance in emotions and feelings linked to focal lesions to structures outlined earlier. The following is a review of evidence from neurological patients with focal brain damage, as well as supporting functional neuroimaging evidence, demonstrating the role of these specific neural structures in processing information about emotions, feelings, and social behavior. The specific neural structures addressed are the amygdala, the insular and somatosensory cortex, and the orbitofrontal and anterior cingulate cortex.

Amygdala Damage

Clinical observations of patients with amygdala damage (especially when the damage is bilateral; figure 12.2) reveal that these patients express one form of emotional lopsidedness: Negative emotions

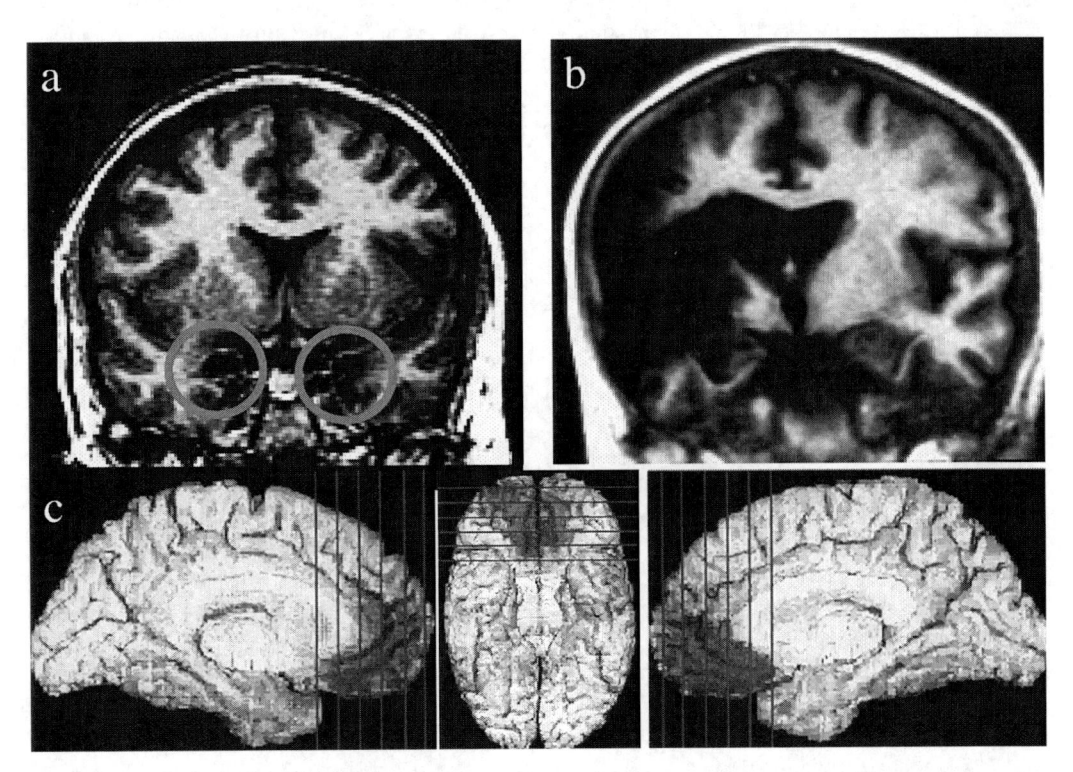

Figure 12.2. (a) Coronal sections through the amygdala taken from the 3-D reconstruction of brains of patients with bilateral amygdala damage. The region showing bilateral amygdala damage is highlighted by circles. (b) Coronal sections through the brain of a patient suffering from anosognosia. These coronoal sections show extensive damage in the right parietal region that include the insula and somatosensory cortices (SII, SI). The left parietal region is intact. (c) Left midsagittal (left), inferior (center), and right midsagittal (right) views of the brain of a patient with bilateral damage to the ventromedial region of the prefrontal cortex.

such as anger and fear are less frequent and less intense in comparison to positive emotions (Damasio, 1999). Many laboratory experiments have also established problems in these patients with processing emotional information, especially in relation to fear (Adolphs, Tranel, & Damasio, 1998; Adolphs, Tranel, Damasio, & Damasio, 1995; LaBar, LeDoux, Spencer, & Phelps, 1995; Phelps et al., 1998). Noteworthy, at least when the damage occurs earlier in life, is that these patients grow up to have many abnormal social behaviors and functions (Adolphs et al., 1995; Tranel & Hyman, 1990).

Laboratory experiments suggest that the amygdala is a critical substrate in the neural system necessary for the triggering of emotional states from what we have termed *primary inducers* (Bechara et al., 2003). Primary inducers are stimuli or entities that are innate or learned to be pleasant or aversive. Once they are present in the immediate environment, they automatically, quickly, and obligatorily elicit an emotional response. Examples of primary inducers include the encounter of a feared object such as a snake. Winning or losing a large sum of money, as in the case of being told that you won the lottery, is a type of learned information, but it has the property of instantly, automatically, and obligatorily eliciting an emotional response. This is also an example of a primary inducer. *Secondary inducers*, on the other hand, are entities generated by the recall of a personal or hypothetical emotional event (i.e., thoughts and memories about the primary inducer), which, when they are brought to working memory, slowly and gradually begin to elicit an emotional response. Examples of secondary inducers include the emotional response elicited by the memory of encountering or being bitten by a snake, the memory or the imagination of winning the lottery, and the recall or imagination of the death of a loved one.

Several lines of study suggest that the amygdala is a critical substrate in the neural system necessary for the triggering of emotional states from primary inducers. Patients with bilateral amygdala lesions have reduced, but not completely blocked, autonomic reactivity to startlingly aversive loud sounds (Bechara, Damasio, Damasio, & Lee, 1999). These patients also do not acquire conditioned autonomic responses to the same aversive loud sounds, even when the damage is unilateral (Bechara et al., 1995; LaBar et al., 1995). Amygdala lesions in humans have also been shown to reduce autonomic reactivity to a variety of stressful stimuli (Lee et al., 1988, 1998).

Bilateral amygdala damage in humans also interferes with the emotional response to cognitive information that through learning has acquired properties that automatically and obligatorily elicit emotional responses. Examples of this cognitive information are learned concepts such as winning or losing. The announcement that you have won a Nobel Prize, an Oscar award, or the lottery can instantly, automatically, involuntarily, and obligatorily elicit an emotional response. Emotional reactions to gains and losses of money, for example, are learned responses because we were not born with them. However, through development and learning, these reactions become automatic. We do not know how this transfer occurs. However, we have presented evidence showing that patients with bilateral amygdala lesions failed to trigger emotional responses in reaction to the winning or losing of various amounts of money (Bechara et al., 1999).

The results of functional neuroimaging studies corroborate those from lesion studies. For instance, activation of the amygdala has been shown in classical conditioning experiments (LaBar, Gatenby, Gore, LeDoux, & Phelps, 1998). Other functional neuroimaging studies have revealed amygdala activation in reaction to winning and losing money (Zalla et al., 2000). Also interesting is that humans tend to automatically, involuntarily, and obligatorily elicit a pleasure response when they solve a puzzle or uncover a solution to a logical problem. In functional neuroimaging experiments involving asking human subjects to find solutions to a series of logical problems, amygdala activations were associated with the "aha" in reaction to finding the solution to a given logical problem (Parsons & Oshercon, 2001).

In essence, the amygdala links the features of a stimulus with the expressed emotional or affective value of that stimulus (Malkova, Gaffan, & Murray, 1997). However, the amygdala appears to respond only when the stimuli are actually present in the environment (Whalen, 1998).

Damage to the Insular or Somatosensory Cortex

The classical clinical condition of patients with parietal damage (involving the insular, somatosensory, and adjacent cortex), especially on the right side, which demonstrates alterations in emotional experience, is called *anosognosia*. Anosognosia means denial of illness or failure to recognize an illness (see figure 12.2). The condition is characterized by apathy and placidity. It is most commonly seen in association with right-hemisphere lesions (as opposed to left).

The classical example of this condition is that the patient is paralyzed in the left side of the body, unable to move hand, arm, and leg, and unable to stand or walk. When asked how they feel, patients with anosognosia report that they feel fine, and they seem oblivious to the entire problem. In stroke patients, the unawareness is typically most profound during the first few days after onset. In a few days or a week, patients will begin to acknowledge that they have suffered a stroke and that they are weak or numb, but they minimize the implications of the impairment. In the chronic epoch (3 months or more after onset), the patients may provide a more accurate account of their physical disabilities. However, defects in the appreciation of acquired cognitive limitations may persist for months or years. Patients with similar damage on the left side of their brain are usually cognizant of their deficit and often feel depressed.

Many laboratory experiments have established problems with processing emotional information in these patients, such as empathy and recognition of emotions in facial expressions (Adolphs, Damasio, Tranel, & Damasio, 1996). Furthermore, although the paralysis and neurological handicaps of these patients limit their social interactions and mask potential abnormal social behaviors, instances in which these patients were allowed extensive social interactions revealed that patients with this condition exhibit severe impairments in judgment and failure to observe social convention. One illustrative example is the case of the Supreme Court Justice William O. Douglas described by Damasio in his book *Descartes' Error* (Damasio, 1994).

Support for the idea that the insular and somatosensory cortex are parts of a neural system that subserves emotions and feelings also comes from numerous experiments using functional neuroimaging methods (Dolan, 2002). In addition, evidence from functional neuroimaging studies suggests that beside the insular and somatosensory cortex, neighboring regions that include the posterior cingulate cortex are consistently activated in experiments involving the generation of feeling states (Damasio et al., 2000; Maddock, 1999), which suggests that the whole region plays a role in the generation of feelings from autobiographical memory.

Lesions of the Orbitofrontal and Anterior Cingulate Cortex

Patients with orbitofrontal cortex damage exhibit varying degrees of disturbance in emotional experience, depending on the location and extent of the damage (figure 12.2). If the damage is localized, especially in the more anterior sector of the orbitofrontal region (i.e., toward the front of the brain), the patients exhibit many manifestations including alterations of emotional experience and social functioning. Previously well-adapted individuals become unable to observe social conventions and decide advantageously on personal matters, and their ability to express emotion and to experience feelings appropriately in social situations becomes compromised (Bechara, Damasio, & Damasio, 2000; Bechara, Tranel, & Damasio, 2002). If the damage is more extensive, especially when it involves parts of the anterior cingulate, the patients exhibit additional problems in impulse control, disinhibition, and antisocial behavior. For instance, such patients may utter obscene words, make improper sexual advances, or say the first thing that comes to mind, without considering the social correctness of what they say or do. As an example, some of these patients may urinate in a completely inappropriate social setting, when the urge arises, without any regard to the social rules of decency.

With more extensive damage, the patient may suffer a condition known as akinetic mutism, especially when the damage involves most of the anterior cingulate cortex and a surrounding region called the supplementary motor area. The condition is a combination of mutism and akinesia. The lesions may result from strokes related to impairment of blood supply in the anterior cerebral artery territories and, in some cases, from rupture of aneurysms of the anterior communicating artery or anterior cerebral artery. It may also result from parasagittal tumors (e.g., meningiomas of the falx cerebri). The lesion can be unilateral or bilateral. There is no difference between left- and right-side lesions in terms of causing the condition. The difference between unilateral and bilateral lesions appears to be only in relation to course of recovery: With unilateral lesions, the condition persists for 1 to 2 weeks; with bilateral lesions, the condition may persist for many months. The patient with akinetic mutism makes no effort to communicate verbally or by gesture. Movements are limited to the eyes (tracking moving targets) and to body or arm movements connected with daily necessities (eating, pulling bed sheets, getting up to go to the bathroom). Speech is exceptionally sparse, with only rare isolated utterances, but linguistically correct and well articulated (although generally hypophonic). With extensive prompting, the patient may repeat words and short sentences.

Provided that the amygdala, insular and somatosensory cortices were normal during development, emotional states associated with secondary inducers develop normally. Generating emotional states from secondary inducers depends on cortical circuitry in which the orbitofrontal cortex plays a central role. Evidence suggests that the orbitofrontal region is a critical substrate in the neural system necessary for the triggering of emotional states from secondary inducers, that is, from recalling or imagining an emotional event (Bechara et al., 2003).

Development of the Neural Systems Subserving Emotions and Feelings

While the amygdala is engaged in emotional situations requiring a rapid response, that is, low-order emotional reactions arising from relatively automatic processes (LeDoux, 1996), the orbitofrontal cortex is engaged in emotional situations driven by thoughts and reason. Once this initial amygdala emotional response is over, high-order emotional reactions begin to arise from relatively more controlled, higher-order processes involved in thinking, reasoning, and consciousness. Unlike the amygdala response, which is sudden and habituates quickly, the orbitofrontal cortex response is deliberate and slow, and lasts for a long time.

Thus the orbitofrontal cortex helps predict the emotion of the future, thereby forecasting the consequences of one's own actions. However, it is important to note that the amygdala system is a priori a necessary step for the normal development of the orbitofrontal system for triggering emotional states from secondary inducers (i.e., from thoughts and reflections). The normal acquisition of secondary inducers requires the integrity of the amygdala, and also the insular and somatosensory cortex. When the amygdala, or critical components of the insular and somatosensory cortex, is damaged, then primary inducers cannot induce emotional states. Furthermore, signals from triggered emotions cannot be transmitted to the insular and somatosensory cortex and then get translated into conscious feelings. For instance, individuals with a congenital absence of a specific type of neurons specialized for transmitting pain signals from the skin, called C fibers, do not feel pain, and they are unable to construct feeling representations related to pain. It follows that such individuals are unable to fear situations that lead to pain, or empathize in contexts related to pain; that is, they lack the brain representations of what it feels like to be in pain (Damasio, 1994, 1999). Thus the normal development of the orbitofrontal system (which is important for triggering emotions from secondary inducers) is contingent upon integrity of the amygdala system, which is critical for triggering emotions from primary inducers.

Given this neural framework, it follows that there may be a fundamental difference between two types of abnormalities that lead to distorted brain representations of emotional and feeling states, which in turn lead to abnormal cognition and behavior, especially in the area of judgment and decision making. The following paragraphs outline of the nature of these potential abnormalities.

One abnormality is neurobiological in nature and may relate to (1) abnormal receptors or cells concerned with the triggering or detection of emotional signals at the level of the viscera and internal milieu; (2) abnormal peripheral neural and endocrine systems concerned with transmission of emotional signals from the viscera and internal milieu to the brain stem, that is, the spinal cord, the vagus nerve, and the circumventricular organs (the brain areas that lack a blood-brain barrier); or (3) abnormal neural systems involved in the triggering (e.g., the amygdala, orbitofrontal cortex, and effector structures in the brain stem) or building of representations of emotional or feeling states (e.g., sensory nuclei in the brain stem, and insular or somatosensory cortex).

The other abnormality is environmental in nature and relates to social learning. For instance, growing up in a social environment where, say, killing another individual is glorified and encouraged leads to abnormal development of the representations of the emotional or feeling states associated with the act of killing. Although both abnormalities may be difficult to distinguish from each other at a behavioral level, the two are distinguishable at a physiological level.

We argue that individuals with abnormal social learning are capable of triggering emotional states under a variety of laboratory conditions. These individuals have the capacity to empathize, feel remorse, and fear negative consequences. In contrast, individuals with neurobiological abnormalities demonstrate failure to trigger emotional states under the same laboratory conditions. Such individuals cannot express emotions, empathize, or fear negative consequences. The distinction between the two abnormalities has important social and legal implications. Individuals whose abnormal neural representations of emotional or feeling states relate to faulty social learning can reverse this abnormality and unlearn the antisocial behavior once they are exposed to proper learning contingencies. In other words, these individuals are likely to benefit from cognitive and behavioral rehabilitation. In contrast, individuals with underlying neurobiological abnormalities do not have the capacity to reverse these emotional or feeling abnormalities. Consequently, these individuals demonstrate repeated and persistent failures to learn from previous mistakes, even in the face of rising and severe punishment. It follows that these individuals are unlikely to benefit from rehabilitation.

The Interplay between Emotions, Feelings, and Decision Making

Situations involving personal and social matters are strongly associated with positive and negative emotions. Reward or punishment, pleasure or pain, happiness or sadness all produce changes in bodily states, and these changes are expressed as emotions. We argue that such prior emotional experiences often come into play when we are deliberating a decision. Whether these emotions remain unconscious or are perceived consciously in the form of feelings, they provide the go, stop, and turn signals needed for making advantageous decisions. In other words, the activation of these brain representations of emotional and body states provides biasing signals that covertly or overtly mark various options and scenarios with a value. Accordingly, these biases assist in the selection of advantageous responses from among an array of available options. Deprived of these biases, response options become more or less equalized, and decisions become dependent on a slow reasoned cost-benefit analysis of numerous and often conflicting options. At the end, the result is an inadequate selection of a response.

Phineas Gage: A Brief History

Phineas Gage was a dynamite worker who survived an explosion that blasted an iron-tamping bar through the front of his head. Before the accident, Phineas Gage was a man of normal intelligence, responsible, sociable, and popular among peers and friends. After the accident, his recovery was remarkable. He survived this accident with normal intelligence, memory, speech, sensation, and movement. However, his behavior changed completely: He became irresponsible and untrustworthy, impatient of restraint or advice when it conflicted with his desires (Damasio, 1994). Phineas Gage died and an autopsy was not performed to determine the location of his brain lesion. However, the skull of Phineas Gage was preserved and kept at a museum at Harvard University. Using modern neuroimaging techniques, Hanna Damasio and colleagues at the University of Iowa reconstructed the brain of Phineas Gage. Based on measures taken from his skull, they reconstituted the path of the iron bar and determined the most likely location of his brain lesion (Damasio, Grabowski, Frank, Galburda, & Damasio, 1994). The key finding of this neuroimaging study was that the most likely placement of Gage's lesion was the ventromedial (VM) region of the prefrontal cortex on both sides. The damage was relatively extensive and involved considerable portions of the anterior cingulate.

Over the years, we have studied numerous patients with VM lesions. Such patients develop severe impairments in personal and social decision making, in spite of otherwise largely preserved intellectual abilities. These patients were intelligent and creative before their brain damage. After the damage, they had difficulties planning their workday and future and difficulties in choosing friends, partners, and activities. The actions they elect to pursue often lead to diverse losses, such as financial losses, losses in social standing, losses of family and friends. The choices they make are no longer advantageous and are remarkably different from the kinds of choices they were known to make before their brain injuries. These patients often decide against their best interests. They are unable to learn from previous mistakes, as reflected by repeated engagement in decisions that lead to negative consequences. In striking contrast to this real-life decision-making impairment, the patients perform normally in most laboratory tests of problem solving. Their intellect remains normal, as measured by conventional clinical neuropsychological tests.

The Somatic Marker Hypothesis

While these VM patients were intact on most neuropsychological tests, there were abnormalities in emotion and feeling, along with the abnormalities in decision making. Based on these observations, the *somatic marker hypothesis* was proposed (Damasio, 1994), which posits that the neural basis of the decision-making impairment characteristic of patients with VM prefrontal lobe damage is defective activation of somatic states (emotional signals) that attach value to given options and scenarios. These emotional signals function as covert, or overt, biases for guiding decisions. Deprived of these emotional signals, patients must rely on slow cost-benefit analyses of various conflicting options. These options may be too numerous, and their analysis may be too lengthy to permit rapid, online decisions to take place appropriately. Patients may resort to deciding based on the immediate reward of an option, or may fail to decide altogether if many options have the same basic value.

In essence, when we make decisions, mechanisms of arousal, attention, and memory are necessary to evoke and display the representations of various options and scenarios in our mind's eye. However, another mechanism is necessary for weighing these various options and for selecting the most advantageous response. This mechanism for selecting good from bad is what we call decision-making, and the physiological changes occurring in association with the behavioral selection are part of what we call somatic states (or somatic signaling).

Evidence That Emotion Guides Decisions

Situations involving personal and social matters are strongly associated with positive and negative emotions. Reward or punishment, pleasure or pain, happiness or sadness all produce changes in bodily states, and these changes are expressed as emotions. We believe that such prior emotional experiences often come into play when we are deliberating a decision. Whether these emotions remain unconscious or are perceived consciously in the form of feelings, they provide the go, stop, and turn signals needed for making advantageous decisions. In other words, the activation of these somatic states provides biasing signals that covertly or overtly mark various options and scenarios with a value. Accordingly, these biases assist in the selection of advantageous responses from among an array of available options. Deprived of these biases or somatic markers, response options become more or less equalized and decisions become dependent on a slow reasoned cost-benefit analysis of numerous and often conflicting options. At the end, the result is an inadequate selection of a response. We conducted several studies that support the idea that decision making is a process guided by emotions.

The Iowa Gambling Task

For many years, these VM patients presented a puzzling defect. Although the decision-making impairment was obvious in the real-world behavioral lives of these patients, there was no effective laboratory probe to detect and measure this impairment. For this reason, we developed what became known as the Iowa gambling task, which enabled us to detect these patients' elusive impairment in the laboratory, measure it, and investigate its possible causes (Bechara, Damasio, Damasio, & Anderson, 1994). The gambling task mimics real-life decisions closely. The task is carried out in real time and it resembles real-world contingencies. It factors reward and punishment (i.e., winning and losing money) in such a way that it creates a conflict between an immediate, luring reward and a delayed, probabilistic punishment. Therefore, the task engages the subject in a quest to make advantageous choices. As in real-life choices, the task offers choices that may be risky, and there is no obvious explanation of how, when, or what to choose.

Each choice is full of uncertainty because a precise calculation or prediction of the outcome of a given choice is not possible. The way that one can do well on this task is to follow one's hunches and gut feelings.

More specifically, this task involves four decks of cards. The goal in the task is to maximize profit on a loan of play money. Subjects are required to make a series of 100 card selections. However, they are not told ahead of time how many card selections they are going to make. Subjects can select one card at a time from any deck they choose, and they are free to switch from any deck to another at any time, and as often as they wish. However, the subject's decision to select from one deck versus another is largely influenced by various schedules of immediate reward and future punishment. These schedules are preprogrammed and known to the examiner, but not to the subject, and they entail the following principles: Every time the subject selects a card from two decks (A and B), the subject gets $100. Every time the subject selects a card from the two other decks (C or D), the subject gets $50. However, in each of the four decks, subjects encounter unpredictable punishments (money loss). The punishment is set to be higher in the high-paying Decks A and B, and lower in the low-paying Decks C and D. For example, if 10 cards were picked from Deck A, one would earn $1,000. However, in those 10 card picks, 5 unpredictable punishments would be encountered, ranging from $150 to $350, bringing a total cost of $1,250. Deck B is similar: Every 10 cards that were picked from Deck B would earn $1,000; however, these 10 card picks would encounter one high punishment of $1,250. On the other hand, every 10 cards from Decks C or D earn only $500, but only cost $250 in punishment. Hence, Decks A and B are disadvantageous because they cost more in the long run; that is, one loses $250 every 10 cards. Decks C and D are advantageous because they result in an overall gain in the long run; that is, one wins $250 every 10 cards.

We investigated the performance of normal controls and patients with VM prefrontal cortex lesions on this task. Normal subjects avoided the bad decks A and B and preferred the good decks C and D. In sharp contrast, the VM patients did not avoid the bad decks A and B; indeed, they preferred decks A and B. From these results, we suggested that the patients' performance profile is comparable to their real-life inability to decide advantageously. This is especially true in personal and social matters, a domain for which in life, as in the task, an exact calculation of future outcomes is not possible and choices must be based on hunches and gut feelings.

Emotional Signals Guide Decisions

In light of the finding that the gambling task is an instrument that detects the decision-making impairment of VM patients in the laboratory, we went on to address the next question of whether the impairment is linked to a failure in somatic (emotional) signaling (Bechara, Tranel, Damasio, & Damasio, 1996).

To address this question, we added a physiological measure to the gambling task. The goal was to assess somatic state activation (or generation of emotional signals) while subjects were making decisions during the gambling task. We studied two groups: normal subjects and VM patients. We had them perform the gambling task while we recorded their electrodermal activity (skin conductance response, SCR). As the body begins to change after a thought, and as a given emotion begins to be enacted, the autonomic nervous system begins to increase the activity in the skin's sweat glands. Although this sweating activity is relatively small and not observable by the naked eye, it can be amplified and recorded by a polygraph as a wave. The amplitude of this wave can be measured and thus provide an indirect measure of the emotion experienced by the subject.

Both normal subjects and VM patients generated SCRs after they had picked a card and were told that they won or lost money. The most important difference, however, was that normal subjects, as they became experienced with the task, began to generate SCRs prior to the selection of any cards, that is, during the time when they were pondering from which deck to choose. These anticipatory SCRs were more pronounced before picking a card from the risky decks A and B, when compared to the safe decks C and D. In other words, these anticipatory SCRs were like gut feelings that warned the subject against picking from the bad decks. Frontal patients failed to generate such SCRs before picking a card. This failure to generate anticipatory SCRs before picking cards from the bad decks correlates with their failure to avoid these bad decks and choose advantageously in this task. These results provide strong support for the notion that decision making is guided by emotional signals (gut feelings) that are generated in anticipation of future events.

Emotional Signals Do Not Need to Be Conscious

Further experiments revealed that these biasing somatic signals (gut feelings) do not need to be perceived consciously. We carried out an experiment similar to the previous one, in which we tested normal subjects and VM patients on the gambling task while recording their SCRs. However, every time the subject picked 10 cards from the decks, we would stop the game briefly and ask subjects to declare whatever they knew about what was going on in the game (Bechara, Damasio, Tranel, & Damasio, 1997). From the answers to the questions, we were able to distinguish four periods as subjects went from the first to the last trial in the task. The first was a prepunishment period, when subjects sampled the decks, and before they had yet encountered any punishment. The second was a prehunch period, when subjects began to encounter punishment, but when asked about what was going on in the game, they had no clue. The third was a hunch period, when subjects began to express a hunch about which decks were riskier but were not sure. The fourth was a conceptual period, when subjects knew very well the contingencies in the task, and which decks were the good ones and which were bad ones, and why this was so.

When examining the anticipatory SCRs from each period, we found that there was no significant activity during the prepunishment period. These were expected results because, at this stage, the subjects were picking cards and gaining money, and had not encountered any losses yet. Then there was a substantial rise in anticipatory responses during the prehunch period, that is, after encountering some money losses, but still before the subject had any clue about what was going on in the game. This SCR activity was sustained for the remaining periods, that is, during the hunch and then during the conceptual period. When examining the behavior during each period, we found that there was a preference for the high-paying decks (A and B) during the prepunishment period. Then there was a hint of a shift in the pattern of card selection, away from the bad decks, even in the prehunch period. This shift in preference for the good decks became more pronounced during the hunch and conceptual periods. The VM patients, on the other hand, never reported a hunch about which of the decks were good or bad. Furthermore, they never developed anticipatory SCRs, and they continued to choose more cards from the bad decks A and B relative to the good decks C and D.

An especially intriguing observation was that not all the normal control subjects were able to figure out the task, explicitly, in the sense that they did not reach the conceptual period. Only 70% of them were able to do so. Although 30% of controls did not reach the conceptual period, they still performed advantageously. On the other hand, 50% of the VM patients were able to reach the conceptual period and state explicitly which decks were good and which ones were bad and why. Although 50% of the VM patients did reach the conceptual period, they still performed disadvantageously. After the experiment, these VM patients were confronted with the question, Why did you continue to pick from the decks you thought were bad? The patients would resort to excuses such as "I was trying to figure out what happens if I kept playing the $100 decks" or "I wanted to recover my losses fast, and the $50 decks are too slow."

These results show that VM patients continue to choose disadvantageously in the gambling task, even after realizing explicitly the consequences of their action. This suggests that the anticipatory SCRs represent unconscious biases derived from prior experiences with reward and punishment. These biases (or gut feelings) help deter the normal subject from pursuing a course of action that is disadvantageous in the future. This occurs even before subjects become aware of the goodness or badness of the choice they are about to make. Without these biases, the knowledge of what is right and what is wrong may still become available. However, by itself, this knowledge is not sufficient to ensure an advantageous behavior. Therefore, although VM patients may manifest declarative knowledge of what is right and what is wrong, they fail to act accordingly. The VM patients may say the right thing, but they do the wrong thing.

Thus, knowledge without emotion or somatic signaling leads to dissociation between what one knows or says and how one decides to act. This dissociation is not restricted to neurological patients but also applies to neuropsychiatric conditions with suspected pathology in the VM cortex or other components of the neural circuitry that process emotion. Addiction is one example, where patients know the consequences of their drug-seeking behavior but still take the drug. Psychopathy is another example, where psychopaths can be fully aware of the consequences of their actions but still go ahead and plan the killing or rape of a victim.

A Brain-Based Model of Robot Decisions

There are many implications for neuroergonomics of the findings on emotion and decision making discussed in this chapter. One area that is also discussed elsewhere in this volume is affective robotics. Breazeal and Picard (chapter 18, this volume) provide a compelling robot model that incorporates emotions and social interactions. This is a remarkable advance, since their approach is based on evolutionary and psychological studies of emotions, and the model yields exciting results. I propose a brain-based model of robot decisions based on what we know about brain mechanisms of emotions as reviewed in this chapter. Obviously, some of the steps of my proposed model overlap with those of Breazeal and Picard's model. However, some differ, especially in the process that Breazeal and Picard referr to as the cognitive-affective control system.

Although most researchers on emotions are preoccupied with understanding subtle differences between states such as anger, sadness, fear, and other different shades of emotions, I think that the most crucial information about emotions that we need at this stage is the following: (1) is the emotion positive or negative; and (2) is it mild, moderate, or strong? This is a classification that is similar and agrees with that of Breazeal and Picard.

The first question we need to know is how humans (and in this case robots) assign value to options. Here the term *value* is interchangeable with the terms *emotion*, *feeling*, *motivation*, *affect*, or *somatic marker*. The idea is that the whole purpose of emotion is to assign value for every option we have. For example, the value of a drink of water to a person stranded in a desert is different from its value to a person who just had a super-size Pepsi.

Other factors that affect the value of a choice include time, which explains a phenomenon named *temporal discounting*, that is, why information conveying immediacy (e.g., having a heart attack tomorrow) exerts a stronger influence on decisions than information conveying delayed outcomes (e.g., having heart attack 20 years from now). Bechara and Damasio (2005) described elsewhere in more detail how such factors may be implemented in the brain. Neuroscientists have been able to address the question of how the brain can encode the value of various options on a common scale (Montague & Berns, 2002), thus suggesting that there may be a common neural currency that encodes the value of different options. This may, for example, allow the reward value of money to be compared to that of food, sex or other goods.

The Primacy of Emotion during Development

A key to emotional development is that the amygdala system (primary induction) and the insular or somatosensory system (which holds records of every emotional experience triggered by the amygdala) must be healthy, in the first place, in order for the prefrontal system to develop and function normally, that is, assign appropriate value to a given option. In terms of a robot, we can build this system that holds records of values linked to categories of actions using one of two methods. First, build a robot with some values that are innately assigned, that is, a system that simulates the amygdala system, which responds to primary inducers, and then let the robot execute a whole bunch of behaviors and decisions, with each one being followed by a consequence (i.e., reward or punishment of mild, moderate, or strong magnitude). In this way, the robot can build a registry of values connected to particular actions, that is, equivalent to the insular or somatosensory and prefrontal systems for secondary induction. In other words, here we have to assume that the robot is like a newly born child exposed to the world. Learning begins from scratch, and building a value system is a process similar to that of raising a child, especially in terms of reinforcing good behaviors and punishing bad ones. The other method is to bypass this learning process and install into a robot a registry of values based on what we already know from normal human development. The latter method is somewhat inaccurate because it does not take into account human individuality, which differs from person to person because of their different developmental histories. At best, this method may account for an average, rather than an individual, human behavior.

The Willpower to Endure Sacrifices and Resist Temptations

Willpower, as defined by the *Encarta World English Dictionary*, is a combination of determination and self-discipline that enables somebody to do something despite the difficulties involved. This is the mechanism that enables one to endure sacrifices now in order to obtain benefits later. Otherwise, how would one accept the pain of surgery? Why would someone resist the temptation to have something irresistible, or delay the gratification from something that is appealing?

I propose that these complex and apparently indeterminist behaviors are the product of a complex cognitive process subserved by two separate, but interacting, neural systems that were discussed earlier: (1) an impulsive, amygdala-dependent, neural system for signaling the pain or pleasure of the immediate prospects of an option; and (2) a reflective, prefrontal-dependent, neural system for signaling the pain or pleasure of the future prospects of an option. The final decision is determined by the relative strengths of the pain or pleasure signals associated with immediate or future prospects. When the immediate prospect is unpleasant, but the future is more pleasant, then the positive signal of future prospects forms the basis for enduring the unpleasantness of immediate prospects. This also occurs when the future prospect is even more pleasant than the immediate one. Otherwise, the immediate prospects predominate and decisions shift toward short-term horizons. The following is a proposal of how a robot should be built in order to manifest the properties of willpower, a key characteristic of human decisions:

Input of Information

Once a registry of values linked to categories of options or scenarios has been acquired, the robot should access and trigger the value assigned to each of these options and scenarios whenever they (or closely related ones) are encountered. The confronting entities and events may have many conflicting values, some of which are triggered impulsively through the amygdala system, and some of them reflectively through the prefrontal system. Another process modulates the strength of these values by factors such as time (i.e., immediate versus delayed), probability (the outcome is certain or very probable), deprivation (e.g., hungry or not), and so on. These modulation effects are mediated by the prefrontal system, as explained by Bechara and Damasio (2005).

Emotional Evaluation

Although the input of information may trigger numerous somatic responses that conflict with each other, the end result is that an overall positive or negative somatic state (or value) emerges. We have proposed that the mechanisms that determine the nature of this overall somatic state (i.e., being positive or negative) are consistent with the principles of natural selection, that is, survival of the fittest (Bechara & Damasio, 2005). In other words, numerous and often conflicting somatic states may be triggered at the same time, but stronger ones gain selective advantage over weaker ones. With each piece of information brought by cognition, the strength of the somatic state triggered by that information determines whether that same information is likely to be kept (i.e., brought back to cognition so that it triggers another somatic state that reinforces the previous one) or is likely to be discarded. Thus over the course of pondering a decision, positive and negative somatic markers that are strong are reinforced, while weak ones are eliminated. This process of elimination can be very fast.

Decision Output

Ultimately, a winner takes all; an overall, more dominant somatic state emerges (a gut feeling or a hunch, so to speak), which then provides signals to cognition that modulate activity in neural structures involved in biasing behavioral decisions. Thus the more dominant an available option over the others, the quicker the decision output; the more equal and similar the available options, the slower the decision output.

Conclusion

Emotions are not, as some might feel, simply a nuisance. Nor can we can ignore emotions on the grounds that the greatest evolutionary development of the human brain is in relation to the cortex and its "cold" cognition, as opposed to the more primitive limbic brain, the seat of emotions. Rather, emotions are a major factor in the interaction between environmental conditions and human cognitive processes, with these emotional systems (underlying somatic state activation) providing valuable implicit or explicit signals that recruit cognitive processes that are most adaptive and advantageous for survival. Therefore, understanding the neural mechanisms underlying emotions, feelings, and their regulation is crucial for many aspects of human behaviors and their disorders.

Main Points

1. Decision making is a process guided by emotional signals (gut feelings) that are generated in anticipation of future events, which help bias decisions away from courses of action that are disadvantageous to the organism, or toward actions that are advantageous.
2. Human factors evaluation of decision making in everyday and work situations must take these emotional signals into account.
3. Impairments of decision making and inappropriate social behaviors are often observed after damage to neural regions that overlap considerably with those subserving the expression of emotions and the experience of feelings.
4. These biasing emotional signals (gut feelings) do not need to be perceived consciously. Emotional signals may begin to bias decisions and guide behavior in the advantageous direction before conscious knowledge does.
5. Knowledge without emotional signaling leads to dissociation between what one knows or says and how one decides to act. Patients who have this disconnection may manifest declarative knowledge of what is right and what is wrong, but they fail to act accordingly. Such patients may say the right thing, but they do the wrong thing.
6. Abnormalities in emotions, feelings, decision making, and social behavior may be biological, that is, caused by damage to specific neural structures, but they can also be learned, such as acquiring or attaching the wrong value to actions and behaviors during childhood.

Acknowledgment. Most of the decision-making studies described in this chapter were supported by NIDA grants DA11779-02, DA12487-03, and DA16708, and by NINDS grant NS19632-23.

Key Readings

Bechara, A., Damasio, H., & Damasio, A. R. (2000). Emotion, decision-making, and the orbitofrontal cortex. *Cerebral Cortex, 10*(3), 295–307.
Craig, A. D. (2002). How do you feel? Interoception: The sense of the physiological condition of the body. *Nature Reviews Neuroscience, 3*, 655–666.
Damasio, A. R. (1994). *Descartes' error: Emotion, reason, and the human brain.* New York: Grosset/Putnam.

References

Adolphs, R., Damasio, H., Tranel, D., & Damasio, A. R. (1996). Cortical systems for the recognition of emotion in facial expressions. *Journal of Neuroscience, 16*, 7678–7687.
Adolphs, R., Tranel, D., & Damasio, A. R. (1998). The human amygdala in social judgment. *Nature, 393*, 470–474.
Adolphs, R., Tranel, D., Damasio, H., & Damasio, A. R. (1995). Fear and the human amygdala. *Journal of Neuroscience, 15*, 5879–5892.
Bechara, A., & Damasio, A. R. (2005). The somatic marker hypothesis: A neural theory of economic decision. *Games and Economic Behavior, 52*(2), 336–372.
Bechara, A., Damasio, A. R., Damasio, H., & Anderson, S. W. (1994). Insensitivity to future consequences following damage to human prefrontal cortex. *Cognition, 50*, 7–15.
Bechara, A., Damasio, H., & Damasio, A. R. (2000). Emotion, decision-making, and the orbitofrontal cortex. *Cerebral Cortex, 10*(3), 295–307.
Bechara, A., Damasio, H., & Damasio, A. (2003). The role of the amygdala in decision-making. *Annals of the New York Academy of Sciences, 985*, 356–369.
Bechara, A., Damasio, H., Damasio, A. R., & Lee, G. P. (1999). Different contributions of the human amygdala and ventromedial prefrontal cortex to decision-making. *Journal of Neuroscience, 19*, 5473–5481.

Bechara, A., Damasio, H., Tranel, D., & Damasio, A. R. (1997). Deciding advantageously before knowing the advantageous strategy. *Science, 275,* 1293–1295.

Bechara, A., Tranel, D., & Damasio, A. R. (2002). The somatic marker hypothesis and decision-making. In F. Boller & J. Grafman (Eds.), *Handbook of neuropsychology: Frontal lobes* (2nd ed., Vol. 7, pp. 117–143). Amsterdam: Elsevier.

Bechara, A., Tranel, D., Damasio, H., Adolphs, R., Rockland, C., & Damasio, A. R. (1995). Double dissociation of conditioning and declarative knowledge relative to the amygdala and hippocampus in humans. *Science, 269,* 1115–1118.

Bechara, A., Tranel, D., Damasio, H., & Damasio, A. R. (1996). Failure to respond autonomically to anticipated future outcomes following damage to prefrontal cortex. *Cerebral Cortex, 6,* 215–225.

Craig, A. D. (2002). How do you feel? Interoception: The sense of the physiological condition of the body. *Nature Reviews Neuroscience, 3,* 655–666.

Damasio, A. R. (1994). *Descartes' error: Emotion, reason, and the human brain.* New York: Grosset/Putnam.

Damasio, A. R. (1999). *The feeling of what happens: Body and emotion in the making of consciousness.* New York: Harcourt Brace and Company.

Damasio, A. R. (2003). *Looking for Spinoza: Joy, sorrow, and the feeling brain.* New York: Harcourt.

Damasio, A. R., Grabowski, T. G., Bechara, A., Damasio, H., Ponto, L. L. B., Parvizi, J., et al. (2000). Subcortical and cortical brain activity during the feeling of self-generated emotions. *Nature Neuroscience, 3,* 1049–1056.

Damasio, H., Grabowski, T., Frank, R., Galburda, A. M., & Damasio, A. R. (1994). The return of Phineas Gage: Clues about the brain from the skull of a famous patient. *Science, 264,* 1102–1104.

Dolan, R. J. (2002). Emotion, cognition, and behavior. *Science, 298,* 1191–1194.

James, W. (1884). What is an emotion? *Mind, 9,* 188–205.

Kahneman, D., & Tversky, A. (1979). Prospect theory: An analysis of decision under risk. *Econometrica, 47,* 263–291.

LaBar, K. S., Gatenby, J. C., Gore, J. C., LeDoux, J. E., & Phelps, E. A. (1998). Human amygdala activation during conditioned fear acquisition and extinction: A mixed-trial fMRI study. *Neuron, 20,* 937–945.

LaBar, K. S., LeDoux, J. E., Spencer, D. D., & Phelps, E. A. (1995). Impaired fear conditioning following unilateral temporal lobectomy in humans. *Journal of Neuroscience, 15,* 6846–6855.

LeDoux, J. (1996). *The emotional brain: The mysterious underpinnings of emotional life.* New York: Simon and Schuster.

Lee, G. P., Arena, J. G., Meador, K. J., Smith, J. R., Loring, D. W., & Flanigin, H. F. (1988). Changes in autonomic responsiveness following bilateral amygdalotomy in humans. *Neuropsychiatry, Neuropsychology, and Behavioral Neurology, 1,* 119–129.

Lee, G. P., Bechara, A., Adolphs, R., Arena, J., Meador, K. J., Loring, D. W., et al. (1998). Clinical and physiological effects of stereotaxic bilateral amygdalotomy for intractable aggression. *Journal of Neuropsychiatry and Clinical Neurosciences, 10,* 413–420.

Maddock, R. J. (1999). The retrosplenial cortex and emotion: New insights from functional neuroimaging of the human brain. *Trends in Neurosciences, 22,* 310–320.

Malkova, L., Gaffan, D., & Murray, E. A. (1997). Excitotoxic lesions of the amygdala fail to produce impairment in visual learning for auditory secondary reinforcement but interfere with reinforcer devaluation effects in rhesus monkeys. *Journal of Neuroscience, 17,* 6011–6020.

Montague, P. R., & Berns, G. S. (2002). Neural economics and the biological substrates of valuation. *Neuron, 36*(2), 265–284.

Parsons, L., & Oshercon, D. (2001). New evidence for distinct right and left brain systems for deductive versus probabilistic reasoning. *Cerebral Cortex, 11,* 954–965.

Phelps, E. A., LaBar, K. S., Anderson, A. K., O'Connor, K. J., Fulbright, R. K., & Spencer, D. D. (1998). Specifying the contributions of the human amygdala to emotional memory: A case study. *Neurocase, 4*(6), 527–540.

Tranel, D., & Hyman, B. T. (1990). Neuropsychological correlates of bilateral amygdala damage. *Archives of Neurology, 47,* 349–355.

Whalen, P. J. (1998). Fear, vigilance, and ambiguity: Initial neuroimaging studies of the human amygdala. *Current Directions in Psychological Science, 7*(6), 177–188.

Zalla, T., Koechlin, E., Pietrini, P., Basso, G., Aquino, P., Sirigu, A., et al. (2000). Differential amygdala responses to winning and losing: A functional magnetic resonance imaging study in humans. *European Journal of Neuroscience, 12,* 1764–1770.

IV

Stress, Fatigue, and Physical Work

Peter A. Hancock and James L. Szalma

Stress and Neuroergonomics

Neuroergonomics has been defined as "the study of human brain function in relation to work and technology" (Parasuraman, 2003, p. 5). Given that stress is an aspect of many forms of work, it is natural that investigations of stress and its neurobehavioral aspects form a core topic in neuroergonomics. We begin this chapter with a brief overview of the major conceptual approaches to the study of stress, including a short historical account and a précis of the more recent stress theories (e.g., Hancock & Desmond, 2001). We discuss how an understanding of stress helps shape and direct new, emerging concepts in neuroergonomics. We then consider the issue of individual differences and the effects on stress in individuals, and ethical issues related to monitoring and mitigating stress in the workplace.

Concepts of Stress

In behavioral research, stress itself has traditionally been viewed as a source of disturbance arising from an individual's physical or social environment. However, since individuals do not react in exactly the same way to common conditions, it is now considered more appropriate to view stress in terms of each individual's response to his or her environment. This is the so-called transactional perspective (see Hockey, 1984, 1986; Lazarus & Folkman, 1984; Matthews, 2001; Wickens & Hollands, 2000). This has led to both continuing debate and some confusion, since some researchers continue to define stress in terms of the external stimuli involved (e.g., noise, temperature, vibration; see Elliott & Eisdorfer, 1982; Jones, 1984; Pilcher, Nadler, & Busch, 2002). Defining stress only in terms of the physical stimulus does not account for why the same stimulus induces different stress responses across individuals or within the same individual on different occasions (Hockey, 1986; Hockey & Hamilton, 1983; Matthews, 2001). Consideration of stimulus properties therefore provides an important but nevertheless incomplete understanding of stress effects. To more fully capture all the dimensions of stress, we have developed a "trinity of stress" model (Hancock & Warm, 1989), which includes (1) environmental stimulation as the input dimension; (2) the transactional perspective, emphasizing the individual response as the adaptation facet; and (3) most critically, an output performance level. We return to this description after a review of some theoretical background.

Arousal Theory

In contrast to the stimulus-driven view of stress, which might be considered a physics-based approach, the biologically based approaches define stress in terms of the physiological response patterns of the organism (Selye, 1976). In this vein, Cannon (1932) conceptualized stress in terms of autonomic nervous system activation triggered by a homeostatic challenge. According to Cannon, external conditions (e.g., temperature, noise, etc.) or internal deficiencies (e.g., low blood sugar, etc.) trigger deviations from homeostatic equilibrium. Such threats to equilibrium result in physiological responses aimed at countering that threat. These responses involve sympathetic activation of the adrenal medulla and the release of several hormones (see Asterita, 1985; Frankenhaeuser, 1986; for an early review see Dunbar, 1954). The response-based approach of Cannon was also championed by Selye (1976), who defined stress in terms of an orchestrated set of these bodily defense reactions against any form of noxious stimulation. This set of physiological reactions and processes he referred to as the *general adaptation syndrome*. Environmental objects or events that give rise to such responses were referred to as *stressors*. Within Selye's theory, physiological responses to stressors are general in character, since the set of responses is similar across different stressors and contexts.

Arousal theory is one of the most widely applied physiological response-based theories of stress and performance (Hebb, 1955). *Arousal level* is a hypothetical construct representing a nonspecific (general) indicator of the level of stimulation of the organism as a whole (Hockey, 1984). Arousal may be assessed using techniques such as electroencephalography (EEG) or indicators of autonomic nervous system activity such as the galvanic skin response (GSR) and heart rate. As a person becomes more aroused, the waveforms of the EEG increase in frequency and decrease in amplitude (see also chapter 2, this volume), and skin conductance and heart rate increase (see also chapter 14, this volume).

Within this framework, stress effects are observed under conditions that either overarouse (e.g., noise) or underarouse the individual (e.g., sleep deprivation; Hockey, 1984; McGrath, 1970). This approach assumes an inverted-U relationship between arousal and performance—the Yekes-Dodson law—such that the optimal level of performance is observed for midrange levels of arousal (Hebb, 1955). Stressors, such as noise or sleep loss, act by either increasing or decreasing the arousal level of the individual relative to the optimum level for a given task (Hockey & Hamilton, 1983). The optimum level is also postulated to be inversely related to the difficulty of the task (Hockey, 1984, 1986). A potential mechanism to account for this relation was first postulated by Easterbrook (1959), who indicated that emotional arousal restricts the utilization of the range of peripheral visual cues from the sensory environment, so that, under conditions of chronic or acute stress, peripheral stimuli are less likely to be processed than more centrally located cues. As attention narrows (i.e., as the number of cues attended to is reduced), performance capacity is preserved by the retention of focus on salient cues. Eventually, performance fails as stress increases and even salient cues become excluded. Hancock and Dirkin (1983) showed that the narrowing phenomenon was attentional rather than sensory in nature, since individuals narrowed to the source of greatest perceived salience wherever it appeared in the visual field. More recently, Hancock and Weaver (2005) showed that the narrowing phenomenon demonstrated in spatial perception is also evident in the temporal domain, as in the effects of stress on time perception (see also Szalma & Hancock, 2002).

There are several problems with the traditional arousal explanation of stress and performance. First, the different physiological indices of arousal often do not correlate well. In the performance of a vigilance task, for instance, muscle tension, as measured by electromyogram, and catecholamine levels can indicate a highly aroused state, but skin conductance might indicate that the observer is de-aroused (Hovanitz, Chin, & Warm, 1989; Parasuraman, 1984). Second, it has proven difficult to define the effects of stressors on arousal independent of effects on performance (Hockey, 1986). Third, the theory can accommodate almost any results, making it a post hoc explanation that is difficult to falsify (i.e., test empirically; Hancock & Ganey, 2003; Hockey, 1984; Holland & Hancock, 1991). Finally, arousal theory assumes that a stressor (or set of stressors) affects overall processing efficiency and that differences in task demands (i.e., difficulty) are reflected only in the position of the optimal level of performance. Hockey and Hamilton

(1983) noted, however, that environmental stressors can have differential effects on the pattern of cognitive activity, and a single dimension, as posited by arousal theory, cannot account for such differences among stressors. Hence, a multidimensional approach is necessary in order to understand stress effects, with arousal mechanisms representing only one facet of this complex construct.

Appraisal and Regulatory Theories

Most modern theories of stress and performance have two central themes. They either explicitly include or implicitly assume an appraisal mechanism by which individuals assess their environments and select coping strategies to deal with those environments (see Hancock & Warm, 1989; Hockey, 1997; Lazarus & Folkman, 1984). Indeed, Lazarus and Folkman defined psychological stress itself as the result of an individual's appraisal of his or her environment as being taxing or exceeding his or her resources or endangering his or her well-being. The negative effects of stress are most likely to occur when individuals view an event as a threat (primary appraisal) and when they assess their coping skills as inadequate for handling the stressor (secondary appraisal; see Lazarus & Folkman, 1984). Both the person-environment interactions and the appraisal processes are likely organized at multiple levels (Matthews, 2001; Teasdale, 1999).

A second central theme of current stress theories is that individuals regulate their internal states and adapt to perturbations resulting from external stressors, including social stressors and task-based stress. Individuals respond to appraised threats (including task load) by exerting compensatory effort to either regulate their internal cognitive-affective state or to recruit resources necessary to maintain task performance. Thus, individuals are often able to maintain performance levels, particularly in real-world settings, but only at a psychological and physiological cost (Hancock & Warm, 1989; Hockey, 1997). Two current models of stress and performance that emphasize these regulatory mechanisms and adaptation are those of Hockey (1997) and Hancock and Warm (1989).

Hockey's (1997) theory is based on assumptions that behavior is goal directed and controlled by self-regulatory processes that have energetic costs associated with them. He distinguished between effort as controlled processing (e.g., working memory capacity) and effort as compensatory control (e.g., arousal level). Hockey proposed that mental resources and effort are allocated and controlled via two levels of negative feedback regulation by an effort monitor that compares current activity to goal-based performance standards. Simple, well-learned tasks are controlled at a lower level that requires very little effort (i.e., very low resource allocation) to maintain performance goals. When demands are placed on the cognitive system, via increased task demand or other forms of stress, control is exerted by a supervisory controller at the higher level. At this level, resources are recruited to compensate for goal discrepancies created by the increased demands, and information processing is more controlled and effortful. Note, however, that this represents only one potential response to stress. A second possibility would be to alter the task goals to maintain low effort or to reduce effort in the face of prolonged exposure to stress. The effortful coping response was referred to as a *strain mode*, and the reduction of effort or performance goals is *passive coping mode*. Hockey's model provides a flexible structure and an energetic mechanism by which the effects that environmental demands (i.e., stress) place on the cognitive system can be understood. As we shall see, however, this view rests on concepts that are difficult to test empirically. Specifically, the "resource concept" employed in Hockey's model, which came to dominate stress theory after the fall of the unitary arousal explanation, presents a problem in that it is ill defined and difficult to quantify. It is our contention that neuroergonomics has a significant contribution to make in improving the resource concept, since such "mental energy" must be specifically understood if neuroergonomic methods are to be effective.

The approach presented by Hancock and Warm (1989) also adopts as a fundamental tenet the idea that in many stressful situations humans adapt to their environments. This adaptation is manifested in an extended inverted-U function that describes performance change as a function of stress level, as shown in figure 13.1. Stress here can take the form of both overstimulation (hyperstress), in which the sensory systems experience an elevated level of stimulation, and understimulation (hypostress), in which the sensory systems receive disturbingly little stimulation. Note that for a wide

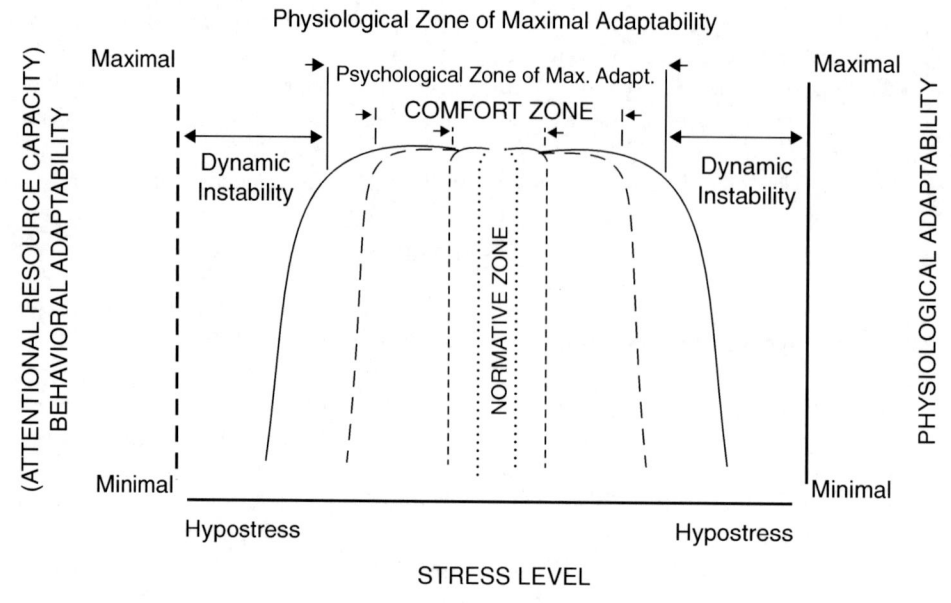

Figure 13.1. The stress-adaptation model of Hancock and Warm (1989).

range of stimulation or task demand, individuals maintain behavioral and physiological stability. There are multiple levels of adaptation that are each nested within the general, extended-U function. Thus subjective state (e.g., the normative and comfort zones in figure 13.1) is altered by relatively minor levels of disturbance, whereas it takes a greater degree of disturbance to affect behavioral performance, which is itself less robust than physiological response capacity. These different facets of response are linked: Incipient failure at one level represents the beginning of stress disturbance at the next level. While each level retains the same extended-U shape, the nesting represents the progressive fragility across levels. In the same way, there are comparable divisions within levels. For example, within the physiological level, there are progressive functions for cell, organelle, organ, and so on.

Hancock and Warm's model explicitly recognizes that tasks themselves often represent the proximal and most salient source of stress. Thus, the adaptation level of an individual will heavily depend upon the characteristics of the tasks facing that individual. Hancock and Warm (1989) proposed two fundamental dimensions along which tasks vary, these being information rate (the speed with which demands are made) and information structure (the complexity of that demand). Combined variations in task and environmental demand impose considerable stress on operators, to which they adapt via various coping efforts. Breakdown of performance under stress, and its inverse, efficient behavioral adaptability, occurs at both psychological and physiological levels, with psychological adaptability failing before comparable physiological adaptability (for a related view, see Matthews, 2001). A representation of the adaptive space in the context of the extended U is shown in figure 13.2. To locate an individual's level of adaptation to a set of stressors in an environment, one defines vectors for the level of stress and cognitive and physiological state, as well as the position of task performance along the space-time axes. Although further work is required to quantify the theoretical propositions in the Hancock and Warm (1989) model, we argue that such quantification will result in a rubric under which neuroergonomic measures of stress can be developed in coordination with performance and subjective measures. For instance, if subjective state or comfort declines prior to task performance, this should be observable not only via self-report but also using well-defined neural measures with well-validated links to cognitive processes. If the task dimensions can be specified precisely, predictions can be made regarding the level of adaptation under different task and arousal conditions. Neuroergonomic methods can thereby facilitate tests of theoretical models of stress such as that discussed here.

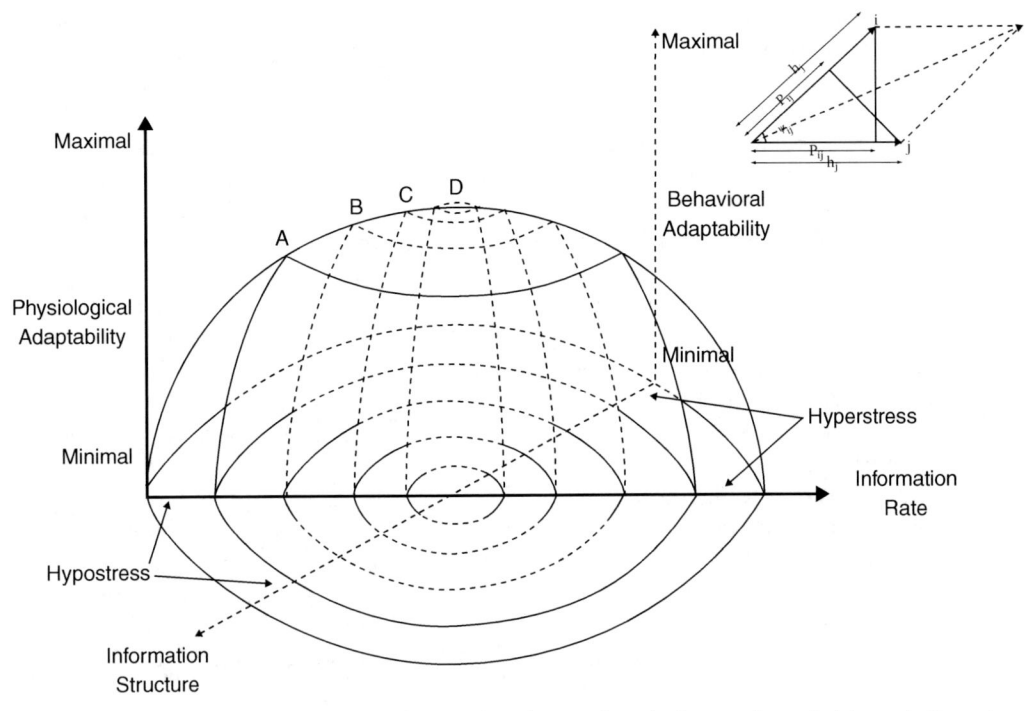

Figure 13.2. The stress-adaptation model of Hancock and Warm (1989), showing the underlying task dimensions that influence behavioral and physiological response to stress. Letters A, B, C, and D represent boundaries (tolerance limits) for physiological function (A), behavior/performance (B), subjective comfort (C), and the normative zone (D). The vector diagram at upper right illustrates how the multiple variables (i.e., information rate, information structure, and stress level) can be combined into a single vector representing cognitive and physiological state.

Stress and Neuroergonomics Research

The above considerations show that stress, like many psychological constructs, is difficult to define precisely. This represents a challenge for neuroergonomics because to identify a particular neurological state as behaviorally stressful requires a well-defined stress concept. It is also an opportunity because consideration of specific neural mechanisms for stress responses can serve to inform neuroergonomic actions at the most critical operational times. Indeed, if neuroergonomics fulfills its potential (see Hancock & Szalma, 2003; Parasuraman, 2003), it may transform the concept of stress itself. If, ultimately, cognitive states can be strongly tied to specific neurological processes, then components of stress that are currently defined psychologically (e.g., appraisal, coping, worry, distress, task engagement, etc.) may be defined in terms of their underlying neural structure and function. A note of caution is in order, however, since this reductionistic aspiration is unlikely to be fulfilled completely.

While the mechanisms by which appraisals occur are likely to be universal and nomothetic, there are almost inevitably individual differences in how thorough an appraisal is, how long it takes, and what aspects of the environment are attended to (Scherer, 1999). These may not have common neurological mechanisms. Thus, there are individual differences that occur spatially (what part of the environment has drawn one's attention, and what is its relevance to the individual) and temporally (what is the time course of appraisal, how long does it take, and is more time spent appraising some criteria over others?). Neurological activity associated with appraisal mechanisms will very probably not be identical across these two dimensions. For a successful neuroergonomic approach to stress, we will need to have the capacity to examine the neural correlates of specific cognitive patterns (cf. Hockey & Hamilton, 1983) and distinguish among varieties of appraisal and coping mechanisms.

The multidimensionality of stress (Matthews, 2001) and the likely hierarchical organization of appraisal mechanisms (Teasdale, 1999) imply that neuroergonomic measures will have to be sufficiently sensitive to delineate these dimensions of stress. For instance, considering the state "Big 3" (worry, distress, and task engagement; see Matthews et al., 1999, 2002), neuroergonomic tools would need to have the capacity to differentiate the neural processes underlying worry (cognitive), distress (mood, cognitive), and task engagement (mood, energetic, cognitive). Further, each of these dimensions has its specific components. For instance, task engagement consists of motivation, concentration, and energetic arousal (see Matthews et al., 2002). For a neuroergonomic system to adapt to operator stress, it will need to be sensitive to such facets of operator state. In addition, a useful neuroergonomic study of stress would be to develop valid neurological indices of the progressive "shoulders of failure" depicted in the Hancock and Warm (1989) model (see figure 13.1). For instance, indices that can track and predict the conditions under which an individual will transition from one curve to the next (e.g., from the comfort zone to failure of behavioral adaptation) would be very useful in aiding operator adaptation to stressful environments.

A core problem in development of stress theory is in the use of resource theory (Wickens, 1984) as an explanatory framework. With the fall of the unitary arousal concept, resource models emerged as a primary intervening variable to account for performance effects (see Hockey, Gaillard, & Coles, 1986). While resource theory has found some support and has been used in theoretical models of stress (e.g., Gopher, 1986; Hockey, 1997), it has been criticized as inadequate to the task of explaining attention allocation and human performance (e.g., Navon, 1984). In addition, the structure of resources and the mechanisms of resource control and allocation may not be common across individuals (Thropp, Szalma, & Hancock, 2004). The problem for stress theory is that a vague construct (resources) was employed to explain mechanisms by which another vague construct (stress) impacts information processing and performance. A significant contribution of neuroergonomics lies in its demand for precision, which only then permits computer and engineering-mediated changes in the effects of stress on information processing and performance. However, one must avoid the temptation to overly simplistic reductionism and recognize that neuroscience can enhance our understanding of resources (as both energetic states and processing capacity) but not necessarily replace psychological constructs with purely neurophysiological mechanisms. Fundamentally, the question for the neuroergonomics approach to stress is the same as for other approaches: How is it that individuals generally adapt to stress and maintain performance and what are the cognitive and perceptual mechanisms by which this occurs? The challenge for neuroergonomic efforts to answer this question will be development of neurological measures that are not merely outputs of information processing but reflect the processing itself and provide a direct measure of brain state.

To the degree that neuroergonomics can illuminate the above issues, it will also facilitate the development of more precise theory regarding the associations, dissociations, and insensitivities between performance and workload (Hancock, 1996; Oron-Gilad, Szalma, Stafford, & Hancock, 2005; Parasuraman & Hancock, 2001; Yeh & Wickens, 1988), as well as distinguishing between effortful and relatively effortless performance (cf. processing efficiency theory; see Eysenck & Calvo, 1992; and see also Hockey, 1997). Eventually, neuroergonomic approaches to stress research can lead to improving not only physiological measurement of workload but also in relating such measures to other forms of workload measurement (i.e., performance and subjective measures; see O'Donnell & Eggemeier, 1986).

Validation of Neuroergonomic Stress Measures

A fundamental problem facing those pursuing research in neuroergonomics is measurement (Hancock & Szalma, 2003). How does one connect mental processes to overt behavior (or physiological outputs) in a valid and reliable fashion? Driving this development of sound measures is the necessity for sound theory. If one adopts a monistic, reductionistic approach, then the ultimate result for neuroscience and neuroergonomics is the attempt to replace psychological models of stress with neurological models that specify the brain mechanisms that produce particular classes of appraisal and coping responses.

The alternative position that we posit here is a functionalist approach (see also Parasuraman, 2003) in which one postulates distinct physical and psychological constructs for a complete understanding of stress and cognition. From this perspective, neuroscience provides another vista into understanding stress and cognition that complements psychological evidence. Whichever position is taken, one must still base neuroergonomic principles on sound theoretical models of psychological and neurological function. As we have indicated in our earlier work, understanding multidimensional concepts, including stress, requires a multimethod assessment so that a more complete picture of cognition and cognitive state can be revealed (Oron-Gilad et al., 2005).

Commonalities and Differences between Individuals

Neuroergonomics is a logical extension of the critical need for stress researchers to consider individual variation in stress response. Because stress response varies as a function of task demand and the physical, social, and organizational context, it is also likely that neuroergonomic stress profiles will vary between and within individuals. Indeed, it is already known that individuals vary in cortical arousal and limbic activation and that indices of these covary with personality traits (Eysenck, 1967; Eysenck & Eysenck, 1985), although the evidence for a causal link is mixed (e.g., see Matthews & Amelang, 1993; for a review, see Matthews, Deary, & Whiteman, 2003). Application of emerging neuroergonomic technologies to adaptive systems will provide information, combined with performance differences associated with specific traits and states, that can be used to adjust systems to particular operators and to adapt as operator states change over time. Further, application of neuroscience techniques might enhance our theoretical understating of individual differences in stress response. Although many theories of stress have been developed, comprehensive theory on individual differences in stress and coping in ergonomic domains has been lacking. Neuroergonomics offers a new set of tools for individual differences researchers for both empirical investigation and theory development.

Individual differences in cognition, personality, and other aspects of behavior have traditionally been examined using the psychometric approach. Understanding the sources of such differences has also been illuminated by behavioral genetic studies of psychometric test performance. For example, this method has been used to show that general intelligence, or g, is highly heritable (Plomin & Crabbe, 2000). However, conventional behavioral genetics cannot identify the particular genes involved in intelligence or personality. The spectacular advances in molecular genetics now allow a complementary approach to behavioral genetics—*allelic association*. In this approach, normal variations in single genes, identified using DNA genotyping, are associated with individual differences in performance on cognitive tests. This method has been applied to the study of individual differences in cognition in healthy individuals, revealing evidence of modulation of attention and working memory by specific genes (Parasuraman, Greenwood, Kumar, & Fossella, 2005). Parasuraman and Caggiano (2005) have incorporated this approach into their neuroergonomic framework and discussed how molecular genetics can pinpoint the sources of individual differences and thereby provide new insights into traditional issues of selection and training. This approach could also be applied to examining individual differences in stress response.

Hedonomics and Positive Psychology

Thus far we have been concerned primarily with implications of neuroergonomics for stress. There is also an opportunity to explore the application of neuroergonomics to the antithesis of stressful conditions. The latter study has been termed *hedonomics*, which is defined as "that branch of science which facilitates the pleasant or enjoyable aspects of human-technology interaction" (Hancock, Pepe, & Murphy, 2005, p. 8). This represents an effort not simply to alleviate the bad but also to promote the good.

Studies of the level of pleasure experienced by individuals are rooted in classic effects of limbic stimulation on behavior (Olds & Milner, 1954). Indeed, neuroergonomic indices of attention might clarify the processes that occur for maladaptive versus adaptive attentional narrowing, recovery from stress and performance degradation (i.e., hysteresis). In addition, neuroergonomics can contribute to the emerging positive psychology trend (Seligman & Csikszentmihalyi, 2000) by identifying the neurological processes underlying flow states (Csikszentmihalyi, 1990) in which individuals are fully engaged in a task and information processing is more automatic and fluid rather than controlled and effortful.

Ethical Issues in Neuroergonomics and Stress

There is a pervasive question that must attend the development of all new human-machine technologies, and that is the issue of purpose (Hancock, 1997). Subsumed under purpose are issues such as morality, benefit, ethics, aesthetics, cost, and the like. As we have merely introduced elsewhere (Hancock & Szalma, 2003), there are issues of privacy, information ownership, and freedom that will also have to be considered as cognitive neuroscience develops and is applied to the design of work. In regard to stress, there is a particular danger that those who are naturally prone to specific patterns of stress response (e.g., those high in neuroticism or trait anxiety) may be excluded from opportunities or even punished by controlling authorities for their personality characteristics. If we assume, however, that the rights of individuals have been secured, application of neuroergonomic technologies to monitor emotional states of individuals could provide useful information for adjusting system status to the characteristics of the individual. Thus, those who are prone to trait anxiety could be assisted by an automated system when it is detected that their state anxiety is increasing to levels that make errors or other performance failures more likely. In such an application, the technology could serve to increase the performance of anxious individuals to a level comparable to that of individuals low in trait anxiety. More generally, neuroergonomics offers a new avenue for research and application in individual differences in coping and responses to stress, as a technology of inclusion rather than a technology of exclusion. In many modern systems, the environment is sufficiently flexible that it can be adapted to the characteristics and current cognitive state of the individual operator. This can improve performance, reduce the workload and stress associated with the task, and perhaps even render the task more enjoyable.

As neuroergonomic interventions emerge, consideration of intention and privacy will be crucial. In the age of ever-increasing technology, the opportunity for any individual to preserve his or her own private experience is vastly diminished. The proliferation of video technology alone means that events that have previously remained hidden behind a barrier of institutional silence now become open to public inspection. The modern threat of terrorism has also engendered new categories of surveillance technologies which use sophisticated software that seeks to process the nominal intent of individuals. Added to these developments are the evolutions in detection technologies that use evidence such as DNA traces to track the presence of specific individuals. Such developments indicate that overt actions are now detectable and recordable and thus potentially punishable. Further, this invasion is moving to the realm of speech and communication. Recorded by sophisticated technologies, utterances regarding (allegedly) maleficent intent now become culpable evidence. Despite a nominal freedom of expression, one cannot offer threats, conspire to harm, or engage in certain forms of conversation without the threat of detection and punishment. Many would argue that destructive acts should be punished, and that voicing violent intent is also culpable, whether the intent is fulfilled or not (hence the ambivalence regarding jokes concerning exploding devices at airport screening facilities). However, neuroergonomics, if the vision is fully or even partly fulfilled, now promises to extend these trends further. It will not only be actions and language that could be considered culpable but, more dangerously, the thought itself. As Marcus Aurelius rightly noted, we are the sole observer and arbiter of our own personal private experience. We may safely and unimpeachably think unpleasant thoughts about anyone or anything, content in the knowledge that our thoughts are private and impenetrable to any agency or individual. Neuroergonomics could threaten this fundamental right and indeed, in so doing, threaten what it is to be an individual human being.

We paint this picture not to discourage the pursuit of neuroergonomics, but rather to sound a cautionary alarm. While neuroergonomic interventions have the potential to provide early warning that an individual is overstressed and therefore could be used to mitigate the negative effects of such stress, the imposition of the neuroergonomic intervention could itself impose significant stress on the individual, particularly if those in authority (e.g., company management, governmental agencies) have access to the information provided by neurological and physiological measures, and individuals appraise this situation as a potential threat with which they cannot effectively cope.

Conclusion

Clearly, one application of neuroergonomics to stress will be a set of additional tools to monitor operator state for markers predictive of later performance failure and, based on such information, adjusting system behavior to mitigate the stress. However, to reduce stress effectively requires an understanding of its etiology and the mechanisms by which it occurs and affects performance. To address these issues requires development of good theoretical models, and we see the neuroergonomic approach to stress research making a significant contribution toward this development. As a multidimensional construct, stress requires multidimensional assessment, and improvement of theory requires that the relations among the dimensions be well articulated. Neuroergonomics will facilitate this process. The key will be to develop reliable and valid neurological metrics relating to cognitive and energetic states and the processes underlying the recruitment, allocation, and depletion of cognitive resources. Our optimism is not unguarded, however. First, neurological measures should be viewed as one piece of a complex puzzle rather than as a reductive replacement for other indices of stress. Omission of psychological constructs and measures would weaken the positive impact of neuroscience for stress research and stress mitigation efforts. Second, in implementing stress research, we must ensure that the application of neuroergonomics to stress mitigation does not increase stress by imposing violations of security and privacy for the individual. If these concerns are adequately addressed, neuroergonomics could not only revolutionize stress as a psychological construct but could also serve to transform the experience of stress itself.

On a more general front, neuroergonomics could be used to mitigate stress completely. With sufficient understanding of neural stress states and their precursors in both the environment and the operator's appraisal of that environment, it would be possible, in theory, to develop a servomechanistic system whereby sources of stress were ablated as soon as they arose. But would this be a good thing? It is not that we enjoy certain adverse situations, but it may be that it is the stimulation of such adverse conditions that spur us to higher achievement. This brings us to our final observations on the very thorny issue of the purpose of technology itself. Is the purpose of technology to eradicate all human need and as a corollary to this, to instantly and effortlessly grant all human physical desires? We suggest not. Indeed, such a state of apparent *dolce far niente* (life without care) might well prove the equivalent of the medieval view of hell! Further, it is currently unclear how to respond to mistakes in neuroergonomics if the purpose is to facilitate the immediate transition from an intention to an action. Complex error recovery processes are built into the current human motor system—will such effective guards be embedded into future neuroergonomics systems? Some scientists have opined that all technology is value neutral in that it can be employed for both laudable and detestable purposes. But this is a flawed argument because the creation of each new embodiment of technology is an intentional act that itself expresses an intrinsic value-based decision (see Hancock, 1997). To whatever degree that value is apparently hidden in the process of conception, design, and fabrication, and to whatever degree others choose to pervert that original intention, the act itself implies value. Thus, we need at the present stage of development to consider not whether we can develop neuroergonomics technologies but rather, whether we should. Needless to say, this is liable to be a somewhat futile discussion since rarely, if ever, in human history have we refrained from doing what is conceived as being possible, whether we should or not. Indeed, it is this very motivation that may well be the demise of the species. To end on a more hopeful note: perhaps not. Perhaps the present capitalist-driven global structure will consider the greater good of all individuals (and indeed all life) and refrain from the crass, materialistic exploitation of whatever innovations are realized. Then again—perhaps not. And that was a hopeful note.

Main Points

1. Stress is a multidimensional construct that requires multidimensional assessment. Neuroergonomics promises to provide valuable tools for this effort.

2. A major problem for stress research is the difficulty in precisely defining the concepts of stress and cognitive resources. Neuroergonomic efforts should be directed toward elucidating these constructs.
3. Neuroergonomics can improve the state of stress theory via programmatic research toward establishing the links between neurological and cognitive states. This is critical, since the validity of neuroergonomic measures depends heavily on sound psychological theory.
4. A potential practical application for stress mitigation will be the ability to monitor operator state in real time so that systems can adapt to those states as operators experience stress. Such efforts, in the context of neuroergonomics, are already underway.
5. Neuroergonomics can also be useful for studying individual differences in stress and coping and establishing general theoretical framework for individual differences in performance, stress, and workload.
6. While the promise for neuroergonomics is high, we must ensure that individuals' privacy and well-being are preserved so that the cure does not become worse than the disease.

Key Readings

Hancock, P. A., & Desmond, P. A. (Eds.). (2001). *Stress, workload, and fatigue*. Mahwah, NJ: Erlbaum.

Hancock, P. A., & Warm, J. S. (1989). A dynamic model of stress and sustained attention. *Human Factors, 31*, 519–537.

Hockey, G. R. J., Gaillard, A. W. K., & Coles, M. G. H. (Eds.). (1986). *Energetics and human information processing*. Dordrecht: Martinus Nijhoff.

Hockey, R., & Hamilton, P. (1983). The cognitive patterning of stress states. In: G. R. J. Hockey (Ed.), *Stress and fatigue in human performance* (pp. 331–362). Chichester: Wiley.

Lazarus, R. S., & Folkman, S. (1984). *Stress, appraisal, and coping*. New York: Springer-Verlag.

References

Asterita, M. F. (1985). *The physiology of stress*. New York: Human Sciences Press.

Cannon, W. (1932). *The wisdom of the body*. New York: Norton.

Csikszentmihalyi, M. (1990). *Flow: The psychology of optimal experience*. New York: Harper.

Dunbar, F. (1954). *Emotion and bodily changes*. New York: Columbia University Press.

Easterbrook, J. A. (1959). The effect of emotion on cue utilization and the organization of behavior. *Psychological Review, 66*, 183–201.

Elliot, G. R., & Eisdorfer, C. (1982). *Stress and human health*. New York: Springer.

Eysenck, H. J. (1967). *The biological basis of personality*. Springfield, IL: Charles Thomas.

Eysenck, H. J., & Eysenck, M. W. (1985). *Personality and individual differences: A natural science approach*. New York: Plenum.

Eysenck, M. W., & Calvo, M. (1992). Anxiety and performance: The processing efficiency theory. *Cognition and Emotion, 6*, 409–434.

Frankenhaeuser, M. (1986). A psychobiological framework for research on human stress and coping. In M. H. Appley & R. Trumball (Eds.), *Dynamics of stress: Physiological, psychological, and social perspectives* (pp. 101–116). New York: Plenum.

Gopher, D. (1986). In defence of resources: On structure, energies, pools, and the allocation of attention. In G. R. J. Hockey, A. W. K. Gaillard, & M. G. H. Coles (Eds.), *Energetics and human information processing* (pp. 353–371). Dordrecht: Martinus Nijhoff.

Hancock, P. A. (1996). Effects of control order, augmented feedback, input device and practice on tracking performance and perceived workload. *Ergonomics, 39*, 1146–1162.

Hancock, P. A. (1997). *Essays on the future of human-machine systems*. Eden Prairie, MN: Banta.

Hancock, P. A., & Desmond, P. A. (Eds.). (2001). *Stress, workload, and fatigue*. Mahwah, NJ: Erlbaum.

Hancock, P. A., & Dirkin, G. R. (1983). Stressor induced attentional narrowing: Implications for design and operation of person-machine systems. *Proceedings of the Human Factors Association of Canada, 16*, 19–21.

Hancock, P. A., & Ganey, H. C. N. (2003). From the inverted-U to the extended-U: The evolution of a law of psychology. *Human Performance in Extreme Environments, 7*(1), 5–14.

Hancock, P. A., Pepe, A., & Murphy, L. L. (2005). Hedonomics: The power of positive and pleasurable ergonomics. *Ergonomics in Design, 13*, 8–14.

Hancock, P. A., & Szalma, J. L. (2003). The future of neuroergonomics. *Theoretical Issues in Ergonomics Science, 4*, 238–249.

Hancock, P. A., & Warm, J. S. (1989). A dynamic model of stress and sustained attention. *Human Factors, 31*, 519–537.

Hancock, P. A., & Weaver, J. L. (2005). On time distortion under stress. *Theoretical Issues in Ergonomics Science, 6,* 193–211.

Hebb, D. O. (1955). Drives and the CNS (conceptual nervous system). *Psychological Review, 62,* 243–254.

Hockey, G. R. J. (1986). Changes in operator efficiency as a function of environmental stress, fatigue, and circadian rhythms. In K. R. Boff, L. Kaufman, & J. P. Thomas (Eds.), *Handbook of human perception and performance: Vol. II. Cognitive processes and performance* (pp. 1–49). New York: Wiley.

Hockey, G. R. J. (1997). Compensatory control in the regulation of human performance under stress and high workload: A cognitive-energetical framework. *Biological Psychology, 45,* 73–93.

Hockey, G. R. J., Gaillard, A. W. K., & Coles, M. G. H. (Eds.). (1986). *Energetics and human information processing.* Dordrecht: Martinus Nijhoff.

Hockey, R. (1984). Varieties of attentional state: The effects of environment. In R. Parasuraman & D. R. Davies (Eds.), *Varieties of attention* (pp. 449–483). New York: Academic Press.

Hockey, R., & Hamilton, P. (1983). The cognitive patterning of stress states. In G. R. J. Hockey (Ed.), *Stress and fatigue in human performance* (pp. 331–362). Chichester: Wiley.

Holland, F. G., & Hancock, P. A. (1991, June). *The inverted-U: A paradigm in chaos.* Paper presented at the annual meeting of the North American Society for the Psychology of Sport and Physical Activity, Asilomar, CA.

Hovanitz, C. A., Chin, K., & Warm, J. S. (1989). Complexities in life stress-dysfunction relationships: A case in point—tension headache. *Journal of Behavioral Medicine, 12,* 55–75.

Jones, D. (1984). Performance effects. In D. M. Jones & A. J. Chapman (Eds.), *Noise and society* (pp. 155–184) Chichester: Wiley.

Lazarus, R. S., & Folkman, S. (1984). *Stress, appraisal, and coping.* New York: Springer-Verlag.

Matthews, G. (2001). Levels of transaction: A cognitive sciences framework for operator stress. In P. A. Hancock & P. A. Desmond (Eds.), *Stress, workload, and fatigue* (pp. 5–33). Mahwah, NJ: Erlbaum.

Matthews, G., & Amelang, M. (1993). Extraversion, arousal theory, and performance: A study of individual differences in the EEG. *Personality and Individual Differences, 14,* 347–364.

Matthews, G., Campbell, S. E., Falconer, S., Joyner, J. A., Huggins, J., Gilliland, K., et al. (2002). Fundamental dimensions of subjective state in performance settings: Task engagement, distress, and worry. *Emotion, 2,* 315–340.

Matthews, G., Deary, I. J., & Whiteman, M. C. (2003). *Personality traits* (2nd ed.). Cambridge, UK: Cambridge University Press.

Matthews, G., Joyner, L., Gilliland, K., Campbell, S., Falconer, S., & Huggins, J. (1999). Validation of a comprehensive stress state questionnaire: Towards a state "big three"? In I. Mervielde, I. J. Deary, F. DeFruyt, & F. Ostendorf (Eds.), *Personality psychology in Europe* (Vol. 7, pp. 335–350). Tilburg: Tilburg University Press.

McGrath, J. J. (1970). A conceptual framework for research on stress. In J. J. McGrath (Ed.), *Social and psychological factors in stress* (pp. 10–21). New York: Holt, Rinehart, and Winston.

Navon, D. (1984). Resources: A theoretical soup stone? *Psychological Review, 91,* 216–234.

O'Donnell, R. D. & Eggemeier, F. T. (1986). Workload assessment methodology. In K. R. Boff, L. Kaufman, & J. P. Thomas (Eds.), *Handbook of human performance: Vol. 2. Cognitive processes and performance* (pp. 1–49). New York: Wiley.

Olds, J., & Milner, P. (1954). Positive reinforcement produced by electrical stimulation of septal area and other regions in the rat brain. *Journal of Comparative and Physiological Psychology, 49,* 281–285.

Oron-Gilad, T., Szalma, J. L., Stafford, S. C., & Hancock, P. A. (2005). *On the relationship between workload and performance.* Manuscript submitted for publication.

Parasuraman, R. (1984). The psychobiology of sustained attention. In J. S. Warm (Ed.), *Sustained attention in human performance* (pp. 61–101). Chichester: Wiley.

Parasuraman, R. (2003). Neuroergonomics: Research and practice. *Theoretical Issues in Ergonomics Science, 1–2,* 5–20.

Parasuraman, R., & Caggiano, D. (2005). Neural and genetic assays of mental workload. In D. McBride & D. Schmorrow (Eds.), *Quantifying human information processing* (pp. 123–155). Lanham, MD: Rowman and Littlefield.

Parasuraman, R., Greenwood, P. M., Kumar, R., & Fossella, J. (2005). Beyond heritability: Neurotransmitter genes differentially modulate visuospatial attention and working memory. *Psychological Science, 16*(3), 200–207.

Parasuraman, R., & Hancock, P. A. (2001). Adaptive control of mental workload. In P. A. Hancock & P. A. Desmond (Eds.), *Stress, workload, and fatigue* (pp. 305–320). Mahwah, NJ: Erlbaum.

Pilcher, J. J., Nadler, E., & Busch, C. (2002). Effects of hot and cold temperature exposure on performance: A meta-analytic review. *Ergonomics, 45,* 682–698.

Plomin, R., & Crabbe, J. (2000). DNA. *Psychological Bulletin, 126,* 806–828.

Scherer, K. R. (1999). Appraisal theory. In T. Dalgleish & M. J. Power (Eds.), *Handbook of cognition and emotion* (pp. 637–663). Chichester: Wiley.

Seligman, M. E. P., & Csikszentmihalyi, M. (2000). Positive psychology: An introduction. *American Psychologist, 55*, 5–14.

Selye, H. (1976). *The stress of life* (Rev. ed.). New York: McGraw-Hill.

Szalma, J. L., & Hancock, P. A. (2002). *On mental resources and performance under stress*. White Paper, MIT^2 Laboratory, University of Central Florida. Available at www.mit.ucf.edu.

Teasdale, J. D. (1999). Multi-level theories of cognition-emotion relations. In T. Dalgleish & M. J. Power (Eds.), *Handbook of cognition and emotion* (pp. 665–681). Chichester: Wiley.

Thropp, J. E., Szalma, J. L., & Hancock, P. A. (2004). Performance operating characteristics for spatial and temporal discriminations: Common or separate capacities? *Proceedings of the Human Factors and Ergonomics Society, 48*, 1880–1884.

Wickens, C. D. (1984). Processing resources in attention. In R. Parasuraman & D. R. Davies (Eds.) *Varieties of attention* (pp. 63–102). San Diego: Academic Press.

Wickens, C. D., & Hollands, J. G. (2000). *Engineering psychology and human performance* (3rd ed.). Upper Saddle River, NJ: Prentice Hall.

Yeh, Y. Y., & Wickens, C. D. (1988). Dissociations of performance and subjective measures of workload. *Human Factors, 30*, 111–120.

Melissa M. Mallis, Siobhan Banks, and David F. Dinges

Sleep and Circadian Control of Neurobehavioral Functions

Overview of Sleep and Circadian Rhythms

Most organisms show daily changes in their behavior and physiology that are not simply controlled by external stimuli in the environment. In mammals, these 24-hour cycles, otherwise knows as circadian rhythms, are primarily controlled by an internal clock called the *suprachiasmatic nucleus* (SCN), located in the hypothalamus (Moore, 1999). These cycles can be synchronized to external time signals, but they can also persist in the absence of such signals. In the absence of time cues, in humans, the SCN shows an average "free running" intrinsic period of 24.18 hours (Czeisler et al., 1999). However, the SCN is entrained to the 24-hour day via *zeitgebers*, or time givers, of which the strongest is light. This endogenous circadian pacemaker affects many physiological functions, including core body temperature, plasma cortisol, plasma melatonin, alertness, and sleep patterns. The nadir of the circadian component of the endogenous core body temperature rhythm is associated with an increased sleep propensity (Dijk & Czeisler, 1995).

For most animals, the timing of sleep and wakefulness under natural conditions is in synchrony with the circadian control of the sleep cycle and all other circadian-controlled rhythms. Humans, however, have the unique ability to cognitively override their internal biological clock and its rhythmic outputs. When the sleep-wake cycle is out of phase with the endogenous rhythms that are controlled by the circadian clock (e.g., during night shift work or rapid travel across time zones), adverse effects will result (Dijk & Czeisler, 1995). The synchrony of an organism with both its external and internal environments is critical to the organism's well-being and survival. A disruption of this synchrony can result in a range of difficulties, such as impaired cognitive function, sleepiness, altered hormonal function, and gastrointestinal complaints.

In addition to the circadian component, another fundamental regulatory process is involved in programming sleep and alertness. The less (good-quality) sleep one obtains daily, the greater will be the homeostatic drive for sleep. The neurobiological mechanisms underlying homeostatic sleep pressure are beginning to be identified (Mignot, Taheri, & Nishino, 2002; Saper, Chou, & Scammell, 2001). The mechanisms promoting wakefulness and sleep appear to involve a number of neurotransmitters and nuclei located in the forebrain, midbrain, and brain stem (Mignot, Taheri, & Nishino, 2002; Saper,

Chou, & Scammell, 2001). For example, one theory posits that adenosine tracks lost sleep and may induce sleep when the homeostat is elevated due to being awake for too long a period, or when sleep quality or quantity are chronically inadequate (Basheer, Strecker, Thakkar, & McCarley, 2004). The homeostatic regulation of sleep interacts nonlinearly with the circadian cycle to produce dynamic changes in the propensity and stability of sleep and waking across each 24-hour period and across days (Van Dongen & Dinges, 2000).

Slow-wave sleep—and especially slow-wave brain activity during sleep—are currently considered the classical markers of the sleep homeostatic process. There is evidence that the time course of the sleep homeostatic process can also be monitored during wakefulness in humans using oculomotor measures (see below). In addition, electroencephalographic (EEG) studies have suggested that certain slow-frequency components of brain activity increase as the duration of wakefulness increases. Scheduling multiple naps during the day attenuates this increase in activity during wakefulness (Cajochen, Knoblauch, Krauchi, Renz, & Wirz-Justice, 2001). It thus appears that low-frequency components of EEG during wakefulness are closely associated with sleep homeostasis. Together, homeostatic and circadian processes determine degree and timing of daytime alertness and cognitive performance (Czeisler & Khalsa, 2000; Durmer & Dinges, 2005; Van Dongen & Dinges, 2000).

Types of Sleep Deprivation

Sleep deprivation can result from either partial or total loss of sleep, which can be either voluntary or involuntary, and it can range in duration from acute to chronic. *Partial sleep deprivation* occurs when an individual is prevented from obtaining a portion of the sleep needed to produce normal waking alertness during the daytime. Typically, this occurs when an individual's sleep time is restricted in duration or is fragmented for environmental or medical reasons. The effects of both acute and chronic partial sleep deprivation on a range of neurobehavioral and physiological variables have been examined over the years using a variety of protocols. These have included restricting time in bed for sleep opportunities in continuous and distributed schedules, gradual reductions in sleep duration over time, selective deprivation of specific sleep stages, and situations where the time in bed is reduced to a percentage of the individual's habitual time in bed.

Many early experiments on the effects of chronic partial sleep restriction concluded that voluntarily reducing nightly sleep duration to between 4 and 6 hours had little adverse effect. This led to the belief that one could adapt to reduced sleep. However, most of these early studies lacked key experimental controls and relied on small sample sizes and limited measurements (Dinges, Baynard, & Rogers, 2005). More recent controlled experiments on the cognitive and neurobehavioral effects of chronic partial sleep restriction corrected for these methodological weaknesses and found that the loss of alertness and performance capability from chronic sleep restriction got worse across days and showed dose-response relationships to the amount of sleep obtained (Belenky et al., 2003; Dinges et al., 1997; Van Dongen, Maislin, Mullington, & Dinges, 2003). When sleep duration was reduced below 7 hours per night, daytime functions deteriorated.

In contrast to partial sleep deprivation, *total sleep deprivation* occurs when no sleep is obtained and the waking period exceeds 16 hours in a healthy adult. Total sleep deprivation that extends beyond 24 hours—such as occurs in sustained or critical operations—reveals the nonlinear interaction of the escalating sleep homeostat and the endogenous circadian clock (Van Dongen & Dinges, 2000). This interaction manifests a counterintuitive outcome—namely, an individual who remains awake for 40 hours (i.e., a day, a night, and a second day) is less impaired from sleepiness at 36–38 hours of wakefulness than at 22–24 hours of wakefulness. Figure 14.1 displays this interaction manifesting in a number of cognitive functions during acute total sleep deprivation.

There is a third condition that can interact with sleep deprivation and endogenous circadian phase, called *sleep inertia*. Sleep inertia describes the grogginess and disorientation that a person feels for minutes to hours after awakening from sleep. It is a transitional brain state between sleep and waking, which can have severe effects on cognitive functions involving attention and memory. Sleep inertia interacts with sleep homeostatic drive and circadian phase. If prior sleep duration is normal (e.g., more than 7 hours), sleep inertia is typically modest and short-lived (Achermann, Werth, Dijk, & Borbely, 1995). The sleep stage prior to awakening is also an important factor in amplifying sleep inertia in that waking up from slow-wave sleep (SWS) is worse than waking up from REM (rapid eye movement) sleep (Ferrara, De Gennaro, & Bertini, 1999). Additionally, it has been found that the existence of prior sleep deprivation increases the intensity and duration of sleep inertia, as does awakening from sleep near the circadian nadir (Naitoh, Kelly, & Babkoff, 1993). Interestingly, there is evidence that caffeine can block sleep inertia (Doran, Van Dongen, & Dinges, 2001), which may explain why this common stimulant is much sought after in the morning, after a night of sleep.

Neurobehavioral and Neurocognitive Consequences of Inadequate Sleep

Cognitive performance degrades with sleep loss, which in operational environments is often referred to as the effect of fatigue—however, this latter term has historically also referred to performance decrements as a function of time on task (i.e., work time). There is now extensive evidence to support the view that a vast majority of instances in which fatigue affects performance are due directly to inadequate sleep or to functioning at a nonoptimal circadian phase. The effects of sleep loss on cognitive performance are primarily manifested as increasing variability in cognitive speed and accuracy. Thus, behavioral responses become unpredictable with increasing amounts of fatigue. When this increased performance variability occurs as a result of sleep deprivation (acute or chronic, partial or total), it is thought to reflect state instability (Doran, Van Dongen, & Dinges, 2001). *State instability* refers to moment-to-moment shifts in the relationship between neurobiological systems mediating wake maintenance and those

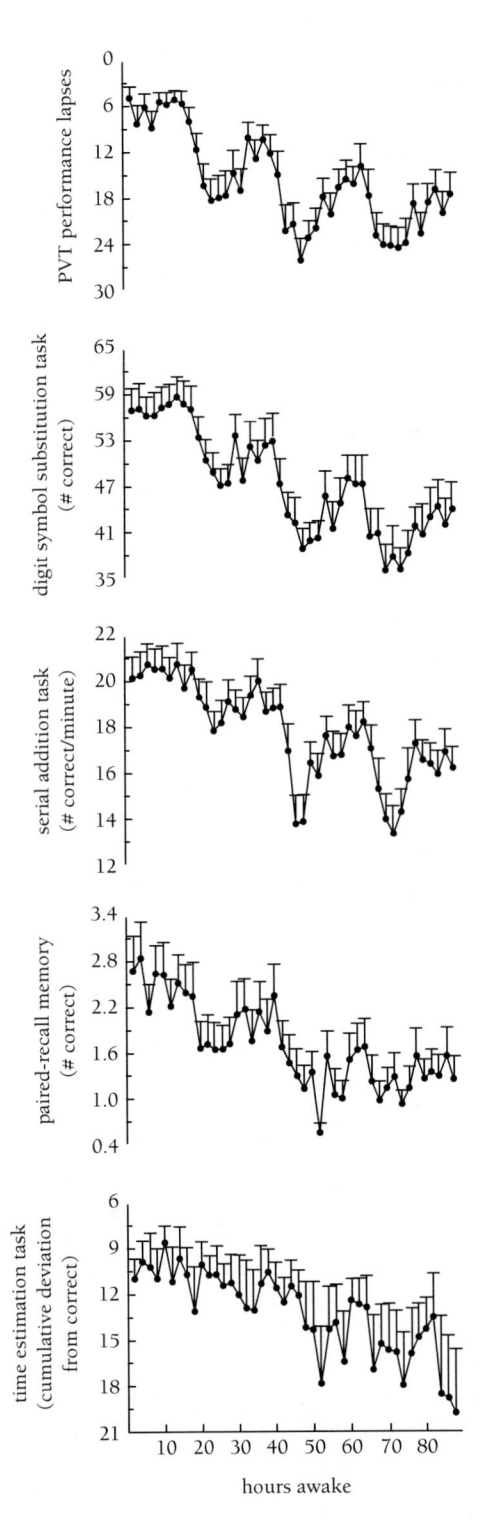

Figure 14.1. Performance decrements on five cognitive tasks during 88 hours of continual wakefulness. In this experiment, subjects ($N = 24$) were tested every 2 hours on a 30-minute performance battery, beginning at 8 a.m. on the first day. The panels show performance profiles for lapses of sustained attention performance on the psychomotor vigilance task (PVT; Dorrian et al., 2005); decreases in performance on working memory and cognitive throughput tasks (digit symbol substitution task, serial addition task); short-term memory (paired-recall memory); and subjective accuracy (time estimation). Each point represents the mean (*SEM*) for a test bout during the 88 hours of total sleep deprivation.

mediating sleep initiation (Mignot et al., 2002; Saper et al., 2001). The increased propensity for sleep as well as the tendency for performance to show behavioral lapses, response slowing, time-on-task decrements, and errors of commission (Doran et al., 2001) are signs that sleep-initiating mechanisms deep in the brain are activating during wakefulness. Thus, the cognitive performance variability that is the hallmark of sleep deprivation (Dinges & Kribbs, 1991) appears to reflect state instability (Dorrian, Rogers, & Baynard, 2005; Durmer & Dinges, 2005). Sleep-initiating mechanisms repeatedly interfere with wakefulness, making cognitive performance increasingly variable and dependent on compensatory measures, such as motivation, which cannot override elevated sleep pressure without consequences (e.g., errors of commission increase as subjects try to avoid errors of omission—lapses; Durmer & Dinges, 2005; Doran et al., 2001).

Intrusions of sleep into goal-directed performance are evident in increases in a variety of neurobehavioral phenomena: lapses of attention, sleep attacks (i.e., involuntary naps), increased frequency of voluntary naps, shortened sleep latency, slow eyelid closures and slow-rolling eye movements, and intrusive daydreaming while engaged in cognitive work (Dinges & Kribbs, 1991; Kleitman, 1963). Remarkably, these phenomena can occur even in healthy sleep-deprived people engaged in potentially dangerous activities such as driving. Sleepiness-related motor vehicle crashes have a fatality rate and injury severity level similar to alcohol-related crashes. Sleep deprivation has been shown to produce psychomotor impairments equivalent to those induced by alcohol consumption at or above the legal limit (Durmer & Dinges, 2005).

Sleep deprivation degrades many aspects of neurocognitive performance (Dinges & Kribbs, 1991; Dorrian & Dinges, 2005; Durmer & Dinges, 2005; Harrison & Horne, 2000). In addition, there are hundreds of published studies on the cognitive effects of sleep deprivation showing that all forms of sleep deprivation result in increased negative mood states, especially feelings of fatigue, loss of vigor, sleepiness, and confusion. Although feelings of irritability, anxiety, and depression are believed to result from inadequate sleep, most experimental settings do not find such changes owing to the comfortable and predictable environment in which subjects are studied. On the other hand, increased negative mood states have been observed often when sleep deprivation occurs in complex real-world conditions.

Table 14.1. Summary of Cognitive Performance Effects of Sleep Deprivation

Involuntary microsleeps occur, which can lead to increasingly long involuntary naps.

Performance on attention-demanding tasks, such as vigilance, is unstable with increased errors of omission and commission.

Reaction time slows.

Time pressure increases cognitive errors, and cognitive slowing occurs in subject-paced tasks.

Working memory and short-term recall decline.

Reduced learning (acquisition) of cognitive tasks.

Performance requiring divergent thinking (e.g., multitasking) deteriorates.

Response perseveration on ineffective solutions is more likely.

Increased compensatory effort is required to remain behaviorally effective.

Task performance deteriorates as task duration increases (e.g., vigilance).

Neglect of activities considered nonessential (i.e., loss of situational awareness).

Sleep deprivation induces a wide range of effects on cognitive performance (see table 14.1). In general, cognitive performance becomes progressively worse when time on task is extended—this is the classic fatigue effect that is exacerbated by sleep loss. However, performance on even very brief cognitive tasks that require speed of cognitive throughput, working memory, and other aspects of attention have been found to be sensitive to sleep deprivation. Cognitive work involving the prefrontal cortex (e.g., divergent thinking) is adversely affected by sleep loss. Divergent skills involved in decision making that decrease with sleep loss include assimilation of changing information, updating strategies based on new information, lateral thinking, innovation, risk assessment, maintaining interest in outcomes, mood-appropriate behavior, insight, communication, and temporal memory skills (Harrison & Horne, 2000). Implicit to divergent thinking abilities is a heavy reliance on executive functions—especially working memory and control of attention (see also chapter 11, this volume). Working memory and executive attention involve the ability to hold and manipulate information and can involve multiple sensory-motor modalities.

Therefore, deficits in neurocognitive performance due to sleep loss can compromise the following functions that depend on good working memory: assessment of the scope of a problem due to changing or distracting information, remembering the temporal order of information, maintaining focus on relevant cues, maintaining flexible thinking, avoiding inappropriate risks, gaining insight into performance deficits, avoiding perseveration on ineffective thoughts and actions, and making behavioral modifications based on new information (Durmer & Dinges, 2005). Although executive functions compromised by sleep deprivation are mediated by changes in prefrontal and related cortical areas, the effects of sleep loss likely originate in subcortical systems (hypothalamus, thalamus, and brain stem). This may explain why the most sensitive performance measure of sleep deprivation is a simple sustained attention task, such as the psychomotor vigilance task (PVT; Dorrian, Rogers, & Dinges, 2005).

Rest Time and the Effects of Chronic Partial Sleep Deprivation

Among the most problematic issues in work-rest regulations is the question of how much rest (time off work) should be mandated to ensure that workers avoid sleep deprivation. Instantiated in many federal regulatory work rules (e.g., in each major transportation modality) is the assumption that allowing 8 hours for rest between work periods will result in adequate recovery sleep to avoid deprivation. Virtually all studies of sleep during varying work-rest schemes reveal that actual physiological sleep accounts for only about 50–75% of rest time, which means that people allowed 8 hours to recover actually sleep 4–6 hours at most. Early experiments on chronic restriction of sleep to 4–6 hours per night found few adverse effects on performance measures, but these studies lacked key experimental controls and relied on small sample sizes and limited measurements (Dinges, Baynard, & Rogers, 2005). More recent carefully controlled experiments on healthy adults found clear and dramatic evidence that behavioral alertness and a range of cognitive performance functions involving sustained attention, working memory, and cognitive throughput deteriorated systematically across days when nightly sleep duration was between 4 and 7 hours (Belenky et al., 2003; Van Dongen et al., 2003). In contrast, when time in bed for sleep was 8 hours (Van Dongen et al., 2003) or 9 hours (Belenky et al., 2003), no cumulative cognitive performance deficits were found across days. Figure 14.2 shows comparable data from each of these studies. In one study, truck drivers were randomized to 7 nights of 3, 5, 7 or 9 hours of time in bed for sleep per night (Belenky et al., 2003). Subjects in the 3- and 5-hour time-in-bed groups experienced a decrease in performance across days of the sleep restriction protocol, with an increase in the mean reaction time, number of lapses, and fastest reaction times on the PVT. In the subjects allowed 7 hours of time in bed per night, a significant decrease in mean response speed was also evident, although no effect on lapses was evident. Performance in the group allowed 9 hours of time in bed was stable across the 7 days.

In an equally large experiment (Van Dongen et al., 2003), adults (mean age 28 years) had their sleep duration restricted to 4, 6, or 8 hours of time in bed per night for 14 consecutive nights. Cumulative daytime deficits in cognitive function were observed for lapses on the PVT, for a memory task, and for a cognitive throughput task. These deficits worsened over days of sleep restriction at a faster rate for subjects in the 4- and 6-hour sleep periods relative to subjects in the 8-hour control condition, which showed no cumulative performance deficits. In order to quantify the magnitude of cognitive deficits experienced during 14 days of restricted sleep, the findings from the 4-, 6-, or 8-hour sleep periods were compared with cognitive effects after 1, 2, and 3 nights of total sleep deprivation (Van Dongen et al., 2003). This comparison revealed that both 4- and 6-hour sleep periods resulted in cumulative cognitive impairments that quickly increased to levels found after 1, 2, and even 3 nights of total sleep deprivation.

These studies suggest that when the nightly recovery sleep period is routinely restricted to 7 hours or less, the majority of motivated healthy adults develop cognitive performance impairments that systematically increase across days, until a longer duration (recovery) sleep period is provided. When nightly sleep periods in these two major experiments were 8–9 hours, no cognitive deficits were found. These data indicate that the basic principle of providing 8 hours for rest between work bouts in regulated industries is inadequate, unless people actually sleep 90% of the time allowed.

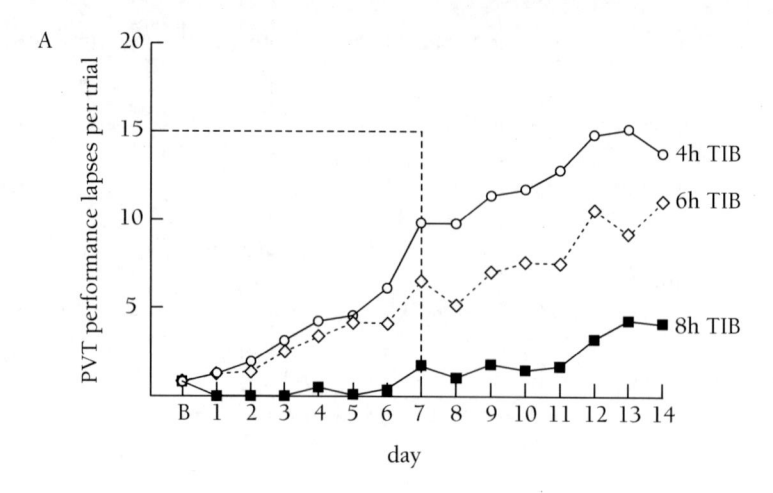

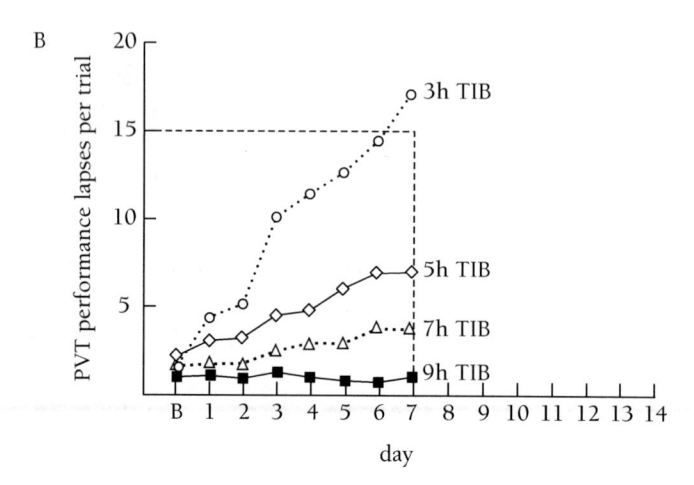

Figure 14.2. Results from two dose-response studies of chronic sleep restriction. Panel A is from Van Dongen et al. (2003). In this experiment, sleep was restricted for 14 consecutive nights in 36 healthy adults (mean age 30 years). Subjects were randomized to 4-hour ($n = 13$), 6-hour ($n = 13$) or 8-hour ($n = 9$) time in bed (TIB) at night. Performance was assessed every 2 hours (9 times each day) from 7:30 a.m. to 11:30 p.m. The graph shows cumulative increases in lapses of attention during the psychomotor vigilance task (PVT; Dorrian et al., 2005) per test bout across days within the 4-hour and 6-hour groups ($p = .001$), with sleep-dose differences between groups ($p = .036$). The horizontal dotted line shows the level of lapsing found in a separate experiment when subjects had been awake continuously for 64–88 hours. For example, by Day 7, subjects in the 6-hour TIB condition averaged 54 total lapses for the 9 test trials that day, while those in the 4-hour TIB averaged 70 lapses per day. Panel B shows data from Belenky et al. (2003). In this experiment, sleep was restricted for 7 consecutive nights in 66 healthy adults (mean age 48 years). Subjects were randomized to 3-hour ($n = 13$), 5-hour ($n = 13$), 7-hour ($n = 13$), or 9-hour ($n = 16$) TIB at night. Performance was assessed 4 times each day from 9 a.m. to 9 p.m. As in (A), the graphs show cumulative increases in PVT lapses per test bout across days within the 3-hour and 5-hour groups ($p = .001$). The horizontal dotted line shows the level of lapsing found in a separate experiment by Van Dongen et al. (2003) when subjects had been awake continuously for 64–88 hours. For example, by Day 7, subjects in the 5-hour TIB averaged 24 total lapses for the four test trials that day, while those in the 3-hour TIB averaged 68 lapses that day. Reprinted with permission of Blackwell Publishing from Belenky et al. (2003).

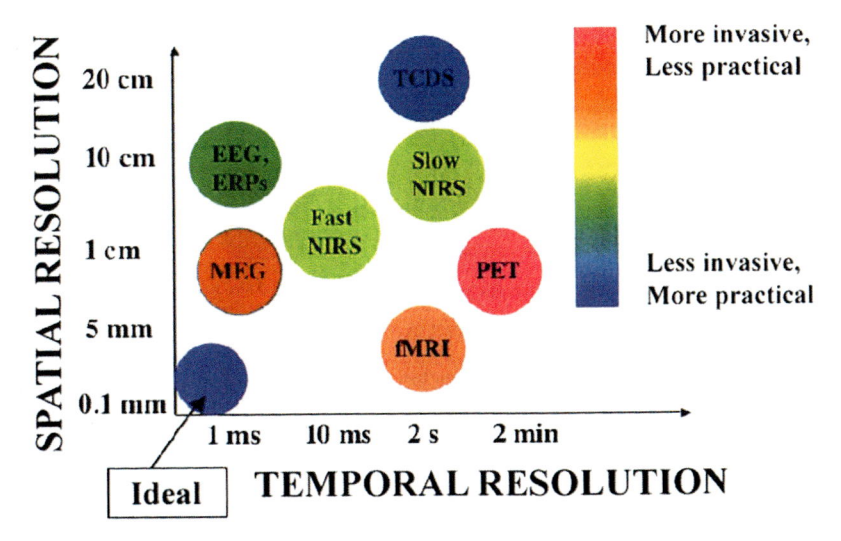

Figure 1.1. Resolution space of brain imaging techniques for ergonomic applications. Trade-offs between the criteria of the spatial resolution (y-axis) and temporal resolution (x-axis) of neuroimaging methods in measuring neuronal activity, as well as the relative noninvasiveness and ease of use of these methods in ergonomic applications (color code). EEG = electronencephalography; ERPs = event-related potentials; fMRI = functional magnetic resonance imaging; MEG = magnetoencenphalography; NIRS = near-infrared spectroscopy; PET = positron emission tomography; TCDS = transcranial doppler sonography.

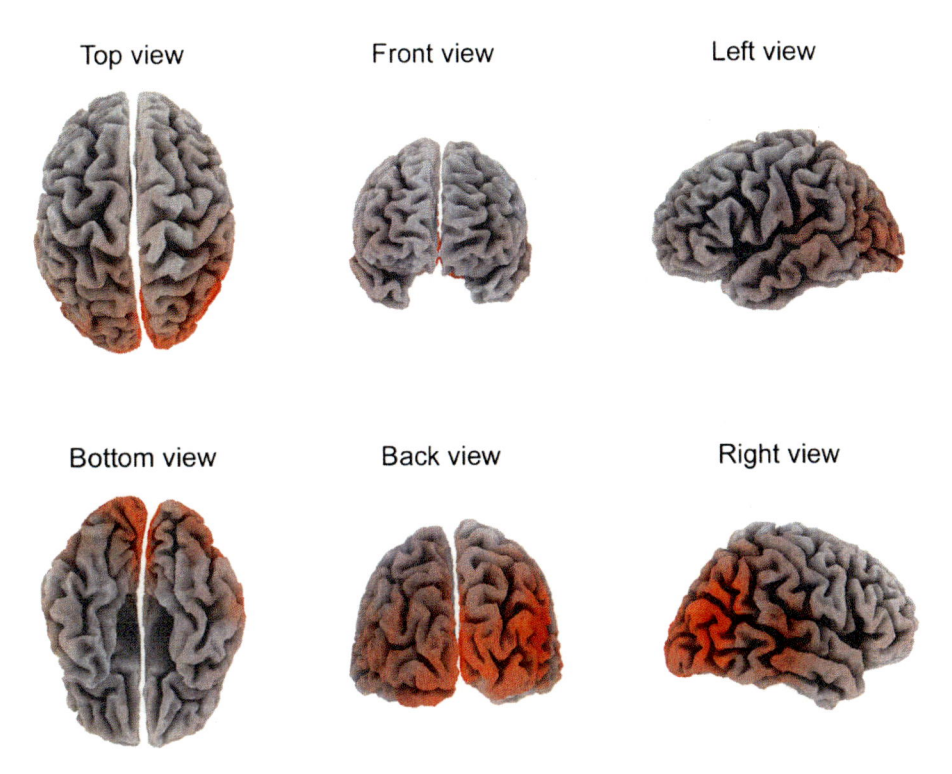

Figure 3.3. Illustration of results obtained from LORETA (Pascual-Marqui, 1999), a distributed localization method. Data were for a left visual field target array under valid and invalid cue conditions, averaged across large and small cue conditions. The attentional effects on the P1 and N1 range were obtained by subtracting ERPs of invalid trials from ERPs of valid trials. The brain activations in response to these attentional effects are displayed at 154 ms, when the global field power is maximal. It is clear that these attentional effects are distributed on the posterior brain regions contralateral to the stimulus visual field. That is, for the left visual field (LVF) stimulus, attentional effects are more pronounced in right posterior brain regions.

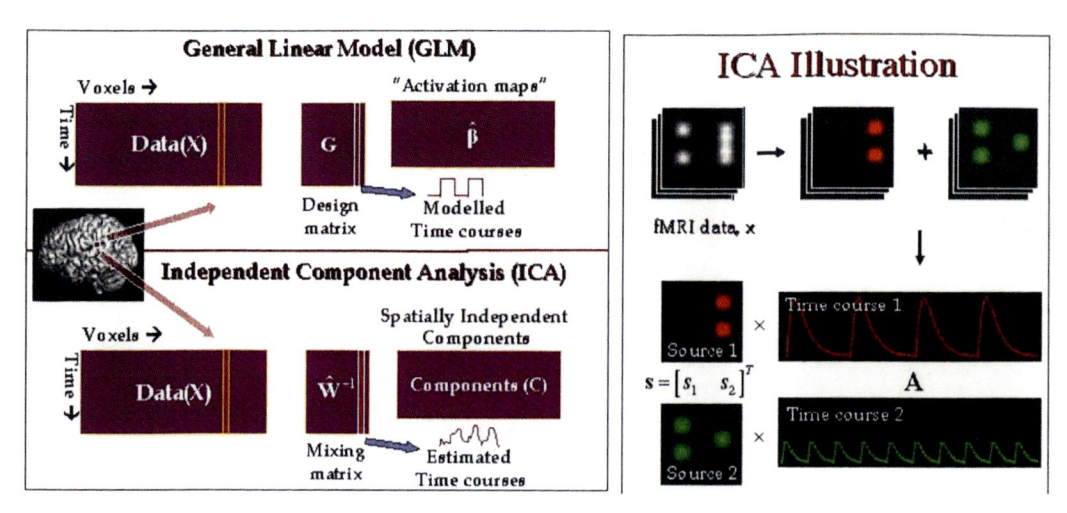

Figure 4.1. Comparison of general linear model (GLM) and independent component analysis (ICA; left) and ICA illustration (right). The GLM (top left) is by far the most common approach to analyzing fMRI data, and to use this approach, one needs a model for the fMRI time course, whereas in spatial ICA (bottom left), there is no explicit temporal model for the fMRI time course. This is estimated along with the hemodynamic source locations (right). The ICA model assumes the fMRI data, x, is a linear mixture of statistically independent sources s, and the goal of ICA is to separate the sources given the mixed data and thus determine the s and A matrices.

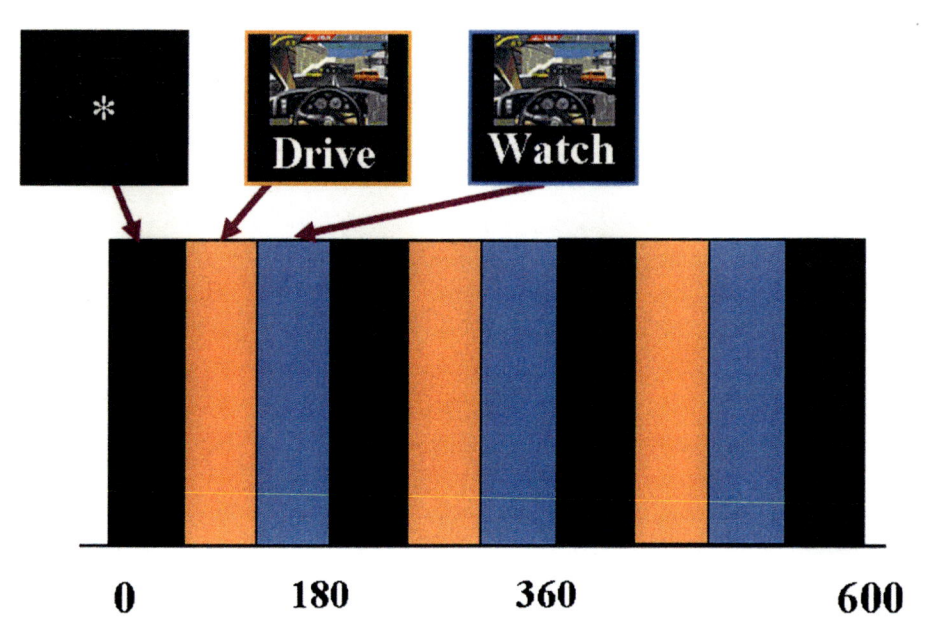

Figure 4.2. fMRI simulated driving paradigm. The paradigm consisted of ten 1-minute epochs of (a) a fixation target, (b) driving the simulator, and (c) watching a simulation while randomly moving fingers over the controller. The paradigm was presented twice, changing the order of the (b) and (c) epochs and counterbalancing the first order across subjects.

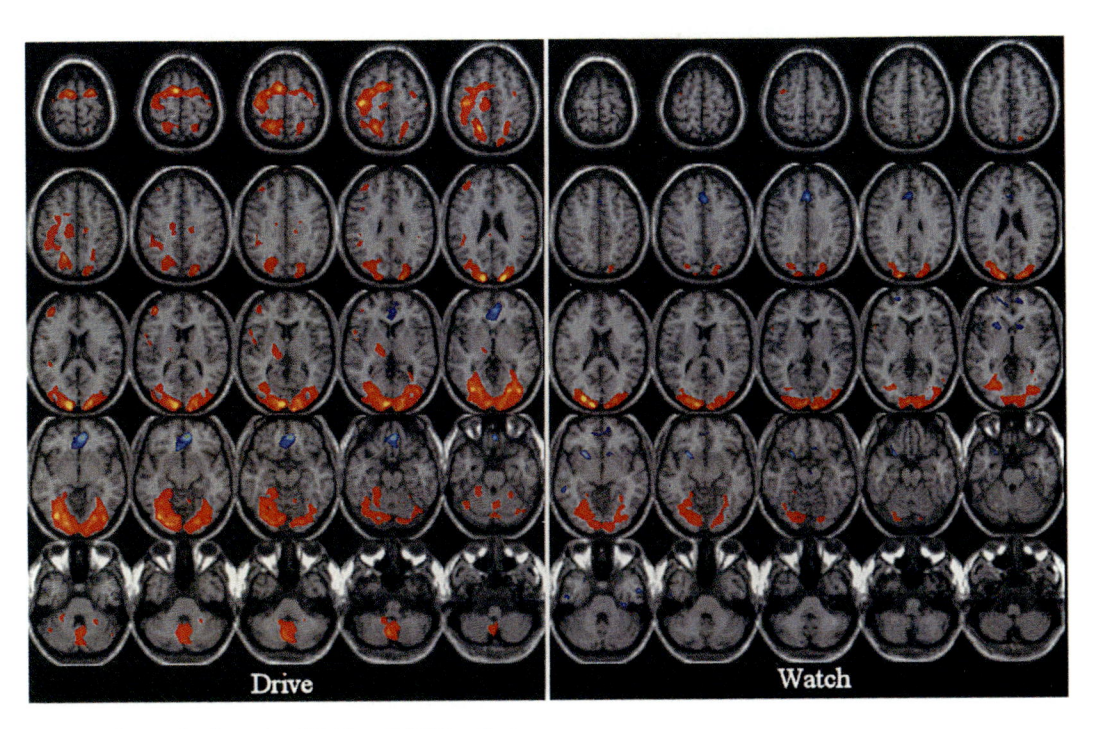

Figure 4.3. Results from the GLM-based fMRI analysis.

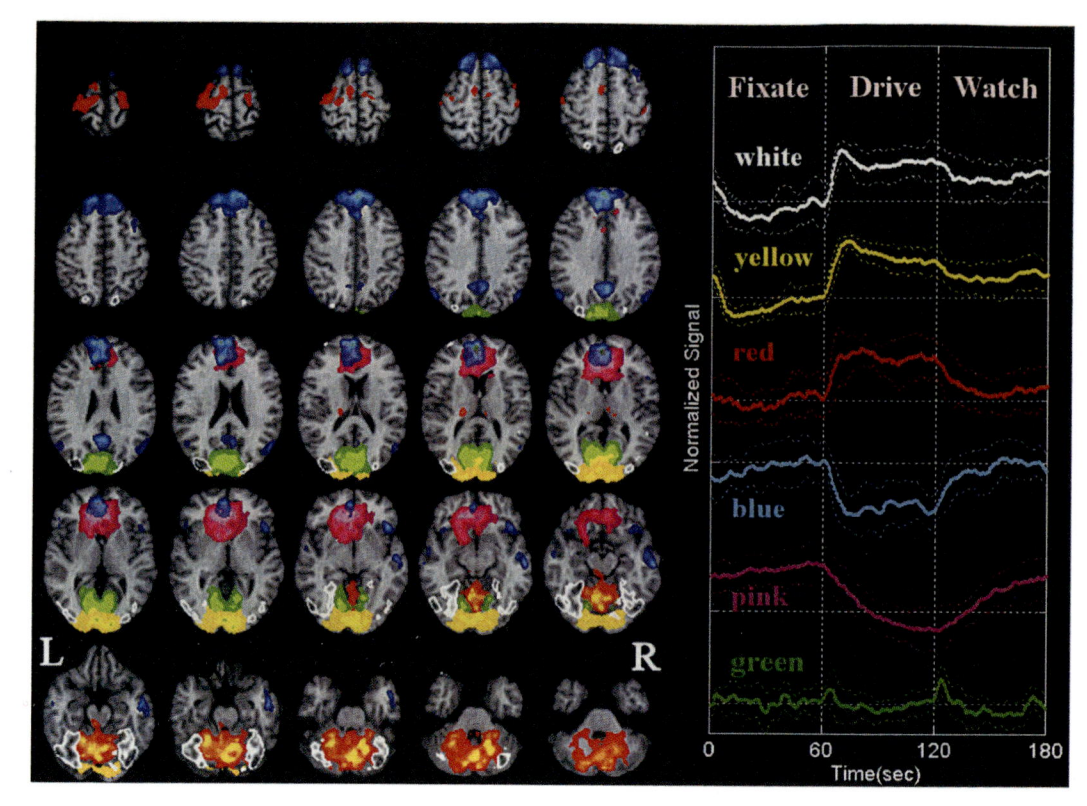

Figure 4.4. Independent component analysis results. Random effects group fMRI maps are thresholded at $p < 0.00025$ ($t = 4.5$, $df = 14$). A total of six components are presented. A green component extends on both sides of the parieto-occipital sulcus including portions of cuneus, precuneus, and the lingual gyrus. A yellow component contains mostly occipital areas. A white component contains bilateral visual association and parietal areas, and a component consisting of cerebellar and motor areas is depicted in red. Orbitofrontal and anterior cingulate areas identified are depicted in pink. Finally, a component including medial frontal, parietal, and posterior cingulated regions is depicted in blue. Group averaged time courses (right) for the fixate-drive-watch order are also depicted with similar colors. Standard deviation across the group of 15 scans is indicated for each time course with dotted lines. The epochs are averaged and presented as fixation, drive, and watch.

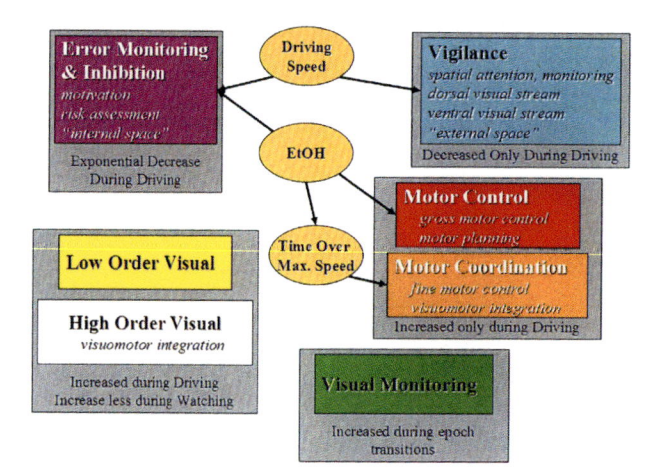

Figure 4.6. Interpretation of imaging results involved in driving. The colors correspond to those used in figure 2.4. Components are grouped according to the averaged pattern demonstrated by their time courses. The speed-modulated components are indicated with arrows.

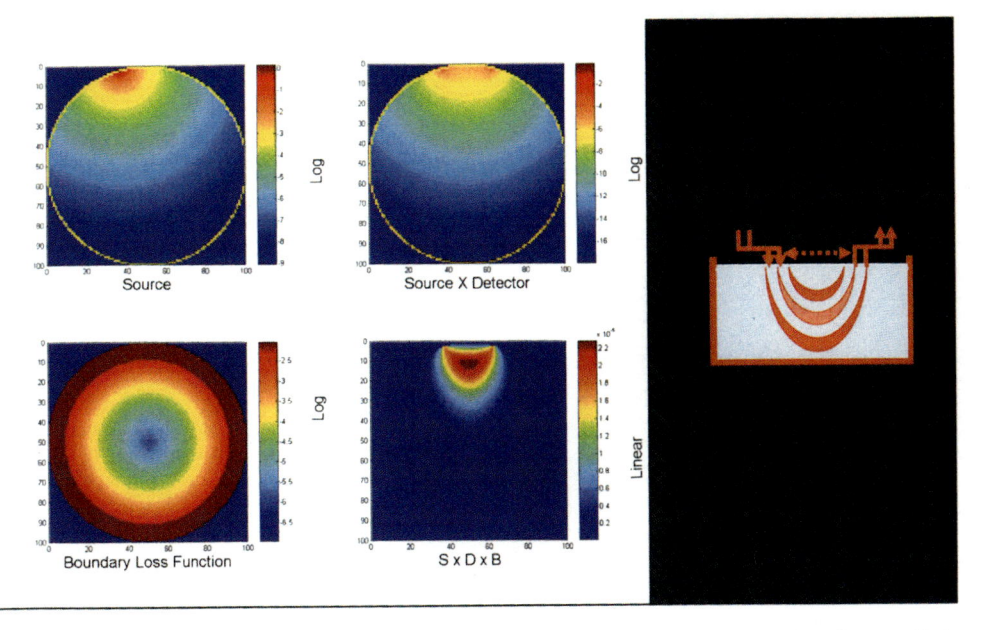

Figure 5.1. Mathematical model of the propagation of light into a scattering medium. Top left, diffusion of light in the scattering medium from a surface source; top middle, area of diffusion for photons produced by a surface source and reaching a source detector; bottom left, effect of a boundary between the scattering medium (such as the head) and a nonscattering medium (such as air); bottom middle, area of diffusion for photons produced by a surface source and reaching a source detector, including effects due to boundary between scattering and nonscattering medium, simulating the actual situation in human head; right, schematic representation of variations in the depth of the diffusion spindle as a function of variations in source-detector distance.

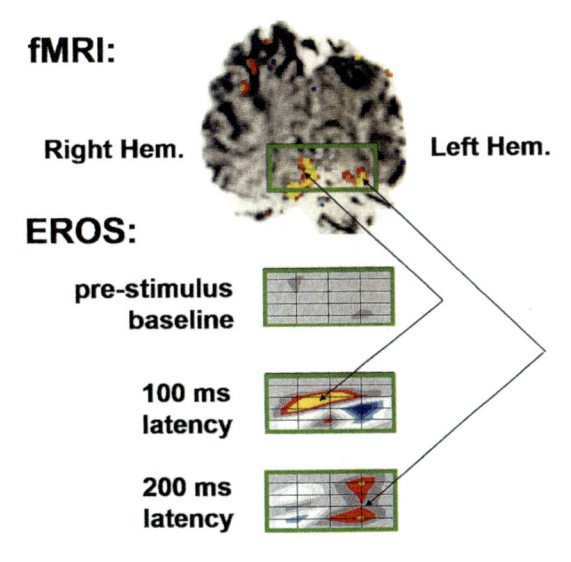

Figure 5.3. Comparison of fMRI and EROS responses in a visual stimulation experiment. The data refer to stimulation of the upper left quadrant of the visual field in one subject. The upper panel reports the BOLD-fMRI slice with maximum activity in this condition. Voxels in yellow are those exhibiting the most significant BOLD response. The green rectangle corresponds to the area explored with EROS. Maps of the EROS response before stimulation and 100 ms and 200 ms after grid reversal are shown at the bottom of the figure, with areas in yellow indicating locations of maximum activity. Note that the location of maximum EROS response varies with latency: The earlier response (100 ms) corresponds in location to the more medial fMRI response, whereas the later response (200 ms) corresponds in location to the more lateral fMRI response. The data are presented following the radiological convention (i.e., the brain is represented as if it were seen from the front).

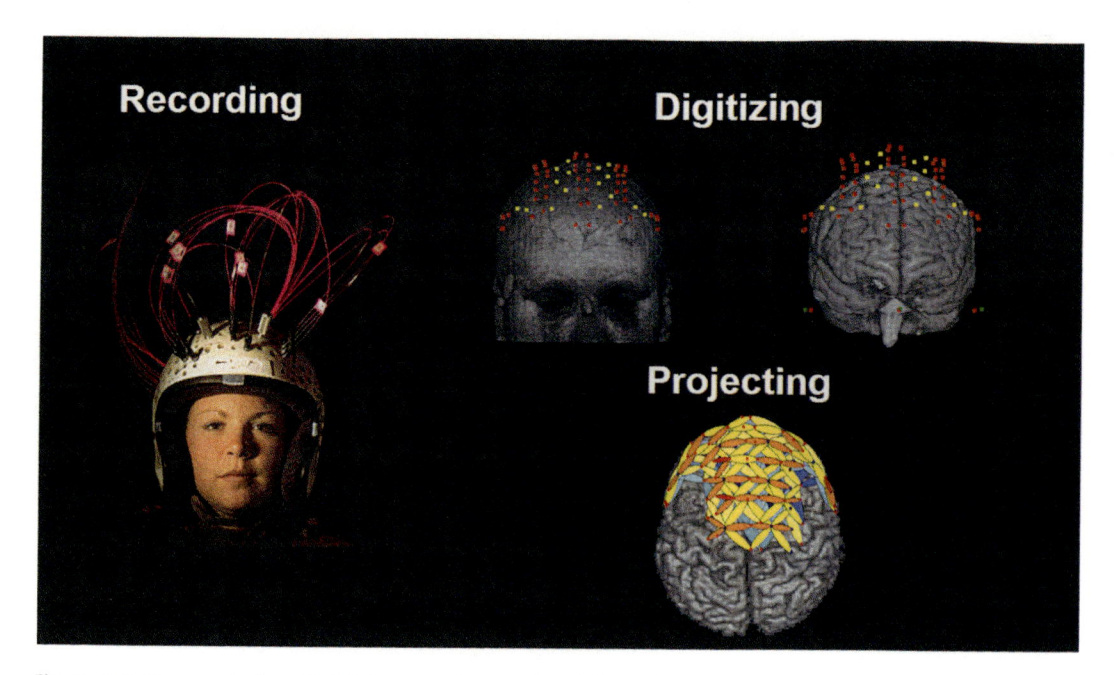

Figure 5.6. Photograph of modified helmet used for recording (left panel); examples of digitized montage (top right) and surface channel projections (bottom right). From Griffin, R. (1999). Illuminating the brain: The love and science of EROS. *Illumination*, 2(2), 20-23. Used with permission from photographer Rob Hill and the University of Missouri-Columbia.

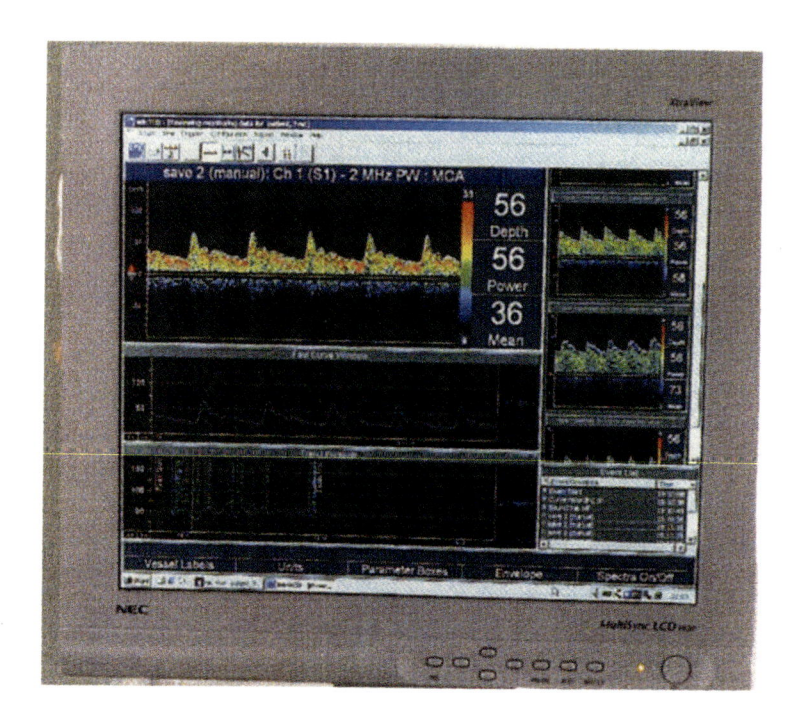

Figure 6.2. Transcranial Doppler display showing flow velocity waveforms, mean cerebral blood flow velocity, and depth of the ultrasound signal.

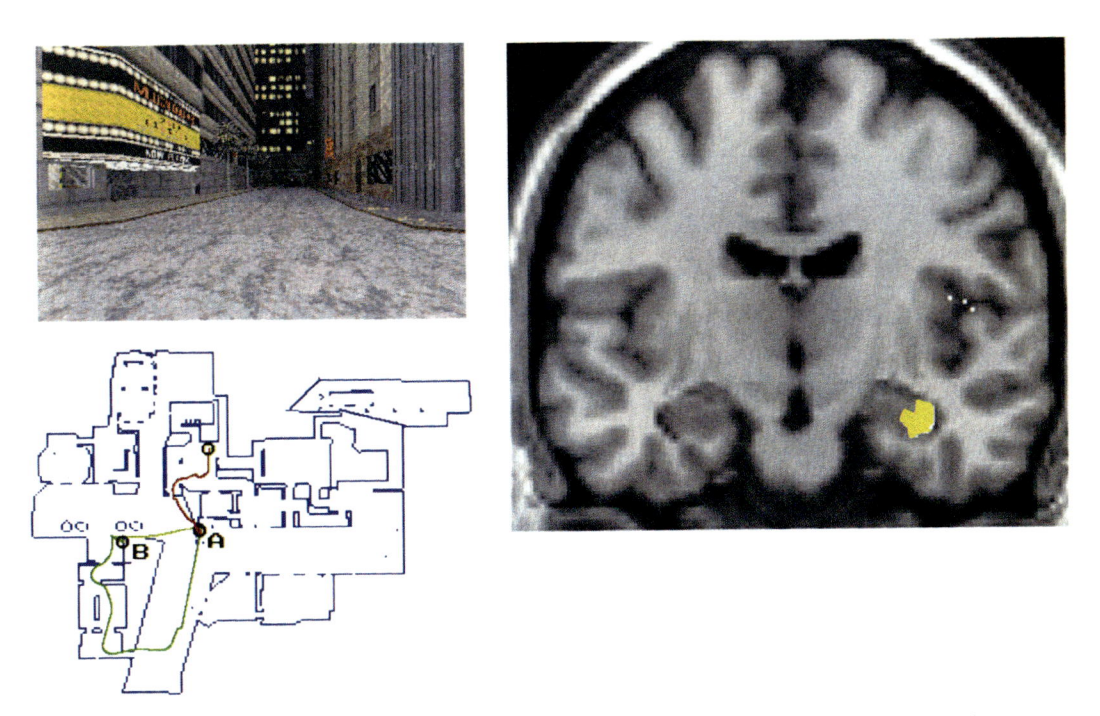

Figure 9.1. Activity in the hippocampus correlates with accuracy of path taken in a virtual reality town (see Maguire, Burgess, Donnett, Frackowiak, et al., 1998). The more direct the path (in yellow on the aerial view), the more active the hippocampus.

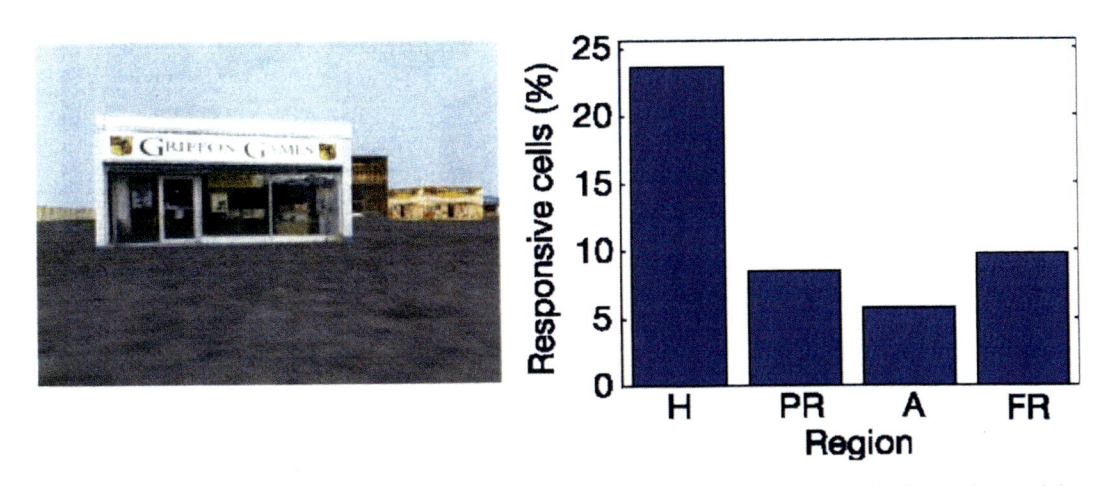

Figure 9.2. Place-responsive cells (see Ekstrom et al., 2003) from intracranial electrodes in the human brain while patients navigated in a VR townlike environment. These neurons were clustered in the hippocampus (H) compared with the amygdala (A), parahippocampal region (PR), and frontal lobes (FR). From Ekstrom et al. (2003). Reprinted by permission from Macmillan Publishers Ltd.: *Nature*, Arne D. Ekstrom et al., Cellular networks underlying human spatial navigation, volume 425, issue 6954, pages 184–188, 2003.

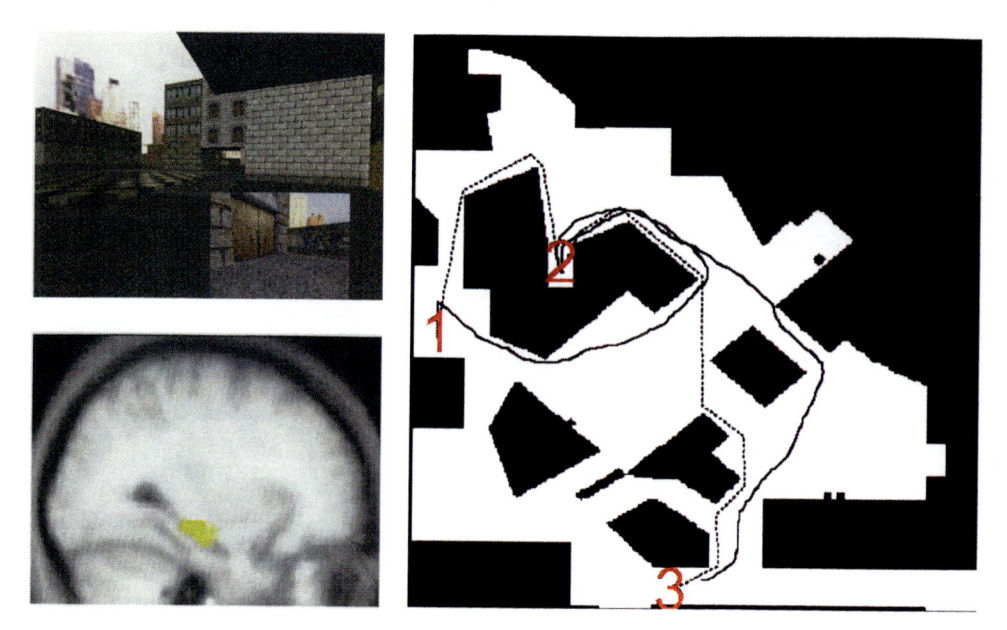

Figure 9.3. Town 1 from Hartley et al. (2003). In the wayfinding task, the current target location is indicated in the lower right corner of the VR display. The map shows an example path followed by a subject (solid line) between the first three target locations. The corresponding ideal path is shown as a dotted line. Accuracy of performance was correlated with activity in the hippocampus.

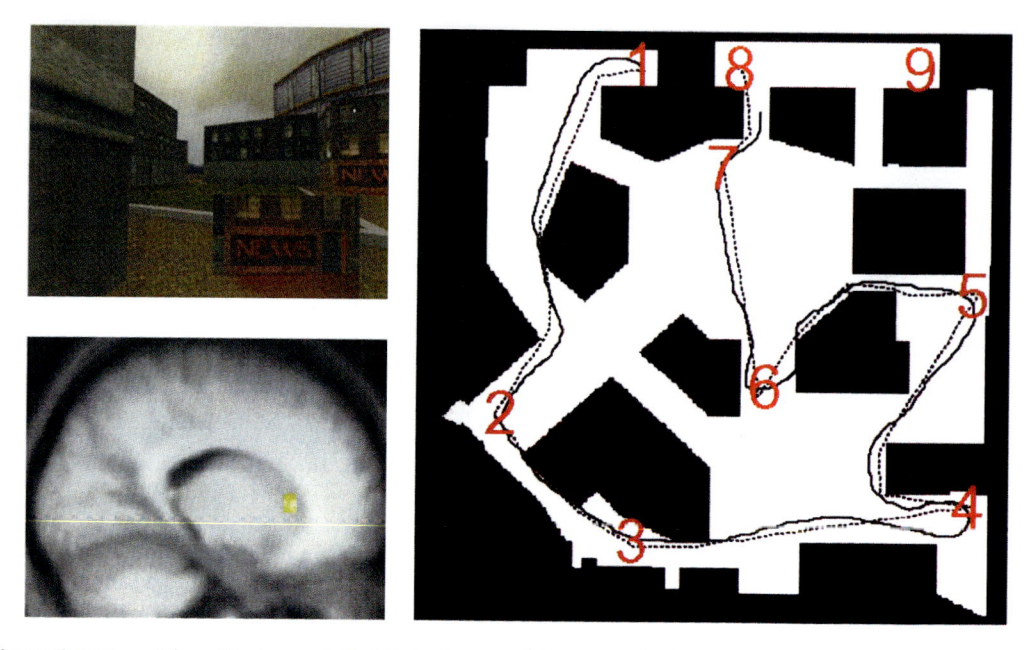

Figure 9.4. Town 2 from Hartley et al. (2003). In the route-following task, the current target location is indicated in the lower right corner of the VR display. The map shows the well-worn route followed by a subject (solid line). Accuracy of performance was correlated with activity in the caudate nucleus.

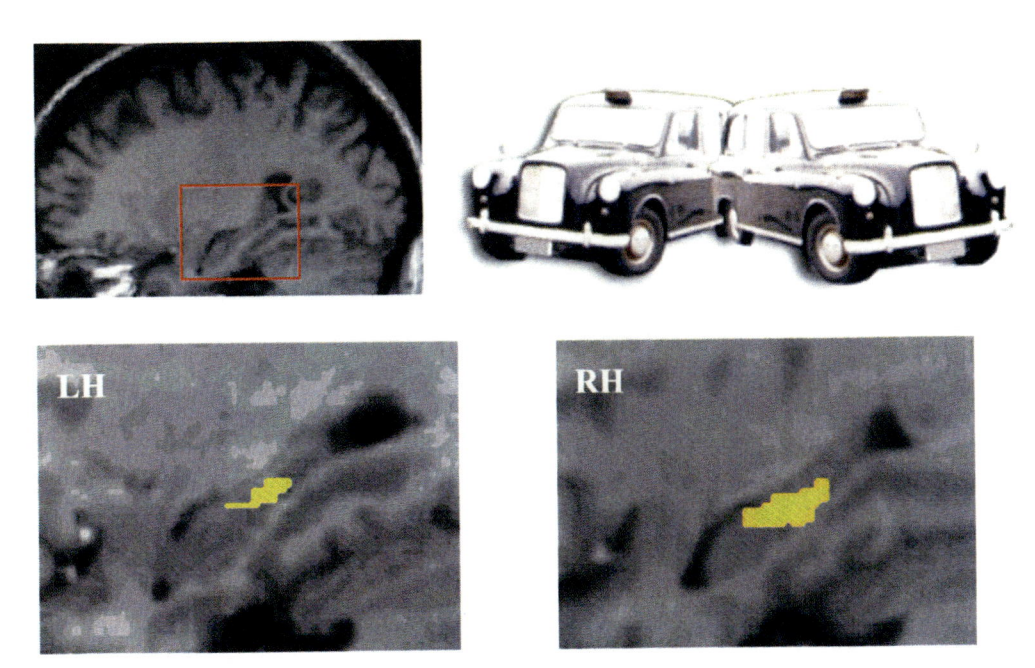

Figure 9.5. MRI sagittal brain sections. Yellow areas indicate where there was increased gray matter density in the left (LH) and right (RH) hippocampi of licensed London taxi drivers compared with non–taxi driver control subjects (see Maguire et al., 2000).

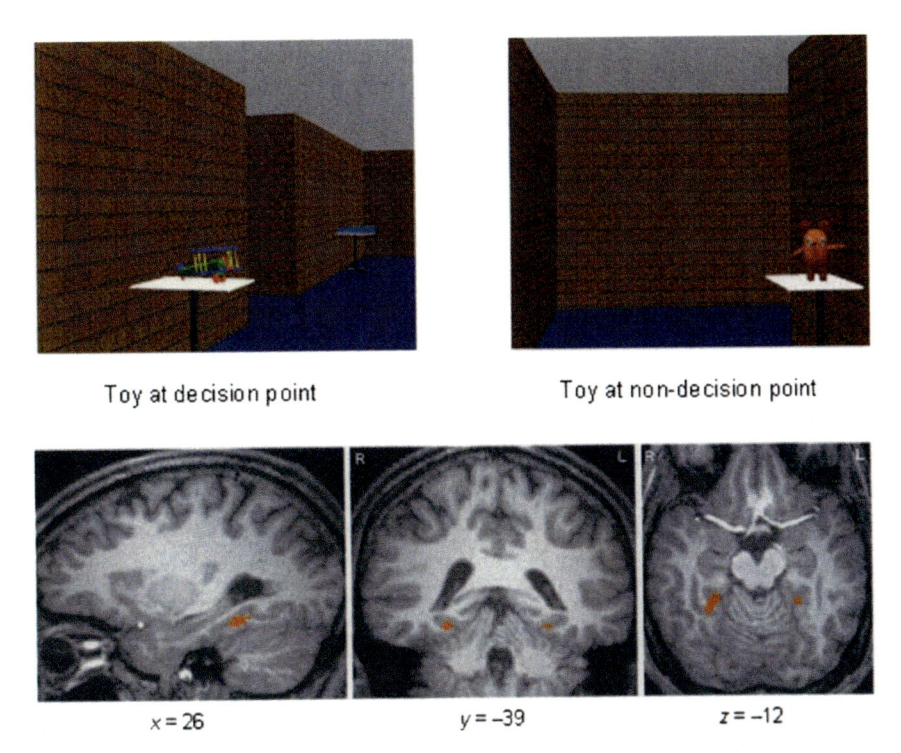

Figure 9.6. Views from the virtual museum from Janzen and van Turennout (2004). The left panel shows a toy placed at a decision point, and the right scene a toy at a nondecision point. A parahippocampal region was more active for toys at decision points, that is, that had navigational relevance. From Janzen and van Turennout (2004). Reprinted by permission from Macmillan Publishers Ltd.: *Nature Neuroscience*, Gabriele Janzen and Miranda van Turennout, Selective neural representation of objects relevant for navigation, volume 7, issue 6, pages 673–677, 2004.

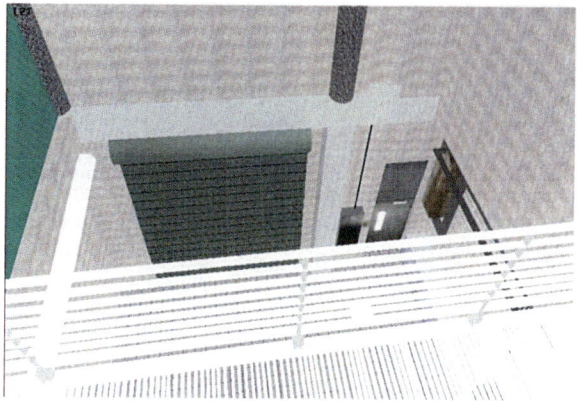

Figure 17.3. Height effects in real and virtual reality environments. Virtual reality environments have been used to study height effects and spatial cognition in neuroergonomic research aimed at reducing falls and injuries in work environments. Simeonov et al. (2005) compared psychological and physiological responses to height in corresponding real and virtual environments. (Courtesy of NIOSH, Division of Safety Research.)

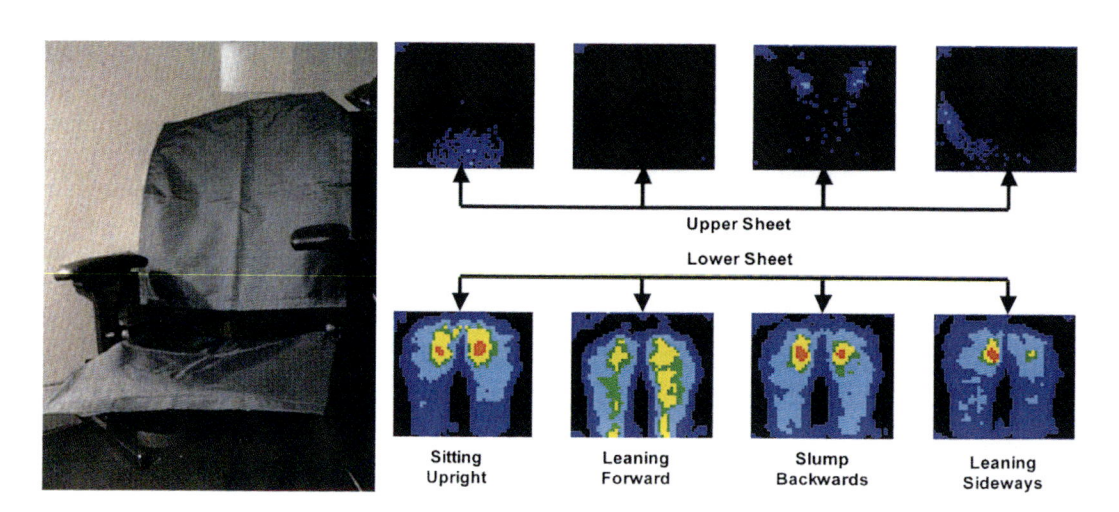

Figure 18.2. The Smart Chair (left) and data (right) showing characteristic pressure patterns used to recognize postural shifts characteristic of high interest, low interest, and taking a break.

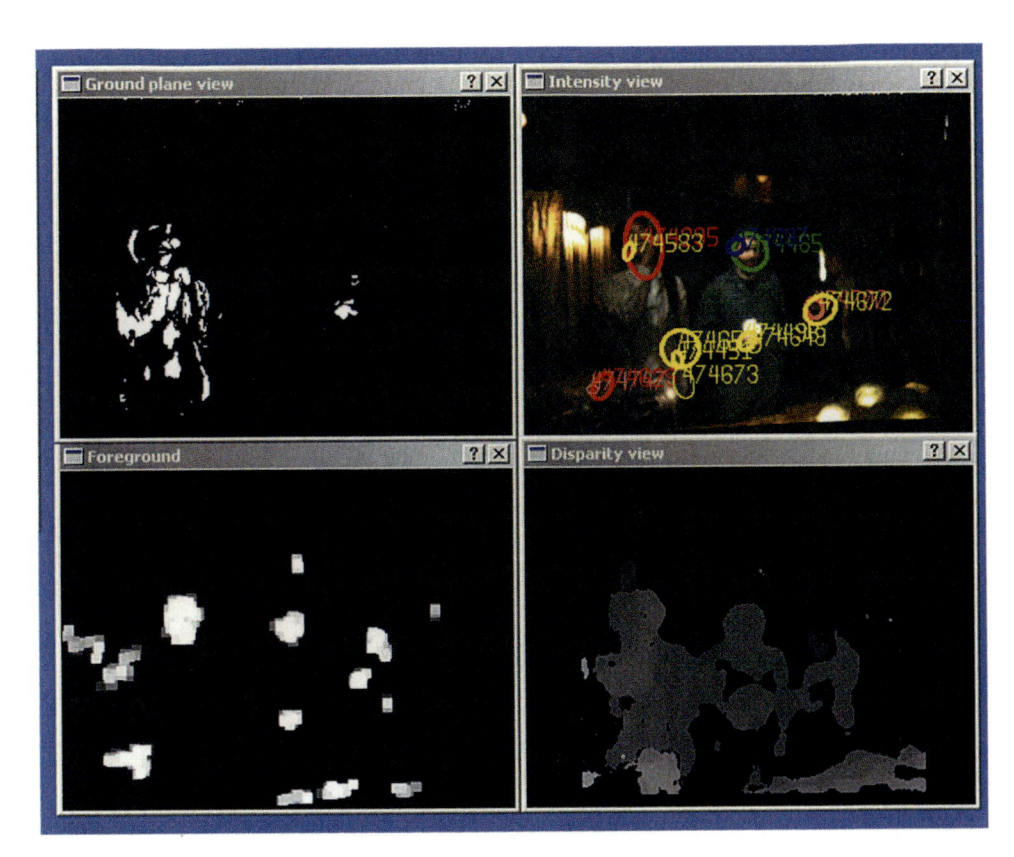

Figure 18.3. A snapshot of the stereo vision system that is mounted in the base of the computer (developed in collaboration with the Vision Interfaces Group at MIT CSAIL). Motion is detected in the upper left frame; human skin chromaticity is extracted in the lower left frame; a foreground depth map is computed in the lower right frame; and the faces and hands of audience participants are tracked in the upper right frame.

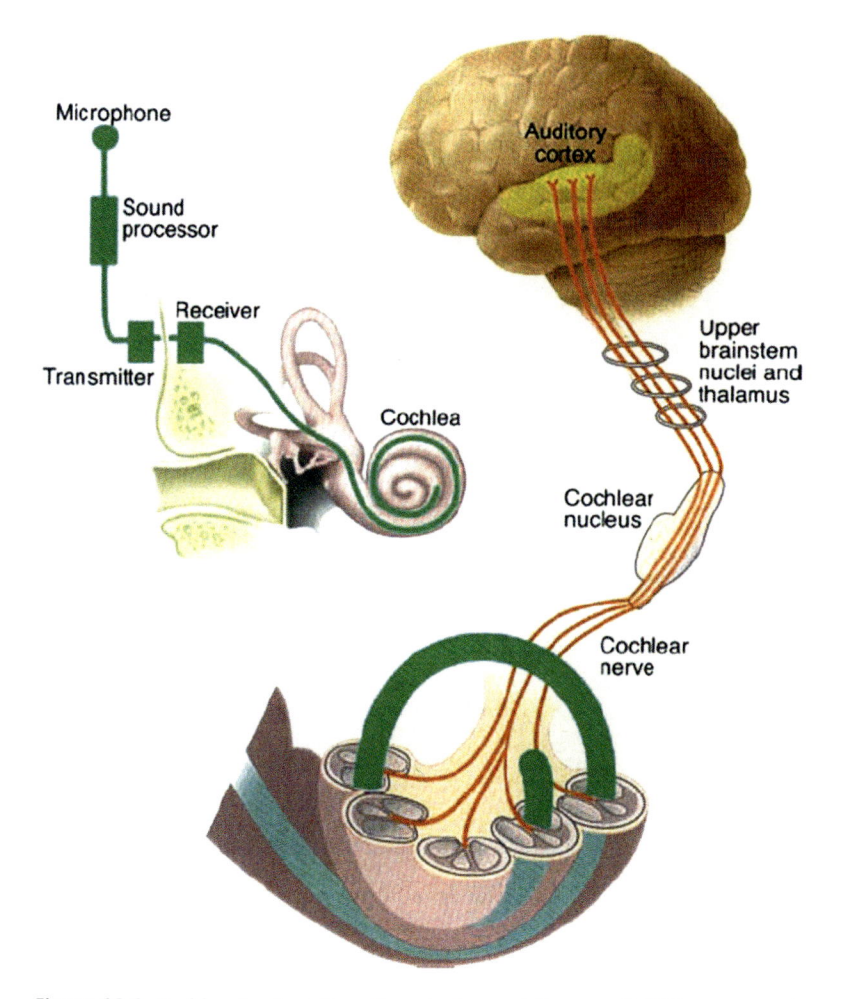

Figure 19.1. Cochlear implant. From Rauschecker and Shannon (2002).

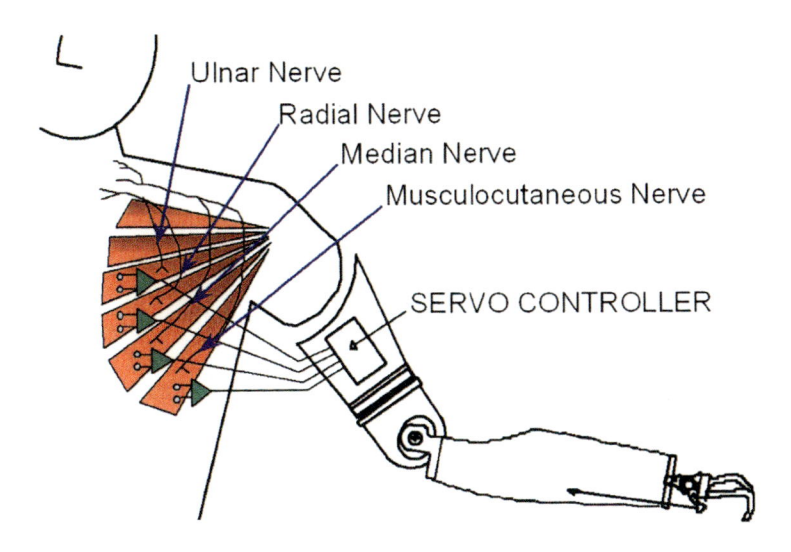

Figure 19.5. Schematic of neuromuscular reinnervation showing how the pectoralis muscle is split into four separate sections and reinnervated with the four main arm nerve bundles.

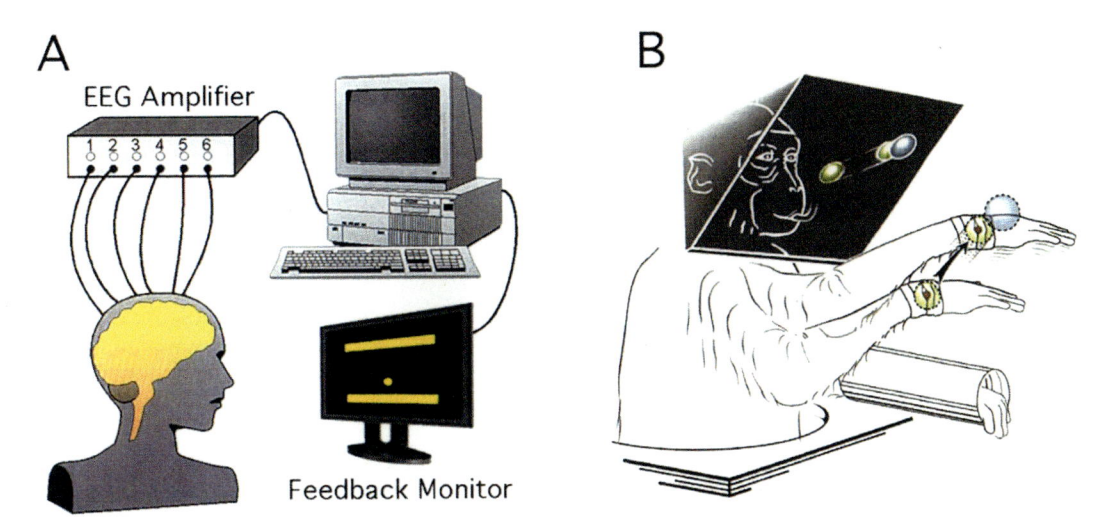

Figure 19.6. Noninvasive and invasive brain-computer interfaces. (A) Electroencephalographic (EEG) signals recorded from the scalp have been used to provide communication or other environmental controls to locked-in patients, who are completely paralyzed due to brain stem stroke or neurodegenerative disease. The signals are amplified and processed by a computer such that the patient can learn to control the position of a cursor on a screen in one or two dimensions. Among other options, this technique can be used to select letters from a menu in order to spell words. From Kubler et al. (2001). (B) Monkeys have learned to control the 3-D location of a cursor (yellow sphere) in a virtual environment. The cursor and fixed targets are projected onto a mirror in front of the monkey. The cursor position can be controlled either by movements of the monkey's hand or by the hand movement predicted in real time on the basis of neuronal discharge recorded from electrodes implanted in the cerebral cortex. From Taylor et al. (2002).

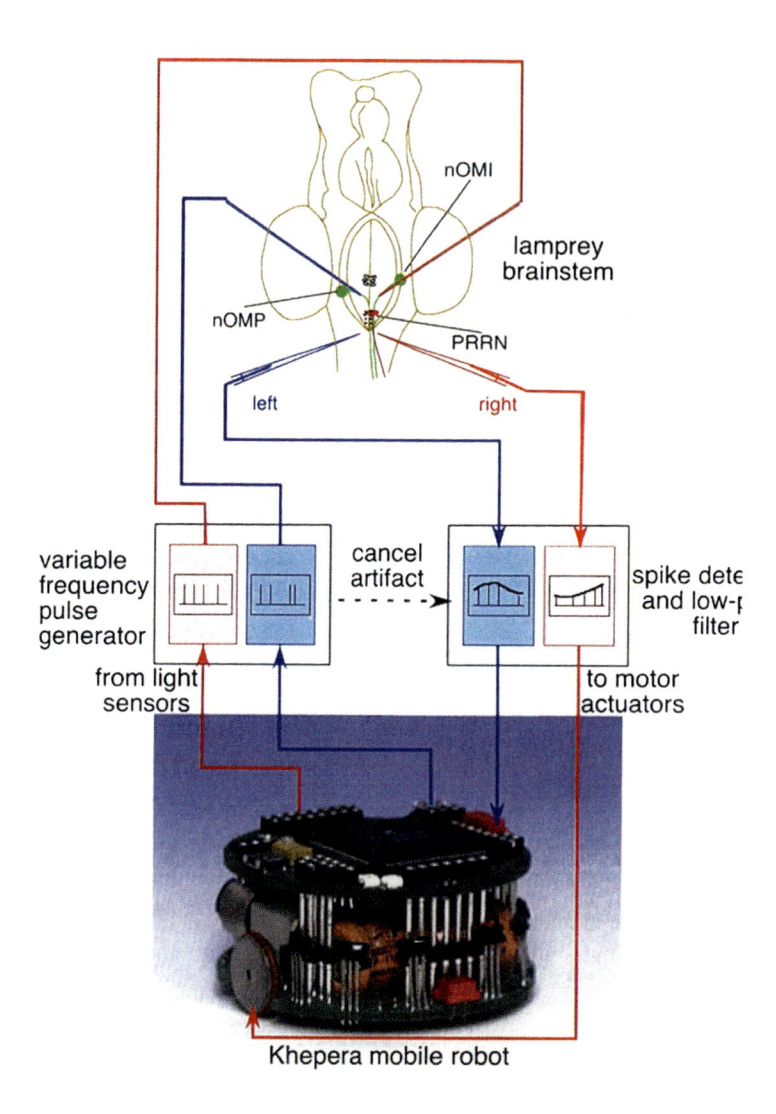

Figure 19.8. A hybrid neurorobotic system. Signals from the optical sensors of Khepera (K-team) mobile robot (bottom) are encoded by the interface into electrical stimulations whose frequency depends linearly upon the light intensity. These stimuli are delivered by tungsten microelectrodes to the right and left vestibular pathways of a lamprey's brain stem (top) immersed in artificial cerebrospinal fluid within a recording chamber. Glass microelectrodes record extracellular responses to the stimuli from the posterior rhomboencephalic neurons (PRRN). Recorded signals from right and left PRRNs are decoded by the interface, which generates the commands to the robot's wheels. These commands are set to be proportional to the estimated average firing rate on the corresponding side of the lamprey's brain stem.

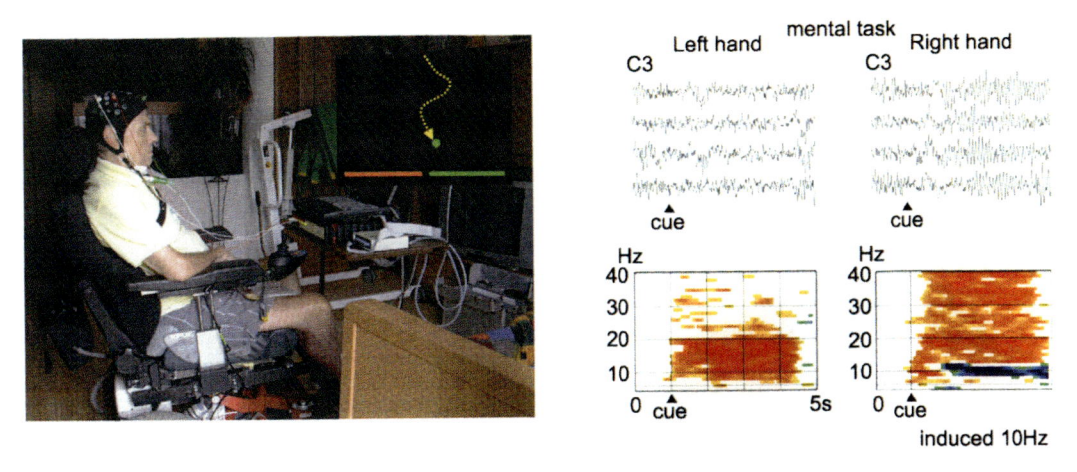

Figure 20.4. Picture of a patient suffering from ALS during feedback training. In the upper right corner, the basket paradigm is shown (left). Examples of trials and event-related desynchronization and event-related synchronization time/frequency maps display ipsilateral ERS (right).

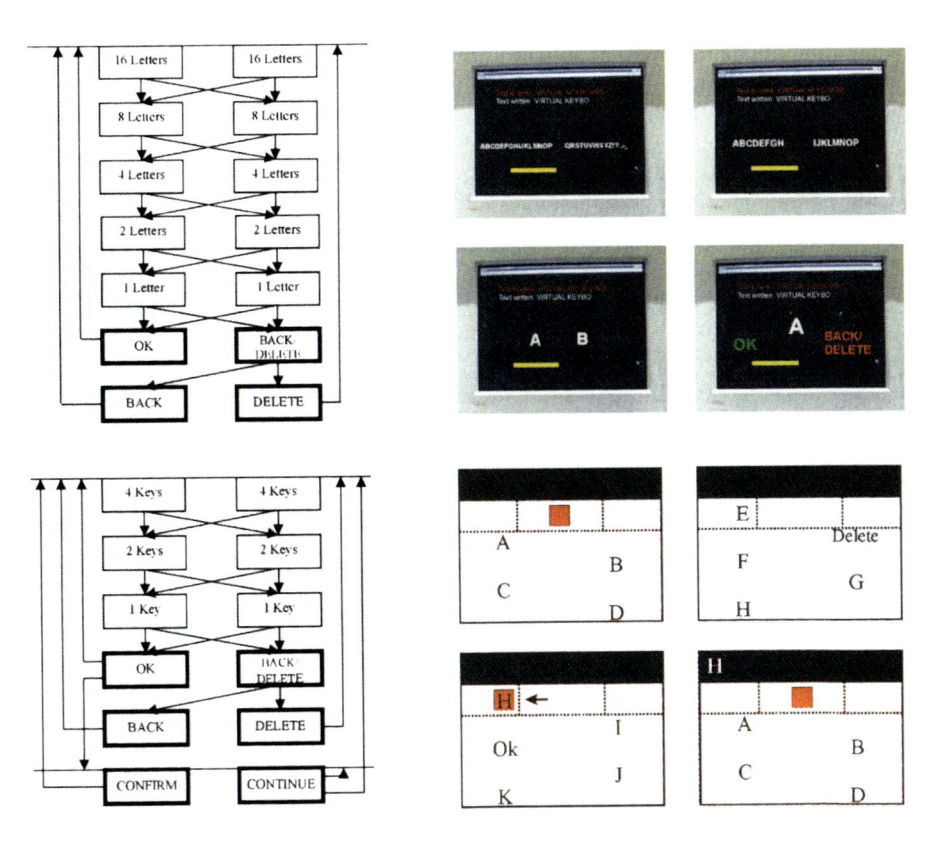

Figure 20.5. Dichotomous letter selection structure (top left). Screen shots of the basic two-class virtual keyboard. Four of the six steps to perform to select a letter are shown (top right). Letter selection structure based on T8 (bottom left). Three-class asynchronously controlled virtual keyboard. Example: Selection of the letter H by virtual keyboard (bottom right).

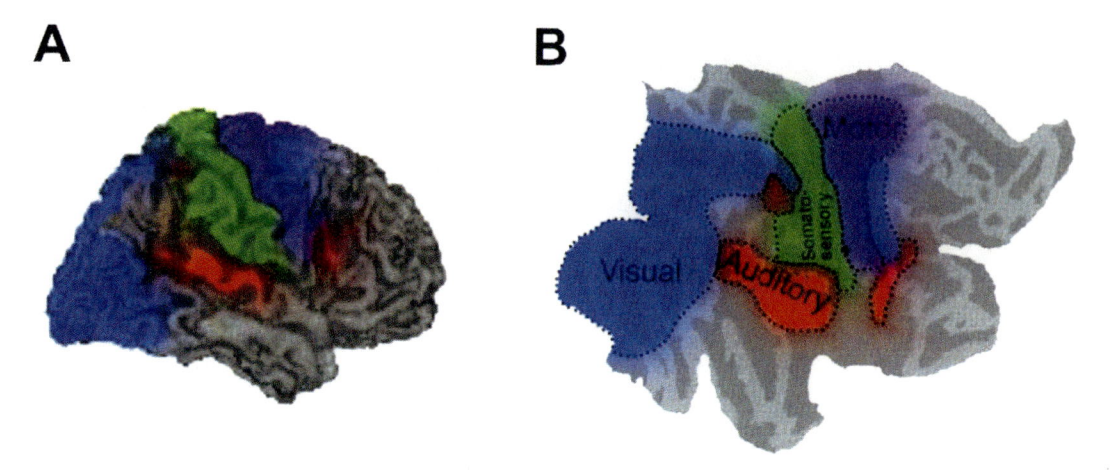

Figure 21.4. Organization of sensory and motor areas shown on a three-dimensional volume-rendered brain (A) and flattened projection (B). For simplicity, only the right hemisphere is shown. Modified from van Essen (2004) with permission from MIT Press.

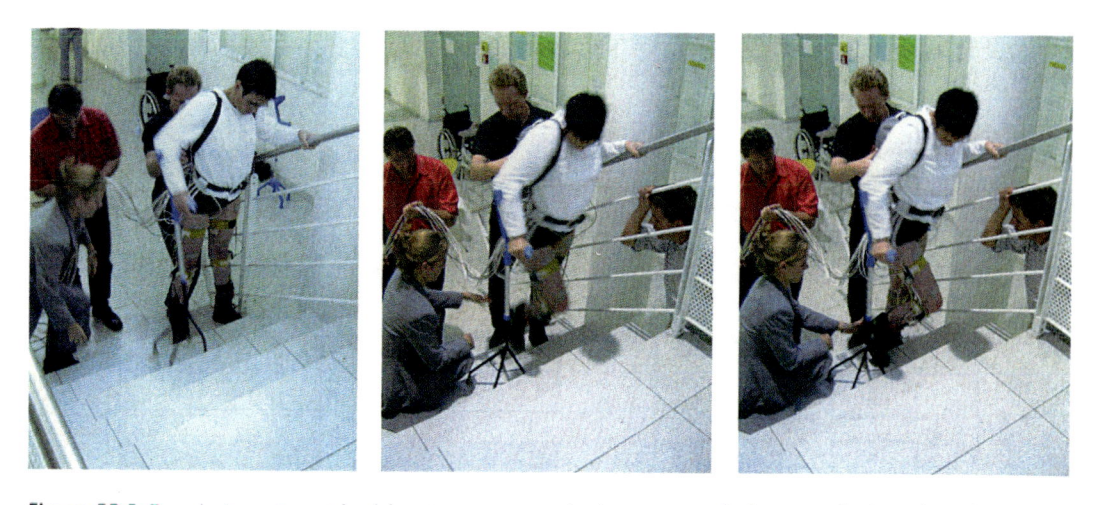

Figure 22.5. Paraplegic patient with a laboratory neuroprosthesis system applied to stair climbing (T. Fuhr, TU München).

Since studies consistently show that people use only 50–75% of rest periods to sleep, it would be prudent to provide longer rest breaks (e.g., 10–14 hours) to ensure that adequate recovery sleep is obtained. Work-rest rules are not the only area in which these new experimental data are relevant. Recent epidemiological studies have found an increased incidence of sleep-related crashes in drivers reporting 6 or fewer hours sleep per night on average (Stutts, Wilkins, Scott Osberg, & Vaughn, 2003).

Perception of Sleepiness Versus Performance during Sleep Deprivation

It is commonly assumed that people know when they are tired from inadequate sleep and hence can actively avoid sleep deprivation and the risks it poses to performance and safety. That has not proven to be the case, however. In contrast to the continuing accumulation of cognitive performance deficits associated with nightly restriction of sleep to below 8 hours, ratings of sleepiness, fatigue, and alertness made by the subjects in the sleep restriction experiments did not parallel performance deficits (Belenky et al., 2003; Van Dongen et al., 2003). Instead, subjects' perceptions of their fatigue and sleepiness showed little change after the first few days of sleep restriction. While perceptions of sleepiness and fatigue did not show systematic increases over days of sleep restriction, cognitive performance functions were steadily deteriorating across days of restriction in a near-linear manner (Belenky et al., 2003; Van Dongen et al., 2003). As a consequence, after a week or two of sleep restriction, subjects were markedly impaired and less alert, but they felt subjectively that they had adjusted to the reduced sleep durations. This suggests that people frequently underestimate the cognitive impact of sleep restriction and overestimate their performance readiness when sleep deprived. Other experiments using driving simulators have found comparable results (Banks, Catcheside, Lack, Grunstein, & McEvoy, 2004)—people often do not accurately identify their performance risks when sleep deprived.

Individual Differences in Response to Sleep Deprivation

Although restriction of sleep periods to below 7 hours duration results in cumulative cognitive performance deficits in a majority of healthy adults, not everyone is affected to the same degree. In fact, sleep deprivation not only increases performance variability within subjects (i.e., state instability) but also reveals marked performance differences between subjects. That is, as sleep deprivation becomes worse over time, intersubject differences also increase markedly. While the majority of people suffer neurobehavioral deficits when sleep deprived, there are individuals at opposite ends of this spectrum—those who experience very severe impairments even with modest sleep loss, and those who show few if any cognitive deficits until sleep deprivation is very severe. Recently, it has been shown that these responses are stable and reliable across subjects. Cognitive performance changes following sleep loss were traitlike, with intraclass correlations accounting for a very high percentage of the variance (Van Dongen, Maislin, & Dinges, 2004). However, as with chronic sleep restriction, subjects were not really aware of their differential vulnerability to sleep loss. The biological basis of the differential responses to sleep loss is not known. Consequently, until objective markers for differential vulnerability to sleep deprivation can be found, it will not be possible to use such information in a manner that reduces the risk posed by fatigue in a given individual.

Operational Causes of Sleep Loss

Work and related operational demands can affect the magnitude of sleep loss and fatigue. It is estimated that more than one third of the population suffers from chronic sleep loss (Walsh, Dement, & Dinges, 2005). This is partially due to society's requirement for around-the-clock operations. Individuals are expected to be able to adjust to any schedule, independent of time of day, and continue to remain alert and vigilant. Often this forces people to extend their waking hours and reduce their sleep time.

Night Shift Work

Night shift work is particularly disruptive to sleep. Many of the 6 million full-time employees in the United States who work at night on a permanent or rotating basis experience daytime sleep disruption leading to sleep loss and nighttime sleepiness on the job from circadian misalignment (Akerstedt, 2003). More than 50% of shift workers complain of shortened and disrupted sleep and overall tiredness, with total amounts of sleep loss ranging anywhere from 2 to 4 hours per night (Akerstedt, 2003).

Irregular and prolonged work schedules, shift work, and night work are not unique to a single operational environment but exist in many work sectors. Such schedules create physiological disruption of sleep and waking because of misalignment of the endogenous circadian clock and imposed work-rest schedules. Individuals are exposed to competing time cues from the day-night cycle and day-oriented society and are usually inadequately adapted to their temporally displaced work-rest schedule.

Fatigue and Drowsy Driving

Individuals working irregular schedules are also more likely to have higher exposure to nighttime driving, increasing the chances of drowsy driving and decreasing the ability to effectively respond to stimuli or emergency situations (Braver, Preusser, Preusser, Baum, Beilock, & Ulmer, 1992; Stutts et al., 2003). Drowsy driving is particularly challenging in the truck-driving environment. Fatigue is considered to be a causal factor in 20–40% of heavy truck crashes. Operational demands often force night driving in an attempt to avoid traffic and meet time-sensitive schedules. Such night work is a double-edged sword, requiring both working when the body is programmed to be asleep and sleeping when the body is programmed to be awake. In recognition of the safety issues surrounding drowsy driving, the U.S. Department of Transportation is actively involved in developing programs that would provide fatigue-tracking technologies to the human operator to help manage drowsy driving and fatigue as part of the development of an "intelligent vehicle" (Mallis et al., 2000).

Transmeridian Travel

Operator fatigue associated with jet lag is a concern in aviation. Although air travel over multiple time zones is considered a technological advance, it poses substantial physiological challenges to human endurance. Crew members can be required to work any schedule, regardless of their geographical location, as well as crossing multiple time zones. As a result, flight crews can experience disrupted circadian rhythms and sleep loss. Studies have physiologically documented episodes of fatigue and the occurrence of uncontrolled sleep periods or microsleeps in pilots (Wright & McGown, 2001). Flight crew members, unlike most passengers, remain at their destination for a short period of time and never have the opportunity to adjust physiologically to the new time zone or work schedule.

Transiting time zones and remaining at the new destination for days with exposure to the new light-dark cycle and social cues does not guarantee a rapid realignment (phase shift) of the sleep-wake cycle and circadian system to the new time zone. A typical jet lag experience involves arriving at a destination (new time zone) with an accumulated sleep debt (i.e., elevated homeostatic sleep drive). This ensures that the first night of sleep in the new time zone will occur—even if it is abbreviated due to a wake-up signal from the endogenous circadian clock—but on the second, third, and fourth nights the person will most likely find it more difficult to obtain consolidated sleep because of the circadian disruption. As a result, the individual's sleep is not maximally restorative for a number of nights in a row, leading to increasing difficulty being alert during the daytime. These cumulative effects (see figure 14.1) can be very incapacitating and can take 1–3 weeks to fully dissipate through full circadian reentrainment to the new time zone.

The effects of jet lag are also partly dependent on the direction of travel. Eastward travel tends to be more difficult for physiological adjustment than westward travel because eastward transit seeks to impose a phase advance on the circadian clock, while westward transit imposes a phase delay. Lengthening a day by a few hours is somewhat easier to adjust to physiologically and behaviorally than advancing a day by the same amount of time, although adjustment to either eastward or westward phase shifts of more than a couple of hours is a slow process, often requiring at least a 24-hour period (day) for each time zone crossed (e.g., transiting six time zones can require 5–7 days) and resulting in a substantial but still incomplete adjustment for most people, assuming they get daily exposure to the light-dark cycle and social rhythms of the new environment. Moreover, the direction of flight does not ensure the direction of circadian phase adjustment—some people physiologically phase delay to an eastward (phase advanced) flight, which can require many more days for physiological adjustment to occur. The reasons for the direction of physiological shift are not well understood, but likely involve individual differences in circadian dynamics and light entrainment responses.

Prolonged Work Hours and Errors: Medical Resident Duty Hours as an Example

Other transportation modalities—such as maritime, rail, and mass transit—as well as many non-transportation industries must manage fatigue from the demands of 24-hour operations. Any occupation that requires individuals to maintain high levels of alertness over extended periods of time is vulnerable to the neurobehavioral and work performance consequences of sleep loss and circadian disruption. The resulting performance effects have the potential of compromising safety (Dinges, 1995). However, the hazards associated with extended duty schedules are not always apparent to the public. There is a lack of public as well as professional awareness regarding the importance of obtaining adequate amounts of sleep (Walsh, Dement, & Dinges 2005). For example, providing acute medical care 24 hours a day, year round, results in physicians, nurses, and allied health care providers being awake at night and often working for durations well in excess of 12 hours. Chronic partial sleep deprivation is an inherent consequence of such schedules, especially in physicians in training (Weinger & Ancoli-Isreal, 2002). Human error also increases with such prolonged work schedules (Landrigan et al., 2004; Rogers, Hwang, Scott, Aiken, & Dinges, 2004).

To address the risks of performance errors posed by sleep loss in resident physicians, the Accreditation Council for Graduate Medical Education (ACGME, 2003) imposed duty hour limits for resident physicians. These duty limits were intended to reduce the risks of performance errors due to both acute and chronic sleep loss by limiting residents to 80 hours work per week and by limiting a continuous duty period to 24–30 hours. They also mandated 1 day in 7 free from duty averaged over 4 weeks, and 10-hour rest opportunities between duty periods (ACGME, 2003). Recent objective studies of residents operating under these duty hour limits reveal significant numbers of medical errors and motor vehicle crashes (Barger et al., 2005; Landrigan et al., 2004; see also chapter 23, this volume). It appears that work schedules that permit extended duty days to well beyond 16 hours result in sleep deprivation and substantial operational errors, consistent with laboratory studies.

Although increased fatigue and sleepiness on the job are more common for those involved in erratic or irregular shifts, when a portion of sleep or the total sleep period is occurring at a time not conducive to sleep, sleep loss can also be problematic for individuals on a "normal" schedule with sleep nocturnally placed. As the number of hours awake increases, levels of fatigue increase, especially with extended duty shifts (Rosa & Bonnet, 1993). Therefore, increased fatigue and sleepiness associated with long shifts should be carefully considered prior to implementing extended schedules. However, individuals often prefer a 12-hour shift to an 8-hour shift (Johnson & Sharit, 2001) because it allows for more consecutive days off and more opportunities for social activities and family time, even though the result can be increased fatigue on the job and significant decrements in performance.

Prediction and Detection of the Effects of Sleep Loss in Operational Environments

The growth of continuous operations in government and industry has resulted in considerable efforts to either predict human performance based on knowledge of sleep-wake temporal dynamics or detect sleepiness and hypovigilance while on the job. Both approaches have come under intensive development and scrutiny in recent years, as governments and industries struggle to manage fatigue.

Biomathematical Models to Predict Performance Capability

One approach to predicting the effects of sleep loss is through the development of computer algorithms and models that reflect the biologically dynamic changes in alertness and performance due to sleep and circadian neurobiology. These models and algorithms are being developed as scheduling tools to quantify the impact of underlying interaction of sleep and circadian physiology on neurobehavioral functioning. A number of major efforts are underway internationally that focus on the application of biomathematical modeling in a software package in order to: (1) predict the times that neurobehavioral functioning will be maintained; (2) establish time periods for maximal recovery sleep; and (3) determine the cumulative effects of different work-rest schedules on overall performance (Mallis, Mejdal, Nguyen, & Dinges, 2004).

These biomathematical models of alertness and performance are founded in part on the two-process model of sleep regulation (Achermann, 2004). The two-process model describes the temporal relationship between the sleep homeostatic process and the endogenous circadian pacemaker in the brain. Although the two-process model was originally designed to be a model of sleep regulation, its application has been extended to describe and predict temporal changes in waking alertness. When used in this manner, the model predicts that performance decreases progressively with prolonged wakefulness and simultaneously varies over time in a circadian pattern.

As a result of U.S. Department of Defense, U.S. Department of Transportation, and NASA interest in the deployment of scheduling software in real-world environments, a workshop on fatigue and performance modeling was conducted that reviewed seven models commonly cited in scientific literature or funded by government funding (Mallis, Mejdal, Nguyen, et al., 2004). Although these biomathematical models were based on the same sleep-wake physiological dynamics, there was considerable diversity among them in the number and type of input and output variables, and their stated goals and capabilities (Mallis, Mejdal, Nguyen, et al., 2004). It is widely believed that such models can help identify less fatiguing work-rest schedules and manage fatigue-related risk. However, it is critical that they validly and reliably predict the effects of sleep loss and circadian desynchrony to help minimize fatigue-related accidents and incidents (Dinges, 2004). It is clear that biomathematical models that instantiate sleep-wake dynamics based in neurobiology are not yet ready for applicability. Current models failed to reliably predict the adverse effects of chronic sleep restriction on performance when evaluated in a double-blind test (Van Dongen, 2004). While biomathematical models hold much promise as work-rest scheduling tools, they should not be transitioned to real-world environments without substantial evidence of their scientific validity. Similarly, their ecological validity relative to different real-world work scenarios also needs to be established. This includes the identification of inputs and outputs that are both relevant and accurate for the specific operational environment.

Technologies for Detecting Operator Hypovigilance

Mathematical models of fatigue seek to predict performance capability based on sleep and circadian dynamics. In contrast, online real-time monitoring technologies for fatigue seek to detect sleepiness as it begins to occur during cognitive work. Many technologies are being developed to detect the effects of sleep loss and night work on attention (i.e., hypovigilance). The goal is to have a real-time system that can alert or warn an operator of increasing drowsiness before a serious adverse event occurs. It is believed that the earlier hypovigilance is detected, the sooner an effective countermeasure can be implemented, thereby reducing the chance of serious human error. There is a need therefore to be able to continuously and unobtrusively monitor an individual's attention to task in an automated fashion. Technologies that purport to be effective in fatigue and drowsiness detection must be shown to meet or exceed a number of criteria (Dinges & Mallis, 1998).

Any technology developed for drowsiness detection in real-world environments must be unobtrusive to the user and capable of calculating real-time measurements. Both its hardware and software must be reliable and its method of drowsiness and hypovigilance detection must be accurate. As far as algorithm development is concerned, it must reliably and validly detect fatigue in all individuals (i.e., reflect individual differences) and it must require as little calibration as possible, both within and between subjects. Overall, the drowsiness devices must meet all scientific standards and be unobtrusive, economical, practical, and easy to use. A great deal of harm can be done if invalid or unreliable technologies are quickly and uncritically implemented.

Initial validation should be tested in a controlled laboratory environment. Although validation may be possible in some field-based studies, there remains a challenge of error variance from extraneous sources. Additionally, field studies that do not allow for complete manipulation of the independent variable (e.g., a range of sleep loss) can mask the validity of a technology or create an apparent validity that is artificial and therefore could not be generalized to other contexts.

Fatigue and Drowsiness Detection: Slow Eyelid Closures as an Example

The first problem to confront in the online detection of fatigue in an operational environment is determining what biological or behavioral (or biobehavioral) parameter to detect. What are the objective early warning signs of hypovigilance, fatigue, sleepiness, or drowsiness? This question is not yet resolved, but research sponsored by the U.S. Department of Transportation has helped identify a likely candidate measure of fatigue from sleep loss and night work (Dinges, Mallis, Maislin, & Powell, 1998). Researchers experimentally tested the scientific validity of six online driver-based alertness-drowsiness detection technologies (e.g., EEG algorithms, eye blink detectors, head position sensor arrays). The criterion variable against which these technologies were evaluated was the frequency of lapses on the PVT—a test well validated to be sensitive to sleep loss and night work (Dorrian, Rogers, & Dinges, 2005). Results showed that only Perclos (a measure of the proportion of time subjects had slow eye closures; Weirwille, Wreggit, Kirn, Ellsworth, & Fairbanks, 1994) was more accurate in the detection of the frequency of drowsiness-induced PVT performance lapses than were other approaches. In these double-blind experiments, Perclos was also superior to subjects' own ratings of their sleepiness when it came to detecting PVT lapses of attention. Perclos has been evaluated in an over-the-road study of technologies for fatigue management in professional trucking operations (Dinges, Maislin, Brewster, Krueger, & Carroll, 2005). This study reveals the critical role operator acceptance plays in both driver-based and vehicle-based measures of occupational fatigue.

The human factors and ergonomic aspects of an operational context in which fatigue detection is undertaken can have a major impact on the validity and utility of the fatigue-detection system. This is illustrated by the application of Perclos to different transportation modalities. Results from implementation research conducted in a high-fidelity truck simulator demonstrated that it was possible to interface an online automated drowsiness-detection Perclos system into the driving environment (Grace, Guzman, Staszewski, Mallis, & Dinges, 1998). Both auditory and visual feedback from the Perclos device improved alertness and driving performance, especially when drivers were drowsy at night (Mallis et al., 2000), suggesting the system may promote user alertness and safety during the drowsiest portions of night driving. However, when the same automated Perclos system was tested in a Boeing 747-400 flight simulator to determine the effects of Perclos feedback on pilot alertness and performance during a night flight, the results were quite different. Unlike the truck-driving environment, the automated Perclos system with feedback did not significantly counteract decrements in performance, physiological sleepiness, or mood (Mallis, Neri, Colletti, et al., 2004). This was largely due to the Perclos system having a limited field of view that could not capture the pilot's eyes at all times during the flight, due to operational requirements of constant visual scanning and head movements. This research demonstrates that implementation obstacles can emerge and must be overcome when transitioning scientifically valid drowsiness-monitoring technologies to an operational environment.

Conclusion

Fatigue and sleepiness on the job are common occurrences in today's society and result from circadian displacement of sleep-wake schedules, and acute and chronic sleep loss. Extensive neurobiological and neurobehavioral research have established that waking neurocognitive functions on the job depend upon stable alertness from adequate daily recovery sleep. Operational demands in 24-hour industries inevitably result in fatigue from sleep loss and circadian displacement, which contribute to increased cognitive errors and risk of adverse events—although the magnitude of the effects can depend on the individual. Understanding and mitigating the risks posed by physiologically based variations in sleepiness and alertness in the workplace should be an essential function of neuroergonomics. The emergence of biomathematical models of temporally dynamic influences on performance capability from sleep-wake and circadian biology and the development of unobtrusive online technologies for detection of fatigue while working are two cogent examples of neuroergonomics in action.

Main Points

1. Neurobiologically based circadian and sleep homeostatic systems interact to regulate changes in alertness, performance, and timing of sleep.
2. Reduced sleep time results in neurobehavioral decrements that include increased reaction times, memory difficulties, cognitive slowing, and lapses of attention.
3. Night work, time zone transit, prolonged work, and work environments that include irregular schedules contribute to fatigue and the risk it poses to safe operations.
4. Efforts to manage fatigue and sleepiness in operational environments include prediction through biomathematical models of alertness and online fatigue-detection technologies.
5. Understanding and mitigating the risks posed by physiologically based variations in sleepiness and alertness in the workplace should be an essential function of neuroergonomics.

Acknowledgments. Supported by NASA cooperative agreement 9-58 with the National Space Biomedical Research Institute; by AFOSR F49620-95-1-0388 and F-49620-00-1-0266; and by NIH NR-04281 and RR-00040. We thank Nick Price for his assistance with the figures.

Key Readings

Durmer, J. S., & Dinges, D. F. (2005). Neurocognitive consequences of sleep deprivation. *Seminars in Neurology, 25*(1), 117–129.

Folkard, S., & Akerstedt, T. (2004). Trends in the risk of accidents and injuries and their implications for models of fatigue and performance. *Aviation, Space, and Environmental Medicine, 75*(3, Suppl.), A161–A167.

Saper, C. B., Chou, T. C., & Scammell, T. E. (2001). The sleep switch: Hypothalamic control of sleep and wakefulness. *Trends in Neuroscience, 24*, 726–731.

Van Dongen, H. P. A., & Dinges, D. F. (2005). Circadian rhythms in sleepiness, alertness and performance. In M. H. Kryger, T. Roth, & W. C. Dement (Eds.), *Principles and practice of sleep medicine* (4th ed.). Philadelphia: W.B. Saunders.

References

Accreditation Council for Graduate Medical Education. (2003). Report of the Work Group on Resident Duty Hours and the Learning Environment, June 11, 2002. In *The ACGME's approach to limit resident duty hours 12 months after implementation: A summary of achievements.*

Achermann, P. (2004). The two-process model of sleep regulation revisited. *Aviation Space and Environmental Medicine, 75*(3, Suppl.), A37–A43.

Achermann, P., Werth, E., Dijk, D. J., & Borbely, A. A. (1995). Time course of sleep inertia after nighttime and daytime sleep episodes. *Archives Italiennes de Biologie, 134*, 109–119.

Akerstedt, T. (2003). Shift work and disturbed sleep/wakefulness. *Occupational Medicine (London), 53*(2), 89–94.

Banks, S., Catcheside, P., Lack, L., Grunstein, R. R., & McEvoy, R. D. (2004). Low levels of alcohol impair driving simulator performance and reduce perception of crash risk in partially sleep deprived subjects. *Sleep, 27*, 1063–1067.

Barger, L. K., Cade, B. E., Ayas, N. T., Cronin, J. W., Rosner, B., Speizer, F. E., et al. (2005). Extended work shifts and the risk of motor vehicle crashes among interns. *New England Journal of Medicine, 352*, 125–134.

Basheer, R., Strecker, R. E., Thakkar, M. M., & McCarley, R. W. (2004). Adenosine and sleep-wake regulation. *Progress in Neurobiology, 73*, 379–396.

Belenky, G., Wesensten, N. J., Thorne, D. R., Thomas, M. L., Sing, H. C., Redmond, D. P., et al. (2003). Patterns of performance degradation and restoration during sleep restriction and subsequent recovery: A sleep dose-response study. *Journal of Sleep Research, 12*(1), 1–12.

Braver, E. R., Preusser, C. W., Preusser, D. F., Baum, H. M., Beilock, R., & Ulmer, R. (1992). Long hours and fatigue: A survey of tractor-trailer drivers. *Journal of Public Health Policy, 13*(3), 341–366.

Cajochen, C., Knoblauch, V., Krauchi, K., Renz, C., & Wirz-Justice, A. (2001). Dynamics of frontal EEG activity, sleepiness and body temperature under high and low sleep pressure. *Neuroreport, 12,* 2277–2281.

Czeisler, C. A., Duffy, J. F., Shanahan, T. L., Brown, E. N., Mitchell, J. F., Rimmer, D. W., et al. (1999). Stability, precision, and near-24-hour period of the human circadian pacemaker. *Science, 284,* 2177–2181.

Czeisler, C. A., & Khalsa, S. B. S. (2002). The human circadian timing system and sleep-wake regulation. In M. H. Kryger, T. Roth, & W. C. Dement (Eds.), *Principles and practice of sleep medicine* (pp. 353–376). Philadelphia: W. B. Saunders.

Dijk, D. J., & Czeisler, C. A. (1995). Contribution of the circadian pacemaker and the sleep homeostat to sleep propensity, sleep structure, electroencephalographic slow waves, and sleep spindle activity in humans. *Journal of Neuroscience, 15,* 3526–3538.

Dinges, D. F. (1995). An overview of sleepiness and accidents. *Journal of Sleep Research, 4*(S2), 4–14.

Dinges, D. F. (2004). Critical research issues in development of biomathematical models of fatigue and performance. *Aviation, Space, and Environmental Medicine, 75*(3, Suppl.), A181–A191.

Dinges, D. F., & Kribbs, N. B. (1991). Performing while sleepy: Effects of experimentally induced sleepiness. In T. H. Monk (Ed.), *Sleep, sleepiness and performance* (pp. 97–128). Winchester, UK: John Wiley.

Dinges, D. F., & Mallis, M. M. (1998). Managing fatigue by drowsiness detection: Can technological promises be realized? In L. Hartley (Ed.), *Managing fatigue in transportation* (pp. 209–229). Oxford, UK: Pergamon.

Dinges, D. F., Pack, F., Williams, K., Gillen, K. A., Powell, J. W., Ott, G. E., et al. (1997). Cumulative sleepiness, mood disturbance, and psychomotor vigilance performance decrements during a week of sleep restricted to 4–5 hours per night. *Sleep, 20*(4), 267–277.

Dinges, D. F., Mallis, M., Maislin, G., & Powell, J. W. (1998). *Evaluation of techniques for ocular measurement as an index of fatigue and the basis for alertness management*. Final report for the U.S. Department of Transportation (pp. 1–112). Washington, DC: National Highway Traffic Safety Administration.

Dinges, D. F., Maislin, G., Brewster, R. M., Krueger, G. P., & Carroll, R. J. (2005). *Pilot test of fatigue management technologies*. Journal of the Transportation Research Board No. 1922 (pp. 175–182). Washington, DC: Transportation Research Board of the National Academies.

Dinges, D. F., Baynard, M., & Rogers, N. L. (2005): Chronic sleep restriction. In: M. H. Kryger,

T. Roth, & W. C. Dement, (Eds.), *Principles and practice of sleep medicine* (4th ed., pp. 67–76). Philadelphia: W.B. Saunders.

Doran, S. M., Van Dongen, H. P. A., & Dinges, D. F. (2001). Sustained attention performance during sleep deprivation: Evidence of state instability. *Archives Italiennes de Biologie, 139,* 253–267.

Dorrian, J., & Dinges, D. F. (2006). Sleep deprivation and its effects on cognitive performance. In: T. Lee-Chiong (Ed.), *Sleep: A comprehensive handbook* (pp. 139–143). Hoboken, NJ: John Wiley & Sons.

Dorrian, J., Rogers, N. L., & Dinges, D. F. (2005). Psychomotor vigilance performance: Neurocognitive assay sensitive to sleep loss. In C. Kushida (Ed.), *Sleep deprivation: Clinical issues, pharmacology and sleep loss effects* (pp. 39–70). New York: Marcel Dekker, Inc.

Durmer, J. S., & Dinges, D. F. (2005). Neurocognitive consequences of sleep deprivation. *Seminars in Neurology, 25*(1), 117–129.

Ferrara, M., De Gennaro, L., & Bertini, M. (1999). The effects of slow-wave sleep (SWS) deprivation and time of night on behavioral performance upon awakening. *Physiology and Behavior, 68*(1–2), 55–61.

Grace, R., Guzman, A., Staszewski, J., Dinges, D. F., Mallis, M., & Peters, B. A. (1998). The Carnegie Mellon truck simulator, a tool to improve driving safety. Society of Automotive Engineers International: Truck and Bus Safety Issues SP1400 (pp. 1–6).

Harrison, Y., & Horne, J. A. (2000). The impact of sleep deprivation on decision making: A review. *Journal of Experimental Psychology: Applied, 6*(3), 236–249.

Johnson, M. D., & Sharit, J. (2001). Impact of a change from an 8hr to a 12hr shift schedule on workers and occupational injury rates. *International Journal of Industrial Ergonomics, 27,* 303–319.

Kleitman, N. (1963). *Sleep and wakefulness* (2nd ed.). Chicago: University of Chicago Press.

Landrigan, C. P., Rothschild, J. M., Cronin, J. W., Kaushal, R., Burdick, E., Katz, J. T., et al. (2004). Effect of reducing interns' work hours on serious medical errors in intensive care units. *New England Journal of Medicine, 351,* 1838–1848.

Mallis, M. M., Mejdal, S., Nguyen, T. T., & Dinges, D. F. (2004). Summary of the key features of seven biomathematical models of human fatigue and performance. *Aviation, Space, and Environmental Medicine, 75*(3), A4–A14.

Mallis, M., Maislin, G., Konowal, N., Byrne, V., Bierman, D., Davis, R., Grace, R., & Dinges, D. F. (1998). *Biobehavioral responses to drowsy driving alarms and alerting stimuli*. Final report for the U.S. Department of Transportation (pp. 1–127).

Washington, DC: National Highway Traffic Safety Administration.

Mallis, M. M., Neri, D. F., Colletti, L. M., Oyung, R. L., Reduta, D. D., Van Dongen, H., & Dinges, D. F. (2004). Feasibility of an automated drowsiness monitoring device on the flight deck. *Sleep* (Suppl. 27), A167.

Mignot, E., Taheri, S., & Nishino, S. (2002). Sleeping with the hypothalamus: Emerging therapeutic targets for sleep disorders. *Nature Neuroscience*, 5(Suppl.), 1071–1075.

Moore, R. Y. (1999). A clock for the ages. *Science, 284*, 2102–2103.

Naitoh, P., Kelly, T., & Babkoff, H. (1993). Sleep inertia: Best time not to wake up? *Chronobiology International, 10*(2), 109–118.

Rogers, A. E., Hwang, W. T., Scott, L. D., Aiken, L. H., & Dinges, D. F. (2004). The working hours of hospital staff nurses and patient safety. *Health Affairs (Millwood), 23*(4), 202–212.

Rosa, R. R., & Bonnet, M. H. (1993). Performance and alertness on 8 h and 12 h rotating shifts at a natural gas utility. *Ergonomics, 36*, 1177–1193.

Saper, C. B., Chou, T. C., & Scammell, T. E. (2001). The sleep switch: Hypothalamic control of sleep and wakefulness. *Trends in Neuroscience, 24*, 726–731.

Stutts, J. C., Wilkins, J. W., Scott Osberg, J., & Vaughn, B. V. (2003). Driver risk factors for sleep–related crashes. *Accident Analysis and Prevention, 35*, 321–331.

Van Dongen, H. P. A. (2004). Comparison of mathematical model predictions to experimental data of fatigue and performance. *Aviation, Space, and Environmental Medicine, 75*(3, Suppl.), A122–A124.

Van Dongen, H. P. A., & Dinges, D. F. (2000). Circadian rhythms in fatigue, alertness, and performance. In M. H. Kyyger, T. Roth, & W. C. Dement (Eds.), *Principles and practice of sleep medicine* (pp. 391–399). Philadelphia: W.B. Saunders.

Van Dongen, H. P., Maislin, G., & Dinges, D. F. (2004). Dealing with inter-individual differences in the temporal dynamics of fatigue and performance: Importance and techniques. *Aviation, Space, and Environmental Medicine, 75*(3), A147–A154.

Van Dongen, H. P., Maislin, G., Mullington, J. M., & Dinges, D. F. (2003). The cumulative cost of additional wakefulness: Dose-response effects on neurobehavioral functions and sleep physiology from chronic sleep restriction and total sleep deprivation. *Sleep, 26*, 117–126.

Walsh, J. K., Dement, W. C., & Dinges, D. F. (2005). Sleep medicine, public policy, and public health. In M. H. Kryger, T. Roth, & W. C. Dement (Eds.), *Principles and practice of sleep medicine* (4th ed., pp. 648–656) Philadelphia: W.B. Saunders.

Weinger, M. B., & Ancoli-Israel, S. (2002). Sleep deprivation and clinical performance. *Journal of the American Medical Association, 287*, 955–957.

Wierwille, W. W., Ellsworth, L. A., Wreggit, S. S., Fairbanks, R. J., & Kirn, C. L. (1994). *Research on vehicle-based driver status/performance monitoring: Development, validation, and refinement of algorithms for detection of driver drowsiness.* Final report for the U.S. Department of Transportation (NHTSA Technical Report No. DOT-HS-808-247). Washington, DC: National Highway Traffic Safety Administration.

Wright, N., & McGown, A. (2001). Vigilance on the civil flight deck: Incidence of sleepiness and sleep during long-haul flights and associated changes in physiological parameters. *Ergonomics, 44*, 82–106.

15

Waldemar Karwowski, Bohdana Sherehiy,
Wlodzimierz Siemionow, and
Krystyna Gielo-Perczak

Physical Neuroergonomics

Over the last 50 years, ergonomics (or human factors) has been maturing and evolving as a unique and independent discipline that focuses on the nature of human-artifact interactions, viewed from the unified perspective of science, engineering, design, technology, and the management of human-compatible systems, including a variety of natural and artificial products, processes, and living environments (Karwowski, 2005). According to the International Ergonomics Association (2002), ergonomics is a systems-oriented discipline that extends across all aspects of human activity. The traditional domains of specialization within ergonomics include physical ergonomics, cognitive ergonomics, and organizational ergonomics. Physical ergonomics is concerned with human anatomical, anthropometric, physiological, and biomechanical characteristics as they relate to human physical activity. Cognitive ergonomics is concerned with mental processes, such as perception, memory, reasoning, and motor response, as they affect interactions among humans and other elements of a system. Organizational ergonomics is concerned with the optimization of sociotechnical systems, including their organizational structures, policies, and processes.

The discipline of ergonomics has witnessed rapid growth and its scope has continually expanded toward new knowledge about humans that can be useful in design (Karwowski, Siemionow, & Gielo-Perczak, 2003). The expansion in scope has roughly followed the sequence from physical (motor) to cognitive, to esthetic, and most recently to affective (emotional) factors. This in turn has made it necessary to consider human brain functioning and the ultimate supreme role of the brain in exercising control over human behavior in relation to the affordances of the environment (Gibson, 1986). Most recently, the above developments have led to the onset of neuroergonomics, or the study of brain and behavior at work (Parasuraman, 2003). This chapter introduces *physical neuroergonomics* as the emerging field of study focusing on the knowledge of human brain activities in relation to the control and design of physical tasks (Karwowski et al., 2003). We provide an introduction to this topic in separate sections on the human brain in control of muscular performance in the work environment and in motor control tasks. We discuss these issues in conditions of health, fatigue, and disease states.

The Human Brain in Control of Muscular Performance

The insights offered by neuroscience (Zigmond et al., 1999) are essential to our understanding of the human operator functioning in complex systems (Parasuraman, 2000). Knowledge of human motor control is also critical to further advances in occupational biomechanics in general, and to prevention of musculoskeletal disorders in industry due to repetitive manual tasks and material handling in particular (Karwowski et al., 2003). One of the important functions of the human brain is the control of motor activities, combined with perceptual, cognitive, and affective processes. Half a century ago, Sperry (1952) proposed that the main function of the central nervous system is the coordinated innervation of the musculature, and its fundamental structure and mechanisms can be understood only on these terms. Sperry also argued that even for the highest human cognitive activities, which do not require motor output, there exist certain essential motoric neural events.

Recent brain imaging studies have provided strong support for Sperry's supposition, which has gained broader acceptance today. For example, according to Malmo and Malmo (2000), the extremely wide diversity of situations yielding electromyographical (EMG) gradients suggests the possibility that these gradients may be universal accompaniments of the organized goal-directed behavioral sequences. Both motor and cognitive tasks, without any requirements for motor output, were found to produce EMG gradients. EMG gradients were not observed during simple, repetitive exercises. On the efferent side, Malmo and Malmo (2000) proposed a dual model for the production of EMG gradients, which is based on empirical findings that reflect the complex relations between EMG gradient steepness and mental effort. The authors provided the evidence for movement-related brain activity generated by proprioceptive input, in relation to different types of feedback to the central nervous system during tasks that produce EMG gradients.

The Human Motor System

The human motor system consists of two interacting parts, peripheral and central (Wise & Shadmehr, 2002; see also chapter 22, this volume). The peripheral motor system includes muscles and both motor and sensory nerve fibers. The central motor system has components throughout the central nervous system (CNS), including the cerebral cortex, basal ganglia, cerebellum, brain stem, and spinal cord. Wise and Shadmehr (2002) proposed that the various components of the motor system work as an integrated neural network and not as isolated motor centers. According to Thach (1999), the spinal cord serves as the central pattern generator for reflexes and locomotion. As opposed to a sensory system, interruption of a motor system causes two abnormal functions: the inability to make an intended movement, and the spontaneous production of an unintended posture or movement. One distinct characteristic of the motor system is that many of its parts are capable of independently generating movement when cut off from other parts. In this way, each part of the system is a central pattern generator of movement.

Human motor systems can also be classified into three interrelated subsystems: skeletal, autonomic, and neuroendocrine, which are hierarchically organized (Swanson, Lufkin, & Colman, 1999). For example, the lowest level of skeletal motor systems consists of the alpha motor neurons that synapse on skeletal muscle fibers. The next higher level consists of the motor pattern generators (MPGs), which form the circuitry of interneurons that innervate unique sets of motor neuron pools. The highest level consists of motor pattern initiators (MPIs), which recognize specific input patterns and project to unique sets of MPGs. Swanson et al. (1999) also argued that a hierarchical organization exists between MPGs and MPIs. In this hierarchical model, at the lowest level (1), pools of motor neurons (MNs) innervate individual muscles that generate individual components of behavior. Pools of interneurons as MPGs at the next higher level (2), innervate specific sets of motor neuron pools. At the highest level (3), MPIs innervate specific sets of MPGs. Complex behaviors are produced when MPIs receive specific patterns of sensory, intrinsic, and cognitive inputs.

A Hierarchical Model of the Human Motor System

The main functions of the human nervous system can also be understood by analyzing the structural organization of the functional subsystems. This

Physical Neuroergonomics 223

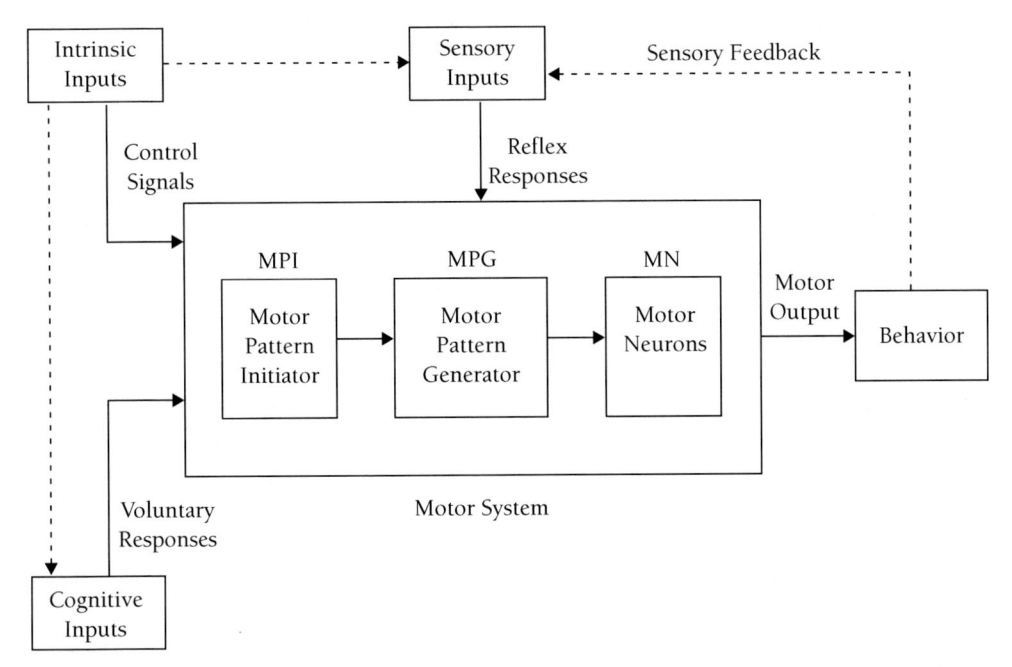

Figure 15.1. Basic information processing in the hierarchical motor system (after Swanson et al., 1999).

approach also provides the circuit diagram of information processing in the nervous system. Swanson et al. (1999) proposed a model of basic information processing (figure 15.1) which assumes that behavior is determined by the motor output of the CNS and that motor output is a function of three inputs: sensory, cognitive, and intrinsic. According to Swanson et al. (1999), the relative importance of these inputs in controlling motor output varies from species to species and from individual to individual. This model postulates that human behavior (B) is determined by the motor system (M), which is influenced by three neural inputs: sensory (S), intrinsic (I), and cognitive (C). Sensory inputs lead to reflex responses (r); cognitive inputs produce voluntary responses (v); and intrinsic inputs act as control signals (c) to regulate the behavioral state. Motor system outputs (1) produce behaviors whose consequences are monitored by sensory feedback (2). The above model also shows that the cognitive, sensory, and intrinsic systems are interconnected.

For example, motor control functioning and the utilization of processes traditionally considered as strictly cognitive can be illustrated with the model of neuronal control of precision grip. This model is based on extensive research of the neuronal implications of precision grip and describes the control of grip and load forces in a lifting task (Johanson, 1998). Adaptation of the fingertip forces to the physical characteristics of the lifted object involves two control mechanisms: (1) anticipatory parameter control, and (2) discrete-event sensory-driven control. Anticipatory parameter control determines appropriate motor programs that generate distributed muscle commands to the muscle exerting the fingertip forces. The specification of the parameters for motor commands is based on the important properties of the object that are stored in sensorimotor memory representation and were acquired during the previous experience. The discrete-event sensory-driven control uses somatosensory mechanisms that monitor the progress of the task. When a movement is made, the motor system predicts the sensory input and it is compared to the actual sensory input produced by the movement (Schmitz, Jenmalm, Ehrosson, & Forssberg, 2005). If there is a mismatch between the predicted and the actual sensory input, the somatosensory system triggers reactive preprogrammed patterns of corrective responses that modify the forces in the ongoing lift. The information from mismatch is used to update the sensorimotor representations of the parameters for specific object (see figure 15.2).

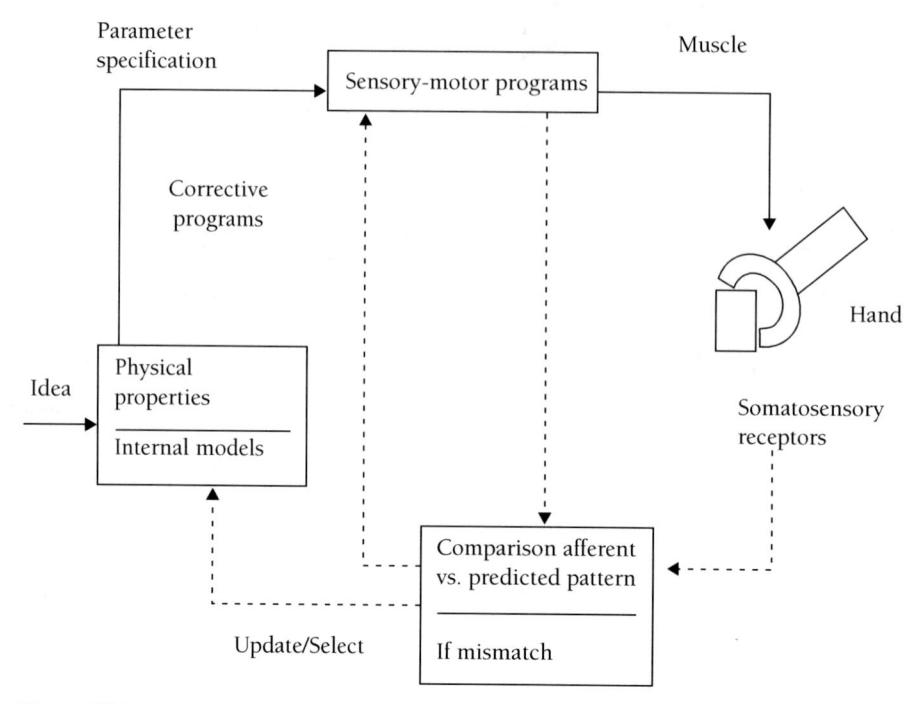

Figure 15.2. Anticipatory parameter control of fingertip forces (after Schmitz et al., 2005).

The Human Brain and the Work Environment

Contemporary work environments often demand effective human control, predictions, and decisions in the presence of uncertainties and unforeseen changes in work system parameters (Karwowski, 1991, 2001). The description of human operators who actively participate in purposeful work tasks in a given environment and their performance on such tasks should reflect the complexity of brain activity, which includes cognition and the dynamic processes of knowing. Bateson (2000) suggested that the processes of knowing were related to perception, communication, coding, and translation. However, he also provided a differentiation of logical levels, including the relationship between the knower and the known, knowledge looping back as knowledge of an expanded self.

Many control problems at work arise from a lack of attention to the interactions among different human system components in relation to affordances of the environment (Karwowski et al., 2003). Affordances, as opportunities for action for a particular organism, can offer both positive (benefits) and negative (injury) effects (Gibson, 1986). Thus, affordances are objective in the sense that they are fully specified by externally observable physical reality, but are subjective in the sense of being dependent on the behavior of a particular kind of organism. Gibson suggested that perception of the world is based upon perception of affordances, or recognizing the features of the environment that specify behaviorally relevant interactions.

Human consciousness at work is manifested in several brain activities, including thought, perception, emotion, will, memory, and imagination (Parasuraman, 2000). Therefore, tools for predicting human performance that take into account human emotions, imagination, and intuition with reference to affordances of the environment are needed (Gielo-Perczak & Karwowski, 2003). For example, Picard (2000) pointed out that in the human brain, a critical part of our ability to see and perceive is not logical but emotional. Emotions are integrated in a hierarchy of neurological processes that influence brain perceptual functions. There is behavioral and physiological evidence for the integration of perception and cognition with emotions (see also chapters 12 and 18, this volume). Emotions define the organism's dynamic structural pattern and their interactions may lead to specific responses in a work system (environment). Furthermore, the findings reported by Gevins, Smith, Mcenvoy, and Yu (1997), based on high-density electroencephalograph (EEG) mapping, suggest that the anterior cingulate cortex is a specialized area of the neocortex devoted to the regulation of emotional and cognitive behavior. It was also noted that when environmental regularities are allowed to take part in human behavior, they can give it coherence without the need of explicit internal mechanisms for binding perceptual entities.

Human Brain Activity in Motor Control Tasks

Motor control studies originated in 1950s when motor activity-related cortical potential (MRCP) was first described as EEG-derived brain potential associated with voluntary movements. Bates (1951) first recorded MRCP from the human scalp during voluntary hand movements using a crude photographic superimposition technique. Subsequent successful studies were reported by Kornhuber and Deecke (1965) who, based on the characteristics of the MRCP recording, described a slowly rising negative potential, known as the *readiness potential*, that precedes a more sharply rising negative potential, known as the *negative slope*. The onset of both the readiness potential and negative slope occurs prior to the onset of the voluntary movement and hence, both are considered to indicate involvement of the underlying brain cortical fields in preparing for the desired movement (Kornhuber & Deecke, 1965).

Recently, there have been numerous studies showing relevance of human motor control in relation to human performance on physical tasks. Explanation of the neural mechanisms underlying voluntary movements using MRCP requires an understanding of the relationship between the magnitude of MRCP and muscle activation. The relationship between magnitude of MRCP and rate of force development has not been well explored. A systematic study of these relationships (MRCP versus force, MRCP versus rate of force development) is still needed. Due to the advancements in brain imaging technology in recent years and the noninvasive feature of the surface EEG and MEG recordings, the number of studies of brain function (especially motor function) involving MRCP measurements is rapidly increasing.

Studies of Muscle Activation

In general, many motor actions are accomplished without moving the body or a body part (e.g., isometric contractions). Siemionow, Yue, Ranganathan, Liu, & Sahgal (2000) investigated the relationship between EEG-derived MRCP and voluntary muscle activation during isometric elbow-flexion contractions. Thus, MRCP in this study represents motor activity-related cortical potential as opposed to movement-related cortical potential. In one session, subjects performed isometric elbow-flexion contractions at four intensity levels (10%, 35%, 60%, and 85% maximal voluntary contraction or MVC). In another session, a given elbow-flexion force (35% MVC) was generated at three different rates (slow, intermediate, and fast). EEG signals were recorded from the scalp overlying the supplementary motor area (SMA) and contralateral sensorimotor cortex, and EMG signals were recorded from the skin surface overlying the belly of the biceps brachii and brachioradialis muscles during all contractions. The study results showed that the magnitude of MRCP from both EEG recording locations (sensorimotor cortex and SMA) was highly correlated with elbow-flexion force, rate of rising of force, and muscle EMG signals. Figure 15.3 illustrates the relationship between MRCP and muscle EMG across four levels of force. In figure 15.3A, MRCP from the SMA site was compared with the EMG recorded from the biceps brachii muscle; in (B), the MRCP (from SMA) was compared with the brachioradialis EMG; in (C), MRCP from the motor cortex site was compared with the EMG of the biceps brachii; and in (D), the MRCP (from motor cortex) was compared with the brachioradialis EMG. Data were recorded at the four force levels, but the EMG data at each force level are expressed as actual percentage of MVC EMG. These results suggest that MRCP represents cortical motor commands that scale the level of muscle activation.

Eccentric and Concentric Muscle Activities

Since different nervous system control strategies may exist for human concentric and eccentric muscle contractions, Fang, Siemionow, Sahgal, Xiong, and Yue (2001) used EEG-derived MRCP to determine whether the level of cortical activation differs between these two types of muscle

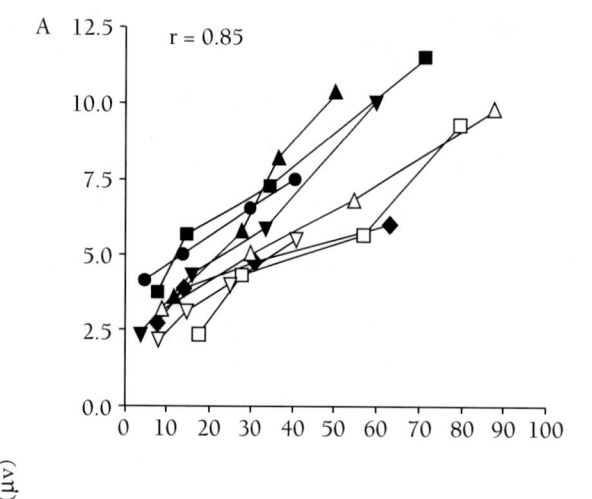

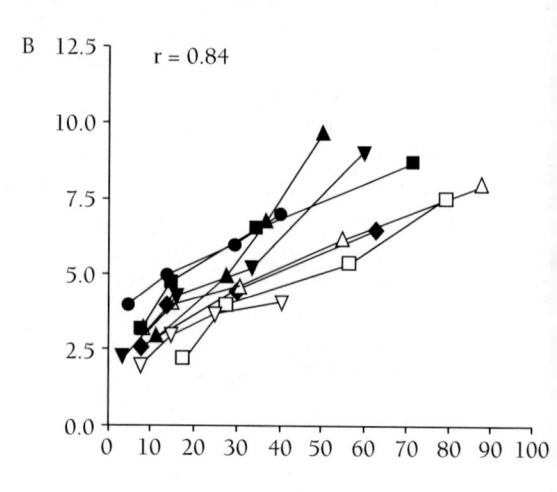

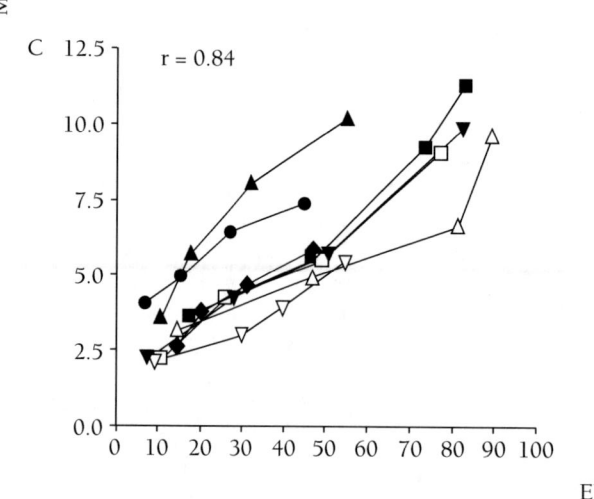

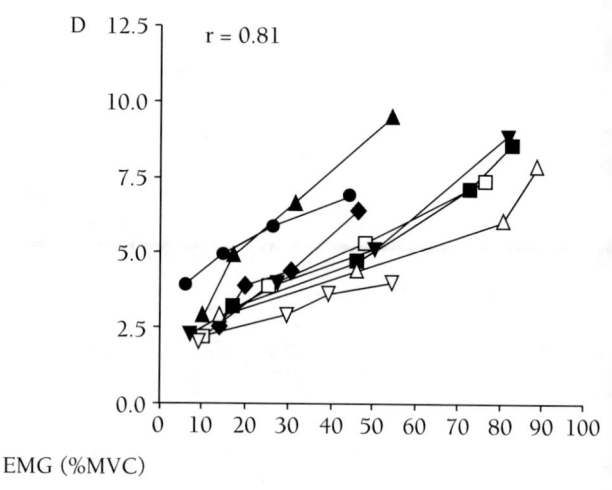

Figure 15.3. Relationship between motor activity-related cortical potential (MRCP) and muscle electromyograph (EMG) across four levels of force (after Siemionow et al., 2000). Each symbol represents a subject ($n = 8$).

activities. Subjects performed voluntary eccentric and voluntary concentric elbow flexor contractions against a load equal to 10% body weight. Surface EEG signals from four scalp locations overlying sensorimotor-related cortical areas in the frontal and parietal lobes were measured along with kinetic and kinematic information from the muscle and joint. The MRCP was derived from the EEG signals of the eccentric and concentric muscle contractions. Although the load supported by the subject was the same between the two tasks, the force increased during concentric and decreased during eccentric contractions from the baseline (isometric) force. The results showed that although the elbow flexor muscle activation (EMG) was lower during eccentric than concentric actions, the amplitude of two major MRCP components—one related to movement planning and execution and the other associated with feedback signals from the peripheral systems—was significantly greater for eccentric than for concentric actions. The MRCP onset time for the eccentric task occurred earlier than that for the concentric task. The authors concluded that the greater cortical signal for eccentric muscle actions suggests that the brain probably plans and programs eccentric movements differently from the concentric muscle tasks.

Mechanism of Muscular Fatigue

A limited number of studies regarding cortical modulation of muscle fatigue have been reported. Liu et al. (2001) investigated brain activity during muscle fatigue using the EEG system. Subjects performed intermittent handgrip contractions at 30% (300 trials) and 100% (150 trials) MVC levels. Each 30% contraction lasted 5 seconds and each 100% contraction 2 seconds, with a 5-second rest period between adjacent contractions. EEG data were recorded from the scalp during all contractions along with handgrip force and muscle EMG signals. MRCP was derived by force-triggered averaging of EEG data from each channel over each block (30 trials) of contractions with MRCP amplitude quantified. Thus, for each channel there were 5 MRCP data points for the 100% level task and 10 such data points for the 30% level task after averaging. Each data point represented cortical activity corresponding to a unique fatigue status of the subject or time frame (e.g., the first and last points corresponded respectively to conditions of least and most fatigue). Fatigue was determined by evaluating changes in force (100% level task) and EMG (30% level task), which declined to about 40%, while EMG of the flexor muscles (i.e., FDP and FDS) declined to about 45% of the maximal levels in 400 seconds. The results showed that the handgrip force and EMG decreased in parallel for the 100% level task. For the 30% level task, the EMG of the finger flexors increased progressively while the force was maintained. The MRCP data, however, did not couple closely with the force and EMG signals, especially for the 100% level task. It was concluded that the uncoupling of brain and muscle signals may indicate cortical and sensory feedback modulation of muscle fatigue.

Chronic fatigue syndrome (CFS) is a controversial condition in which people report malaise, fatigue, and reduction in daily activities, despite few or no physiological or laboratory findings. To examine the possibility that CFS is a biological illness involving pathology of the CNS, Siemionow et al. (2001) investigated whether brain activity of CFS patients during voluntary motor activities differs from that of healthy individuals. Eight CFS patients and eight age- and sex-matched healthy volunteers performed isometric handgrip contractions at 50% MVC level. In the first experiment, they performed 60 contractions with a 10-second rest between adjacent trials—the nonfatigue (NF) task. In the second experiment, the same number of contractions was performed with only a 5-second rest period—the fatigue (FT) task; 64 channels of surface EEG were recorded simultaneously from the scalp. Depicted data were recorded during fatiguing tasks from a CFS patient and a control subject (Ctrl). The amplitude of MRCP for the NF task was greater for the patient group than the control group. Similarly, MRCP for the FT task was greater for the patients than for the healthy subjects. Spectrum analysis of the EEG signals indicated that there were substantial differences at the delta and theta frequency bands between the two groups. The study results support the notion that CFS involves alterations of the CNS system.

Motor Control in Human Movements

Control of Extension and Flexion Movements

Corticospinal projections to the motor neuron pool of upper-limb extensor muscles have been reported to differ from those of the flexor muscles in humans and other primates. The influence of this difference on the CNS control for extension and flexion movements was studied by Yue, Siemionow, Liu, Ranganathan, and Sahgal (2000). Cortical activation during thumb extension and flexion movements of eight human volunteers was measured using functional magnetic resonance imaging (fMRI), which detects signal changes caused by an alteration in the local blood oxygenation level. The amplitude of MRCP was recorded during the thumb flexion and extension. The amplitude of MRCP recorded during thumb extension was significantly higher than that during flexion at both the motor cortex and SMA ($p < .05$, paired t test). Although the relative activity of the extensor and flexor muscles of the thumb was similar, the brain volume activated during extension was substantially larger than that during flexion. These fMRI results were confirmed by measurements of EEG-derived MRCP. It was concluded that the higher brain activity during thumb extension movement may be a result of differential corticospinal and possibly other pathway projections to the motor neuron pools of extensor and flexor muscles of the upper extremities.

Power and Precision Grip

Over the last decade, a large number of neurophysiological studies have focused on investigating the neural mechanisms controlling the precision grip (Fagergren, Ekeberg, & Forssberg, 2000; Kinoshita, Oku, Hashikawa, & Nishimura, 2002; Schmitz et al., 2005). These studies analyzed how a human manipulates an object with the tips of the index finger and thumb, or compensates for sudden perturbation when holding the object. The studies, using cell recording and different brain imaging methods, established a close relationship between primary motor cortex and precision grip (Ehrsson et al., 2000; Lemon, Johanson, & Westling, 1995; Salimi, Brochier, & Smith, 1999). The activation of brain areas during performance of precision grip in humans was also investigated in reference to several factors, including the following: different force size applied to object during the grip (Ehrsson, Fagergren, & Forssberg, 2001), object weight changes (Kinoshita et al., 2000; Schmitz et al., 2005), different types of object surface texture and friction characteristics (Salimi, 1999), and different types of grip (precision vs. power grip; Ehrsson et al., 2000).

Kilner, Baker, Salenius, Hari, and Lemon (2001) investigated task-dependent modulation incoherence between motor cortex and hand muscles during precision grip tasks. Twelve right-handed subjects used index finger and thumb to grip two levers that were under robotic control. Each lever was fitted with a sensitive force gauge. Subjects received visual feedback of lever force levels and were instructed to keep them within target boxes throughout each trial. Surface EMGs were recorded from four hand and forearm muscles, and magnetoencephalography (MEG) was recorded using a 306-channel neuromagnetometer. Overall, all subjects showed significant levels of coherence (0.086–0.599) between MEG and muscle in the 15–30 Hz range. Coherence was significantly smaller when the task was performed under an isometric condition (levers fixed) compared with a compliant condition in which subjects moved the levers against a springlike load. Furthermore, there was a positive, significant relationship between the level of coherence and the degree of lever compliance. These results argue in favor of coherence between cortex and muscle being related to specific parameters of hand motor function.

Ehrsson et al. (2000) used fMRI to analyze human brain activity during performance of two different kinds of manual grip: power grip and precision grip. The power grip is a palmar opposition grasp in which all digits are fixed around the object. Results showed that application of the precision grip was associated with a different pattern of brain activity in comparison to the power grip. When the power grip was performed, the activity of the primary sensory and motor cortex contralateral to the operating hand was higher than when subjects performed the precision grip. The precision grip task was more strongly associated with activity in the ipsilateral ventral premotor area, the rostral cingulate motor area, and at several locations in the posterior parietal and prefrontal cortices. Besides, it was also found that while the precision grip involved extensive activation in both hemispheres, the power grip was mainly related to activity in the left hemisphere.

Kinoshita (2000) investigated regional cortical and subcortical activation induced by repetitive lifting using a precision grip between index finger and thumb with positron emission tomography (PET). Results revealed significant activation of the primary motor (M1), primary sensory (S1), dorsocaudal premotor (PM), caudal SMA, and cingulate motor (CMA) cortices contralateral to the hand used during the object lifting. Behavioral adaptations to a heavier object weight were reflected in a nearly proportional increase of grip and lift forces, prolonged force application period, and a higher level of hand and arm muscle activity. An increase of regional cerebral blood flow (rCBF) that can be associated with these changes was found in several cortical and subcortical areas. However, consistent object weight-dependent activation was noted only in the M1/S1 contralateral to the hand used.

Ehrsson et al. (2001) conducted an fMRI study on human subjects which investigated if the cortical control of small precision grip forces differs from control of the large grip forces when the same grasp is applied to a stationary object. The research revealed that several sensory- and motor-related frontoparietal areas were more strongly activated when subjects applied a small force in comparison to when they generated a larger force. The fMRI showed brain regions with significant increased activity when the subjects applied small fingertip forces as compared with large forces. It was concluded that secondary sensorimotor areas in the frontal and parietal lobes of the human brain play an important role in the control of fine precision-grip forces in the small range. The stronger activation of those areas during small force application may reflect the involvement of additional sensorimotor mechanism control and somatosensory information processing due to more demanding control of fine object manipulation.

Reaching Movements

Neuroimaging studies (Grossman et al., 2000) point to the existence of neural mechanisms specialized for analysis of the kinematics defining motion. In order to reach an object, the spatial information about a target from the sensory system should be transformed into the motor plan. Studies have shown that the transformation of this information is performed in the posterior parietal cortex (PPC), since it is placed between visual areas that encode spatial information and motor cortical areas (Cloweer et al., 1996; Kalaska, Scott, Cisek, & Sergio, 1997; Kerztman, Schwarz, Zeffiro, & Hallt, 1997; Sabes, 2000; Snyder, Batista, & Andersen, 1997).

Snyder et al. (1997) analyzed neuronal activity in PPC during the early planning phases of the reaching task in the monkey and during saccades to the single remembered location. The pattern of neuronal activities before the movement depended on the type of movement being planned. The activity of 373 out of 652 recorded neurons was significantly modulated by direction of movement during the tasks. Also, 68% of these neurons were motor-intention specific: 21% were significantly modulated before eye but not arm movement, and 47% were significantly modulated before arm but not eye movement. Thus it has been suggested that there are two separate pathways in the PPC area, one related to reach intention and another related to saccade intentions.

Kerztman et al. (1997) conducted a PET study on visual guided reaching with the left or right arm to the targets presented in either the right or left visual fields. Two separate effects on neural activity were analyzed: (1) *hand effect*, the hand used to reach irrespective of field reach; and (2) the *field effect*, the effect of the target field of reach irrespective of the hand used. Significant rCBF increases in the hand and field conditions occurred bilaterally in the supplementary motor area, premotor cortex, cuneus, lingual gyrus, superior temporal cortex, insular cortex, thalamus, and putamen. Based on the results, Kerztman et al. (1997) suggested that the visual and motor components of reaching have different functional organizations and many brain areas represent both reaching limb and target field. The authors also observed a close correspondence in identified activation areas for both effects. The comparison of results for both conditions did not show separated regions related either to hand or visual field. Besides, the posterior parietal cortex is related to all areas identified in the study; thus it has been concluded that it plays a main role in the integration of the limb and visual field information.

The control of grasping and manipulations relies on the distributed processes in the CNS that involve most areas related to the sensorimotor control (Lemon et al., 1995; Schmitz et al., 2005). The motor cortex, via descending influences over the spinal cord, modulates activity in all motoneuron pools involved in reach and grasp. It is believed that the cortico-motoneuron is particularly important for the execution of skilled hand tasks (Lemon et al., 1995). The influence of the corticospinal system on motor output during the reach, grasp, and lift of the object in human subjects was investigated with application of transcranial magnetic stimulation (TMS; Lemon et al., 1995). It was hypothesized that the influence of the motor cortex on the hand muscles should appear during the movement phases that place high demands on the sensorimotor control, that is, during the positioning of the fingers prior to touch and during the early lifting phases. TMS was directed at the hand area of the motor cortex and was delivered during the six phases: reaching, at grip close, object manipulation, parallel increase in grip and load forces, lifting movement, and holding of the object. It was concluded that observed modulation in the EMG responses across different phases of the task evoked by the TMS reflect phasic changes in corticospinal excitability. This conclusion was based on the observation that for each muscle clear dissociations were found between the changes in the amplitude of the response to the TMS and the variations of EMG related to the different phases of the task. The obtained results suggest that the cortical representations of the intrinsic muscles that control the hand and fingertip orientation were highly activated when the digits closed around the object and just after the subject touched the object. The extrinsic muscles that control orientation of the hand and fingertips received the largest cortical input during the reaching phase of the task.

Postural Adjustments and Control

One of the unresolved problems in motor control research is whether maintaining fixed limb posture and movement between postures involves single robust or different specialized neural control processes (Kurtzer, 2005). Both approaches have their own advantages and disadvantages. The single control process may allow generality and conserved neural circuitry, which can be illustrated with equilibrium point control. The specialized control processes allow for context dependency and flexibility. Kurtzer analyzed this problem by investigating motor cortex neuron activity representation of mechanical loads during posture and reaching tasks. It was observed that the activity of approximately half of the neurons that reflected load-related activity was related specifically to either posture or movement only. Those neurons that were activated during both tasks randomly switched their response magnitude between tasks. Kurtzer suggested that the observed random changes in load representation of neuronal activity provide evidence for different specialized control processes of posture and movement. The existence of a switch in neural processing of the motor cortex just before the transition from stationary posture to body movement provides support for task-dependent control strategies.

Kazennikov, Solopova, Talis, Grishin, and Ioffe (2005) investigated the role of the motor cortex in the anticipatory postural adjustments for forearm unloading. In this study, motor evoked potentials (MEPs) evoked by TMS were analyzed in a forearm flexor at the time of bimanual unloading. MEPs were recorded in the forearm flexor during different tasks, including active and passive unloading, static forearm loading, and static loading of one hand with simultaneous lifting of the same weight by the other hand. Anticipatory postural adjustments consisted of changes in the activity of a forearm flexor muscle prior to active unloading of the limb and acted to stabilize the forearm position. It was found that during active and passive unloading, MEP amplitude decreased with the decrease of muscle activity. During stationary forearm loading, the change in MEP corresponded to the degree of loading. If during static loading the contralateral arm has lifted a separate, equivalent weight, the amplitude of MEP decreased. No specific changes in cortical excitability were found during the anticipatory postural adjustment. Kazennikov et al. suggested a possibility of direct corticospinal volley and the motor command mediated by subcortical structures in anticipatory postural adjustments.

Also, based on the results of several previous studies, it was suggested that the motor cortex plays a predominant role in learning a new pattern of postural adjustments, but not in physical performance.

Motor Control Task Difficulty

Winstein, Grafton, and Pohl (1997) analyzed brain activity related to motor task difficulty during performance of goal-directed arm aiming using PET. The study used and applied Fitts continuous aiming paradigm with three levels of difficulty and two aiming types (transport vs. targeting). Kinematics characteristics and movement time were analyzed along with the magnitude of brain activity in order to determine the brain areas related to the task and movement variables. The results showed significant differences in rCBF in reference to different task conditions. Reciprocal aiming compared with no-movement conditions resulted in significant differences in brain activity in the following areas: the cortical areas in the left sensorimotor, dorsal premotor, and ventral premotor cortices, caudal SMA proper, and parietal cortex, and subcortical areas of the left putamen, globus pallidus, red nucleus, thalamus, and anterior cerebellum. These brain areas are associated with the planning and execution of goal-directed movements. The increase of task difficulty gave the increase of rCBF in areas associated with planning complex movements requiring greater visuomotor processing. The decrease of task difficulty resulted in significant increases of brain activity in the areas related to high motor execution demands and minimal demands for precise trajectory planning.

This study presented an interesting testing of Fitts' law. According to Fitts' law, movement time increases as the width of the target decreases or when the distance to the target increases (Fitts, 1954). This time dependency describes the speed-accuracy trade-off for rapid, goal-directed aiming movements. The increase in movement time was explained as result of increased motor planning demands and the use of visual feedback to maintain accuracy with increase task difficulty. Winstein et al. (1997) stated that obtained brain activation patterns in this study support both arguments. The observed activation increase in bilateral fusiform gyrus suggests increase in visual information processing. The results also showed that with an increase in task difficulty, the activation of three areas of frontal cortex increases. These areas are associated with motor planning. When the aiming task required shorter-amplitude movements with precise endpoint constraints, the results showed increased activity in the dorsal parietal area and left central sulcus. According to Winstein et al. (1997), these results suggest that contrary to Fitts' law predictions, the brain areas' activity pattern reflects increased targeting demands.

Load Expectation

Schmitz et al. (2005) investigated brain activity related to the sensorimotor memory representation during predictable and unpredictable weight changes in the lifting task. During repetitive object lifting, the fingertip forces are targeted to the weight of the objects. The anticipatory programming depends on the sensorimotor memory representation that possesses information on the object weight. The unpredicted changes of object weight lead to mismatch between the predicted and actual sensory output related to the object weight. This in turn triggers corrective mechanisms and updating of the sensory memory. In this study, 12 subjects lifted an object with the right index finger and thumb in three conditions: constant weight, regular weight change, and irregular weight change. Results obtained with fMRI showed that some of the cerebral and subcortical brain areas related to the precision grip lift were more active during irregular and regular lifting than during constant weight conditions. The larger activation of the three cortical regions (left parietal operculum, bilateral supramarginal gyrus, and bilateral inferior frontal cortex) may be related to the fact that weight changing conditions caused errors in motor programming that triggered corrective reactions and updating of sensory memory representation. Some of these areas were more activated during the irregular than regular change of weight. It was suggested that larger activation during the irregular weight change may be due to larger errors occurring in the programming of the fingertip forces during unexpected weight changes.

Internal Models and Motor Learning

Research on motor control shows the existence of representations of sensorimotor transformations within the CNS and close covariation of motor cortex with arm postures, output force, and muscle activity. This evidence was used to support the theoretical framework, claiming that a nervous system uses internal models in controlling complex movement (Kalaska et al., 1997; Kawato, Furukawa, & Suzuki, 1987; Sabes, 2000). Internal models of the motor system are hypothetical computations within the nervous system that simulate the naturally occurring transformations between the sensory signals and motor commands (Witney, 2004). Two main types of internal models are distinguished: forward models that predict the outcome of some motor event and inverse models or feedforward controllers that calculate the motor command required to achieve some desired state. In the case of visually guided movement, it is believed that the early stages of a reaching movement are under control of the forward internal model, whereas the later stages are under control of a feedforward model (Sabes, 2000).

Neuroimaging studies have revealed that learning a new motor or cognitive task causes an increase of regional blood flow in the human cerebellum at the beginning of the learning process, which decreases as the learning process progresses. Nezafat, Shadmehr, and Holcomb (2001) used PET to examine changes in the cerebellum as subjects learned to make movements with their right arm while holding the handle of a robot that produced a force field. They observed that motor errors decreased from the control condition to the learning condition. With practice, initially distorted movements became smooth, straight-line trajectories. Shadmehr and Mussa-Ivaldi (1994) hypothesized that this improvement is due to formation of an internal model in the brain. Finally, Shadmehr and Brashers-Krug (1997) observed that a single session of training was sufficient to allow the subject to form an accurate internal model and maintain this model for up to 5 months.

Discussion

One of the most important functions of the human brain is the control of motor activities, combined with perceptual, cognitive, and affective processes. The studies discussed above demonstrate the complexity of the relationships between muscular performance and the nature of human brain control over physical activities. In addition to enhancing our understanding of human motor performance, the results of these studies provide many important insights into workplace design. For example, it has been shown that during the isometric elbow-flexion contractions, the cerebral-cortex system controls the extent of muscle activation and is responsible for smoothing out high-speed motor control processes (Siemionow et al., 2001). Surprisingly, brain activation during thumb extension was substantially larger than that observed during thumb flexion (Yue, Siemonow, Ranganatham, & Sahgal, 2000). It was also suggested that the higher brain activity might be a result of differential corticospinal projections to the motor neuron pool of upper-limb extensor muscles. Research on the precision grip demonstrated that the relationships between muscular force and human brain activity patterns are more complex than the proportional increase of brain activity with increase of force applied (Kinoshita et al., 2000). The activation in some brain areas significantly increased when small forces were applied at the fingertip-object interface (Ehrsson et al., 2001). The investigation of differences between brain activation patterns during different grip types (small force vs. large force, precision grip vs. power grip) demonstrated that precise and fine finger-object manipulations require additional sensorimotor control mechanisms to control force, and are more demanding in terms of neural control (Ehrsson et al., 2000, 2001). The results of eccentric versus concentric muscle contractions revealed that brain areas can be activated by motor imagery of those actions (Fang et al., 2001). The influence of mental training on strength of muscles was also analyzed (Ranganathan, Siemionow, Sahgal, & Yue, 2001). The range of synchronization of beta range EEG-EMG revealed a state of the corticomuscular network when attention was directed toward the motor task (Kristeva-Feige, Fritsch, Timmer, & Lucking, 2002), stipulating higher synchronization effects with higher precision of the motor task. The magnetoencephalography study revealed the coherence between motor cortex and muscle activity, with smaller coherence for isometric tasks compared with the dynamic hand activities (Kilner et al., 2001). A brain study of the affected patients versus healthy group confirmed that CFS involves impairments of the CNS (Siemionow et al., 2001).

The neurophysiological research on the motor control of different types of complex movement may provide important and very specific information for ergonomic design concerning what important parameters of the object to be grasped or lifted are used by central neuronal processes to plan body movement. For example, knowledge about how sensory spatial information is used by the brain to plan the goal-directed voluntary movements may be helpful in planning and design of the manual work task and work place. Furthermore, knowledge about how the neural control mechanism adapts the hand movement and force parameters to the physical properties of manipulated objects may help in developing optimal and specific strategies for manual handling tasks or other skilled hand tasks. The results of research on brain activity during reaching movement revealed the role of visual feedback at the early stages of reach planning and the ability to precisely control the complex dynamics of multijoint movements. It has also provided information on the role of learning in the maintenance of internal models and the way in which intrinsic (e.g., joint, muscle) and extrinsic (e.g., perceptual, task specific) information is used and combined to form a motor action plan. The investigation of how the CNS utilizes spatial information about target locations and limb position for specification of movement parameters, such as arm position or work posture, can help in providing specific visual cues to guide optimal and safe physical performance. Finally, mathematical models based on the concepts of cerebellum structures and neural control mechanisms can provide a wide range of possibilities to simulate and model human musculoskeletal responses during performance under a variety of working conditions, in order to validate the design of tools, workplaces, and physically demanding tasks.

Conclusion

As discussed by Karwowski et al. (2003), the selective approaches taken by researchers who work in different domains of ergonomics and the critical lack of knowledge synthesis may significantly slow down growth of the human factors discipline. For example, physical ergonomics, which focuses on musculoskeletal system performance (biomechanics), provides very little, and inadequate at best, consideration of cognitive design issues. On the other hand, cognitive ergonomics mostly disregards the muscular aspects of performance requirements of the variety of human-machine interaction tasks (Karwowski, 1991; Karwowski, Lee, Jamaldin, Gaddie, & Jang, 1999). The physical neuroergonomics approach offers a new methodological perspective for the study of human-compatible systems in relation to working environments and work task design. In this approach, the human brain is exerting control over its environment by constructing behavioral control networks, which functionally extend outside of the body, making use of consistent properties of the environment. These networks and the control they allow are the very reason for having a brain. In this context, human performance can be modeled as a dynamic, nonlinear process taking place over the interactions between the human brain and the environment (Gielo-Perczak & Karwowski, 2003).

The focus on the human brain in control of physical activities can be a very useful tool for minimizing incompatibilities between the capacities of workers and demands of their jobs in the context of affordances of the working environment. Such an approach can also help in assessing the suitability of human-machine system design solutions and determining the most effective improvements at the workplace. The exemplary results of neuroscience studies discussed in this chapter point out the critical importance of our understanding of brain functioning in control of human tasks. Such knowledge is also very much needed to understand the mechanisms for causation of musculoskeletal injuries, including work-related low back or upper extremity disorders, and to stimulate the development of new theories and applications to prevent such disorders.

Consideration of human motor activities separately from the related sensation and perception of our environment, information processing, decision making, learning, emotions, and intuition leads to system failures, industrial disasters, workplace injuries, and many failed ergonomics interventions. Neuroergonomics design rejects the traditional perspective of divisions between physical, perceptual, or cognitive activities, as they all are controlled by the brain. From this perspective, the functioning of the brain must be reflected in development of design principles and operational requirements for human-compatible systems.

Main Points

1. One of the most important functions of the human brain is the control of motor activities, combined with perceptual, cognitive, and affective processes.
2. The studies discussed in this chapter demonstrate the complexity of the relationships between muscular performance and the nature of human brain control over physical activities.
3. Specifically, knowledge about how the neural control mechanism adapts the hand movement and force parameters to the physical properties of manipulated objects may help in developing optimal and specific strategies for manual handling tasks or other skilled hand tasks.
4. Mathematical models based on the concepts of cerebellum structures and neural control mechanisms allow for modeling human musculoskeletal responses under a variety of working conditions in order to validate the design of tools, workplaces, and physically demanding tasks.
5. Knowledge of human motor control is critical to further advances in occupational biomechanics in general and prevention of musculoskeletal disorders in industry due to repetitive tasks and manual handling of loads in particular.
6. The physical neuroergonomics approach offers a new methodological perspective for the study of human-compatible systems in relation to working environments and work task design.

Key Readings

Grossman, E., Donnelly, M., Price, R., Pickens, D., Morgan, V., Neighbor, G., et al. (2000). Brain areas involved in perception of biological motion. *Journal of Cognitive Neuroscience, 12*, 711–720.

Karwowski, W., Siemionow, W., & Gielo-Perczak, K. (2003). Physical neuroergonomics: The human brain in control of physical work activities. *Theoretical Issues in Ergonomics Science, 4*(1–2), 175–199.

Lemon, R. N., Johanson, R. S., & Westling, G. (1995). Corticospinal control during reach, grasp, and precision lift in man. *Journal of Neuroscience, 15,* 6145–6156.

Thach, W. T. (1999). Fundaments of motor systems. In M. J. Zigmond, F. E. Bloom, S. C. Landis, J. L. Roberts, & L. R. Squire (Eds), *Fundamental neuroscience* (pp. 855–861). San Diego, CA: Academic Press.

References

Bates, A. V. (1951). Electrical activity of the cortex accompanying movement. *Journal of Physiology, 113,* 240–254.

Bateson, G. (2000). *Steps to an ecology of mind.* Chicago: University of Chicago Press.

Cloweer, D. M., Hoffman J. M., Votaw, J. R., Faber, T. L., Woods, R. P., & Alexander, G. E. (1996). Role of the posterior parietal cortex in the recalibration of visually guided reaching. *Nature, 383,* 618–621.

Ehrsson, H. H., Fagergren, E., & Forssberg, H. (2001). Differential fronto-parietal activation depending on force used in a precision grip task: An fMRI study. *Journal of Neurophysiology, 85,* 2613–2623.

Ehrsson, H. H., Fagergren, A., Jonsson, T., Westling, G., Johansson, R. S., & Forssberg, H. (2000). Cortical activity in precision- versus power-grip tasks: An fMRI study. *Journal of Neurophysiology, 83,* 528–536.

Fagergren, A., Ekeberg, O., & Forssberg, H. (2000). Precision grip force dynamics: A system identification approach. *IEEE Transactions on Biomedical Engineering, 47,* 1366–1375.

Fang, Y., Siemionow, V., Sahgal, V., Xiong, F., & Yue, H. G. (2001). Greater movement-related cortical potential during human eccentric and concentric muscle contractions. *Journal of Neurophysiology, 86,* 1764–1772.

Fitts, P. M. (1954). The information capacity of the human motor system controlling the amplitude of movement. *Journal of Experimental Psychology, 47*(6), 381–391.

Gevins, A., Smith, M. E., Mcenvoy, L., & Yu, D. (1997). High-resolution EEG mapping of cortical activation related to working memory: Difficulty, types of processing, and practice. *Cerebral Cortex, 7,* 374–385.

Gibson, J. J. (1986). *The ecological approach to visual perception.* Hillsdale, NJ: Erlbaum.

Gielo-Perczak, K., & Karwowski, W. (2003). Ecological models of human performance based on affordance, emotion and intuition. *Ergonomics, 46*(1–3), 310–326.

Grossman, E., Donnelly, M., Price, R., Pickens, D., Morgan, V., Neighbor, G., et al. (2000). Brain areas involved in perception of biological motion. *Journal of Cognitive Neuroscience, 12,* 711–720.

IEA. (2002). International Ergonomics Association website. http://www.iea.cc/ergonomics/.

Johansson, R. S. (1998). Sensory input and control of grip. *Novartis Foundation Symposium, 218,* 45–59.

Kalaska, J. F., Scott, S. H., Cisek, P., & Sergio, L. E. (1997). Cortical control of reaching movements. *Current Opinion in Neurobiology, 7,* 849–859.

Karwowski, W. (1991). Complexity, fuzziness and ergonomic incompatibility issues in the control of dynamic work environments. *Ergonomics, 34,* 671–686.

Karwowski, W. (Ed.). (2001). *International encyclopedia of ergonomics and human factors.* London: Taylor and Francis.

Karwowski, W. (2005). Ergonomics and human factors: The paradigms for science, engineering, design, technology, and management of human-compatible systems. *Ergonomics, 48,* 436–463.

Karwowski, W., Grobelny, J., & Yang Yang W. G. L. (1999). Applications of fuzzy systems in human factors. In H. Zimmermman (Ed.), *Handbook of fuzzy sets and possibility theory* (pp. 589–620). Boston: Kluwer.

Karwowski, W., Lee, W. G., Jamaldin, B., Gaddie, P., & Jang, R. (1999). Beyond psychophysics: A need for a cognitive modeling approach to setting limits in manual lifting tasks. *Ergonomics, 42*(1), 40–60.

Karwowski, W., Siemionow, W., & Gielo-Perczak, K. (2003). Physical neuroergonomics: The human brain in control of physical work activities. *Theoretical Issues in Ergonomics Science, 4*(1–2), 175–199.

Kawato, M., Furukawa, K., & Suzuki, R. (1987). A hierarchical neural-network model for control and learning of voluntary movement. *Biological Cybernetics, 57,* 169–187.

Kazennikov, O., Solopova, I., Talis, V., Grishin, A., & Ioffe, M. (2005). TMS-responses during anticipatory postural adjustment in bimanuaul unloading in humans. *Neuroscience Letters, 383,* 246–250.

Kertzman, C., Schwarz, U., Zeffiro, T. A., & Hallt, M. (1997). The role of posterior parietal cortex in visually guided reaching movements in humans. *Experimental Brain Research, 114,* 170–163.

Kilner, J. M., Baker, S. N., Salenius, S., Hari, R., & Lemon, R. N. (2001). Human cortical muscle coherence is directly related to specific motor parameters. *Journal of Neuroscience, 20,* 8838–8845.

Kinoshita, H., Oku, N., Hashikawa, K., & Nishimura, T. (2000). Functional brain areas used for the lifting of objects using a precision grip: A PET study. *Brain Research, 857*(1–2), 119–130.

Kornhuber, H. H., & Deecke, L. (1965). Hirnpotentialänderungen bei Willkürbewegungen und passiven Bewegungen des Menschen: Bereitschaftspotential und reafferente Potentiale. *Pflüger's Arch, 284,* 1–17.

Kristeva-Feige, R., Fritsch, C., Timmer, J., & Lucking, C. H. (2002). Effects of attention and precision of exerted force on beta range EEG-EMG synchronization during a maintained motor contraction task. *Clinical Neurophysiology, 113*(1), 124–131.

Kurtzer, I., Herter, T. M., & Scott, S. H. (2005). Random change in cortical load representation suggests distinct control of posture and movement. *Nature Neuroscience, 8*(4), 498–504.

Lemon, R. N., Johanson, R. S., & Westling, G. (1995). Corticospinal control during reach, grasp, and precision lift in man. *Journal of Neuroscience, 15,* 6145–6156.

Liu, J., Bing, Y., Zhang, L. D., Siemionow, V., Sahgal, V., & Yue, G. H. (2001). Motor activity-related cortical potential during muscle fatigue. *Society for Neuroscience Abstracts, 27,* 401.6.

Malmo, R. B., & Malmo, H. P. (2000). On electromyographic EMG gradients and movement-related brain activity: Significance for motor control, cognitive functions, and certain psychopathologies. *International Journal of Psychophysiology, 38,* 143–207.

Nezafat, R., Shadmehr, R., & Holcomb, H. H. (2001). Long-term adaptation to dynamics of reaching movements: A PET study. *Experimental Brain Research, 140,* 66–76.

Parasuraman, R. (2000). *The attentive brain.* Cambridge, MA: MIT Press.

Parasuraman, R. (2003). Neuroergonomics: Research and practice. *Theoretical Issues in Ergonomics Science, 1–2,* 5–20.

Picard, R. W. (2000). *Affective computing.* Cambridge, MA: MIT Press.

Ranganathan, V. K., Siemionow, V., Sahgal, V., & Yue, G. H. (2001). Increasing muscle strength by training the central nervous system without physical exercise. *Society for Neuroscience Press Book, 2,* 518–520.

Sabes, P. N. (2000). The planning and control of reaching movements. *Current Opinion in Neurobiology, 10,* 740–746.

Salimi, I., Brochier, T., & Smith, A. M. (1999). Neuronal activity in somatosensory cortex of monkeys using a precision grip: III. Responses to altered friction perturbations. *Journal of Neurophysiology, 81,* 845–857.

Schmitz, C., Jenmalm, P., Ehrosson, H. H., & Forssberg, H. (2005). Brain activity during predictible and unpredictible weight changes when lifting objects. *Journal of Neurphysiology, 93,* 1498–1509.

Shadmehr, R., & Brashers-Krug, T. (1997). Functional stages in the formation of human long-term motor memory. *Journal of Neuroscience, 17,* 409–419.

Shadmehr, R., & Mussa-Ivaldi, F. A. (1994). Adaptive representation of dynamics during learning of a motor task. *Journal of Neuroscience, 14,* 3208–3224.

Siemionow, V., Fang, Y., Nair, P., Sahgal, V., Calabrese, L., & Yue, G. H. (2001). Brain activity during voluntary motor activities in chronic fatigue syndrome. *Society for Neuroscience Press Book, 2,* 725–727.

Siemionow, W., Yue, G. H., Ranganathan, V. K., Liu, J. Z., & Sahgal, V. (2000). Relationship between motor-activity related potential and voluntary muscle activation. *Experimental Brain Research, 133,* 3030–3311.

Snyder, L. H., Batista, A. P., & Andersen, R. A. (1997). Coding of intention in the posterior parietal cortex. *Nature, 386,* 167–170.

Sperry, R. W. (1952). Neurology and the mind-brain problem. *American Scientist, 40,* 291–312.

Swanson, L. W., Lufkin, T., & Colman, D. R. (1999). Organization of nervous systems. In M. J. Zigmond, F. E. Bloom, S. C. Landis, J. L. Roberts, & L. R. Squire (Eds.), *Fundamental neuroscience* (pp. 9–37). San Diego, CA: Academic Press.

Thach, W. T. (1999). Fundaments of motor systems. In M. J. Zigmond, F. E. Bloom, S. C. Landis, J. L. Roberts, & L. R.Squire (Eds.), *Fundamental neuroscience* (pp. 855–861). San Diego, CA: Academic Press.

Winstein, C. J., Grafton, S. T., & Pohl, P. S. (1997). Motor task difficulty and brain activity: Investigation of goal-directed reciprocal aiming using positron emission tomography. *Journal of Neurophysiology, 77,* 1581–1594.

Witney, A. G. (2004). Internal models for bi-manual tasks. *Human Movement Science, 23*(5), 747–770.

Yue, G. H., Ranganathan, V., Siemionow, W., Liu, J. Z., & Sahgal, V. (2000). Evidence of inability to fully activate human limb muscle. *Muscle and Nerve, 23,* 376–384.

Yue, G. H., Siemionow, W., Liu, J. Z., Ranganathan, V., & Sahgal, V. (2000). Brain activation during human finger extension and flexion movements. *Brain Research, 856,* 291–300.

Zigmond, M. J., Bloom, F. E., Landis, S. C., Roberts, J. L., & Squire, L. R. (Eds.). (1999). *Fundamental neuroscience.* San Diego, CA: Academic Press.

V

Technology Applications

Mark W. Scerbo

Adaptive Automation

> We humans have always been adept at dovetailing our minds and skills to the shape of our current tools and aids. But when those tools and aids start dovetailing back—when our technologies actively, automatically, and continually tailor themselves to us just as we do to them—then the line between tool and human becomes flimsy indeed.
>
> Andy Clark, *Natural-Born Cyborgs: Minds, Technologies and the Future of Human Intelligence* (p. 7)

Neuroergonomics has been described as the study of brain and behavior at work (Parasuraman, 2003). This emerging area focuses on current research and developments in the neuroscience of information processing and how that knowledge can be used to improve performance in real-world environments. Parasuraman (2003) argued that an understanding of how the brain processes perceptual and cognitive information can lead to better designs for equipment, systems, and tasks by enabling a tighter match between task demands and the underlying brain processes. Ultimately, research in neuroergonomics can lead to safer and more efficient working conditions.

Ironically, interest in neuroergonomics evolved from research surrounding how operators interact with a form of technology designed to make work and our lives easier—automation. In general, automation can be thought of as a machine agent capable of carrying out functions normally performed by a human (Parasuraman & Riley, 1997). For example, the automatic transmission in an automobile allocates the tasks of depressing the clutch, shifting gears, and releasing the clutch to the vehicle. Automated machines and systems are intended and designed to reduce task demands and workload. Further, they allow individuals to increase their span of operation or control, perform functions that are beyond their normal abilities, maintain performance for longer periods of time, and perform fewer mundane activities. Automation can also help reduce human error and increase safety. The irony behind automation arises from a growing body of research demonstrating that automated systems often actually increase workload and create unsafe working conditions.

In his book, *Taming HAL: Designing Interfaces Beyond 2001*, Degani (2004) relates the story of an airline captain and crew performing the last test flight with a new aircraft. This was to be the second such test that day and the captain, feeling rather tired, requested that the copilot fly the aircraft. The test plan required a rapid takeoff, followed by engaging the autopilot, simulating an engine failure by reducing power to the left engine, and then turning off the left hydraulic system. The test flight started out just fine. Four seconds into the flight, however, the aircraft was pitched about 4 degrees higher than normal, but the captain continued with the test plan and attempted to engage the autopilot. Unfortunately, the autopilot did not engage. After a few more presses of the autopilot button, the control panel display indicated that the system had engaged (although in reality, the autopilot had not assumed control). The aircraft was still pitched too high and was beginning to lose speed. The captain apparently did not notice these conditions and continued with the next steps, requiring power reduction to the left engine and shutting down the hydraulic system.

The aircraft was now flying on one engine with increasing attitude and decreasing speed. Moreover, the attitude was so steep that the system intentionally withdrew autopilot mode information from its display. Suddenly, the autopilot engaged and assumed the altitude capture mode to take the aircraft to the preprogrammed setting of 2,000 feet, but this information was not presented on the autopilot display. The autopilot initially began lowering the nose, but then reversed course. The attitude began to pitch up again and airspeed continued to fall. When the captain finally turned his attention from the hydraulic system back to the instrument panel, the aircraft was less than 1,500 feet above ground, pitched up 30 degrees, with airspeed dropping to about 100 knots. The captain then had to compete with the envelope protection system for control of the aircraft. He attempted to bring the nose down and then realized he had to reduce power to the right engine in order to undo a worsening corkscrew effect produced by the simulated left engine failure initiated earlier. Although he was able to bring the attitude back down to zero, the loss of airspeed coupled with the simulated left engine failure had the aircraft in a 90-degree roll. The airspeed soon picked up and the captain managed to raise the left wing, but by this time the aircraft was only 600 feet above ground. Four seconds later the aircraft crashed into the ground, killing all on board.

Degani (2004) discussed several factors that contributed to this crash. First, no one knows why the autopilot's altitude was preprogrammed for 2,000 feet, but it is possible that the pilot never entered the correct value of 10,000 feet. Second, although the pilot tried several times to engage the autopilot, he did not realize that the system's logic would override his requests because his copilot's attempts to bring the nose down were canceling his requests. Third, there was a previously undetected flaw in the autopilot's logic. The autopilot calculated the rate of climb needed to reach 2,000 feet when both engines were powered up, but did not recalculate the rate after the left engine had been powered down. Thus, the autopilot continued to demand the power it needed to reach the preprogrammed altitude despite that the aircraft was losing speed. Last, no one knows why the pilot did not disengage the autopilot when the aircraft continued to increase its attitude. Degani suggested that pilots who have substantial experience with autopilot systems may place too much trust in them. Thus, it is possible that assumptions regarding the reliability of the autopilot coupled with the absence of mode information on the display left the captain without any information or reason to question the status of the autopilot.

This incident clearly highlights the complexity and problems that can be introduced by automation. Unfortunately, it is not a unique occurrence. Degani (2004) described similar accounts of difficulties encountered with other automated systems including cruise control in automobiles and blood pressure devices.

Research on human interaction with automation has shown that it does not always make the job easier. Instead, it changes the nature of work. More specifically, automation changes the way activities are distributed or carried out and can therefore introduce new and different types of problems (Woods, 1996). Automation can also lead to different types of errors because operator goals may be incongruent with the goals of systems and subsystems (Sarter & Woods, 1995; Wiener, 1989). Woods (1996) argued further that in systems where subcomponents are tightly coupled, problems may propagate more quickly and be more difficult to isolate. In addition, highly automated systems leave fewer activities for individuals to perform. Consequently, the operator becomes a more passive monitor instead of an active participant. Parasuraman, Mouloua, Molloy, and Hilburn (1996) have shown that this shift from performing tasks to monitoring automated systems can actually inhibit one's ability to detect critical signals or warning conditions. Further, an operator's manual skills can begin to deteriorate in the presence of long periods of automation (Wickens, 1992).

Adaptive Automation

Given the problems associated with automation noted above, researchers and developers have begun to turn their attention to alternative methods for implementing automated systems. *Adaptive automation* is one such method that has been proposed to address some of the shortcomings of traditional automation. In adaptive automation, the level of automation or the number of systems operating under automation can be modified in real time. In addition, changes in the state of automation can be initiated by either the human or the system (Hancock & Chignell, 1987; Rouse, 1976; Scerbo, 1996). Consequently, adaptive automation enables the level or modes of automation to be tied more closely to operator needs at any given moment (Parasuraman, Bahri, Deaton, Morrison, & Barnes, 1992).

Adaptive automation systems can be described as either adaptable or adaptive. Scerbo (2001) described a taxonomy of adaptive technology. One dimension of this taxonomy concerns the underlying source of flexibility in the system, that is, whether the information displayed or the functions themselves are flexible. A second dimension addresses how the changes are invoked. In adaptable systems, changes among presentation modes or in the allocation of functions are initiated by the user. By contrast, in adaptive systems both the user and the system can initiate changes in the state of the system.

The distinction between adaptable and adaptive technology can also be described with respect to authority and autonomy. Sheridan and Verplank (1978) described several levels of automation that range from completely manual to semiautomatic to fully automatic. As the level of automation increases, systems take on more authority and autonomy. At the lower levels of automation, systems may offer suggestions to the user. The user can either veto or accept the suggestions and then implement the action. At moderate levels, the system may have the autonomy to carry out the suggested actions once accepted by the user. At higher levels, the system may decide on a course of action, implement the decision, and merely inform the user. With respect to Scerbo's (2001) taxonomy, adaptable systems are those in which the operator maintains authority over invoking changes in the state of the automation (i.e., they reflect a superordinate-subordinate relationship between the operator and the system). In adaptive systems, on the other hand, authority over invocation is shared. Both the operator and the system can initiate changes in state of the automation.

There has been some debate over who should have control over changes among modes of operation. Some argue that operators should always have authority over the system because they are ultimately responsible for the behavior of the system. In addition, it is possible that operators may be more efficient at managing resources when they can control changes in the state of automation (Billings & Woods, 1994; Malin & Schreckenghost, 1992). Many of these arguments are based on work with life-critical systems in which safe operation is of the utmost concern. However, it is not clear that strict operator authority over changes among automation modes is always warranted. There may be times when the operator is not the best judge of when automation is needed. For example, changes in automation may be needed at the precise moment when the operator is too busy to make those changes (Wiener, 1989). Further, Inagaki, Takae, and Moray (1999) have shown mathematically that the best piloting decisions concerning whether to abort a takeoff are not those where either the human or the avionics maintain full control. Instead, the best decisions are made when the pilot and the automation share control.

Scerbo (1996) has argued that in some hazardous situations where the operator is vulnerable, it would be extremely important for the system to have authority to invoke automation. If lives are at stake or the system is in jeopardy, allowing the system to intervene and circumvent the threat or minimize the potential damage would be paramount. For example, it is not uncommon for many of today's fighter pilots to sustain G forces high enough to render them unconscious for periods of up to 12 seconds. Conditions such as these make a strong case for system-initiated invocation of automation. An example of one such adaptive automation system is the Ground Collision-Avoidance System (GCAS) developed and tested on the F-16D (Scott, 1999). The system assesses both internal and external sources of information and calculates the time it will take until the aircraft breaks through a pilot-determined minimum altitude. The system issues a warning to the pilot. If no action is taken, an audio "fly up" warning is then presented and the system takes control of the aircraft. When the system has maneuvered the aircraft out of the way of the terrain, it returns control of the aircraft to the pilot with the message, "You got it." The intervention is designed to right the aircraft quicker than any human pilot can respond. Indeed, test pilots who were given the authority to override GCAS eventually conceded control to the adaptive system.

Adaptive Strategies

There are several strategies by which adaptive automation can be implemented (Morrison & Gluckman, 1994; Rouse & Rouse, 1983; Parasuraman et al., 1992). One set of strategies addresses system functionality. For instance, entire tasks can be allocated to either the system or the operator, or a specific task can be partitioned so that the system and operator each share responsibility for unique portions of the task. Alternatively, a task could be transformed to a different format to make it easier (or more challenging) for the operator to perform.

A second set of strategies concerns the triggering mechanism for shifting among modes or levels of automation (Parasuraman et al., 1992; Scerbo, Freeman, & Mikulka, 2003). One approach relies on goal-based strategies. Specifically, changes among modes or levels of automation are triggered by a set of criteria or external events. Thus, the system might invoke the automatic mode only during specific tasks or if it detects an emergency situation. Another approach would be to use real-time measures of operator performance to invoke the changes in automation. A third approach uses models of operator performance or workload to drive the adaptive logic (Hancock & Chignell, 1987; Rouse, Geddes, & Curry, 1987–1988). For example, a system could estimate current and future states of an operator's activities, intentions, resources, and performance. Information about the operator, the system, and the outside world could then be interpreted with respect to the operator's goals and current actions to determine the need for adaptive aiding. Finally, psychophysiological measures that reflect operator workload can also be used to trigger changes among modes.

Examples of Adaptive Automation Systems

Adaptive automation has its beginnings in artificial intelligence. In the 1970s, efforts were directed toward developing adaptive aids to help allocate tasks between humans and computers. By the 1980s, researchers began developing adaptive interfaces. For instance, Wilensky, Arens, and Chin (1984) developed the UNIX Consultant (UC) to provide general information about UNIX, procedural information about executing UNIX commands, and debugging information. The UC could analyze user queries, deduce user goals, monitor the user's interaction history, and present the system's response.

Associate Systems

Adaptive aiding concepts were applied in a more comprehensive manner in the Defense Advanced Research Projects Agency (DARPA) Pilot's Associate program (Hammer & Small, 1995). The goal of the program was to use intelligent systems to provide pilots with the appropriate information, in the proper format, at the right time. The Pilot's Associate could monitor and assess the status of its own systems as well as events in the external environment. The information could then be evaluated and presented to the pilot. The Pilot's Associate could also suggest actions for the pilot to take. Thus, the system was designed to function as an assistant for the pilot.

In the 1990s, the U.S. Army attempted to take this associate concept further in its Rotorcraft Pilot's Associate (RPA) program. The goal was to create an associate that could serve as a "junior crew member" (Miller & Hannen, 1999). A major component of the RPA is the Cognitive Decision Aiding System (CDAS), which is responsible for detecting and organizing incoming data, assessing the internal information regarding the status of the aircraft, assessing external information about target and mission status, and feeding this information into a series of planning and decision-making modules. The Cockpit Information Manager (CIM) is the adaptive automation system for the CDAS. The CIM is designed to make inferences about current and impending activities for the crew, allocate tasks among crew members as well as to the aircraft, and reconfigure cockpit displays to support the ability of the "crew-automation team" to execute those activities (see figure 16.1). The CIM monitors crew activities and external events and matches them against a database of tasks to generate inferences about crew intentions. The CIM uses this information to make decisions about allocating tasks, prioritizing information to be presented on limited

Figure 16.1. The Rotorcraft Pilot's Associate cockpit in a simulated environment. Reprinted from *Knowledge-Based Systems*, 12, Miller, C. A., and Hansen, M. D., The Rotorcraft Pilot's Associate: Design and evaluation of an intelligent user interface for cockpit information management, pp. 443–456, copyright 1999, with permission from Elsevier.

display spaces, locating pop-up windows, adding or removing appropriate symbology from displays, and adjusting the amount of detail to be presented in displays. Perhaps most important, the CIM includes a separate display that allows crew members and the system to coordinate the task allocation process and communicate their intentions (located above the center display in figure 16.1). The need for communication among members is strong for highly functioning human teams and, as it turned out, was essential for user acceptance of the RPA. Evaluations from a sample of pilots indicated that the RPA often provided the right information at the right time. Miller and Hannen reported that in the initial tests, no pilot chose to turn off the RPA.

The RPA was an ambitious attempt to create an adaptive automation system that would function as a team member. Several characteristics of this effort are particularly noteworthy. First, a great deal of the intelligence inherent in the system was designed to anticipate user needs and be proactive about reconfiguring displays and allocating tasks. Second, both the users and the system could communicate their plans and intentions, thereby reducing the need to decipher what the system was doing and why it was doing it. Third, unlike many other adaptive automation systems, the RPA was designed to support the simultaneous activities of multiple users.

Although the RPA is a significant demonstration of adaptive automation, it was not designed from the neuroergonomics perspective. It is true that a good deal of knowledge about cognitive processing related to decision making, information representation, task scheduling, and task sharing was needed to create the RPA, but the system was not built around knowledge of brain functioning.

Brain-Based Systems

An example of adaptive automation that follows the neuroergonomics approach can be found in systems that use psychophysiological indices to trigger changes in the automation. There are many psychophysiological indices that reflect underlying cognitive activity, arousal levels, and external task demands. Some of these include cardiovascular measures (e.g., heart rate, heart rate variability), respiration, galvanic skin response (GSR), ocular motor activity, and speech, as well as those that reflect cortical activity such as the electroencephalogram (EEG; see also chapter 2, this volume) and event-related potentials (ERPs; see also chapter 3, this volume) derived from EEG signals to stimulus presentations. Additional cortical measures include functional magnetic resonance imaging (fMRI; see also chapter 4, this volume), and near-infrared spectrometry (NIRS; see also chapter 5, this volume) that measures changes in oxygenated and deoxygenated hemoglobin (see Byrne & Parasuraman, 1996, for a review of the use of psychophysiological measures in adaptive systems). One of the most important advantages to brain-based systems for adaptive automation is that they provide a continuous measure of activity in the presence or absence of overt behavioral responses (Byrne & Parasuraman, 1996; Scerbo et al., 2001).

The first brain-based adaptive system was developed by Pope, Bogart, and Bartolome (1995). Their system uses an index of task engagement based upon ratios of EEG power bands (alpha, beta, theta, etc.). The EEG signals are recorded from several locations on the scalp and are sent to a LabView Virtual Instrument that determines the power in each band for all recording sites and then calculates the engagement index used to change a tracking task between automatic and manual modes. The system recalculates the engagement index every 2 seconds and changes the task mode if necessary. Pope and his colleagues studied several different engagement indices under both negative and positive feedback contingencies. They argued that under negative feedback the system should switch modes more frequently in order to maintain a stable level of engagement. By contrast, under positive feedback the system should be driven to extreme levels and remain there longer (i.e., fewer switches between modes). Moreover, differences in the frequency of task mode switches obtained under positive and negative feedback conditions should provide information about the sensitivity of various engagement indices. Pope et al. found that the engagement index based on the ratio of beta/(alpha + theta) proved to be the most sensitive to differences between positive and negative feedback.

The study by Pope et al. (1995) showed that their system could be used to evaluate candidate engagement indices. Freeman, Mikulka, Prinzel, and Scerbo (1999) expanded upon this work and studied the operation of the system in an adaptive context. They asked individuals to perform the compensatory tracking, resource management, and system monitoring tasks from the Multi-Task Attribute Battery (MAT; Comstock & Arnegard, 1991). Figure 16.2 shows a participant performing the MAT task while EEG signals are being recorded. In their study, all tasks remained in automatic mode except the tracking task, which shifted between automatic and manual modes. They also examined performance under both negative and positive feedback conditions. Under negative feedback, the

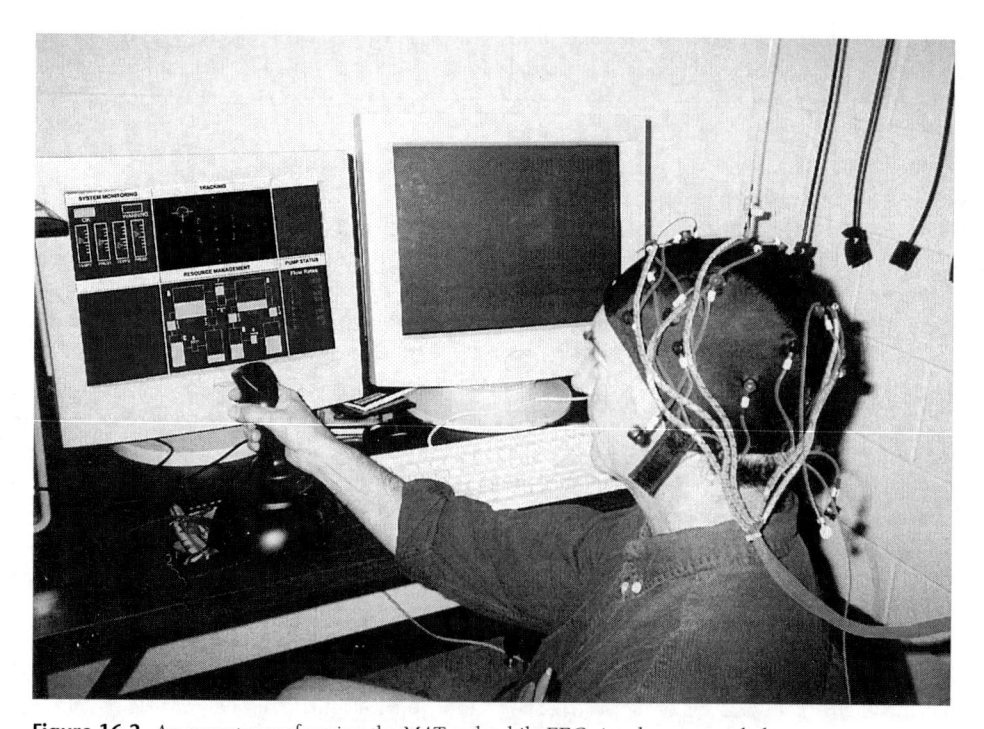

Figure 16.2. An operator performing the MAT task while EEG signals are recorded.

tracking task was switched to or maintained in automatic mode when the index increased above a preestablished baseline, reflecting high engagement. By contrast, the tracking task was switched to or maintained in manual mode when the index decreased below the baseline, reflecting low engagement. The opposite schedule of task changes occurred under the positive feedback conditions. Freeman and his colleagues argued that if the system could moderate workload, better tracking performance should be observed under negative as compared to positive feedback conditions. Their results confirmed this prediction. In subsequent studies, similar findings resulted when individuals performed the task over much longer intervals and under conditions of high and low task load (see Scerbo et al., 2003).

More recently, St. John, Kobus, Morrison, and Schmorrow (2004) described a new DARPA program aimed at enhancing an operator's effectiveness by managing the presentation of information and cognitive processing capacity through cognitive augmentation derived from psychophysiological measures. The goal of the program is to develop systems that can detect an individual's cognitive state and then manipulate task parameters to overcome perceptual, attentional, and working memory bottlenecks. Unlike the system described by Pope et al. (1995) that relies on a single psychophysiological measure, EEG, the augmented cognition systems use multiple measures including NIRS, GSR, body posture, and EEG. The physiological measures are integrated to form gauges that reflect constructs such as effort, arousal, attention, and workload. Performance thresholds are established for each gauge to trigger mitigation strategies for modifying the task. Some of these mitigation strategies include switching between verbal and spatial information formats, reprioritizing or rescheduling tasks, or changing the level of display detail.

Wilson and Russell (2003, 2004) reported on their experiences with an augmented cognitive system designed to moderate workload. Operators were asked to perform a target identification task with a simulated uninhabited combat air vehicle under different levels of workload. They recorded EEG from six sites as well as heart, blink, and respiration rates. In their system, the physiological data were analyzed by an artificial neural network (ANN). The ANN was trained to distinguish between high and low levels of operator workload in real time. The output from the ANN was used to trigger changes in the task to moderate workload. Comparisons among adaptive aiding, no aid, or random aiding revealed some performance benefits and lower ratings of subjective workload for the adaptive aiding condition under the more difficult condition.

Taken together, the findings from these studies suggest that it is indeed possible to obtain indices of one's brain activity and use that information to drive an adaptive automation system to improve performance and moderate workload. There are, however, still many critical conceptual and technical issues (e.g., making the recording equipment less obtrusive and obtaining reliable signals in noisy environments) that must be overcome before systems such as these can move from the laboratory to the field (Scerbo et al., 2001).

Further, many issues still remain surrounding the sensitivity and diagnosticity of psychophysiological measures in general. There is a fundamental assumption that psychophysiological measures provide a reliable and valid index of underlying constructs such as arousal or attention. In addition, variations in task parameters that affect those constructs must also be reflected in the measures (Scerbo et al., 2001). In fact, Veltman and Jansen (2004) have argued that there is no direct relation between information load and physiological measures or state estimators because an increase in task difficulty does not necessarily result in a physiological response. According to their model, perceptions of actual performance are compared to performance requirements. If attempts to eliminate the difference between perceived and required levels of performance are unsuccessful, one may need to increase mental effort or change the task goals. Both actions have consequences. Investing more effort can be fatiguing and result in poorer performance. Likewise, changing task goals (e.g., slowing down, skipping low-priority tasks, etc.) can also result in poorer performance. They suggest that in laboratory experiments, it is not unusual for individuals to compensate for increases in demand by changing task goals because there are no serious consequences to this strategy. However, in operational environments, where the consequences are real and operators are highly motivated, changing task goals may not be an option. Thus, they are much more likely to invest the effort needed to meet the required levels of performance. Consequently, Veltman and Jansen contend that physiological measures can only be valid and reliable in an adaptive automation environment if they are sensitive to information about task difficulty, operator output, the environmental context, and stressors.

Another criticism of current brain-based adaptive automation systems is that they are primarily reactive. Changes in external events or brain activity must be recorded and analyzed before any instructions can be sent to modify the automation. All of this takes time and even with short delays, the system must still wait for a change in events to react. Recently, however, Forsythe, Kruse, and Schmorrow (2005) described a brain-based system that also incorporates a cognitive model of the operator. The system is being developed by DaimlerChrysler through the DARPA Augmented Cognition program to support driver behavior. Information is recorded from the automobile (e.g., steering wheel angle, lateral acceleration) as well as the operator (e.g., head turning, postural adjustments, and vocalizations) and combined with EEG signals to generate inferences about workload levels corresponding to different driving situations. In this regard, the system is a hybrid of brain-based and operator modeling approaches to adaptive automation and can be more proactive than current adaptive systems that rely solely on psychophysiological measures.

Workload and Situation Awareness

Workload

One of the arguments for developing adaptive automation is that this approach can moderate operator workload. Most of the research to date has assessed workload through primary task performance or physiological indices (see above). Kaber and Riley (1999), however, conducted an experiment using both primary and secondary task measures. They had their participants perform a simulated radar task where the object was to eliminate targets before they reached the center of the display or collided with one another. During manual control, the participants were required to assess the situation on the display, make decisions about which targets to eliminate, and implement those decisions. During a shared condition, the participant and the computer could each perform the situation assessment task. The computer scheduled and implemented the actions, but the operator had the ability to override the computer's plans. The participants were also asked to perform a secondary task requiring them to monitor the movements of a pointer and correct any deviations outside of an ideal range. Performance on the secondary task was used to invoke the automation on the primary task. For half of the participants, the computer suggested changes between automatic or manual operation of the primary task, and for the remaining participants, those changes were mandated.

Kaber and Riley (1999) found that shared control resulted in better performance than manual control on the primary task. However, the results showed that mandating the use of automation also bolstered performance during periods of manual operation. Regarding the secondary task, when use of automation was mandated, workload was lower during periods of automation; however, under periods of manual control, workload levels actually increased and were similar to those seen when its use was suggested. These results show that authority over invoking changes between modes had differential effects on workload during periods of manual and automated operation. Specifically, Kaber and Riley found that the requirement to consider computer suggestions to invoke automation led to higher levels of workload during periods of shared or automated control than when those decisions were dictated by the computer.

Situation Awareness

Thus far, there have been few attempts to study the effects of adaptive automation on situation awareness (SA). Endsley (1995) described SA as the ability to perceive elements in the environment, understand their meaning, and make projections about their status in the near future. One might assume that efforts to moderate workload through adaptive automation would lead to enhanced SA; however, that relationship has yet to be demonstrated empirically. In fact, within an adaptive paradigm, periods of high automation could lead to poor SA and make returning to manual operations more difficult. The findings of Kaber and Riley (1999) regarding secondary task performance described above support this notion.

Bailey, Scerbo, Freeman, Mikulka, and Scott (2003) examined the effects of a brain-based adaptive automation system on SA. The participants were given a self-assessment measure of complacency toward automation (i.e., the propensity to become reliant on automation; see Singh, Molloy, & Parasuraman, 1993) and separated into groups who scored either high or low on the measure. The participants performed a modified version of the MAT battery that included a number of digital and analog displays (e.g., vertical speed indicator, GPS heading, oil pressure, and autopilot on/off) used to assess SA. Participants were asked to perform the compensatory tracking task during manual mode and to monitor that display during automatic mode. Half of the participants in each complacency potential group were assigned to either an adaptive or yoke control condition. In the adaptive condition, Bailey et al. used the system modified by Freeman et al. (1999) to derive an EEG-based engagement index to control the task mode switches. In the other condition, each participant was yoked to one of the individuals in the adaptive condition and received the same pattern of task mode switches; however, their own EEG had no effect on system operation. All participants performed three 15-minute trials. At the end of each trial, the computer monitor went blank and the experimenter asked the participants to report the current values for a sample of five displays. Participants' reports for each display were then compared to the actual values to provide a measure of SA (Endsley, 2000).

Bailey and his colleagues (2003) found that the effects of the adaptive and yoke conditions were moderated by complacency potential. Specifically, for individuals in the yoke control conditions, those who were high as compared to low in complacency potential had much lower levels of SA. On the other hand, there was no difference in SA scores for high- and low-complacency individuals in the adaptive conditions. More important, the SA scores for both high- and low-complacency individuals were significantly higher than those of the low-complacency participants in the yoke control condition. The authors argued that a brain-based adaptive automation system could ameliorate the effects of complacency by increasing available attentional capacity and in turn, improving SA.

Human-Computer Etiquette

Interest has been shown in the merits of an etiquette for human-computer interaction. Miller (2002) described etiquette as a set of prescribed and proscribed behaviors that permit meaning and intent to be ascribed to actions. Etiquette serves to make social interactions more cooperative and polite. Importantly, rules of etiquette allow one to form expectations regarding the behaviors of others. In fact, Nass, Moon, and Carney (1999) have shown that people adopt many of the same social conventions used in human-human interactions when they interact with computers. Moreover, they also expect computers to adhere to those same conventions when computers interact with users.

Miller (2004) argued that when humans interact with systems that incorporate intelligent agents, they may expect those agents to conform to accepted rules of etiquette. However, the norms may be implicit and contextually dependent: What is acceptable for one application may violate expectations in another. Thus, there may be a need to understand the rules under which computers should behave and be more polite.

Miller (2004) also claimed that users ascribe expectations regarding human etiquette to their interactions with adaptive automation. In their work with the RPA, Miller and Hannen (1999) observed that much of the dialogue between team members in a two-seat aircraft was focused on communicating plans and intentions. They reasoned that any automated assistant would need to communicate in a similar manner to be accepted as a team player. Consequently, the CIM described earlier was designed to allow users and the system to communicate in a conventionally accepted manner.

The benefits of adopting a human-computer etiquette are described by Parasuraman and Miller (2004) in a study of human-automation interactions. In particular, they focused on interruptions. In their study, participants were asked to perform the tracking and fuel resource management tasks from the MAT battery. A third task required participants to interact with an automated system that monitored engine parameters, detected potential failures, and offered advice on how to diagnose faults. The automation support was implemented in two ways. Under the "patient" condition, the automated system would withhold advice if the user was in the act of diagnosing the engines, or provide a warning, wait 5 seconds, and then offer advice if it determined the user was not interacting with the engines. By contrast, under the "impatient" condition the automated system offered its advice without warning while the user was performing the diagnosis. Parasuraman and Miller referred to the patient and impatient automation as examples of good and poor etiquette, respectively. In addition, they examined two levels of system reliability. Under low and high reliability, the advice was correct 60% and 80% of the time, respectively.

As might be expected, performance was better under high as opposed to low reliability. Further, Parasuraman and Miller (2004) found that when the automated system functioned under the good etiquette condition, operators were better able to diagnose engine faults regardless of reliability level. In addition, overall levels of trust in the automated system were much higher under good etiquette within the same reliability conditions. Thus, "rude" behavior made the system seem less trustworthy irrespective of reliability level. Several participants commented that they disliked being interrupted. The authors argued that systems designed to conform to rules of etiquette may enhance performance beyond what might be expected from system reliability and may even compensate for lower levels of reliability.

Parasuraman and Miller's (2004) findings were obtained with a high-criticality simulated system; however, the rules of etiquette (or interruptions) may be equally important for business or home applications. Bubb-Lewis and Scerbo (2002) examined the effects of different levels of communication on task performance with a simulated adaptive interface. Specifically, participants worked with a computer "partner" to solve problems (e.g., determining the shortest mileage between two cites or estimating gasoline consumption for a trip) using a commercial travel-planning software package. In their study, the computer partner was actually a confederate in another room who followed a strict set of rules regarding how and when to intervene to help complete a task for the participant. In addition, they studied four different modes of communication that differed in the level of restriction ranging from context-sensitive natural language to no communication at all. The results showed that as restrictions on communication increased, participants were less able to complete their tasks, which in turn caused the computer to intervene more often to complete the tasks. This increase in interventions also led the participants to rate their interactions with the computer partner more negatively. Thus, these findings suggest that even for less critical systems, poor etiquette makes a poor impression. Apparently, no one likes a show-off, even if it is the computer.

Living with Adaptive Automation

Adaptive automation is also beginning to find its way into commercial and more common technologies. Some examples include the adaptive cruise control found on several high-end automobiles and "smart homes" that control electrical and heating systems to conform to user preferences.

Mozer (2004) has described his experiences living in an adaptive home of his own creation. The home was designed to regulate air and water temperature and lighting. The automation monitors the inhabitant's activities and makes inferences about the inhabitant's behavior, predicts future needs, and adjusts the temperature or lighting accordingly. When the automation fails to meet the user's expectations, the user can set the controls manually.

The heart of the adaptive home is the adaptive control of home environment (ACHE), which functions to balance two goals: user desires and energy conservation. Because these two goals can conflict with one another, the system uses a reinforcement learning algorithm to establish an optimal control policy. With respect to lighting, the ACHE controls multiple independent light fixtures, each with multiple levels of intensity (see figure 16.3). The ACHE encompasses a learning controller that selects light settings based on current states. The controller receives information about an event change that is moderated by a cost evaluator. A state estimator generates high-level information about inhabitant patterns and integrates it with output from an occupancy model as well as information regarding levels of natural light available to make decisions about changes in the control settings. The state estimator also receives input from an anticipator module that uses neural nets to predict which zones are likely to be inhabited within the next 2 seconds. Thus, if the inhabitant is moving within the home, the ACHE can anticipate the route and adjust the lights before he arrives at his destination. Mozer (2004) recorded the energy costs as well as costs of discomfort (i.e., incorrect predictions and control settings) for a month and found that both decreased and converged within about 24 days.

Mozer (2004) had some intriguing observations about his experiences living in the adaptive house.

Figure 16.3. Michael Mozer's adaptive house. An interior photo of the great room is shown on the left. On the right is a photo of the data collection room where sensor information terminates in a telephone punch panel and is routed to a PC. A speaker control board and a microcontroller for the lights, electric outlets, and fans are also shown here. (Photos courtesy of Michael Mozer).

First, he found that he generated a mental model of the ACHE's model of his activities. Thus, he knew that if he were to work late at the office, the house would be "expecting" him home at the usual time, and he often felt compelled to return home. Further, he admitted that he made a conscious effort to be more consistent in his activities. He developed a meta-awareness of his occupancy patterns and recognized that as he made his behavior more regular, it facilitated the operation of the ACHE, which in turn helped it to save energy and maximize his comfort. In fact, Mozer claimed, "the ACHE trains the inhabitant, just as the inhabitant trains the ACHE" (p. 293).

Mozer (2004) also discovered the value of communication. At one point, he noticed a bug in the hardware and modified the system to broadcast a warning message throughout the house to reset the system. After the hardware problem had been addressed, however, he retained the warning message because it provided useful information about how his time was being spent. He argued that there were other situations where the user could benefit from being told about consequences of manual overrides.

Conclusion

The development of adaptive automation represents a qualitative leap in the evolution of technology. Users of adaptive automation will be faced with systems that differ significantly from the automated technology of today. These systems will be much more complex from both the users' and designers' perspectives. Scerbo (1996) argued that adaptive automation systems will need time to learn about users and users will need time to understand the automation. In Mozer's (2004) case, he and his home needed almost a month to adjust to one another. Further, users may find that adaptive systems are less predictable due to the variability and inconsistencies of their own behavior. Thus, users are less likely to think of these systems as tools, machines, or even traditional computer programs. As Mozer indicated, he soon began to think about how his adaptive home would respond to his behavior. Others have suggested that interacting with adaptive systems is more like interacting with a teammate or coworker (Hammer & Small, 1995; Miller & Hannen, 1999; Scerbo, 1994).

The challenges facing designers of adaptive systems are significant. Current methods in system analysis, design, and evaluation fall short of what is needed to create systems that have the authority and autonomy to swap tasks and information with their users. These systems require developers to be knowledgeable about task sharing, methods for communicating goals and intentions, and even assessment of operator states of mind. In fact, Scerbo

(1996) has argued that researchers and designers of adaptive technology need to understand the social, organizational, and personality issues that impact communication and teamwork among humans to create more effective adaptive systems. In this regard, Miller's (2004) ideas regarding human-computer etiquette may be paramount to the development of successful adaptive systems.

Thus far, most of the adaptive automation systems that have been developed address life-critical activities where the key concerns surround the safety of the operator, the system itself, and recipients of the system's services. However, the technology has also been applied in other contexts where the consequences of human error are less severe (e.g., Mozer's adaptive house). Other potential applications might include a personal assistant, butler, tutor, secretary, or receptionist. Moreover, adaptive automation could be particularly useful when incorporated into systems aimed at training and skill development as well as entertainment.

To date, most of the adaptive automation systems that have been developed were designed to maximize the user-system performance of a single user. Thus, they are user independent (i.e., designed to improve the performance of any operator). However, overall user-system performance is likely to be improved further if the system is capable of learning and adjusting to the behavioral patterns of its user, as was shown by Mozer (2004). This would indeed make the line between human and machine more flimsy, as Clark (2003) has suggested. Although building systems capable of becoming more user specific might seem like a logical next step, that approach would introduce a new and significant challenge for designers of adaptive automation—addressing the unique needs of multiple users. The ability of Mozer's house to successfully adapt to his routines is due in large part to his being the only inhabitant. One can imagine the challenge faced by an adaptive system trying to accommodate the wishes of two people who want the temperature set at different levels.

The problem of accommodating multiple users is not unique to adaptive automation. In fact, the challenge arises from a fundamental aspect of humanity. People are social creatures and, as such, they work in teams, groups, and organizations. Moreover, they can be colocated or distributed around the world and networked together. Developers of collaborative meeting and engineering software realize that one cannot optimize the individual human-computer interface at the expense of interfaces that support team and collaborative activities. Consequently, even systems designed to work more efficiently based on knowledge of brain functions must ultimately take into consideration groups of people. Thus, the next great challenge for the neuroergonomics approach may lie with an understanding of how brain activity of multiple operators in social situations can improve the organizational work environment.

Main Points

1. One source of interest in neuroergonomics stems from research showing that automated systems do not always reduce workload and create safer working conditions.
2. Adaptive automation refers to systems in which the level of automation or the number of subsystems operating under automation can be modified in real time. In addition, changes in the state of automation can be initiated by either the operator or the system.
3. Adaptive automation has traditionally been implemented in associate systems that use models of operator behavior and workload, but recent brain-based systems that follow the neuroergonomics approach show much promise.
4. User interactions with adaptive automation may be improved through an understanding of human-computer etiquette.
5. Working with adaptive automation can be like working with a teammate. Further, living with adaptive automation can modify both the system and the user's behavior.
6. The next great challenge for neuroergonomics may be to address the individual needs of multiple users.

Key Readings

Degani, A. (2004). *Taming HAL: Designing interfaces beyond 2001*. New York: Palgrave Macmillan.

Parasuraman, R. (Ed.). (2003). Neuroergonomics [Special issue]. *Theoretical Issues in Ergonomics Science, 4*(1–2).

Parasuraman, R., & Mouloua, M. (1996). *Automation and human performance: Theory and applications*. Mahwah, NJ: Erlbaum.

Schmorrow, D., & McBride, D. (Eds.). (2004). Augmented cognition [Special issue]. *International Journal of Human-Computer Interaction, 17*(2).

References

Bailey, N. R., Scerbo, M. W., Freeman, F. G., Mikulka, P. J., & Scott, L. A. (2003). The effects of a brain-based adaptive automation system on situation awareness: The role of complacency potential. In *Proceedings of the Human Factors and Ergonomics Society 47th annual meeting* (pp. 1048–1052). Santa Monica, CA: Human Factors and Ergonomics Society.

Billings, C. E., & Woods, D. D. (1994). Concerns about adaptive automation in aviation systems. In M. Mouloua & R. Parasuraman (Eds.), *Human performance in automated systems: current research and trends* (pp. 264–269). Hillsdale, NJ: Erlbaum.

Bubb-Lewis, C., & Scerbo, M. W. (2002). The effects of communication modes on performance and discourse organization with an adaptive interface. *Applied Ergonomics, 33*, 15–26.

Byrne, E. A., & Parasuraman, R. (1996). Psychophysiology and adaptive automation. *Biological Psychology, 42*, 249–268.

Clark, A. (2003). *Natural-born cyborgs: Minds, technologies and the future of human intelligence*. New York: Oxford University Press.

Comstock, J. R., & Arnegard, R. J. (1991). *The multi-attribute task battery for human operator workload and strategic behavior research* (NASA Technical Memorandum No. 104174). Hampton, VA: Langley Research Center.

Degani, A. (2004). *Taming HAL: Designing interfaces beyond 2001*. New York: Palgrave Macmillan.

Endsley, M. R. (1995). Toward a theory of situation awareness in dynamic systems. *Human Factors, 37*, 32–64.

Endsley, M. R. (2000). Theoretical underpinnings of situation awareness: A critical review. In M. Endsley & D. Garland (Eds.), *Situation awareness analysis and measurement* (pp. 3–32). Mahwah, NJ: Erlbaum.

Freeman, F. G., Mikulka, P. J., Prinzel, L. J., & Scerbo, M. W. (1999). Evaluation of an adaptive automation system using three EEG indices with a visual tracking task. *Biological Psychology, 50*, 61–76.

Forsythe, C., Kruse, A., & Schmorrow, D. (2005). Augmented cognition. In C. Forsythe, M. L. Bernard, & T. E. Goldstein (Eds.), *Cognitive systems: Human cognitive models in system design* (pp. 97–117). Mahwah, NJ: Erlbaum.

Hammer, J. M., & Small, R. L. (1995). An intelligent interface in an associate system. In W. B. Rouse (Ed.), *Human/technology interaction in complex systems* (Vol. 7, pp. 1–44). Greenwich, CT: JAI Press.

Hancock, P. A., & Chignell, M. H. (1987). Adaptive control in human-machine systems. In P. A. Hancock (Ed.), *Human factors psychology* (pp. 305–345). North Holland: Elsevier Science.

Inagaki, T., Takae, Y., & Moray, N. (1999). Automation and human-interface for takeoff safety. *Proceedings of the 10th International Symposium on Aviation Psychology*, 402–407.

Kaber, D. B., & Riley, J. M. (1999). Adaptive automation of a dynamic control task based on secondary task workload measurement. *International Journal of Cognitive Ergonomics, 3*, 169–187.

Malin, J. T., & Schreckenghost, D. L. (1992). *Making intelligent systems team players: Overview for designers* (NASA Technical Memorandum 104751). Houston, TX: Johnson Space Center.

Miller, C. A. (2002). *Definitions and dimensions of etiquette: The AAAI Fall Symposium on Etiquette for Human-Computer Work* (Technical Report FS-02-02, pp. 1–7). Menlo Park, CA: AAAI.

Miller, C. A. (2004). Human-computer etiquette: Managing expectations with intentional agents. *Communications of the ACM, 47*(4), 31–34.

Miller, C. A., & Hannen, M. D. (1999). The Rotorcraft Pilot's Associate: Design and evaluation of an intelligent user interface for cockpit information management. *Knowledge-Based Systems, 12*, 443–456.

Morrison, J. G., & Gluckman, J. P. (1994). Definitions and prospective guidelines for the application of adaptive automation. In M. Mouloua & R. Parasuraman (Eds.), *Human performance in automated systems: Current research and trends* (pp. 256–263). Hillsdale, NJ: Erlbaum.

Mozer, M. C. (2004). Lessons from an adaptive house. In D. Cook & R. Das (Eds.), *Smart environments: Technologies, protocols, and applications* (pp. 273–294). New York: Wiley.

Nass, C., Moon, Y., & Carney, P. (1999). Are respondents polite to computers? Social desirability and direct responses to computers. *Journal of Applied Social Psychology, 29*, 1093–1110.

Parasuraman, R. (2003). Neuroergonomics: Research and practice. *Theoretical Issues in Ergonomics Science, 4*, 5–20.

Parasuraman, R., Bahri, T., Deaton, J. E., Morrison, J. G., & Barnes, M. (1992). *Theory and design of adaptive automation in aviation systems* (Technical Report No. NAWCADWAR-92033-60). Warminster, PA: Naval Air Warfare Center, Aircraft Division.

Parasuraman, R., & Miller, C. A. (2004). Trust and etiquette in high-criticality automated systems. *Communications of the ACM, 47*(4), 51–55.

Parasuraman, R., Mouloua, M., Molloy, R., & Hilburn, B. (1996). Monitoring of automated systems. In R. Parasuraman & M. Mouloua (Eds.), *Automation and human performance: Theory and applications* (pp. 91–115). Mahwah, NJ: Erlbaum.

Parasuraman, R., & Riley, V. (1997). Humans and automation: Use, misuse, disuse, abuse. *Human Factors, 39,* 230–253.

Pope, A. T., Bogart, E. H., & Bartolome, D. (1995). Biocybernetic system evaluates indices of operator engagement. *Biological Psychology, 40,* 187–196.

Rouse, W. B. (1976). Adaptive allocation of decision making responsibility between supervisor and computer. In T. B. Sheridan & G. Johannsen (Eds.), *Monitoring behavior and supervisory control* (pp. 295–306). New York: Plenum.

Rouse, W. B., Geddes, N. D., & Curry, R. E. (1987–1988). An architecture for intelligent interfaces: Outline of an approach to supporting operators of complex systems. *Human-Computer Interaction, 3,* 87–122.

Rouse, W. B., & Rouse, S. H. (1983). *A framework for research on adaptive decision aids* (Technical Report AFAMRL-TR-83-082). Wright-Patterson Air Force Base, OH: Air Force Aerospace Medical Research Laboratory.

Sarter, N. B., & Woods, D. D. (1995). How in the world did we ever get into that mode? Mode errors and awareness in supervisory control. *Human Factors, 37,* 5–19.

Scerbo, M. W. (1994). Implementing adaptive automation in aviation: The pilot-cockpit team. In M. Mouloua & R. Parasuraman (Eds.), *Human performance in automated systems: current research and trends* (pp. 249–255). Hillsdale, NJ: Erlbaum.

Scerbo, M. W. (1996). Theoretical perspectives on adaptive automation. In R. Parasuraman & M. Mouloua (Eds.), *Automation and human performance: Theory and applications* (pp. 37–63). Mahwah, NJ: Erlbaum.

Scerbo, M. W. (2001). Adaptive automation. In W. Karwowski (Ed.), *International encyclopedia of ergonomics and human factors* (pp. 1077–1079). London: Taylor and Francis.

Scerbo, M. W., Freeman, F. G., & Mikulka, P. J. (2003). A brain-based system for adaptive automation. *Theoretical Issues in Ergonomic Science, 4,* 200–219.

Scerbo, M. W., Freeman, F. G., Mikulka, P. J., Parasuraman, R., Di Nocero, F., & Prinzel, L. J. (2001). *The efficacy of psychophysiological measures for implementing adaptive technology* (NASA TP-2001-211018). Hampton, VA: NASA Langley Research Center.

Scott, W. B. (1999, February). Automatic GCAS: "You can't fly any lower." *Aviation Week and Space Technology,* 76–79.

Sheridan, T. B., & Verplank, W. L. (1978). *Human and computer control of undersea teleoperators.* Cambridge, MA: MIT Man-Machine Systems Laboratory.

Singh, I. L., Molloy, R., & Parasuraman, R. (1993). Automation-induced "complacency": Development of the complacency-potential rating scale. *International Journal of Aviation Psychology, 3,* 111–122.

St. John, M., Kobus, D. A., Morrison, J. G., & Schmorrow, D. (2004). Overview of the DARPA augmented cognition technical integration experiment. *International Journal of Human-Computer Interaction, 17,* 131–149.

Veltman, H. J. A., & Jansen, C. (2004). The adaptive operator. In D. A. Vincenzi, M. Mouloua, & P. A. Hancock (Eds.), *Human performance, situation awareness, and automation: Current research and trends* (Vol. II, pp. 7–10). Mahwah, NJ: Erlbaum.

Wickens, C. D. (1992). *Engineering psychology and human performance* (2nd ed.). New York: Harper Collins.

Wiener, E. L. (1989). *Human factors of advanced technology ("glass cockpit") transport aircraft* (Technical report 117528). Moffett Field, CA: NASA Ames Research Center.

Wilensky, R., Arens, Y., & Chin, D. N. (1984). Talking to Unix in English: An overview of UC. *Communications of the ACM, 27,* 574–593.

Wilson, G. F., & Russell, C. A. (2003). Real-time assessment of mental workload using psychophysiological measures and artificial neural networks. *Human Factors, 45,* 635–643.

Wilson, G. F., & Russell, C. A. (2004). Psychophysiologically determined adaptive aiding in a simulated UCAV task. In D. A. Vincenzi, M. Mouloua, & P. A. Hancock (Eds.), *Human performance, situation awareness, and automation: Current research and trends* (pp. 200–204). Mahwah, NJ: Erlbaum.

Woods, D. D. (1996). Decomposing automation: Apparent simplicity, real complexity. In R. Parasuraman & M. Mouloua (Eds.), *Automation and human performance: Theory and applications* (pp. 3–17). Mahwah, NJ: Erlbaum.

17

Joseph K. Kearney, Matthew Rizzo, and Joan Severson

Virtual Reality and Neuroergonomics

Virtual reality (VR) describes the use of computer-generated stimuli and interactive devices to situate participants in simulated surroundings that resemble real or fantasy worlds. As early as 1965, Ivan Sutherland foresaw the potential of computer graphics to create a window into a virtual world. In its very first episode, the late 1980s television show *Star Trek: The Next Generation* introduced the concept of a virtual environment into the public imagination with the spaceship's "holodeck," which persists today as the archetype for a virtual world. The usage of the term *virtual reality* is attributed to Jaron Lanier, who defined it as "a computer-generated, interactive, three-dimensional environment in which a person is immersed" (Tsirliganis, 2001). Brooks (1999) defined a VR experience as "any in which the user is effectively immersed in a responsive virtual world." The Lanier and Brooks definitions both identify the two elements most commonly associated with VR: immersion and interaction.

With VR, a computer-controlled setup creates an illusory environment that relies on visual, auditory, somatosensory (e.g., haptic [tactile], vibratory, and proprioceptive), and even olfactory displays. The term *presence* describes the degree to which a person feels a part of, or engaged in, the VR world (figure 17.1). It is sometimes described as a sense of "being there." Users may navigate within the virtual world and interact with objects, obstacles, and inhabitants such as avatars, the digital representations of real humans in virtual worlds. Optimally, participants will lose awareness of their physical surroundings and any sense of artificiality and will feel and behave in a virtual environment (VE) as they would in the real world. This allows presentation of useful tasks or diversions in many circumstances for entertaining, training, testing, or treating normal or impaired users.

Ellis (1991, 1993, 1995), Rheingold (1991), and Heim (1993) have reviewed the philosophical underpinnings, history, and early implementations of VR. Brooks (1999) reviewed the state of the art for VR through 1999. Continued gains in computer power and display technologies since then have made VR better, more economical, and accessible to a wider range of users and disciplines. Availability of high-quality personal computer (PC) graphics-rendering hardware (largely due to the commercial demands of computer gaming applications) and steady improvements in the quality and cost of video projectors (driven by presentation applications and, more recently, by home entertainment) have dramatically reduced the cost of VR

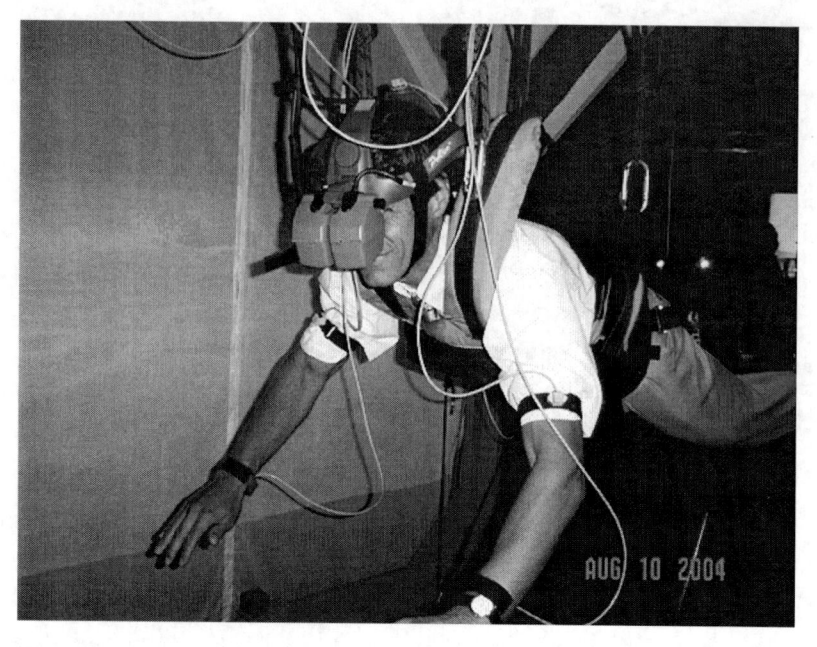

Figure 17.1. Immersion in virtual reality (VR). A VR user in a head-mounted display and harness is swimming across the Pacific. The lack of wetness or fatigue, the harness, and noisy spectators spoil the immersion.

equipment. Coupled with technical advancements in graphics and sound rendering, gesture and speech recognition, modeling of biomechanics, motion tracking, motion platforms, force feedback devices, and improvements in software for creating complex VE databases, it is now possible to pursue research using VR with a modest budget and limited technical expertise.

It is difficult to precisely define what it takes to make a simulated experience VR. It is generally accepted that VR refers to 3-D environments in which participants perceive that they are situated and can interact with the environment in some way, typically by navigating through it or manipulating objects in it. The sensory experience should be immersive in some respects, and participants should have a feeling of presence. The factors that influence the degree and cognitive underpinnings of presence are under investigation (Meehan, Insko, Whitton, & Brooks, 2002; Sanchez-Vives & Slater, 2005). Panoramic movies are typically not considered VR (and neither is watching "reality TV") because the participants are not able to interact with the objects or entities in the environment.

VR is highly relevant to neuroergonomics because VR can replicate situations with greater control than is possible in the real world, allowing behavioral and neurophysiological observations of the mind and brain at work in a wide range of situations that are impractical or impossible to observe in the real world. VR can be used to study the performance of hazardous tasks without putting subjects at risk of injury. For example, VR has been used to study the influence of disease, drugs, and disabilities on driving, to understand how the introduction of in-vehicle technologies such as cell phones and heads-up displays influence crash risk, to investigate underlying causes of bicycle crashes (Plumert, Kearney, & Cremer, 2004), and to examine how to reduce the risk of falling from roofs and scaffolding (Simeonov, Hsiao, Dotson, & Ammons, 2005).

VR can also help train students in areas where novice misjudgments and errors can have devastating outcomes, such as learning to do medical procedures, to fly an airplane, or to operate heavy machinery. VR can provide a cost-effective means to train users in tasks that are by their nature destructive, such as learning to fire shoulder-held missiles. VR can provide training for environments that are dangerous or inaccessible, such as mission rehearsal for fire training (St. Julien & Shaw, 2003), mining (Foster & Burton, 2003), or hostage rescue. VR is also proving to be useful for therapy and rehabilitation of persons with motor, cognitive, or psychiatric impairments. VR is especially useful when the job requires spatial awareness, complex motor skills, or decision making to choose among possible actions appropriate to changing circumstances.

As we shall see, detailed observations in these immersive VR environments offer a window to mechanisms of the human brain and cognition and have applications to public health and safety, education, training, communication, entertainment, science, engineering, psychology, health care, and the military (e.g., Barrett, Eubank, & Smith, 2005; Brown, 2005; Thacker, 2003).

The Physiology of a VR Experience

VR setups provide a closely controlled environment for examining links between human physiology and behavior. Relevant measurements in different VR tasks can assess body temperature, heart rate, respiratory rate, galvanic skin response, electroencephalographic (EEG) activity, electromyographic activity, eyelid closure (as an index of alertness/arousal), eye movements (as an index of cognitive processing), and even brain metabolic activity.

Physiological sensors can be fastened to a subject or positioned remotely to record physiological data from a subject immersed in the VE. These instruments differ on spatial and timing accuracy, ease and reliability of calibration, ability to accommodate appliances (such as spectacles in the case of eye-recording systems), susceptibility to vibration and lighting conditions, variability and reliability, ability to connect with auxiliary devices, synchronization with the performance measures collected, and the ability to automate analysis of large data files and to visualize and analyze these data in commercial software and statistical packages.

Physiological measurements are generally easier to make in VR than in the real world where movement, lighting, and other artifacts can be a nuisance and a source of confounding variables. Synchronous measures of operator physiology and performance in VR scenarios can illuminate relationships between cognition and performance in operators with differing levels of self-awareness (e.g., drivers, pilots, factory workers, medical personnel, patients) under effects of fatigue, distraction, drugs, or brain disorders. Some physiological indices that can be measured in VE are listed in table 17.1.

Common technical difficulties facing researchers who record physiological measures include how specific measures are analyzed and reported. Common problems include the spatial and timing precision of the physiological recording hardware, postprocessing of recorded measurements to find physiologically meaningful events (e.g., localization of fixation points from raw eye movement data or identification of microsleeps from EEG recordings), and synchronization of the stream of data on subject performance with the timing of simulation events and activities. Measurement devices such as those that record galvanic skin response or eye movements are notoriously temperamental and require frequent calibration, and dependent measures must be clearly defined (see figure 17.2).

The Honeywell Laboratories AugCog team developed a closed-loop, adaptive system that monitors neurophysiological measures—EEG, pupilometric, and physiological measures such as electrooculography and electrocariography—as a comprehensive set of cognitive gauges for assessing the cognitive state of a subject. The Closed-Loop Integrated Prototype of a military communications scheduler is designed to schedule a soldier's tasks, communications rates, and modalities to prevent cognitive overload and provide higher situational awareness and message comprehension in high-workload conditions (Dorneich, Whitlow, Ververs, Carciofini, & Creaser, 2003).

The HASTE (Human Machine Interface and the Safety of Traffic in Europe) group proposed several physiological indices to assess user workload in VR for driving. Mandatory workload measures (which had to be collected at all sites) included a rating scale of mental effort, eye glance measures (glance frequency, glance duration), measures of vehicle control while using vehicle information system performance, and a measure of situation awareness. Optional workload measures (which could be collected at sites with appropriate facilities) included heart rate and heart rate variability, respiration, and skin conductance.

Note that physiological indices can be used to make inferences about cognitive and emotional activity during VR tasks. Rainville, Bechara, Naqvi, and Damasio (2006) investigated profiles of cardiorespiratory activity during the experience of

Table 17.1. Physiological Measures in Virtual Environments

Test	Description
Electroencephalography (EEG)	Routinely used to determine sleep onset and microsleeps (Harrison & Horne, 1996; Risser et al., 2000). EEG power spectral analysis may detect subtle shifts into lower frequency bands (theta, delta) that would be associated with developing drowsiness (Horne & Reyner, 1996; Kecklund & Akerstedt, 1993; see also chapter 2, this volume).
Heart rate	Variability can be assessed by subjecting continuous EKG data and may be more sensitive than absolute heart rate in studies of fatigued operators (Egelund, 1982).
Respiratory frequency	Changes in respirations/minute and relative amplitude of abdominal respiratory effort can continuously be recorded. Sleep onset is typically associated with decreasing abdominal respiratory effort, while thoracic effort remains stable (Ogilvie et al., 1989).
Near-infrared spectroscopy	Has the potential to passively monitor the effects of cognition on oxygenation of blood in the brain during challenges posed in VR environments. Near-infrared light incident on the skin diffuses through the skull into the brain and then diffuses back out to the skin where it can be detected as an index of brain activity (see also chapter 2, this volume).
Eye movements	Can index attention and cognition in virtual environments. They can provide an index of information processing and depend on the stimulus, context for search, and scenario. For example, in driving on straight roads in low traffic, drivers tend to fixate around the focus of expansion, in the direction of forward travel in virtual environments. Experienced drivers may fixate farther away from the car than novices do (e.g., Mourant & Rockwell, 1972). Cognitive load affects eye movements. Eye movements generally precede driver actions over vehicle controls, so prediction of driver behavior may be possible. For example, driver eye and head checks may precede lane position and steering changes. Relevant analyses can assess fixation duration, distance, location (in regions of interest in the virtual environment), scan path length, and transitions between fixations (see chapter 7). Eye movements can also provide an index of drowsiness during VR (Wierwille et al., 1995; Hakkanen et al., 1999; Dinges & Mallis, 2000; see also chapter 7, this volume)
Eye blinks	Frequency is reported in blinks/minute. Blink duration is expressed as PERCLOS (PERcent eyelid CLOSure) in consecutive, 3-minute epochs. PERCLOS can predict performance degradation in studies of sleep-deprived individuals (Dinges & Mallis, 2000).
Cervical paraspinal electromyographic recordings	Reduction in activity has correlated with poor performance in fatigued operators (Dureman & Boden, 1972).
Electrogastrogram	Can be recorded from electrodes placed over the abdomen. Tachygastria (acute increased stomach activity of 4–9 cycles per minute) correlates with feelings of nausea and excitement.
Skin electrical resistance: galvanic skin response, a.k.a. electrodermal response	Increases with drowsiness (Dureman & Boden, 1972; Johns et al., 1969) and peaks in response to an emotionally charged stimulus (see figure 17.4).

fear, anger, sadness, and happiness. They recorded electrocardiographic and respiratory activity in 43 volunteers during recall and experiential reliving of potent emotional episodes or during a neutral episode. Multiple linear and spectral indices of cardiorespiratory activity were reduced to five physiologically meaningful factors using principal component analysis. Multivariate analyses of variance and effect size estimates calculated based on those factors confirmed the differences between the four emotion conditions. The results are compatible with the proposal of William James (1894) that afferent signals felt by the viscera are essential for the unique experience associated with distinct emotions.

Physiological measures can provide a quantitative index of a user's feeling of presence in VR. Meehan et al. (2002) examined reactions to stressful situations presented in VR by observing physiological measures that correlate with stress in real environments. They found that heart rate and skin

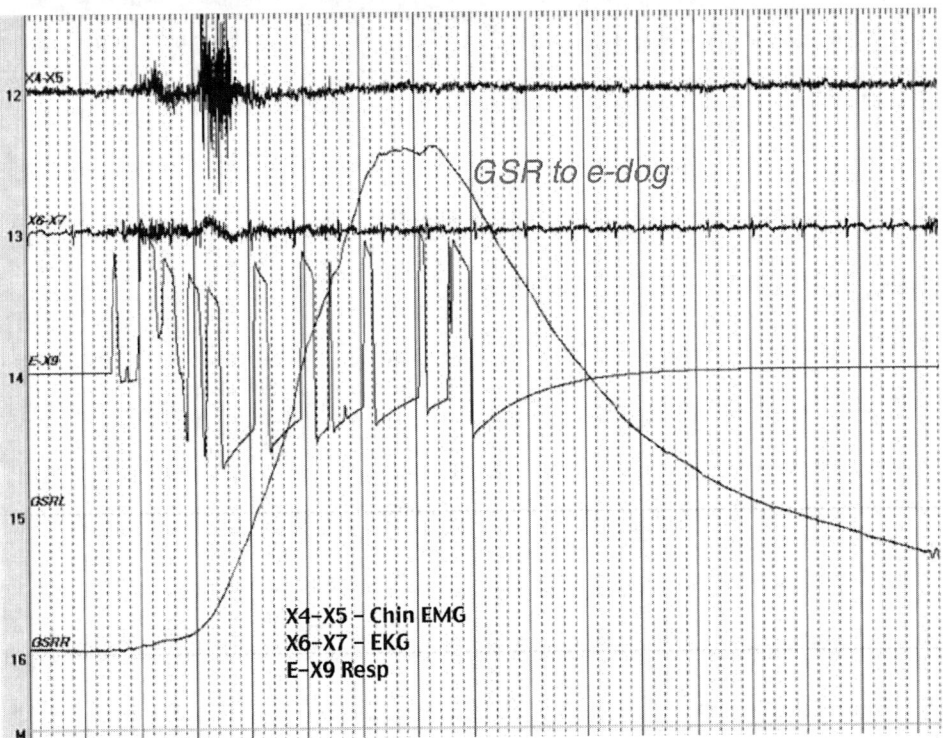

Figure 17.2. Physiology in virtual reality. Physiology in virtual reality. Top. Quad view (Top) shows a surprised driver (upper left panel) who is braking (lower left panel) in response to an e-dog running across the driver's path (lower right panel) in a virtual driving environment. Bottom. Galvanic skin response activity (GSR) accompanies the surprised driver's response.

Figure 17.3. Height effects in real and virtual reality environments. Virtual reality environments have been used to study height effects and spatial cognition in neuroergonomic research aimed at reducing falls and injuries in work environments. Simeonov et al. (2005) compared psychological and physiological responses to height in corresponding real and virtual environments. (Courtesy of NIOSH, Division of Safety Research.) See also color insert.

conductance significantly increased and skin temperature significantly decreased when subjects were exposed to a virtual pit—a room with a large hole in the floor that dropped 20 feet to the ground below. The most consistent correlate was heart rate. Simeonov et al. (2005) compared psychological and physiological responses to height in corresponding real and virtual environments, pictured in figure 17.3. They found similar levels of anxiety and perceived risk of falling in the real and virtual environments. Subjects had similar skin conductance responses and comparable postural instabilities at real and virtual heights. However, while heart rates were elevated in real environments, heart rates were not stable across heights in the virtual environment.

From a cognitive neuroscience perspective, Sanchez-Vives and Slater (2005) noted that VEs can break the everyday connection between where our senses tell us we are and where we actually are located and whom we are with. They argue that studies of presence, the phenomenon of behaving and feeling as if we are in the virtual world created by computer displays, may aid the study of perception and consciousness.

Augmented Reality

The term *augmented reality* (AR) refers to the combining of real and artificial stimuli, generally with the aim of improving human performance and creativity. This typically involves overlaying computer-generated graphics on a stream of video images so that these virtual objects appear to be embedded in the real world. Sounds, somatosensory cues, and even olfactory cues can also be added to the sensory stream from the real world. The augmentation may highlight important objects or regions, superimpose informative annotations, or supplement a real environment.

AR has the potential to help normal individuals accomplish difficult tasks by enhancing their normal perceptual abilities with information that is typically unavailable. Perhaps the most visible and commercially successful example of AR is the yellow first-down line shown on television broadcasts of college and professional football games. Using precise information on camera position and orientation and a geometric model of the shape of the football field, an imaginary yellow line is electroni-

Figure 17.4. Augmented cognition. The Rockwell Collins Synthetic Vision System overlays wire-frame terrain imagery rendered from an onboard database onto a heads-up display along with guidance symbology. This improves the pilot's awareness of terrain under poor visibility conditions without obstructing the real-world view as visibility improves (courtesy of Rockwell Collins, Inc.).

cally superimposed on the image of the football field to indicate the first-down line.

Imagine a surgeon, about to begin an incision, who can see a patient's internal organs through that patient's skin, or a landscape architect who can see property lines overlaid on the ground and the underground pipes and conduits (Spohrer, 1999). Such AR applications can use semitransparent head-mounted displays (HMDs) to superimpose computer-generated images over the images of the real world. They could help aircraft pilots maintain situational awareness of weather, other air traffic, aircraft state, and tactical operations using an HMD that enhances salient features (e.g., the horizon, occluded runway markings) or attaches labels to those features to identify runways, buildings, or nearby aircraft (http://hci.rsc.rockwell.com/AugmentedReality). Such augmented information could also be presented using a "heads-up display" located or projected on a window along the pilot or driver's line of sight to the external world (see figure 17.4).

AR might also help a cognitively impaired older walker to navigate better by superimposing visual labels for landmarks and turning points on images of the terrain viewed through instrumented spectacles. Sensors on the navigator would transmit personal position, orientation, and movements that enable accurate AR overlays (see chapter 8, this volume).

The key challenges in creating effective AR systems are (1) modeling the virtual objects to be embedded in the image, (2) precisely registering the real and virtual coordinate systems, (3) producing realistic lighting and shading, (4) generating images quickly enough to avoid disconcerting lag when there is relative movement, and (5) building portable devices that do not encumber the wearer.

Augmented Cognition

The information presented to the user using an AR setup can also be used in systems for augmented cognition (AC). AC research aims to enhance user performance and cognitive capabilities through adaptive assistance. AC systems can employ physiological sensors to monitor a user's performance and regulate the information presented to the user to minimize stress, fatigue, and information overload. When an AC system identifies a state of cognitive overload, it modifies the presentation or pace of information for the user, offloads information, or alerts others about the state of the user (Dorneich, Whitlow, Ververs, Rogers., 2003).

Honeywell used a gaming environment to test AC for subjects engaged in a military scenario. Subjects played the role of a platoon leader charged with leading soldiers to an objective in a hostile urban setting, shooting enemy soldiers and not their own (Dorneich, Whitlow, Ververs, Carciofini, & Creaser, 2004). Participants also received communications, some of which required a response. This included status and mission updates and requests for information and reports. This exercise showed that physiological sensors could provide inputs to regulate the information presented so that the user is not overwhelmed with data or inappropriately distracted when full attention should be directed to the matters at hand.

Finally, note that AC might also be achieved by very invasive means, such as implantable brain devices to enhance memory or visual processing (see chapter 21, this volume).

Fidelity

In VR, fidelity refers to the accuracy with which real sensory cues are reproduced. Visual cue reproduction is crucial; Ackerman (1990) highlighted the key role of somatosensory information in human experience in her book *A Natural History of the Senses*, and Rheingold (1991) underscored its compelling importance in VR. Yet, despite relentless gains in image generation technology, display resolution (measured by numbers of pixels, levels of intensity, and range of colors), and processing power of real-time graphics rendering hardware and software, VR visuals still fall short of the resolution of the human visual system. Three-dimensional spatial sound, motion generation, and haptic (tactile) interfaces in VR are likewise imperfect.

The fidelity of a computer-generated stimulus depends on the entire process of stimulus generation. This includes the fidelity with which the rendering computations capture the properties of the physical stimulus. At present, images can be generated in real time for only simple lighting models. Diffuse illumination, reflection of light between objects, and caustic effects (caused by uneven refraction of light) that are now commonplace in computer animation cannot be computed at rates necessary for real-time applications. As a result, VR images typically appear cartoonish and flat compared to real images, computer-generated still images, and Hollywood production animations, which can require hours of computation to render a single image.

Because fidelity costs money (and money is scarce), researchers and practitioners must often make difficult choices on how to allocate equipment budgets and trade off fidelity in one area for another. For example, stereo imaging doubles the requirement for image generation because two images must be rendered (one for each eye). Other systems, which use liquid crystal shutter glasses to alternately expose the two eyes to a single display, require projectors with rapid refresh rates and can greatly increase projector cost. Stereo imaging has been identified as an important factor in near-field perception of 3-D, particularly for judging depth of objects within arm's reach. However, for applications where the user is principally focused on objects at a distance (such as flight or driving simulation), most researchers have chosen to present a single image to both eyes, thereby halving the cost of image generation and permitting the use of relatively low-cost LCD projectors.

The choice of whether or not to use stereo image presentation is one of many choices that influence the fidelity of a VR device. The complexity of the 3-D geometric models determines the resolution of shape. Fine details are typically captured in textures overlaid on simple polygonal models, in the manner of trompe l'oeil architectural techniques of painting false facades on building surfaces. The properties of these textures are important in determining the quality of rendered images. In addition, the resolution and level of anti-aliasing of generated images, the brightness, contrast, and resolution of the display device, and even the screen material onto which images are projected all influence the fidelity of the viewed images. Similar issues concern the fidelity of haptic displays and motion systems. There is little information to guide the architects of VR systems when deciding what matters most and what level of fidelity is adequate for their purpose.

Notwithstanding their infidelities, VR worlds have been described as the "reality on your retina." However, VR sensory worlds differ from the real world at several levels, no matter how much money is spent on the VR setup. First, computer display technology is imperfect. Second, even if one could present sensory cues at spatial and temporal resolution rates equal to the human visual system's, cognitive and sensory scientists do not yet know all the perceptual and contextual cues needed to accurately recreate the experience of the real world. Because we do not know what all these cues are, we cannot represent the world as accurately as we might like using software, hardware, and displays.

A dozen cues to 2-D surface representation and 3-D structure have been identified since Fechner's initial investigations of human sensory psychophysics over 150 years ago (Palmer, 1999). However, others likely remain to be discovered. Similar problems exist for representing auditory cues, inertial and somatosensory cues, and olfactory cues, and with seamlessly synchronizing these cues. In many VR implementations, binocular stereopsis cues to 3-D structure and depth are absent, and motion parallax cues to structure and depth are inaccurate.

Another signal of unreality in simulation and VR is the need for the viewer's eyes to accommodate and converge on displays that simulate faraway objects that should not require ocular convergence, which may occur with implementations of HMDs, surround-screen "caves," flat-panel displays, and other displays that do not compensate for a closer-than-expected focal point. The use of collimated displays can help mitigate this problem by displacing the apparent location of the image into the expected depth plane.

At a different level, the shape and fluid motion of clothing and living bodies remain difficult to represent. The artificial surface appearances of skin texture, body surface geometry, and shading, and inaccurate deformation of body surfaces, such as muscle and skin, and inaccurate models of biological movement and facial expressions on an otherwise faithful reproduction of a moving human figure can seem disturbing and takes getting used to (Garau, Slater, Pertaub, & Razzaque, 2005).

Independent of the physical appearance of a VR scenario represented by multisensory cues, the logic of the VR scenario and the behavior of autonomous agents such as other vehicles in a driving scenario may not act like real-world entities. Audiovisual fidelity notwithstanding, consider how easy it is to detect the awkward responses of a human voice spoken by a computer agent designed to depict a human being. Current artificial beings or synthetic actors would likely fail Turing's test of interrogation for intelligence (Turing, 1950) and be distinguished as not human, or at least not normal. At best they might seem strange, mechanical, or autistic.

Notwithstanding these shortcomings, VR can be used to probe the brain at work, and user reactions to current setups provide a window to how humans use a flood of sensory and cognitive cues, how these cues should be reproduced, how they interact, and what is the essence of real-world experience.

Simulator Adaptation and Discomfort

Because VR differs from reality, it can take time to adapt. For example, steering behavior in a real automobile depends on sensory, perceptual, cognitive, and motor factors. In driving simulators, visual cues and steering feedback may differ from corresponding feedback in a real car (Dumas & Klee, 1996). For simulator experiments to be useful, drivers must adapt quickly to differences between the simulator and reality (Gopher, Weil, & Bareket, 1994; Nilsson, 1993). Objective measurements of adaptation and training are needed in the VR environment.

A study assessed time required for 80 experienced drivers to adapt to driving on simulated two-lane rural highways (McGehee, Lee, Rizzo, Dawson, & Bateman, 2004). The drivers' steering behavior stabilized within just 2 minutes (figure 17.5). Fourier analyses provided additional information on adaptation with respect to different frequency components of steering variability. Older drivers' steering behavior is more variable than younger drivers', but both groups adapted at similar rates. Steering is only one of many factors that can index adaptation to driving simulation, and there are many other factors to adapt to in other types of VR applications.

There can be a mismatch between cues associated with the simulator or VR experience. For example, in a driving simulator, rich visual cues of self-motion (heading) without corresponding inertial (vestibular) cues of self-movement can create discomfort ("cybersickness" or simulator adaptation syndrome), analogous to the discomfort passengers feel in a windowless elevator or ship cabin when they receive strong inertial cues without corresponding visual movement cues. Related symptoms (Brandt & Daroff, 1979) have the potential to adversely affect a subject's performance, causing distraction, somatic preoccupation, and dropout.

Adaptation and discomfort in VR environments can be studied using physiological recordings of the user, including electrodermal response, electrogas-

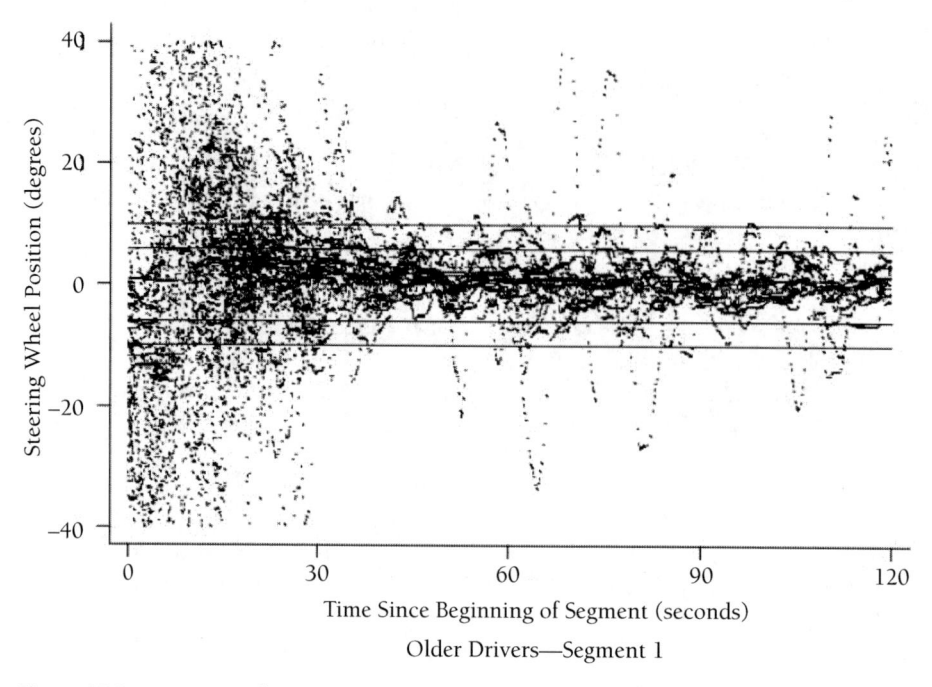

Figure 17.5. Time series of steering wheel position shows adaptation of 52 older drivers in the first 120 seconds of the training drive in the driving simulator SIREN.

trogram, and heart rate (see The Physiology of a VR Experience above and table 17.1). Subjective report is also important and can be measured with questionnaire tools that rate the user's visual, perceptual, and physiological experience in the VR environment (Kellogg, Kennedy, & Graybiel, 1965; Kennedy, Lane, Berbaum, & Lilienthal, 1993). Such tools can be administered before and after exposure to a VR environment to determine how exposure increases discomfort relative to baseline, and how it affects performance and dropout. The Simulator Adaptation Questionnaire (SAQ) is a brief tool (Rizzo, Sheffield, Stierman, & Dawson, 2003) that incorporates items from the Simulator Sickness Questionnaire (SSQ; Kennedy et al., 1993) and Motion Sickness Questionnaire (MSQ; Kellogg et al., 1965), avoids redundancy, and excludes less helpful items (e.g., "desire to move the bowels" from the MSQ).

One way to mitigate user discomfort is to design VEs that minimize exposure to events that can trigger discomfort, such as left turns and abrupt braking in a driving simulator. However, this risks eliminating key scenarios for study. Better understanding of the signs, symptoms, and factors associated with adaptation to VR environments is needed, including the role of presence, immersion, visual aftereffects, vestibular adaptation, display configuration, and movement cues. The efficacy of scopolamine patches, mild electrical stimulation of the median nerve "to make stomach rhythms return to normal," and the extent to which movement cues can reduce discomfort are under investigation (Mollenhauer, 2004).

Greenberg (2004) pointed out that the goal of characterizing a simulator's fidelity is to identify the major artifacts and how they matter to a given VR experiment. A VR study should be undertaken only after the investigator understands the artifacts and can control or adjust adequately for them. Most neuroergonomics research questions cannot be addressed using VR unless there is a useful level of accuracy. In certain settings, the accuracy need not be high and a useful simulation may even be surrealistic (see below).

Understanding the level of accuracy of VR depends on knowledge of the most important characteristics of physical fidelity in the VR cues, including visual, auditory, somatosensory, vestibular (inertial), and even olfactory cuing systems. This depends on a clear understanding of the system being replicated (e.g., driving, flying, phobias) and on the accuracy of reproduction of appliances in the laboratory set up to interact with the user, such as the cab in a driving simulator or the cockpit in a flight simulator or the HMD or glove. Another factor affecting the fidelity (and acceptance) of the VR setup is the intrusiveness of devices used to present the stimuli and measure user responses, including physiological (such as EEG, eye movement recording systems, and galvanic skin response; see below). Greenberg (2004) summarized relevant factors for the assessment of the fidelity of driving simulation along 21 dimensions in three main domains: the visual world, terrain and roadways, and vehicle models.

How Low Can You Go?

In VR or simulation, a real-life task is reduced to a lower-dimension approximation. One theory is that transfer between the simulated and real-life tasks will occur to the extent that they share common components. More relevant, perhaps, is the level of psychological fidelity or functional equivalence of the simulation (Alessi, 1988; Lintern, 1991; Lintern, Roscoe, Koonce, & Segal, 1990). Replicating key portions of a task convincingly and with enough fidelity to immerse, engage, interest, and provide presence may matter more than reproducing exactly what is out there in the real world.

A common approach to VR strives for computer-generated photorealistic representations (Brooks, 1999) using multiple large display screens or HMDs yielding 150- to 360-degree fields of view and providing optical flow and peripheral vision cues not easily achieved on a single small display. Yet the high cost and technical complexity of operating and maintaining these systems, including software provenance, legacy, and updates, limits their use to large university, government, or corporate research settings, and they cannot be practically deployed in physicians' or psychologists' offices for clinical applications (e.g., the use of VR to assess drivers who are at risk for crashes due to cognitive impairments). Moreover, scenario design in VR settings has been ad hoc and unfocused.

Early driving simulators created video game–like scenarios, and some operators felt discomfort, possibly because low microprocessor speeds introduced coupling delays between visual motion and driver performance (Frank, 1988). Yet, despite modest degrees of realism, such simulations successfully showed how operators in the loop, such as pilots and drivers running hand and foot controls, are affected by secondary task loads, fatigue, alcohol intoxication, aging, and cognitive impairments (Brouwer, Ponds, Van Wolffelaar, & Van Zomeren, 1989; de Waard & Rooijers, 1994; Dingus, Hardee, & Wierwille, 1987; Guerrier, Manivannan, Pacheco, & Wilkie, 1995; Haraldsson, Carenfelt, Diderichsen, Nygren, & Tingvall, 1990; Katz et al., 1990; McMillen & Wells-Parker, 1987; Rizzo, McGehee, Dawson, & Anderson, 2001). Even environments with very impoverished sensory cues can provide a useful basis for studying perception and action. For example, a study of the perception of direction of heading examined the role of motion cues in driving (Ni, Andersen, McEvoy, & Rizzo, 2005). A variety of cues provided crucial inputs for moving through the environment. Optical flow is the apparent motion of object points across the visual field as an observer moves through an environment. It is a well-studied cue for the perception and control of locomotion (Gibson, 1966, 1979). The optical flow field can be conceptualized as an image of vectors that represent how points are moved from image to image in an image sequence. Visual landmarks also provide key information for the perception and control of locomotion (Hahn, Andersen, & Saidpour, 2003) and are important for movement perception in real environments (see chapter 9) and virtual environments.

Ni et al. (2005) used a tracking task in which younger and older drivers viewed optical flow fields comprising achromatic dots that were randomly positioned in 3-D space, as with a star field (see figure 17.6). The task simulated motion along the subject's line of sight displaced laterally by a forcing function (resembling unpredictable vehicle sidewind on a windy road). The driver had to compensate for the sidewind using a steering wheel so that the path of locomotion was always straight. On half of the trials, landmark information was given by a few colored dots within the flow field of achromatic dots. Steering control was indexed by tracking error (root mean square) and coherency (the squared correlation between the input and response at a particular frequency). Independent measures were number of dots in the optical flow field, landmark information (present or not), and frequency of lateral perturbation. Results suggested greater accuracy and less steering control error for younger subjects (Ni et al., 2005). Both groups showed improvement with an increase in optical flow information (i.e., an increase in the density of dots in the flow field). Younger subjects were more efficient at using optical flow information and could use landmark information to improve steering control. This study illustrates how a perceptually

Figure 17.6. Heading from optical flow. In this schematic illustration, each vector represents the change in position of a texture element in the display during forward motion. The subject's task was to judge the direction of observer motion relative to the vertical bar. In the illustration, the correct response is the direction to the right of the bar.

impoverished environment can replicate key elements involved in real-world navigation.

Although VR environments differ from natural environments, not all differences matter. Different physical stimuli known as *metamers* produce indistinguishable neural or psychological states (Brindley, 1970). Humans "complete" images across the normal physiological blind spot (caused by the lack of retinal receptors where the optic nerve meets the back of the eye). They may fill in gaps and missing details from the area of an acquired visual field defect and fail to recognize that they have a deficit (Rizzo & Barton, 1998). Normal observers may seem to be blind to glaring details that defy everyday expectations, such as the famous change-blindness demonstration showing that observers with normal vision failed to notice a gorilla walking through a group of people tossing a ball (Simons & Chabris, 1999). Consequently, they may not be disturbed by some missing details or low fidelity in VR environments.

Low fidelity can be a problem if it reduces immersion in a simulated task. One approach to increase immersion is to place real objects in VEs as props. The touch of a physically real surface or simply feeling a ledge can do much to heighten the sense of immersion (Insko, 2001; Lok, Naik, Whitton, & Brooks, 2003). The use of simple objects as props to augment high-fidelity visual representations compellingly demonstrates the power that small cues have to give participants a sense of realness and presence. For example, a wooden plank can be used to create an apparent ledge at the edge of a virtual precipice.

A key question is, is it fair to cheat (i.e., reduce, enhance, or augment some cues) to increase users' immersion and decrease discomfort? For example, is it acceptable to substantially enlarge road signs in a driving simulator to compensate for poor acuity caused by low-resolution displays? The fidelity-related question of how closely a simulated world should match the real world depends on the goals of the simulation and requires multilevel considerations of and comparisons between simulated cues (e.g., visual, auditory, haptic, and movement cues) and tasks and corresponding real-world cues and tasks. Potential reasons for exploring the lower boundaries of fidelity besides economy are that increasing levels of fidelity may limit data collection, dilute training effects, undermine experimental generalizations, and increase the likelihood of simulator discomfort.

Nonrealistic Virtual Environments

VEs need not be highly realistic to be effective for experimental and training applications in neuroergonomics. A number of researchers have examined the use of deliberately unreal VR environments as surrogates for real environments. This line of research parallels developments in nonphotorealistic rendering (NPR) techniques for static and animated graphics. NPRs abstract the essential content in an image or image stream and often can be produced very efficiently. Simple drawings are often more informative than photographs. For example, instruction manuals frequently make use of schematics to illustrate assembly procedures, and medical textbooks are filled with hand-drawn illustrations of anatomical structures (Finkelstein & Markosian, 2003).

Fischer, Bartz, and Strasser (2005) examined the use of NPR to improve the blending of real and synthetic objects in AR by applying a painterly filter to produce stylized images from a video camera. The filtered video stream was combined with computer-generated objects rendered with a nonphotorealistic method. The goal was to blur the distinction between real and virtual objects. The resultant images are similar to hand-drawn sketches (Fischer et al., 2005). Sketches and line drawings can convey a sense of volume and depth while reducing distracting imperfections and artifacts of more realistic renderings. Gooch and Willemsen (2002) examined distance judgments for VR environments rendered with line drawings on an HMD. Subjects were asked to view a target on the ground located 2–5 meters away, close their eyes, and walk to the target. They undershot distances relative to the same walking task in a real environment, which corresponded with undershoot of similar experiments conducted with more realistically rendered VR environments in HMDs.

Severson and colleagues (Rizzo, Severson, Cremer, & Price, 2003; Rizzo, 2004) developed an abstract VR environment tool that captures essential elements of driving-related decision making and is deployable on single-screen PC systems (figure 17.7). This tool tests go/no-go decision making. It is based on a surreal environment and provides performance measures along the lines of a neuropsychological test. The nonconventional design approach was motivated by results of synthetic vision display system research for navigation and situational awareness. Designers focused on what was needed for the intended assessment rather than assuming visual realism was necessary and used abstract, rather than photorealistic, representations in the VE (Ellis, 1993, 2000). This approach draws from perceptual psychology, computer graphics, art, and human factors, and provides sufficient pictorial and motion cues relevant to the task of perceiving the spatial relationships of objects and the orientation of the user in the VE (Cutting, 1997; Palmer, 1999; Wanger, Ferwerda, & Greenberg., 1992). Similar to high-fidelity driving simulators and avionics synthetic vision systems, the go/no-go tool provided motion parallax, optical flow, shading, texture gradients, relative size, perspective, occlusion, convergence of parallel lines, and position of objects relative to the horizon. Application of the go/no-go tool showed significant differences in errors between 16 subjects with neurological impairments (14 with focal brain lesions, 2 with Alzheimer's disease) and 16 neurologically normal subjects. The finding of a shallower learning curve across go/no-go trials in brain-damaged drivers suggests a failure of response selection criteria based on prior experience, as previously reported in brain-damaged individuals with decision-making impairments on the Gambling Task (e.g., Bechara, Damasio, Tranel, & Damasio, 1997; see above).

These results suggest that VR can be used to develop new neuropsychological tests. VR tasks can provide a window on the brain at work in strategic and tactical situations encountered in the real world that are not tested by typical neuropsychological tests. A PC-based VR tool can distinguish decision-making-impaired people, which traditional neuropsychological test batteries may not (Damasio & Anderson, 2003).

Validation and Cross-Platform Comparisons

The results of experiments conducted in VR are valid only to the extent that people behave the same in the real world as they do in the virtual world. Likewise, VR training will be effective only if learned behaviors transfer to the real world. The most direct way to test the validity of a study conducted in a VR environment is to compare the results to performance under the same conditions in the real world. However, in many cases it is not possible to conduct comparable tests in the real world because to do so would expose subjects to dangerous circumstances, posing a risk of injury. Even if a methodology is validated for a particular VE, the question that remains is how that methodology transfers to a different VE. VE installations vary in a multitude of ways, including the display technology, the rendering software, the interaction devices, and the content presented (i.e., object models and scenarios). In order to be assured that results generalize to a variety of platforms, it is important to conduct cross-platform comparative studies to determine what differences influence behavior.

A recent body of research has examined whether or not fundamental aspects of visual perception and

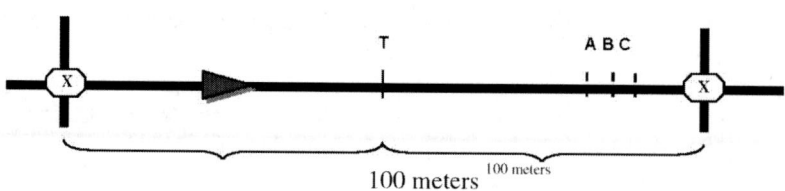

Figure 17.7. (Top) The go/no-go tool used a nonphotorealistic representation of a 3-D virtual space. Visual and pictorial cues provided situational awareness in a small field of view, similar to the design constraints placed on aviation research display system researchers (Theunissen, 1997). (Bottom) At point T, gate-closing trigger point A (easy), B (medium), or C (difficult) is computed, based on current speed, deceleration limit, and other parameters. Each subject drove through a series of intersections (marked by Xs), which had gates that opened and closed. When the driver reached point T, 100 meteres before an intersection, a green Go or red Stop signal appeared at the bottom of the display and a gate-closing trigger point (A, B, or C) was computed based on a deceleration constant, gate closure animation parameters, driver speed, and amount of time allotted to the driver to make a decision.

action are preserved in VR environments. For example, do people judge distances and motions the same in VEs as they do in real environments? A number of studies have found that people underestimate distances in VR by as much as 50% relative to comparable judgments in real environments (Loomis & Knapp, 2003; Thompson et al., 2004; Willemsen & Gooch, 2002). For example, Loomis and Knapp compared distance estimates in the real world to distance estimates in VR presented on an HMD. Subjects viewed a target on the ground. They then closed their eyes, turned 90 degrees, took a small number of steps, and then pointed to where they thought the target would be. On average, subjects underestimated the true angle by a factor of 2 in VR relative to the same task in a corresponding real environment.

Contradicting these results, a number of studies have found distance estimates to be similar in real and virtual environments (Interrante, Anderson, & Ries, 2004; Plumert, Kearney, & Cremer, 2005; Witmer & Sadowski, 1998). Plumert et al. (2005) examined distance estimation in large-screen virtual environments. Subjects judged the distance of a person standing on a grassy lawn in a real environment and an image of a person placed on a textured model of the outdoor environment presented on a panoramic display. Perceived distance was estimated by the time it took subjects to perform an imagined walk to the target. Subjects started a stopwatch as they initiated their imagined walk and stopped it when (in their imagination) they arrived at the target. They found that time-to-walk estimates were very similar in real and virtual environments. In general, subjects were accurate in judging the distance to targets at 20 feet and underestimated the distance to targets at 40 to 120 feet.

If distance perception is significantly distorted in VR, then performance on tasks that require judgment of distance is likely to be very different in VR than in the real world. This could profoundly influence the outcomes of experiments in VR. However, the conditions under which distance is underestimated in VR and the cause of this distortion remain unclear and are under investigation.

The validity of a simulation can be assessed by comparing results in VR to a variety of gold standard real-world outcomes. For example, the behaviors of drivers in a ground vehicle simulator could be compared to state records of crashes and moving violations of real people driving real cars on real roads. This epidemiological record can be assessed prospectively, but there are few direct observations, and records are reconstructed from imperfect and incomplete accounts by untrained observers in chaotic conditions. It is now also possible to measure real behavior in the real world (see chapter 8). Results make it clear that the epidemiological record may not accurately reflect the truth. In one remarkable example, an innocent driver was charged with an at-fault, rear-end collision after his car was struck by an oncoming car that spun 180 degrees and veered into his lane (Dingus et al., 2005).

Can a standard VR scenario that is run using different hardware and software be expected to give similar results in similar populations of subjects? This question is relevant to developing standard VR scenarios and performing clinical trials across multiple research sites. In this vein, the HASTE group considered several specific criteria for cross-platform comparisons in VR scenarios implemented in a range of driving simulators. Key comparison measures included minimum speed and speed variation, lateral position and lateral position variation, lane exceedances, time to line crossing, steering wheel reversal rate, time to collision, time headway, distance headway, and brake reaction time. Optional comparison measures were high-frequency component of steering wheel angle variation, steering entropy, high-risk overtakings, abrupt onsets of brakes, and postencroachment time (HASTE, 2005).

Guidelines, Standards, and Clinical Trials

A current problem in using VR in neuroergonomics research is lack of guidelines and standards. Guidelines are recommendations of good practice that rely on their authors' credibility and reputation. Standards are formal documents published by standards-making bodies that develop through extensive consultation, consensus, and formal voting. Guidelines and standards help establish best practices. They provide a common frame of reference across VR (simulator and scenario) designs over time to avoid misunderstanding and to allow different user groups to connect. They allow reviewers to assess research quality, to compare results across studies for quality and consistency, and to help identify inconsistencies and the need for further guidelines and standards.

The lack of guidelines and standards is hindering progress in human factors and neuroergonomics research and has emerged as a major issue in VR setups for driving simulation. VR can be an important means for safe and objective assessment of performance capabilities in normal and impaired automobile drivers. Yet it remains a cottage industry of homegrown devices with few guidelines and standards.

Having guidelines and standards for simulation and VR scenarios can facilitate comparisons of operator performance across different VR platforms, including the collection of large amounts of data at different institutions with greater power for addressing worldwide public health issues conducive to research using VR. Comparisons against standards can clearly reveal missing descriptors and other weaknesses in ever-mounting numbers of research reports on driving simulation and other applications of VR.

In driving simulation, the scenario might focus on key situations that convey a potential high crash risk, such as intersection incursion avoidance and rear-end collision avoidance scenarios involving older drivers and scenarios that address behavioral effects of using in-vehicle devices such as cell phones and information displays. In a training application, the scenario might focus on risk acceptance

Table 17.2. Sample Driving Scenario Description

Script	The driver is traveling on a road with two lanes of traffic, each moving at different speeds. At different times, one lane of traffic is moving more advantageously (faster), although overall this may be the slower lane. The driver's task is to pass through traffic as quickly as possible. This task would be similar to the gambling task (Bechara et al, 1997) in which an individual has to overcome an immediate reward to ensure long-term benefit. Variations can be made on this script: the two lanes of traffic could average the same speed or could even be moving at the same speed. The driver's perception that one lane is moving faster may be a visual and cognitive illusion.
Cognitive constructs stressed	Attention, perception, and decision making.
Dependent measures	Time it takes the driver to maneuver through the traffic to arrive at a destination, number of navigation errors, and number of moving violations or safety errors (e.g., excessive speed, near misses, and collisions) could be recorded.
Independent variables	The number of vehicles involved, number of lanes, speed of the different lanes and final destination, and travel contingencies could be manipulated (e.g., the driver could be instructed to get off at a specific exit for a hospital).
Implementation challenges	It may be difficult to create a realistic feel. The problem may be lessened by giving the driver external instructions, thus altering driver expectations and rewards or incentives. For instance, the instructions could be to drive as if you were taking a pregnant woman or a critically sick person to the emergency room. Is it possible for the driver to be able to change lanes when desired? How will the surrounding cars respond to the driver?
Measurement challenges	Some drivers will not try to get into the faster lane, if they think it makes little difference in the long run. A questionnaire following the task may be helpful in assessing how fast the driver felt each lane was moving and whether or not the driver would have changed lanes if given the opportunity. Variations in personalities would have to be considered.
Testing validity of scenario	An instrumented vehicle could be used to study lane-change behavior during times of high-density traffic, yet the environmental variables could not be easily controlled.

in teenage drivers. In a medical application, the scenario might focus on a high-risk situation in anesthesia. In a psychological application, the scenario might be tailored to desensitize a patient to a certain phobia (see below).

Task analysis (Hackos & Redish, 1998) can deconstruct and help clarify the logical structure of complex VR scenarios, the detailed flow of actions a subject has to take in a VR task, and related issues such as operational definitions and measurements of variables. This can facilitate cross-platform comparisons, help create a body of shared VR scenarios for multicenter research trials, and contribute to an infrastructure for worldwide research on a variety of global public health issues.

A proposed format for describing a VR scenario for neuroergonomics research in traffic safety is shown in table 17.2. An example specification is shown for a lane-change scenario that includes a script, the cognitive constructs stressed, dependent measures, independent variables, implementation challenges, plan for testing the validity of the scenario, and a bibliography of related literature. Similar strategies could be considered for other VR applications.

Key topics for guidelines and standards in VR are (1) scenario design; (2) selection and definition of performance measures; (3) standards for reporting experimental setup and results; (4) subject selection criteria; (5) reporting of independent measures; (6) standards for physiological recording; (7) graphics, audio, and movement; (8) training criteria; (9) subject adaptation and comfort; and (10) criteria for validation.

Making guidelines and standards involves science, logic, policy, and politics. Standards may seem to favor one group over another. Key persons may seem to exert undue influence over the final product. The structure and formality of standards might seem to hinder innovation. It may be easier to propose standards than to apply them in practice. For example, how well can most driving simulator scenarios be described within a common framework? Should researchers be expected to implement a key VR scenario just one way? Can levels of fidelity be adequately specified for cross-platform comparisons? Can we overcome vested interests and entrenched opinions? These are questions that demand attention and a call for greater action.

Having VR standards would facilitate multicenter clinical trials in a variety of settings in which VR has shown promise, such as driving simulation, treatment of phobias, and so on. A clinical trial has "some formal structure of an experiment, particularly control over treatment assignment by the investigator" (Piantadosi, 1997). According to the National Institutes of Health (2005), "A clinical trial . . . is a research study in human volunteers to answer specific health questions. [Clinical trials] are the fastest and safest way to find treatments that . . . improve health. Interventional trials determine whether experimental treatments . . . are safe and effective under controlled environments."

For instance, driving simulator scenarios are becoming key elements in state and federal efforts to probe (1) the efficacy of novice and older driver training programs; (2) fitness to drive in patients with performance declines due to aging and mild cognitive impairment, traumatic brain injury, neurodegenerative disorders such as Alzheimer's and Parkinson's diseases, and sleep disturbances; and (3) acute and chronic effects of many medications such as analgesics, antidepressants, psychostimulants, antidepressants, and cancer chemotherapy agents. To make meaningful comparisons between research studies conducted on differing platforms, it is essential that the scenarios used be specified in sufficient detail to implement them on simulators that depend on very different hardware, visual database and scenario development tools, image rendering software, and data collection systems.

Future Directions

VR has many potential applications in neuroergonomic assessments of healthy and impaired operators. VR has been used to assess the effects of introducing new technology to the human operator in the loop: cell phones, GPS, infotainment, automated highways, heads-up displays, and aircraft displays. Human performance measurements in VEs include motor learning and rehabilitation, and information visualization in VR (glass cockpit, information-gathering behaviors of military or homeland security analysts, and such). VR can be used in applications to train physicians effectively, resulting in fewer medical errors. To assess the efficacy of applications that use VEs as training simulators, we must measure transfer of training effects.

VR applications are being proposed more and more as a tool for assessing and treating a variety of neuropsychological, psychiatric, and physical disorders. These applications pertain to planning and decision making, spatial orientation and navigation, mood disorders, anxiety, panic disorder, agoraphobia, and eating disorders. The feeling of being somewhere else in VR may help people suffering under drug addiction, dental extractions, chemotherapy, dialysis, physical therapy, or other discomforts.

There are numerous research groups providing insights and therapeutic use of VR as a tool for therapy and rehabilitation of panic disorders and post-traumatic stress disorders, and as an analgesic during therapeutic and rehabilitation treatment (Hodges et al., 2001; Hoffman Garcia-Palacios, Carlin, & Botella-Arbona, 2003; Hoffman, Richards, Coda, Richards, & Sharar, 2003; Hoffman et al., 2004; Rothbaum & Schwartz, 2002; Rothbaum et al., 1995; Wiederhold & Wiederhold, 2004). VR therapy is an ongoing area of exploration, currently with limited clinical use, but has been embraced by a small community of providers. Beyond the challenges of utilizing high-tech computing technology in a therapy setting, there are operational challenges in the use of VR therapy: with customization of environments for the broad range of scenarios required to meet a subject's therapeutic needs; and with the limited educational resources for the design and development of scenarios, environments, and protocols and the training of therapists to utilize the techniques. For example, in traditional therapy sessions, subjects would utilize in vivo exposure or be required to imagine a scenario and setting in which their phobia would take place (Hodges, Rothbaum, Watson, Kessler, & Opdyke, 1996). Utilizing VR therapy, an appropriate scenario and environment would need to be available to the therapist and patient for treatment of the disorder. For each scenario and environment, the efficacy and usability of the VR scenario would need to be validated for clinical use. As a therapeutic modality, VR therapy appears to manifest a number of the challenges noted throughout this chapter.

Clinical application challenges notwithstanding, there are particularly compelling examples of research into the use of VR therapy. The University of Washington HITLab is exploring VR therapy as a nonpharmacological analgesic for patients receiving burn treatment therapy, and it is conducting exploratory research utilizing synchronized VR exposure with fMRI brain scans to study the neural correlates of psychological disorders and the impact of therapy on patterns of fear-related brain activity (Hoffman et al., 2004).

Patients with anxiety disorders, post-traumatic stress disorders or phobias (such agoraphobia (fear of crowds), arachnophobia (fear of spiders), acrophobia (fear of heights), and aviophobia (fear of flying) might be desensitized by exposure to virtual threats. Burn victims may be distracted from their pain by moving through icy VR worlds with snowmen, igloos, and penguins in association with modulation of brain activity (Hoffman et al., 2004). Patients with eating disorders might be treated by exposure to differing body images or renditions of virtual dining experiences.

Advances in the technology such as HMDs, interactive gloves, and the Internet will facilitate penetration of VR in these applications. Wireless movement trackers (see chapter 8) will lead to new and better interfaces for users to manipulate objects in VR. Improvements in cost and fidelity may be driven by the resources of the gaming industry. VR users may be connected to distributed virtual worlds for collaborative work and play over the Internet.

Honeywell's use of a gaming platform for assessing the performance of research tools and user performance provided a visually rich VE and complex scenario capabilities that would be challenging for most researchers to create from scratch (Dorneich et al., 2004). The marriage of VR and gaming has been a growing collaboration fostered by military funding of computer-based simulation and training and the popularity and resources for development of first-person-shooter and distributed gaming (Zyda & Sheehan, 1997).

One study suggested that action video game playing enhances visual attention skills in habitual video game players versus non–video game players (Green & Bavalier, 2003). Nonplayers trained on an action video game showed improvement from their pretraining abilities, establishing the role of playing in this effect. Further work is needed to investigate this effect and address whether immersion in VR through video game playing can enhance other aspects of sensation, memory, and other cognitive processes.

Avatars and simulated humans will assume more realistic traits, including movement, speech, and language, and will seem to understand and respond appropriately to what we say and do. This may result in virtual friends, advisors, and therapists (e.g., on long space flights, in lonely outposts).

Conclusion

Multidisciplinary vision and collaboration has established VR as a powerful tool to advance neuroergonomics. Researchers are creatively exploring and expanding VR applications using systems that vary from high-fidelity, motion-based simulation and fully immersive, head-mounted VR to PC-based VR gaming platforms. These systems are providing computer-based, synthetic environments to train, treat, and augment human behavior and cognitive performance in simulacra of the real world and to investigate the relationships between perception, action, and consciousness.

Challenges remain. Although the term VR generally describes computer-based environments, there are no clearly defined standards for what the systems should provide in terms of fidelity, display technology, or interaction devices. Standards are needed for assessing research outcomes and replicating experiments in multi-institution research. The current vacuum of standards can be at least partly attributed to the rapid pace of change. VR is an evolving, fluid, and progressing medium that benefits from the advances in PC-based computing and graphics-rendering capabilities.

Main Points

1. VR applications show promise for studying the behavior of people at work, for training workers, and for improving worker performance through augmented reality and augmented cognition environments.
2. The results of simulator scenarios in different labs with different VR environments should be compared with each other, as well as with the results in comparable tasks in real-world

settings, in order to test the replicability and validity of the results.
3. VR environments can be useful even if they depart significantly from reality. Even surreal VR environments can provide useful information on the brain and behavior at work.

Key Readings

Nof, S. Y. (1999). *Handbook of industrial robotics* (2nd ed.). New York: John Wiley & Sons.

Rheingold, H. (1991). *Virtual reality*. New York: Simon and Schuster.

Sherman, W. R., & Craig, A. (2003) *Understanding virtual reality: Interface application, and design (The Morgan Kaufmann Series in Computer Graphics)*. New York: Morgan Kaufmann.

Stanney, K. (Ed.). (2002). *Handbook of virtual environments*. Mahwah, NJ: Erlbaum.

References

Ackerman, D. (1990). *A natural history of the senses*. New York: Random House.

Alessi, S. M. (1988). Fidelity in the design of instructional simulations. *Journal of Computer-Based Instruction, 15*(2), 40–47.

Barrett, C., Eubank, S., & Smith, G. (2005). If small pox strikes Portland. *Scientific American, 292*, 53–61.

Bechara, A., Damasio, H., Tranel, D., & Damasio, A. R. (1997). Deciding advantageously before knowing the advantageous strategy. *Science, 172*, 1293–1295.

Brandt, T., & Daroff, R. (1979). The multisensory physiological and pathological vertigo syndromes. *Annals of Neurology, 7*, 195–203.

Brindley, G. S. (1970). *Physiology of the retina and the visual pathway* (2nd ed.). Baltimore, MD: Williams and Wilkins.

Brooks, F. (1999). What's real about virtual reality. *IEEE Computer Graphics and Applications, 19*(6), 16–27.

Brouwer, W. H., Ponds, R. W., Van Wolffelaar, P. C., & VanZomeren, A. H. (1989). Divided attention 5 to 10 years after severe closed head injury. *Cortex, 25*, 219–230.

Brown, J. (2005). Race against reality. *Popular Science, 266*, 46–54.

Cutting, J. E. (1997). How the eye measures reality and virtual reality. *Behavior Research Methods, Instrumentation, and Computers, 29*, 29–36.

Damasio, A. R., & Anderson, S. (2003). The frontal lobes. In K. Heilman & E. Valenstein (Eds.), *Clinical neuropsychology* (4th ed., pp. 402–446). New York: Oxford University Press.

de Waard, D., & Rooijers, T. (1994). An experimental study to evaluate the effectiveness of different methods and intensities of law enforcement on driving speed on motorways. *Accident Analysis and Prevention, 26*(6), 751–765.

Dinges, D., & Mallis, M. (2000). Alertness monitors. In T. Akerstad & P.-O. Haraldsson (Eds.), *The sleepy driver and pilot* (pp. 31–32). Stockholm: National Institute for Psychosocial Factors and Health.

Dingus, T. A., Hardee, H. L., & Wierwille, W. W. (1987). Development of models for on-board detection of driver impairment. *Accident Analysis and Prevention, 19*, 271–283.

Dingus, T. A., Klauer, S. G., Neale, V. L., Petersen, A., Lee, S. E., Sudweeks, J., et al. (2005). *The 100-Car Naturalistic Driving Study: Phase II—Results of the 100-car field experiment* (Project Report for DTNH22-00-C-07007, Task Order 6; Report No. TBD). Washington, DC: National Highway Traffic Safety Administration.

Dorneich, M., Whitlow, S., Ververs, P., Rogers, W. (2003). Mitigating cognitive bottlenecks via an Augmented Cognition Adaptive System, *Proceedings of the 2003 IEEE International Conference on Systems, Man, and Cybernetics*, 937–944.

Dorneich, M., Whitlow, S., Ververs, P., Carciofini, J., & Creaser, J. (2004, September). Closing the loop of an adaptive system with cognitive state. *Proceedings of the 48th Annual meeting of the Human Factors and Ergonomics Society Conference*, New Orleans, 590–594.

Dumas, J. D., & Klee, H. I. (1996). Design, simulation, and experiments on the delay compensation for a vehicle simulator. *Transactions of the Society for Computer Simulation, 13*(3), 155–167.

Dureman, E., & Boden, C. (1972). Fatigue in simulated car driving. *Ergonomics, 15*, 299–308.

Egelund, N. (1982). Spectral analysis of the heart rate variability as an indicator of driver fatigue. *Ergonomics, 25*, 663–672.

Ellis, S. R. (1991). Nature and origin of virtual environments: A bibliographical essay. *Computer Systems in Engineering, 2*(4), 321–347.

Ellis, S. R. (1993). Pictorial communication: Pictures and the synthetic universe, *Pictorial communication in virtual and real environments* (2nd ed.). Bristol, PA: Taylor and Francis.

Ellis, S. R. (1995). Origins and elements of virtual environments. In W. Barfield & T. Furness (Eds.), *Virtual environments and advanced interface design* (pp. 14–57). Oxford, UK: Oxford University Press.

Ellis, S. R. (2000). On the design of perspective displays. In *Proceedings: 15th Triennial Conference, International Ergonomics Association/44th annual meeting, Human Factors and Ergonomics Society*. Santa Monica, CA: HFES, 411–414.

Finkelstein, A., & Markosian, L. (2003). Nonphotorealistic rendering. *IEEE Computer Graphics and Applications, 23*(4), 26–27.

Fischer, J., Bartz, D., & Strasser, W. (2005). Stylized augmented reality for improved immersion. *Proceedings of the IEEE Virtual Reality Conference (VR '05)*, IEEE Computer Society, 195–202.

Foster, P. J., & Burton, A. (2003). Virtual reality in improving mining ergonomics. In *Applications of computers and operations research in the minerals industry* (pp. 35–39). Symposium Series S31. Johannesburg: South African Institute of Mining and Metallurgy.

Frank, J. F. (1988, November). *Further laboratory testing of in-vehicle alcohol test devices.* Washington, DC: National Highway Traffic Safety Administration, report DOT-HS-807-333.

Garau, M., Slater, M., Pertaub, D., & Razzaque, S. (2005). The responses of people to virtual humans in an immersive virtual environment. *Presence, 14*(1), 104–116.

Gibson, J. J. (1966). *The senses considered as perceptual systems.* Boston: Houghton Mifflin.

Gibson, J. J. (1979). *The ecological approach to visual perception.* Boston: Houghton Mifflin.

Gooch, A., & Willemsen, P. (2002). Evaluating space perception in NPR immersive environments. *Proceedings of the 2nd International Symposium on Non-Photorealistic Animation and Rendering*, ACM Press, 105–110.

Gopher, D., Weil, M., & Bareket, T. (1994). Transfer of skill from a computer game trainer to flight. *Human Factors, 36*(3), 387–405.

Green, C., & Bavelier, D. (2003). Action video game modifies visual attention. *Nature, 423*(6939), 534–537.

Greenberg, J. (2004, January 13). *Issues in simulator fidelity.* The Simulator Users Group. TRB meeting in Washington, DC. Safe Mobility of Older Persons Committee (ANB60). Retrieved from http://www.uiowa.edu/neuroerg/index.html.

Guerrier, J. H., Manivannan, P., Pacheco, A., & Wilkie, F. L. (1995). The relationship of age and cognitive characteristics of drivers to performance of driving tasks on an interactive driving simulator. In *Proceedings of the 39th Annual Meeting of the Human Factors and Ergonomics Society* (pp. 172–176). San Diego: Human Factors and Ergonomics Society.

Hackos, J., & Redish, J. (1998). *User and task analysis for interface design.* Chichester: Wiley.

Hahn, S., Andersen, G. J., & Saidpour, A. (2003). Static scene analysis for the perception of heading. *Psychological Science, 14*, 543–548.

Hakkanen, H., Summala, H., Partinen, M., Tihonen, M., & Silvo, J. (1999). Blink duration as the indicator of driver sleepiness in professional bus drivers. *Sleep, 22*, 798–802.

Haraldsson, P.-O., Carenfelt, C., Diderichsen, F., Nygren, A., & Tingvall, C. (1990). Clinical symptoms of sleep apnea syndrome and automobile accidents. *Journal of Oto-Rhino-Laryngology and Its Related Specialties, 52*, 57–62.

Harrison, Y., & Horne, J. A. (1996). Long-term extension to sleep—are we chronically sleep deprived? *Psychophysiology, 33*, 22–30.

HASTE. (2005, March 22). *Human-machine interface and the safety of traffic in Europe.* Project GRD1/2000/25361 S12.319626. HMI and Safety-Related Driver Performance Workshop, Brussels.

Heim, M. (1993). *The metaphysics of virtual reality.* New York: Oxford University Press.

Hodges, L., Anderson, P., Burdea, G., Hoffman, H., & Rothbaum, B. (2001). VR as a tool in the treatment of psychological and physical disorders. *IEEE Computer Graphics and Applications, 21*(6), 25–33.

Hodges, L., Rothbaum, B., Watson, B., Kessler, G., & Opdyke, D. (1996). Virtually conquering fear of flying. *IEEE Computer Graphics and Applications, 16*, 42–49.

Hoffman, H., Garcia-Palacios, A., Carlin, C., III, T. F., & Botella-Arbona, C. (2003). Interfaces that heal: Coupling real and virtual objects to cure spider phobia. *International Journal of Human-Computer Interaction, 16*, 283–300.

Hoffman, H., Richards, T., Coda, B., Bills, A. R., Blough, D., Richards, A. L., et al. (2004). Modulation of thermal pain-related brain activity with virtual reality: Evidence from fMRI. *Neuroreport, 15*, 1245–1248.

Hoffman, H., Richards, T., Coda, B., Richards, A., & Sharar, S. R. (2003). The illusion of presence in immersive virtual reality during an fMRI brain scan. *CyberPsychology and Behavior, 6*, 127–132.

Horne, J. A., & Reyner, L. A. (1996). Driver sleepiness: Comparisons of practical countermeasures—caffeine and nap. *Psychophysiology, 33*, 306–309.

Insko, B. (2001). *Passive haptics significantly enhances virtual environments.* Doctoral dissertation, University of North Carolina.

Interrante, V., Anderson, L., & Ries, B. (2004). *An experimental investigation of distance perception in real vs. immersive virtual environments via direct blind walking in a high-fidelity model of the same room.* Paper presented at the 1st Symposium on

Applied Perception in Graphics and Visualization, Los Angeles, August 7–8.

James, W. (1894). The physical basis of emotion. *Psychological Review, 1*, 516–529.

Johns, M. W., Cornell, B. A., & Masterton, J. P. (1969). Monitoring sleep of hospital patients by measurement of electrical resistance of skin. *Journal of Applied Physiology, 27*, 898–910.

Katz, R. T., Golden, R. S., Butter, J., Tepper, D., Rothke, S., Holmes, J., et al. (1990). Driving safely after brain damage: Follow up of twenty-two patients with matched controls. *Archives of Physical Medicine and Rehabilitation, 71*, 133–137.

Kecklund, G., & Akerstedt, T. (1993). Sleepiness in long distance truck driving: An ambulatory EEG study of night driving. *Ergonomics, 36*, 1007–1017.

Kellogg, R. S., Kennedy, R. S., & Graybiel, A. (1965). Motion sickness symptomatology of labyrinthine defective and normal subjects during zero gravity maneuvers. *Aerospace Medicine, 36*, 315–318.

Kennedy, R. S., Lane, N. E., Berbaum, K. S., & Lilienthal, M. G. (1993). Simulator sickness questionnaire: An enhanced method for quantifying simulator sickness. *International Journal of Aviation Psychology, 3*(3), 203–220.

Lintern, G. (1991). An informational perspective on skill transfer in human-machine systems. *Human Factors, 33*, 251–266.

Lintern, G., Roscoe, S. N., Koonce, J. M., & Segal, L. D. (1990). Transfer of landing skills in beginning flight training. *Human Factors, 32*, 319–327.

Lok, B., Naik, S., Whitton, M., & Brooks, F. (2003). Effects of handling real objects and self-avatar fidelity on cognitive task performance and sense of presence in virtual environments. *Presence, 12*, 615–628.

Loomis, J., & Knapp, J. (2003). *Visual perception of egocentric distance in real and virtual environments.* In L. Hettinger & M. Haas (Eds.), *Virtual and adaptive environments* (pp. 21–46). Mahwah, NJ: Erlbaum.

McGehee, D. V., Lee, J. D., Rizzo, M., Dawson, J., & Bateman, K. (2004). Quantitative analysis of steering adaptation on a high performance fixed-base driving simulator. *Transportation Research, Part F: Traffic Psychology and Behavior, 7*, 181–196.

McMillen, D. L., & Wells-Parker, E. (1987). The effect of alcohol consumption on risk-taking while driving. *Addictive Behaviors, 12*, 241–247.

Meehan, M., Insko, B., Whitton, M., & Brooks, F. (2002). Physiological measures of presence in virtual environments. *ACM Transactions on Graphics, 21*(3), 645–652.

Mollenhauer, M. (2004). Simulator adaptation syndrome literature review. *Realtime Technologies Technical Report.* Retrieved from http://www.simcreator.com/documents/techreports.htm.

Mourant, R. R., & Rockwell, T. H. (1972). Strategies of visual search by novice and experienced drivers. *Human Factors, 14*, 325–335.

National Institutes of Health. (2005). *An introduction to clinical trials.* Retrieved from http://www.clinicaltrials.gov/ct/info/whatis#whatis.

Ni, R., Andersen, G. J., McEvoy, S., & Rizzo, M. (2005). *Age-related decrements in steering control the effects of landmark and optical flow information.* Paper presented at Driving Assessment 2005, 3rd International Driving Symposium on Human Factors in Driver Assessment, Training, and Vehicle Design, Rockport, ME, June 27–30.

Nilsson, L. (1993). Contributions and limitations of simulator studies to driver behavior research. In A. Parkes & S. Franzen (Eds.), *Driving future vehicles* (pp. 401–407). London: Taylor and Frances.

Ogilvie, R., Wilkinson, R., & Allison, S. (1989). The detection of sleep onset: Behavioral, physiological, and subjective convergence. *Sleep, 21*, 458–474.

Palmer, S. E. (1999). *Vision science: Photons to phenomenology.* Cambridge, MA: MIT Press.

Piantadosi, S. (1997). *Clinical trials: A methodologic perspective.* New York: John Wiley.

Plumert, J., Kearney, J., & Cremer, J. (2004). Children's perception of gap affordances: Bicycling across traffic-filled intersections in an immersive virtual environment. *Child Development, 75*, 1243–1253.

Plumert, J., Kearney, J., & Cremer, J. (2005). Distance perception in real and virtual environments. *ACM Transactions on Applied Perception, 2*(3), 1–18.

Rainville, P., Bechara, A., Naqvi, N., & Damasio, A. R. (2006). Basic emotions are associated with distinct patterns of cardiorespiratory activity. *International Journal of Psychophysiology.*

Rheingold, H. (1991). *Virtual reality.* New York: Simon and Schuster.

Risser, M., Ware, J., & Freeman, F. (2000). Driving simulation with EEG monitoring in normal and obstructive sleep apnea patients. *Sleep, 23*, 393–398.

Rizzo, M. (2004). Safe and unsafe driving. In M. Rizzo & P. Eslinger (Eds.), *Principles and practice of behavioral neurology and neuropsychology* (pp. 197–222). Philadelphia, PA: W.B. Saunders.

Rizzo, M., & Barton, J. J. (1998). *Central disorders of visual function.* In N. Miller & N. Newman (Eds.), *Walsh and Hoyt's clinical neuro-ophthalmology* (Vol. 1, 5th ed., pp. 387–482). Baltimore: Williams and Wilkins.

Rizzo, M., McGehee, D., Dawson, J., & Anderson, S. (2001). Simulated car crashes at intersections in drivers with Alzheimer's disease. *Alzheimer Disease and Associated Disorders, 15,* 10–20.

Rizzo, M., Severson, J., Cremer, J., & Price, K. (2003). An abstract virtual environment tool to assess decision-making in impaired drivers. *Proceedings 2nd International Driving Symposium on Human Factors in Driver Assessment, Training and Vehicle Design,* Park City, UT, 40–47.

Rizzo, M., Sheffield, R., Stierman, L., & Dawson, J. (2003). *Demographic and driving performance factors in simulator adaptation syndrome.* Paper presented at the Second International Driving Symposium on Human Factors in Driver Assessment, Training and Vehicle Design, Park City, Utah, July 21–24, pp. 201–208.

Rothbaum, B., Hodges, L., Kooper, R., Opdyke, D., Williford, J., & North, M. (1995). Effectiveness of computer-generated (virtual reality) graded exposure in the treatment of acrophobia. *American Journal of Psychiatry, 152,* 626–628.

Rothbaum, B., & Schwartz, A. (2002). Exposure therapy for posttraumatic stress disorder. *American Journal of Psychotherapy, 56,* 59–75.

Sanchez-Vives, M., & Slater, M. (2005, April). From presence to consciousness through virtual reality. *Nature Reviews Neuroscience, 6*(4), 332–339.

Simeonov, P., Hsiao, H., Dotson, B., & Ammons, D. (2005). Height effects in real and virtual environments. *Human Factors, 47*(2), 430–438.

Simons, D., & Chabris, C. (1999). Gorillas in our midst: Sustained inattentional blindness for dynamic events. *Perception, 28,* 1059–1074.

Spohrer, J. (1999). Information in places. *IBM Systems Journal, 38*(4), 602–628.

St. Julien, T., & Shaw, C. (2003). *Firefighter command training virtual environment.* Proceedings of the 2003 Conference on Diversity in Computing TAPIA '02, ACM Press, pp. 30–33.

Sutherland, I. (1965). The ultimate display. In *Proceedings of IFIP 65, 2,* 506–508.

Thacker, P. (2003). Fake worlds offer real medicine. *Journal of the American Medical Association, 290,* 2107–2112.

Theunissen, E. R. (1997). *Integrated design of a man-machine interface for 4-D navigation.* Delft, Netherlands: Delft University Press.

Thompson, W., Willemsen, P., Gooch, A., Creem-Regehr, S., Loomis, J., & Beall, A. (2004). Does the quality of computer graphics matter when judging distance in visually immersive environments. *Presence: Teleoperators and Virtual Environments, 13,* 560–571.

Tsirliganis, N., Pavlidis, G., Tsompanopoulos, A., Papadopoulou, D., Loukou, Z., Politou, E., et al. (2001, November). *Integrated documentation of cultural heritage through 3D imaging and multimedia database, VAST 2001.* Paper presented at Virtual Reality, Archaeology, and Cultural Heritage, Glyfada, Athens, Greece.

Turing, A. M. (1950). Computing machinery and intelligence. *Mind, 49,* 433–460.

Wanger, L., Ferwerda, J., & Greenberg, D. (1992). Perceiving spatial relationships in computer-generated images. *IEEE Computer Graphics and Applications, 12*(3), 44–51, 54–58.

Wiederhold, B. K., & Wiederhold, M. D. (2004). *Virtual-reality therapy for anxiety disorders: Advances in education and treatment.* New York: American Psychological Association Press.

Wierwille, W., Fayhey, S., Fairbanks, R., & Kirn, C. (1995). *Research on vehicle-based driver status/performance monitoring development* (Seventh Semiannual Report No. DOT HS 808 299). Washington, DC: National Highway Traffic Safety Administration.

Willemsen, P., & Gooch, A. (2002). *Perceived egocentric distances in real, image-based and traditional virtual environments.* Proceedings of the IEEE Virtual Reality Conference (VR'02), IEEE Computer Society, pp. 89–90.

Witmer, B., & Sadowski, W. (1998). Nonvisually guided locomotion to a previously viewed target in real and virtual environments. *Human Factors, 40,* 478–488.

Zyda, M., & Sheehan, J. (Eds.). (1997, September). *Modeling and simulation: Linking entertainment and defense.* Washington, DC: National Academy Press.

Cynthia Breazeal and Rosalind Picard

The Role of Emotion-Inspired Abilities in Relational Robots

This chapter presents our motivation and a snapshot of our work to date in building robots with social-emotional skills, inspired by findings in neuroscience, cognitive science, and human behavior. Our primary motivation for building robots with social-emotional-inspired capabilities is to develop "relational robots" and their associated applications in diverse areas such as health, education, or work productivity where the human user derives performance benefit from establishing a kind of social rapport with the robot. We describe some of the future applications for such robots, provide a brief summary of the current capabilities of state-of-the-art socially interactive robots, present recent findings in human-computer interaction, and conclude with a few challenges that we would like to see addressed in future research. Much of what we describe in this chapter cannot be done by most machines today, but we are working to bring about research breakthroughs that will make such things possible.

To many people, the idea of an emotional robot is nonsensical if not impossible. How could a machine ever have emotions? And why should it? After all, robots today are largely automated machines that people use to perform useful work in various settings (see also chapter 16, this volume). Manufacturing robots, for instance, tirelessly perform simple repetitive tasks. They are large and potentially dangerous, and therefore are located safely apart from people. Autonomous and semiautonomous mobile robots have had successes as probes for exploring remote and hazardous environments, such as planetary rovers, oil well inspection robots, or urban search and rescue robots. These successes have all come without explicitly building emotion into any of these machines. One might even argue that the fact that these machines lack emotion is a benefit. For example, a state like boredom could interfere with the ability to perform the simple repetitive tasks that are, for many machines, their raison d'etre. And who really needs an emotional toaster?

The above views, while quite reasonable, reveal only part of the story. A new understanding of emotion from neuroscience discoveries has led to significant rethinking of the role of emotion in intelligent processes. While the popular view of emotion is still generally associated with undesirable episodes of "being emotional," implying a state of imbalance and reduced rationality in thinking, the new scientific view, rooted in growing evidence, is a more balanced one: Emotion facilitates flexible, rational, useful processing, especially when an entity faces complex unpredictable inputs and has to respond with limited resources in an intelligent way (e.g., Bechara, Damasio, Tranel, & Damasio, 1997; Damasio, 1994; Le Doux, 1996; Panksepp, 1998; see also chapter 12, this volume). Not only is too much emotion detrimental, but so too is too little emotion. Emotion mechanisms are now seen as regulatory, biasing cognition, perception, decision making, memory, and action in useful ways that are adaptive and lead to intelligent real-time behavior. Mechanisms of emotion, when they are functioning appropriately, do not make people appear emotional in the negative sense of the word.

Intelligence in Uncertain Environments

With this new understanding of the valuable role of emotion in people and animals comes a chance to reexamine the utility it might provide in machines. If emotion is as integral to intelligent real-time functioning as the evidence suggests, then more intelligent machines are likely to need more sophisticated emotion-like mechanisms, even if we never want the machines to appear emotional. Much like an animal, an autonomous mobile robot must apply its limited resources to address multiple concerns (performing tasks, self-preservation, etc.) while faced with complex, unpredictable, and often dangerous situations. For instance, balancing emotion-inspired mechanisms associated with interest and fear could produce a focused yet safe searching behavior for a routine surveillance robot. For this application, one could take inspiration from the classic example of Lorenz regarding the exploratory behavior of a raven when investigating an object on the ground starting from a perch high up in a tree. For the robot, just as for the raven, interest encourages exploration and sustains focus on the target, while recurring low levels of fear motivate it to retreat to safe distances, thereby keeping its exploration within safe bounds. Thus, an analogous exploratory pattern for a surveillance robot would consist of several iterative passes toward the target: On each pass, move closer to investigate the object in question and return to a launching point that is successively closer to the target.

In fact, mechanisms of emotion, in the form of regulatory signals and biasing functions, are already present in some low-level ways in many machine architectures today (e.g., emergency interrupt cycles, which essentially hijack a machine's routine processing, function analogously to the fear system in an animal). Moreover, many of the problems today's unintelligent machines have difficulty with bear some similarity to problems faced by people with impaired emotional systems (Picard, 1997). There is a growing number of reasons why scientists might wish to consider emotion-like mechanisms in the design of future intelligent machines.

Social Interaction with Humans

There is another domain where a need for social-emotional skills in machines is becoming of immediate significance, and where machines will probably need not only internal emotion-like regulatory mechanisms, but also the ability to sense and respond appropriately to your emotions. We are beginning to witness a new breed of autonomous robots called *personal service robots*. Rather than operating far from people in hazardous environments, these robots function in the human environment. In particular, consumer robots (e.g., robot toys and autonomous vacuum cleaners) are already successful products. Reports by UNEC and IFR (2002) predict that the personal robot market—robots designed to assist, protect, educate, or entertain in the home—is on the verge of dramatic growth. In particular, it is anticipated that the needs of a growing aging population will generate a substantial demand for personal robots that can assist them in the home. Such a robot should be persuasive in ways that are sensitive to people, such as reminding them when to take medication, without being annoying or upsetting. It should understand what the person's changing needs are and the urgency for satisfying them so that it can set appropriate priorities. It needs to understand when the person is distressed or in trouble so that it can get help if needed. Furthermore, people should enjoy having the robot in their life because it is useful and pleasant to have around. The issue of enduring consumer appeal, long after the novelty wears off, is important not only for the success of robotic products, but also for the well-being of the people for whom they will serve as companions.

Yet the majority of robots today treat people either as other objects in the environment to be navigated around or at best in a manner characteristic of socially impaired people. Robots do not understand or interact with you any differently than if you were a television. They are not aware of your goals and intentions. As a result, they do not know how to appropriately adjust their behavior to help you as your goals and needs change. They do not flexibly draw their attention to what you currently find of interest in order to coordinate their behavior with yours. They do not realize that perceiving a given situation from different visual perspectives impacts what you know and believe to be true about it. Consequently, they do not bring important information to your attention that is not easily accessible to you when you need it. They are not aware of your emotions, feelings, or attitudes. As a result, they cannot prioritize what is the most important to do for you according to what pleases you or to what you find to be most urgent, relevant, or significant. Although there are initial strides in giving robots these social-emotive abilities (see Breazeal, 2002; Fong, Nourbakshsh, & Dautenhahn, 2002; Picard, 1997), there remains quite a lot of work to be done before robots are socially and emotionally intelligent entities.

Our interaction with technology evokes strong emotional states within us, unfortunately often negative. Many people have experienced the frustration of a computer program crashing in the middle of work, trying to customize our electronic gadgets through confusing menus, or waiting through countless phone options as we try to book an airline flight. Recent human-computer interaction studies have found that the ways in which our interaction with technology influences our cognitive and affective states can either help or hinder our performance. For instance, Nass et al. (2005) found that during a simulated driving task, it is very important to have the affective quality of the car's voice match the driver's own affective state (e.g., a subdued car voice when the driver is in a negative affective state, and a cheerful car voice when in a positive state). When these conditions were crossed, the drivers had twice as many accidents. One likely explanation is that the mismatched condition required the driver to use more cognitive and attentive effort to interpret what the car was saying, distracting the driver from the changing road conditions. This study also points to the importance of enabling a computational system to sense and respond to a person's affect in real time: Adaptation to a person's emotion has a direct, measurable impact on safety. It is important to understand how to design our interaction with technology in ways that help us rather than hinder us. Whereas past effort in human-computer interaction has focused on the cognitive aspects, it is growing increasingly clear that it is also important to understand the affective aspects and how they influence cognitive processing.

A New Scientific Tool

While the potential applications (and consumer markets) for robotic companions demand substantial progress in this research area, it is also true that this applied research will advance basic scientific theories and understanding. The challenge in neuroergonomics, to bring together new findings about the brain and cognition with new understanding of human behavior and human factors, requires advances in tools. New tools are needed to sense emotion and behavior in natural environments and to provide controlled stimuli and responses in repeatable, measurable ways. While a natural interaction between two people is rarely, if ever, repeatable and controllable, and even scripted ones have a lot of variation (witness the differences in performance from evening to evening of two professional actors on stage in any long-running scripted production), the interactions made by a robot can be designed to be repeatable and measurable, quite precisely.

The robot can be designed to sense and measure a variety of channels of information from you (e.g., your eye and body movements associated with level and direction of attention) and can be built to respond in preprogrammed, precise ways (such as mirroring your postures). For example, La France's (1982) theory that mirroring of body postures between people builds rapport and liking is based on observations of people by video. On the other hand, robots could be designed to respond in very precise and controlled ways to specific human states or behaviors, thus adding repeatability and a new measure of control in interaction studies. Hence, robots can be viewed not only as future companions for a variety of useful tasks, but also as research platforms for furthering basic scientific understanding of how emotion, cognition, and behavior interact in natural and social situations. Our work involves answering a myriad of basic science questions, as well as long-term goals of building companions that people would look forward to having around.

Design of a Relational Robot for Education

RoCo is our most recent effort in developing a robotic desktop computer with elements of social and emotional intelligence. Our primary motivation for building a physically animated system is the development of applications that benefit from establishing a kind of social rapport between human and computer. RoCo is an actuated version of a desktop computer where its monitor has the ability to physically move in subtly expressive ways that respond to its user (see figure 18.1). The movement of the machine is inspired by natural human-human interaction: When people work together, they move in a variety of reciprocal ways, such as shifting posture at conversational boundaries and leaning forward when interested (Argyle, 1988). Physical movement plays an important role in aiding communication, building rapport, and even in facilitating the health of the human body, which was not designed to sit motionless for long periods of time. We have developed a suite of perceptual technologies that sense and interpret multimodal affective cues from the user via custom sensors (including facial and postural expression sensing) using machine learning algorithms that have been designed to recognize patterns of human behavior from multiple modes. RoCo is designed to respond to the user's cues with carefully crafted subtle mechanical movements and occasional auditory feedback, using principles derived from natural human-human interaction.

The novelty of introducing ergonomic physical movement to human-computer interaction enables us to explore several interesting questions in the interaction of the human body with cognition and affect, and how this coupling could be applied to foster back health as well as task performance and learning gains of the user. For instance, one important open question we are exploring is whether the reciprocal movement of human and computer can be designed to promote back health, without being distracting or annoying to the user. Another intriguing question is whether reciprocal physical movement of human and computer, and its interaction with affect and cognition, could be designed to improve the human's efficacy of computer use. We believe this is possible in light of new theories that link physical posture and its influence on cognition and affect. An example is the theory in

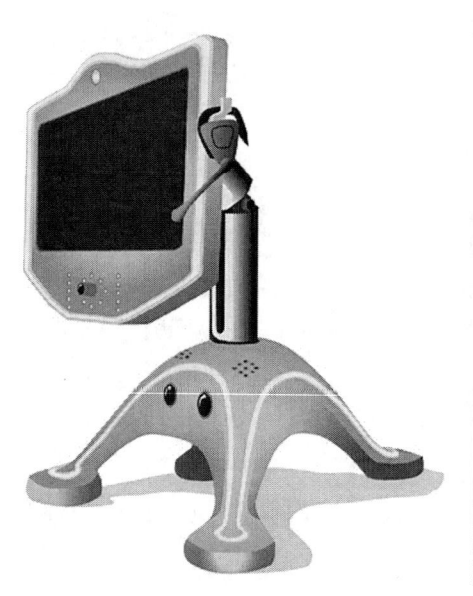

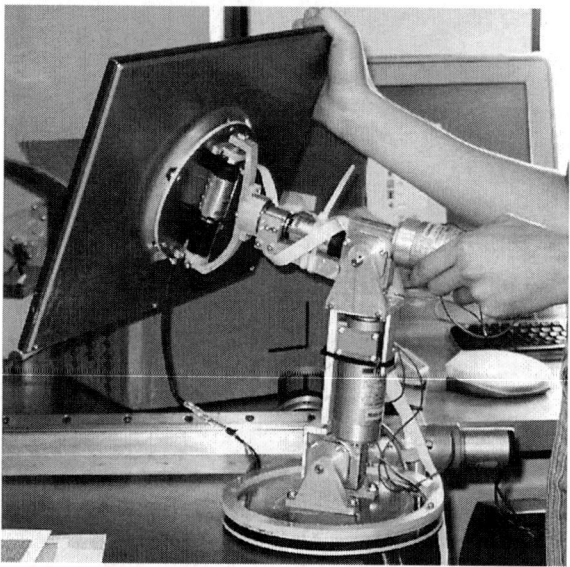

Figure 18.1. Concept drawing of the physically animated computer (left), and a mechanical prototype (right). RoCo can move its monitor "head" and its mechanical "neck" using 5 degrees of freedom. There are two axes of rotation at the head, an elbow joint, and a swivel and a lean degree of freedom at the base. It does not have explicit facial features, but this does not preclude the ability to display these graphically on the LCD screen.

the "stoop to conquer" research, where it was found that slumping following a failure in problem solving, and sitting up proudly following an achievement, led to significantly better performance outcomes than crossing those conditions (Riskind, 1984).

Other theories, such as the role of subtle postural mirroring for building rapport and liking between humans (La France, 1982), could be examined in human-computer interaction by using the proposed system to induce and measure subtle postural responses. In particular, nonverbal immediacy behaviors alone when displayed by a teacher (such as close conversational distance, direct body orientation, forward lean, and postural openness) have been shown to increase learning outcomes (Christensen & Menzel, 1998). Growing interest in the use of computers as pedagogical aids raises the question of the role of these postural immediacy behaviors in influencing the effectiveness of the learning experience provided by computers. It also motivates development of systems that can recognize and respond to the affective states of a human learner (e.g., interest, boredom, frustration, pleasure, etc.) to keep the learner engaged and to help motivate the learner to persevere through challenges in order to ultimately experience success (Aist, Kort, Reilly, Mostow, & Picard, 2002).

RoCo as a Robotic Learning Companion

Inspired by these scientific findings, we have been developing theory and tools that can be applied to RoCo so that it may serve as a robotic learning companion (Picard et al., 2004) that can help a child to persist and stay focused on a learning task, and mirrors some of the child's affective states to increase awareness of the role that these states play in propelling the learning experience. For example, if the child's face and posture show signs of intense interest in what is on the screen, this computer should hold very still so as to not distract the child. If the child shifts her posture and glances in a way that shows she is taking a break, the computer could do the same and may note that moment as a good time to interrupt the child and provide scaffolding (encouragement, tips, etc.) to help the learning progress. In doing so, the system would not only acknowledge the presence of the child and show respect for her level of attentiveness, but also would show subtle expressions that, in human-human interaction, are believed to help build rapport and liking (La France, 1982).

By increasing likeability, we aim to make the robotic computer more enjoyable to work with and potentially facilitate measurable task outcomes, such as how long the child perseveres with the learning task. Within K-6 education, there is evidence that relationships between students are important in peer learning situations, including peer tutoring and peer collaborative learning methodologies (Damon & Phelps, 1989). Collaborations between friends involved in these exercises have been shown to provide a more effective learning experience than collaboration between acquaintances (Hartup, 1996). Friends have been shown to engage in more extensive discourse with one another during problem solving, to offer suggestions more readily, to be more supportive and more critical, and to work longer on tasks and remember more about them afterward than nonfriends.

Building Social Rapport

In our experience, most computer agents in existence are, at best, charming for a short interaction, but rapidly grow tiring or annoying with longer-term use. We believe that in order to build a robot or computer that people will enjoy interacting with over time, it must be able to establish and maintain a good social rapport with the user. This requires the robot to be able to accumulate memory of ongoing interactions with the user and exhibit basic social-emotional skills. By doing so, it can respond in ways that appear socially intelligent given the user's affective state and expressions. Major progress has already been made within our groups in defining, designing, and testing relational agents, computer and robotic agents capable of building long-term social-emotional relationships with people (Bickmore, 2003; Breazeal, 2002).

We believe that relational and immediacy behaviors (see Richmond & McCroskey, 1995)—especially close proximity, direct orientation, animation, and postural mirroring to demonstrate liking of the user and engagement in the interaction—should be expressible by a computer learning companion to help build rapport with the child. In support of this view, immediacy and relational behaviors have been implemented in a virtual exercise advisor application and shown to successfully build social rapport with the user (Bickmore, 2003). In a

99-person 1-month test of the exercise advisor agent, where subjects were split into three groups (one third with task but no agent, one third with task plus nonrelational agent, one third with task plus relational agent), we found task outcome always improved, while a "bond" rating toward the agent was significantly higher ($p < .05$) in the relational case. This includes people reporting that the agent cared more about them, was more likeable, showed more respect, and earned more of their trust than the nonrelational agent. People interacting with the relational agent were also significantly more likely to want to continue interacting with that agent. Most of these measures were part of a standard instrument from clinical psychotherapy—the Working Alliance Inventory—that measures the trust and belief that the therapist and patient have in each other as team members in achieving a desired outcome (Horvath & Greenberg, 1989). The significant difference in people's reports held at both times of evaluation: after 1 week and after 4 weeks, showing that the improvement was sustained over time. These findings confirm an important role for research in giving agents social-emotional and other relational skills for successful long-term interactions.

Social Presence

Why should the computer screen move, rather than simply moving a graphical character on a static screen? Does physical movement, especially drawing nearer or pulling further from the user, impact the social presence of a character? To answer this question, we carried out a series of human-robot interaction studies to explore how the medium through which an interaction takes place affects a person's perception of social presence of a character (Kidd & Breazeal, 2004). In these studies, the robot had static facial features with movable eyes mounted upon a neck mechanism (and therefore had degrees of freedom comparable to those proposed for our physically animate computer). The study involved naive subjects ($n = 32$) interacting with a physical robot, an on-screen animated character, and a person in a simple task. Each subject interacted with each of the characters, one at a time, in a preassigned order of presentation. The character made requests of the subject to move simple physical objects (three colored blocks). All requests were presented in a prerecorded female voice to minimize the effects of different voices, and each character made these requests in a different order. At the conclusion of the interaction, subjects were asked to complete a questionnaire on their experiences based on the Lombard and Ditton scale for measuring social presence (Lombard et al., 2000). Subjects were asked to read and evaluate a series of statements and questions about engagement with the character on a seven-point scale. All data were evaluated using a single-factor ANOVA and paired two-sample t tests for comparisons between the robot and animated character. The data presented were found to be statistically significant to $p < .05$.

Our data show that the robot consistently scored higher on measures of social presence than the animated character (and both below that of the human). Overall, people found the robot character to be easier to read, more engaging of their senses and emotions, and more interested in them than the animated character. Subjects also rated the robot as more convincing, compelling, and entertaining than the animated character. These findings suggest that in situations where a high level of motivational and attentional arousal is desired, a physically copresent and animated computer may be a preferred medium for the task over an animated character trapped within a screen. We are continuing this work by looking at more specific measures of the interaction (e.g., trust, reliability, and immediacy) when subjects interact with characters in different types of tasks, such as a cooperative task or a learning task.

Sensing and Responding to Human Affect

In this section, we briefly present three major components of our robotic computer system:

- Expressive behavior of the robotic computer
- Perceptual systems for passively sensing subtle movement and expression by the user
- The cognitive-affective control system

Expressive Behavior

Character animators appreciate the importance of body posture and movement (i.e., the principle axes of movement) to convincingly portray life and convey expression in inanimate objects (Thomas & Johnston, 1981). Hence, the primary use of the LCD screen for our purposes is to display task-relevant information, where physical movement is used for emotive and social expression. For instance, the mechanical expressions include postural shifts like moving closer to the user, and looking around in a curious sort of way. However, it is possible to graphically render facial features on the screen if it makes sense to do so.

We have developed expressive anthropomorphic robots in the past that convey emotive states through facial expression and posture (Breazeal, 2000). We have found that the scientific basis for how emotion correlates to facial expression or vocal expression is very useful in mapping the robot's emotive states to its face actuators (Breazeal, 2003a), and to its articulatory-based speech synthesizer (Breazeal, 2003b). In human-robot interaction studies (Breazeal, 2003d), we have found that these expressive cues are effective in regulating affective or intersubjective interactions (Trevarthen, 1979) and proto-dialogs (Tronick, Als, & Adamson, 1979) between the human and the robot that resemble their natural correlates during infant-caregiver exchanges. The same techniques are applied to give RoCo its ability to express itself through posture.

With respect to communicating emotion through the face, psychologists of the componential theory of facial expression posit that these expressions have a systematic, coherent, and meaningful structure that can be mapped to affective dimensions that span the relationship between different emotions (Smith & Scott, 1997). Some of the individual features of expression have inherent signal value. The raised brows, for instance, convey attentional activity for the expression of both fear and surprise. For RoCo, this same state can be communicated by an erect body posture. By considering the individual facial action components that contribute to the overall facial display, it is possible to infer much about the underlying properties of the emotion being expressed. This promotes a signaling system that is robust, flexible, and resilient. It allows for the mixing of these components to convey a wide range of affective messages, instead of being restricted to a fixed pattern for each emotion.

Inspired by this theory, RoCo's facial expressions and body postures can be generated using an interpolation-based technique over a three-dimensional affect space that we have used with past robots. The three dimensions correspond to arousal (high/low), valence (good/bad), and stance (advance/withdraw)—the same three attributes that our robots have used to affectively assess the myriad of environmental and internal factors that contribute to their overall affective state. There are nine basic postures that collectively span this space of emotive expressions.

The current affective state of the robot (as defined by the net values of arousal, valence, and stance) occupies a single point in this space at a time. As the robot's affective state changes, this point moves around this space and the robot's facial expression and body posture change to mirror this. As positive valence increases, the robot may move with a more buoyant quality. If facial expressions are shown on the screen, than lips would curve upward. However, as valence decreases, the posture would droop, conveying a heavy, sagging quality. If eyebrows are shown on the screen, they would furrow. Along the arousal dimension, the robot moves quicker, with a more dartlike quality (positive), or with greater lethargy as arousal decreases. Along the stance dimension, the robot leans toward (increasing) or shrinking away (decreasing) from the user. These expressed movements become more intense as the affective state moves to more extreme values in the affect space.

The computer can also express itself through auditory channels. The auditory expressions, designed to be similar in spirit to the fictional *Star Wars* robot R2D2, are nonlinguistic but aim to complement the movements, such as electronic sounds of surprise. This does not preclude the use of speech synthesis if the task demands it, but our initial emphasis is on using modes of communication that do not explicitly evoke high-level human capabilities.

Preliminary studies have been carried out to evaluate the readability of the physically animated computer's subtle expressions (Liu & Picard, 2003). In this preliminary study, 19 subjects watched 15 different video clips of an animated version of the computer or heard audio sequences to convey certain expressive behaviors (e.g., welcoming, sorrow, curious, confused, and surprised). The expressive behaviors were conveyed through body movement only—the LCD screen was blank. The subjects were asked to rate the strength of each of these expressions on a seven-point scale for video only, audio only, or video with audio. In a two-tailed *t* test, there was significant recognition of the behavior sequences designed to express curiosity, sadness, and surprise. These initial studies are encouraging, and we are continuing to refine expressive movements and sounds.

Perceptual Systems

We are interested in sensing those affective and attentive states that play an important role in extended learning tasks. Detecting affective states such as interest, boredom, confusion, and excitement are important. Our goal is to sense these emotional and cognitive aspects in an unobtrusive way. Cues like the learner's posture, gesture, eye gaze, facial expression, and so on help expert teachers recognize whether a learner is on-task or off-task. For instance, Rich et al. (1994) have defined symbolic postures that convey a specific meaning about the actions of a user sitting in an office, such as interested, bored, thinking, seated, relaxed, defensive, and confident. Leaning forward toward a computer screen might be a sign of attention (on-task), while slumping on the chair or fidgeting suggests frustration or boredom (off-task).

We have developed a postural sensing system with custom pattern recognition that watches the posture of children engaged in computer learning tasks and learns associations between their postural movements and their level of interest—high, low, or "taking a break" (a state of shifting forward and backward, sometimes with hands stretched above the head, which tended to occur frequently before teachers labeled a child as bored). The system attained 82% recognition accuracy training and testing on different episodes within a group of eight children, and performed at 77% accuracy recognizing these states in two children that the system had not seen before (Mota, 2002; Mota & Picard, 2003).

A computer might thus use postural cues to decide whether it is a good time to encourage the user to take a break and stretch or move around. Specifically, we have focused on identifying the surface-level behaviors (indicative of both attention and affect) that suggest a transition from an on-goal to off-goal state or vice versa. Toward this aim, we are further developing and integrating multimodal perceptual systems that allow RoCo to extract this information about the user. The systems under development are described in the following sections.

The Sensor Chair

The user sits in a sensor chair that provides posture data from an array of force-sensitive resistors—similar to the Smart Chair used by Tan, Ifung, and Pentland (1997). It consists of two 0.10 mm thick sensor sheets, with an array of 42 × 48 sensing units. Each unit outputs an 8-bit pressure reading. One of the sheets is placed on the backrest and one on the seat (see figure 18.2). The pressure distribution map (2 of 42 × 48

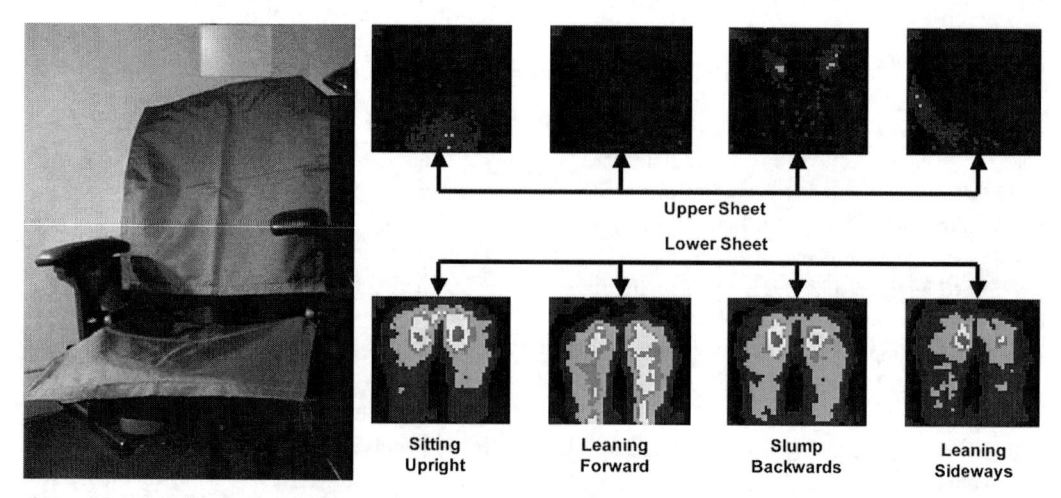

Figure 18.2. The Smart Chair (left) and data (right) showing characteristic pressure patterns used to recognize postural shifts characteristic of high interest, low interest, and taking a break. See also color insert.

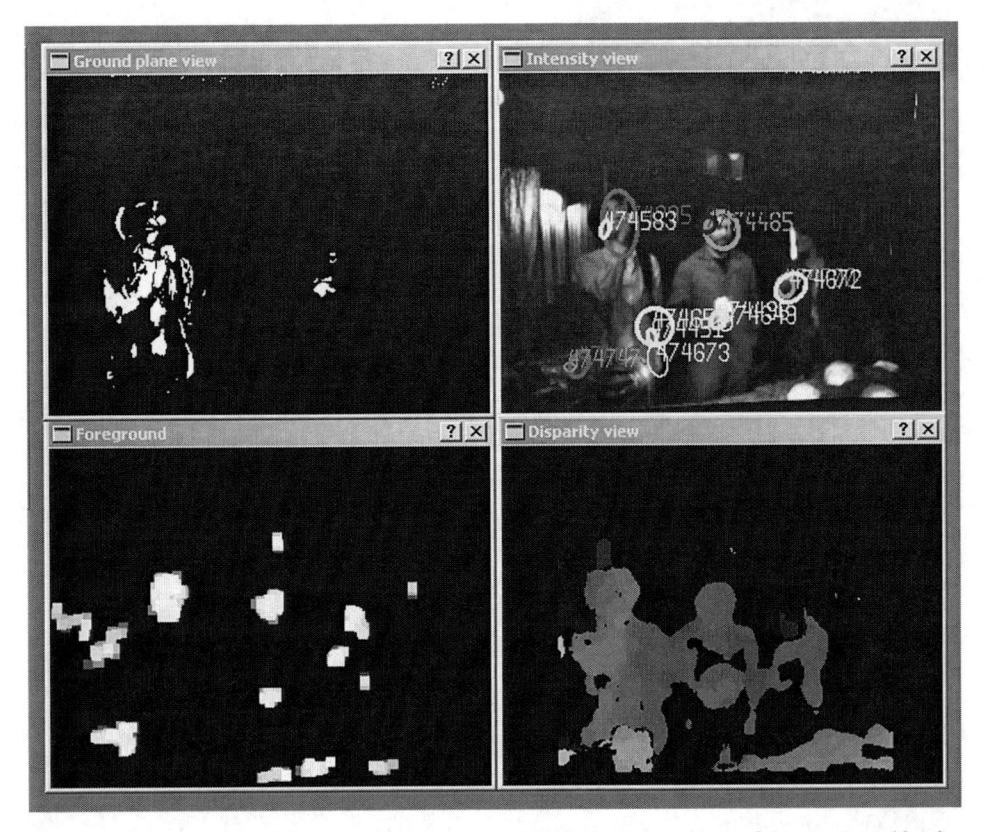

Figure 18.3. A snapshot of the stereo vision system that is mounted in the base of the computer (developed in collaboration with the Vision Interfaces Group at MIT CSAIL). Motion is detected in the upper left frame; human skin chromaticity is extracted in the lower left frame; a foreground depth map is computed in the lower right frame; and the faces and hands of audience participants are tracked in the upper right frame. See also color insert.

points) sensed at a sampling frequency of 50 Hz. A custom pattern recognition system was developed to distinguish a set of 9 static postures that occur frequently during computer learning tasks (e.g., lean forward, lean back, etc.), and to analyze patterns of these postures over time in order to distinguish affective states of high interest, low interest, and taking a break. These postures and associated affective states were recognized with significantly greater than random accuracy, indicating that the postural pressure cues carry significant information related to the student's interest level (Mota & Picard, 2003).

Color Stereo Vision

A color Mega-D stereo vision system (manufactured by Small Vision Systems, Inc.) is mounted inside the computer's base. This is a compact, integrated megapixel system that uses advanced CMOS imagers from PixelCam and fast 1394 bus interface electronics from Vitana to deliver high-resolution, real-time stereo imagery to any PC equipped with a 1394 interface card. We use it to detect and track the movement and orientation of the user's hands and face (Breazeal et al., 2003; Darrell, Gordon, & Woodfill, 1998). This vision system segments people from the background and tracks the face and hands of the user who interacts with it. We have implemented relatively cheap algorithms for performing certain kinds of model-free visual feature extraction. A stereo correlation engine compares the two images for stereo correspondence, computing a 3-D depth (i.e., disparity map) at about 15 frames per second. This is compared with a background depth estimate to produce a foreground depth map. The color images are simultaneously normalized and analyzed with a probabilistic model of human skin chromaticity to segment out areas of likely correspondence to human flesh. The foreground depth map and the skin probability map are then filtered and combined, and positive regions extracted. An optimal bounding ellipse is computed for each region. For the camera behind the machine facing the user, a Viola-Jones face detector (Viola & Jones, 2001) runs on each to determine whether or not the region corresponds to a face. The regions are then tracked over time, based on their position, size, orientation, and velocity. Connected components are examined to match hands and faces to a single owner (see figure 18.3).

Blue Eyes Camera

An IBM Blue Eyes camera system is mounted on the lower edge of the LCD screen (http://www.almaden.ibm.com/cs/blueeyes), as shown in figure 18.4. It uses a combination of off-axis and on-axis infrared LEDs and an infrared camera to track the user's pupils unobtrusively by producing the red eye effect (Haro, Essa, & Flickner, 2000). We have developed real-time techniques using the pupil tracking system to automatically detect the users facial features (e.g., eyes, brows, etc.), gaze direction, and head gestures such shaking or nodding (Kapoor & Picard, 2002; Kapoor, Qi, & Picard, 2003). Also physiological parameters like pupillary dilation, eye-blink rate, and so on can be extracted to infer information about arousal and cognitive load. The direction of eye gaze is an important signal to assess the focus of attention of the learner. In an on-task state, the focus of attention is mainly toward the problem the student is working on, whereas in an off-task state the eye gaze might wander off from it.

Mouse Pressure Sensor

We have modified a computer mouse to sense not only the usual information (where the mouse is placed and when it is clicked) but also how it is clicked, the adverbs of its use (see figure 18.5). It has been observed that users apply significantly more pressure to the mouse when a task is frustrating to them than when it is not (Dennerlein, Becker, Johnson, Reynolds, & Picard, 2003). Use of this sensor is new and much is still to be learned about how its signals change during normal use. We propose to replace the traditional computer mouse in our setup with this pressure mouse, and combine its data with that from the other sensors, to assess if it is also helpful in learning about the user state.

In addition to the physical synchronization of data from the sensors described above, we have also developed algorithms that integrate these multiple channels of data for making joint inference about the human state. Our approach refines recent techniques in machine learning based on a Bayesian combination of experts. These techniques use statistical machine learning to learn multiple classifiers and to learn how they tend to perform given certain observations. This information is then combined to learn how to best combine their outputs to make a joint decision. The techniques we have been developing here generalize those of Miller and Yan (1999) and so far appear to perform better than classifier combination methods such as the product rule, sum rule, vote, max, min, and so forth (Kapoor et al., 2003). We are continuing to improve upon these methods using better approximation techniques for the critics and their performance, as well as integrating new techniques that combine learning from both labeled and unlabeled data (semisupervised, with the human in the loop).

Cognitive-Affective Control System

Robotics researchers continue to be inspired by scientific findings that reveal the reciprocally interrelated roles that cognition and emotion play in intelligent decision making, planning, learning, attention, communication, social interaction, memory, and more (Lewis & Haviland-Jones, 2000). Several early works in developing computational models of emotions for robots were inspired by theories of basic emotions and the role they play in promoting a creature's survival in unpredictable environments given limited resources (Ekman, 1992; Izard, 1977). Later works explored different models (such as cognitive appraisal models) and their theorized role in intelligent decision making, while other works have explored the role of affect in reinforcement learning (see Picard, 1997, for a review).

Our own work has explored various social-emotional factors that benefit a robot's ability to achieve its goals (Breazeal, 2002) and to learn new skills (Lockerd & Breazeal, 2004) in partnership with people. For instance, the design of our first

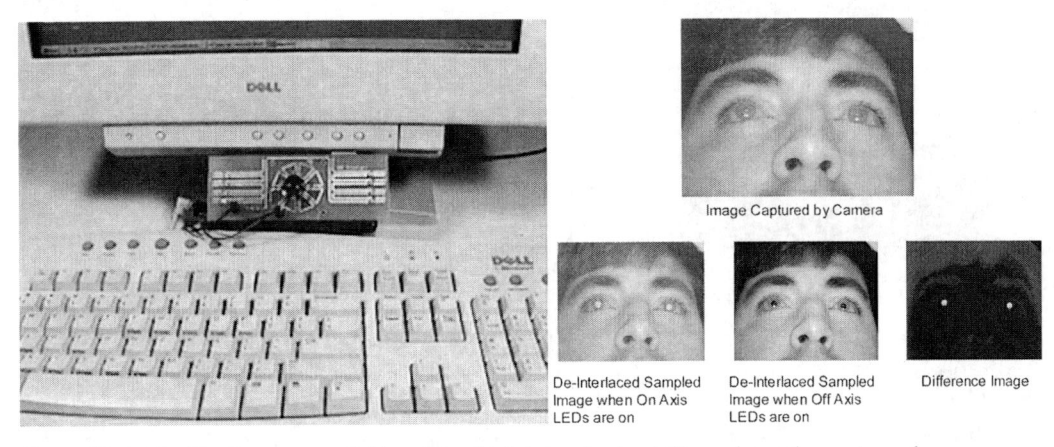

Figure 18.4. The Blue Eyes camera (left) and data (right) showing its ability to detect the user's pupils.

socially interactive robot, called Kismet, was inspired by mother-infant communicative interactions. Any parent can attest to the power that infants' social-emotional responses have in luring adults to help satisfy their infant's needs and goals. Infants socially shape their parents' behavior by exhibiting a repertoire of appropriate negative responses to undesirable or inappropriate interactions (e.g., fear, disgust, boredom, frustration) and a repertoire of positive responses (e.g., happiness, interest) to those that help satisfy their goals. Similarly, Kismet's emotive responses were designed and have been demonstrated in human-robot interaction studies to do the same (Breazeal, 2002; Turkle, Breazeal, Detast, & Scassellati, 2004).

Our cognitive-affective architecture features both a cognitive system and an emotion system (see figure 18.6). It is being constructed using the C5 behavior architecture system developed at the MIT Media Lab (Burke, Isla, Downie, Ivanov, & Blumberg, 2001) to create autonomous, animated, interactive characters that interact with each other as well as the user. Whereas the cognitive system is responsible for interpreting and making sense of the world, the emotion system is responsible for evaluating and judging events to assess their overall value with respect to the creature, for example, positive or negative, desirable or undesirable, and so on. These parallel systems work in concert to address the competing goals and motives of the robot given its limited resources and the ever-changing demands of interacting with people.

The cognitive system (the light-gray modules shown in figure 18.6) is responsible for perceiving and interpreting events, and for arbitrating among the robot's goal-achieving behaviors to address competing motivations. To achieve this, we based the design of the cognitive system on classic ethological models of animal behavior (heavily inspired by those models proposed by Tinbergen [1951] and Lorenz [1950]). The computational subsystems and mechanisms that comprise the cognitive system (i.e., perceptual releasers, visual attention, homeostatic drives, goal-oriented behavior and arbitration, and motor acts) work in concert to decide which behavior to activate, at what time, and for how long to service the appropriate objective. Overall, the robot's behavior must exhibit an appropriate degree of relevance, persistence, flexibility, and robustness.

The robot's emotion system implements the style and personality of the robot, encoding and conveying its attitudes and behavioral inclinations toward the events it encounters. Furthermore, these cognitive mechanisms are enhanced by emotion-inspired mechanisms (the white modules shown in figure 18.6), informed by basic emotion theory, that further improve the robot's communicative effectiveness via emotively expressive behavior, its ability to focus its attention on relevant stimuli despite distractions, and its ability to prioritize goals to promote flexible behavior that is suitably opportunistic when it can afford to be, yet persistent when it needs to be. The emotion system achieves this by assessing and signaling the value of immediate events in order to appropriately regulate and bias the cognitive system to help focus attention, prioritize goals, and to

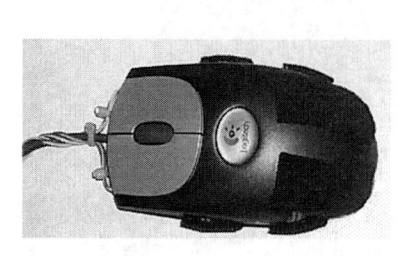

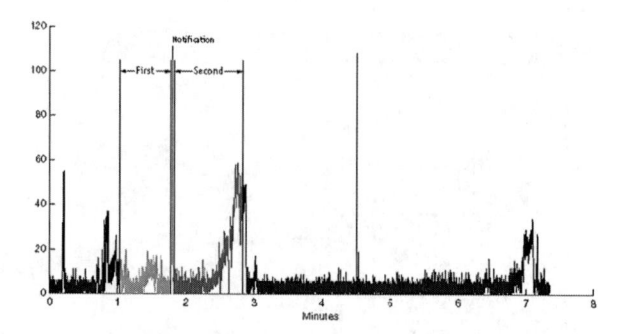

Figure 18.5. Pressure-sensitive mouse collects data on how you handle it—the waveform is higher when more pressure is applied.

pursue the current goal with an appropriate degree of persistence and opportunism. Furthermore, the emotive responses protect the robot from intense interactions that may be potentially harmful, and help the robot to sustain interactions that are beneficial for the robot.

Broadly speaking, each emotive response consists of the four factors described in the following sections.

A Precipitating Event

Precipitating events are detected by perceptual processes. Environmental stimuli influence the robot's behavior through perceptual releasers (inspired by Tinbergen's innate releasing mechanisms). Each releaser is modeled as a simple elicitor of behavior that combines lower-level perceptual features into behaviorally significant perceptual categories.

An Affective Appraisal of That Event

Behavioral responses can be pragmatically elicited by rewards and punishments, where a reward is something for which an animal (or robot) will work and a punishment is something it will work to escape or avoid. The robot's affective appraisal process assesses whether an internal or external event is rewarding, punishing, or not relevant, and tags it with an affective value that reflects its expected benefit or harm to the robot. This mechanism is inspired by Damasio's somatic marker hypothesis (Damasio, 1994). For instance, internal and external events are appraised in relation to the quality of the stimulus (e.g., the intensity is too low, too high, or just right), or whether it relates to the robot's current goals or motivations.

There are three classes of tags used within the robot to affectively characterize a given event. Each tag has an associated intensity that scales its contribution to the overall affective state. The arousal tag (A) specifies how arousing (or intense) this factor is and very roughly corresponds to the activity of the autonomic nervous system. Positive values correspond to high arousal whereas negative values correspond to low arousal. The valence tag (V) specifies how favorable or unfavorable the event is to the robot and varies with the robot's current goals. Positive values correspond to a beneficial (desirable) event, whereas negative values correspond to an event that is not beneficial (not desirable). The stance tag (S) specifies how approachable the event is—positive values encourage the robot to approach whereas negative values correspond to retreat. It also varies with the robot's current goals.

These affective tags serve as the common currency for the inputs to the behavioral response selection mechanism, allowing the influences of externally elicited perceptual states and a rich repertoire of internal states to be combined. Perceptual states contribute to the net affective state of the robot based on their relevance to the robot's current goals or their intrinsic affective value according to the robot's programmed preferences. The ability of the robot to keep its drives satiated also influences the robot's affective state: positive when a drive is in balance, and negative when out of balance. The ability of the robot to achieve its goals is another factor. Forward progress culminating in success is tagged with positive valence. In contrast, prolonged delay in achieving a goal is tagged with negative valence.

A Characteristic Display

Characteristic display can be expressed through facial expression, vocal quality, or body posture, as described earlier.

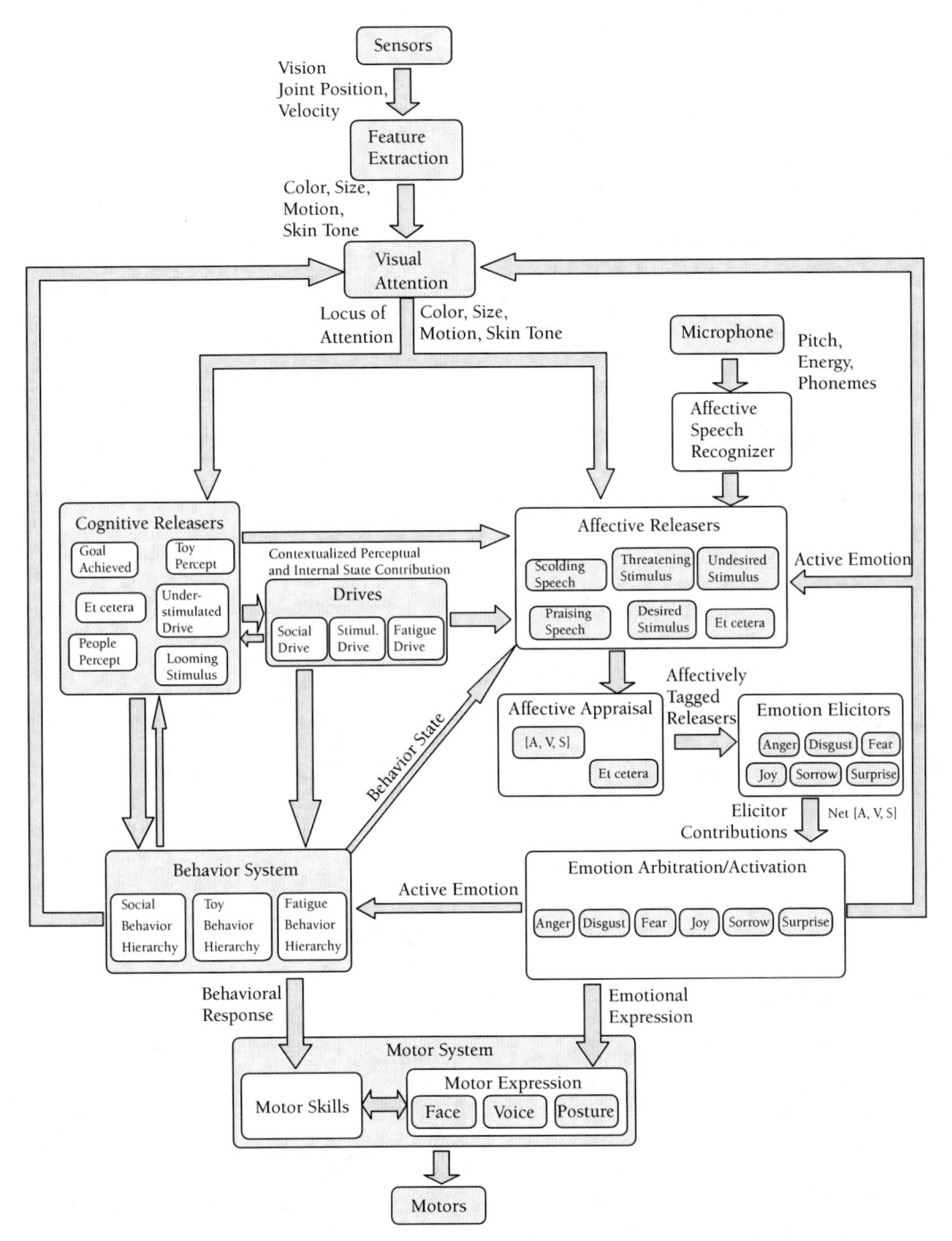

Figure 18.6. An overview of the cognitive-affective architecture showing the tight integration of the cognitive system (shown in light gray), the emotion system (shown in white), and the motor system. The cognitive system comprises perception, attention, drives, and behavior systems. The emotion system comprises affective releasers, appraisals, elicitors, and gateway processes that orchestrate emotive responses. A, arousal tag; V, valence tag; S, stance tag.

Modulation of the Cognitive and Motor Systems to Motivate a Behavioral Response

The emotive elicitor system combines the myriad of affectively tagged inputs (i.e., perceptual elicitors, cognitive drives, behavioral progress, and other internal states) to compute the net affective state of the robot. This affective state characterizes whether the robot's homeostatic needs are being met in a timely manner, and whether the present interaction is beneficial to its current goals. When this is the case, the robot is in a mildly positive, aroused, and open state, and its behavior is pleasant and engaging. When the robot's internal state diverges from this desired internal relationship, it will work to restore the balance—to acquire desired stimuli, to avoid undesired stimuli, and to escape dangerous stimuli. Each emotive response carries this out in a distinct fashion by modulating the cognitive system to redirect attention, to evoke a corrective behavioral pattern, to emotively communicate that the interaction is not appropriate, and to socially cue others how they might respond to correct the problem (e.g., a fearful response to a threatening stimulus, an angry response to something blocking the robot's goal, a disgusted response to a undesirable stimuli, a joyful response to achieving its goal, etc.). In a process of behavioral homeostasis, the emotive response maintains activity through external and internal feedback until the correct relation of robot to environment is established (Plutchik, 1991).

Social Learning

We have been adapting this architecture to explore various forms of social learning on robots, such as tutelage (Lockerd & Breazeal, 2004), imitation (Buschbaum, Blumberg, & Breazeal, 2004), and social referencing (Breazeal, Buchsbaum, Gray, Gatenby, & Blumberg, 2004). For instance, social referencing is an important form of socially guided learning in which one person utilizes another person's interpretation of a given situation to formulate his or her own interpretation of it and to determine how to interact with it (Feinman, 1982; Klinnert, Campos, Source, Emde, & Svejda, 1983). Given the large number of novel situations, objects, or people that infants encounter (as well as robots), social referencing is extremely useful in forming early appraisals and coping responses toward unfamiliar stimuli with the help of others.

Referencing behavior operates primarily under conditions of uncertainty—if the situation has low ambiguity, then intrinsic appraisal processes are used (Campos & Stenberg, 1981). In particular, emotional referencing is viewed as a process of emotional communication whereby the infant learns how to feel about a given situation, and then responds to the situation based on his or her emotional state (Feinman, Roberts, Hsieh, Sawyer, & Swanson, 1992). For example, the infant might approach a toy and kiss it upon receiving a joy message from the adult, or swat the toy aside upon receiving a fear message (Hornik & Gunnar, 1988).

In our model, shown in figure 18.7 (for our expressive humanoid robot, Leonardo), a perception-production coupling of the robot's facial imitation abilities based on the active-intermodal mapping hypothesis of Meltzoff and Moore (1997) is leveraged to allow the robot to make simple inferences about the emotional state of others. Namely, imitating the expression of another as they look upon the novel object induces the appropriate affective state within the robot. The robot applies this induced affective state as the appraisal for the novel object via the robot's joint attention and emotion-based mechanisms. This allows Leonardo to use the emotionally communicated assessment of others to form its own appraisals of the same situations and to guide its own subsequent responses.

Conclusion

This chapter summarizes our ongoing work in developing and embedding affective technologies in learning interactions with automated systems such robotic learning companions. These technologies include a broad repertoire of perceptual, expressive, and behavioral capabilities. Importantly, we are also beginning to develop computational models and learning systems that interact with people to elucidate the role that social, cognitive, and affective mechanisms play in learning (see Picard et al., 2004, for a review). By doing so, we hope to better answer such questions as: What affective

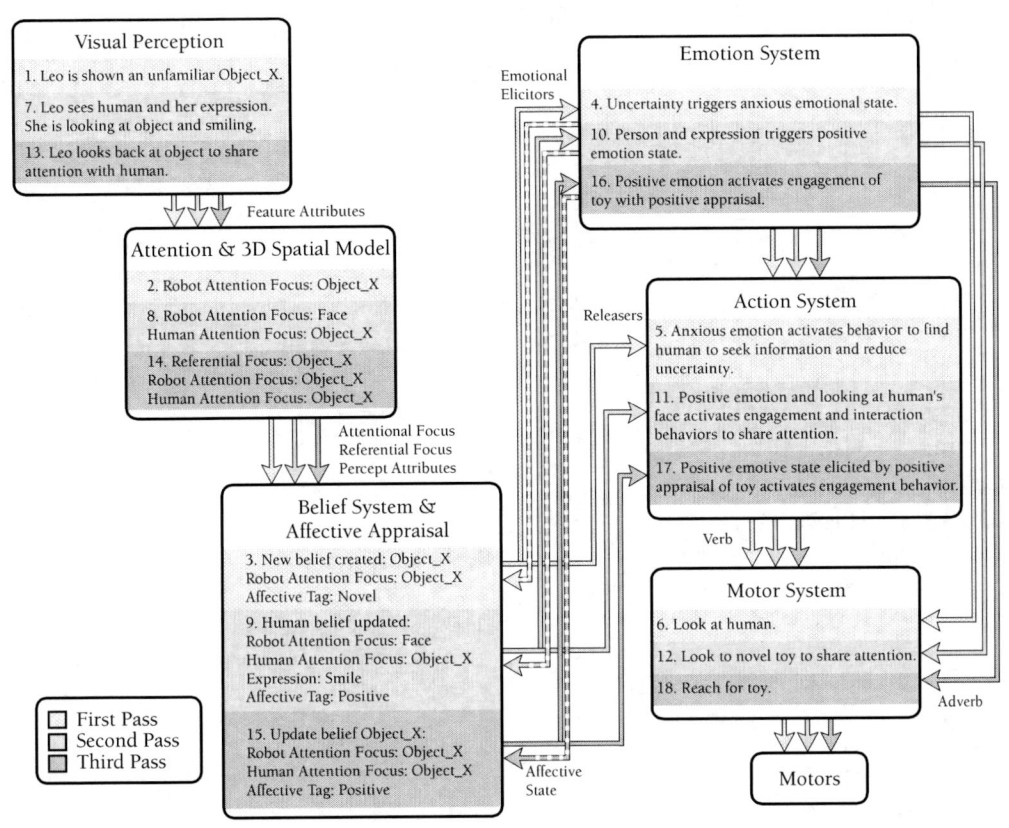

Emotion-Inspired Abilities in Relational Robots

Figure 18.7. Model of social referencing. This schematic shows how social referencing is implemented within Leonardo's extended cognitive-affective architecture. Significant additions include a perception-production system for facial imitation, an attention system that models the attentive and referential state of the human and the robot, and a belief system that bundles visual features with attentional state to represent coherent entities in the 3-D space around the robot. The social referencing behavior executes in three passes through the architecture, each pass shown by a different colored band. The numbers represent steps in processing as information flows through the architecture. In the first pass, the robot encounters a novel object. In the second pass, the robot references the human to see his or her reaction to the novel object. On the third pass, the robot uses the human's assessment as a basis to form its own affective appraisal of the object (step 15) and interacts with the object accordingly (step 18).

states are most important to learning and how do these states change with various kinds of pedagogy? How does knowledge of one's affective state influence outcomes in the learning experience? Additionally, these technologies form the basis for building systems that will interact with learners in more natural ways, even bootstrapping the machine's own ability to learn from people.

In the grand picture, we hope to realize three equally important goals. First, we wish to advance the state of the art in machine learning to develop systems that can learn far more quickly, broadly, and continuously from natural human instruction and interaction than they could alone. The ability of personal service robots of the future to quickly learn new skills from humans while on the job will be important for their success. Second, we aspire to achieve a deeper understanding of human learning and development by creating integrated models that permit an in-depth investigation into the social, emotional, behavioral, and cognitive factors that play an important role in human learning. Third, we want to use these models and insights to create engaging technologies that help people learn better. We believe that such models can provide new insights into numerous cognitive-affective mechanisms and shape the design of learning tools and environments, even if they do not compare to the marvelous nature of those that make children tick.

Main Points

1. Endowing robots with social-emotional skills will be important where robots are expected to interact with people, and will be especially critical when the interactions involve supporting various requests and needs of humans.
2. New tools are needed, and are under development, for real-time multimodal perception of human affective states in natural environments and for providing controlled stimuli in repeatable and measurable ways. This requires new technological developments to create a broad repertoire of perceptual, expressive, and behavioral capabilities. It also requires the development of computational models to elucidate the role that physical, social, cognitive, and affective mechanisms play in intelligent behavior.
3. We are developing a physically animate desktop computer system (called RoCo) to explore the interaction of the human body with cognition and affect, and how this coupling could be applied to foster back health as well as task performance and learning gains for the human user. It also enables us to explore scientific questions such as "What affective states are most important to learning?" "How awareness of one's own affective states might influence learning outcomes?" and so on.
4. Successful human-computer interaction over the long term remains a significant challenge. The ability for a computer to build and maintain social rapport with its user (via immediacy behaviors and conveying liking or caring) could not only make computers more engaging, but also provide performance benefits for the human user by establishing an effective working alliance for behavior change goals or learning gains.

Acknowledgments. The authors gratefully acknowledge the MIT Media Lab corporate sponsors of the Things That Think and Digital Life consortia for supporting their work and that of their students in Breazeal's Robotic Life Group and Picard's Affective Computing Group. Tim Bickmore of the Boston University Medical School provided valuable discussions on the topic of building social rapport with relational agents. We developed the stereo vision system in collaboration with Trevor Darrell and David Demirdjian of MIT CSAIL, building upon their earlier system. The C5M code base was developed by Bruce Blumberg and the Synthetic Characters Group at the MIT Media Lab. Kismet was developed at MIT Artificial Intelligence Lab and funded by NTT and DARPA contract DABT 63-99-1-0012. The development of RoCo is funded by a NSF SGER award IIS-0533703.

Key Readings

Bickmore, T. (2003). *Relational agents: Effecting change through human-computer relationships.* PhD thesis, MIT, Cambridge, MA.

Bickmore, T., & Picard, R. W. (2005). Establishing and maintaining long-term human-computer relationships. *Transactions on Computer-Human Interaction, 12*(2), 293–327.

Breazeal, C. (2002). *Designing sociable robots.* Cambridge, MA: MIT Press.

Breazeal, C. (2003). Function meets style: Insights from emotion theory applied to HRI. *IEEE Transactions in Systems, Man, and Cybernetics, Part C, 34*(2), 187–194.

La France, M. (1982). Posture mirroring and rapport. In M. Davis (Ed.), *Interaction rhythms: Periodicity in communicative behavior* (pp. 279–298). New York: Human Sciences Press.

Picard, R. W. (1997) *Affective computing.* Cambridge, MA: MIT Press.

Riskind, J. H. (1984). They stoop to conquer: Guiding and self-regulatory functions of physical posture after success and failure. *Journal of Personality and Social Psychology, 47,* 479–493.

References

Aist, G., Kort, B., Reilly, R., Mostow, J., & Picard, R. (2002). Analytical models of emotions, learning, and relationships: Towards an affective-sensitive cognitive machine. In *Proceedings of the Intelligent Tutoring Systems Conference (ITS2002)* (pp. 955–962). Biarritz, France.

Argyle, M. (1988). *Bodily communication.* New York: Methuen.

Bechara, A., Damasio, H., Tranel, D., & Damasio, A. R. (1997). Deciding advantageously before knowing the advantageous strategy. *Science, 275*, 1293–1295.

Bickmore, T. (2003). *Relational agents: Effecting change through human-computer relationships*. PhD thesis, MIT, Cambridge, MA.

Breazeal, C. (2000). Believability and readability of robot faces. In J. Ferryman, & A. Worrall (Eds.), *Proceedings of the Eighth International Symposium on Intelligent Robotic Systems (SIRS2000)* (pp. 247–256). Reading, UK: University of Reading.

Breazeal, C. (2002). *Designing sociable robots*. Cambridge, MA: MIT Press.

Breazeal, C. (2003a). Emotion and sociable humanoid robots. *International Journal of Human Computer Interaction, 59*, 119–155.

Breazeal, C. (2003b). Emotive qualities in lip synchronized robot speech. *Advanced Robotics, 17*(2), 97–113.

Breazeal, C. (2003c). Function meets style: Insights from emotion theory applied to HRI. *IEEE Transactions in Systems, Man, and Cybernetics, Part C, 34*(2), 187–194.

Breazeal, C. (2003d). Regulation and entrainment for human-robot interaction. *International Journal of Experimental Robotics, 21*, 883–902.

Breazeal, C., Brooks, A., Gray, J., Hancher, M., McBean, J., Stiehl, W. D., et al. (2003). Interactive robot theatre. *Communications of the ACM, 46*(7), 76–85.

Breazeal, C., Buchsbaum, D., Gray, J., Gatenby, D., & Blumberg, B. (2004). Learning from and about others: Towards using imitation to bootstrap the social understanding of others by robots. *Artificial life, 11*(1–2), 1–32.

Burke, R., Isla, D., Downie, M., Ivanov, Y., & Blumberg, B. (2001). CreatureSmarts: The art and architecture of a virtual brain. In *Proceedings of the Game Developers Conference* (pp. 147–166). San Jose, CA.

Buschbaum, D., Blumberg, B., & Breazeal, C. (2004). Social learning in humans, animals, and agents. In Alan Schultz (Ed.)., *Papers from the 2004 Fall Symposium*, (pp. 9–16). Technical Report FS-04-05. American Association for Artificial Intelligence, Menlo Park, CA.

Campos, J., & Stenberg, C. (1981). Perception, appraisal, and emotion: The onset of social referencing. In M. Lamb & L. Sherrod (Eds.), *Infant social cognition* (pp. 273–314). Hillsdale, NJ: Erlbaum.

Christensen, L., & Menzel, K. (1998). The linear relationship between student reports of teacher immediacy behaviors and perception of state motivation, and of cognitive, affective, and behavioral learning. *Communication Education, 47*, 82–90.

Damasio, A. (1994). *Descartes' Error: Emotion, reason, and the human brain*. New York: Putnam.

Damon, W., & Phelps, E. (1989). Strategic uses of peer learning in children's education. In T. Berndt & G. Ladd (Eds.), *Peer relationships in child development* (pp. 135–157). New York: Wiley.

Darrell, T., Gordon, G., & Woodfill, J. (2000). Integrated person tracking using stereo, color, and pattern detection. *International Journal of Computer Vision, 37*(2), 175–185.

Dennerlein, J. T., Becker, T., Johnson, T., Reynolds, C., & Picard, R. W. (2003). Frustrating computer users increases exposure to physical risk factors. *Proceedings of International Ergonomics Association*, Seoul, Korea.

Ekman, P. (1992). Are there basic emotions? *Psychological Review, 99*, 550–553.

Feinman, S. (1982). Social referencing in infancy. *Merrill-Palmer Quarterly, 28*, 445–470.

Feinman, S., Roberts, D., Hsieh, K.-F., Sawyer, D., & Swanson, K. (1992). A critical review of social referencing in infancy. In S. Feinman (Ed.), *Social referencing and the social construction of reality in infancy* (pp. 15–54). New York: Plenum.

Fong, T., Nourbakshsh, I., & Dautenhahn, K. (2002). A survey of social robots. *Robotics and Autonomous Systems, 42*, 143–166.

Haro, A., Essa, I., & Flickner, M. (2000). A non-invasive computer vision system for reliable eye tracking. In *Proceedings of ACM CHI 2000 Conference* (pp. 167–168). The Hague, Netherlands. Published by ACM Press, New York.

Hartup, W. (1996). Cooperation, close relationships, and cognitive development. In W. Bukowski, A. Newcomb, & W. Hartup (Eds.), *The company they keep: Friendship in childhood and adolescence* (pp. 213–237). Cambridge, UK: Cambridge University Press.

Hornik, R., & Gunnar, M. (1988). A descriptive analysis of infant social referencing. *Child Development, 59*, 626–634.

Horvath, A., & Greenberg, L. (1989). Development and validation of the working alliance inventory. *Journal of Counseling Psychology, 36*(2), 223–233.

Izard, C. (1977). *Human emotions*. New York: Plenum.

Kapoor, A., & Picard, R. (2002). Real-time, fully automatic upper facial feature tracking. In *Proceedings of the 5th IEEE International Conference on Automatic Face and Gesture Recognition* (Washington, DC, May 20–21) (p. 0010). Published by IEEE Computer Society, Washington, DC.

Kapoor, A., Qi, Y., & Picard, R. W. (2003). *Fully automatic upper facial action recognition*. Paper presented at the Workshop on IEEE International Workshop on Analysis and Modeling of Faces and Gestures (AMFG) 2003, held in conjunction with International Conference on Computer Vision (ICCV-2003), Nice, France, October.

Kidd, C., & Breazeal, C. (2004). *"Effect of a Robot on Engagement and User Perceptions."* Paper presented at IEEE/RSJ International Conference on Intelligent Robots and Systems (IROS04). Sendai, Japan.

Klinnert, M., Campos, J., Source, J., Emde, R., & Svejda, M. (1983). Emotions as behavior regulators: Social referencing in infancy. In R. Plutchik & H. Kellerman (Eds.), *The emotions* (Vol. 2, pp. 57–86). New York: Academic Press.

La France, M. (1982). Posture mirroring and rapport. In M. Davis (Ed.), *Interaction rhythms: Periodicity in communicative behavior* (pp. 279–298). New York: Human Sciences Press.

Le Doux, J. (1996). *The emotional brain.* New York: Simon and Schuster.

Lewis, M., & Haviland-Jones, J. (2000). *Handbook of emotions* (2nd ed.). New York: Guilford.

Liu, K., & Picard, R. W. (2003). *Subtle expressivity in a robotic computer.* Paper presented at CHI 2003 Workshop on Subtle Expressiveness in Characters and Robots, Ft. Lauderdale, FL, April 7, 2003.

Lockerd, A., & Breazeal, C. (2004). *Tutelage and socially guided robot learning.* Paper presented at IEEE/RSJ International Conference on Intelligent Robots and Systems (IROS04). Sendai, Japan.

Lombard, M., Ditton, T. B., Crane, D., Davis, B., Gil-Egul, G., Horvath, K., et al. (2000). *Measuring presence: A literature-based approach to the development of a standardized paper-and-pencil instrument.* Paper presented at Presence 2000: The Third International Workshop on Presence, Delft, The Netherlands.

Lorenz, C. (1950). *Foundations of ethology.* New York: Springer-Verlag.

Meltzoff, M., & Moore, M. K. (1997). Explaining facial imitation: A theoretical model. *Early Development and Parenting, 6,* 179–192.

Miller, D. J., & Yan, L. (1999). Critic-driven ensemble classification. *Signal Processing, 47,* 2833–2844.

Mota, S. (2002). *Automated posture analysis for detecting learners' affective state.* Master's thesis, MIT Media Lab, Cambridge, MA.

Mota, S., & Picard, R. W. (2003). *Automated posture analysis for detecting learner's interest level.* Paper presented at First IEEE Workshop on Computer Vision and Pattern Recognition, CVPR HCI 2003.

Nass, C., Jonsson, I., Harris, H., Reaves, B., Endo, J., Brave, S., et al. (2005). Improving automotive safety by pairing driver emotion and car voice emotion (pp. 1973–1976). In *Proceedings of the Conference on Human Factors in Computing Systems, CHI'05,* Portland, OR. Published by ACM Press, New York.

Panksepp, J. (1998). *Affective neuroscience: The foundations of human and animal emotions.* New York: Oxford University Press.

Picard, R. W. (1997). *Affective computing.* Cambridge, MA: MIT Press.

Picard, R. W., Papert, S., Bender, W., Blumberg, B., Breazeal, C., Cavallo, D., et al. (2004). Affective learning—a manifesto. *British Telecom Journal,* http://www.media.mit.edu/publications/bttj/Paper26Pages253–269.pdf.

Plutchik, R. (1991). *The emotions.* Lanham, MD: University Press of America.

Rich, C., Waters, R. C., Strohecker, C., Schabes, Y., Freeman, W. T., Torrance, M. C., et al. (1994). *A prototype interactive environment for collaboration and learning* (Technical Report TR-94-06). Retrieved from http://www.merl.com/projects/emp/index.html.

Richmond, V., & McCroskey, J. (1995). *Immediacy, nonverbal behavior in interpersonal relations.* Boston: Allyn and Bacon.

Riskind, J. H. (1984). They stoop to conquer: Guiding and self-regulatory functions of physical posture after success and failure. *Journal of Personality and Social Psychology, 47,* 479–493.

Smith, C., & Scott, H. (1997). A componential approach to the meaning of facial expressions. In J. Russell & J. Fernandez-Dolls (Eds.), *The psychology of facial expression* (pp. 229–254). Cambridge, UK: Cambridge University Press.

Tan, H. Z., Ifung, L., & Pentland, A. (1997). *The chair as a novel haptic user interface.* Paper presented at Workshop on Perceptual User Interfaces, Banff, Alberta, Canada, October.

Thomas, F., & Johnson, O. (1981). *The illusion of life: Disney animation.* New York: Hyperion.

Tinbergen, N. (1951). *The study of instinct.* New York: Oxford University Press.

Trevarthen, C. (1979). Communication and cooperation in early infancy: A description of primary intersubjectivity. In M. Bullowa (Ed.), *Before speech: The beginning of interpersonal communication* (pp. 321–348). Cambridge, UK: Cambridge University Press.

Tronick, E., Als, H., & Adamson, L. (1979). Structure of early face-to-face communicative interactions. In M. Bullowa (Ed.), *Before speech: The beginning of interpersonal communication* (pp. 349–370).. Cambridge, UK: Cambridge University Press.

Turkle, S., Breazeal, C., Detast, O., & Scassellati, B. (2004). *Encounters with Kismet and Cog: Children's relationship with humanoid robots.* Paper presented at Humanoids 2004, Los Angeles, CA.

UNEC & IFR. (2002). *United Nations Economic Commission and the International Federation of Robotics: World Robotics 2002.* New York and Geneva: United Nations.

Viola, P., & Jones, M. (2001). Rapid object detection using a boosted cascade of simple features. In *Proceedings of the IEEE Conference on Computer Vision and Pattern Recognition* (pp. 511–518). Kauai, HI.

19

Ferdinando A. Mussa-Ivaldi, Lee E. Miller,
W. Zev Rymer, and Richard Weir

Neural Engineering

The desire to re-create or at least emulate the functions of the nervous system has been one of the prime movers of computer science (Von Neumann, 1958) and artificial intelligence (Marr, 1977). In pursuit of this objective, the interaction between engineering and basic research in neuroscience has produced important advances in both camps. Key examples include the development of artificial neural networks for automatic pattern recognition (Bishop, 1996) and the creation of the field of computational neuroscience (Sejnowski, Koch, & Churchland, 1988), which investigates the mechanisms for information processing in the nervous system. At the beginning of this new millennium, neural engineering has emerged as a rapidly growing research field from the marriage of engineering and neuroscience.

As this is a new discipline, a firm definition of neural engineering is not yet available. And perhaps it would not be a good idea to constrain the growing entity within narrow boundaries. Yet a distinctive focus of neural engineering appears to be in the establishment of direct interactions between the nervous system and artificial devices. As pointed out by Buettener (1995), this focus leads to a partition of the field into two reciprocal application domains: the application of technology to improve biological function and the application of biological function to improve technology.

In the first case, the physical connection between neural and artificial systems aims primarily at restoring functions lost to disabling conditions, such as brain injury, stroke, and a large number of chronic impairments. In the second case, one can visualize the visionary goal of tapping into the as-yet unparalleled computational power of the biological brain and of its constituents to build intelligent, adaptable, and self-replicating machines.

Both cases involve the combination of science and technology from fields as diverse as cellular and molecular biology, material science, signal processing, artificial intelligence, neurology, rehabilitation medicine, and robotics. But perhaps most remarkably, in both cases the focus on establishing information exchange between artificial and neural systems is likely to lead to deeper advances of our basic understanding of the nervous system itself.

While neural engineering appears to be a new field, emerging at the beginning of the new millennium, its roots reach quite far into the past. This chapter begins with a brief review of the earlier ideas and advances in the interaction between neural and artificial systems. A significant impulse to this field is contributed by the need to develop artificial spare parts for our bodies, such as arms and legs. In a subsequent section, we present ideas concerning the interaction between design and neural control of such prosthetic devices. The perspective of using neural control for artificial devices has been enhanced by advances in our ability to extract information from brain signals. We then review recent advances in the use of signals extracted by electroencephalography (EEG) and by electrodes implanted in cortical areas for generating commands to artificial devices. Subsequently, we discuss the role of sensory feedback and the problem of creating an artificial version of it by transmitting sensory information to the nervous system via electrical stimulation. The next section offers a view on the clinical impact of neural engineering, both in its current state and in the foreseeable future.

A second perspective on neural engineering concerns the development of novel technology based on the computational power of the nervous system. One issue associated with this perspective is the development of biomimetic and hybrid technologies, which we consider in this chapter. Others, considered in subsequent sections, include the analysis of neural plasticity as a biological programming language and the use of interactions between brain and machines as a means to investigate neural computation.

From Fiction to Reality: Cochlear Implants

The marriage of biological and artificial systems has captured the dreams and nightmares of humanity since biblical times. One of the earliest ideas in this regard is the golem, the automaton made of clay first mentioned in the Talmud, resurfacing in the 16th century's legend of a creature created by Rabbi Loews of Prague to protect and support the city's Jewish community. However, not everything worked as planned, because the golem became a threat to innocent lives and the rabbi had to deprive it of its spiritual force. Interfering with nature in such potentially hazardous ways has been of great concern to literary figures since at least the 19th century. We see these themes emerge in Mary Shelley's *Frankenstein* and in the reappearance of the golem in the work of Isaac Singer and in the classic film by Paul Wegener, on the eve of World War II.

Years later, Manfred Clynes and Nathan Kline (1960) coined the term *cyborg* to define the intimate interaction of human and machine. This term originated in the semifictional context of space exploration, from the idea of augmenting human sensory and motor capabilities through artificial motion and sensing organs. As in the case of the golem and Frankenstein, cyborgs rapidly became part of a fearful view of a future in which we create monstrous creatures that threaten our own existence. This rather dark view was particularly intense at a time of global wars, when millions of lives were destroyed. And, indeed, all technologies are potential sources of mortal danger. However, one can look at them with a different perspective (Clark, 2003). This is particularly the case for human-machine interactions, which are contributing in a variety of ways to enhance the quality of life of the disabled population.

Cochlear implants (Loeb, 1990; Rubinstein, 2004) are among the earliest and most successful interfaces between brain and artificial systems (figure 19.1). Currently, some 40,000 patients worldwide have such implants, which allow deaf adult recipients to speak on the phone and children to attend regular school classes (Rauschecker & Shannon, 2002). Cochlear implants are based upon the simple concept of transforming the signals picked up by a microphone into electrical stimulations, which are delivered to the basilar membrane of the cochlea.

In normal conditions, the cochlea transforms the sound vibrations generated by the eardrums into neural impulses that are transmitted via the acoustic nerve to the cochlear nucleus of the brain stem and then to the auditory cortex. An important characteristic of this signal transduction is the separation of spectral components that are topographically organized in regions sensitive to specific frequency bands (Merzenich & Brugge, 1973). A key step in the development of cochlear implants has been the understanding of the critical value of this spectral organization or *tonotopy* (Loeb, 1990; Rauschecker & Shannon, 2002), which has led the pioneers of this technology to decompose the acoustic signal into bands of increasing median frequency and to deliver the corresponding

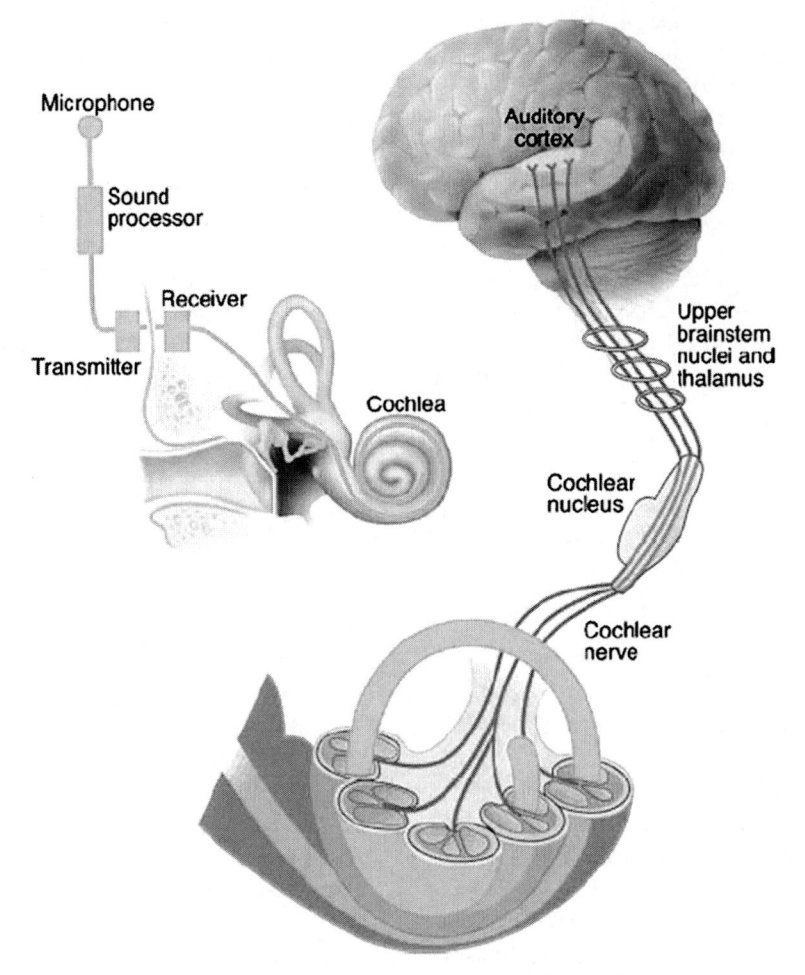

Figure 19.1. Cochlear implant. From Rauschecker and Shannon (2002). See also color insert.

electrical stimuli to contiguous portions of the cochlear membrane.

While these stimuli bear some resemblance to natural sound stimuli, patients typically do not reacquire new hearing ability right after the implant. At first, they most often hear only noise. The recovery of hearing involves a reorganization of neural information processing, particularly in the auditory cortex. Reorganization may take place thanks to the plastic properties of the nervous system.

Interestingly, this learning and reorganization process has highlighted a significant difference between children that receive the implant before the onset of language and children that receive it afterward. The latter group shows a much faster adaptation to the implant and greater ability to extract sound-related information (Balkany, Hodges, & Goodman, 1996; Rauschecker, 1999). However, there is also a trend toward the reduction of plasticity in the auditory and visual system as development progresses toward adulthood. Therefore, younger postlingual patients have generally better results than older adults. The possibility of success with cochlear implants is contingent upon the conditions of the cochlea and of the pathways joining the cochlea to the central auditory system. These conditions are compromised in pathological conditions such as type 2 neurofibromatosis, which is treated by surgical transection of the auditory nerve and causes total deafness.

Auditory–brain stem interfaces work on the same signal-processing principles as cochlear implants, but the stimulating electrodes are placed in the brain stem, over the cochlear nucleus. This offers a potential alternative when the peripheral auditory system is compromised or disconnected from the central auditory system. The outcomes of these interfaces are promising (Kanowitz, 2004), but not yet as impressive as those of cochlear implants. One of the possible causes is the less accessible distribution of frequencies inside the cochlear nucleus compared to the tonotopic organization of the cochlea. To address this issue, researchers are developing electrodes that deliver at different depths signals encoding different frequencies (McCreery, Shannon, Moore, & Chatterjee, 1998). Success in dealing with the transmission of sensory information to central brain structures will have critical implications for the use of artificial electrical stimuli as a source of feedback to the motor system.

Current Challenges in Motor Prosthetics

There have been many attempts to design fully articulated arms and hands in an effort to re-create the full function of the hand. While the ultimate goal of upper-extremity prosthetics research is the meaningful subconscious control of a multifunctional prosthetic arm or hand—a true replacement for the lost limb—the current state-of-the-art electric prosthetic hands are generally single degree-of-freedom (opening and closing) devices usually implemented with myoelectric control (electric signals generated as a by-product of normal muscle contraction). Current prosthetic arms requiring multiple-degree-of-freedom control most often use sequential control. Locking mechanisms or special switch signals are used to change control from one degree of freedom to another. As currently implemented, sequential control of multiple motions is slow; consequently, transradial prostheses are generally limited to just opening and closing of the hand, greatly limiting the function of these devices (figure 19.2). Persons with recent hand amputations expect modern hand prostheses to be like hands, similar to depictions of artificial hands in stories like *The Six Million Dollar Man* or *Star Wars*. Because these devices fail to meet some users' expectations, they are frequently rejected (Weir, 2003).

A major factor limiting the development of more sophisticated hand and arm prostheses is the difficulty of finding sufficient control sources to control the many degrees of freedom required to replace a physiological hand or arm. In addition, most multiple-degree-of-freedom prosthetic hands are doomed by practicality, even before the control interface becomes an issue. Most mechanisms fail because of poor durability, lack of performance, and complicated control. No device will be clinically successful if it breaks down frequently. A multifunctional design is by its nature more complex than a single-degree-of-freedom counterpart. However, some robustness and simplicity must be traded if the increase in performance possible with a multiple-degree-of-freedom hand is ever to be realized.

Neuroelectric control in its broadest context arises where a descending neural command is readily interpreted by a peripheral apparatus, and an ascending command returned, holds the allure of being able to provide multiple-channel control and multiple-channel sensing. There has been much research into interfacing prosthesis connections directly to nerves and neurons (Andrews et al., 2001; Edell, 1986; Horch, 2005; Kovacs, Storment, James, Hentz, & Rosen, 1988) but the practicality of human-machine interconnections of this kind is still problematic. Nervous tissue is sensitive to mechanical stresses, and sectioned nerves are more sensitive still, in addition to which, this form of control also requires the use of implanted systems.

Edell (1986) attempted to use nerve cuffs to generate motor control signals. Kovacs et al. (1988) tried to encourage nerve fibers to grow through arrays of holes in silicon integrated circuits that had been coated with nerve growth factor. Andrews et al. (2001) reported on their progress in developing a multipoint microelectrode peripheral nerve implant. Horch (2005) has demonstrated the control of a prosthetic arm with sensory feedback using needle electrodes in peripheral nerves, but these are short-term experiments with the electrodes being removed after 3 weeks.

The development of BIONs (Loeb et al., 1998) for functional electrical stimulation (FES) is a promising new implant technology that may have a far more immediate effect on prosthesis control. These devices are hermetically encapsulated, leadless electrical devices that are small enough to be

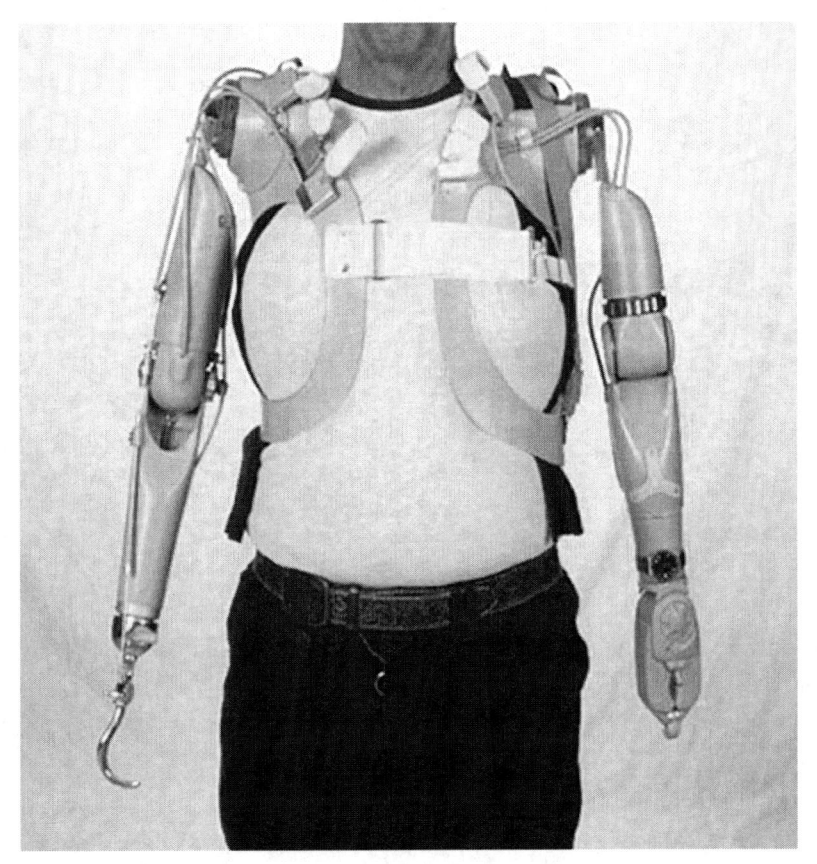

Figure 19.2. Current state-of-the-art prosthesis for a man with bilateral shoulder disarticulation amputations.

injected percutaneously into muscles (2 mm diameter × 15 mm long). They receive their power, digital addressing, and command signals from an external transmitter coil worn by the patient. The hermetically sealed capsule and electrodes necessary for long-term survival in the body have received investigational device approval from the FDA for experimental use in people with functional electrical stimulation systems. As such, these BIONs represent an enabling technology for a prosthesis control system based on implanted myoelectric sensors.

Weir, Troyk, DeMichele, and Kuiken (2003) are involved in an effort to revisit (Reilly, 1973) the idea of implantable myoelectric sensors. Implantable myoelectric sensors (IMES) have been designed that will be implanted into the muscles of the forearm and will transcutaneously couple via a magnetic link to an external exciter/r data telemetry reader. Each IMES will be packaged in a BION II (Arcos et al., 2002) hermetic ceramic capsule (figure 19.3). The external exciter/data telemetry reader consists of an antenna coil laminated into a prosthetic interface so that the coil encircles the IMES. No percutaneous wires will cross the skin. The prosthesis controller will take the output of an exciter/data telemetry reader and use this output to decipher user intent. While it is possible to locate three, possibly four, independent (free of cross talk) surface electromyographic (EMG) sites on the residual limb, it will be feasible to create many more independent EMG sites in the same residual limb using implanted sensors. There are 18 muscles in the forearm that are involved in the control of the hand and wrist. Using intramuscular signals in this manner means the muscles are effectively acting as biological amplifiers for the descending neural commands (figure 19.4). Neural signals are on the order of microvolts while muscle or EMG signals are on the order of millivolts. Intramuscular EMG signals from multiple residual muscles offer a means of providing simultaneous control of

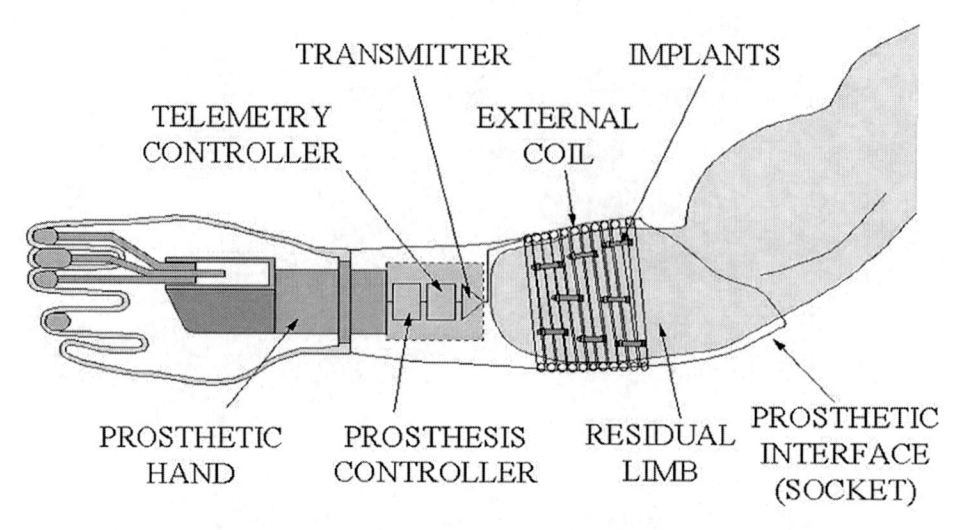

Figure 19.3. Schematic of how the implantable myoelectric sensors will be located within the forearm and encircled by the telemetry coil when the prosthesis is donned.

multiple degrees of freedom in a multifunction prosthetic hand.

At levels of amputation above the elbow, the work of Kuiken (2003) offers a promising surgical technique to create physiologically appropriate control sites. He advocates the use of "targeted muscle reinnervation" or neuromuscular reinnervation to improve the control of artificial arms. Kuiken observed that although the limb is lost in an amputation, the control signals to the limb remain in the residual peripheral nerves of the amputated limb.

The potential exists to tap into these lost control signals using nerve-muscle grafts. As first suggested by Hoffer and Loeb (1980), it is possible to denervate expendable regions of muscle in or near an amputated limb and graft the residual peripheral nerve stumps to these muscles. The peripheral nerves then reinnervate the muscles and these nerve-muscle grafts would provide additional EMG control sites for an externally powered prosthesis. Furthermore, these signals relate directly to the original function of the limb.

In the case of the high-level amputee, the median, ulnar, radial, and musculocutaneous nerves are usually still present (figure 19.5). The musculocutaneous nerve controls elbow flexion, while the radial nerve controls extension. Pronation of the

Figure 19.4. A muscle as a biological amplifier of the descending neural command. CNS, central nervous system.

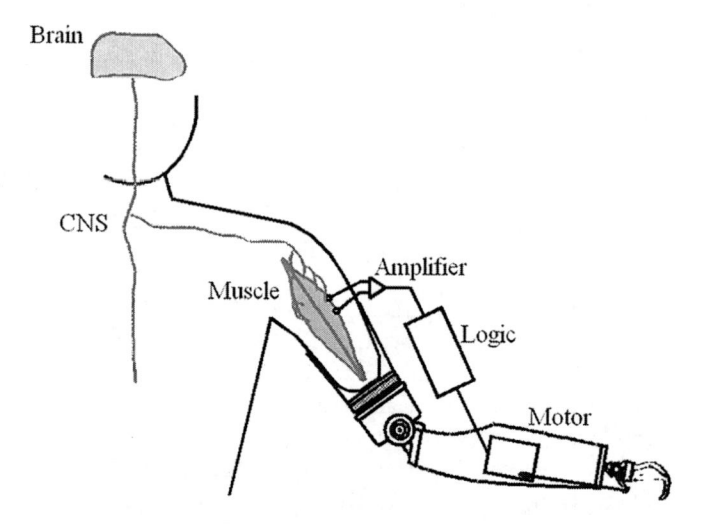

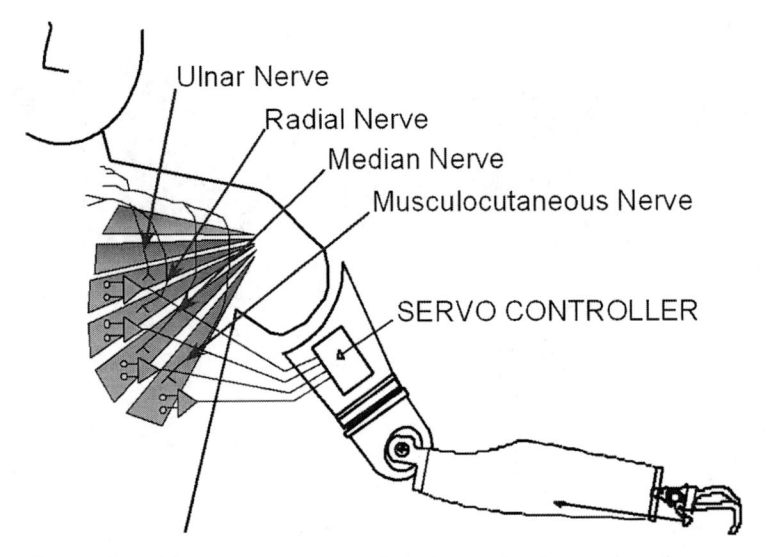

Figure 19.5. Schematic of neuromuscular reinnervation showing how the pectoralis muscle is split into four separate sections and reinnervated with the four main arm nerve bundles. See also color insert.

forearm is directed by the median nerve and supination by the radial and musculocutaneous nerve. Extension of the hand is governed by the radial nerve and flexion by the median and ulnar nerves. Since each of these nerves innervates muscles that control the motion of different degrees of freedom, they should theoretically supply at least four independent control signals. The nerves are controlling functions in the prosthesis that are directly related to their normal anatomical function.

IMES located at the time of initial surgery would complement this procedure nicely by providing focal recording of sites that may be located physically close together. Targeted muscle reinnervation has been successfully applied in two transhumeral amputees and one bilateral, high-level subject to control a shoulder disarticulation prosthesis. In the shoulder disarticulation case, each of the residual brachial plexus nerves wsa grafted to different regions of denervated pectoralis muscle, and in the transhumeral cases the nerves were grafted to different regions of denervated biceps brachii.

Kuiken's neuromuscular reorganization technique is being performed experimentally in people. IMES are due to be soon available for human testing. However, the problem is not solved. Now that multiple control signals are available for control, we need to figure out how to use the information they provide to control a multifunction prosthesis in a meaningful way.

The CNS as a Source of Control Signals: The Brain-Computer Interface

The area of brain-computer interfaces (BCI) has attracted considerable attention in the last few years, with various demonstrations of brain-derived signals used to control external devices in both animals and humans. Potential applications include communications systems for "locked-in" patients suffering from complete paralysis due to brain stem stroke or ALS, and environmental controls, computer cursor controls, assistive robotics, and FES systems for spinal cord injury patients (see chapters 20 and 22, this volume). The field encompasses invasive approaches that require the surgical implantation of recording electrodes either beneath the skull or actually within the cortex, and noninvasive approaches based on potentials recorded over the scalp (figure 19.6). The latter approach has much less spatial and temporal resolution and consequently less potential bandwidth than the signals available from intracortical or even epidural electrodes. As a result, there is a trade-off between the quality of the signals and the risk to which a patient would be exposed. Issues of convenience

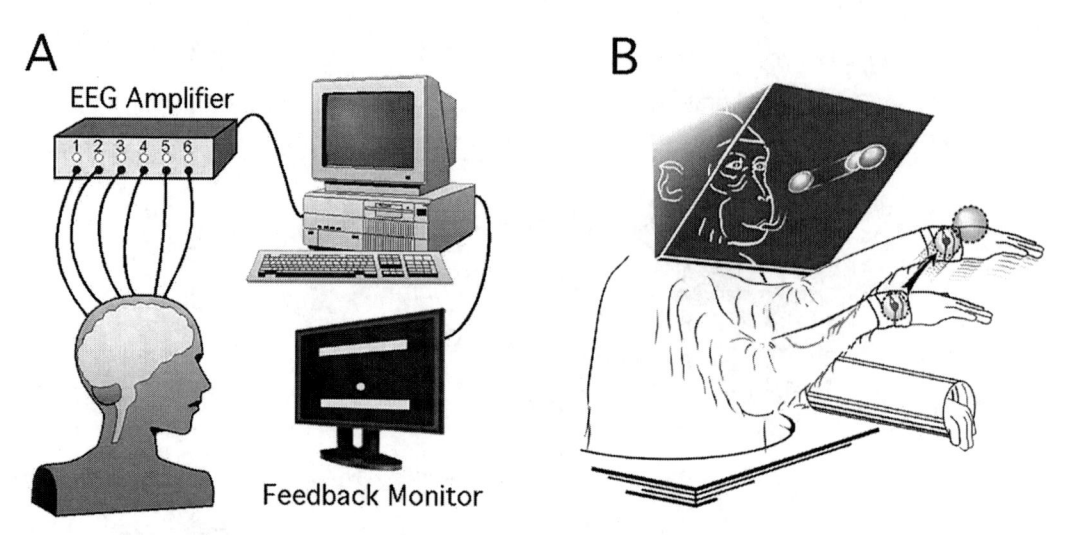

Figure 19.6. Noninvasive and invasive brain-computer interfaces. (A) Electroencephalographic (EEG) signals recorded from the scalp have been used to provide communication or other environmental controls to locked-in patients, who are completely paralyzed due to brain stem stroke or neurodegenerative disease. The signals are amplified and processed by a computer such that the patient can learn to control the position of a cursor on a screen in one or two dimensions. Among other options, this technique can be used to select letters from a menu in order to spell words. From Kubler et al. (2001). (B) Monkeys have learned to control the 3-D location of a cursor (yellow sphere) in a virtual environment. The cursor and fixed targets are projected onto a mirror in front of the monkey. The cursor position can be controlled either by movements of the monkey's hand or by the hand movement predicted in real time on the basis of neuronal discharge recorded from electrodes implanted in the cerebral cortex. From Taylor et al. (2002). See also color insert.

and patient acceptability also come into play. External recording electrodes can be unsightly and uncomfortable and require much caregiver assistance to don and maintain. The state of the field is that a range of useful technologies are being developed to accommodate the priorities and concerns of a varied group of potential users.

EEG Recordings

For more than 20 years, the most systematic attempts at clinical application of BCIs to the sensory-motor system have used specific components of EEG signals (Niedermeyer & Lopes Da Sylva, 1998). A few BCI systems use involuntary neural responses to stimuli to provide some rudimentary access to the thoughts and wishes of the most severely paralyzed individuals (Donchin, Spencer, & Wijesinghe, 2000). However, for paralyzed individuals with a relatively intact motor cortex, there is evidence that volitional command signals can be produced in association with attempted or imagined movements. Characteristic changes in the EEG in the μ (8–12 Hz) and β (18–26 Hz) bands usually accompany movement preparation, execution, and termination (Babiloni et al., 1999; Pregenzer & Pfurtscheller, 1999; Wolpaw, Birbaumer, McFarland, Pfurtscheller, & Vaughan, 2002) and can still be observed years after injury (Lacourse, Cohen, Lawrence, & Romero, 1999). However, fine details about the intended movement, such as direction, speed, limb configuration, and so on, are much more difficult to discern from EEG than from intracortical unit recordings.

There are two primary ways these sensorimotor rhythms have been utilized in BCIs. One option is to use discrete classifiers that recognize spatiotemporal patterns associated with specific attempted movements (Birch, Mason, & Borisoff, 2003; Blankertz et al., 2003; Cincotti et al., 2003; Garrett, Peterson, Anderson, & Thaut, 2003; Graimann, Huggins, Levine, & Pfurtscheller, 2004; Millan & Mourino, 2003; Obermaier, Guger, & Pfurtscheller, 1999). This approach might allow selection between several different options or states of a device, but it cannot provide continuous, proportional control. The second option is to translate sensorimotor rhythms continuously into one or more proportional command signals. In 1994, Wolpaw et al. demonstrated x- and y-axis cursor control using the sum and difference respectively, of the μ-range power recorded from left and right sensorimotor areas (Wolpaw & McFarland, 1994). More recently, this group has improved this paradigm by adaptively refining the EEG decoder during the training process. The decoder's coefficients were regularly adjusted in order to make use of learning-induced changes in the person's EEG modulation capabilities (Wolpaw & McFarland, 2004). In that study, four subjects, including two spinal injury patients, learned to make fairly accurate 2-D cursor movements. This result with injured patients is particularly important, as it demonstrates that the ability to modulate activity in primary motor cortex (M1) voluntarily apparently survives spinal cord injury. Additional evidence from both EEG and functional magnetic resonance imaging indicates that the motor cortex can still be activated by imagined movements even years after a spinal cord injury (Lacourse et al., 1999; Shoham, Halgren, Maynard, & Normann, 2001).

Intracortical Recordings

While EEG signals show increasing promise, it seems reasonable to anticipate that the more invasive intracortical recordings offer more degrees of freedom and more natural control than does EEG. This conclusion is supported by the commonsense observation that EEG signals are heavily filtered products of cortical activities. However, a direct comparison of the two approaches has yet to be made. The information transmission rate calculated in a limited number of separate experiments allows an indirect comparison to be made. A rate of 0.5 bits per second (bps) has been achieved from EEG recordings (Blankertz et al., 2003), and 1.5 bps through intracortical recordings (Taylor, Tillery, & Schwartz, 2003).

Understanding the relation between time-varying neuronal discharge of single neurons and voluntary movement has been a mainstay of motor systems studies since the pioneering work of Evarts (1966). Within 4 years of that research, Humphrey, Schmidt, and Thompson (1970) showed that linear combinations of three to five simultaneously recorded cells could significantly improve the estimates afforded by single-neuron recordings. About the same time, in one of the earliest examples of intracortical control, monkeys learned to control the firing rate of individual neurons in M1 on the basis of simple visual feedback (Fetz, 1969). A number of years later, a simple two- or three-electrode system intended to provide rudimentary communication was implanted in several locked-in human patients (Kennedy & Bakay, 1998).

Within the past 5 years, technologies for microelectrode recording and signal processing have developed to the point that it is now feasible to use simultaneous recordings from nearly one or two orders of magnitude more cells. High-density arrays of electrodes implanted in M1 or premotor areas have been used to control both virtual and real robotic devices. In the first such system, Chapin and coworkers trained rats to retrieve drops of water by pressing a lever controlling the rotation of a robotic arm (Chapin, Moxon, Markowitz, & Nicolelis, 1999). They used the activities of 21 to 46 neurons, recorded with microwires implanted in M1, as input to a computer program which controlled the motion of the robot. Several rats learned to operate the arm using the neural signals, without actually moving their own limbs.

More recently, monkeys have controlled multijoint physical robotic arms (Carmena et al., 2003; Taylor et al., 2003; Wessberg et al., 2000) and virtual devices in two and three dimensions (Musallam, Corneil, Greger, Scherberger, & Andersen, 2004; Serruya, Hatsopoulos, Paninski, Fellows, & Donoghue, 2002; Taylor, Tillery, & Schwartz, 2002). In these experiments, as many as 100 electrodes were implanted into the cerebral cortex and control was based on activities of 10 to 100 neurons. In early 2004, Cyberkinetics Inc. received FDA approval to implant chronic intracortical microelectrode arrays in human patients with high-level spinal cord injuries. The first patient was implanted several months later, several years after the original injury. After several months of practice and testing, the patient achieved sufficient control to play a very simple video game using the interface (Serruya, Caplan, Saleh, Morris, & Donoghue, 2004).

As with EEG signals, it is possible to use intracortical recordings for either classification or continuous control. Andersen and coworkers (Shenoy et al., 2003) implanted electrodes in the posterior parietal cortex, a region that is believed to participate in movement planning. Monkeys were required to reach toward one of two targets displayed on a

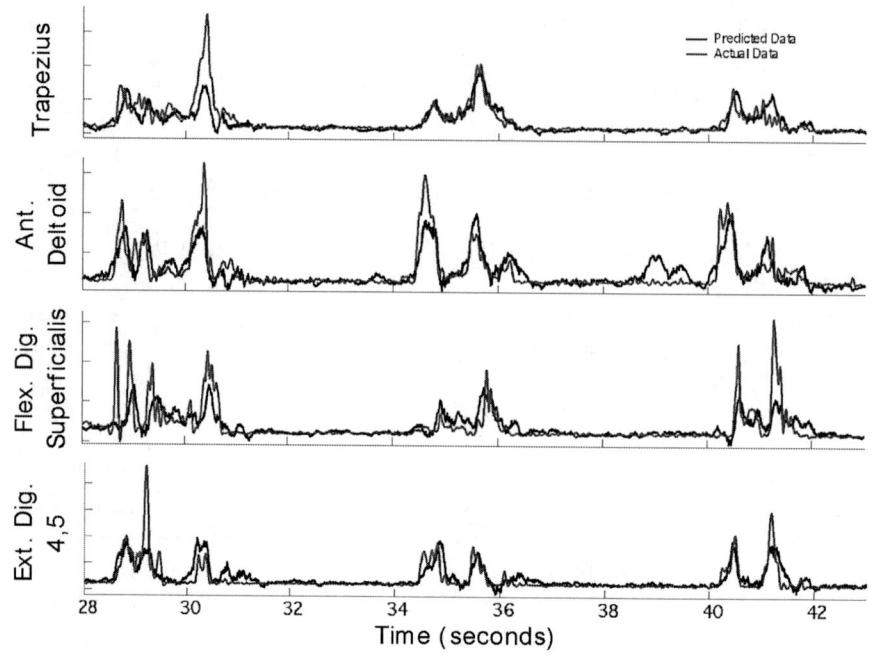

Figure 19.7. Reconstruction of electromyographic (EMG) activity based on motor cortical signals. The blue traces are EMG signals recorded from four arm muscles of a monkey during the execution of a multitarget button-pressing task. The green traces were obtained from neural signals recorded in the primary motor cortex (M1). The relationship between recorded M1 activity and EMGs was estimated using multiple input least-squares system identification techniques, resulting in a set of optimal nonparametric linear filters. Principal component analysis was first used to reduce the complexity of the inputs to these models. To make the filters more robust, a singular value decomposition pseudo-inverse of the input autocorrelation was employed. The resulting filters were then used to predict EMG activity in separate data sets to test the stability and generalization of the correlation between M1 and EMG.

touch screen, and a probabilistic algorithm predicted the preferred target based on discharge recorded during the delay period preceding movement. Within 50 trials, the monkeys learned to modulate the discharge in order to indicate the intended target in the absence of any limb movement.

Notably, however, none of the BCI-based control systems implemented to date has reflected the dynamics of the arm. Instead they have used a position or velocity controller to drive a reference point whose location is instantaneously derived from neuronal activity. Yet there is abundant evidence that the activity of most M1 neurons is affected by forces exerted at the hand (Ashe, 1997; Evarts, 1969; Kalaska, Cohon, Hyde, & Prud'homme, 1989) and by the posture of the limb (Caminiti, Johnson, & Urbano, 1990; Kakei, Hoffman, & Strick, 1999; Scott & Kalaska, 1995). One report did describe the control of a single-degree-of-freedom force gripper though M1 recordings from a rhesus monkey (Carmena et al., 2003). In those experiments, M1 discharge typically accounted for more of the variance within the force signal than it did for either hand position or velocity.

Beyond the prediction of a single-degree-of-freedom force signal, we have shown that rectified and filtered EMG signals can be predicted on the basis of 40 single- and multiunit intracortical recordings (Pohlmeyer, Miller, Mussa-Ivaldi, Perreault, & Solla, 2003). In those experiments, the activity of as many as four muscles of the arm and hand was predicted simultaneously, accounting for between 60% and 75% of the total variance during a series of button presses. A typical example of the correspondence between actual and predicted EMG is shown in figure 19.7. In another series of very recent experiments, we estimated the torques generated during a series of planar, random-walk movements. We calculated predictions of Cartesian hand position, as well as shoulder and elbow joint torque using the discharge of 88 single neurons recorded from M1 and dorsal premotor cortex (PMd). On average, these data accounted for 78% of the torque variance but only 53% of the hand position variance. The implications for BCI control are significant. If the motor commands expressed by M1 neurons have a substantial kinetic component, using them in a system that emulates the dynamics of the limb might provide more natural control.

Feedback Is Needed for Learning and for Control

Not surprisingly, the success of the experiments described above was substantially improved by real-time feedback of performance. Feedback normally allows for two corrective mechanisms. One is the online correction of errors; the other is the gradual adaptation of motor commands across trials. The latter mechanism has been extensively studied in both humans and monkeys (Scheidt, Dingwell, & Mussa-Ivaldi, 2001; Shadmehr & Mussa-Ivaldi, 1994; Thoroughman & Shadmehr, 2000).

The long delays intrinsic to the visual system (100–200 milliseconds) make it unsuitable for the online correction of errors in a complex dynamic system like the human arm, except during movements that are much slower than normal. This problem has been partially sidestepped in most brain-machine interface experiments by eliminating plant dynamics—for example, by controlling a cursor or virtual limb on a computer monitor or by means of a servomechanism enforcing a commanded trajectory. While these approaches eliminate mechanical dynamics, they do not eliminate the neural dynamics responsible for the generation of the open-loop signals. As a consequence, computational errors can only be corrected after long latencies. This is likely to be one of the causes of tracking inaccuracy. If the muscles of a paralyzed patient were to be activated through a BCI, the dynamics of the musculoskeletal system would need to be considered.

While visual feedback plays an important role in the planning and adaptation of movement, other sensory systems ordinarily supply more timely information. Recordings from peripheral sensory nerves have been used as a source of feedback for a closed-loop FES system for controlling hand grip in a patient (Inmann, Haugland, Haase, Biering-Sorensen, & Sinkjaer, 2001). Signals derived from a controlled robot might instead be used to stimulate these nerves as a means of approximating the natural feedback during reaching. However, in most of the situations in which such systems would be clinically useful, conduction through these nerves to the central nervous system is not present. The auditory system might provide another rapid pathway into the brain. Much effort has also been expended in the development of a visual prosthesis, including attempts to stimulate both the visual cortex (Bak et al., 1990; Normann, Maynard, Rousche, & Warren, 1999) and the retina (Zrenner, 2002). The latter methods, while in several ways more promising for blind patients, are unlikely to provide useful feedback for a motor prosthesis, as they would suffer nearly as long a latency as the normal visual system. Direct cortical stimulation might substantially decrease the delay, but mimicking the sophisticated visual signal processing of the peripheral nervous system is a daunting prospect.

The somatosensory system, including proprioception, offers a more natural modality for movement-related feedback. Limited experimental efforts have been made to investigate the perceptual effects of electrical stimulation in the somatosensory cortex. Monkeys have proven capable of distinguishing different frequencies of stimulation, whether applied mechanically to the fingertip or electrically to the cortex (Romo, Hernandez, Zainos, Brody, & Lemus, 2000). In one demonstration, the temporal association of electrical stimuli to somatosensory cortex (cue) with stimuli to the medial forebrain bundle (reward) conditioned freely roaming rats to execute remotely controlled turns (Talwar et al., 2002). It is not yet known whether cortical stimulation could provide adequate feedback to guide movement in the absence of normal proprioception.

The Clinical Impact of Neural Engineering

Cochlear Implants

As outlined earlier in this chapter, the cochlear implant is perhaps the best modern-day example of a successful application of neural engineering design and technology that has clearly generated strong benefits for its recipients. The impact of the cochlear implant has been revolutionary in that a relatively simple design has generated substantial enhancement in hearing for its recipients. Predictably, this device has also raised expectations for other clinical applications of neural engineering; however, these have not yet been fulfilled. Nonetheless, our experience with cochlear implants is salutary in many respects.

First, our experience has verified that implantable neural stimulators could function effectively for long periods of time. Although implanted battery-powered stimulators have been available for a variety of applications for many years (most notably for cardiac pacing in complete heart block or other cardiac arrhythmias), the cochlear implant is the first fully implantable neural system shown to operate effectively without failure or significant disruption of function.

Second, the sophistication of the electrode designs, placement, and stimulus protocols that were utilized in the early stages of this technology were quite primitive, especially when compared with the elegant design and intrinsic function of the natural biological organ (i.e., the cochlea) itself. Nonetheless, the clinical success of the implant illustrated that relatively unsophisticated technologies could still prove to be very useful for treating both congenital and acquired hearing loss. From this, we can deduce that neural plasticity and adaptation appear to be extremely important in generating this success, and that the inherent neural plasticity present in most nervous system elements is extremely important and can readily compensate for considerable technical or design inadequacies.

The third lesson is that the devices have proved to be relatively inexpensive and readily accepted by clinicians and consumers, providing a framework for successful design and implementation of other neural engineering devices in the future.

EEG-Based BCIs

EEG-based BCIs have been a focus of intensive research for several decades (see also chapter 20, this volume). In the beginning, relatively simple measures of EEG root-mean squared (RMS) amplitude or of spectral content were utilized to drive a cursor on a computer screen or to generate binary choices for controlling electrical or mechanical systems or devices. The signal processing time was often quite long and the error rates high, so that many patients were reluctant to utilize these approaches because these factors caused considerable frustration.

To optimize the utility of these approaches, a number of time-efficient and elegant signal processing and classification algorithms have been advanced, and many appear to be able to accelerate the speed with which binary or ternary choices can be made. There is, however, no well-documented example in which surface EEG signals have been shown to have appropriate input-output properties in that signal power or some other measured magnitude component could be linked to the amplitude or speed of the desired outcome measure (such as speed of motion or muscle force). Such improvements are likely to be necessary if natural and spontaneous scaling of motion is to occur.

In spite of these difficulties, certain impairments such as high quadriplegia or widespread neural or muscular loss with amyotrophic lateral sclerosis or Guillian-Barré syndrome may require large-scale substitution of neuromuscular function. Under these conditions, the time delays that are characteristic of many BCI systems are potentially less onerous, and many patients appear willing to deal with these inadequacies because they have few acceptable alternatives.

A related but somewhat different approach is to use EEG signals to control implanted functional neuromuscular stimulation systems, either for upper-extremity hand control or for lower-extremity function in standing up. When real-time control approaches are attempted, the time constraints and scaling nonlinearities become more pressing, and there is as yet no widespread application of these techniques.

Implanted EEG Recording Systems

There is an understandable reluctance to insert electrodes chronically into undamaged human cerebral cortical tissues because of the potential for bleeding, infection, and motion-related damage of cortical neurons that could lead to scarring. As an alternative, recording through transcalvarial EEG electrodes has considerable appeal, in that the dura remains intact, yet signal amplitude rises substantially, making efficient signal processing tasks more feasible.

Implanted Unitary Recording Systems

As outlined earlier, promising examples of implantable human cortical electrode systems are also now emerging that focus either on unitary recordings from the motor cortex (Serruya et al., 2004) or on multiunit recordings and field responses from parietal areas of the cortex (Musallam et al., 2004; Shenoy et al., 2003). Although these approaches are promising in that they may allow the physician to tap the sources of the command signals directly, this work is still in its infancy, and we know little about the benefits and potential complications of such approaches.

Implanted Nerve Cuffs

Other long-standing neural engineering approaches that have promise include implantation of nerve cuffs around cutaneous nerves such as the sural nerve to detect foot contact during gait, a useful measure for controlling functional electrical stimulation. Additional novel peripheral recording systems, described earlier, include the use of EMG recordings from muscles that have received implanted peripheral nerves after an amputation (Kuiken, 2003). The muscles receiving such nerves act as biological amplifiers, providing useful signals for prosthesis control.

Implanted Stimulation Systems

It is also clear that electrical stimulation may be combined with intensive physical training to augment improvements in performance such as arise in Nudo's work on stroke recovery in animal models and in parallel studies in human brain stimulation using the Northstar Neuroscience system (Nudo, Plautz, & Frost, 2001; Nudo, Wise, Sifuentes, & Milliken, 1996; Plautz et al., 2002).

It has become evident that significant improvement in performance may rely on remapping and reorganization of cortical tissues located in the neighborhood of the stroke and that these cortical systems can be retrained optimally using a variety of interventional methods.

The Brain-Computer Metaphor: Biomimetic and Hybrid Technologies

During the past century, studies of computers and of the brain have evolved in a reciprocal metaphor: The brain is seen as an organ that processes information and computers are developed in imitation of the brain. Despite the speed with which today's computers execute billions of operations, their biological counterparts have unsurpassed performance when it comes to recognizing a familiar face or controlling the complex dynamics of the arm. Hence, the computational power of biological systems has sparked intense activity aimed at mimicking neurobiological processes in artificial systems.

Two distinguishing features of biological systems are guiding the development of biomimetic research: (a) their ability to adapt to novel conditions without the need for being "reprogrammed" by some external agent, and (b) their ability to regenerate and reproduce themselves. In addition, biological systems use massively parallel computation systems. The first feature is particularly relevant to the creation of machines that can explore remote and dangerous environments, such as the surface of other planets or deep undersea environments.

The ability to reproduce and regenerate is perhaps the most visible defining feature of living organisms. And indeed, while one may have the impression that materials like metals and plastics are superior to flesh and bones, there is hardly any human artifact that can continuously operate for several decades without irreparably breaking down. The durability of biological tissue is ensured by the complex molecular mechanisms that underlie their continuous renewal. And the biological mechanisms of reproduction are an object of growing interest in robotics, where the theme of self-assembling devices is making promising strides (Murata, Kurokawa, & Kokaji, 1994).

While mimicking the nervous system has led to the creation of artificial neural networks, a different idea has begun to take shape: to construct hybrid computers in which neurons are grown over a semiconductor substrate (Fromherz, Offenhausser, Vetter, & J., 1991; Fusi, Annunziato, Badoni, Salamon, & Amit, 2000; Grattarola & Martinoia, 1993; Zeck & Fromherz, 2001). Fromhertz's team has developed a simple prototype, in which electrical signals are delivered by the substrate to a nerve cell. The responses are transmitted via an electrical synapse to a second cell, and the activity of the second cell is read out by the semiconductor substrate. Chiappalone et al. (2003) have grown cell cultures from chick embryos over microelectrode arrays and have succeeded inducing plastic changes by applying drugs that acted on glutamate receptors.

Neural Plasticity as a Programming Language

A neurobiological basis for programming brain-machine interactions is offered by the different mechanisms of neural plasticity, such as long-term potentiation (LTP; Bliss & Collingridge, 1993), long-term depression (LTD; Ito, 1989), short-term memory (Zucker, 1989), and habituation (Castellucci, Pinsker, Kupfermann, & Kandel, 1970). A common feature of the different forms of plasticity is their history-dependence: In addition to what is known as *phyletic memory* (Fuster, 1995; i.e., instructions "written" in the genetic code), neural systems are programmed by their individual experience. A clear example of this is offered by Hebbian mechanisms that relate the strengthening or weakening of synaptic connections to the temporal correlation of pre- and postsynaptic activations: If the firing of a presynaptic neuron is followed by firing of the postsynaptic neuron, then the efficacy of the synapse increases (LTP). Conversely, if the firing of the postsynaptic neuron occurs when the presynaptic neuron is silent, the connection efficacy is decreased (LTD). From an operational standpoint, these forms of plasticity can be regarded as an assembly language for the neural component in a brain-machine interaction.

To gain access to this "programming language," it is not sufficient to observe that some form of plasticity must take place, for example, in the operant conditioning of motor-cortical signals (Fetz, 1969; Wessberg et al., 2000). A challenge that has yet to be met is to acquire the means for obtaining the desired efficacy of the synaptic connections between neurons and neuronal populations at specific locations of the nervous system. This would correspond to acquiring the ability to design the behavior of biological neural network, as we know how to design the behavior of artificial neural networks.

Brain-Machine Interactions for Understanding Neural Computation

Some investigators have used the closed-loop interaction between nerve cells and external devices as a means to study neural information processing (Kositsky, Karniel, Alford, Fleming, & Mussa-Ivaldi, 2003; Potter, 2001; Reger, Fleming, Sanguineti, Alford, & Mussa-Ivaldi, 2000; Zelenin, Deliagina, Grillner, & Orlovsky, 2000). Mussa-Ivaldi and coworkers (Reger et al., 2000) have established bidirectional connections between a robotic device and a lamprey's brain stem (figure 19.8) with the goal of understanding the operations carried out by a single layer of connections in the reticular formation. Signals generated by the left and right optical sensors of a small mobile robot were translated into electrical stimuli with frequency proportional to the light intensity. The stimuli were applied to the pathways connecting the lamprey's right and left vestibular organs to two populations of reticular neurons. A simple interface translated the resultant discharge frequency of the reticular neurons into motor commands to the robot's right and left wheels. In this simple arrangement, the reticular neurons acted as a processing element that determined the closed-loop response of the neuro-robotic system to a source of light. These studies revealed that (a) different behavior can be generated by different electrode locations; (b) the input-output relation of the reticular synapses are well approximated by simple linear models with a recurrent dynamical component; and (c) the prolonged suppression of one input channel leads to altered responsiveness well after it has been restored.

In a similar experiment, Deliagina, Grillner, and coworkers (Zelenin et al., 2000) used the activity of reticulo-spinal neurons recorded from a swimming lamprey to rotate the platform supporting the fish tank. The lamprey was able to stabilize the hybrid system, and this compensatory effect was most efficient in combination with undulatory swimming motions. These studies demonstrate the feasibility of closed-loop interactions between a specific region of the nervous system and an artificial device. Closed-loop BCIs offer the unparalleled possibility to replace the neural system with a computational model having the same input-output structure, thus providing a direct means for testing the predictions of specific hypotheses about neural information processing.

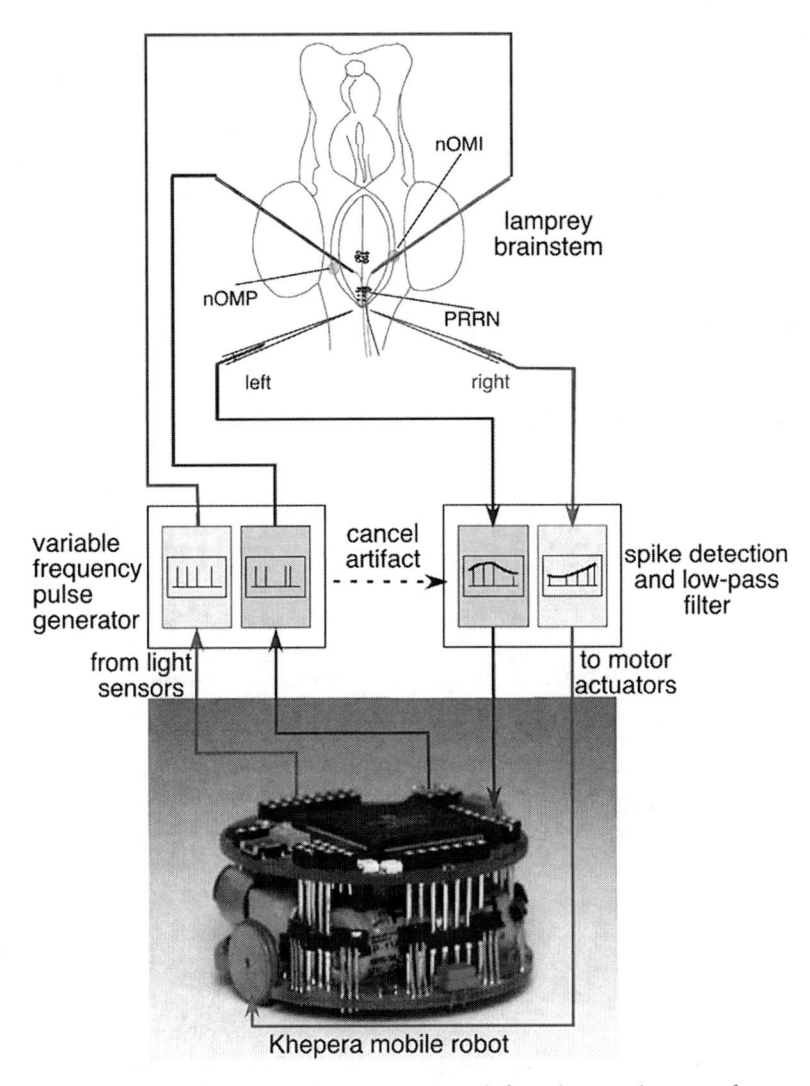

Figure 19.8. A hybrid neurorobotic system. Signals from the optical sensors of Khepera (K-team) mobile robot (bottom) are encoded by the interface into electrical stimulations whose frequency depends linearly upon the light intensity. These stimuli are delivered by tungsten microelectrodes to the right and left vestibular pathways of a lamprey's brain stem (top) immersed in artificial cerebrospinal fluid within a recording chamber. Glass microelectrodes record extracellular responses to the stimuli from the posterior rhomboencephalic neurons (PRRN). Recorded signals from right and left PRRNs are decoded by the interface, which generates the commands to the robot's wheels. These commands are set to be proportional to the estimated average firing rate on the corresponding side of the lamprey's brain stem. See also color insert.

Conclusion

It is self-evident that neural engineering is a multidisciplinary endeavor. Perhaps what makes it even more so is the fact that both components of it—neuroscience and engineering—arise in their own right from a combination of disciplines, such as molecular biology, electrophysiology, mathematics, signal processing, physics, and psychology. In this chapter, we have attempted to paint a portrait of neural engineering based on its dual character.

On one hand, we are seeing a rapid development of approaches that involve the use of brain-machine interaction to improve the living conditions of severely disabled patients. Remarkably, this type of intervention is not merely aimed at restoring functions. Brain-machine and brain-computer interfaces can actually open the door to functions that are not naturally available. For example, neural signals can be used in combination with telecommunication technologies for controlling remote devices or for transmitting (and receiving) information across geographical distances. This will impact not only patients' lives but also the activities of health care providers. Surgical procedures have been already carried out via radio link connecting physician and patient across the Atlantic Ocean (Marescaux et al., 2001). We may expect that the possibility of monitoring the actions of a surgeon to a high level of detail—eye movements, hand movements, hand forces, and so on—will bring new advances in telemedicine.

On the other hand, we are witnessing attempts at interfacing with the nervous system with the intention of capturing its power. As Steve DeWeerth puts it, one may consider "neurobiology as a technology" and neural engineering as an attempt at accessing this technology by creating hybrid systems (personal communication). Can we possibly succeed? Perhaps this is not the right question. Better questions are: To what extent we may succeed, and what may failures teach us about both technology and biology?

Failures along this difficult way will certainly occur because our knowledge and models of brain functions are almost certainly flawed and unquestionably limited. For example, on one hand computational neuroscientists describe synaptic connections using real valued numbers or weights-sto describe the efficacy of transmission. But even accepting this view, we do not really know what ranges of values are possible. Can synaptic connections be approximated by continuous variables? On the other hand, neurobiologists describe plasticity as a basis for learning and memory and find that postsynaptic potentials may be altered by particular patterns of stimuli. But how are these changes reflected in the spiking activities of neurons? And how is it possible to consistently drive a cell or a population to a desired level of responsiveness? By exposing the difficulties in achieving controlled brain-machine interactions and by challenging artificial divisions of labor between theoretical and experimental work, neural engineering may become a driving force in neuroergonomics, neurobiology, and computational neuroscience.

Main Points

1. Neural engineering is a rapidly growing research field that arises from the marriage of engineering and neuroscience. A specific goal of neural engineering is to establish direct and operational interactions between the nervous system and artificial devices.
2. The area of brain-computer interfaces has attracted considerable attention, with various demonstrations of brain-derived signals used to control external devices in both animals and humans. High-density arrays of electrodes implanted in the cerebral cortex have been used to control both virtual and real robotic devices. Potential applications include communications systems for patients suffering from complete paralysis due to brain stem stroke and environmental controls, computer cursor controls, and assistive robotics for spinal cord injury patients.
3. Success in dealing with the transmission of sensory information to central brain structures will have critical implications for the use of artificial electrical stimuli as a source of feedback to the motor system. Sensory feedback allows for two corrective mechanisms. One is the online correction of errors; the other is the gradual adaptation of motor commands across trials. Signals derived from a controlled robot might be used to stimulate these nerves as a means of approximating the natural feedback.
4. Neural plasticity and adaptation are of critical importance for the success of BCIs. The inherent neural plasticity present in most nervous system elements can compensate for considerable technical or design inadequacies. From an operational standpoint, these forms of plasticity can be regarded as an assembly language for the neural component in a brain-machine interaction.
5. As we see a rapid development of approaches that involve the use of brain-machine interactions to improve the living conditions of severely disabled patients, this is not the only value of BCIs. One may consider

neurobiology as a technology and neural engineering as an attempt at accessing this technology by creating hybrid systems. In this endeavor, brain-machine interfaces are a new and unparalleled tool for investigating neural information processing.

Acknowledgments. This work was supported by ONR grant N00014-99-1-0881 and NINDS grant NS048845.

Key Readings

Mussa-Ivaldi, F. A., & Miller, L. E. (2003). Brain–machine interfaces: Computational demands and clinical needs meet basic neuroscience. *Trends in Neurosciences, 26*, 329–334.

Nudo, R. J., Wise, B. M., Sifuentes, F., & Milliken, G. W. (1996). Neural substrates for the effects of rehabilitative training on motor recovery following ischemic infarct. *Science, 272*, 1791–1794.

Serruya, M. D., Hatsopoulos, N. G., Paninski, L., Fellows, M. R., & Donoghue, J. P. (2002). Instant neural control of a movement signal. *Nature, 416*, 141–142.

Taylor, D. M., Tillery, S. I., & Schwartz, A. B. (2002). Direct cortical control of 3D neuroprosthetic devices. *Science, 296*, 1829–1832.

Wolpaw, J. R., & McFarland, D. J. (2004). Control of a two-dimensional movement signal by a noninvasive brain-computer interface in humans. *Proceedings of the National Academy of Sciences, USA, 101*, 17849–17854.

References

Andrews, B., Warwick, K., Jamous, A., Gasson, M., Harwin, W., & Kyberd, P. (2001). Development of an implanted neural control interface for artificial limbs. In *Proceedings of the 10th World Congress of the International Society for Prosthetics and Orthotics (ISPO)*, p. TO8.6. Glasgow, Scotland: ISPO Publications.

Arcos, I., David, R., Fey, K., Mishler, D., Sanderson, D., Tanacs, C. et al. (2002). Second-generation microstimulator. *Artificial Organs, 26*, 228–231.

Ashe, J. (1997). Force and the motor cortex. *Behavioral Brain Research, 86*, 1–15.

Babiloni, C., Carducci, F., Cincotti, F., Rossini, P. M., Neuper, C., Pfurtscheller, G., et al. (1999). Human movement-related potentials vs. desynchronization of EEG alpha rhythm: A high-resolution EEG study. *Neuroimage, 10*, 658–665.

Bak, M., Girvin, J. P., Hambrecht, F. T., Kufta, C. V., Loeb, G. E., & Schmidt, E. M. (1990). Visual sensations produced by intracortical microstimulation of the human occipital cortex. *Medical and Biological Engineering and Computing, 28*(3), 257–259.

Balkany, T., Hodges, A. V., & Goodman, K. W. (1996). Ethics of cochlear implantation in young children. *Otolaryngology—Head and Neck Surgery, 114*, 748–755.

Birch, G. E., Mason, S. G., & Borisoff, J. F. (2003). Current trends in brain-computer interface research at the Neil Squire foundation. *IEEE Transactions on Neural Systems and Rehabilitation Engineering, 11*(2), 123–126.

Bishop, C. (1996). *Neural networks for pattern recognition.*: New York: Oxford University Press.

Blankertz, B., Dornhege, G., Schafer, C., Krepki, R., Kohlmorgen, J., Muller, K. R., et al. (2003). Boosting bit rates and error detection for the classification of fast-paced motor commands based on single-trial EEG analysis. *IEEE Transactions on Neural Systems and Rehabilitation Engineering, 11*(2), 127–131.

Bliss, T. V. P., & Collingridge, G. L. (1993). A synaptic model of memory: Long-term potentiation in the hippocampus. *Nature, 361*, 31–39.

Buettener, H. M. (1995). Neuroengineering in biological and biosynthetic systems. *Current Opinion in Biotechnology, 6*, 225–229.

Caminiti, R., Johnson, P. B., & Urbano, A. (1990). Making arm movements within different parts of space: Dynamic aspects in the primate motor cortex. *Journal of Neuroscience, 10*, 2039–2058.

Carmena, J. M., Lebedev, M. A., Crist, R. E., O'Doherty, J. E., Santucci, D. M., Dimitrov, D. F., et al. (2003). Learning to control a brain–machine interface for reaching and grasping by primates. *PLoS Biology, 1*, 1–16.

Castellucci, V., Pinsker, H., Kupfermann, I., & Kandel, E. R. (1970). Neuronal mechanisms of habituation and dishabituation of the gill-withdrawal reflex in Aplysia. *Science, 167*, 1745–1748.

Chapin, J. K., Moxon, K. A., Markowitz, R. S., & Nicolelis, M. A. (1999). Real-time control of a robot arm using simultaneously recorded neurons in the motor cortex. *Nature Neuroscience, 2*(7), 664–670.

Chiappalone, M., Vato, A., Tedesco, M. T., Marcoli, M., Davide, F. A., & Martinoia, S. (2003). Networks of neurons coupled to microelectrode arrays: A neuronal sensory system for pharmacological applications. *Biosensors and Bioelectronics, 18*, 627–634.

Cincotti, F., Mattia, D., Babiloni, C., Carducci, F., Salinari, S., Bianchi, L., et al. (2003). The use of EEG modifications due to motor imagery for brain-computer interfaces. *IEEE Transactions on Neural Systems and Rehabilitation Engineering, 11*(2), 131–133.

Clark, A. (2003). *Natural-born cyborgs: Minds, technologies and the future of human intelligence.* Oxford, UK: Oxford University Press.

Clynes, M. E., & Kline, N. S. (1960). *Astronautics.* New York: American Rocket Society.

Donchin, E., Spencer, K. M., & Wijesinghe, R. (2000). The mental prosthesis: Assessing the speed of a P300-based brain-computer interface. *IEEE Transactions on Rehabilitation Engineering, 8,* 174–179.

Edell, D. J. (1986). A peripheral nerve information transducer for amputees: Long-term multichannel recordings from rabbit peripheral nerves. *IEEE Transactions on Biomedical Engineering, BME-33,* 203–214.

Evarts, E. V. (1966). Pyramidal tract activity associated with a conditioned hand movement in the monkey. *Journal of Neurophysiology, 29,* 1011–1027.

Evarts, E. V. (1969). Activity of pyramidal tract neurons during postural fixation. *Journal of Neurophysiology, 32,* 375–385.

Fetz, E. E. (1969). Operant conditioning of cortical unit activity. *Science, 163,* 955–958.

Fromherz, P., Offenhausser, A., Vetter, T., & J., W. (1991). A neuron-silicon junction: A retzius cell of the leech on an insulated-gate field effect transistor. *Science, 252,* 1290–1292.

Fusi, S., Annunziato, M., Badoni, D., Salamon, A., & Amit, D. J. (2000). Spike-driven synaptic plasticity: Theory, simulation, VLSI implementation. *Neural Computation, 12,* 2227–2258.

Fuster, J. M. (1995). *Memory in the cerebral cortex.* Cambridge, MA: MIT Press.

Garrett, D., Peterson, D. A., Anderson, C. W., & Thaut, M. H. (2003). Comparison of linear, nonlinear, and feature selection methods for EEG signal classification. *IEEE Transactions on Neural Systems and Rehabilitation Engineering, 11*(2), 141–144.

Graimann, B., Huggins, J. E., Levine, S. P., & Pfurtscheller, G. (2004). Toward a direct brain interface based on human subdural recordings and wavelet-packet analysis. *IEEE Transactions on Neural Systems and Rehabilitation Engineering, 51,* 954–962.

Grattarola, M., & Martinoia, S. (1993). Modeling the neuronmicrotransducer junction: From extracellular to patch recording. *IEEE Transactions on Biomedical Engineering, 40,* 35–41.

Hoffer, J. A., & Loeb, G. E. (1980). Implantable electrical and mechanical interfaces with nerve and muscle. *Annals of Biomedical Engineering, 8,* 351–360.

Horch, K. (2005). *Neural control.* Paper presented at the Speech to Advisory Panel, DARPA Advanced Prosthesis Workshop, Maryland, January 10–11.

Humphrey, D. R., Schmidt, E. M., & Thompson, W. D. (1970). Predicting measures of motor performance from multiple cortical spike trains. *Science, 170,* 758–761.

Inmann, A., Haugland, M., Haase, J., Biering-Sorensen, F., & Sinkjaer, T. (2001). Signals from skin mechanoreceptors used in control of a hand grasp neuroprosthesis. *Neuroreport, 12,* 2817–2820.

Ito, M. (1989). Long-term depression. *Annual Review of Neuroscience, 12,* 85–102.

Kakei, S., Hoffman, D. S., & Strick, P. L. (1999). Muscle and movement representations in the primary motor cortex. *Science, 285,* 2136–2139.

Kanowitz, S. J., Shapiro, W. H., Golfinos, J. G., Cohen, N. L., & Roland, J. T., Jr. (2004). Auditory brainstem implantation in patients with neurofibromatosis type 2. *Laryngoscope, 114,* 2135–2146.

Kalaska, J. F., Cohon, D. A. D., Hyde, M. L., & Prud'homme, M. (1989). A comparison of movement direction-related versus load direction-related activity in primate motor cortex, using a two-dimensional reaching task. *Journal of Neuroscience, 9,* 2080–2102.

Kennedy, P. R., & Bakay, R. A. (1998). Restoration of neural output from a paralyzed patient by a direct brain connection. *Neuroreport, 9,* 1707–1711.

Kositsky, M., Karniel, A., Alford, S., Fleming, K. M., & Mussa-Ivaldi, F. A. (2003). Dynamical dimension of a hybrid neuro-robotic system. *IEEE Transactions on Neural Systems and Rehabilitation, 11,* 155–159.

Kovacs, G. T., Storment, C. W., James, B., Hentz, V. R., & Rosen, J. M. (1988). Design and Implementation of two-dimensional neural interfaces. *IEEE Engineering in Medicine and Biology Society, Proceedings of the 10th Annual Conference,* New Orleans.

Kanowitz, S. J., Shapiro, W. H., Golfinos, J. G., Cohen, N. L., & Roland, J. T., Jr. (2004). Auditory brainstem implantation in patients with neurofibromatosis type 2. *Laryngoscope, 114,* 2135–2146.

Kuiken, T. A. (2003). Consideration of nerve-muscle grafts to improve the control of artificial arms. *Technology and Disability, 15,* 105–111.

Lacourse, M. G., Cohen, M. J., Lawrence, K. E., & Romero, D. H. (1999). Cortical potentials during imagined movements in individuals with chronic spinal cord injuries. *Behavioral Brain Research, 104,* 73–88.

Loeb, G. E. (1990). Cochlear prosthetics. *Annual Review of Neuroscience, 13,* 357–371.

Loeb, G. E., Richmond, F. J. R., Olney, S., Cameron, T., Dupont, A. C., Hood, K., et al. (1998). Bionic neurons for functional and therapeutic electrical stimulation. *Proceedings of the IEEE-EMBS, 20*, 2305–2309.

Marescaux, J., Leroy, J., Gagner, M., Rubino, F., Mutter, D., Vix, M., et al. (2001). Transatlantic robot-assisted telesurgery.[erratum, Nature, 414, 710]. *Nature, 413*, 379–380.

Marr, D. (1977). Artificial intelligence— personal view. *Artificial Intelligence, 9*, 37–48.

McCreery, D. B., Shannon, R. V., Moore, J. K., & Chatterjee, M. (1998). Accessing the tonotopic organization of the ventral cochlear nucleus by intranuclear microstimulation. *IEEE Transactions on Rehabilitation Engineering, 6*, 391–399.

Merzenich, M. M., & Brugge, J. F. (1973). Representation of the cochlear partition of the superior temporal plane of the macaque monkey. *Brain Research, 50*, 275–296.

Millan, J. R., & Mourino, J. (2003). Asynchronous BCI and local neural classifiers: An overview of the adaptive brain interface project. *Neural Systems and Rehabilitation Engineering, IEEE Transactions on, 11*(2), 159–161.

Murata, S., Kurokawa, H., & Kokaji, S. (1994). Self-assembling machine. In *Proceedings of the 1994 IEEE International Conference on Robotics and Automation* (pp. 441–448).IEEE Publications, Los Alamitos, CA.

Musallam, S., Corneil, B. D., Greger, B., Scherberger, H., & Andersen, R. A. (2004). Cognitive control signals for neural prosthetics. *Science, 305*, 258–262.

Niedermeyer, E., & Lopes Da Sylva, F. (1998). *Electroencephalography: Basic principles, clinical applications, and related fields.* Baltimore, MD: Williams and Wilkins.

Normann, R. A., Maynard, E. M., Rousche, P. J., & Warren, D. J. (1999). A neural interface for a cortical vision prosthesis. *Vision Research, 39*, 2577–2587.

Nudo, R. J., Plautz, E. J., & Frost, S. B. (2001). Role of adaptive plasticity in recovery of function after damage to motor cortex. *Muscle and Nerve, 24*, 1000–1019.

Nudo, R. J., Wise, B. M., Sifuentes, F., & Milliken, G. W. (1996). Neural substrates for the effects of rehabilitative training on motor recovery following ischemic infarct. *Science, 272*, 1791–1794.

Obermaier, B., Guger, C., & Pfurtscheller, G. (1999). Hidden Markov models used for the offline classification of EEG data. *Biomedical Technology (Berlin), 44*(6), 158–162.

Plautz, E. J., Barbay, S., Frost, S. B., Friel, K. M., Dancause, N., Zoubina, E. V., et al. (2002). Induction of novel forelimb representations in peri-infarct motor cortex and motor performance produced by concurrent electrical and behavioral therapy. Program No. 662.2. 2002 Abstract Viewer/Itinerary Planner. Washington, DC: Society for Neuroscience, CD-ROM.

Pohlmeyer, E. A., Miller, L. E., Mussa-Ivaldi, F. A., Perreault, E. J., & Solla, S. A. (2003). Prediction of EMG from Multiple Electrode Recordings in M1. Presented at Abstracts of Neural Control of Movement 13th Annual Meeting.

Potter, S. (2001). Distributed processing in cultured neuronal networks. *Progress in Brain Research, 130*, 49–62.

Pregenzer, M., & Pfurtscheller, G. (1999). Frequency component selection for an EEG-based brain to computer interface. *IEEE Transactions on Rehabilitation Engineering, 7*(4), 413–419.

Rauschecker, J. P. (1999). Making brain circuits listen. *Science, 285*, 1686–1687.

Rauschecker, J. P., & Shannon, R. V. (2002). Sending sound to the brain. *Science,* 1025–1029.

Reger, B. D., Fleming, K. M., Sanguineti, V., Alford, S., & Mussa-Ivaldi, F. A. (2000). Connecting brains to robots: An artificial body for studying the computational properties of neural tissue. *Artificial Life, 6*, 307–324.

Reilly, R. E. (1973). Implantable devices for myoelectric control. In P. Herberts, R. Kadefors, R. I. Magnusson, & I. Petersén (Eds.), *Proceedings of the Conference on the Control of Upper-Extremity Prostheses and Orthoses* (pp. 23–33). Springfield, IL: Charles C. Thomas.

Romo, R., Hernandez, A., Zainos, A., Brody, C. D., & Lemus, L. (2000). Sensing without touching: Psychophysical performance based on cortical microstimulation. *Neuron, 26*(1), 273–278.

Rubinstein, J. T. (2004). How cochlear implants encode speech. *Current Opinion in Otolaryngology and Head and Neck Surgery, 12*(5), 444–448.

Scheidt, R. A., Dingwell, J. B., & Mussa-Ivaldi, F. A. (2001). Learning to move amid uncertainty. *Journal of Neurophysiology, 86*, 971–985.

Scott, S. H., & Kalaska, J. F. (1995). Changes in motor cortex activity during reaching movements with similar hand paths but different arm postures. *Journal of Neurophysiology, 73*, 2563–2567.

Sejnowski, T. J., Koch, C., & Churchland, P. S. (1988). Computational neuroscience. *Science, 241*, 1299–1306.

Serruya, M. D., Caplan, A. H., Saleh, M., Morris, D. S., & Donoghue, J. P. (2004). The Braingate pilot trial: Building and testing novel direct neural output for patients with severe motor impairment (pp. 190–222). Washington, DC: Society for Neuroscience.

Serruya, M. D., Hatsopoulos, N. G., Paninski, L., Fellows, M. R., & Donoghue, J. P. (2002). Instant neural control of a movement signal. *Nature, 416,* 141–142.

Shadmehr, R., & Mussa-Ivaldi, F. A. (1994). Adaptive representation of dynamics during learning of a motor task. *Journal of Neuroscience, 14,* 3208–3224.

Shenoy, K. V., Meeker, D., Cao, S., Kureshi, S. A., Pesaran, B., Buneo, C. A., et al. (2003). Neural prosthetic control signals from plan activity. *Neuroreport, 14,* 591–596.

Shoham, S., Halgren, E., Maynard, E. M., & Normann, R. A. (2001). Motor-cortical activity in tetraplegics. *Nature, 413,* 793.

Talwar, S. K., Xu, S., Hawley, E. S., Weiss, S. A., Moxon, K. A., & Chapin, J. K. (2002). Rat navigation guided by remote control. *Nature, 417,* 37–38.

Taylor, D. M., Tillery, S. I., & Schwartz, A. B. (2002). Direct cortical control of 3D neuroprosthetic devices. *Science, 296,* 1829–1832.

Taylor, D. M., Tillery, S. I., & Schwartz, A. B. (2003). Information conveyed through brain-control: Cursor versus robot. *IEEE Tranactions on Neural Systems and Rehabilitation Engineering, 11,* 195–199.

Thoroughman, K. A., & Shadmehr, R. (2000). Learning of action through adaptive combination of motor primitives. *Nature, 407,* 742–747.

Von Neumann, J. (1958). *The computer and the brain.* New Haven, CT: Yale University Press.

Weir, R. F. f. (2003). Design of artificial arms and hands for prosthetic applications. In K. Myer (Ed.), *Standard handbook of biomedical engineering and design* (pp. 32.31–32.61). New York: McGraw-Hill.

Weir, R. F. f., Troyk, P. R., DeMichele, G., & Kuiken, T. (2003). Implantable myoelectric sensors (IMES) for upper-extremity prosthesis control—preliminary work. In *Proceedings of the 25th Silver Anniversary International Conference of the IEEE Engineering in Medicine and Biology Society (EMBS)* (pp. 1562–1565)..

Wessberg, J., Stambaugh, C. R., Kralik, J. D., Beck, P. D., Laubach, M., Chapin, J. K., et al. (2000). Real-time prediction of hand trajectory by ensembles of cortical neurons in primates. *Nature, 408,* 361–365.

Wolpaw, J. R., Birbaumer, N., McFarland, D. J., Pfurtscheller, G., & Vaughan, T. M. (2002). Brain-computer interfaces for communication and control. *Clinical Neurophysiology, 113,* 767–791.

Wolpaw, J. R., & McFarland, D. J. (1994). Multichannel EEG-based brain-computer communication. *Electroencephalography and Clinical Neurophysiology, 90,* 444–449.

Wolpaw, J. R., & McFarland, D. J. (2004). Control of a two-dimensional movement signal by a non-invasive brain-computer interface in humans. *Proceedings of the National Academy of Sciences, USA, 101,* 17849–17854.

Zeck, G., & Fromherz, P. (2001). Noninvasive neuroelectronic interfacing with synaptically connected snail neurons immobilized on a semiconductor chip. *Proceedings of the National Academy of Sciences, USA, 98,* 10457–10462.

Zelenin, P. V., Deliagina, T. G., Grillner, S., & Orlovsky, G. N. (2000). Postural control in the lamprey: A study with a neuro-mechanical model. *Journal of Neurophysiology, 84,* 2880–2887.

Zrenner, E. (2002). Will retinal implants restore vision? *Science, 295,* 1022–1025.

Zucker, R. S. (1989). Short-term synaptic plasticity. *Annual Review of Neuroscience, 12,* 13–31.

VI

Special Populations

20

Gert Pfurtscheller, Reinhold Scherer, and Christa Neuper

EEG-Based Brain-Computer Interface

A brain-computer interface (BCI) is a system that allows its user to interact with his environment without the use of muscular activity as, for example, hand, foot, or mouth movement (Wolpaw et al., 2002). This would mean that a specific type of mental activity and strategy are necessary to modify brain signals in a predictable way. These signals have to be recorded, analyzed, classified, and transformed into a control signal at the output of the BCI. From the technical point of view, a BCI has to classify brain activity patterns online and in real time. This process is highly subject specific and requires a number of training or learning sessions.

The current and most important application of a BCI is to assist patients who have highly limited motor functions, such as completely paralyzed patients with amyotrophic lateral sclerosis (ALS) or high-level spinal cord injury. In the first case, the BCI can help to realize a spelling device to facilitate communication (Birbaumer et al., 2000), and in the second case the BCI can bypass the damaged neuromuscular channels to control a neuroprosthesis (Pfurtscheller, Müller, Pfurtscheller, Gerner, & Rupp, 2003). Further applications occur in neurofeedback therapy and the promising field of multimedia and virtual reality applications.

Components Defining a BCI

A BCI system is, in general, defined by the following components: type of signal recording, feature of the brain signal used for control, mental strategy, mode of operation, and feedback (see figure 20.1).

The brain signal can be measured in the form of electrical potentials or as blood oxygen level-dependent (BOLD) response using real-time functional magnetic resonance imaging (fMRI; Weiskopf et al., 2003; see also chapter 4, this volume). The electrical potential can be obtained by direct recording from cortical neurons in the form of local field potentials and multiunit activity, electrocorticogram (ECoG), or electroencephalogram (EEG). Intracortical electrode arrays are used for direct brain potential recording (Nicolelis et al., 2003), while electrode arrays and strips are utilized for subdural recording (Levine et al., 2000). The main difference between EEG and subdural or intracortical recordings is that the former is a noninvasive method without any risk (with a poor signal-to-noise ratio) and the latter is an invasive method (but results in a very good signal-to-noise ratio).

A variety of features can be extracted from electrical signals. In the case of intracortical recordings,

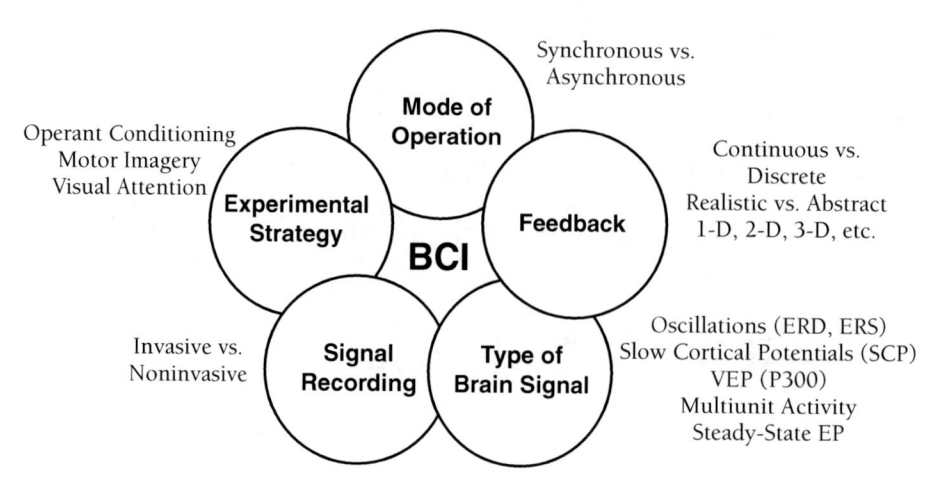

Figure 20.1. Components of a brain-computer interface (BCI).

they include, for example, the firing patterns of cortical neurons. In EEG and ECoG recordings, either slow cortical potential shifts, components of visual evoked potentials, amplitudes of steady-state evoked potentials, or dynamic changes of oscillatory activity are of importance and have to be analyzed and classified (Wolpaw et al., 2000).

One important mental strategy to operate a BCI is motor imagery. Others are focused attention or operant conditioning. In the case of operant conditioning, the subject has to learn (by feedback) to produce, for example, negative or positive slow cortical potential shifts (Birbaumer et al., 1999). Focused attention on a certain visually presented item modifies the P300 component of the visual evoked potential (VEP; Donchin, Spencer, & Wijesinghe, 2000) or enhances the amplitude of the steady-state visual evoked potential (SSVEP; Middendorf, McMillan, Calhoun, & Jones, 2000). Motor imagery changes central mu and beta rhythms in a way similar to that observed during execution of a real movement (Pfurtscheller & Neuper, 2001).

The mode of operation determines when the user performs, for example, a mental task, thereby intending to transmit a message. In principle, there are two distinct modes of operation, the first being externally paced or cue-based (computer-driven, synchronous BCI) and the second internally paced or uncued (user-driven, asynchronous BCI). In the case of a synchronous BCI, a fixed, predefined time window is used. After a visual or auditory cue stimulus, the subject has to produce a specific brain pattern within a predefined time window (of not more than a few seconds) which is simultaneously analyzed. An asynchronous protocol requires the continuous analysis and feature extraction of the recorded brain signal because the user acts at will. Thus, such an asynchronous BCI is in general even more demanding and more complex than a BCI operating with a fixed timing scheme.

Feedback is usually presented in the form of a visualization of the classifier output or an auditory or tactile signal. It is an integral part of the BCI system since the users observe the intended action (e.g., a certain movement of a neuroprosthesis) as they produce the required brain responses.

Focused attention to both a visually presented character (letter) in a P300 paradigm and to a flickering item with evaluation of SSVEPs needs gaze control. Patients in a late stage of ALS have lost such conscious control of eye muscles, and consequently P300 or SSVEP are not suited for communication. We focus therefore on motor imagery as mental strategy because no eye control is necessary.

Motor Imagery Used as Control Strategy for a BCI

Over the last decade, reports based on fMRI demonstrated consistently that primary motor and premotor areas are involved not only in the execution of limb movement but also in the imagination thereof (de Charms et al., 2004; Dechent, Merboldt, & Frahm, 2004; Lotze et al., 1999; Porro et al., 1996). It is difficult, however, to determine the dynamics of these activities based on such imaging techniques that rely on physiological phenomena which lack the necessary fast time response. In contrast, EEG and MEG offer this possibility, though at the cost of reduced spatial resolution.

Two types of changes in the electrical activity of the cortex may occur during motor imagery: One change is time-locked and phase-locked in the form of a slow cortical potential shift (Birbaumer et al., 1999); the other is time-locked but not phase-locked, showing either a desynchronization or synchronization of specific frequency components. The term event-related desynchronization (ERD) means that an ongoing signal presenting a rhythmic component may undergo a decrease in its amount of synchrony, reflected by the disappearance of the spectral peak. Similarly, the term event-related synchronization (ERS) only has meaning if the event (e.g., motor imagery) is followed by an enhanced rhythmicity or spectral peak that was initially not detectable (Pfurtscheller & Lopes da Silva, 1999).

By means of quantification of temporal-spatial ERD (amplitude decrease) and ERS (amplitude increase) patterns, it can be shown that motor imagery can induce different types of activation patterns, as, for example:

1. Desynchronization (ERD) of sensorimotor rhythms (mu rhythm and central beta oscillations; Pfurtscheller & Neuper, 1997)
2. Synchronization (ERS) of the mu rhythm (Neuper & Pfurtscheller, 2001)
3. Short-lasting synchronization (ERS) of central beta oscillations after termination of motor imagery (Pfurtscheller, Neuper, Brunner, & Lopes da Silva, 2005)

Desynchronization and Synchronization of Sensorimotor Rhythms

It is important to note that motor imagery can modify sensorimotor rhythms in a way very similar to that observed during the preparatory phase of actual movement (Neuper & Pfurtscheller, 1999). This means, for example, that imagination of one-sided hand or finger movement results in a mu ERD localized at the contralateral hand representation area. Because motor imagery results in somatotropically organized activation patterns, mental imagination of different movements (e.g., hand, foot, tongue) represent an efficient strategy to operate a BCI.

There is, however, an important point to consider. When a naive subject starts to practice hand motor imagery, generally only a desynchronization pattern is found. Desynchronization of brain rhythms is a relatively unspecific phenomenon and characteristic of most cognitive tasks (Klimesch, 1999). This is extremely detrimental for a BCI operating in an uncued or asynchronous mode, because many false positive decisions will result during resting or idling periods. Training sessions in which the subject receives feedback about the performed mental task are therefore very important in BCI research. Furthermore, in the case of a simple 2-class motor imagery task with imagination of right- versus left-hand movement, an ipsilateral localized ERS often develops as the number of training sessions increases (Pfurtscheller & Neuper, 1997). This is in addition to the usual contralateral localized ERD. Generally, such a contralateral ERD or ipsilateral ERS pattern is associated with an increase in the classification accuracy of single EEG trials. The example in figure 20.2 displays the comparison of band power (11–13 Hz) time courses (ERD/ERS curves) at two electrode positions (left sensorimotor hand area C3 and right sensorimotor hand area C4) between an initial session without feedback and a session with feedback. It can be clearly seen that initially one-sided hand motor imagery revealed only ERD patterns with a clear dominance over the contralateral hemisphere. After feedback training, however, an ipsilateral ERS became apparent. The classification accuracy achieved in the training without feedback session, computed by means of Fisher's linear discriminant analysis (LDA), was 87%. After feedback training, the brain patterns could be classified with 100% accuracy.

This example documents, first, the plasticity of the brain and the dynamics of brain oscillations and, second, the importance of induced brain oscillations or ERS for a high classification accuracy in a BCI.

This interesting observation of a contralateral ERD together with an ipsilateral ERS during hand motor imagery is the manifestation of a phenomenon known as focal ERD/surround ERS (Suffczynski, Kalitzin, Pfurtscheller, & Lopes da Silva, 2001). While the ERD can be seen as a correlate of an activated cortical network, the ERS in the upper alpha band, at least under certain circumstances, can be interpreted as a correlate of a deactivated or

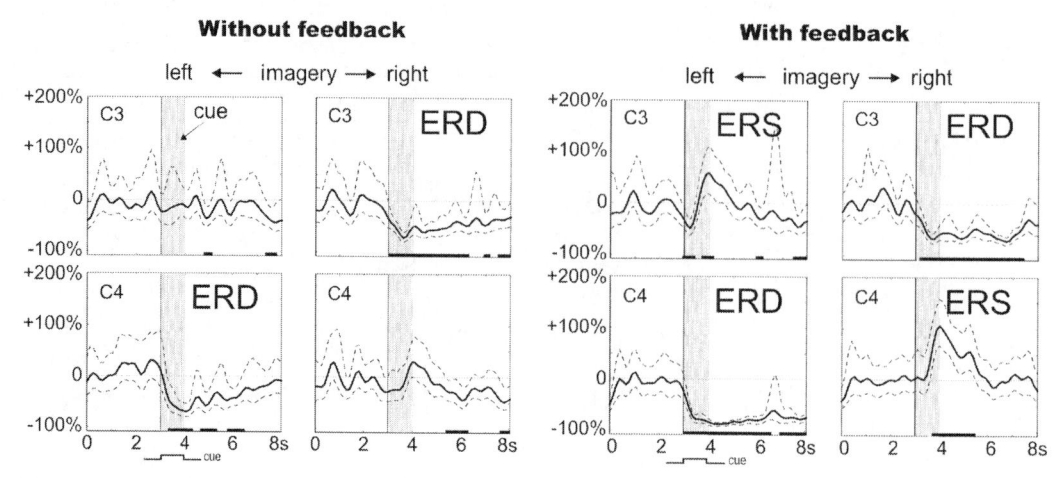

Figure 20.2. Band power (11–13 Hz) time courses ±95% confidence interval displaying event-related desynchronization (ERD) and event-related synchronization (ERS) from training session without feedback (left) and session with feedback (right). Data from one able-bodied subject during imagination of left- and right-hand movement. Gray areas indicate the time of cue presentation.

even inhibited network in an anatomically distinct area. This interaction between different cortical areas can be found not only within the same modality, but also between different modalities (Pfurtscheller & Lopes da Silva, 1999). The combination of the focal ERD and the surround ERS may form a suitable neural mechanism for increasing neural efficiency, thereby optimizing mental strategies for BCI control.

Another example of an intramodal focal ERD/surround ERS is found with the imagination of foot movement. In this case, neural structures are activated in the foot representation area and, as a result, a midcentral focused ERD can be found. Such an ERD pattern is, however, relatively rare since the location of the foot representation area is in the mesial brain surface (Ikeda, Lüders, Burgess, & Shibasaki, 1992) and the potentials of which are not easily accessible by EEG electrodes. In contrast to this rare midcentral ERD, lateralized ERS (in both hand representation areas) is frequently observed (Neuper & Pfurtscheller, 2001). To quantify, five out of nine (trained) able-bodied subjects exhibited hand-area mu ERS during a foot motor imagery task. One typical example is shown on the right hand side of figure 20.3. Foot motor imagery desynchronized the foot-area mu rhythm and enhanced the hand-area mu rhythm in both hemispheres. This can be interpreted such that foot motor imagery activates not only the corresponding representation area but simultaneously deactivates

(inhibits) networks in the hand representation area and synchronizes the hand-area mu rhythm. This is very important in BCI research, because a motor-imagery-induced synchronization of sensorimotor rhythms is an important feature that is used to obtain high classification accuracy. Further support for this idea of intramodal interaction is derived from PET experiments, where a decrease in blood flow has been observed in the somatosensory cortical representation area of one body part (e.g., hand area), whenever attention is diverted to a distant body part (e.g., foot area; Drevets et al., 1995).

For comparison, the data obtained during hand motor imagery in the same subject are also displayed on the left side of figure 20.3. As expected, hand motor imagery induced a marked desynchronization (ERD) at the electrodes overlaying the hand representation area and a synchronization (ERS) at parieto-occipital electrodes. This pattern, central ERD and parieto-occipital ERS, can be seen as an example of an intermodal interaction between motor and extrastriate visual areas, characteristic of an activation of hand-area networks and deactivation or inhibition of parieto-occipital networks. In this respect, it is of interest to refer to the work of Foxe, Simpson, and Ahlfors (1998). They reported an increase of parieto-occipital alpha band activity when the subject was engaged in an auditory attention task. This would indicate an active mechanism for suppressing stimulation in the visual processing areas.

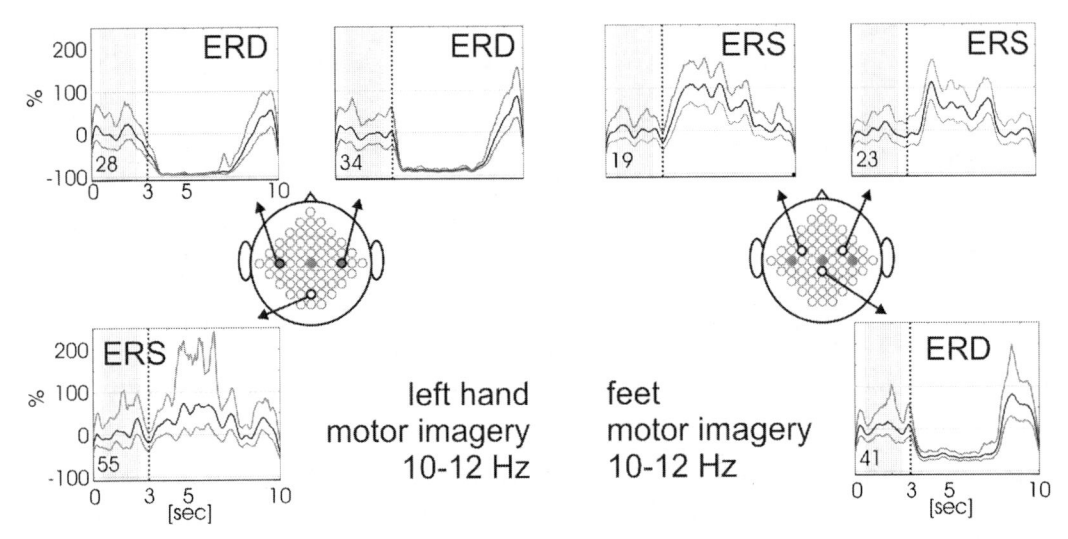

Figure 20.3. Intermodal focal ERD/surround ERS during hand motor imagery (left) and intramodal focal ERD/surround ERS during foot motor imagery (right). Displayed are band power time courses (ERD/ERS) ±95% confidence intervals for selected electrode positions. The dashed line corresponds to the cue onset. ERD, event-related desynchronization; ERS, event-related synchronization.

Synchronization of Central Beta Oscillations

Imagination of foot movement not only synchronizes the hand-area mu rhythm but also produces a beta rebound at the vertex in the majority of subjects. Of nine able-bodied subjects, seven were found to exhibit such a beta rebound (second and third columns, table 20.1). Both a strict localization at the vertex and the fact that the most reactive frequency components were found in the 25–35 Hz band characterized this beta rebound. Such a mid-centrally induced beta rebound after hand motor imagery had been reported earlier (Pfurtscheller & Lopes da Silva, 1999), but it was found to be sporadic by comparison to the beta rebound seen after both feet imagery.

We hypothesize that the occurrence of motor cortex activity, independent of whether it follows the actual execution or just imagination of a movement, may involve at least two networks, one corresponding to the primary motor area and another one in the supplementary motor area (Pfurtscheller et al., 2005). Imagination of both feet movements may involve both the supplementary motor area and the two cortical foot representation areas. Taking into consideration the close proximity of these cortical areas (Ikeda et al., 1992), along with the fact that the response of the corresponding networks in both areas may be synchronized, it is likely that a large-amplitude beta rebound occurs after foot motor imagery. Because of its limited frequency band (between 20 and 35 Hz) and its large magnitude at the vertex, this beta rebound feature is a good candidate for obtaining a high classification accuracy in single-trial EEG classification. In a

Table 20.1. Percentage Band Power Increase and Beta Rebound

Subject	Average Band Power Increase %	Hz	Single-Trial Beta Classification %	Hz
s1	190	29	62	29–31
s2	1059	26	89	26–28
s3	491	25	86	25–27
s4	94	25	66	24–26
s5	320	25	82	24–26
s6	377	27	75	28–30
s7	150	23	59	18–20
Mean ± SD	383 ± 329		74 ± 12	

Note. Percentage band power increase and reactive frequency band after termination of foot motor imagery referred to a 1-second time interval before the motor imagery task (left side). Results of single-trial classification of the beta rebound using Distinction Sensitive Learning Vector Quantization (Pregenzer & Pfurtscheller, 1999) and frequency band with the highest classification accuracy (right side). Only one EEG channel was analyzed, recorded at electrode position Cz. Modified from Pfurtscheller et al., 2005.

pilot study, classification accuracies between 59% and 89% in seven out of nine healthy subjects were achieved when only one EEG channel (electrode position Cz) was classified against rest (fourth and fifth columns, table 20.1).

BCI Training

BCI Training with Feedback

The enhancement of oscillatory EEG activity (ERS) during motor imagery is a very important aspect of BCI research and presumably requires positive reinforcement. Feedback regulation of the sensorimotor oscillatory activity was originally derived from animal experiments, where cats were rewarded for producing increases of the sensorimotor rhythm (SMR; Sterman, 2000). It has been documented over many years that human subjects can also learn to enhance or to suppress rhythmic EEG activity when they are provided with information regarding the EEG changes (e.g., Mulholland, 1995, Neuper, Schlögl, & Pfurtscheller, 1999; Sterman, 1996; Wolpaw & McFarland, 1994; Wolpaw, McFarland, Neat, & Forneris, 1991). The process of acquiring control of brain activity (i.e., to deliberately enhance patterns of oscillatory activity) can therefore be conceptualized as an implicit learning mechanism involving, among other processes, operant learning.

The main rationale of (classifier-based) BCI training is, however, to take advantage of both the learning progress of the individual user and, simultaneously, the learning capability of the system (Pfurtscheller & Neuper, 2001). This implies that two systems (human being and machine) have to be adapted to each other simultaneously to achieve an optimal outcome. Initially, the computer has to learn to recognize EEG patterns associated with one or more states of mental imagery. This implies that the computer has to be adapted to the brain activity of a specific user. After this phase of machine learning, when an appropriate classifier is available, the online classification of single EEG trials can start and feedback can be provided to enable the user's learning, thereby enhancing the target EEG patterns. As a result of feedback training, the EEG patterns usually change, but not necessarily in the desired direction (i.e., divergence may occur). For this reason, the generation of appropriate EEG feedback requires dynamic adjustment of the classifier and of the feedback parameters.

To keep the training period as short as possible, a well thought-out training paradigm is necessary. In this context, two aspects are crucial: (a) the exact manner of how the brain signal is translated into the feedback signal (i.e., information content of the feedback; for advantages of providing continuous or discrete feedback, see McFarland, McCane, & Wolpaw, 1998; Neuper et al., 1999); and (b) the type of feedback presentation (i.e., visual versus auditory feedback; abstract versus concrete or realistic feedback). In any case, the influence of the feedback on the capacity for attention, concentration, and motivation of the user, all aspects that are closely related to the learning process, should be considered. In general, it is important to design an attractive and motivating feedback paradigm. One such example, the so-called basket paradigm, is described in the following section.

BCI Training with a Basket Paradigm

Motivation is crucial for the learning process. The same applies to BCI feedback training. Users have to learn to control their own brain activity by reliably generating different brain patterns. For this purpose, a simple gamelike feedback paradigm was implemented. The object of the game is to move a falling ball horizontally to hit a highlighted target (basket) at the base of a computer monitor. The horizontal position of the ball is controlled by the user's mental activity (BCI classification output), whereas the velocity of fall (defined by the trial length) is constant. Even though the graphical representation of the paradigm is deliberately simple, like those of the first computer games, users achieved good BCI control within a short period of time. The results of a study including four young paraplegic patients showed that three out of the four had satisfying results after a few runs using the basket paradigm (Krausz, Scherer, Korisek, & Pfurtscheller, 2003). Predefined electrode positions (C3 and C4 according to the international 10/20 system) and standard alpha (10–12 Hz) and beta (16–24 Hz) frequency bands were used to generate the feedback. A Butterworth filter and simple amplitude squaring was applied for the band power estimation of the acquired EEG signal (analog band pass between 0.5 and 30 Hz, sample rate 128 Hz). The feature values used for classification were calculated by averaging across a

1-second time window that was shifted sample-by-sample along the band power estimates. The patients had the task of landing the ball on the basket (situated on either the left or right half of the base of the screen). For this two-class problem, an LDA classifier was used. The classifier was computed by analyzing (10 × 10 cross-validation) cue-based motor-imagery-related EEG patterns collected during a motor imagery assessment session at the beginning of the study: The patients were required to imagine the execution of movements corresponding to each of a sequence of visual cues presented. By analysis of the resulting brain patterns for each subject, the most suited (discriminable) motor imagery tasks were found, and the classifier output (position of the ball) was weighted to adjust the mean deflection to the middle of the target basket. In this way, the BCI output was uniquely adapted to each patient. It was found that online classification accuracies of 85% and above could be achieved in a short time (3 days, 1.5 hours per day).

BCI Application for Severely Paralyzed Patients

Completely paralyzed patients without any conscious control of muscle activity can only communicate with their environment when, through the use of EEG signals, an electronic spelling system is controlled. There is evidence that a BCI based on oscillatory EEG changes, induced by motor imagery, can be utilized to restore communication in severely disabled people (Neuper, Müller, Kübler, Birbaumer, & Pfurtscheller, 2003). Even completely paralyzed patients, who had lost all voluntary muscle control, learned to control how to enhance or suppress specific frequency components of the sensorimotor EEG by using a motor imagery strategy. In order for such a patient to obtain control over brain oscillations, BCI training sessions have to be conducted regularly (i.e., 2 times a week) over a period of several months.

Here we report, as an example, the case of a male patient (60 years old) who suffered from ALS for more than 5 years, being artificially ventilated. This patient was totally paralyzed and had almost lost his ability to communicate altogether. Initially, the patient was trained to produce two distinct EEG patterns by using an imagery strategy. For this, the so-called basket paradigm, as described above (for details, see Krausz et al., 2003), was employed. It requires continuous (one-dimensional) cursor control to direct a falling ball on a computer monitor into one of two baskets (the target is indicated in each trial). The EEG signal (band pass 5–30 Hz, sampling rate 128 Hz) used for classification and feedback was initially recorded from three bipolar channels over the left and right sensorimotor areas and close to the vertex. To generate the feedback, two frequency bands were used (8–12 Hz and 18–30 Hz). The feedback was calculated by a linear discriminant classifier, which was developed to discriminate between two brain states (Pfurtscheller, Neuper, Flotzinger, & Pregenzer, 1997). In 17 training days, the patient performed 82 runs with the basket paradigm. The effectiveness of the training is suggested by the significant increase of the classification accuracy from random level at the beginning (mean over 23 runs, performed on the first two training days) to an average classification rate of 83% (mean over 22 runs, performed on the last six training days). At the end of the reported training period, this patient was able to voluntarily produce two distinct EEG patterns, one being characterized by a broad-banded desynchronization and the other by a synchronization in the form of induced oscillations (ERS) in the alpha band (see figure 20.4). This EEG control enabled the patient to later use the so-called virtual keyboard (VK; Obermaier, Müller, & Pfurtscheller, 2003) for copy spelling of presented words and, finally, for free spelling of a short message. Since the design of a convenient BCI spelling system is still a matter of current research, the following section provides a short overview of different approaches.

BCI-Based Control of Spelling Systems

The VK spelling system presented in this section is an important application of the Graz-BCI (Guger et al., 2001; Pfurtscheller & Neuper, 2001; Scherer, Schlögl, Müller-Putz, & Pfurtscheller, 2004). As described above, it is based on the detection and classification of motor imagery-related EEG patterns.

The basic VK allows the selection of letters from an alphabet by making a series of binary decisions (Birbaumer et al., 1999; Obermaier et al., 2003). This means that the user has to learn to reliably reproduce two different EEG patterns (classes).

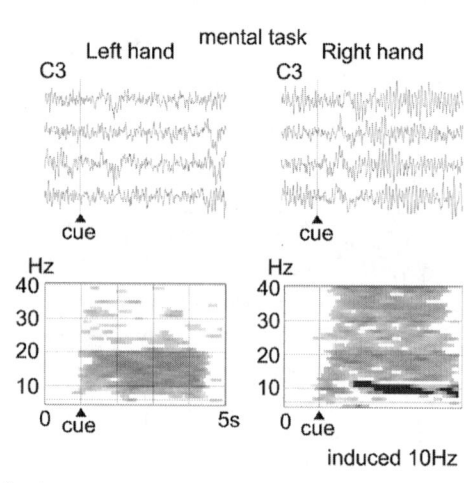

Figure 20.4. Picture of a patient suffering from ALS during feedback training. In the upper right corner, the basket paradigm is shown (left). Examples of trials and event-related desynchronization and event-related synchronization time/frequency maps display ipsilateral ERS (right). See also color insert.

Starting with the complete alphabet displayed on a screen, subsets of decreasing size are successively selected until the desired letter is one of two options. A dichotomous structure with five consecutive levels of selection (corresponds to $2^5 = 32$ different letters) and two further levels of confirmation (OK) and correction (BACK/DEL) is implemented (see figure 20.5, top left). BACK allows a cancellation of the previous subset selection by returning to the higher selection level, while choosing DEL results in deletion of the last written letter. A measure for the communication performance is the spelling rate σ, given as correctly selected letters per minute. A study using healthy subjects showed that spelling rates between 0.67 and 1.02 letters per minute could be achieved when using a trial length of 8 seconds. Six correct selections are required per letter, resulting in a maximum $\sigma = (60 \text{ sec}/6.8 \text{ sec}) = 1.25$ letters per minute (Obermaier et al., 2003). A bar, either on the left- or right-hand side of the screen, was presented as feedback (figure 20.5, top right). Its purpose was to indicate either the first or the second subset. The subjects were required to spell predefined words by selecting the appropriate letter subgroup by motor imagery. Hidden Markov models (Rabiner & Juang, 1986) were used to characterize the motor imagery-related EEG and classification based on maximum selection of likelihood.

The dichotomous selection strategy is a simple but efficient method to make a single choice (letter) from a number of items (alphabet). The selection of 1 item from a set of 32 items requires only five binary selections. Although the achieved spelling rates are low and do not reflect an appropriate communication speed, nonneuromuscular interaction with one's environment is possible. A major problem of this method, however, comes with a misclassification. One false selection requires, depending on the selection level, up to 13 correct selections in order to cancel the mistake. Given the fact that subjects usually start with a classification rate of 70% (which means that 3 out of 10 EEG patterns are misclassified, thereby greatly decreasing the communication rate), it is clear that training to achieve a high classification accuracy and consequently a good adaptation between the subject and the BCI is necessary. For the majority of the subjects, a longer training period results in better BCI performance.

There is also potential to increase the spelling rate in other ways. One possibility is the use of word prediction. The prediction can be based on considering language-specific probability occurrences of letters (e.g., the letter *e* has the highest probability of occurrence in the German and English languages) and their intra- and interword position. The prediction can also be based on a predefined dictionary such as the T9 (Tegic Communications, Inc.) predictive text input system (currently implemented in many mobile phones). A modified version based on eight keys was implemented for the two-class cue-based VK. Again, a dichotomous structure such as that described above is used (figure 20.5, bottom left). With this system, we demonstrated that healthy subjects could achieve spelling rates (σ) of up to

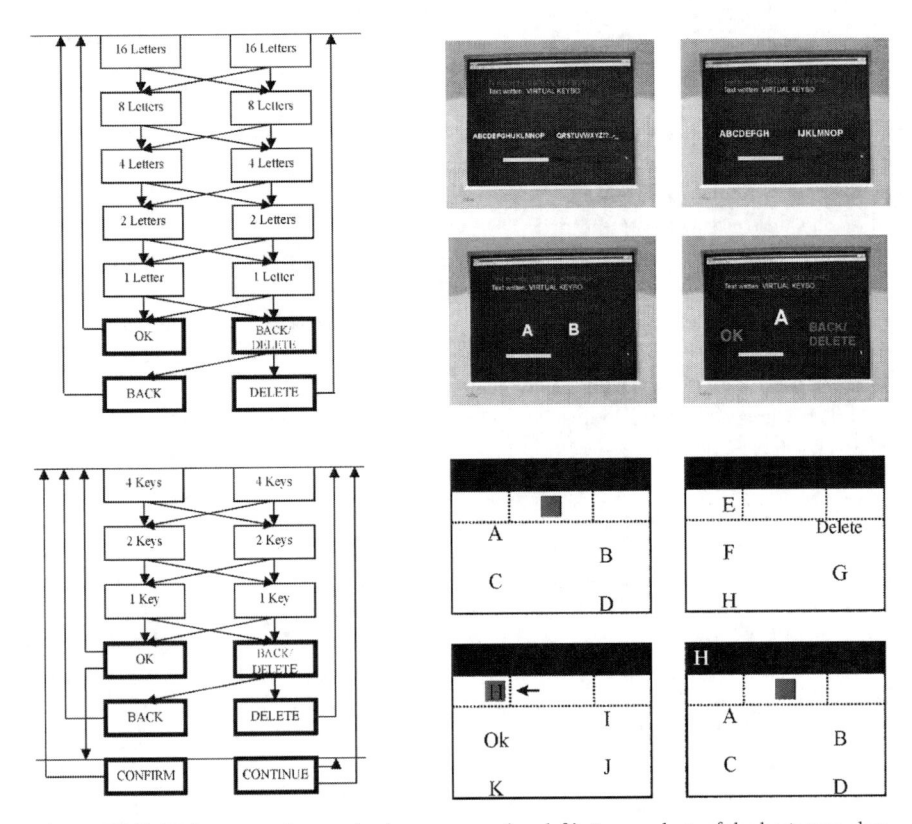

Figure 20.5. Dichotomous letter selection structure (top left). Screen shots of the basic two-class virtual keyboard. Four of the six steps to perform to select a letter are shown (top right). Letter selection structure based on T8 (bottom left). Three-class asynchronously controlled virtual keyboard. Example: Selection of the letter H by virtual keyboard (bottom right). See also color insert.

4.24 letters per minute when a dictionary containing 145 words was used.

A further possibility to speed up communication performance is to use more than two reliably detectable brain patterns. If the two-class procedure described above is divided into three instead of two parts, fewer selection steps are necessary. A decrease of the trial length would also increase the communication performance.

Another important factor is classification accuracy. The information transfer rate

$$\text{ITR [bit/min]} = \left(\log_2 N + P \log_2 P + (1-P) \log_2 \frac{1-P}{N-1} \right) \frac{60}{\text{trial length}}$$

describes the relationship between number of classes (N), classification accuracy (P), and trial length (Wolpaw et al., 2000). A large N, a high P, and a short trial length result in an increased ITR.

A VK based on three classes that additionally uses the asynchronous operating mode has been introduced (Scherer, Müller, Neuper, Graimann, & Pfurtscheller, 2004). The asynchronous mode frees subjects from the constraints of a fixed timing scheme. Whenever the subject is ready to generate a motor imagery-related brain pattern, the BCI is able to detect and process it. The new operation mode also allows a new VK design. The screen is divided into two parts: A small upper part displays the selected letters and words, while the letter selection process and the visual biofeedback (moving cursor) take up the remaining lower part of the screen. The letters of the alphabet are arranged alphabetically on two vertically moving assembly lines on the left and right half of the screen. A vertical displacement between the letters on both sides should avoid the sensation of competition between the single letters. Every five letters, a control command can be selected: DEL, used to delete the last spelled letter, and OK to confirm the spelled word.

At all times, five items are visible on each side. As long as the first of the three brain patterns is detected, the items on both sides of the screen scroll from the bottom to the top. If an item reaches the top of the selection area, it disappears and the next one appears from the bottom. In order to avoid disturbing influences and give subjects the opportunity to concentrate on the moving objects, the feedback cursor was hidden during the scrolling process. The item at the topmost position can be selected by moving the feedback cursor toward the desired left or right direction by performing the second or third motor imagery-related brain pattern. An item is selected if the cursor exceeds a subject-specific left- or right-side threshold for a subject-specific time period. Selected letters are presented in the upper part of the screen and the spelling procedure starts over again. If none of the trained brain patterns is detected, the VK goes into the standby mode (see figure 20.5, bottom right). A study with healthy subjects revealed spelling rates (σ) of up to 3.38 letters per minute (mean 1.99 letters per minute). Compared to the standard two-class VK, double in performance could be achieved. Another outcome of the study was the conclusion that not every subject was able to control the application, even if the classification accuracy was higher than 90%. A possible explanation for this may be the short period available for the training. Three classes and asynchronous control seems to be more demanding than the synchronous two-class control mode. Also, the dynamic scenario of many moving objects may have disturbing influences and cause changes to, or even a deterioration of, the EEG. The BCI analyzes the ongoing EEG by estimating the band power in subject-specific bands (optimized by means of a genetic algorithm; Holland, 1975).

Three LDA classifiers were trained to discriminate in each case between two out of the three classes. The overall classification output was generated by combining the results of the three LDA classifiers (majority voting).

BCI-Based Control of Functional Electrical Stimulation in Tetraplegic Patients

For patients with high spinal cord injury, the restoration of grasp function by means of an orthosis (Pfurtscheller et al., 2000) or neuroprostheses (Heasman et al., 2002; Pfurtscheller et al., 2003) has developed into an important area of application.

We had the opportunity to work with a tetraplegic patient (complete motor and sensory lesion below C5) for several months. During this time, the patient learned to reliably induce 17 Hz oscillations in EEG (modulated by foot motor imagery) recorded from electrode position Cz. By estimating the band power in the frequency band between 15 and 19 Hz and by applying simple threshold classification, the foot motor imagery brain pattern can be detected with an accuracy of 100% (see figure 20.6). Since the patient learned to generate the brain pattern at will, the asynchronous operating mode could be used to control a neuroprosthesis based on surface functional electrical stimulation (Pfurtscheller et al., 2003). Every time the patient would like to grasp an object, he can consecutively initiate ("brain-switch") phases of the grasp. In the case of a lateral grasp (e.g., to use a spoon), three phases have to be executed in this order: (1) finger and thumb extension (hand opens); (2) finger flexion and thumb extension (fingers close); and (3) finger and

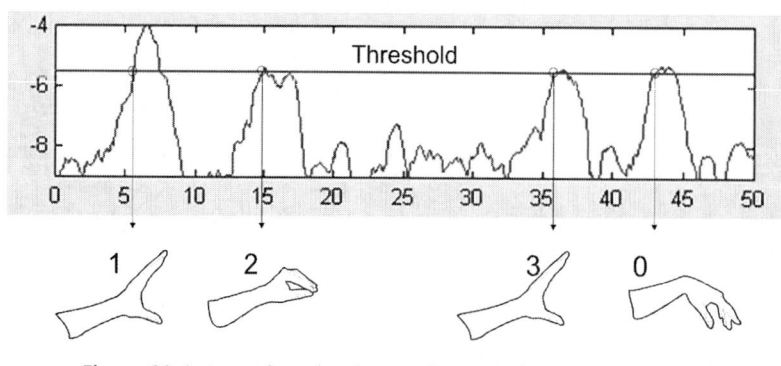

Figure 20.6. Logarithmic band power (15–19 Hz) with threshold used for detection and the phases of the lateral grasp (left). Picture of the patient with surface functional electrical stimulation (right).

thumb flexion (thumb moves against closed fingers). In this way, the quality of life of the patient improved dramatically because of the independence gained (Pfurtscheller et al., 2005). Another patient (complete motor and sensory lesion below C5) with whom we had the opportunity to work had a neuroprosthesis (Freehand system; Keith et al., 1989) implanted in his right arm. During a three-day training program, the patient learned to generate two different brain patterns. First the patterns were reinforced using the cue-based basket paradigm, and then during free training the patient learned to produce the patterns voluntarily (asynchronous mode). In the latter case, the feedback was presented in the form of a ball, placed in the middle of the screen, which could be moved either to the left or to the right side. The EEG signals (C3 and Cz) were analyzed continuously and fed back to the patient. After the short training period, the patient was able to operate the neuroprosthesis by mental activity alone. These results show that a BCI may also be an alternative approach for clinical purposes.

Perspectives for the Future

User acceptance of a BCI depends on its reliability in obtaining a high information transfer rate (ITS) within a minimum number of training sessions and its practicability of application (e.g., reduction of number of electrodes required). The ITS can be increased by a corresponding increase in the number of recognizable mental tasks and also by achieving BCI operation in an asynchronous mode. A realistic projection for the future would see three or four mental tasks with noninvasive EEG recordings being utilized, where the type and optimal combination of mental tasks would have to be selected in separate sessions for each user. To obtain high accuracy in single trial classifications, it is important not only to use powerful feature extraction methods but also to search for spatiotemporal EEG patterns displaying task-related synchronization or ERS. Recognition of such ERS phenomena is a prerequisite for a high hit rate and a low false positive detection rate in an asynchronous BCI.

The ITS can be increased when subdural ECoG recordings are available. In this case, the signal-to-noise ratio is much higher than that of the EEG, and high-frequency components such as gamma band oscillations can be used for classification. ECoG recordings in patients therefore hold great promise, and their utilization is an immediate challenge for the near future.

When the EEG is used as the input signal for a BCI system, multichannel recordings and special methods of preprocessing (i.e., independent component analysis, ICA) are recommended. Sensorimotor rhythms such as mu and central beta are particularly suitable for ICA because both are spatially stable and can therefore be separated easily from other sources (Makeig et al., 2000). More extensive research is also needed in order to extract as many possible features from the EEG, from which a small number may be selected to optimize the quality of the classification system. For this feature selection, different algorithms have been proposed such as distinction sensitive learning vector quantification, an extension of Kohonen's learning vector quantification (Pregenzer & Pfurtscheller, 1999) and the genetic algorithm (Graimann, Huggins, Levine, & Pfurtscheller, 2004).

More extensive work is also needed to specify the mental task and to optimize the feedback presentations. So, for example, hand motor imagery can be realized either by visualization of one's own or another's hand movement (visuomotor imagery) or by remembering the feeling of hand movement (kinesthetic motor imagery; Curran & Stokes, 2003). Both types of mental strategies result in different activation patterns and have to be investigated in detail. The feedback is important to speed up the training period and to avoid fatigue by the user. The importance of visual feedback and its impact on premotor areas was addressed by Rizzolatti, Fogassi, and Gallese (2001).

EEG recording with subdural electrodes is an invasive method of importance in neurorehabilitation engineering (Levine et al., 2000). The advantage of the ECoG over the EEG is its better signal-to-noise ratio and therewith also the easier detection of, for example, gamma activity. Patient-oriented work with subdural electrodes and ECoG single trial classification have shown the importance of gamma activity in the frequency band of 60–90 Hz (Graimann, Huggins, Schlogl, Levine, & Pfurtscheller, 2003).

A number of studies in monkeys have demonstrated that 3-D control is possible when multiunit activity is recorded in cortical areas and firing patterns are analyzed (Serruya, Hatsopoulos, Paninski, Fellows, & Donoghue, 2002; Wessberg et al., 2000). This technique is highly invasive and based on the recording from multiple electrode arrays, each of which with up to 100 sensors implanted in different cortical motor areas (Maynard et al., 1997).

Acknowledgments These investigations were supported in part by the Allgemeine Unfallversicherungsanstalt (AUVA), the Lorenz Böhler Gesellschaft, and the Ludwig Boltzmann Institute of Medical Informatics and Neuroinformatics, Graz University of Technology.

Main Points

1. The preparatory phase of an actual movement and the imagination of the same movement (motor imagery) can modify sensorimotor rhythms in a very similar way.
2. Motor imagery-induced changes in oscillatory electroencephalogram activity can be quantified (event-related desynchronization, ERD/ERS), classified, and translated into control commands.
3. The mutual adaptation of machine and user, as well as feedback training, is crucial to achieve optimal control.
4. Existing applications show that BCI research is ready to leave the laboratory and prove its suitability in assisting disabled people in everyday life.

Key Readings

Introductory Readings

Pfurtscheller, G., & Neuper, C. (2001). Motor imagery and direct brain-computer communication. *Proceedings of the IEEE, 89*, 1123–1134.

Wolpaw, J. R., Birbaumer, N., McFarland, D. J., Pfurtscheller, G., & Vaughan, T. M. (2002). Brain-computer interfaces for communication and control. *Clinical Neurophysiology, 113*, 767–791.

Applications

Birbaumer, N., Ghanayim, N., Hinterberger, T., Iversen, I., Kotchoubey, B., Kubler, A., et al. (1999). A spelling device for the paralyzed. *Nature, 398*, 297–298.

Krausz, G., Scherer, R., Korisek, G., & Pfurtscheller, G. (2003). Critical decision-speed and information transfer in the "Graz Brain-Computer Interface." *Applied Psychophysiology and Biofeedback, 28*(3), 233–240.

Neuper, C., Müller, G., & Kübler, A. (2003). Clinical application of an EEG-based brain-computer interface: A case study in a patient with severe motor impairment. *Clinical Neurophysiology, 114*, 399–409.

Obermaier, B., Müller, G. R., & Pfurtscheller, G. (2003). "Virtual keyboard" controlled by spontaneous EEG activity. *IEEE Transactions on Neural Systems and Rehabilitation Engineering, 11*, 422–426.

Pfurtscheller, G., Müller, G. R., Pfurtscheller, J., Gerner, H. J., & Rupp, R. (2003). "Thought"-control of functional electrical stimulation to restore hand grasp in a patient with tetraplegia. *Neuroscience Letters, 351*, 33–36.

References

Birbaumer, N., Ghanayim, N., Hinterberger T., Iversen, I., Kotchoubey, B., Kubler, A., et al. (1999). A spelling device for the paralyzed. *Nature, 398*, 297–298.

Birbaumer, N., Kübler, A., Ghanayim, N., Hinterberger, T., Perelmouter, J., Kaiser, J., et al. (2000). The thought translation device (TTD) for completely paralyzed patients. *IEEE Transactions on Rehabilitation Enineering, 8*, 190–193.

Curran, E. A., & Stokes, M. J. (2003). Learning to control brain activity: A review of the production and control of EEG components for driving brain-computer interface (BCI) systems. *Brain and Cognition, 51*, 326–336.

de Charms, R. C., Christoff, K., Glover, G. H., Pauly J. M., Whitfield, S., & Gabrieli, J. D. (2004). Learned regulation of spatially localized brain activation using real-time fMRI. *Neuroimage, 21*, 436–443.

Dechent, P., Merboldt, K. D., & Frahm, J. (2004). Is the human primary motor cortex involved in motor imagery? *Cognitive Brain Research, 19*, 138–144.

Donchin, E., Spencer, K. M., & Wijesinghe, R. (2000) The mental prosthesis: Assessing the speed of a P300-based brain-computer interface. *IEEE Transactions on Rehabilitation Engineering, 8*, 174–179.

Drevets, W. C., Burton, H., Videen, T. O., Snyder, A. Z., Simpson, J. R., Jr, Raichle, M. E. (1995). Blood flow changes in human somatosensory cortex during anticipated stimulation. *Nature, 373*, 249–252.

Foxe, J. J., Simpson, G. V., & Ahlfors, S. P. (1998). Parieto-occipital ~10 Hz activity reflects anticipatory state of visual attention mechanisms. *Neuroreport, 9*, 3929–3922.

Graimann, B., Huggins, J. E., Levine, S. P., & Pfurtscheller, G. (2004). Toward a direct brain interface based on human subdural recordings and wavelet-packet analysis. *IEEE Transactions on Biomedical Engineering, 51*, 954–962.

Graimann, B., Huggins, J. E., Schlogl, A., Levine, S. P., & Pfurtscheller, G. (2003). Detection of movement-related desynchronization patterns in ongoing single-channel electrocorticogram. *IEEE Transactions on Neural Systems Rehabilitation Engineering, 11*, 276–281.

Guger, C., Schlögl, A., Neuper, C., Walterspacher, D., Strein, T., & Pfurtscheller, G. (2001). Rapid prototyping of an EEG-based brain-computer interface (BCI). *IEEE Transactions on Neural Systems Rehabilitation Engineering, 9*, 49–58.

Heasman, J. M., Scott, T. R., Kirkup, L., Flynn, R. Y., Vare, V. A., & Gschwind, C. R. (2002). Control of a hand grasp neuroprosthesis using an electroencephalogram-triggered switch: Demonstration of improvements in performance using wavepacket analysis. *Medical and Biological Engineering and Computation, 40*, 588–593.

Holland, J. H. (1975). *Adaptation: Natural and artificial systems.* Ann Arbor: University of Michigan Press.

Ikeda, A., Lüders, H. O., Burgess, R. C., & Shibasaki, H. (1992). Movement-related potentials recorded from supplementary motor area and primary motor area: Role of supplementary motor area in voluntary movements. *Brain, 115*, 1017–1043.

Keith, M. W., Peckham, P. H., Thrope, G. B., Stroh, K. C., Smith, B., Buckett, J. R., et al. (1989). Implantable functional neuromuscular stimulation in the tetraplegic hand. *Journal of Hand Surgery [Am], 14*, 524–530.

Klimesch, W. (1999). EEG alpha and theta oscillations reflect cognitive and memory performance: A review and analysis. *Brain Research Reviews, 29*, 169–195.

Krausz, G., Scherer, R., Korisek, G., & Pfurtscheller, G. (2003). Critical decision-speed and information transfer in the "Graz Brain-Computer Interface." *Applied Psychophysiology and Biofeedback, 28*, 233–240.

Levine, S. P., Huggins, J. E., BeMent, S. L., Kushwaha, R. K., Schuh, L. A., Rohde, M. M., et al. (2000). A direct brain interface based on event-related potentials. *IEEE Transactions on Rehabilitation Engineering, 8*, 180–185.

Lotze, M., Montoya, P., Erb, M., Hulsmann, E., Flor, H., Klose, U., et al. (1999). Activation of cortical and cerebellar motor areas during executed and imaginated hand movements: A functional MRI study. *Journal of Cognitive Neuroscience, 11*, 491–501.

McFarland, D. J., McCane, L. M., & Wolpaw, J. R. (1998). EEG-based communication and control: Short-term role of feedback. *IEEE Transactions on Rehabilitation Engineering, 6*, 7–11.

Middendorf, M., McMillan, G., Calhoun, G., & Jones, K. S. (2000). Brain-computer interfaces based on the steady-state visual-evoked response. *IEEE Transactions on Rehabilitation Engineering, 8*, 211–214.

Mulholland, T. (1995). Human EEG, behavioral stillness and biofeedback. *International Journal of Psychophysiology, 19*, 263–279.

Neuper, C., Müller, G., Kübler, A., Birbaumer, N., & Pfurtscheller, G. (2003). Clinical application of an EEG-based brain-computer interface: A case study in a patient with severe motor impairment. *Clinical Neurophysiology, 114*, 399–409.

Neuper, C., & Pfurtscheller, G. (1999). Motor imagery and ERD. In G. Pfurtscheller & F. H. Lopes da Silva (Eds.), *Event-related desynchronisation: Handbook of electroencephalography and clinical neurophysiology* (Vol. 6, rev. ed., pp. 303–325). Amsterdam: Elsevier.

Neuper, C., & Pfurtscheller, G. (2001). Event-related dynamics of cortical rhythms: Frequency-specific features and functional correlates. *International Journal of Psychophysiology, 43*, 41–58.

Neuper, C., Schlögl, A., & Pfurtscheller, G. (1999). Enhancement of left-right sensorimotor EEG differences during feedback-regulated motor imagery. *Journal of Clinical Neurophysiology, 16*, 373–382.

Nicolelis, M. A. (2003). Brain-machine interfaces to restore motor function and probe neural circuits. *Nature Reviews Neuroscience, 4*, 417–422.

Obermaier, B., Müller, G. R., & Pfurtscheller, G. (2003). "Virtual keyboard" controlled by spontaneous EEG activity. *IEEE Transactions on Neural Systems Rehabilitation Engineering, 11*, 422–426.

Pfurtscheller, G., Guger, C., Müller, G., Krausz, G., & Neuper, C. (2000). Brain oscillations control hand orthosis in a tetraplegic. *Neuroscience Letters, 292*, 211–214.

Pfurtscheller, G., & Lopes da Silva, F. H. (1999). Functional meaning of event-related desynchronization (ERD) and synchronization (ERS). In G. Pfurtscheller & F. H. Lopes da Silva (Eds.), *Event-related desynchronisation: Handbook of electroencephalography and clinical neurophysiology* (Vol. 6, rev. ed., pp. 51–65). Amsterdam: Elsevier.

Pfurtscheller, G., Müller, G. R., Pfurtscheller, J., Gerner, H. J., & Rupp, R. (2003). "Thought"-control of functional electrical stimulation to restore hand grasp in a patient with tetraplegia. *Neuroscience Letters, 351*, 33–36.

Pfurtscheller, G., & Neuper, C. (1997). Motor imagery activates primary sensorimotor area in humans. *Neuroscience Letters, 239*, 65–68.

Pfurtscheller, G., & Neuper, C. (2001). Motor imagery and direct brain-computer communication. *Proceedings of the IEEE, 89,* 1123–1134.

Pfurtscheller, G., Neuper, C., Brunner, C., & Lopes da Silva, F. H. (2005). Beta rebound after different types of motor imagery in man. *Neuroscience Letters, 378,* 156–159.

Pfurtscheller, G., Neuper, C., Flotzinger, D., & Pregenzer, M. (1997). EEG-based discrimination between imagination of right and left hand movement. *Electroencephalography and Clinical Neurophysiology, 103,* 642–651.

Porro, C. A., Francescato, M. P., Cettolo, V., Diamond, M. E., Baraldi, P., Zuiani, C., et al. (1996). Primary motor and sensory cortex activation during motor performance and motor imagery: A functional magnetic resonance imaging study. *Journal of Neuroscience, 16,* 7688–7698.

Pregenzer, M., & Pfurtscheller, G. (1999). Frequency component selection of an EEG-based brain to computer interface. *IEEE Transactions on Rehabilitation Engineering, 7,* 413–419.

Rabiner, L., & Juang, B. (1986). An introduction to hidden Markov models. *IEEE ASSP Magazine, 3,* 4–16.

Rizzolatti, G., Fogassi, L., & Gallese, V. (2001). Neurophysiological mechanisms underlying the understanding and imitation of action. *Perspectives, 2,* 661–670.

Scherer, R., Müller, G. R., Neuper, C., Graimann, B., & Pfurtscheller, G. (2004). An asynchronously controlled EEG-based virtual keyboard: Improvement of the spelling rate. *IEEE Transactions on Biomedical Engineeering, 51,* 979–984.

Scherer, R., Schlögl, A., Müller-Putz, G. R., & Pfurtscheller, G. (2004). Inside the Graz-BCI. In G. R. Müller-Putz, C. Neuper, A. Schlögl, & G. Pfurtscheller (Eds.), *Biomedizinische technik: Proceedings of the 2nd International Brain-Computer Interface Workshop and Training Course* (pp. 81–82). Graz, Austria.

Serruya, M. D., Hatsopoulos, N. G., Paninski, L., Fellows, M. R., & Donoghue, J. P. (2002). Instant neural control of a movement signal. *Nature, 416,* 141–142.

Sterman, M. B. (1996). Physiological origins and functional correlates of EEG rhythmic activities: Implications for self-regulation. *Biofeedback Self-Regulation, 21,* 3–33.

Sterman, M. B. (2000). Basic concepts and clinical findings in the treatment of seizure disorders with EEG operant conditioning [Review]. *Clinical Electroencephalogram, 31,* 45–55.

Suffczynski, P., Kalitzin, S., Pfurtscheller, G., & Lopes da Silva, F. H. (2001). Computational model of thalamo-cortical networks: Dynamical control of alpha rhythms in relation to focal attention. *International Journal of Psychophysiology, 43,* 25–40.

Weiskopf, N., Veit, R., Erb, M., Mathiak, K., Grodd, W., Goebel, R., et al. (2003). Physiological self-regulation of regional brain activity using real-time functional magnetic resonance imaging (fMRI): Methodology and exemplary data. *Neuroimage, 19,* 577–586.

Wessberg, J., Stambaugh, C. R., Kralik, J. D., Beck, P. D., Laubach, M., Chapin, J. K., et al. (2000). Real-time prediction of hand trajectory by ensembles of cortical neurons in primates. *Nature, 408,* 361–365.

Wolpaw, J. R., Birbaumer, N., Heetderks, W. J., McFarland, D. J., Peckham, P. H., Schalk, G., et al. (2000). Brain-computer interface technology: A review of the first international meeting. *IEEE Transactions on Rehabilitation Engineering, 8,* 164–173.

Wolpaw, J. R., Birbaumer, N., McFarland, D. J., Pfurtscheller, G., & Vaughan, T. M. (2002). Brain-computer interfaces for communication and control. *Clinical Neurophysiology, 113,* 767–791.

Wolpaw, J. R., & McFarland, D. J. (1994). Multichannel EEG-based brain-computer communication. *Electroencephalography and Clinical Neurophysiology, 90,* 444–449.

Wolpaw, J. R., McFarland, D. J., Neat, G. W., & Forneris, C. A. (1991). An EEG-based brain-computer interface for cursor control. *Electroencephalography and Clinical Neurophysiology, 78,* 252–259.

21

Dorothe A. Poggel, Lotfi B. Merabet, and Joseph F. Rizzo III

Artificial Vision

The main guideline of research in neuroergonomics is to find out how the brain copes with the massive amount and variety of information that must be processed to carry out complex tasks of everyday life (Parasuraman, 2003). From the viewpoint of visual neuroscience, our primary attention is addressed to the question of how the brain encodes visual information and how we can learn more about the neural mechanisms underlying vision. Such a pursuit is relevant not only to developing an understanding of visual physiology per se, but also to providing insight into environmental modifications to assist and enhance human performance. Our long-term goal is to learn how to "talk" to the brain—and create vision where the eye and the brain fail, so that we might be able to improve the quality of life for some visually impaired patients.

Based on the dramatic evolution of microtechnology over the past decades, it is now possible to create devices, that is, neuroprostheses, that have the potential to restore functions even though elements of the sensory or motor systems are damaged or lost, for example, due to a disease process like macular degeneration, or due to lesions induced by surgery, brain trauma, or stroke. In the auditory domain, the attempt to artificially stimulate peripheral parts of the pathway in deaf individuals by means of a cochlear implant has proven highly successful. Cochlear implants are capable of restoring hearing and providing language comprehension in patients (Copeland & Pillsbury, 2004). Although knowledge about the structure and function of the visual system is more advanced than information on the auditory system, the goal of creating a comparable prosthesis for blind patients has yielded only moderate success so far (see next section). The challenge in designing retinal implants or visual cortical stimulators is not related to the manufacture or implantation of such a device—this has been done already by several groups. Rather, the most serious challenge relates to the need to construct devices that talk to the visual brain in a way that makes sense.

Thus, visual neuroscience and the neuroergonomics of artificial vision are not confined to matters of engineering and must include inquiries to increase our understanding of the anatomy and physiology of the visual pathway. We are using a systemic approach to make an effort to understand the language of the brain.

In this chapter, we present an overview of research on retinal prostheses and progress in this field and then provide a perspective on artificial vision and the role of modern technologies to support research in this sector. We also review studies on brain plasticity that provide essential information on changes in the brain as a consequence of (partial) blindness, the interaction of visual areas with other sensory modalities, and possible ways of influencing brain plasticity as a basis of visual rehabilitation.

History of Visual Prostheses

The primary impetus to develop a visual prosthesis stems from the fact that blindness affects millions of people worldwide. Further, there are no effective treatments for the most incapacitating causes of blindness. In response to this large unmet need, many groups have pursued means of artificially restoring vision in the blind. It has been known for decades that electrical stimulation delivered to intact visual structures in a blind individual can evoke patterned sensations of light called phosphenes (Gothe, Brandt, Irlbacher, Roricht, Sabel & Meyer, 2002; Marg & Rudiak, 1994). It has therefore been assumed that if electrical stimulation could be somehow delivered in a controlled and reproducible manner, patterns encoding meaningful shapes could be generated in order to potentially restore functional vision. With this foundation and with roughly three decades of work on visual prostheses, significant advances have been made in a relatively short time (for review, see Loewenstein, Montezuma, & Rizzo, 2004; Margalit et al., 2002; Maynard, 2001; Merabet, Rizzo, Amedi, Somers, & Pascual-Leone, 2005; Rizzo et al., 2001; Zrenner, 2002).

One of the earliest attempts to develop a visual prosthetic was made by applying electrical stimulation to the visual cortex (figure 21.1A). In one profoundly blind patient, electrical stimulation delivered to the surface of the brain allowed the patient to report crude phosphenes that were at least in spatial register with the known cortical retinotopic representation of visual space (Brindley & Lewin, 1968). More recent efforts have incorporated a digital video camera mounted onto a pair of glasses interfaced with a cortical-stimulating array that decodes an image into appropriate electrical stimulation patterns (Dobelle, 2000). Although the cortical approach has provided an important foundation in terms of feasibility, several technical challenges and concerns remain, including the sheer complexity of the visual cortex and the invasiveness of surgical implantation.

A. Cortical Approach

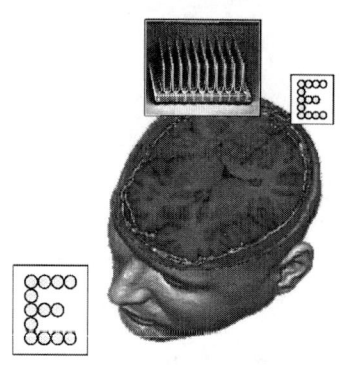

B. Retinal Approach

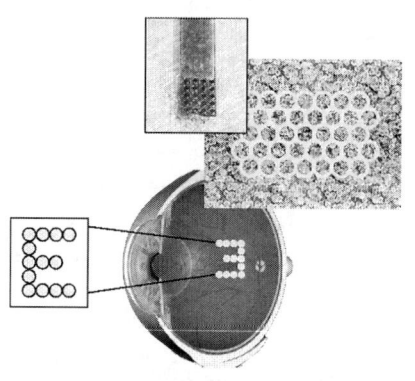

Figure 21.1. Schematic diagram of a cortical (A) and retinal (B) approach to restore vision. In the cortical approach, an image captured by a camera (not shown) is translated into an appropriate pattern of cortical stimulation to generate a visual precept. Inset figure shows a 100 microelectrode array that could be used to stimulate the cortex (Utah Array, modified from Maynard, Nordhausen, & Normann, 1997). In the retinal approach, an electrode array (shown in inset) can be placed directly on the retinal surface (epiretinal) or below (subretinal) to stimulate viable retinal ganglion cells. The retinal prosthesis can stimulate by the power captured by incident light or use signals received from a camera and signal processor mounted externally on a pair of glasses (not shown).

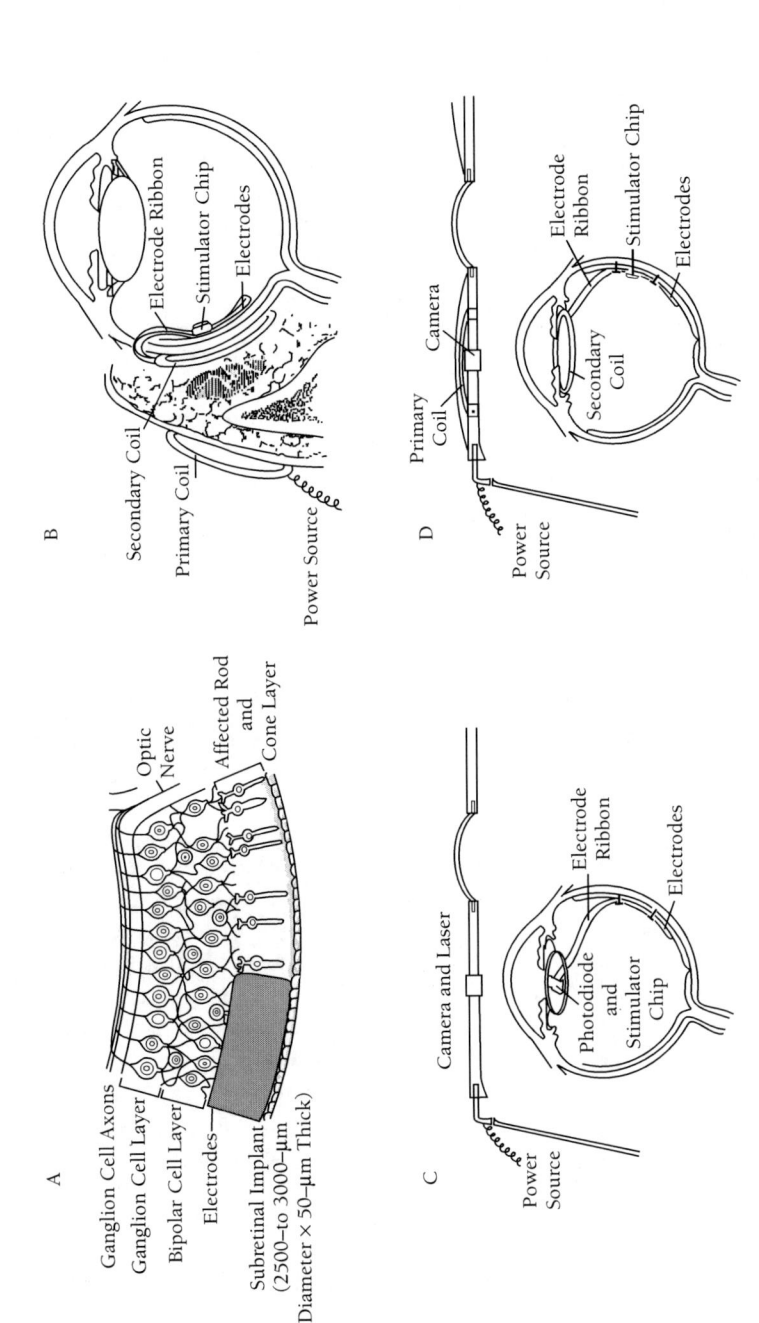

Figure 21.2. Geometries of some subretinal and epiretinal prosthetic devices currently being pursued. (A) Schematic drawing of a subretinal implant placed under a retina with photoreceptor degeneration. The small electrodes on the implant are designed to stimulate inner retinal cells. (B) Epiretinal implant with extraocular components. A primary coil on the temple transmits to a secondary coil on the sclera. A cable carries power and signal to a stimulator chip, which distributes energy appropriately to electrodes on the epiretinal surface. (C) Epiretinal implant with intraocular components. A camera on a spectacle frame provides the signal. A primary coil in the spectacles transmits to a secondary coil in an intraocular lens. A stimulator chip distributes power to multiple electrodes on the epiretinal surface. From Loewenstein, Montezuma, and Rizzo (2004). Copyright 2004 American Medical Association.

An alternative approach is to implant the prosthesis at a more proximal point of the visual pathway, namely the retina (figure 21.1B). In retinitis pigmentosa and age-related macular degeneration, two disorders that contribute greatly to the incidence of inherited blindness and blindness in the elderly (Hims, Diager, & Inglehearn, 2003; Klein, Klein, Jensen, & Meuer, 1997), there is a relatively selective degeneration of the outer retina, where the photoreceptors lie. Ganglion cells, which lie on the opposite side of the retina and connect the eye to the brain, survive in large numbers and are able to respond to electrical simulation even in highly advanced stages of blindness (Humayun et al., 1996; Rizzo, Wyatt, Loewenstein, Kelly, & Shire, 2003a, 2003b; see also Loewenstein et al., 2004, for review). Thus, one could potentially implant a prosthetic device at the level of the retina to stimulate the middle and inner retina (ganglion and bipolar cells) and potentially replace lost photoreceptors. Given that ganglion cells are arranged in topographical fashion throughout the retina, the generation of a visual image can theoretically be made possible by delivering multisite patterns of electrical stimulation. Two methods are being pursued that differ primarily with respect to the location at which the device interfaces with the retina (see figure 21.2). A subretinal implant is placed beneath the degenerated photoreceptors by creating a pocket between the sensory retina and retinal pigment epithelium. Alternatively, an epiretinal implant is attached to the inner surface of the retina in close proximity to the ganglion cells. Both devices require an intact inner retina to send visual information to the brain and thus are not designed to work in conditions where the complete retina or optic nerve have been compromised (e.g., as might occur in diabetic retinopathy, retinal detachment, or glaucoma).

The subretinal implant design is being pursued by several groups of investigators, including our Boston-based project (see also Chow et al., 2004; Zrenner et al., 1999). In one design, natural incident light that falls upon a subretinal photodiode array generates photocurrents that stimulate retinal neurons in a 2-D spatial profile. Such devices can easily be made with hundreds or thousands of neurons. Chow and coworkers have carried out a phase I feasibility trial in six patients with profound vision loss from retinitis pigmentosa. Patients were followed from 6 to 18 months after implantation and reported an improvement in visual function. This was documented by an increase in visual field size and the ability to name more letters using a standardized visual acuity chart (Chow et al., 2004). However, these results have met with controversy. It seems that the purported beneficial outcome might not be the result of direct and patterned electrical stimulation as initially anticipated but instead an indirect "cell rescue" effect from low-level current generated by the device (Pardue et al., 2005).

Concurrently, large efforts have pursued the epiretinal approach (Rizzo et al., 2003a, 2003b; Humayun et al., 2003). Much like the cortical approach (and unlike the subretinal approach), the current epiretinal strategies incorporate a digital camera mounted on a pair of eyeglasses to capture an image that in turn is converted into electrical signals. Both short-term and long-term testing in human volunteers with advanced RP has been carried out (Rizzo et al., 2003a, 2003b; Humayun et al., 1996, 2003). Intraoperative experiments that lasted only minutes to hours while patients remained awake produced visually patterned perceptions that were fairly crude. As predicted, the gross geometric structure of the phosphene patterns could be altered to some extent by varying the position and number of the stimulating electrodes and the strength or duration of the delivered current (Rizzo et al., 2003a, 2003b). Long-term human testing has only been performed by Humayun and coworkers, who have permanently implanted an epiretinal prosthesis in roughly six blind patients. The implanted device included an intraocular electrode array of platinum electrodes arranged in a 4 × 4 matrix. The array is designed to interface with the retina and camera connected to image-processing electronics. Their first subject has reported seeing perceptions of light (phosphenes) following stimulation of any of the 16 electrodes of the array. In addition, the subject is able to use images captured with the camera to detect the presence or absence of ambient light, to detect motion, and to recognize simple shapes (Humayun et al., 2003).

The results from these research efforts are encouraging and demonstrate (at least in principle) that patterned electrical stimulation can evoke patterned light perceptions. However, the perceptual pattern often does not match the stimulation pattern (Rizzo et al., 2003a, 2003b). The inconsistencies and limitations of the results may relate to the fact that stimulation is carried out on severely degenerated and therefore physiologically compromised retinas. To date, a key milestone has yet to be achieved: the demonstration that a visual neuroprosthesis can improve the quality of life of these individuals by allowing truly functional vision, such as the recognition of objects or even skillful navigation in an unfamiliar environment.

We believe that these engineering and surgical issues no longer represent the greatest barriers to future progress. Rather, the greatest limitation to the effective implementation of this technology is our own ignorance about how to communicate visual information to the brain in a meaningful way (Merabet et al., 2005). Simple generation of crude patterns of light will not likely be very helpful to blind patients when they ambulate in an unfamiliar environment. However, we appreciate that even this meager success may provide useful cues in a familiar environment, like sitting at one's own kitchen table. Furthermore, increasing image resolution (e.g., by increasing the number of stimulating electrodes) with the goal of generating more complex perceptions may or may not provide more useful visual percepts—the potential for enhanced success depends upon our ability to communicate with the brain. We propose that a deeper understanding of how the brain adapts to the loss of sight and how the other remaining senses process information within the visually deprived brain is necessary to make it more likely that we will be able to restore vision with a neuroprosthetic device.

The Systems Perspective and the Role of New Technologies

Given the above comments, it appears that delivering simple patterns of electrical stimulation may not be adequate to create useful vision. This is not surprising given that visual information processing is a very complex process. Our approach to this complexity is systemic, including a very broad perspective on the investigation of visual functions. Our approach incorporates not only engineering, information technology, single-cell physiology, ophthalmology, surgery, and material science, but also aims at improving our understanding of the mechanisms of encoding visual information to create alternatives to the natural pathways. Incorporating this broad approach, we believe, will make it more likely that we will be able to create useful prostheses for blind patients. By *useful*, we mean to imply that our assistive technology will be able to, at the least, help blind patients navigate in unfamiliar environments.

The particular tools used for investigating visual information processing are crucial for the implementation of such an approach. Over the past decades, the development of imaging and stimulation methods has provided a toolbox for neuroscientists that allows the noninvasive monitoring of brain functions and their changes over time. Using these techniques, we can identify not only where in the brain a process takes place (functional Magnetic Resonance Imaging, (fMRI), but also when or in which sequence it occurs (Electroencephalography, EEG, or Magnetoencephalography, MEG), and whether a part of the brain is causally related to the process in question (Transcranial Magnetic Stimulation, TMS; lesion studies, Kobayashi & Pascual-Leone, 2003; Merabet, Theoret, & Pascual-Leone, 2003; Rossini & Dal Forno, 2004). In this section, we give a short introduction to these technologies and describe how they can be applied to the systems approach to artificial vision (see figure 21.3). (These technologies are also described in more detail in part II of this volume.)

Technologies

fMRI

Positron Emission Tomography (PET) and related methods were the first techniques that yielded three-dimensional pictures of brain activity (Haxby, Grady, Ungerleider, & Horwitz, 1991). However, those images provided only low spatial and temporal resolution, and the method was too invasive in that radioactive agents were required. When Magnetic Resonance Imaging (MRI) was modified to allow scanning of brain function (fMRI) in addition to anatomical imaging, neuroscience had found the low-risk method that could be applied to a wide range of populations and research questions and that yielded astonishingly detailed information on brain activity (figure 21.3A; see chapter 4, this volume, for more details on fMRI).

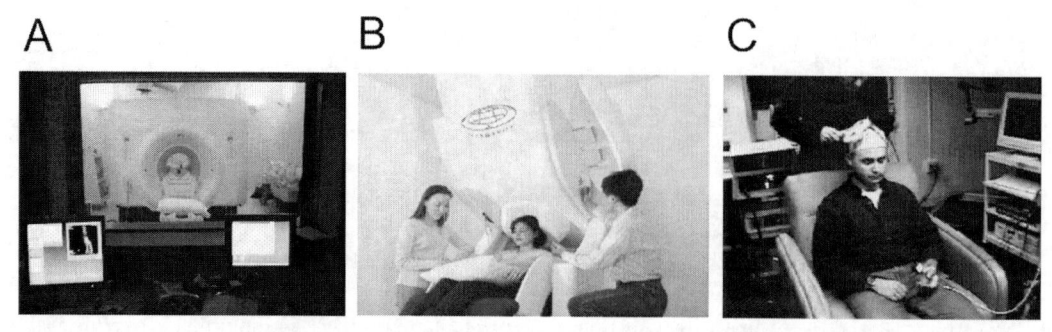

Figure 21.3. Techniques used to study brain function. (A) Example of a functional Magnetic Resonance Imaging scanner. (B) Magnetic encephalography Vectorview; Elekta Neuromag Oy, Finland. (C) Transcranial Magnetic Stimulation.

fMRI has yielded important information on mechanisms of visual processing (Wandell, 1999). In combination with other methods (e.g., EEG or MEG) with a higher temporal resolution, many behavior-based theories regarding the normal response pattern of the visual system have been confirmed by this type of imaging. New studies have explored the observation of changes of brain activity patterns over time, including over the relatively short time required for learning new tasks and over long periods of time to track neural development.

EEG and MEG

The EEG represents one of the earliest and simplest methods to noninvasively record human brain activity. By characterizing different patterns of brain wave activity, an EEG can identify a person's state of alertness and can measure the time it takes the brain to process various stimuli (Malmivuo, Suihko, & Eskola, 1997). While EEG has great temporal sensitivity, a major drawback of this technique is that it cannot reveal the anatomy of the brain nor identify with certainty the specific regions of the brain that are implicated with task performance. This issue, referred to as the *inverse problem*, defines that for a given recorded pattern of electrical activity, there are essentially an infinite number of configurations of current sources that could account for the recorded signal. To localize bioelectric sources within the brain more accurately, highly complex mathematical models and combined imaging techniques (e.g., fMRI) are used (Dale & Halgren, 2001). EEG methods are described in more detail in chapter 2, this volume.

Another related technology that provides greater temporal resolution of brain activity is MEG; figure 21.3B). MEG is a complex, noninvasive technique that attempts to detect, measure, record, and analyze the extracranial magnetic fields induced by the electrical activity of the brain. The skull is transparent to magnetic fields and the measurement sensitivity of MEG in the brain is theoretically more concentrated compared to EEG, since the skull has low conductivity to electric current (Malmivuo et al., 1997). Over 100 magnetic detection coils (superconducting quantum interference devices or SQUIDs) are positioned over the subject's head to detect bioelectrical activity. MEG has the advantage of not needing electrodes to be attached to the skin, and it provides both superior temporal and spatial resolution of measured brain activity. The raw data are filtered and processed by mathematical models to estimate the location, strength, and orientation of the current sources in the brain. Of all the brain-scanning methods, MEG provides the most accurate resolution of the timing of neural activity, but this technique is expensive to purchase, operate, and maintain (Lounasmaa, Hamalainen, Hari, & Salmelin, 1996).

Lesion Studies

To those not aware of basic visual system organization, it might seem odd that we see with the brain rather than with our eyes and that the part of the brain primarily receiving visual information is located at the back of the head, that is, at a maximal distance from the eyes. In fact, 27% of the cortex contributes in one way or another to visual perception, and even more areas link visual functions to other perceptual modalities (Grill-Spector, 2003;

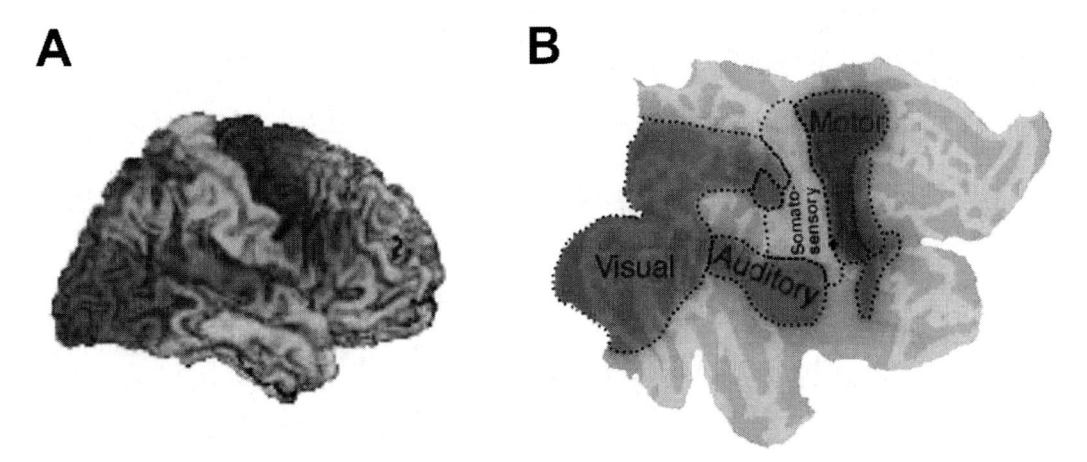

Figure 21.4. Organization of sensory and motor areas shown on a three-dimensional volume-rendered brain (A) and flattened projection (B). For simplicity, only the right hemisphere is shown. Modified from van Essen (2004) with permission from MIT Press. See also color insert.

Van Essen, 2004) that are distributed over even more extensive cortical and subcortical areas (see figure 21.4). Therefore, it is not surprising that any brain lesion bears a high risk of inducing some visual impairment (Kasten et al., 1999).

Most of the knowledge about the function of a particular brain area in humans is derived indirectly, for example, from brain imaging or electrophysiological studies (see below). However, such experiments usually cannot provide causal evidence that a specific function is supported by a specific brain area because there are many alternative explanations for any given data set (e.g., the interaction of several brain areas or inhibitory processes that modulate brain activity). Clear conclusions on causal connections between a brain function and its neuronal substrate have historically been obtained by lesion studies in which the effect of damaging a specific area of the brain on function is studied or by naturally occurring disease. The role of posterior parts of the brain in the generation of visual perception has been slowly uncovered by way of "natural experiments"—that is, by observing correlations between specific brain lesions (most obvious in traumatic injury) and the loss of visual functions (Finger, 1994). Lesion studies in animals with homologues of particular human brain regions, especially primates, have provided information and confirmed hypotheses on the behavior of neuronal networks of vision in the brain. Remarkable single case studies, such as that of a patient with a selective impairment of perceiving visual motion (Zihl,

von Cramon, Mai, & Schmid, 1991), have shed light on the high degree of specialization of visual areas; that is, local brain damage affecting the motion areas in extrastriate cortex results in a precisely defined loss of perceptual function, that is, perception of visual motion. The global organization of the visual system in pathways for parallel processing of "where" and "what" information (Ungerleider & Haxby, 1994) was supported by predictable functional loss in patients with selective lesions to one of the pathways (Stasheff & Barton, 2001).

The major drawback of this lesion approach is that many subjects have chronic lesions and information about the reorganization of the brain cannot be gained. Further, it is difficult to conclude that the findings from such patient studies actually reveal insight into the normal function of the visual brain in a healthy person. From the experimental standpoint, a disadvantage is that the lesions are not under control of the experimenter (e.g., the timing, size, and location). Better control can be achieved in animal studies, but structural homologies between animals and humans do not necessarily imply functional similarity. Moreover, the introspection of patients, which is a useful tool of investigation in neuropsychology, cannot be extracted from animals. New techniques make it possible to temporarily block brain activity in specific locations, like, for instance, TMS, which permits the study of temporary and nonharmful lesion effects in humans (see "Transcranial Magnetic Stimulation" below). As such, TMS enables performance of nondestructive lesion studies in volunteers to learn more about brain function.

Transcranial Magnetic Stimulation

TMS is a noninvasive neurophysiological technique that can be used to transiently disrupt the function of the targeted brain area, functionally map cortical areas, assess the excitability of the stimulated cortex, and even modify its level of activity for a short period of time (figure 21.3C). TMS is widely used to investigate complex aspects of human brain function such as memory, language, attention, learning, and motor function, and is even being studied as a potential therapy for depression (Kobayashi & Pascual-Leone, 2003; Merabet et al., 2003; Walsh & Pascual-Leone, 2003).

The physical basis of TMS rests on the principles of electromagnetic induction originally discovered by Faraday in 1831. Simply stated, a brief electrical pulse of rapidly alternating current passed through a coil of wire generates a strong, transient magnetic field (on the order of 1 to 2 Tesla). By placing a stimulating coil near a subject's scalp, the induced magnetic field pulse penetrates the scalp and skull virtually unattenuated (decaying only with distance) to reach the underlying brain tissue. The rate of change of this magnetic field induces a secondary electrical current in any nearby conducting medium, such as the brain tissue. Unlike direct electrical cortical stimulation, TMS does not require the underlying cortex to be exposed. In essence, TMS represents a method of electrical stimulation without electrodes, whereby the magnetic component is responsible for bridging the transition between current passed in the coil and current induced in the brain.

The effects of this induced current on brain tissue are dependent on numerous stimulation parameters (such as stimulation frequency, intensity, and coil configuration and orientation). The voltage of the primary current and the geometry of the stimulating coil determine the strength and shape of the magnetic field and thus the density and spatial resolution of the secondary current induced in the tissue. There are two main coil designs: (1) circular coils, which generate large currents and are useful for stimulating relatively large cortical areas; and (2) figure eight coils, which are more focal (owing to a maximal current being generated at the intersection of the two round coil wings). Current evidence suggests that the locus of TMS stimulation has a spatial resolution of 1 cm^2 with a penetration depth of approximately 2 cm using commercially available, figure eight stimulation coils. To stimulate at greater depths, larger amounts of current are required, but this comes at the expense of spatial resolution.

Typically, a single pulse lasts 200 ms; however, stimulation can also be delivered as a train of repetitive pulses (referred to as repetitive TMS or rTMS). Unlike the single pulse, repetitive TMS can modulate cortical excitability with effects lasting far beyond the period of stimulation itself (Pascual-Leone et al., 1998). While significant interindividual differences in responsiveness exist, it is generally accepted that motor excitability can be enhanced by stimulating at high frequencies (i.e., 5–25 Hz). Conversely, low frequency rTMS (at 1 Hz) lowers cortical excitability and is thought to result in cortical inhibition (Pascual-Leone et al., 1998).

From a safety standpoint, the great concern of using TMS is the risk of inducing a seizure (particularly using repetitive TMS and in subjects with a known prior history of epilepsy or a predisposing factor). Since the institution of safety guidelines defining stimulation parameters and the monitoring and screening of appropriate subjects, no TMS-induced seizures have been reported (Wassermann, 1998).

TMS has also been used extensively in the mapping of other, nonmotor cortical targets including frontal, temporal, and occipital areas in relation to studies involving memory, attention, language, visual perception, and mental imagery. A TMS pulse delivered with appropriate parameters can temporally disrupt cortical function in a given area. More recently, TMS has been used to supplement and refine conclusions drawn from neuroimaging studies (such as PET and fMRI). By establishing a causal link between brain activity and behavior, TMS can be used to identify which parts of a given network are necessary for performing a behavioral task. For example, when a TMS pulse is delivered at a specific time to the visual cortex of the brain, the detection of a presented visual stimulus can be blocked. Using this paradigm, the timing between the presentation of a visual stimulus and TMS pulse has been determined by varying the chronometry of visual perception within primary and extrastriate visual cortical areas (Amassian et al., 1998).

As mentioned earlier, TMS can be used to establish functional significance of data obtained through neuroimaging. For example, by combining PET imaging and TMS, the common neurophysiological substrate between visual perception and mental visual imagery can be investigated. Data collected from PET studies have suggested that similar visual areas are activated whether patients mentally imagine a visual scene or view it directly. TMS delivered to primary visual cortex disrupts not only visual perception but visual imagery, which suggests that occipital cortex is necessary in both these processes (Kosslyn et al., 1999).

The Top-Down or Systems Approach

In what way does the recent development of brain imaging and stimulation methods described above change the approach of creating a retinal prosthesis? To understand how the visual system works, the higher stages of processing beyond the level of the retina and the eye have to be considered, because many crucial aspects of the perceptual process are shaped at the thalamus and various cortical locations. Consideration of all stages of processing along the visual pathway is essential to develop a retinal prosthesis, which must in some way talk to the brain. Such "smart" devices must be developed with this knowledge about brain processes, because the visual cortex is not only a passive receptor of visual information from lower stages of processing, but—guided by influences from higher-level visual and cognitive areas of the brain—it actively changes the input signals. For example, what we see is not just determined by the pattern of photons or different wavelengths of light striking the retina—the final percept depends on previous experience, expectation, current intentions, and motivation, as well as attentional and other inner states. Thus, a wide range of brain areas contribute to the perception of sight. As has been described above, early experiments with retinal implants in our group (Rizzo et al., 2003a, 2003b) showed that electrical stimulation of the inner retina induces percepts, but that a scoreboard approach of conveying stimulation patterns to higher visual areas was not tenable. Similar results were later found in patients who had been chronically implanted with a retinal prosthesis (Humayun et al., 2003). For creating a smart stimulation device, we first have to understand the relationship of lower-level signals with high-level brain activity during visual perception. But we must also learn more about the interaction between bottom-up signals from the retina and top-down signals from frontal and parietal areas of the brain and higher areas of the visual pathway.

As a first step, we should also consider that blindness changes the brain and the way perceptual information is processed by the remaining, intact parts of the visual system (Grüsser & Landis, 1991). The brain and the sequences of signal processing are not static, but flexibly adapt to transient alterations of internal or external aspects of perception and to long-term changes such as occur with lesions of the visual pathway that create blindness. The techniques of brain imaging and stimulation described above are useful for investigating such processes of brain plasticity and provide relevant information that will influence the functional design of our retinal prostheses. Moreover, with the assistance of fMRI, TMS, and other imaging and stimulation methods, we should also be able to learn more about strategies to develop a retinal prostheses and its effects on the activity of the visual system and the brain in general.

Brain Plasticity

Visual System Plasticity and Influence of Cognitive Factors

The earliest neuropsychological theories about brain function were strongly influenced by the idea that any given capacity was represented in a localized brain area and that each region was assigned a specific function that was immutable. More specifically, by the 1940s and 1950s, the highly specific processing of information and the strict topographical organization of the visual system had been discovered (see Grüsser & Landis, 1991). These discoveries advanced the belief that brain plasticity and functional recovery are impossible, or at least extremely limited. Until recently, the visual system was mainly regarded as hard-wired. The potential of the visual brain to heal itself was not appreciated (see Pambakian & Kennard, 1997; Sabel, 1999), other than within a critical period early in life. Even then, selective deprivation of certain aspects of visual experience in young animals was shown to produce a massive effect on the architecture of the visual system that led to permanent functional impairment (Sabel, 1999; Wiesel & Hubel, 1965).

However, more recent animal experiments as well as human research have shown that visual deficits induced by lesions or deprivation can either recover spontaneously or that function can be regained by systematic training even beyond the critical period (Daas, 1997; Freund et al., 1997; Kasten et al., 1999; Poggel, Kasten, Müller-Oehring, Sabel, & Brandt, 2001; Sabel, 1999). These newer findings have documented a previously unappreciated plasticity of the visual system that can provide improved function, albeit the exact mechanisms of reorganization are not yet clear. Notwithstanding this potential for plasticity, in a recent debate, the outcomes from human training trials are criticized as not being relevant to the everyday life of patients and possibly being due to artifacts (e.g., a shift of the patient's fixation toward the blind field; see Horton, 2005; Plant, 2005, for a critical review). Obviously, more work must be performed in this emerging area to better define the capacity of the visual system to recover after injury, and to judge how this capacity is affected by age, and the extent and location of a lesion.

Earlier behavioral and electrophysiological animal experiments indicated that retinal and cortical lesions are followed by widespread reorganization of receptive field structures (Chino, Smith, Kaas, Sasaki, & Cheng, 1995; Eysel et al., 1999). In these experiments, cortical activity was initially suppressed within cortical areas that had previously received input from a lesioned area of retina or the border zones of cortical scotoma. Even minutes after a lesion of the retina or cortex, the receptive fields of neurons adjacent to a lesioned area increased considerably in size (up to three times the original size) and expanded into the region originally represented by the lesioned part (Eysel et al., 1999; Gilbert, 1998; Kasten et al., 1999; Sabel, 1997, 1999). Hence, the representation of the lesioned area in the cortex shrinks over time, and neural activation recovers. This process of recovery seems to be triggered by the areas surrounding the lesion, which show increased spontaneous activation and excitability by visual stimulation. In addition to the quick increase of receptive field size immediately after the lesion, there are also slower processes of receptive field plasticity that take place over weeks to months. Thus, retinal and cortical defects of the visual system have immense effects on the topographical maps in the visual cortex (see Eysel et al., 1999, for a review). These processes take place in the mature visual system and may be based on molecular and morphological mechanisms similar to those that have been found to occur during the normal process of brain development.

In the beginning of the 20th century, the first systematic observations of spontaneous recovery of visual functions in humans were made (e.g., Poppelreuter, 1917, cited in Poggel, 2002), mainly in soldiers surviving brain lesions during the two world wars. Since then, many investigators have contributed findings in favor of some degree of spontaneous recovery in patients with visual field loss. However, the results concerning the duration of the period of recovery, the size of the visual field regained, and predictors of recovery were often contradictory (see Poggel, 2002, for an overview). Still, the observation that vision can improve even months after a lesion of the visual system prompts the question as to whether processes of recovery can be actively manipulated to enhance the process of visual rehabilitation. Notwithstanding these insights, for several decades, treatment approaches for patients with partial blindness concentrated on the compensation for the visual field defect and did not aim at a restoration of the lost function (Kasten et al., 1999; Kerkhoff, 1999).

In the 1960s and 1970s, training-induced recovery of visual function was first shown in animals (Cowey, 1967; Mohler & Wurtz, 1977), which was followed by attempts at restoring visual function in patients (Zihl & von Cramon, 1979, 1985). In the meantime, many studies supported the initial evidence of vision restoration (see Kasten et al., 1999, for an overview). The benefit of training has been asserted for patients with optic nerve lesions and postgeniculate lesions in one prospective, randomized, placebo-controlled clinical trial (Kasten, Wüst, Behrens-Baumann, & Sabel, 1998). In this study, patients who were trained with a computer-based program that stimulated the border regions of the defective visual area showed a larger visual field and a higher number of detected light stimuli in high-resolution visual field testing after 6 months of training, versus no change in a control group who had performed a placebo fixation training. Then,

Poggel, Kasten, Müller-Oehring, Sabel, and Brandt (2001) compared processes of spontaneous and training-induced recovery of vision in a patient with a shotgun lesion. The phenomenology and topography of training-induced and spontaneous recovery were very similar, which supports a hypothesis that similar mechanisms may underlie both processes of recovery.

The amount of residual vision at the border of the blinded area of the visual field is the main predictor of training success (Kasten et al., 1998; Poggel, Kasten, & Sabel, 2004). Presumably, partially defective brain regions, at the border of the lesion, represent cortical areas where vision is not completely lost but is quite impaired. In border areas, perimetric thresholds are increased, discrimination of forms and colors is impossible, reaction times are prolonged, and the subjective quality of perception is reduced. However, systematic stimulation of these areas of residual vision seems to be able to reactivate the partially defective regions, possibly by co-activating neurons connected to the defective regions via long-range horizontal connections in the cortex. These factors may increase receptive field size and reduce the size of the blind field (Kasten et al., 1999; Poggel, 2002).

In a more recent study, an attentional cue was used to help the patients focus attention at the visual field border, specifically to the areas of residual vision that are crucial for regaining visual functions. This strategy increased the training effect in patients who received the cue versus the results obtained from a group of participants who performed the conventional visual field training without attentional cueing (Poggel, 2002; Poggel et al., 2004). This beneficial effect of the cue might be due to a systematic combination of bottom-up stimulation (i.e., the light stimuli presented during the training) with top-down attentional activation.

As alluded to above, the effects of visual field training are highly controversial (Horton, 2005; Pambakian & Kennard, 1997; Plant, 2005; see also Sabel, 2006). While it remains to be proven in future studies that the treatment effects are clinically relevant and cannot be explained by artifacts, there appears to be growing evidence from animal and human studies of a heretofore unrecognized potential for cortical plasticity.

The example of the study by Poggel et al. (2004) on attention effects in visual field training shows that more than just the neural output from the retina should be taken into account when therapeutic measures are planned. Top-down influences may be systematically applied to improve visual performance in patients. Cognitive aspects of vision also have to be included to understand the various effects that bottom-up stimulation can have on perception. A percept cannot be predicted solely on the basis of the bottom-up signal (Loewenstein et al., 2004; Rizzo et al., 2003a, 2003b); factoring in the effects of top-down signals may make the effects of stimulation, such as those induced by a retinal prosthesis, more predictable. The situation becomes even more complex because the visual cortex cannot be looked at in isolation, given that cortical plasticity takes place in many nonvisual areas that interact with the visual system.

Cross-Modal Plasticity

Neuroplasticity in the Blind and Cross-Modal Sensory Integration

The existence of specialized receptors for different sensory modalities provides the opportunity to process different forms of input and hence capture different views of the world in parallel. While some experiences are uniquely unimodal (e.g., color and tone), the different sensory systems presumably closely collaborate with each other, given that our perceptual experience of the world is richly multimodal and seamlessly integrated (Stein & Meredith, 1990). Despite the overwhelming evidence that we integrate multimodal sensory information to obtain the most accurate representation of our environment, our thinking about the brain is shaped by historical notions of parallel systems specialized for different sensory modalities. For each of those modalities, we assume a hierarchically organized system that begins with specialized receptors that feed unimodal primary cortical areas. A series of secondary areas unimodally integrate different aspects of the processed information. Eventually, multimodal association areas integrate the processed signals with information derived from other sensory modalities.

But what happens when parts of the brain that process a given modality of sensory information are separated from their input? It seems reasonable to presume that the brain would reorganize, so that neuroplasticity would compensate for the loss in concert with the remaining senses. The study of blind individuals provides insight into such brain reorganization and behavioral compensations that occur following sensory deprivation. For instance, Braille reading in the blind is associated with a variety of neuroplastic changes, including development of an expanded sensorimotor representation of the reading finger (Pascual-Leone & Torres, 1993; Pascual-Leone et al., 1993) and recruitment of the occipital cortex during tactile discrimination (Buchel, Price, Frackowiak, & Friston, 1998; Burton et al., 2002; Sadato et al., 1996, 1998). These findings suggest that pathways for tactile discrimination change following blindness. Indeed, somatosensory-to-visual cross-modal plasticity has been shown to be behaviorally relevant, given that interfering with the visual cortex using TMS can disrupt Braille reading in blind patients (Cohen et al., 1997). Converging evidence suggests that cross-modal plasticity following sensory deprivation might be a general attribute of the cerebral cortex. For instance, Weeks et al. (2000) have shown recruitment of occipital areas in congenitally blind individuals during a task that required auditory localization. Interestingly, the opposite relation also holds true. Activation of the auditory cortex was observed in a sample of profoundly deaf subjects in response to purely visual stimuli (Finney, Fine, & Dobkins, 2001) and vibrotactile stimulation using MEG (Levanen, Jousmaki, & Hari, 1998). These reports can be interpreted as meaning that activation of the occipital cortex during processing of nonvisual information may not necessarily represent the establishment of new connections, but rather the unmasking of latent pathways that participate in the multisensory perception. This functional recruitment of cortical areas may be achieved by way of rapid neuroplastic changes. These observations support the belief that unimodal sensory areas are part of a network of areas subserving multisensory integration and are not merely processors of a single modality.

In parallel with development of visual prostheses, great strides are being made in other areas of neuroprosthetic research. Cochlear implant research has enjoyed the most progress and success to date (see Loeb, 1990) and in many respects has served as an impetus and model for visual prosthetic development. Deaf individuals learn to use cochlear implants by establishing new associations between sounds generated by the device and objects in the auditory world. However, there remains great intersubject variability in adapting to cochlear implants, with speech recognition performance ranging from very poor to near perfect among various patients.

In many cases, specific rehabilitation strategies for prospective recipients of a cochlear implant have to be tailored to the profile of the candidate in order to maximize the likelihood of success. For instance, the assessment of the degree of visual-auditory cross-modal plasticity in the deaf has also been extended to patients receiving cochlear implants. Using PET imaging, Lee et al. (2001) found that in prelingually deaf patients (i.e., hearing loss prior to the development of language), the primary auditory cortex was activated by the sound of spoken words following receipt of a cochlear implant device. Even more astounding, this group found that the degree of resting hypometabolism before device implantation was positively correlated with the amount of improvement in hearing capability after the operation. These authors suggest that if cross-modal plasticity restores metabolism in the auditory cortex before implantation, the auditory cortex may no longer be as able to respond to signals from a cochlear implant, and prelingually deaf patients will show no improvement in hearing function (despite concentrated rehabilitation) after implantation of a cochlear prosthesis.

The results of these studies have important implications. First, they suggest that functional neuroimaging may have a prognostic value in selecting those individuals who are ideal candidates for a cochlear implant. It is conceivable that a corollary scenario exists regarding cross-modal sensory interactions within the visual cortex of the blind, and we therefore suspect that such a combined approach (i.e., using information obtained by neuroimaging) may help identify candidates who are most likely to succeed with a visual prosthesis implant. Put another way, simple reintroduction of lost sensory input may not suffice to restore the loss of a sense. We hypothesize that specific strategies will be needed to modulate brain processing and to enhance the extraction of relevant and functionally meaningful information generated from neuroprosthetic inputs.

Is it possible to exploit a blind person's existing senses in order to learn how to see again? Tapping into such mechanisms may be advantageous for enhancing the integration of the encoding of meaningful percepts by a prosthesis. Clearly, there exists a correspondence between how an object appears, how it sounds, and how that same object feels

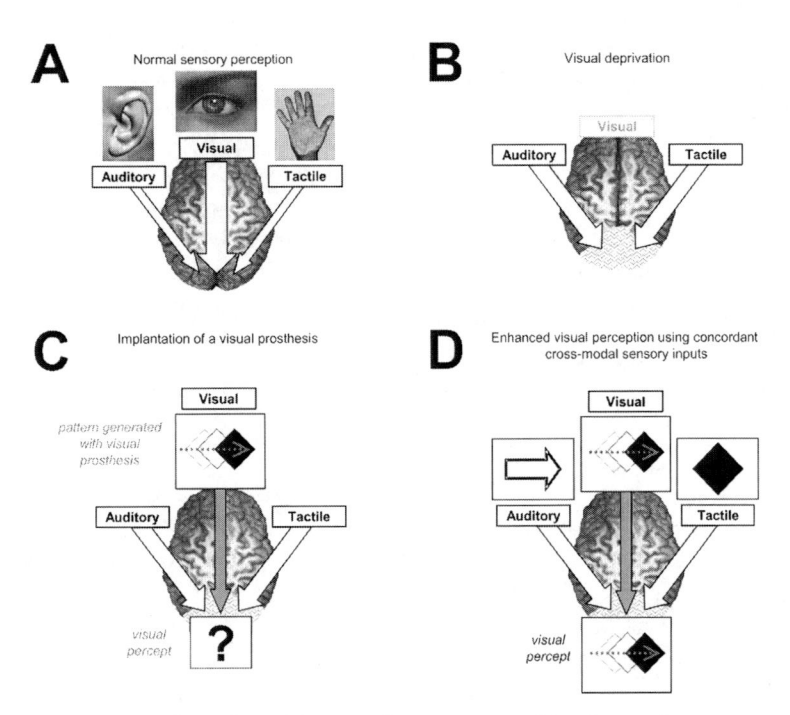

Figure 21.5. The multimodal nature of our sensory world and its implications for implementing a visual prosthesis to restore vision. (A) Under normal conditions, the occipital cortex receives predominantly visual inputs but also from cross-modal sensory areas. (B) Following visual deprivation, neuroplastic changes occur such that the visual cortex is recruited to process sensory information from other senses (illustrated by larger arrows for touch and hearing). (C) After neuroplastic changes associated with vision loss have occurred, the visual cortex is fundamentally altered in terms of its sensory processing, so that simple reintroduction of visual input (by a visual prosthesis; dark gray arrow) is not sufficient to create meaningful vision (in this example, a pattern encoding a moving diamond figure is generated with the prosthesis). (D) To create meaningful visual percepts, a patient who has received an implanted visual prosthesis can incorporate concordant information from remaining sensory sources. In this case, the directionality of a moving visual stimulus can be presented with an appropriately timed directional auditory input, and the shape of the object can be determined by simultaneous haptic exploration. In summary, modification of visual input by a visual neuroprosthesis in conjunction with appropriate auditory and tactile stimulation could potentially maximize the functional significance of restored light perceptions and allow blind individuals to regain behaviorally relevant vision. From Merabet et al. (2005). Reproduced with permission from *Nature Reviews Neuroscience*, copyright 2005 Macmillan Magazines Ltd.).

when explored through touch. Not surprisingly, functional neuroimaging has demonstrated a significant overlap between cortical areas involved in object recognition through sight and touch (e.g., Amedi, Malach, Hendler, Peled, & Zohary, 2001, 2002; Beauchamp, Lee, Argall, & Martin, 2004; Diebert, Kraut, Kremen, & Hartt, 1999; James et al., 2002; Pietrini et al., 2004). It thus appears that visual cortical areas—once considered part of a specialized system—are involved in multiple forms of sensory processing, and they incorporate widely distributed and overlapping object representations in their processing schemes. This notion implies that a crucial aspect of information processing in the brain is not solely dependent upon a specific kind of sensory input, but rather on an integrated computational assimilation across diverse areas of the brain (Pascual-Leone & Hamilton, 2001). We hypothesize that the tactile and auditory inputs (processed in the cross-modally changed brain) can be used to remap restored visual sensations and, in turn, help devise and refine visual stimulation strategies using a common conceptual sensory framework (for discussion, see Merabet et al., 2005; see figure 21.5). Given that sensory representations are shared, appropriate tactile and auditory inputs can assist a patient in using a visual neuroprosthesis to functionally integrate concordant sources of sensory stimuli into meaningful percepts.

Conclusion

This chapter has reviewed broad neuroscience topics and new technologies that may enhance the understanding of how to create useful vision for blind persons and for related neuroergonomic applications. This discussion has emphasized the fact that the visual system does not operate in isolation. What a patient will see with a retinal implant will relate to the processes of plasticity and perhaps other perceptual modalities, as well as top-down influences like attention that play a significant role in shaping the percept. Using that information and the new technologies available for visual neuroscience, it may be possible to develop smarter devices and to make visual rehabilitation more efficient.

Creating a retinal implant is a challenge for ophthalmologists and surgeons, electrical engineers, and material scientists. But the creation of retinal implants will also have an impact on neuroscience disciplines. Once the mechanisms of visual stimulation and processing incoming signals are better understood, another demanding task will be the development of smart algorithms to control the neuroprosthesis and to implement learning procedures that will allow tuning of the implant to the patient's needs. This process will require close cooperation with cognitive scientists, perceptual specialists, and the patients who will hopefully benefit from these efforts.

MAIN POINTS

1. The prevalence of blindness after retinal or cortical lesions is high. No cure exists for most of these conditions.
2. Retinal prostheses may help to restore useful visual performance and improve the quality of life in blind patients.
3. Efforts to create visual prosthetics have been partially successful but are limited mainly by our inability to communicate with the brain.
4. A systems perspective, taking into account state-of-the-art neuroimaging methods and results from cognitive research and neuropsychology, may help to create "smart" visual prosthetics.
5. Visual brain plasticity and the interaction of the visual system with other sensory modalities have to be included in the effort to create a retinal implant that is able to talk to the brain.
6. Advances in microtechnology and the possibility of creating artificial vision create new challenges for the field of rehabilitation.

Key Readings

Chow, A. Y., Chow, V. Y., Packo, K. H., Pollack, J. S., Peyman, G. A., & Schuchard, R. (2004). The artificial silicon retina microchip for the treatment of vision loss from retinitis pigmentosa. *Archives of Opthalmology, 122,* 460–469.

Humayun, M. S., de Juan, E., Jr., Dagnelie, G., Greenberg, R. J., Propst, R. H., & Phillips, D. H. (1996). Visual perception elicited by electrical stimulation of retina in blind humans. *Archives of Opthalmology, 114,* 40–46.

Loewenstein, J. I., Montezuma, S. R., & Rizzo, J. F., 3rd. (2000). Outer retinal degeneration: An electronic retinal prosthesis as a treatment strategy. *Archives of Opthalmology, 122,* 587–596.

Rizzo, J. F., 3rd, Wyatt, J., Humayun, M., de Juan E., Liu, W., Chow, A., et al. (2001). Retinal prosthesis: An encouraging first decade with major challenges ahead. *Opthalmology, 108,* 13–14.

References

Amassian, V. E., Cracco, R. Q., Maccabee, P. J., Cracco, J. B., Rudell, A. P., & Eberle, L. (1998). Transcranial magnetic stimulation in study of the visual pathway. *Journal of Clinical Neurophysiology, 15,* 288–304.

Amedi, A., Jacobson, G., Hendler, T., Malach, R., & Zohary, E. (2002). Convergence of visual and tactile shape processing in the human lateral occipital complex. *Cerebral Cortex, 12,* 1202–1212.

Amedi, A., Malach, R., Hendler, T., Peled, S., & Zohary, E. (2001). Visuo-haptic object-related activation in the ventral visual pathway. *Nature Neuroscience, 4,* 324–330.

Beauchamp, M. S., Lee, K. E., Argall, B. D., & Martin, A. (2004). Integration of auditory and visual information about objects in superior temporal sulcus. *Neuron, 41,* 809–823.

Brindley, G. S., & Lewin, W. S. (1968). The sensations produced by electrical stimulation of the visual cortex. *Journal of Physiology, 196,* 479–493.

Buchel, C., Price, C., Frackowiak, R. S., & Friston, K. (1998). Different activation patterns in the visual cortex of late and congenitally blind subjects. *Brain, 121,* 409–419.

Burton, H., Snyder, A. Z., Conturo, T. E., Akbudak, E., Ollinger, J. M., & Raichle, M. E. (2002). Adaptive changes in early and late blind: A fMRI study of Braille reading. *Journal of Neurophysiology, 87,* 589–607.

Chino, Y. M., Smith, E. G., Kaas, J. H., Sasaki, Y., & Cheng, H. (1995). Receptive field properties of deafferented visual cortical neurons after topographic map reorganization in adult cats. *Journal of Neuroscience, 15,* 2417–2433.

Chow, A. Y., Chow, V. Y., Packo, K. H., Pollack, J. S., Peyman, G. A., & Schuchard, R. (2004). The artificial silicon retina microchip for the treatment of vision loss from retinitis pigmentosa. *Archives of Ophthalmology, 122,* 460–469.

Cohen, L. G., Celnik, P., Pascual-Leone, A., Corwell, B., Falz, L., Dambrosia, J., et al. (1997). Functional relevance of cross-modal plasticity in blind humans. *Nature, 389,* 180–183.

Copeland, B. J., & Pillsbury, H. C., 3rd. (2004). Cochlear implantation for the treatment of deafness. *Annual Review of Medicine, 55,* 157–167.

Cowey, A. (1967). Perimetric study of field defects in monkeys after cortical and retinal ablations. *Quarterly Journal of Experimental Psychology, 19,* 232–245.

Daas, A. (1997). Plasticity in adult sensory cortex: A review. *Network, Computation, and Neural Systems, 8,* R33–R76.

Dale, A. M., & Halgren, E. (2001). Spatiotemporal mapping of brain activity by integration of multiple imaging modalities. *Current Opinion in Neurobiology, 11,* 202–208.

Deibert, E., Kraut, M., Kremen, S., & Hartt, J. (1999). Neural pathways in tactile object recognition. *Neurology, 52,* 1413–1417.

Dobelle, W. H. (2000). Artificial vision for the blind by connecting a television camera to the visual cortex. *ASAIO Journal, 46,* 3–9.

Eysel, U. T., Schweigart, G., Mittmann, T., Eyding, D., Qu, Y., Vandesande, F., et al. (1999). Reorganization in the visual cortex after retinal and cortical damage. *Restorative Neurology and Neuroscience, 15,* 153–164.

Finger, S. (1994). Vision: From antiquity through the renaissance. In S. Finger (Ed.), *Origins of neuroscience: A history of explanations into brain function* (pp. 65–95). Oxford, UK: Oxford University Press.

Finney, E. M., Fine, I., & Dobkins, K. R. (2001). Visual stimuli activate auditory cortex in the deaf. *Nature Neuroscience, 4,* 1171–1173.

Freund, H. J., Sabel, B. A., & Witte, O. (Eds.). (1997). *Brain plasticity.* New York: Lippincott-Raven.

Gilbert, C. D. (1998). Adult cortical dynamics. *Physiological Reviews, 78,* 467–485.

Gothe, J., Brandt, S. A., Irlbacher, K., Roricht, S., Sabel, B. A. & Meyer, B. U. (2002). Changes in visual cortex excitability in blind subjects as demonstrated by transcranial magnetic stimulation. *Brain, 125,* 479–490.

Grill-Spector, K. (2003). The neural basis of object perception. *Current Opinion in Neurobiology, 13,* 159–166.

Grüsser, O. J., & Landis, T. (1991). *Vision and visual dysfunction: Visual agnosias and other disturbances of visual perception and cognition, 12.* Houndmills: Macmillan.

Haxby, J. V., Grady, C. L., Ungerleider, L. G., & Horwitz, B. (1991). Mapping the functional neuroanatomy of the intact human brain with brain work imaging. *Neuropsychologia, 29,* 539–555.

Hims, M. M., Diager, S. P., & Inglehearn, C. F. (2003). Retinitis pigmentosa: Genes, proteins and prospects. *Developments in Ophthalmology, 37,* 109–125.

Horton, J. C. (2005). Disappointing results from Nova Vision's visual restoration therapy. *British Journal of Ophthalmology, 89,* 1–2.

Humayun, M. S., de Juan, E., Jr., Dagnelie, G., Greenberg, R. J., Propst, R. H., & Phillips, D. H. (1996). Visual perception elicited by electrical stimulation of retina in blind humans. *Archives of Ophthalmology, 114,* 40–66.

Humayun, M. S., Weiland, J. D., Fujii, G. Y., Greenberg, R., Williamson, R., Little, J., et al. (2003). Visual perception in a blind subject with a chronic microelectronic retinal prosthesis. *Vision Research, 43,* 2573–2581.

James, T. W., Humphrey, G. K., Gati, J. S., Servos, P., Menon, R. S., Goodale, & M. A. (2002). Haptic study of three-dimensional objects activates extrastriate visual areas. *Neuropsychologia, 40,* 1706–1714.

Kasten, E., Poggel, D. A., Müller-Oehring, E. M., Gothe, J., Schulte, T., & Sabel, B. A. (1999).

Restoration of vision II: Residual functions and training-induced visual field enlargement in brain-damaged patients. *Restorative Neurology and Neuroscience, 15,* 273–287.

Kasten, E., Wüst, S., Behrens-Baumann, W., & Sabel, B. A. (1998). Computer-based training for the treatment of partial blindness. *Nature Medicine, 4,* 1083–1087.

Kerkhoff, G. (1999). Restorative and compensatory therapy approaches in cerebral blindness—a review. *Restorative Neurology and Neuroscience, 1,* 255–271.

Klein, R., Klein, B. E., Jensen, S. C., & Meuer, S. M. (1997). The five-year incidence and progression of age-related maculopathy: The Beaver Dam Eye Study. *Ophthalmology, 104,* 7–21.

Kobayashi, M., & Pascual-Leone, A. (2003). Transcranial magnetic stimulation in neurology. *Lancet Neurology, 2,* 145–156.

Kosslyn, S. M., Pascual-Leone, A., Felician, O., Camposano, S., Keenan, J. P., Thompson, W. L., et al. (1999). The role of area 17 in visual imagery: Convergent evidence from PET and rTMS. *Science, 284,* 167–170.

Lee, D. S., Lee, J. S., Oh, S. H., Kim, S. K., Kim, J. W., Chung, J. K., et al. (2001). Cross-modal plasticity and cochlear implants. *Nature, 409,* 149–150.

Levanen, S., Jousmaki, V., & Hari, R. (1998). Vibration-induced auditory-cortex activation in a congenitally deaf adult. *Current Biology, 8,* 869–872.

Loeb, G. E. (1990). Cochlear prosthetics. *Annual Review of Neuroscience, 13,* 357–371.

Loewenstein, J. I., Montezuma, S. R., & Rizzo, J. F., 3rd. (2004). Outer retinal degeneration: An electronic retinal prosthesis as a treatment strategy. *Archives of Ophthalmology, 122,* 587–596.

Lounasmaa, O. V., Hamalainen, M., Hari, R., & Salmelin, R. (1996). Information processing in the human brain: Magnetoencephalographic approach. *Proceedings of the National Academy of Sciences, USA, 93,* 8809–8815.

Malmivuo, J., Suihko, V., & Eskola, H. (1997). Sensitivity distributions of EEG and MEG measurements. *IEEE Transactions on Biomedical Engineering, 4,* 196–208.

Marg, E., & Rudiak, D. (1994). Phosphenes induced by magnetic stimulation over the occipital brain: description and probable site of stimulation. *Optometry and Vision Science, 71,* 301–311.

Margalit, E., Maia, M., Weiland, J. D., Greenberg, R. J., Fujii, G. Y., Torres, G., et al. (2002). Retinal prosthesis for the blind. *Survey of Ophthalmology, 47,* 335–356.

Maynard, E. M. (2001). Visual prostheses. *Annual Review of Biomedical Engineering, 3,* 145–168.

Maynard, E. M., Nordhausen, C. T., & Normann, R. A. (1997). The Utah intracortical electrode array: A recording structure for potential brain-computer interfaces. *Electroencephalography and Clinical Neurophysiology, 102,* 228–239.

Merabet, L. B., Rizzo, J. F., Amedi, A., Somers, D. C., & Pascual-Leone, A. (2005). What blindness can tell us about seeing again: Merging neuroplasticity and neuroprostheses. *Nature Reviews Neuroscience, 6,* 71–77.

Merabet, L. B., Theoret, H., & Pascual-Leone, A. (2003). Transcranial magnetic stimulation as an investigative tool in the study of visual function. *Optometry and Visual Science, 80,* 356–368.

Mohler, C. W., & Wurtz, R. H. (1977). Role of striate cortex and superior colliculus in visual guidance of saccadic eye movements in monkeys. *Journal of Neurophysiology, 40,* 74–94.

Pambakian, L., & Kennard, C. (1997). Can visual function be restored in patients with homonymous hemianopia? *British Journal of Ophthalmology, 81,* 324–328.

Parasuraman, R. (2003). Neuroergonomics: Research and practice. *Theoretical Issues in Ergonomics Sciences, 4,* 5–20.

Pardue, M. T., Phillips, M. J., Yin, H., Sippy, B. D., Webb-Wood, S., Chow, A. Y., et al. (2005). Neuroprotective effect of subretinal implants in the RCS rat. *Investigative Ophthalmological Visual Science, 46,* 674–682.

Pascual-Leone, A., Cammarota, A., Wassermann, E. M., Brasil-Neto, J. P., Cohen, L. G., & Hallett, M. (1993). Modulation of motor cortical outputs to the reading hand of braille readers. *Annuals of Neurology, 34(1),* 333–337.

Pascual-Leone, A., & Hamilton, R. (2001). The metamodal organization of the brain. *Progress in Brain Research, 134,* 427–445.

Pascual-Leone, A., Tormos, J. M., Keenan, J., Tarazona, F., Canete, C., & Catala, M. D. (1998). Study and modulation of human cortical excitability with transcranial magnetic stimulation. *Journal of Clinical Neurophysiology, 15,* 333–343.

Pascual-Leone, A., & Torres, F. (1993). Plasticity of the sensorimotor cortex representation of the reading finger in braille readers. *Brain, 116,* 39–52.

Pietrini, P., Furey, M. L., Ricciardi, E., Gobbini, M. I., Wu, W. H., Cohen, L., et al. (2004). Beyond sensory images: Object-based representation in the human ventral pathway. *Proceedings of the National Academy of Sciences, USA, 101,* 5658–5663.

Poggel, D. A. (2002). *Effects of visuo-spatial attention on the restitution of visual field defects in patients with cerebral lesions.* Aachen: Shaker.

Poggel, D. A., Kasten, E., Müller-Oehring, E. M., Sabel, B. A., & Brandt, S. A. (2001). Unusual spontaneous and training induced visual field recovery in a patient with a gunshot lesion. *Journal of Neurology, Neurosurgery and Psychiatry, 69,* 236–239.

Poggel, D. A., Kasten, E., & Sabel, B. A. (2004). Attentional cueing improves vision restoration therapy in patients with visual field defects. *Neurology, 63,* 2069–2076.

Plant, G. T. (2005). A work out for hemianopia. *British Journal of Ophthalmology, 89,* 2.

Rizzo, J. F., 3rd, Wyatt, J., Humayun, M., de Juan, E., Liu, W., Chow, A., Eckmiller, R., et al. (2001). Retinal prosthesis: An encouraging first decade with major challenges ahead. *Ophthalmology, 108,* 13–14.

Rizzo, J. F., 3rd, Wyatt, J., Loewenstein, J., Kelly, S., & Shire, D. (2003a). Methods and perceptual thresholds for short-term electrical stimulation of human retina with microelectrode arrays. *Investigative Ophthalmology and Visual Science, 44,* 355–361.

Rizzo, J. F., 3rd, Wyatt, J., Loewenstein, J., Kelly, S., & Shire, D. (2003b). Perceptual efficacy of electrical stimulation of human retina with a microelectrode array during short-term surgical trials. *Investigative Ophthalmology and Visual Science, 12,* 5362–5369.

Rossini, P. M., & Dal Forno, G. (2004). Integrated technology for evaluation of brain function and neural plasticity. *Physical and Medical Rehabilitation Clinics of North America, 15(1),* 263–306.

Sabel, B. A. (1997). Unrecognized potential of surviving neurons: Within-systems plasticity, recovery of function, and the hypothesis of minimal residual structure. *Neuroscientist, 3,* 366–370.

Sabel, B. A. (1999). Restoration of vision I: Neurobiological mechanisms of restoration and plasticity after brain damage—a review. *Restorative Neurology and Neuroscience, 1,* 177–200.

Sabel, B. A. (2006). Vision restoration therapy and raising red flags too early. *British Journal of Ophthalmology, 90,* 659–660.

Sadato, N., Pascual-Leone, A., Grafman, J., Deiber, M. P., Ibanez, V., & Hallett, M. (1998). Neural networks for Braille reading by the blind. *Brain, 12,* 1213–1229.

Sadato, N., Pascual-Leone, A., Grafman, J., Ibanez, V., Deiber, M. P., Dold, G., et al. (1996). Activation of the primary visual cortex by Braille reading in blind subjects. *Nature, 11, 380,* 526–528.

Stasheff, S. F., & Barton, J. J. (2001). Deficits in cortical visual function. *Ophthalmological Clinics of North America, 14(1),* 217–242.

Stein, B. E., & Meredith, M. A. (1990). Multisensory integration: Neural and behavioral solutions for dealing with stimuli from different sensory modalities. *Annals of the New York Academy of Sciences, 608,* 51–65.

Ungerleider, L. G., & Haxby, J. V. (1994). "What" and "where" in the human brain. *Current Opinion in Neurobiology, 4(2),* 157–165.

Van Essen, D. (2004). Organization of visual areas in macaque and human cerebral cortex. In L. Chalupa & J. Werner (Eds.), *The visual neurosciences* (pp. 507–521). Cambridge, MA: MIT Press.

Walsh, V., & Pascual-Leone, A. (2003). *Neurochronometrics of mind: TMS in cognitive science.* Cambridge, MA: MIT Press.

Wandell, B. A. (1999). Computational neuroimaging of human visual cortex. *Annual Review of Neuroscience, 22,* 145–173.

Wassermann, E. M. (1998). Risk and safety of repetitive transcranial magnetic stimulation: Report and suggested guidelines from the International Workshop on the Safety of Repetitive Transcranial Magnetic Stimulation, June 5–7, 1996. *Electroencephalogram and Clinical Neurophysiology, 108,* 1–16.

Weeks, R., Horwitz, B., Aziz-Sultan, A., Tian, B., Wessinger, C. M., Cohen, L. G., et al. (2000). A positron emission tomographic study of auditory localization in the congenitally blind. *Journal of Neuroscience, 20,* 2664–2672.

Wiesel, T. N., & Hubel, D. H. (1965). Extent of recovery from the effects of visual deprivation in kittens. *Journal of Neurophysiology, 28,* 1060–1072.

Zihl, J., & von Cramon, D. (1979). Restitution of visual function in patients with cerebral blindness. *Journal of Neurology, Neurosurgery and Psychiatry, 42,* 312–322.

Zihl, J., & von Cramon, D. (1985). Visual field recovery from scotoma in patients with postgeniculate damage: A review of 55 cases. *Brain, 108,* 335–365.

Zihl, J., von Cramon, D., Mai, N., & Schmid, C. (1991). Disturbance of movement vision after bilateral posterior brain damage. *Brain, 114,* 2235–2252.

Zrenner, E. (2002). Will retinal implants restore vision? *Science, 295,* 1022–1025.

Zrenner, E., Stett, A., Weiss, S., Aramant, R. B., Guenther, E., Kohler, K., et al. (1999). Can subretinal microphotodiodes successfully replace degenerated photoreceptors? *Vision Research, 39,* 2555–2567.

22

Robert Riener

Neurorehabilitation Robotics and Neuroprosthetics

The anatomical structure and function of the human motor system is highly complex (figure 22.1). Billions of central nervous system (CNS) neurons in the cerebral cortex, cerebellum, brain stem, and spinal cord are involved in planning and executing movements. These efferent motor signals propagate via millions of upper motor neurons through the spinal cord to the lower motor neurons of the peripheral nervous system (PNS) that innervate the skeletal muscles. Muscle contractions take place in thousands of independent motor units. Spatial and temporal modulation of motor signals adjusts muscle forces to produce smooth movements. Dozens of muscles span single or multiple skeletal joints to generate specific limb movements or maintain certain body postures. Millions of receptors in joints, muscles, tendons, and the skin sense muscle forces and movements and send this feedback information via afferent (sensory) nerve fibers in the PNS back to the CNS. This information is processed and compared with the planned movement, together with visual inputs and inertial (vestibular) inputs on the position of the limbs and body in space to ensure accuracy. In general, the brain generates and controls voluntary movements, whereas central pattern generators in the spinal cord mediate a variety of stereotypical movements. These movements can be affected by lesions of the CNS, PNS, and the musculoskeletal system (table 22.1). As we shall see, various disorders caused by lesions along these pathways can be mitigated with a variety of new strategies and devices.

Pathologies of the Human Motor System

As mentioned, human movement relies on a variety of neural pathways between the brain, spinal cord, and sensory (afferent) receptors. These lesions can be caused by tumor, trauma, strokes (both hemorrhagic and ischemic), tumors, infections, demyelinative disorders, systemic disorders, and a host of other conditions that are reviewed in standard textbooks of neurology and internal medicine. Briefly, focal brain lesions in the primary motor cortex or other CNS regions responsible for the planning and generation of motor behavior may impair limb movements on the opposite side of the body (known as a hemiparesis when there is some residual function or hemiplegia when there is total paralysis); the opposite arm or leg may be affected alone (known as a monoparesis or monoplegia), depending on the size and location of the lesions.

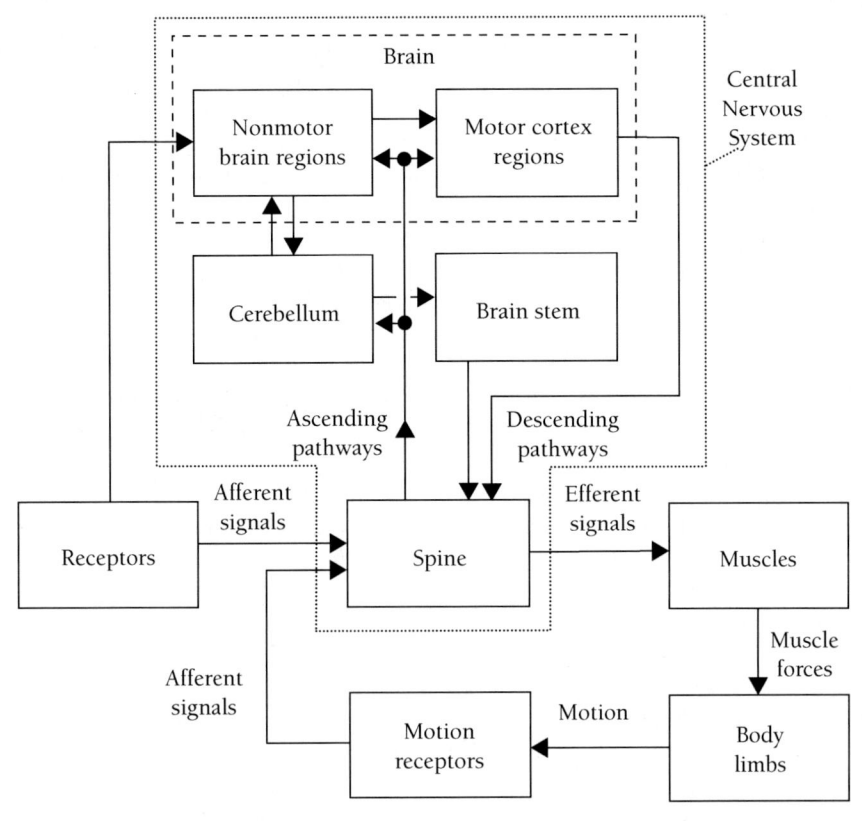

Figure 22.1. Main anatomical components of the human motor system. Adapted from Shumway-Cook & Woollacott (2001).

Table 22.1. Relation Between Injured Region, Physiological Function, Pathology, and Methods of Restoration

Region of Lesion	Functions Involved	Possible Pathologies	Possible Methods of Restoration
Brain	Motion planning Stimulus generation	Stroke, trauma, tumor, demyelinating disease (e.g, multiple sclerosis, Parkinson disease, cerebral palsy)	Surgical intervention Motion therapy Neuroprosthesis Orthosis
Spinal cord	Stimulus generation Stimulus propagation	Spinal cord injury (paraparesis, tetraparesis)	Neuroprosthesis Orthosis
Peripheral nervous system	Stimulus propagation	Neuropathy, neurapraxia, after peripheral nerve injuries	Spontaneous healing Surgical intervention Orthosis
Muscles	Motion execution Body posture	Myopathy, trauma	Spontaneous healing Orthosis Robotic support Exoprosthesis
Skeletal system	Motion execution Body posture	Osteoporosis, arthrosis Fracture Amputation	Spontaneous healing Orthosis Robotic support Exoprosthesis

Lesions of the cerebellum can affect the execution of fine motor tasks and goal-directed motor tasks without producing weakness. This can be characterized by poorly synchronized activities of muscle agonists and antagonists (asynergia) leading to (intention) tremor during movement, uncoordinated limb movements (appendicular ataxia), and unsteady posture and gait (truncal ataxia).

Spinal cord injuries (SCI) interrupt signal transfer between the brain and the periphery (muscles, receptors), causing loss of sensory and motor functions. Depending on lesion location and size, this causes partial or complete loss of tactile and kinesthetic sensations, and loss of control over voluntary motor functions, respiration, bladder, bowel, and sexual functions. Lumbar and thoracic spinal cord lesions impair lower extremity sensation and movement (paraparesis, paraplegia), whereas cervical cord lesions affect lower and upper extremities and the trunk (tetraparesis, tetraplegia). Lesions above the third cervical vertebra (C3) can lead to a loss of breathing function and head movements.

Lesions produce spasticity when upper motor neurons are damaged and lower motor neurons remain intact. Patients with spasticity show pathologically brisk tendon reflexes and an increased muscle tone, which can lead to contractures of a limb. These patients may be treated with electrical stimulation. In contrast, lesions produce atonia (or hypotonia) when lower motor neurons are injured. Compared to lesions that cause spasticity, lesions that produce atonia (hypotonia) reduce reflex activity and muscle tone and produce more striking muscle atrophy.

Finally, musculoskeletal function can be affected by muscle diseases (myopathies), bone diseases (such as osteoporosis), and by tumors and trauma. Musculoskeletal injuries include muscle fiber ruptures, ligament ruptures, joint cartilage damage, and bone fractures. Degenerative joint lesions (osteoarthritis) can result from chronic posture imbalances and inappropriate joint stress. Even entire body limbs can be lost after accidents or surgically removed because of tumors or metabolic diseases (e.g., diabetes). Clearly, these conditions offer many challenges and novel opportunities for treatment.

Natural and Artificial Mechanisms of Movement Restoration

Motion impairments due to neural and musculoskeletal lesions can be restored by natural and artificial mechanisms (table 22.1). Briefly, the body may use three natural mechanisms to restore functions: (1) Areas not affected by the lesion may partially compensate for lost functions; (2) functions of injured brain regions may be transferred to nonaffected brain regions by generation of new synaptic connections due to CNS plasticity; and (3) damaged brain regions may regenerate to some degree. These mechanisms may be enhanced and accelerated by pharmaceutical, physiotherapeutic, or surgical treatments.

Minor peripheral nerve damage (neuropraxies) can heal without any additional treatment. After full nerve transection, an artificial nerve graft can be surgically inserted to support nerve growth. Natural restoration of the musculoskeletal system is limited to healing effects of muscles and bones, for example, after muscle fiber lesions or bone fracture.

If the impairment of the nervous or musculoskeletal system cannot be restored by natural mechanisms, artificial technical support is required. Totally lost functions can be substituted by prostheses, whereas orthoses are used to support remaining (but impaired) body functions. Examples of mechanical prostheses are artificial limbs (exoprostheses) or artificial joints (endoprostheses). Lesions in the CNS can be substituted by neuroprostheses, which generate artificial stimuli in the PNS by functional electrical stimulation. A mechanical orthosis is an orthopedic apparatus used to stabilize, support, and guide body limbs during movements. Typical examples for mechanical orthoses are crutches, shells, gait and stance orthoses, and wheelchairs.

In the following sections, two examples of movement restoration principles are described in more detail. In the first example, it is shown how natural restoration principles of the CNS can be enhanced by robot-aided motion therapy. In the second example, principles and problems of neuroprostheses are clarified.

Neurorehabilitation Robotics

Rationale for Movement Therapy

Task-oriented repetitive movements can improve muscular strength and movement coordination in patients with impairments due to neurological or orthopedic problems. A typical repetitive movement is the human gait. Treadmill training has been shown to improve gait and lower limb motor function in patients with locomotor disorders. Manually assisted treadmill training was first used in the 1990s as a regular therapy for patients with SCI or stroke. Currently, treadmill training is well established at most large neurorehabilitation centers, and its use is steadily increasing. Numerous clinical studies support the effectiveness of the training, particularly in SCI and stroke patients (Barbeau & Rossignol, 1994; Dietz, Colombo, & Jensen, 1994; Hesse et al., 1995).

Similarly, arm therapy is used for patients with paralyzed upper extremities after stroke or SCI. Several studies prove that arm therapy has positive effects on the rehabilitation progress of stroke patients (see Platz, 2003, for review). Besides recovering motor function and improving movement coordination, arm therapy serves also to teach new motion strategies, so-called trick movements to cope with activities of daily living (ADL).

Lower and upper extremity movement therapy also serves to prevent secondary complications such as muscle atrophy, osteoporosis, and spasticity. It was observed that longer training sessions and a longer total training duration have a positive effect on the motor function. In a meta-analysis comprising nine controlled studies with 1,051 stroke patients, Kwakkel, Wagenaar, Koelman, Lankhorst, and Koetsier (1997) showed that increased training intensity yielded positive effects on neuromuscular function and ADLs. This study did not distinguish between upper and lower extremities. The finding that rehabilitation progress depends on training intensity motivates the application of robot-aided arm therapy.

Rationale for Robot-Aided Training

Manually assisted movement training has several major limitations. The training is labor intensive, and therefore training duration is usually limited by personnel shortage and fatigue of the therapist, not by that of the patient. During treadmill training, therapists often suffer from back pain because the training has to be performed in an ergonomically unfavorable posture. The disadvantageous consequence is that the training sessions are shorter than required to gain an optimal therapeutic outcome. Finally, manually assisted movement training lacks repeatability and objective measures of patient performance and progress.

In contrast, with automated (i.e., robot-assisted) gait and arm training, the duration and number of training sessions can be increased, while reducing the number of therapists required per patient. Long-term automated therapy can be an efficient way to make intensive movement training affordable for clinical use. One therapist may be able to train two or more patients in the future. Thus, personnel costs can be significantly reduced. Furthermore, the robot provides quantitative measures, thus allowing the observation and evaluation of the rehabilitation process.

Automated Gait-Training Devices

One commercially available system for locomotion therapy is the Gait Trainer from the German company Reha-Stim in Berlin (Hesse & Uhlenbrock, 2000). Here, the feet of the patient are mounted on two separate plates that move along a trajectory that is similar to a walking trajectory. The device does not control knee and hip joints. Thus, the patient still needs continuous assistance by at least one therapist. Forces or torques can only be measured in the moving plates and not in the leg joints.

Reinkensmeyer, Wynne, and Harkema (2002) developed a different automated gait trainer, which is characterized by several separate actuator units that are spanned between a static frame and the patient.

A third device is the Lokomat (Colombo, Jörg, Schreier, & Dietz, 2000; Colombo, Jörg, & Jezernik, 2002), which is a bilateral robotic orthosis being used in conjunction with a body weight support system to control patient leg movements in the sagittal plane (figure 22.2). The Lokomat's hip and knee joints are actuated by linear drives, which are integrated in an exoskeletal structure. A passive elastic foot lifter induces an ankle dorsiflexion during the swing phase. The legs of the patient are moved with highly repeatable predefined hip and knee joint trajectories on the basis of a position control strategy. Knee and hip joint torques can be measured via force sensors integrated inside the Lokomat (Jezernik, Colombo, & Morari, 2004). The Lokomat is currently being used in more than 30 different clinics and institutes around the world.

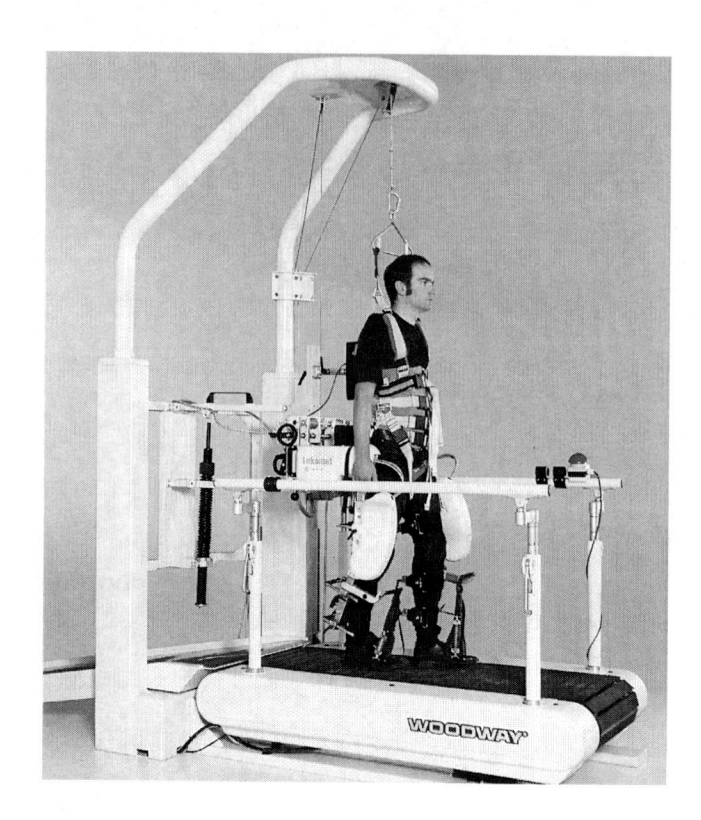

Figure 22.2. Current version of the Lokomat (Hocoma AG).

Automated Training Devices for the Upper Extremities

Hesse, Schulte-Tigges, Konrad, Bardeleben, and Werner (2003) developed an arm trainer for the therapy of wrist and elbow movements. Each hand grasps a handle and can be moved in one degree of freedom (DOF). The device position has to be changed depending on the selected movement. Force and position sensors are used to enable different control modes, including position and impedance control strategies.

Another one-DOF device is the arm robot from Cozens (1999), which acts like an exoskeleton for the elbow joint. Interactive assistance is provided on the basis of position and acceleration signals measured by an electrogoniometer and an accelerometer.

The Haptic Master is a three-DOF robot designed as haptic display by Fokker Control Systems, FCS (Van der Linde, Lammertse, Frederiksen, & Ruiter, 2002). It has formed the basis of the GENTLE/s project supported by the European Union (Harwin et al., 2001). In this project, it was suggested to use the Haptic Master as a rehabilitation device for the training of arm movements by attaching the wrist of the patient to the end-effector of the robot. However, this setup yields an undetermined spatial position for the elbow. Therefore, two ropes of an active weight-lifting system support the arm against gravity. The robot can be extended by a robotic wrist joint, which provides one additional active and two additional passive DOF. Force and position sensors are integrated to enable admittance control strategies for interactive support for patient movements. The system has been designed for the rehabilitation of stroke patients.

One of the most advanced and commonly used arm therapy robots is the MIT-Manus (Hogan, Krebs, Sharon, & Charnnarong, 1995; Krebs, Hogan, Aisen, & Volpe, 1998). It is a planar SCARA module that provides two-dimensional movements of the patient's hand (figure 22.3). Forces and movements are transferred via a robot-mounted handle gripped by the patient. The MIT-Manus was designed to have a low intrinsic end-point impedance (i.e., it is back-drivable) with a low inertia and friction. Force and position sensors are used to feed the impedance controllers. A three-DOF module can be mounted on the end of the planar module, providing additional wrist mo-

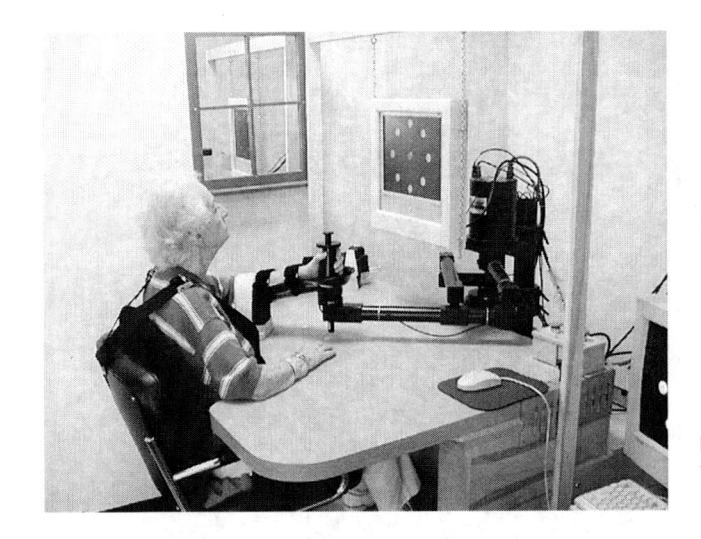

Figure 22.3. Patient using the MIT-Manus (Hogan et al., 1995; Krebs et al., 1998).

tions in three active DOF. Visual movement instructions are given by a graphical display. Clinical results with more then 100 stroke patients have been published so far (Volpe, Ferraro, Krebs, & Hogan, 2002).

Lum, Burgar, Shor, Majmundar, and van der Loos (2002) developed the MIME (Mirror Image Movement Enhancer) arm therapy robot. The key element of the MIME is a six-DOF industrial robot manipulator (Puma 560, Stäubli, Inc.) that applies forces to a patient's hand that is holding a handle connected to the end-effector of the robot. With this setup, the forearm can be positioned within a large range of spatial positions and orientations. The affected arm performs a mirror movement of the movement defined by the intact arm. A six-axis force sensor and position sensors inside the robot allow the realization of four different control modes, including position and impedance control strategies. Clinical results based on 27 subjects have been published so far.

ARMin is another rehabilitation robot system currently being developed at the Swiss Federal University of Technology (ETH) and Balgrist University Hospital, both in Zurich (figure 22.4). The robot is fixed at the wall with the patient sitting beneath. The distal part is characterized by an exoskeleton structure, with the patient's arm placed inside an orthotic shell. The current version comprises four active DOF in order to allow elbow flexion and extension and spatial shoulder movements. A vertically oriented, linear motion module performs shoulder abduction and adduction. Shoulder rotation in the horizontal plane is realized by a conventional rotary drive attached to the slide of the linear motion module. Internal and external shoulder rotation is achieved by a special custom-made drive that is connected to the upper arm via an orthotic shell. Elbow flexion and extension are realized by a conventional rotary drive. Several force and position sensors enable the robot to work with different impedance control strategies. The robot can be extended by one additional DOF to allow also hand pronation and supination—an important DOF to perform ADLs. The robot is designed primarily for the rehabilitation of incomplete tetraplegic and stroke patients. It works in two different main modes. In the zero-impedance mode, the therapist can move the patient's arm together with the robot with almost zero resistance. This movement is recorded and saved so that it can be repeated with any kind of controller, such as a position or cooperative controller.

A review on developments and clinical use of current arm therapy robots has been presented by Riener, Nef, and Colombo (2005).

Cooperative Control Strategies

Many robotic movement trainers do not adapt their movement to the activity of the patient. Even if the patient is passive, that is, unable to intervene, she or he will be moved by the device along a predefined fixed trajectory.

Future projects and studies will focus on so-called patient-cooperative or subject-centered

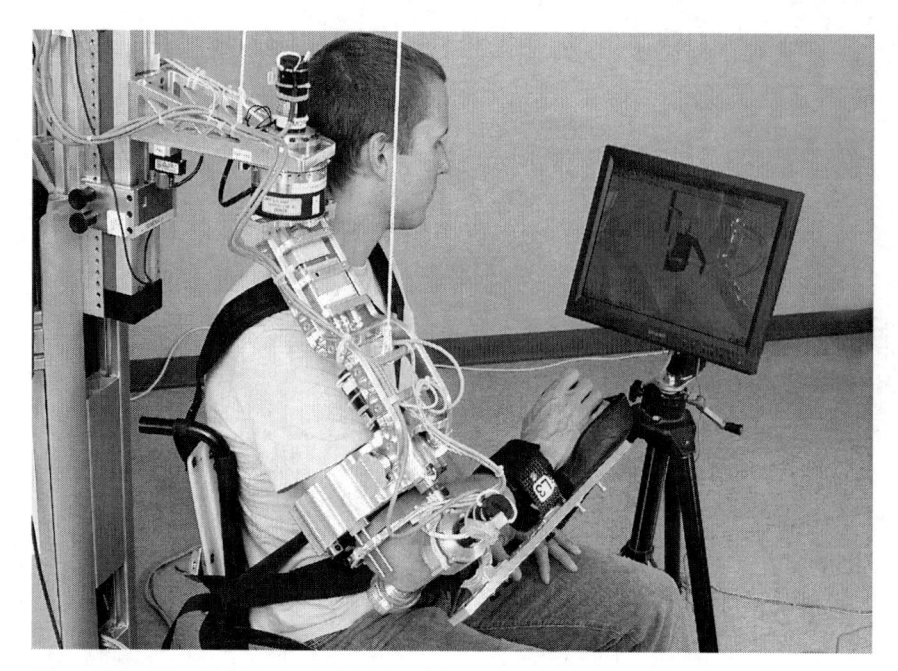

Figure 22.4. The Zurich arm rehabilitation robot ARMin.

strategies that will recognize the patient's movement intention and motor abilities in terms of muscular efforts, feed the information back to the patient, and adapt the robotic assistance to the patient's contribution. The best control and display strategy will do the same as a qualified human therapist—it will assist the patient's movement only as much as necessary. This will allow the patient to actively learn the spatiotemporal patterns of muscle activation associated with normal gait and arm/hand function.

The term *cooperativity* comprises the meanings of *compliant*, because the robot behaves softly and gently and reacts to the patient's muscular effort; *adaptive*, because the robot adapts to the patient's remaining motor abilities and dynamic properties; and *supportive*, because the robot helps the patient and does not impose a predefined movement or behavior. Examples of cooperative control strategies are, first, impedance control methods that make the Lokomat soft and compliant (Hogan 1985; Riener, Burgkart, Frey, & Pröll, 2004); second, adaptive control methods that adjust reference trajectory or controller to the individual subject (Jezernik, Schärer, Colombo, & Morari, 2003; Jezernik et al., 2004); and, third, patient-driven motion reinforcement methods that support patient-induced motions as little as necessary (Riener & Fuhr, 1998).

It is expected that patient-cooperative strategies will stimulate active participation by the patient. They have also the potential to increase the motivation of the patient, because changes in muscle activation will be reflected in the walking pattern, consistently causing a feeling of success. It is assumed that patient-cooperative strategies will maximize the therapeutic outcome. Intensive clinical studies with large patient populations have to be carried out to prove these hypotheses.

Neuroprosthetics

Background

Neuroprostheses on the basis of functional electrical stimulation (FES) may be used to restore motor function in patients with upper motor neuron lesions. The underlying neurophysiological principle is the generation of action potentials in the uninjured lower motor neurons by external electrical stimulation (for review, see Quintern, 1998).

The possibility of evoking involuntary contractions of paralyzed muscles by externally applied

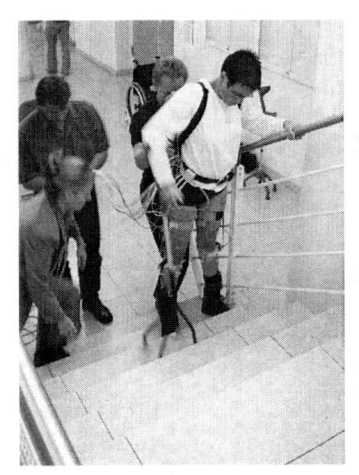

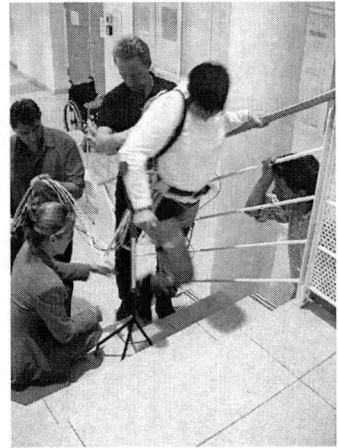

Figure 22.5. Paraplegic patient with a laboratory neuroprosthesis system applied to stair climbing (T. Fuhr, TU München). See also color insert.

electricity was already known in the 18th century (Franklin, 1757). However, after these initial feasibility demonstrations, more than 200 years were to pass until functionally useful movements of paralyzed muscles could be evoked by electrical stimulation. The first demonstration of standing by FES in a spinal cord injury patient was reported by Kantrowitz (1960). He applied electrical stimulation to the quadriceps and gluteus muscles via surface electrodes. The first portable neuroprosthesis for the lower extremities in patients with upper motor neuron lesions was developed by Liberson, Holmquest, Scot, and Dow (1961). They stimulated the peroneal nerve with surface electrodes in hemiplegic patients to prevent foot drop during the swing phase of gait. One decade later, several implantable FES systems for lower extremity applications in hemiplegic patients (Waters, McNeal, & Perry, 1975) and paraplegic patients (Brindley, Polkey, & Rushton, 1978; Cooper, Bunch, & Campa, 1973) were developed and tested. Later, several groups derived multichannel neuroprostheses with more sophisticated stimulation sequences and stimulation via surface electrodes (Kralj, Bajd, & Turk, 1980; Kralj, Bajd, Turk, Krajnik, & Benko, 1983; Malezic et al., 1984) or percutaneous wire electrodes (Marsolais & Kobetic, 1987). In the last 20 years, rapid progress in microprocessor technology has provided the means for computer-controlled FES systems (Petrofsky & Phillips, 1983; Riener & Fuhr, 1998; Thrope, Peckham, & Crago, 1985), which enable flexible programming of stimulation sequences or even the realization of complex feedback (closed-loop) control strategies (figure 22.5).

However, current commercially available neuroprostheses for the lower extremities still work in the same fashion as the first peroneal nerve stimulator (Liberson et al., 1961) or the early multichannel systems (Kralj et al., 1980, 1983). Whereas other neuroprosthetic devices such as the cochlea implant, the phrenic pacemaker, and the sacral anterior root stimulator for bladder control have grown into reliable, functionally useful, commercially available neuroprostheses (Peckham et al., 1996), lower extremity applications are far from this stage of development.

Technical Principles

Although FES is often referred to as muscle stimulation, mainly nerve fibers innervating a muscle are stimulated, irrespective of the type and localization of the electrodes. This, of course, requires that the respective lower motor neurons are preserved. Electrical stimulation activates the motor neurons and not the muscle fibers, because the threshold for electrical stimulation of the motor axons is far below the threshold of the muscle fibers (Mortimer, 1981).

In neuroprostheses, pulsed currents are applied, each pulse releasing a separate action potential in neurons, which are depolarized above threshold. Not only the current amplitude of the externally applied stimulation pulse but also the duration of the pulse, its pulse width, determines if a specific neuron is recruited. The threshold value above which a neuron is recruited depends on its size, the electrical properties of the neuron and electrodes, the position of the electrodes relative to the neuron, and the type of electrodes. When electrical pulses of low intensity (low charge per pulse) are applied, only large low-threshold neurons which are close to the electrodes are recruited. With increasing intensity of the pulses, also small neurons with higher thresholds and neurons that are located further away from the electrodes are recruited (Gorman & Mortimer, 1983).

When FES is applied to the neuromuscular system, muscle force increases with the number of recruited motor units (spatial summation), and therefore modulation of pulse width or pulse amplitude can be used to control muscle force (Crago, Peckham, & Thrope, 1980; Gorman & Mortimer, 1983). Another possible method of controlling muscle force in FES applications is modulation of the stimulation frequency (temporal summation). However, the frequency range is limited, as low-stimulation frequencies produce unfused single twitches rather than a smooth muscular contraction (i.e., tetanus). On the other hand, muscle force saturates when stimulated with frequencies above 30 Hz. With increasing frequencies, the muscle is also subjected to fatigue earlier (Brindley et al., 1978).

Stimulation systems and electrodes can be grouped into external, percutaneous, and implanted systems. In external systems, control unit and stimulator are outside the body. Surface electrodes are used that are attached to the skin above the muscle or peripheral nerve, whereas in percutaneous systems wire electrodes pierce the skin near the motor point of the muscle. In implanted systems, both stimulator and electrodes are inside the body. Different kinds of implanted electrodes are used. They can be inserted into muscle (e.g., on muscle surface: epimysial electrodes), nerve (epineural electrodes), or fascicle (intrafascicular electrodes), or surround the nerve (nerve cuff electrodes).

Challenges in the Development of Neuroprostheses

The development of control systems for neuroprostheses presents challenges at several different levels. First, the physiological system that we are trying to control has many complex features, many of which are poorly understood or poorly characterized. Musculoskeletal geometry, dynamic response properties of muscle, segmental coupling, reflex interactions, joint stiffness properties, and nonlinear system properties have all caused problems for many of the FES control systems that have been tested (Adamczyk & Crago, 1996; Chizeck, 1992; Hatwell, Oderkerk, Sacher, & Inbar, 1991; Quintern, 1998; Veltink, Chizeck, Crago, & El-Bialy, 1992). Perhaps the most important features of the physiological system are the high degree of uncertainty and variability in response properties from person to person and the fact that these properties change over time due to fatigue and other factors. The uncertainty, variability, and time dependence make it extremely difficult to determine a stimulation pattern that will achieve the desired posture or movement.

The second level of challenges includes those that are specific to the implementation of FES control systems. By far the most prominent of these issues has been sensors (Crago, Chizeck, Neuman, & Hambrecht, 1986). Achieving improved control is critically dependent on reliable measurements of neuromotor variables in real time. While this has occasionally been achieved in laboratory environments, it has yet to be achieved in a manner that would be suitable for use on an everyday basis. An exciting approach that has great potential for solving some sensing problems is to record and interpret signals from intact sensory neurons (Hoffer et al., 1996; Yoshida & Horch, 1996). Other implementation-level challenges include input devices, stimulator design, cosmesis, and battery weight.

The third level of challenges includes those related to the interactions between three competing control systems. The neuroprosthesis control system acts via the electrically stimulated muscles; the intact voluntary control system acts via muscles that are not paralyzed; and the spinal reflex control system (mediated by circuits below the level of the lesion) acts via paralyzed muscles. The challenge for the neuroprosthesis control system is to act in a manner that is coordinated with the voluntary control system while effectively exploiting appropriate reflexes and counteracting the effects of inappropriate reflexes and spasticity.

Approaches to Neuroprosthesis Control

Current neuroprostheses for the lower extremities have not found wide acceptance for clinical use. The gain of mobility in terms of walking speed and distance is limited. Complex movements with high coordination requirements, such as ascending and descending stairs, are impossible as yet. The reason for this limited function is that all commercially available systems are open-loop systems, which do not provide sensor feedback to determine the stimulation pattern.

The block diagram in figure 22.6 demonstrates the various types of control that have been used in FES systems (see also Abbas & Riener, 2001). To operate FES systems, users provide inputs that are either discrete selections or continuously variable signals. Discrete input signals can be used to select a task option or to trigger the initiation of a movement pattern. A continuously variable signal can be used to adjust stimulation patterns in a continuous manner as the task is being performed. This signal, sometimes called a command input, can be used to adjust the stimulation to several muscles simultaneously using a nonlinear mapping function (Adamczyk & Crago, 1996; Peckham, Keith, & Freehafer, 1988). Most FES systems use the input signals in an open-loop control system configuration, which means that the action of the controller will be the same each time the user gives the input (Hausdorff & Durfee, 1991; McNeal, Nakai, Meadows, & Tu, 1989). For example, in lower-extremity FES systems for standing up, each time the user gives the discrete input signal to stand up, a prespecified pattern of stimulation is delivered to a set of muscles (Kobetic & Marsolais, 1994). In upper-extremity systems for hand grasp, each time the user changes the level of command input, the stimulation levels delivered to a set of muscles are adjusted using a prespecified mapping function (Peckham et al., 1988). In either case, the relationship between the input signal and the stimulation must be predetermined in a process that can be described as fitting the FES control system to the user. This process is typically time consuming and requires the effort of a trained rehabilitation team.

The effectiveness of an open-loop approach is often limited by the ability of the rehabilitation team to predict or anticipate the response of the musculoskeletal system to the stimulation. If these predetermined stimulation patterns are not appropriate, then the desired posture or movement may not be achieved. Furthermore, as muscles fatigue and system properties change over time, a repetition of the fitting process may be required. In systems that use continuously variable inputs, the user can often observe system performance and make online adjustments as needed. This approach

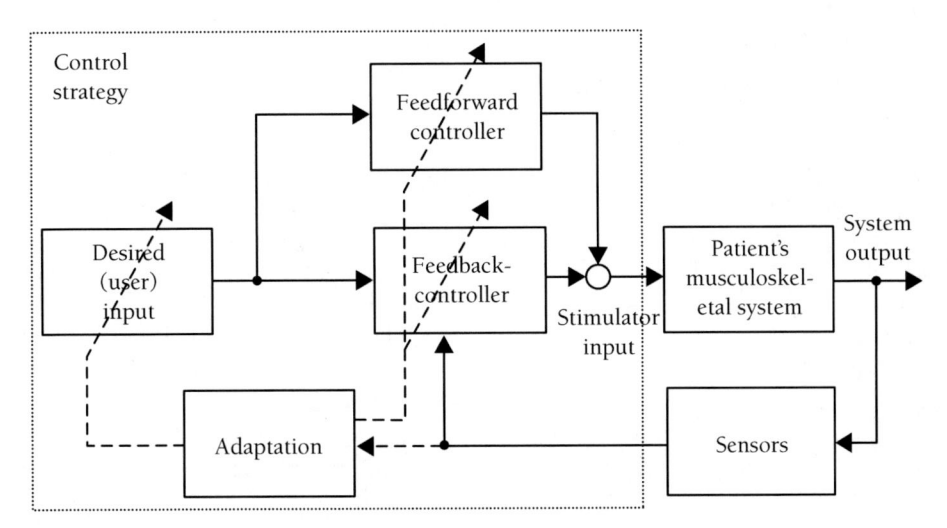

Figure 22.6. Block diagram representation of the neuroprosthesis control components. The control system may include feedforward, feedback, or adaptive components. Note that if feedforward control is used alone, it is usually referred to as open-loop control.

takes advantage of the intact sensory functions of the user (often visual observations), but it may put excessive demands on the attention of the user.

In order to improve the performance of these systems, advanced controllers must be designed that are capable of determining appropriate stimulation levels to accomplish a given task. The most commonly used approach to improving the quality of the control system has been to utilize feedback or closed-loop control strategies (see figure 22.6). Closed-loop control means that information regarding the actual state of the system (e.g., body posture and ground reaction forces) is recorded by sensors and fed back to a controller. Based on the measured signals, the controller then determines the stimulation pattern that is required to achieve a specific movement task. This type of control mimics the online adjustments that can be made by the user but does so in a manner that is automatic and does not require conscious effort on the part of the user. Feedback control can be used to supplement the signals from the open-loop controller and may be able to improve performance by adjusting the stimulation to account for inaccuracies in the open-loop fitting process. In addition, external and internal disturbances can be recognized and the stimulation pattern readjusted to result in a smooth and successful movement. Most closed-loop systems have been evaluated for control of force or angle at single joints (Hatwell et al., 1991; Munih, Donaldson, Hunt, & Barr, 1997; Veltink et al., 1992; Yoshida & Horch, 1996). Little work has been done in the field of closed-loop control of multijoint movements such as standing (Jaeger, 1986), standing up and sitting down (Mulder, Veltink, & Boom, 1992), and walking (Fuhr, Quintern, Riener, & Schmidt, 2001).

One promising approach to improving lower-extremity systems is the use of continuously variable inputs from the user in a manner that is similar to the command input used in upper-extremity systems (Donaldson & Yu, 1996; Riener & Fuhr, 1998; Riener, Ferrarin, Pavan, & Frigo, 2000). Such strategies are called patient-driven or subject-centered strategies, because the person drives the movement, in contrast to controller-centered approaches, where a predefined reference signal is used.

Whenever a continuous input signal is employed, it can be difficult for the user to make adjustments when the system input-output properties are unpredictable. Adaptive control strategies have been developed to adjust the overall system behavior (i.e., the response of the combined controller and system) so that it is more linear, repeatable, and therefore predictable (Adamczyk & Crago, 1996; Kataria & Abbas, 1999). These techniques adjust the parameters of the control system and attempt to self-fit the system to the user in order to make it easier to use and easier to learn to use (Chang et al., 1997; Chizeck, 1992; Crago, Lan, Veltink, Abbas, & Kantor, 1996).

The performance of closed-loop approaches is still not satisfactory in terms of disturbance compensation, upper body incorporation, variable step adjustment, or movement smoothness. This is one of the main reasons why current systems are not applied clinically so far. The use of computational models can significantly enhance the design and test of closed-loop control strategies applied to FES (see Riener, 1999). Time-consuming and perhaps troublesome trial-and-error experimentation can be avoided or at least shortened, and the number of experiments with humans can be reduced, both of which can accelerate the development of neuroprostheses. Furthermore, physiologically based mathematical models can provide significant insight into relevant activation and contraction processes. This insight may help us to better understand and eventually avoid the disadvantageous effects occurring during FES, such as increased muscular fatigue. Eventually, muscle force production and the resulting movement may be optimized to obtain better functionality.

Summary

After a general overview of the principle of human motion generation and related pathologies, two examples of movement restoration were presented in more detail. In the first example, it was shown how robots can be applied to support natural restoration principles of the CNS. In the second example, the technical principles and challenges of neuroprostheses were presented.

Main Points

1. The damaged nervous system can recover by application of natural and artificial restoration principles.

2. Robots can make motion therapy more efficient by increasing training duration and number of training sessions, while reducing the number of therapists required per patient.
3. Patient-cooperative control strategies have the potential to further increase the efficiency of robot-aided motion therapy.
4. Neuroprostheses can restore movement in patients with upper motor neuron lesions.
5. Neuroprosthesis function can be improved by applying closed-loop control components and computational models.

Key Readings

Abbas, J., & Riener, R. (2001). Using mathematical models and advanced control systems techniques to enhance neuroprosthesis function. *Neuromodulation, 4,* 187–195.

Quintern, J. (1998). Application of functional electrical stimulation in paraplegic patients. *Neurorehabilitation, 10,* 205–250.

Riener, R. (1999). Model-based development of neuroprostheses for paraplegic patients. *Royal Philosophical Transactions: Biological Sciences, 354,* 877–894.

Riener, R., Nef, T., & Colombo, G. (2005). Robot-aided neurorehabilitation for the upper extremities. *Medical and Biological Engineering and Computing, 43,* 2–10.

Riener, R., Lünenburger, L., Jezernik, S., Anderschitz, M., Colombo, G. & Dietz, V. (2005). Cooperative subject-centered strategies for robot-aided treadmill training: first experimental results. *IEEE Transactions on Neural Systems and Rehabilitation Engineering 13,* 380–393.

References

Abbas, J., & Riener, R. (2001). Using mathematical models and advanced control systems techniques to enhance neuroprosthesis function. *Neuromodulation, 4,* 187–195.

Adamczyk, M. M., & Crago, P. E. (1996). Input-output nonlinearities and time delays increase tracking errors in hand grasp neuroprostheses. *IEEE Transactions on Rehabilitation Engineering, 4,* 271–279.

Barbeau, H., & Rossignol, S. (1994). Enhancement of locomotor recovery following spinal cord injury. *Current Opinion in Neurology, 7,* 517–524.

Brindley, G. S., Polkey, C. E., & Rushton, D. N. (1978). Electrical splinting of the knee in paraplegia. *Paraplegia, 6,* 428–435.

Chang, G. C., Luh, J. J., Liao, G. D., Lai, J. S., Cheng, C. K., Kuo, B. L., et al. (1997). A neuro-control system for the knee joint position control with quadriceps stimulation. *IEEE Transactions on Rehabilitation Engineering, 5,* 2–11.

Chizeck, H. J. (1992). Adaptive and nonlinear control methods for neural prostheses. In R. B. Stein, P. H. Peckham, & D. B. Popovic (Eds.), *Neural prostheses: Replacing motor function after disease or disability* (pp. 298–328). New York: Oxford University Press.

Colombo, G., Jörg, M., & Jezernik, S. (2002). Automatisiertes Lokomotionstraining auf dem Laufband. *Automatisierungstechnik, 50,* 287–295.

Colombo, G., Jörg, M., Schreier, R., & Dietz, V. (2000). Treadmill training of paraplegic patients using a robotic orthosist. *Journal of Rehabilitation Research and Development, 37,* 693–700.

Cooper, E. B., Bunch, W. H., & Campa, J. H. (1973). Effects of chronic human neuromuscular stimulation. *Surgery Forum, 24,* 477–479.

Cozens, J. A. (1999). Robotic assistance of an active upper limb exercise in neurologically impaired patients. *IEEE Transactions on Rehabilitation Engineering, 7,* 254–256.

Crago, P. E., Chizeck, H. J., Neuman, M. R., & Hambrecht, F. T. (1986). Sensors for use with functional neuromuscular stimulation. *E Transactions on Biomedical Engineering, 33,* 256–268.

Crago, P. E., Lan, N., Veltink, P. H., Abbas, J. J., & Kantor, C. (1996). New control strategies for neuroprosthetic systems. *Journal of Rehabilitation Research and Development, 33,* 158–172.

Crago, P. E., Peckham, P. H., & Thrope, G. B. (1980). Modulation of muscle force by recruitment during intramuscular stimulation. *IEEE Transactions on Biomedical Engineering, 27,* 679–684.

Dietz, V., Colombo, G., & Jensen, L. (1994). Locomotor activity in spinal man. *Lancet, 44,* 1260–1263.

Donaldson, N. de N., & Yu, C.-H. (1996). FES standing control by handle reactions of leg muscle stimulation (CHRELMS). *IEEE Transactions on Rehabilitation Engineering, 4,* 280–284.

Franklin, B. (1757). On the effects of electricity in paralytic cases. *Philosophical Transactions, 50,* 481–483.

Fuhr, T., Quintern, J., Riener, R., & Schmidt, G. (2001). Walk! Experiments with a cooperative neuroprosthetic system for the restoration of gait. In Ron J. Triolo (Ed.), *Proceedings of the IFESS Conference* (pp. 46–47). Cleveland OH, June.. Published by Dept. of Orthopaedics and Biomedical Engineering, Case Western Reserve University, and Louis Stokes Veterans Affairs Medical Center.

Gorman, P. H., & Mortimer, J. T. (1983). The effect of stimulus parameters on the recruitment characteristics of direct nerve stimulation. *IEEE Transactions on Biomedical Engineering, 30,* 301–308.

Harwin, W., Loureiro, R., Amirabdollahian, F., Taylor, M., Johnson, G., Stokes, E., et al. (2001). The GENTLE/s project: A new method for delivering neuro-rehabilitation. In C. Marincek et al. (Eds.), *Assistive technology—added value to the quality of life. AAATE'01* (pp. 36–41). Amsterdam: IOS Press.

Hatwell, M. S., Oderkerk, B. J., Sacher, C. A., & Inbar, G. F. (1991). The development of a model reference adaptive controller to control the knee joint of paraplegics. *IEEE Transactions on Automatic Control, 36,* 683–691.

Hausdorff, J. M., & Durfee, W. K. (1991) Open-loop position control of the knee joint using electrical stimulation of the quadriceps and hamstrings. *Medical and Biological Engineering and Computing, 29,* 269–280.

Hesse, S., Bertelt, C., Jahnke, M. T., Schaffrin, A., Baake, P., Malezic, M., et al. (1995). Treadmill training with partial body weight support compared with physiotherapy in nonambulatory hemiparetic patients. *Stroke, 26,* 976–981.

Hesse, S., Schulte-Tigges, G., Konrad, M., Bardeleben, A., & Werner, C. (2003). Robot-assisted arm trainer for the passive and active practice of bilateral forearm and wrist movements in hemiparetic subjects. *Archives of Physical Medicine and Rehabilitation, 84,* 915–920.

Hesse, S., & Uhlenbrock, D. (2000). A mechanized gait trainer for restoration of gait. *Journal of Rehabilitation Research and Development, 37,* 701–708.

Hoffer, J. A., Stein, R. B., Haugland, M. K., Sinkjaer, T., Durfee, W. K., Schwartz, A. B., et al. (1996). Neural signals or command control and feedback in functional neuromuscular stimulation: A review. *Journal of Rehabilitation Research and Development, 33,* 145–157.

Hogan, N. (1985). Impedance control: An approach to manipulation, Parts I, II, III. *Journal of Dynamic Systems, Measurement, and Control, 107,* 1–23.

Hogan, N., Krebs, H. I., Sharon, A., & Charnnarong, J. (1995). Interactive robotic therapist. U.S. Patent 5466213.

Jaeger, R. J. (1986). Design and simulation of closed-loop electrical stimulation orthoses for restoration of quiet standing in paraplegia. *Journal of Biomechanics, 19,* 825–835.

Jezernik, S., Colombo, G., & Morari, M. (2004). Automatic gait-pattern adaptation algorithms for rehabilitation with a 4 DOF robotic orthosi. *IEEE Transactions on Robotics and Automation, 20,* 574–582.

Jezernik, S., Schärer, R., Colombo, G., & Morari, M. (2003). Adaptive robotic rehabilitation of locomotion: A clinical study in spinally injured individuals. *Spinal Cord, 41*(12).

Kantrowitz, A. (1960). Electronic physiologic aids. In *Report of the Maimonides Hospital* (pp. 4–5). Unpublished report, Maimonides Hospital, Brooklyn, NY.

Kataria, P., & Abbas, J. J. (1999). Adaptive user-specified control of movements with functional neuromuscular stimulation. *Proceedings of the IEEE/BMES Conference* (Atlanta, GA, p. 604).

Kobetic, R., & Marsolais, E. B. (1994). Synthesis of paraplegic gait with multichannel functional neuromuscular stimulation. *IEEE Transactions on Rehabiltation Engineering, 2,* 66–79.

Kralj, A., Bajd, T., & Turk, R. (1980). Electrical stimulation providing functional use of paraplegic patient muscles. *Medical Progress Through Technology, 7,* 3–9.

Kralj, A., Bajd, T., Turk, R., Krajnik, J., & Benko, H. (1983). Gait restoration in paraplegic patients: A feasibility demonstration using multichannel surface electrode FES. *Journal of Rehabilitation Research and Development, 20,* 3–20.

Krebs, H. I., Hogan, N., Aisen, M. L., & Volpe, B. T. (1998). Robot-aided neurorehabilitation. *IEEE Transactions on Rehabilation Engineering, 6,* 75–87.

Kwakkel, G., Wagenaar, R. C., Koelman, T. W., Lankhorst, G. J., & Koetsier, J. C. (1997). Effects of intensity of rehabilitation after stroke: A research synthesis. *Stroke, 28,* 1550–1556.

Liberson, W. T., Holmquest, M. E., Scot, D., & Dow, M. (1961). Functional electrotherapy: Stimulation of the peroneal nerve synchronized with the swing phase of gait of hemiplegic patients. *Archives of Physical Medicine and Rehabilitation, 42,* 101–105.

Lum, P. S., Burgar, C. G., Shor, P. C., Majmundar, M., & van der Loos, M. (2002). Robot-assisted movement training compared with conventional therapy techniques for the rehabilitation of upper-limb motor function after stroke. *Archives of Physical Medicine and Rehabilitation, 83,* 952–959.

Malezic, M., Stanic, U., Kljajic, M., Acimovic, R., Krajnik, J., Gros, N., et al. (1984). Multichannel electrical stimulation of gait in motor disabled patients. *Orthopaedics, 7,* 1187–1195.

Marsolais, E. B., & Kobetic, R. (1987). Functional electrical stimulation for walking in paraplegia. *Journal of Bone and Joint Surgery, 69-A,* 728–733.

McNeal, D. R., Nakai, R. J., Meadows, P., & Tu, W. (1989). Open-loop control of the freely swinging paralyzed leg. *IEEE Transactions on Biomedical Engineering, 36,* 895–905.

Mortimer, J. T. (1981). Motor prostheses. In V. B. Brooks (Ed.), *Handbook of physiology, nervous system*

II (pp. 155–187). Bethesda, MD: American Physiological Society.

Mulder, A. J., Veltink, P. H., & Boom, H. B. K. (1992). On/off control in FES-induced standing up: A model study and experiments. *Medical and Biological Engineering and Computing, 30,* 205–212.

Munih, M., Donaldson, N. de N., Hunt, K. J., & Barr, F. M. D. (1997). Feedback control of unsupported standing in paraplegia—part II: Experimental results. *IEEE Transactions on Rehabilation Engineering, 5,* 341–352.

Peckham, P. H., Keith, M. W., & Freehafer, A. A. (1988). Restoration of functional control by electrical stimulation in the upper extremity of the quadraplegic patient. *Journal of Bone Joint Surgery, 70-A,* 144–148.

Peckham, P. H., Thrope, G., Woloszko, J., Habasevich, R., Scherer, M., & Kantor, C. (1996). Technology transfer of neuroprosthetic devices. *Journal of Rehabilitation Research and Development, 33,* 173–183.

Petrofsky, J. S., & Phillips, C. A. (1983). Computer controlled walking in the paralyzed individual. *Journal of Neurological and Orthopedic. Surgery, 4,* 153–164.

Platz, T. (2003). Evidenzbasierte Armrehabilitation: Eine systematische Literaturübersicht. *Nervenarzt, 74,* 841–849.

Quintern, J. (1998). Application of functional electrical stimulation in paraplegic patients. *Neurorehabilitation, 10,* 205–250.

Reinkensmeyer, D. J., Wynne, J. H., & Harkema, S. J. (2002). A robotic tool for studying locomotor adaptation and rehabilitation. *Second Joint Meeting of the IEEE, EMBS and BMES 2002* (pp. 2353–2354).

Riener, R. (1999). Model-based development of neuroprostheses for paraplegic patients. *Royal Philosophical Transactions: Biological Sciences, 354,* 877–894.

Riener, R., Burgkart, R., Frey, M., & Pröll, T. (2004). Phantom-based multimodal interactions for medical education and training: The Munich Knee Joint Simulator. *IEEE Transactions on Information Technology in Biomedicine, 8,* 208–216.

Riener, R., Ferrarin, M., Pavan, E., & Frigo, C. (2000). Patient-driven control of FES-supported standing up and sitting down: Experimental results. *IEEE Transactions on Rehabilitation Engineering, 8,* 523–529+.

Riener, R., & Fuhr, T. (1998). Patient-driven control of FES-supported standing-up: A simulation study. *IEEE Transactions on Rehabilitation Engineering, 6,* 113–124.

Riener, R., Nef, T., & Colombo, G. (2005). Robot-aided neurorehabilitation for the upper extremities. *Medical and Biological Engineering and Computing, 43,* 2–10.

Shumway-Cook, A., & Woollacott, M. H. (2001). *Motor control: Theory and practical applications* (2nd ed.). Baltimore: Lippincott Williams and Wilkins.

Thrope, G. B., Peckham, P. H., & Crago, P. E. (1985). A computer controlled multichannel stimulation system for laboratory use in functional neuromuscular stimulation. *IEEE Transactions on Biomedical Engineering, 32,* 363–370.

Van der Linde, R. Q., Lammertse, P., Frederiksen, E., & Ruiter, B. (2002). The HapticMaster, a new high-performance haptic interface. In *Proceedings of Eurohaptics, Edinburgh,* UK (pp. 1–5).

Veltink, P. H., Chizeck, H. J., Crago, P. E., & El-Bialy, A. (1992). Nonlinear joint angle control for artificially stimulated muscle. *IEEE Transactions on Biomedical Engineering, 39,* 368–380.

Volpe, B. T., Ferraro, M., Krebs, H. I., & Hogan, N. (2002). Robotics in the rehabilitation treatment of patients with stroke. *Current Atherosclerosis Reports, 4,* 270–276.

Waters, R. L., McNeal, D. R., & Perry, J. (1975). Experimental correction of footdrop by electrical stimulation of the peroneal nerve. *Journal of Bone and Joint Surgery, 57-A,* 1047–1054.

Yoshida, K., & Horch, K. (1996). Closed-loop control of ankle position using muscle afferent feedback with functional neuromuscular stimulation. *IEEE Transactions on Biomedical Engineering, 43,* 167–176.

# 23

Matthew Rizzo, Sean McEvoy, and John Lee

Medical Safety and Neuroergonomics

Primum non noscere: First, do no harm.
                              Physicians' credo

Errare humanum est: To err is human.
                              Seneca

Errors in medicine are an important public health policy issue that can be mitigated by applying principles and techniques of neuroergonomics. The Institute of Medicine (IOM, 2000) issued a publication, *To Err Is Human: Building a Safer Health System*, which asserted that errors in health care are a leading cause of death and injury, killing more people than do car crashes, AIDS, or breast cancer. This IOM report reviewed the frequency, cost, and public perceptions of safety errors and suggested that between 44,000 and 98,000 deaths per year result from medical errors. These figures were extrapolated from a 1984 study of New York and a 1992 study of Colorado and Utah. Whether these limited samples truly reflect what goes on in the huge and variegated U.S. medical population is unclear. Nevertheless, medical errors are clearly a public health problem and systems should be developed to mitigate error-related injuries and death—as they were decades ago in other industries for which safety is of critical importance, such as aviation and nuclear power.

This chapter considers how neuroergonomics, the study of the brain and behavior at work in healthy and impaired states, is relevant to assessments and interventions in patient safety at the levels of individuals and health care systems. For example, knowledge of how the brain processes visual, auditory, and tactile information can provide guidelines and constraints for theories of information presentation and health care task and system designs. It is particularly important to consider alternative approaches because attempts to improve safety in response to the IOM report have not been particularly successful. For instance, computerized physician order entry (CPOE) systems, aimed at reducing errors caused by illegibly written or ill-advised prescriptions, have limited utility in preventing adverse drug effects (Nebeker, Hoffman, Weir, Bennett, & Hurdle, 2005) and may even exacerbate medication errors (Koppel et al., 2005). One widely used CPOE system facilitated 22 types of medication error risks; among these were fragmented CPOE displays that prevented a coherent view of patient medications, pharmacy inventory displays mistaken for dosage guidelines, ignored antibiotic renewal notices placed on paper charts but not in the CPOE system, and inflexible ordering formats generating wrong orders. There are clearly many more opportunities for reducing errors in the delivery of health care and improving patient safety through systemwide assessments of health care delivery processes.

Safe delivery of health care can draw from many of the approaches, strategies, and techniques outlined in other chapters of this book. For example, patient safety efforts can benefit from insights on augmented reality (see chapter 17, this volume) by using interactive projection systems that allow a surgeon to "see" a patient's anatomy projected on a visual model of the patient's skin. Continuous monitoring of a physician's physiological state may enable systems to modulate the amount of information that goes to a physician, say an anesthesiologist, to make sure he or she does not get overloaded or confused, or to alert others to the physician's state. Other systems could help track patients at each step of the health care system to improve situation awareness of medical personnel interacting with patients and each other in complex settings like a busy hospital ward, emergency room, or day-of-surgery lounge. This chapter reviews potential areas for neuroergonomic interventions at the level of individuals and systems, and cultural and legal issues that affect the ability to intervene.

Systems Perspective

Often the health care system is unsafe, not the practitioner or equipment (Bogner, 1994; Woods, 2000; Woods, Johannesen, Cook, & Sarter, 1994). Because many levels of distraction and responsibility affect health care providers, a systems-based approach is needed to locate the precursors of error and identify effective mitigation strategies. Figure 23.1 (Rasmussen, Pejtersen, & Goodstein, 1994) shows how the behavioral sequence that leads to errors depends on a complex context that includes the task situation and mental strategies, as well as the management structure and work preferences of individuals. To date, a systems-based approach to error assessment has been limited by inadequate

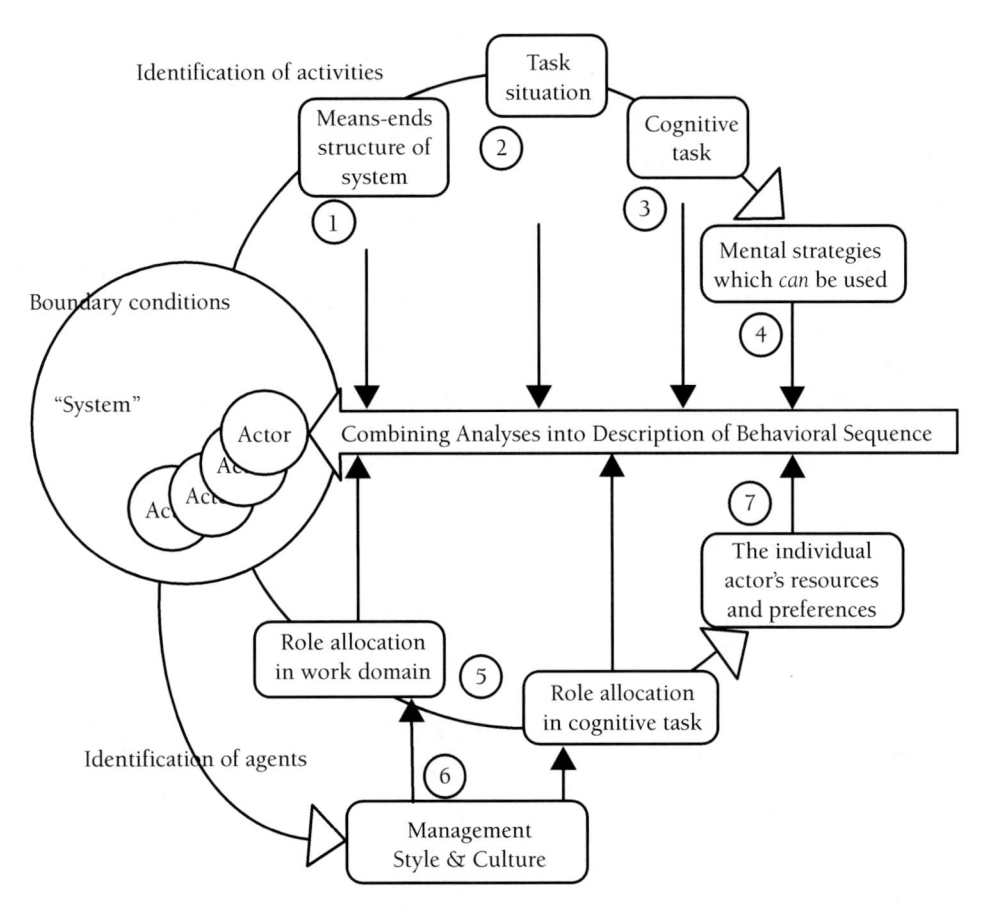

Figure 23.1. A systems-based approach to identifying the behavioral sequences that lead to errors. From Rasmussen et al. (1994).

systems for reporting and classifying errors in health care. Reliable and accurate health care error databases generally do not exist, due in part to culture, fear of reprisal and litigation, and ambiguity on what constitutes an error.

Relationships Between Health Care Delivery and Errors

Relationships between health care delivery and safety errors can be represented by an imaginary triangle (Heinrich, Petersen, & Roos, 1980) or "iceberg" (Maycock, 1997; figure 23.2). The simple model can be applied to errors that lead to injuries in diverse settings, including factories, automobile driving, aviation, nuclear power, and health care. Visible safety errors ("above the waterline") include errors resulting in fatality, serious injury, mild injury, or legal claims. While the number of fatalities and injuries from health care errors may be unacceptably high (IOM, 2000), these events are relatively infrequent. Submerged below the waterline are relatively benign mishaps and near misses that are theoretically related to injury outcome and occur more frequently.

With sufficient numbers of observations, it might be possible to accurately estimate the risk of a fatality (a low-frequency, high-severity event) through the assessment of measurable safety errors (high-frequency, low-severity events). The relationship between these low-frequency, high-severity events that result in reported injuries and high-severity, low-frequency events that are neither systematically observed nor reported might be better defined using a system for naturalistic observations to record continuous behavior and the patterns of performance and circumstances leading to adverse events (as described in chapter 8, this volume).

Cognitive Factors

Risk of human errors in complex systems such as health care increases with distraction, workload, fatigue, licit and illicit drug use, illness, and associated impairments of attention, perception, response selection (which depends on memory, reasoning, and decision making), and implementation (see figure 23.3). Some medical errors can be detected because personnel normally monitor their performance and will often detect discrepancies when feedback on performance fails to match expectations based on correctly formulated intentions.

In this heuristic framework, the health care practitioner (1) perceives and attends stimulus or situation evidence and interprets the situation to arrive at a diagnosis; (2) formulates a plan based on the particular health care situation and relevant previous experience or memory; and (3) executes an action (e.g., by ordering laboratory tests, medications, or making referrals to additional practitioners). The outcome is safe or unsafe due to errors at one or more stages, and depends not only upon the practitioner but also on the engineering of the health care delivery system. The outcome of the behavior provides a source of potential feedback for the practitioner to take subsequent action.

In some cases, the feedback loops of errors in health care are short and direct, such as the sound of an alarm signifying critical alteration of vital signs during a surgical procedure on a patient under general anesthesia. In medical subspecialties such as internal medicine or neurology, however, these feedback loops are often indirect and have a long latency before they return to the operator or operators. In this situation, a system of reporting and classifying errors is especially important. An error database would provide a more global level of understanding

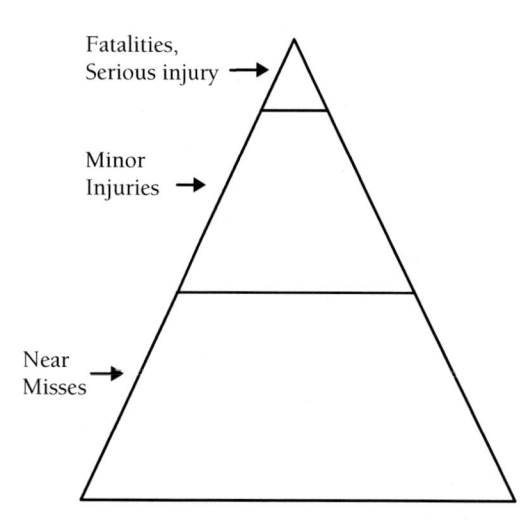

Figure 23.2. Health care errors that lead to fatality or serious injury represent only a small portion of health care errors. The majority of errors lead only to mild consequences or have no direct effect on the patient (near misses).

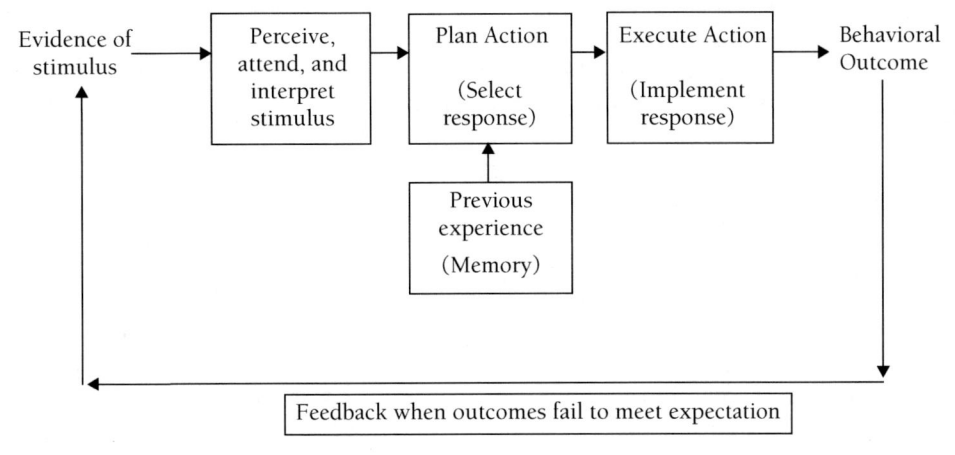

```
Evidence of   →   Perceive,      →   Plan Action    →   Execute Action  →   Behavioral
stimulus          attend, and        (Select            (Implement          Outcome
                  interpret          response)          response)
                  stimulus
                                          ↑
                                     Previous
                                     experience
                                     (Memory)

              ← Feedback when outcomes fail to meet expectation ←
```

Figure 23.3. Information-processing model for understanding practitioner error.

about the types of errors that occur during extended treatment and individual and systematic variables that contribute to errors, and help to focus intervention efforts where they are most needed.

The Need to Track Health Care Errors

Mitigating safety errors in medicine depends on knowing the type, frequency, and severity of the errors that occur, as well as what actions lead to successful outcomes and in what particular circumstances. Error-reporting systems and a taxonomy or lexicon of errors are thus necessary. Analysis of error reports can identify the cognitive or organizational stresses that contributed to the error and suggest mitigation strategies to relieve these stresses.

Systems for reporting and classifying errors in health care are currently inadequate. Consequently, proposed interventions are guided by the best available evidence, which is often limited. Most errors in medical practice are reported at local levels, as with incident reports of nursing or medication error at hospitals, or in morbidity and mortality rounds, in which health care personnel (especially physicians) discuss complications of patient care and how to improve related procedures and practice. These reports are not systematically examined, and the analysis is not disseminated broadly. Any lessons learned from the local analysis of errors are confined to a few people and do not reach the larger organization. Reliable and valid error reporting, analysis, and dissemination systems do not exist in most medical specialties. Frequency of errors is not known and may not be knowable. In the absence of these data, useful evidence for directing health care safety interventions comes from malpractice claims.

Closed Claims Analyses

Malpractice data can serve as a surrogate for identifying severe medical errors at the tip of the iceberg (figure 23.2). Along these lines, Glick (2001) summarized data from available Massachusetts closed malpractice claims involving neurological problems. Errors were classified as failure to diagnose, act, or decide (e.g., delay in ordering studies, failure to perform a proper history or physical examination, and misinterpretation of studies). The basic premise is that malpractice claims are surrogates for poor practices, bad outcomes, errors, and miscommunication and can indicate the need for modifying health care systems and educational programs. Glick proposed that the information provided valuable lessons for neurological teachers on what to teach and whom to teach. Overall, there were approximately 150 cases involving 250 neurological defendants. Findings showed that the main errors were diagnostic failure in one third and treatment failure (especially medication errors and professional behavior and communication problems including improper consent) in one third of all cases. Among the diagnostic failures were failure to diagnose stroke and other vascular problems, spinal cord and nerve root problems, meningitis, encephalitis, head injury, and brain tumors. About two thirds of these problems were acute. A review of malpractice claims, inpatient incident reports and chart reviews, and journal literature (Glick, 2005) reaffirmed that failure of accurate and timely diagnosis was a leading category of neurological health care error.

These data demonstrate the potential utility of error analysis, although the available data are limited. It remains difficult to extract detailed characteristics of individual and systemic performance that led to litigation, because insurers' data collection systems are not designed for this use. Importantly, malpractice claims may not accurately reflect medical error, but may reflect factors unrelated to physician competence, such as tone of voice (Ambady, LaPlante, Nguyen, Rosenthal, Chaumeton, & Levinson, 2002). A need remains for more serviceable sources of error data.

Mandatory Reporting

One obstacle to creating a database of errors for tracking is bias in reporting. One potential source of more reliable data would be a mandatory reporting system that requires health care personnel to report all medical errors. Yet such systems run against the cultural and ethical norms of many Americans and may not lead to mandatory reporting by healthcare personnel. Along these lines, the State of California Health and Safety Code mandates that all physicians report immediately the identity of every driver diagnosed with a lapse of consciousness or an episode of marked confusion due to neurological disorders or senility. The mandatory report triggers an evaluation of the individual, during which driving privileges may be suspended, but may discourage patients with treatable forms of mental impairment to avoid evaluation for fear of losing the license to drive (Drachman, 1990). A practice among many California physicians is to inform the patient and patient's family of concerns about driving with dementia but not to report the names of drivers with dementia to the state. No reliable or fair means of dealing with nonreporters has been devised. Contrary to traditional views of epidemiology and teaching on quality control, mandatory reporting may not generate objective measurements to track.

Voluntary Reporting

To provide an accurate source of health care error information, it may be necessary to establish a health care information safety report system (HISRS) to gather voluntary reports from health care workers, protect the identity of the error reporters, and provide safeguards against the use of the error information during litigation proceedings. This type of protected voluntary reporting system exists in other fields where analysis of safety errors carries high importance.

One such program is the Aviation Safety Reporting System (ASRS). The system depends on attempts to understand the most we can from a smattering of reports. Close analysis of the details of these reports can remove some or all reporter bias. The ASRS is funded by the Federal Aviation Administration (FAA) and operated by NASA. Because the ASRS is operated by an independent agency, it has no licensing or legislative power. ASRS has no authority to direct that action be taken in potentially hazardous conditions or situations, but alerts those (FAA personnel) responsible for correcting the conditions. The ASRS acts directly and indirectly to inform pilots, flight crews, air-traffic controllers, airport managers, and others in the aviation community about potentially unsafe practices. In certain situations (such as "altitude busts"), there is an incentive for pilots to report unsafe practices to ASRS to avoid penalty. Individuals who report these incidents are granted confidentiality and immunity.

Also, reporters are granted use immunity under FAR [Federal Aviation Rule] 91.25 which prohibits the FAA from using reports submitted under ASRS, or information derived from these reports, in any enforcement actions against reporters, provided that criminal offenses and accidents are not involved. FAA's Advisory Circular 00-46 also provides for limited disciplinary immunity for pilots in conjunction with errors resulting from actions or lack of actions, if the certain criteria are met. The transactional immunity of ASRS is a powerful incentive for reporting an unintentional violation of FAA rules because it "inoculates" the reporting person against adverse certificate action or civil penalties.

Along these lines, in March 2002 the NASA ASRS model was adopted by the Veterans Administration (VA) health care system in its Patient Safety Reporting System. The VA can implement such a system with relative ease because of unique legal protections for VA quality assurance activities. The VA model uses reports that are voluntary and confidential to begin with and, in later stages, are deidentified. The analyst can call the reporter back for further information and a better understanding of the mechanisms underlying an unsafe occurrence. The courts so far appear to have ruled that the deidentified database is hearsay and inadmissible, though it may be possible to identify some reports in the database for use in adversarial proceedings.

A significant obstacle to obtaining accurate data on health care errors, even within a voluntary reporting structure, has been the current lack of safe harbors. Voluntary reporting may be preferable to mandatory reporting but could still produce discoverable evidence that fuels litigation without illuminating the fundamental causes of medical errors. Persons who are aware of data on medical errors may not be reporting the data for fear of retribution to themselves or colleagues in terms of professional sanctions, civil or criminal liability, or economic loss. Physicians are reluctant to share data that has traditionally been managed in an adversarial manner. A culture change is required so that reporting is voluntary, viewed as supportive and part of a process of quality improvement.

Legislative actions must be taken to ensure the legal protections required to introduce such an error-tracking program across the whole of the health care system. Accordingly, the Senate passed the Patient Safety and Quality Improvement Act of 2005 (introduced by Senator James Jeffords, Vermont), which was subsequently passed by the House of Representatives and signed into law by the president. The law encourages health care providers to report errors and near misses to patient safety organizations, defined as private or public organizations that analyze reported patient safety data and develop strategic feedback and guidelines to providers on how to improve patient safety and the quality of care. The bill also includes protections for reporter and patient confidentiality, does not mandate a punitive reporting system, and does not alter existing rights or remedies available to injured patients. This development offers enormous promise for systematic study and systemic interventions on behalf of patient safety, starting with reporting tools.

Development of a Reporting Tool

A systems-based approach to patient safety can adapt and integrate several human factors frameworks, including the following:

- The SHEL (software/hardware/environment/liveware) model. First introduced by Edwards (1972) and further developed by Hawkins (1987), this model is an organizational tool used to identify factors and work system elements that can influence human performance (Molloy & O'Boyle, 2005).
- The accident causation and generic error-modeling system framework. Developed by Reason (1990), this guiding tool constructs a sequence of events and circumstances that can help identify the unsafe conditions, actions, and decisions that led to a given incident.
- The taxonomy of error (Hawkins, 1987; Rasmussen, 1987) links skill-, rule-, and knowledge-based error types to appropriate levels of intervention.

After reconciling the legal issues of a voluntary reporting tool, the HISRS framework outlines the collection and analysis of data on errors and the tracking of the results of interventions (see figure 23.4). The reporting addresses multiple issues:

- Information about the reporter.
- In what medical sector are you involved?
- What is your specific job (e.g., physician, administrator, nurse, laboratory technician)?
- In what type of facility do you typically work (e.g., clinic, intensive care unit, pharmacy, outpatient clinic)?
- Type of incident or hazard (e.g., wrong procedure or medicine; fall)
- When did the incident/hazard occur?
- How were you involved in the incident (or in discovering the hazard)?
- Describe the working conditions.
- Where did the incident happen?
- What specific type of care was being rendered at the time of the incident or hazard?
- Describe what happened.
- What do you think caused the incident? Consider decisions, actions, inactions, information overload, communication, fatigue, drugs or alcohol, physical or mental condition, procedures, policies, design of equipment, facility, workers (experience, staffing), equipment failure (reasons), maintenance.
- What went right? How was an accident avoided? Consider corrective actions, contingency plans, emergency procedures, luck.

1) Event
- Who?
- What?
- When?
- Where?
- Why?
- How?

⇩

2) Reporter
- Ability to comprehend, analyze and deconstruct events
- Memory/accuracy
- Verbal ability
- How much does s/he know and *not* know?
- Willingness to divulge (fear of repercussion for self or colleagues)

⇩

3) Form Completion
Form ➤
- Format
- Items
- Detail
- Ease of use

⇩

4) Form Review/Data Entry
Database Structure
- Sufficient detail?
- Well-defined?

Analyst
- Subject matter knowledge (medical and HF)
- Ability to logically analyze
- Willingness to return call and consult
- Validity/reliability

⇩

5) Data Analyses
Database
- Does taxonomy support analysis?
- Search and sort algorithms

Analyst
- Respect for limitations of the data, database
- Knowledge of specific definitions of data items
- Creativity in search techniques
- Willingness to perform complete, accurate analysis

⇩

6) HISRS Products
- Safety product objectives
- Target audience
- Product content (sufficient detail? new info?)
- Product format (easy to use and find info?)
- Web site database

⇩

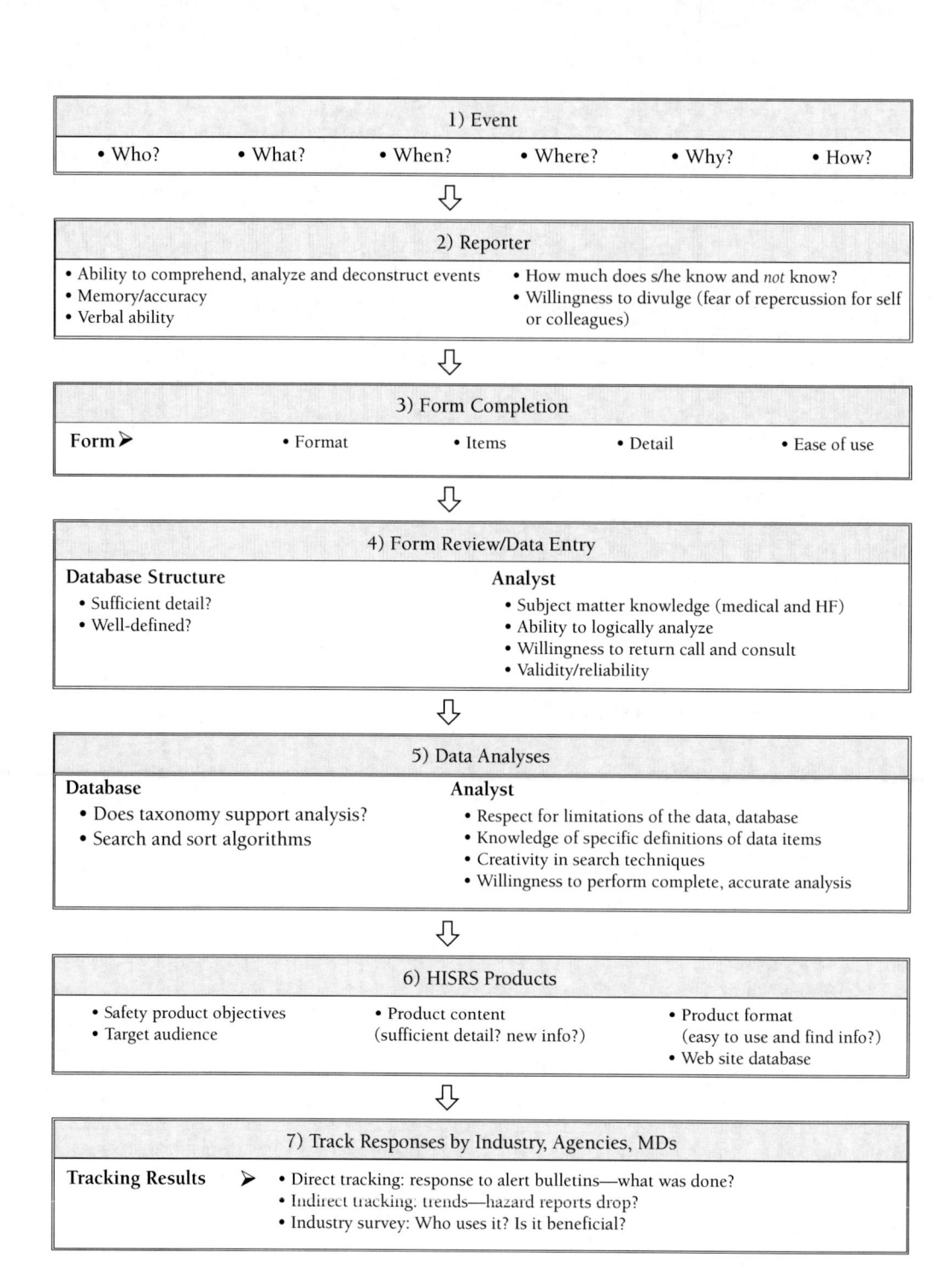

7) Track Responses by Industry, Agencies, MDs
Tracking Results ➤
- Direct tracking: response to alert bulletins—what was done?
- Indirect tracking: trends—hazard reports drop?
- Industry survey: Who uses it? Is it beneficial?

Figure 23.4. Health care information safety report system (HISRS) framework.

- How can we prevent similar incidents (correct the hazard)? What changes need to be made? By whom? Describe lessons learned, safety tips, and suggestions.

Taxonomy of Error

To effectively catalog errors for analysis, a taxonomy of information-processing categories is needed. An example of one system of taxonomy is summarized below (cf. Norman, 1981, 1988; Reason, 1984, 1990).

Knowledge-based mistakes signify inappropriate planning due to failure to comprehend because the operator is overwhelmed by the complexity of a situation and lacks information to interpret it correctly. For example, a specialist may misdiagnose a patient because a primary care provider omitted relevant information when referring the patient.

Rule-based mistakes occur when operators believe they understand a situation and formulate a plan by *if-then* rules, but the *if* conditions are not met, a "bad" rule is applied, or the *then* part of the rule is poorly chosen. For example, a doctor misdiagnoses a patient with an extremely rare disorder because of similarity of symptoms, when the diagnosis of a more common disorder is more probable.

Slips are errors in which an intention is incorrectly carried out because the intended action sequence departs slightly from routine, closely resembles an inappropriate but more frequent action, or is relatively automated. The "reins of action" or perception are captured by a contextually appropriate strong habit due to lack of close monitoring by attention.

Lapses represent failure to carry out an action (omission of a correct action rather than commission of an incorrect action), may be caused by interruption of an ongoing sequence by another task, and give the appearance of forgetfulness. For example, a doctor may intend to write a prescription for a patient in intensive care, but forgets the prescription when his attention is taken away by an emergency situation.

The promise of error taxonomies is to provide an organizing framework for identifying common causes and mitigation strategies from seemingly unrelated instances. Despite their promise, traditional error taxonomies have been generally ineffective in generating useful mitigation strategies.

It remains unclear how these taxonomies of error map onto specific cognitive deficits (e.g., of attention, perception, memory; see figure 23.3) that lead to errors.

Systemic Factors

A more constructive approach to classifying errors may be to focus on collecting narrative descriptions of events and then using a multidisciplinary team to identify the factors that contributed to the mishap. Under this framework, errors may then by classify by level of the health care system and areas of cognition that contributed to the error. An important reason for this is that the factors that shape the outcome of any particular situation have their roots in several levels of system description (figure 23.5; Rasmussen et al., 1994). A level of description that focuses on the information-processing limits of cognition may fail to identify the contribution of management or team deficiencies. A multidisciplinary approach brings multiple viewpoints to the analysis of an error and often identifies a range of contributing factors and system deficiencies that lead to the error. Figure 23.5 shows the range of potential viewpoints that a multidisciplinary team can adopt in understanding the range of factors that contribute to errors and then identifying successful mitigation strategies.

A notable example of a systemic approach to mitigating medical errors is the cognitive work analysis (CWA) framework (Rasmussen et al., 1994; Vicente, 1999). CWA provides a multilevel taxonomy to classify and analyze medical errors, defining several potential levels for neuroergonomic intervention. CWA comprises five layers, each analyzing a different aspect of an application domain (see figure 23.6). The first layer of constraint is the work domain, a map of the environment to be acted upon. The second layer is the set of control tasks that represents what needs to done to the work domain. The third layer is the set of strategies that represents the various processes by which action can be effectively carried out. The fourth layer is the social-organizational structure that represents how the preceding set of demands is allocated among actors, as well as how those actors can productively organize and coordinate themselves. Finally, the fifth layer of constraint is the set of worker competencies that represent the

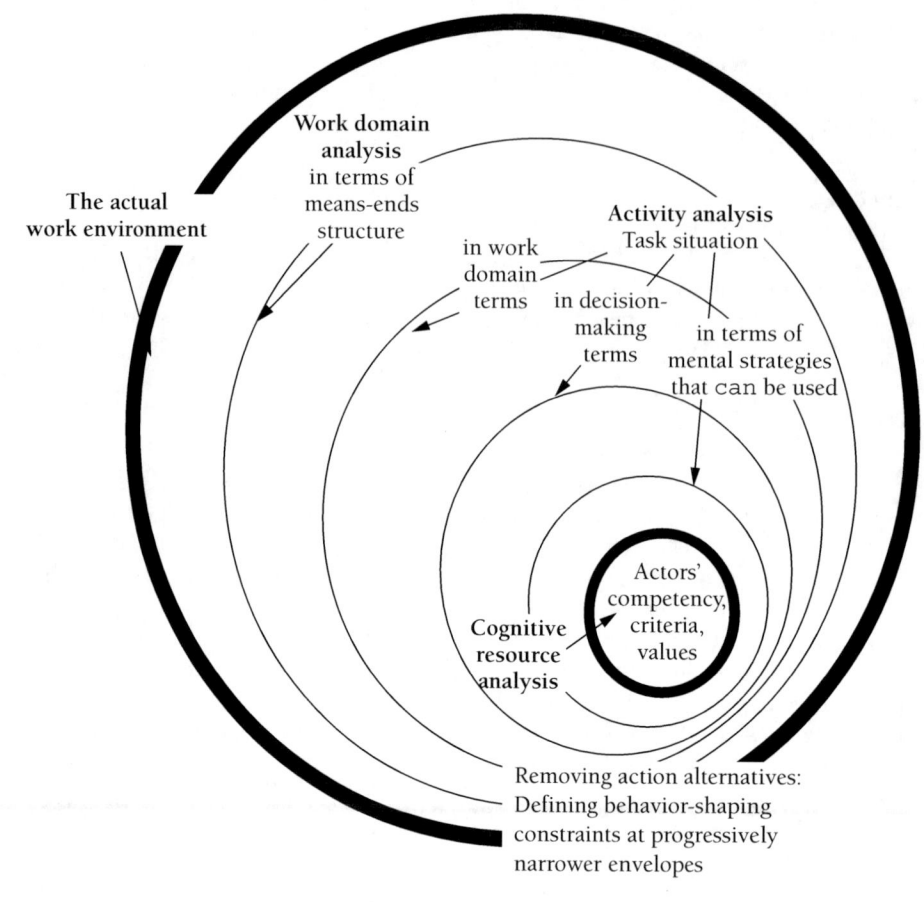

Figure 23.5. Shows a range of perspectives a multidisciplinary approach can adopt in a systematic analysis of errors to identify effective mitigation strategies. From Rasmussen et al. (1994).

Figure 23.6. The cognitive work analysis framework is an example of a constraint-based approach comprising five layers: work domain, control tasks, strategies, social-organizational (soc-org) analysis, and worker competencies. These relationships are logically nested with a progressive reduction of degrees of freedom. Adapted from Vicente (1999).

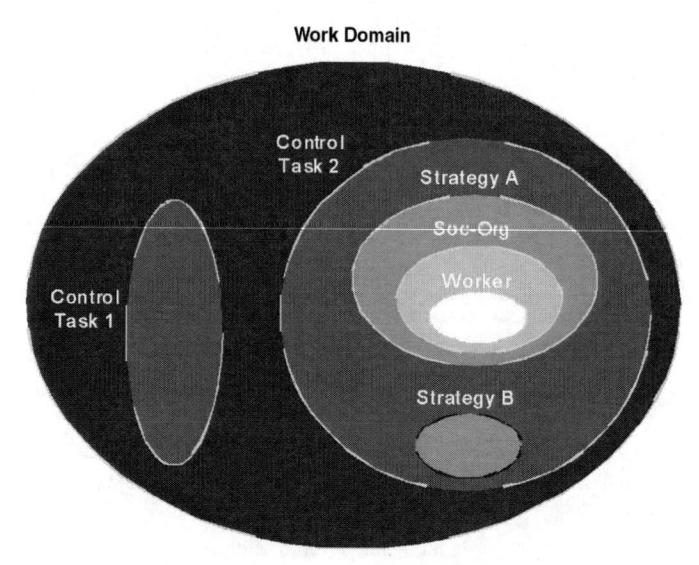

Table 23.1. Relationships Between the Five Phases of Cognitive Work Analysis and Various Classes of System Design Interventions

Phase	Systems Design Intervention
Work Domain	
What information should be measured?	Sensors
What information should be derived?	Models
How should information be organized?	Database
Control Tasks	
What goals must be pursued and what are the constraints on those goals?	Procedures or automation
What information and relations are relevant for particular classes of situations?	Context-sensitive interface
Strategies	
What frames of reference are useful?	Dialogue modes
What control mechanisms are useful?	Process flow
Social-Organizational	
What are the responsibilities of all of the actors?	Role allocation
How should actors communicate with each other?	Organizational structure
Worker Competencies	
What knowledge, rules, and skills do workers need to have?	Selection, training, and interface form

Source: Vicente (1999).

capabilities required for success. Given this breadth, CWA provides an integrated basis for the development of mitigation strategies for medical error in general and neurological misdiagnosis in particular. CWA, described below, has already been successfully applied to aviation safety.

Figure 23.6 illustrates how these five layers of constraint are nested. The size of each set in this diagram represents the productive degrees of freedom for actors, so large sets represent many relevant possibilities for action, whereas small sets represent fewer relevant possibilities for action.

The CWA framework also comprises modeling tools that can be used to identify each layer of constraint (Rasmussen et al., 1994; Vicente, 1999). Table 23.1 shows how each of these models is linked to a particular class of design interventions. The list is illustrative, not definitive or exhaustive, but it shows how CWA is geared toward uncovering implications for systems design that are clearly relevant to neuroergonomic interventions. Beginning with the work domain, analyzing the system being controlled provides insight into what information is required to understand its state. In turn, this analysis has implications for the design of sensors and models. The work domain analysis also reveals the functional structure of the system being controlled. These insights can then be used to design a database that keeps track of the relationships between variables, providing a coherent, integrated, and global representation of the information contained therein. CWA has been successfully applied to aviation and process control. The underpinnings of a single error may have complex roots and may arise from a concatenation of problems across several of the layers.

Safety Management Through Error Reports and Proactive Analysis: A Control Theoretical Approach

Control theory provides a useful framework for describing some of the fundamental challenges to reducing the incidence of medical errors and enhancing patient safety. Control theory is a systems analysis concept that describes the relationship between the current system state, the goal state, environmental disturbances, operating characteristics of the system components, and the control strategies used to achieve the goal state (Jagacinski & Flach, 2002).

CWA evaluates system characteristics to identify organizational and cognitive stress that may induce errors before errors occur. Error reporting and analysis examines the causes of errors after they occur. Both of these approaches generate mitigation strategies that can be introduced into the system to reduce error rates in the future. In the language of control theory, these approaches are examples of feedback and feedforward control.

Feedback control is analogous to safety management through error reporting systems. With feedback control, mitigation strategies are adjusted based on observed differences between the goal and observed levels of safety. This approach requires a comprehensive reporting of errors and introduces a lag between when errors occur and when mitigation strategies can be deployed.

Feedforward control is analogous to safety management through cognitive work analysis. This approach requires an ability to fully describe the medical diagnosis system and catalog all the factors that influence errors.

Feedforward and feedback control have well-known capabilities and limits that can help clarify the requirements for the study of medical error. Specifically, unmeasurable errors and the time lag between error reporting and intervention design provide a rationale for feedforward control. No error-reporting system can capture every important error in a timely manner. Likewise, the rationale for feedback control comes from the difficulty in comprehensively describing the cognitive and organizational stressors of a complex system such as the one that supports neurological diagnosis. No cognitive work analysis will identify all possible error mechanisms. The feedforward approach (cognitive work analysis) and the feedback approach (error reporting and analysis) are complementary strategies that are both required to mitigate medical error.

The control theoretical framework integrates feedback and feedforward approaches to safety management and identifies several other critical requirements associated with the process of identifying and evaluating mitigation strategies:

- *What cannot be measured cannot be controlled.* Safety measurement is stressed, with effective strategies for capturing incident data.
- *Mitigation, not blame.* Reporting mechanisms are defined that go beyond administrative and punitive purposes to address the underlying factors that contributed to the failure. Error reports focus on identifying mitigation strategies, not assigning blame.
- *Getting the word out.* Attention is focused on the need to identify information pathways into which the understanding of failures can be fed so that the lessons learned from errors can be widely disseminated and used to develop effective technological and social interventions.
- *Ongoing evaluation of strategies.* Mitigation strategies are continuously evaluated using contemporary error data.

Figure 23.7 summarizes how this approach integrates the two complementary strategies described in this chapter: a proactive strategy of cognitive work analysis and a reactive strategy of error reporting and analysis (Lee & Vicente, 1998).

Examples of Neuroergonomic Interventions

Having reviewed a framework for understanding human error, taxonomies of error, reporting systems, feedback loops, CWA, and cultural and legal factors surrounding error reporting and patient safety, we now describe a few incipient interventions. These current efforts to advance patient safety aim to improve interactions between health care personnel, systems, products, environments, and tools. These interventions can involve procedural interventions and policy changes, as outlined in National Patient Safety Goals (Joint Commission on Accreditation of Healthcare Organizations, 2005).

A simple example of safety-relevant culture change is limiting the work week of postgraduate physicians-in-training to 80 hours. This intervention by the American College of Graduate Medical Education aimed to (1) curtail the traditional Oslerian abuse of physicians in training, also known

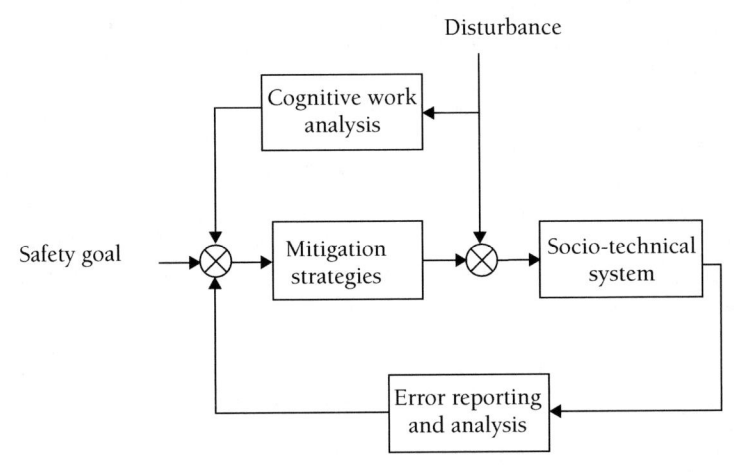

Figure 23.7. A control theoretical framework for safety management in medical systems.

as residents or house staff (perhaps because many of these doctors were always in the house); and (2) improve patient safety by reducing cognitive errors due to physician fatigue (see also chapter 14, this volume).

Neuroergonomic interventions may be initiated at different levels in the CWA framework outlined above, using techniques described throughout parts II–VI of this book (including neuroergonomics methods, perception and cognition, stress, fatigue and physical work, technology applications, and special populations). In general, these interventions are in early phases of planning and development but hold substantial promise for improving patient safety. Relevant efforts can involve monitoring of individual health care providers, patients, and processes, and systems for tracking the ongoing performance of health care workers, teams, and patients (see chapter 8, this volume). These efforts can take advantage of miniature physiological sensors and monitoring devices, improved health care information displays, systems for enhanced communications between individuals, offices, and institutions, and applications from virtual reality (VR).

As outlined earlier, health care areas and processes to assess errors and improve safety are ideally informed by reporting systems that indicate points or levels in the system where problems are localized. Consequently, new opportunities for intervention arise with the passage of the Patient Safety and Quality Improvement Act of 2005, allowing for reporting systems with safe harbors and patient confidentiality protections in the context of patient safety organizations. This new legislation should allow more comprehensive reporting of adverse events that allow CWA analyses.

Accordingly, the chapter describes an interface design system that uses CWA, an HISRS, and principles of ecological interface design to mitigate diagnostic errors in long-loop feedback systems. The chapter also reviews techniques for tracking the patient through the health care system using bar codes, applications of augmented reality in continuous control tasks with short-loop feedback to improve the safety of surgical and endoscopic procedures, and telepresense setups, allowing the mind of the expert to operate at a distance.

Ecological Interface Design: Mapping Information Requirements onto a Computer Interface to Reduce Diagnosis Error

Ecological interface design (EID) is a systems design framework that complements the systems analysis framework comprising CWA. EID takes the information requirements identified by CWA models and turns them into specific interface designs. Specific design principles enable this process to present information in a format that makes it easier for people to understand what they need to get a job done (Vicente & Rasmussen, 1992).

Providing rich feedback in the interface has the potential to minimize errors and facilitate error detection and, thus, error recovery.

The EID framework has been applied to complex systems, including aviation (Dinadis & Vicente, 1999), computer network management (Kuo & Burns, 2000), fossil-fuel power plants (Burns, 2000), information retrieval (Xu, Dainoff, & Mark, 1999), military command and control (Burns et al., 2000), nuclear power plants (Itoh, Sakuma, & Monta, 1995), petrochemical plants (Jamieson & Vicente, in press), and software engineering (Leveson, 2000). Applications of EID to medicine promise to improve patient safety in a variety of settings including intensive care (e.g., Hajdukiewicz, Vicente, Doyle, Milgram, & Burns, 2001; Sharp & Helmicki, 1998). An EID-based computer application could also be developed to mitigate diagnosis error in long-loop feedback settings relevant to community and office-based clinical medical practice settings. The computer application could organize information about each case in a graphical format that combines the history and results of previous medical tests into an easily accessible format.

The evaluation of the computer application could include two approaches. First, the application can be evaluated in a controlled setting using a selection of difficult diagnostic cases culled from reported errors. These cases could be reconstructed and presented to physicians with and without the EID-based computer application. Second, the computer application could be disseminated to the physician community, where it would be evaluated in terms of comments from the physicians and in terms of reductions in errors measured by the error reporting system.

Controlled experiments and a reliable error reporting system allow measurements of the benefits of such a tool in terms of error reduction, compared to base rates of error in current practice. Such EID-based computer application could enhance diagnostic accuracy and lead physicians to be more confident about correct diagnoses and less confident about incorrect diagnoses, compared to the physicians using traditional organizational schemes. Also, physicians using an EID-based computer application would conduct a more thorough diagnosis and examine a broader range of information, compared to the traditional approach.

Tracking the Patient Through the System Using Bar Codes

Bar codes are commonly used in supermarkets to improve efficiency and reduce charge errors at checkout counters. They can also be used to improve patient safety and avoid a variety of error types. For instance, heath care personnel can scan information into a computer terminal at the point of care from bar codes on their ID badges, a patient's ID bracelet, and medicine vials. The computer can track the patient through the health care system and improve situation awareness for the individual caregiver and health care team regarding where a patient is, what procedures are being done, and by whom (Yang, Brown, Trohimovich, Dana, & Kelly, 2002). Bar code-enabled point-of-care software can mitigate transfusion and tissue specimen identification errors (Bar Code Label for Human Drug Products and Blood, 2003) and make certain a patient receives the proper procedure or drug dose, on time, while warning against potential adverse drug interactions or allergic reactions (Patterson, Cook, & Render, 2002). A "smart" drug infusion pump can read drug identity, concentration, and units from the bar code on a drug vial, display these values, and remind the operator to check them against the patient's prescription (Tourville, 2003). It could prevent harm from drug dosing errors by halting drug administration and issuing a warning if it cannot read the bar code on the vial, if the drug dose is out of range, or if it detects a potentially harmful interaction with another drug the patient is taking. Similar technology may help patients avoid errors in self-administration of rapidly acting and potentially dangerous drugs, as in diabetic patients who self-inject insulin to control their blood sugar levels. Such technology can even be combined with a small global positioning system transmitter that can be attached to a patient at risk for getting lost. Such an intervention might help rescue a nursing facility resident with a memory disorder such as Alzheimer's disease who wanders outdoors in thin clothes in midwinter, unattended and unnoticed. Further in the future, galvanic vestibular stimulation (Fitzpatrick, Wardman, & Taylor, 1999) might be used to modulate patient posture and gait (Bent, Inglis, & McFadyen, 2004), perhaps even to steer lost patients back or prevent unsteady patients from falling.

Augmented Reality and Virtual Reality

VR environments can be used to train novice personnel in safety critical care procedures on virtual patients without risk of harm to a patient. Augmented reality (AR) and augmented cognition applications can enhance the planning, conduct, and safety of complex surgeries by skilled surgeons. It is possible to enhance the display of information to improve situation awareness by a health care worker and team (see Ecological Interface Design). Continuously monitored indices of performance and arousal of the worker can inform personnel and their supervisors of impairments or performance decline to avert impending errors and injuries, using optimized alerting and warning signals. These preventive strategies dovetail with applications of crew resource management (CRM) training techniques (Wiener, Kanki, & Helmreich, 1995), first developed to prevent errors in aviation settings where stress levels and workload are high and communications may fail, leading to disaster. CRM may be applied to mitigate errors in emergency rooms and other stressful critical care settings by improving communications and situation awareness among the health care team.

VR also provides neuroergonomic tools to aid an operator's spatial awareness and performance in complex navigational situations, such as performing complex surgeries in small spaces and recognizing and coping with ramifications of anatomical variations (McGovern & McGovern, 1994; Satava, 1993). These systems can use principles from AR, discussed in chapter 17 in terms of combining of real and artificial stimuli, with the aim of improving human performance and insight. This generally involves overlaying computer-generated graphics on a stream of video images so that these virtual images appear to be embedded in the real world.

For example, an AR application—projecting 3-D radiographic information on the location of a patient's organs on surface of a patient's own body—would allow a physician to see through a patient to localize and better understand and treat disease processes. These systems ultimately should reduce spatial errors and navigation problems (see chapter 9, this volume) that may result in procedure-related iatrogenic injuries to patients.

Imagine a system that uses commercially available video products and computer hardware to combine video of an actual patient with 3D computed tomography (CT) scan or magnetic resonance (MR) images of the brain to help in planning surgical operations. The surgeon can view the CT or MR images projected over images of the actual patient. These methods can be applied before and during surgery to locate borders of a lesion. For example, using these techniques, a neurosurgeon could plan the best site for a skin incision, craniotomy, and brain incision, practice the procedure, and limit injury to normal brain tissue (Spicer & Apuzzo, 2003).

In a similar vein, Noser, Stern, and Stucki (2004) assessed a synthetic vision-based application that was designed to find paths through complex virtual anatomical models of the type encountered during endoscopy of colons, aortas, or cerebral blood vessels (Gallagher & Cates, 2004). These spaces typically contain loops, bottlenecks, cul de sacs, and related impasses that can create substantial challenges, even for highly skilled operators. The application found collision-free paths from a starting position to an end-point target in a timely fashion. The results showed the feasibility of automatic path searching for interactive navigation support, capable of mitigating injuries to anatomical structures caused by navigation errors during endoscopic procedures.

A navigational aid might display, in real time, the shape of a flexible endoscope inside the colon (Cao, 2003). Spatial orientation error and workload may be reduced in operators using such a device, which provides additional short-loop feedback to the operator for error correction. Similar VR-based strategies can be applied during teaching, training (Seymour et al., 2002), practicing, and actual surgery to reduce errors in sinus surgery (Satava & Fried, 2002), urological surgery, neurosurgery (Peters, 2000), and breast biopsies.

To see inside patients and guide physicians during internal procedures, neuroergonomists are testing augmented reality systems that combine ultrasound echography imaging, laparoscopic range imaging, a video see-through head-mounted display, and a graphics computer with the live video image of a patient (Rolland & Fuchs, 2000). Figure 23.8 shows the AR setup being applied to improve performance and reduce errors during ultrasound-guided breast biopsy. This procedure requires guiding a biopsy needle to a breast lesion with the aid of ultrasound imagery, requires good

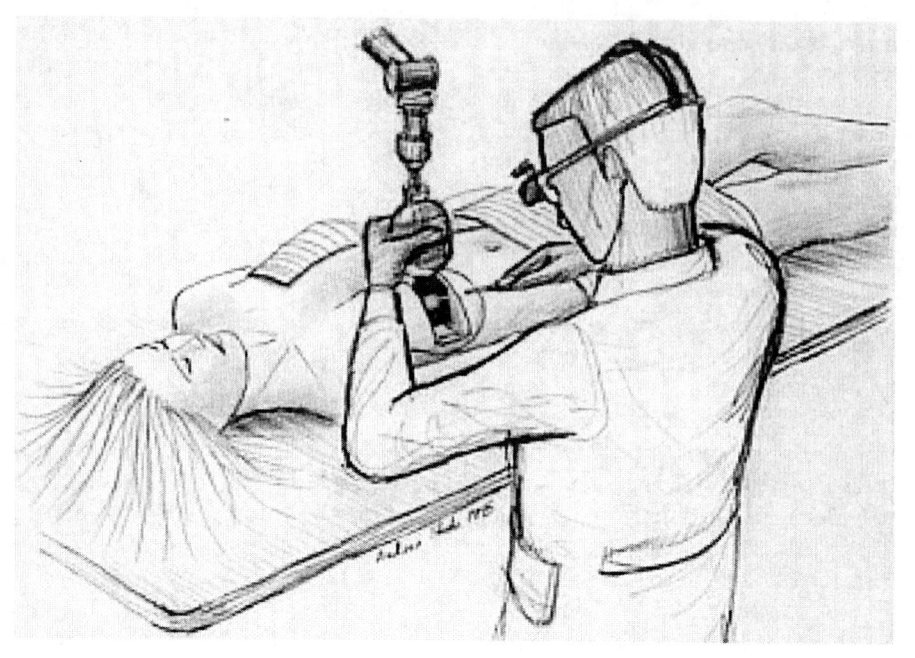

Figure 23.8. Shows the augmented reality setup being applied to improve performance and reduce errors during ultrasound-guided breast biopsy. This procedure requires guiding a biopsy needle to a breast lesion with the aid of ultrasound imagery, requires good 3-D visualization skills and hand-eye coordination, and is difficult to learn and perform.

3-D visualization skills and hand-eye coordination, and is difficult to learn and perform. Figure 23.9 shows an AR image for localizing a lesion during sinus surgery.

Action at a Distance: Extending the Human Mind and Distributing Expertise

Neuroergonomics applications can not only help to guide and focus an operator's mind but can also empower the operator to act at a distance. Preliminary research suggests the feasibility of telepresence operations, allowing a remotely stationed operator to perform invasive procedures on a patient located far beyond arm's length in a different room. Think in terms of an expert using a robotic arm manipulating control of the nuclear fuel rods in a nuclear power plant, the Canadian payload arm of the Space Shuttle, the arms of an unmanned submarine, or a digging tool of the Mars Rover. In the neuroergonomic health care application, the patient is located in a remote procedure room containing a stereoscopic camera for 3-D visualization of the operative field and a robot with tools to implement the operator's commands to perform medical procedures. Such a setup uses some of the same technology that allows the control of unmanned robotic vehicles used in surveillance and in undersea exploration and space exploration. It would extend the human mind, allowing action at a distance in remote locations around the globe for treatment of patients where there is insufficient local expertise, such as in rural areas, battlefields, ships, and perhaps even in extraterrestrial settings aboard space ships, space stations, and in space colonies.

Conclusion

Medical errors depend on multilevel factors from individual performance to systems engineering. Overwork, understaffing, and overly complex procedures may stress the mental resources of health care practitioners. Available information from existing sources (such as closed malpractice claims analyses of visible tip-of-the-iceberg safety errors)

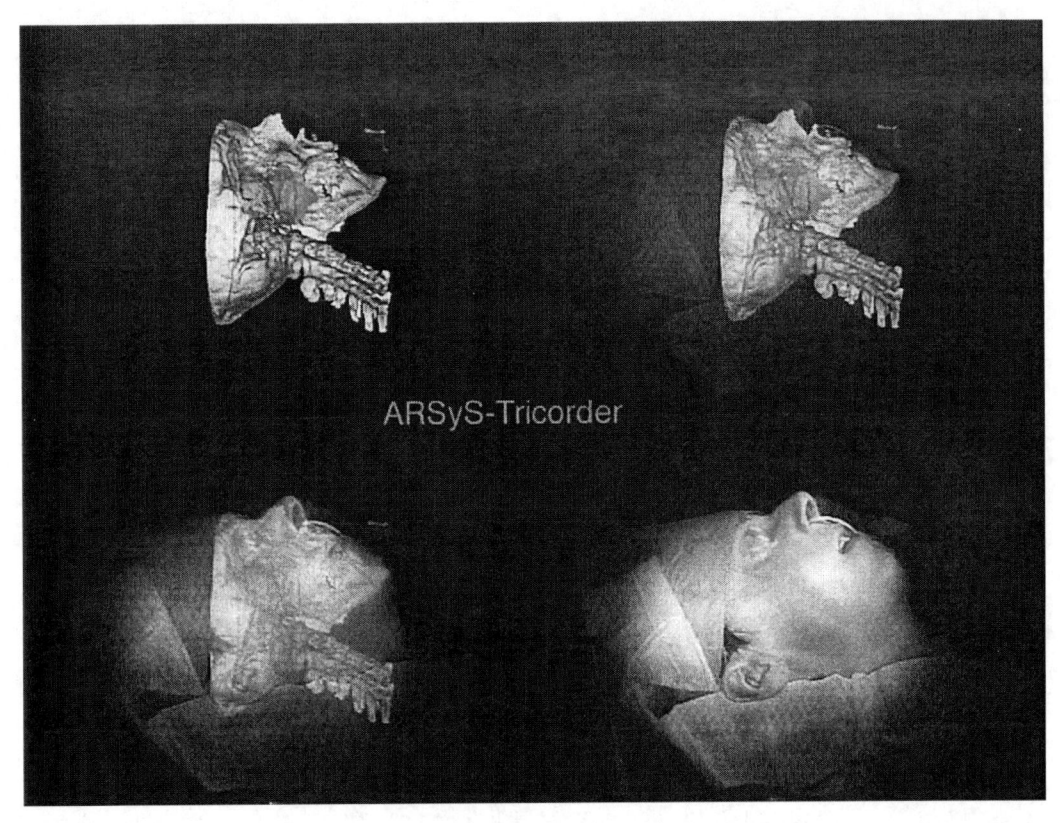

Figure 23.9. Shows an augmented reality image for a localized lesion during sinus surgery.

indicates that errors of diagnosis are an important type of medical error. Taxonomies of error provide a heuristic framework for understanding such errors but have not yet led to concrete improvements in safety. Feedback to the operator is essential to mitigation of errors in health care systems. In medical subspecialties, these feedback loops can be long and indirect. Errors of diagnosis can be reduced by redesigning procedures and systems, using techniques borrowed from other safety-critical industries.

New opportunities arise for reducing errors and improving patient safety following the passage from bill to law of the Patient Safety and Quality Improvement Act of 2005. This development allows for voluntary reporting systems with safe harbors and patient confidentiality protections. It permits the development of more comprehensive reporting systems amenable to cognitive work analysis for localizing and mitigating errors and improving medical care at all levels of the health care industry. These interventions can use modern tools and techniques including ecological interface design, information display technologies, virtual reality environments, and telepresence systems that extend the mind to distribute expertise and allow skilled medical action and procedures at a distance.

Main Points

1. Medical errors depend on multilevel factors from individual performance to systems engineering.
2. Overwork, understaffing, and complex procedures may stress the mental resources of health care personnel and increase the likelihood of error.
3. Errors can be reduced by redesigning procedures and systems, using techniques borrowed from other safety-critical industries such as air transportation and nuclear power.

4. Feedback to the operator is essential to reducing errors in health care systems. Feedback loops can be short and direct in surgical specialties, or long and indirect in medical specialties.
5. These interventions can use modern tools and techniques including ecological interface design, information display technologies, and virtual reality applications.

Key Readings

Akay, M., & Marsh, A. (Eds.). (2001). *Information technologies in medicine: Volume 1. Medical simulation and education.* New York: Wiley.

Bogner, M. S. (Ed.). (1994). *Human error in medicine.* Hillsdale, NJ: Erlbaum.

Institute of Medicine. (2000). *To err is human: Building a safer health system.* Washington, DC: National Academy Press.

Rasmussen, J., Pejtersen, A. M., & Goodstein, L. P. (1994). *Cognitive systems engineering.* New York: Wiley.

Vicente, K. J. (1999). *Cognitive work analysis: Toward safe, productive, and healthy computer-based work.* Mahwah, NJ: Erlbaum.

References

Ambady, N., LaPlante, D., Nguyen, T., Rosenthal, R., Chaumeton, N., & Levinson, W. (2002). Surgeons' tone of voice: A clue to malpractice history. *Surgery, 132*(1), 5–9.

Bar Code Label for Human Drug Products and Blood; Proposed Rule. (2003). Fed. Reg. Department of Health and Human Services, Food and Drug Administration. 21 C.F.R. pts. 201, 606, 610.

Bent, L. R., Inglis, J. T., & McFadyen, B. J. (2004). When is vestibular information important during walking? *Journal of Neurophysiology, 92,* 1269–1275.

Bogner, M. S. (Ed.). (1994). *Human error in medicine.* Hillsdale, NJ: Erlbaum.

Burns, C. M. (2000). Putting it all together: Improving display integration in ecological displays. *Human Factors, 42,* 226–241.

Burns, C. M., Bryant, D. J., & Chalmers, B. A. (2000). A work domain model to support shipboard command and control. In *Proceedings of the 2000 IEEE International Conference on Systems, Man, and Cybernetics* (pp. 2228–2233). Piscataway, NJ: IEEE.

Cao, C. G. L. (2003). *How do endoscopists maintain situation awareness in colonoscopy?* Paper presented at the International Ergonomics Association XVth Triennial Congress, Korea, August 25–29.

Dinadis, N., & Vicente, K. J. (1999). Designing functional visualizations for aircraft system status displays. *International Journal of Aviation Psychology, 9,* 241–269.

Drachman, D. A. (1990). Driving and Alzheimer's disease. *Annals of Neurology, 28,* 591–592.

Edwards, E. (1972). Man and machine: Systems for safety. In *Proceedings of the BALPA Technical Symposium* (pp. 21–36). London: British Airline Pilots Association.

Fitzpatrick, R. C., Wardman, D. L., & Taylor, J. L. (1999). Effects of galvanic vestibular stimulation during human walking. *Journal of Physiology, 517.3,* 931–939.

Gallagher, A. G., & Cates, C. U. (2004). Approval of virtual reality training for carotid stenting: What this means for procedural-based medicine. *Journal of the American Medical Association, 292,* 3024–3026.

Glick, T. (2001). Malpractice claims as outcome markers: Applying evidence to choices in neurologic education. *Neurology, 56,* 1099–1100.

Glick, T. H. (2005). The neurologist and patient safety. *Neurologist, 11,* 140.

Hajdukiewicz, J. R., Vicente, K. J., Doyle, D. J., Milgram, P., & Burns, C. M. (2001). Modeling a medical environment: An ontology for integrated medical informatics design. *International Journal of Medical Informatics, 62,* 79–99.

Hawkins, F. H. (1987). *Human factors in flight.* Aldershot, UK: Gower Technical Press.

Heinrich, H. W., Petersen, D., & Roos, N. (1980). *Industrial accident prevention.* New York: McGraw Hill.

Institute of Medicine. (2000). *To err is human: Building a safer health system.* Washington, DC: National Academy Press.

Itoh, J., Sakuma, A., & Monta, K. (1995). An ecological interface for supervisory control of BWR nuclear power plants. *Control Engineering Practice, 3,* 231–239.

Jagacinski, R., & Flach, J. (2002). *Control theory for humans: Quantitative approaches to modeling performance.* Mahwah, NJ: Erlbaum.

Jamieson, G. A., & Vicente, K. J. (2001). Ecological interface design for petrochemical applications: Supporting operator adaptation, continuous learning, and distributed, collaborative work. *Computers and Chemical Engineering, 25*(7–8), 1055.

Joint Commission on Accreditation of Healthcare Organizations. (2006). *National patient safety goals.* Retrieved from http://www.jcipatientsafety.org/show.asp?durki=10289.

Koppel, R., Metlay, J. P., Cohen, A., Abaluck, B., Localio, A. R., Kimmel, S. E., et al. (2005). Role of computerized physician order entry systems in facilitating medication errors. *Journal of the American Medical Association, 293*, 1261–1263.

Kuo, J., & Burns, C. M. (2000). Work domain analysis for virtual private networks. In *Proceedings of the 2000 IEEE International Conference on Systems, Man, and Cybernetics* (pp. 1972–1977). Piscataway, NJ: IEEE.

Lee, J. D., & Vicente, K. J. (1998). Safety concerns at Ontario Hydro: The need for safety management through incident analysis and safety assessment. In N. Leveson (Ed.). *Proceedings of the second workshop on human error, safety, and system development* (pp. 17–26). Seattle: University of Washington Press.

Leveson, N. G. (2000). Intent specifications: An approach to building human-centered specifications. *IEEE Transactions on Software Engineering, 26*, 15–35.

Maycock, G. (1997). Accident liability—The human perspective. In T. Rothengatter, & E. Vaya Carbonell (Eds.), *Traffic and transport psychology: Theory and application* (pp. 65–76). Amsterdam: Pergamon.

McGovern, K. T., & McGovern, L. T. (1994). The virtual clinic, a virtual reality surgical clinic. *Virtual Reality World*, (March–April), 41–44.

Molloy, G. J., & O'Boyle, C. A. (2005). The SHEL model: A useful tool for analyzing and teaching the contribution of human factors to medical error. *Academic Medicine, 80*(2), 152–155.

Nebeker, J. R., Hoffman, J. M., Weir, C. R., Bennett, C. L., & Hurdle, J. F. (2005). High rates of adverse drug events in a highly computerized hospital. *Archives of Internal Medicine, 165*, 1111–1116.

Norman, D. A. (1981). Categorization of action slips. *Psychological Review, 88*, 1–15.

Norman, D. A. (1988). *The psychology of everyday things*. New York: Harper and Row.

Noser, H., Stern, C., & Stucki, P. (2004). Automatic path searching for interactive navigation support within virtual medical 3-dimensional objects. *Academic Radiology, 11*, 919–930.

Patterson, E. S., Cook, R. I., & Render, M. L. (2002). Improving patient safety by identifying side effects from introducing bar coding in medication administration. *Journal of the American Medical Information Association, 9*, 540–553.

Peters, T. M. (2000). Image-guided surgery: From X-rays to virtual reality. *Computing Methods and Biomechanical and Biomedical Engineering, 4*(1), 27–57.

Rasmussen, J. (1987). The definition of human error and a taxonomy for technical system design. In J. Rasmussen, K. Duncan, & J. Leplat (Eds.), *New technology and human error* (pp. 23–30). Toronto, Canada: Wiley.

Rasmussen, J., Pejtersen, A. M., & Goodstein, L. P. (1994). *Cognitive systems engineering*. New York: Wiley.

Reason, J. (1990). *Human error*. New York: Cambridge University Press.

Reason, J. T. (1984). *Lapses of attention*. In R. Parasuraman & D. R. Davies (Eds.), *Varieties of attention* (pp. 515–549). San Diego: Academic Press.

Satava, R. M., & Fried, M. P. (2002). A methodology for objective assessment of errors: An example using an endoscopic sinus surgery simulator. *Otolaryngology Clinics of North America, 35*, 1289–1301.

Seymour, N. E., Gallagher, A. G., Roman, S. A., O'Brien, M. K., Bansal, V. K., Vipin, K., et al. (2002). Virtual reality training improves operating room performance: Results of a randomized, double-blinded study. *Annals of Surgery, 236*, 458–464.

Sharp, T. D., & Helmicki, A. J. (1998). The application of the ecological interface design approach to neonatal intensive care medicine. In *Proceedings of the Human Factors and Ergonomics Society 42nd Annual Meeting* (pp. 350–354). Santa Monica, CA: HFES.

Spicer, M. A., & Apuzzo, M. L. (2003). Virtual reality surgery: Neurosurgery and the contemporary landscape. *Neurosurgery, 53*, 1010–1011; author reply 1011–1012.

Tourville, J. (2003). Automation and error reduction: How technology is helping Children's Medical Center of Dallas reach zero-error tolerance. *U.S. Pharmacist, 28*, 80–86.

Vicente, K. J. (1999). *Cognitive work analysis: Toward safe, productive, and healthy computer-based work*. Mahwah, NJ: Erlbaum.

Vicente, K. J., & Rasmussen, J. (1992). Ecological interface design: Theoretical foundations. *IEEE Transactions on Systems, Man and Cybernetics, 22*, 589–606.

Wiener, E., Kanki, B., & Helmreich, R. (1995). *Crew resource management*. New York: Academic Press.

Woods, D. (2000, September 11). National Summit on Medical Errors and Patient Safety Research (Panel 2: Broad-based systems approaches) [Testimony]. Washington, DC. Retrieved January 22, 2000, from http://www.quic.gov/summit/wwoods.htm.

Woods, D. D., Johannesen, L. J., Cook, R. I., & Sarter, N. B. (1994). *Behind human error: Cognitive systems, computers, and hindsight*. Wright-Patterson AFB, OH: Crew Systems Ergonomics Information Analysis Center (SOAR/CERIAC).

Xu, W., Dainoff, M. J., & Mark, L. S. (1999). Facilitate complex search tasks in hypertext by externalizing functional properties of a work domain. *International Journal of Human-Computer Interaction, 11,* 201–229.

Yang, M., Brown, M. M., Trohimovich, B., Dana, M., & Kelly, J. (2002). The effect of bar-code enabled point-of-care technology on medication administration errors. In R. Lewis (Ed.), *The impact of information technology on patient safety* (pp. 37–56). Chicago: Healthcare Information and Management Systems Society.

VII

Conclusion

24

Matthew Rizzo and Raja Parasuraman

Future Prospects for Neuroergonomics

The preceding chapters present strong evidence for the growth and development of neuroergonomics since its inception a few years ago (Parasuraman, 2003). The ever-increasing understanding of the brain and behavior at work in the real world, the development of theoretical underpinnings, and the relentless spread of facilitative technology in the West and abroad are inexorably broadening the substrates for this interdisciplinary area of research and practice. Neuroergonomics blends neuroscience and ergonomics to the mutual benefit of both fields and extends the study of brain structure and function beyond the contrived laboratory settings often used in neuropsychological, psychophysical, cognitive science, and other neuroscience-related fields.

Neuroergonomics is providing rich observations of the brain and behavior at work, at home, in transportation, and in other everyday environments in human operators who see, hear, feel, attend, remember, decide, plan, act, move, or manipulate objects among other people and technology in diverse, real-world settings. The neuroergonomics approach is allowing researchers to ask different questions and develop new explanatory frameworks about humans at work in the real world and in relation to modern automated systems and machines, drawing from principles of neuropsychology, psychophysics, neurophysiology, and anatomy at neuronal and systems levels.

Better understanding of brain function, as outlined in the chapters on perception, cognition, and emotion, is leading to the development and refinement of theory in neuroergonomics, which in turn is promoting new insights, hypotheses, and research. For example, research on how the brain processes visual, auditory, and tactile information is providing important guidelines and constraints for theories of information presentation and task design, optimization of alerting and warning signals, development of neural prostheses, mitigation of errors by operators whose physiological profiles indicate poor functioning or fatigue, and development of robots that emulate or are part of human beings.

Specific challenges for this new field, both in the near term and beyond, are outlined in each part of this book (part II, Neuroergonomics Methods; part III, Perception, Cognition, and Emotion; part IV, Stress, Fatigue, and Physical Work; part V, Technology Applications; and part VI, Special Populations). In this closing chapter, we briefly discuss prospects for the future, both in the near term and the longer term. We also address some of the general challenges facing neuroergonomics research and practice. These include issues of privacy and ethics and the development of standards and guidelines to ensure the quality and safety of a host of new procedures and applications.

The Near Future

Simulation and Virtual Reality

An imminent challenge in neuroergonomics will be to disseminate and advance new methods for measuring human performance and physiology in natural and naturalistic settings. It is likely that functional magnetic resonance imaging (fMRI) methods will be further applied to study brain activity in tasks that simulate instrumental activities of daily living, within the constraints of the scanner environment. The ability to image brain activity during complex, dynamic behavior such as driving and navigation tasks will enhance our understanding of the neural correlates of complex behaviors and the performances of neurologically normal and impaired individuals in the real world. Further use of fMRI paradigms involving naturalistic behaviors will depend on improved software design and analytic approaches such as independent component analysis to decompose and understand the complex data sets collected from people engaged in these complex tasks. This exciting future also depends on advances in brain imaging hardware, fMRI experimental design and data analyses, ever-better virtual reality (VR) environments, and stronger evidence to determine the extent to which tasks in these environments are valid surrogates for tasks in the real world.

Multidisciplinary vision and collaboration have established VR as a powerful tool to advance neuroergonomics. VR applications—using systems that vary from surrealistic, PC-based gaming platforms to high-fidelity, motion-based simulation and fully immersive, head-mounted VR—are providing computer-based synthetic environments to train, treat, and augment human behavior and cognitive performance in renditions of the real world. An unanswered question regarding the fidelity of usable VR environments is, "How low can you go?" A range of VR systems will continue to be used to investigate the relationships between perception, action, and consciousness, including how we become immersed in an environment and engaged by a task, and why we believe we are where we think we are.

Augmented reality (AR) systems combine real and artificial stimuli, often by overlaying computer-generated graphics on a stream of video images, so that the virtual objects appear to be embedded in the real world. The augmentation can highlight important objects or regions and superimpose informative annotations to help operators accomplish difficult tasks. These systems may help aircraft pilots maintain situational awareness of weather, air traffic, aircraft state, and tactical operations by using a head-mounted display that displays such information and enhances occluded features (such as the horizon or runway markings); other systems may allow a surgeon to visualize internal organs through a patient's skin. Immediate challenges in creating effective AR systems include modeling the virtual objects to be embedded in the image, precisely registering the real and virtual coordinate systems, generating images quickly enough to avoid any disconcerting lag when there is relative movement, and building portable devices that do not encumber the wearer.

Physiological Monitoring

Advances in physiological measurements will permit additional observations of brain function in simulated and real-world tasks. Electroencephalogram (EEG) data should provide further data about changes in regional functional brain systems activated by ongoing task performance to complement the evidence from fMRI, positron emission tomography (PET), and other techniques. As discussed by Gevins and Smith (chapter 2), such studies are possible because EEG patterns change predictably with changes in task load, mental effort, arousal, and fatigue, and can be assessed using algorithms that combine parameters of the EEG power spectra into multivariate functions. Although these EEG data lack the 3-D spatial resolution of brain fMRI or PET, EEG is more readily applied in tasks that resemble those that an individual might encounter in a real-world environment.

Event-related potentials (ERPs) provide additional insights and applications for neuroergonomics and are computed by averaging EEG epochs time-locked to sensory, motor, and cognitive events. Although ERPs also have lower spatial resolution than fMRI, ERP resolution is improving because of new source localization techniques. Moreover, ERPs have better temporal resolution for evaluating neural activity than other neuroimaging techniques do. As Fu and Parasuraman (chapter 3) discuss, ERP components such as P300, N1, P1, ERN, and LRP will provide additional information relevant to neuroergonomic assessments of mental workload, attention resource allocation, dual-task performance, error detection and prediction, and motor control. Advances in device miniaturization and portability will continue to enhance the utility of EEG and ERP systems. Development of automated systems, in which human operators monitor control system functions, will provide further opportunities to use ERP-based neuroergonomic measures and theories relevant to brain function at work.

Noninvasive optical imaging tools, such as near-infrared spectroscopy (NIRS), assess neuronal activity that occurs directly after stimulus presentation or in preparation for responses and hemodynamic changes that occur a few seconds later, and these add to the neuroergonomics toolbox (which includes fMRI, PET, EEG, and ERP). As Gratton and Fabiani (chapter 5) discuss, optical imaging provides a good combination of spatial and temporal resolution of brain activity, can be deployed in a range of environments and experimental conditions, and costs little compared to magnetoencephalography, PET, and fMRI. Because optical imaging systems are relatively portable, they may be applied more commonly to map the time course of brain function and hemodynamic responses. Further advances will need to overcome the relatively low signal-to-noise ratio that affects the faster neuronal signal and the limited penetration by the technique to within a few centimeters beneath the scalp, which precludes measurements of activity in deeper brain structures (subcortical and brain stem) in adults.

Images of the brain at work will be complemented by transcranial Doppler sonography (TCD), which allows fast and mobile assessment of task-related brain activation and related effects of workload, vigilance, and sustained attention. Tripp and Warm (chapter 6) show that, like NIRS-based measurement of blood oxygenation, TCD offers a noninvasive and relatively inexpensive way to "monitor the monitor." TCD may prove especially useful to assess task difficulty and task engagement and to determine when human operators need rest and whether they may benefit from adaptive automation systems designed to flexibly allocate tasks between the operators and computer systems to mitigate operator workload and fatigue.

Eyelid closure measurements can be used to monitor fatigue and falling asleep on the job or at the wheel. McCarley and Kramer (chapter 7) show that eye movement assessments provide an additional window on perception, cognition, and how we search for meaningful information in displays and real-world scenes. Advances in gaze-contingent control procedures will enable neuroergonomics researchers to infer capability, strategies, and perhaps even the intent of operators who are inspecting the panorama in a variety of complex simulated and real-world environments. Real-time measures of where a user is gazing are already being used to develop attentive user interfaces, namely devices and computers that know where a person is looking and therefore do or do not interrupt the user accordingly (Vertegaal, 2002). Evidence from high-speed measurements of eye movements may also be combined with other measures (such as EEG, ERP optical imaging, heart rate, NIRS) to enhance the design of display devices aboard aircraft, automobiles, and industrial systems and improve real-time workload and performance assessment algorithms.

It is important to recognize what can and cannot be achieved specifically by physiological monitoring in extralaboratory environments. We must also be careful not to overstate what neuroergonomics can achieve given the noisiness of most real-world settings. Some recent programmatic efforts have set very ambitious goals for real-time assessment of operator cognitive state using physiological measures (St. John, Kobus, Morrison, & Schmorrow, 2004). Meeting these goals will require substantial effort focused on the elimination of potential artifacts that may mask the signal of interest, or, more seriously, lead to flawed assessments of operator state. Given the diversity and magnitude of artifacts that are likely to be encountered in natural settings, obtained artifact-free recordings, particularly in real time, will pose a considerable technical challenge. Nevertheless, there have been a number of promising developments in the design of miniaturized recording equipment that can withstand the rigors of operational environments. Automated artifact rejection procedures have also been developed, which, if proven robust, would help considerably in meeting the real-time monitoring challenge (see Gevins et al., 1995; see also chapter 2, this volume).

The Brain in the Wild

In addition to physiological monitoring, neuroergonomic evaluations can also involve assessment of other aspects of human behavior in natural environments, what Rizzo and colleagues call the brain in the wild (chapter 8). New technologies are allowing the development of various implementations of "people trackers" using combinations of accelerometers, global positioning systems, video, and other sensors (e.g., to measure gross and fine body movement patterns, skin temperature, eye movements, heart rate, etc.) to make naturalistic observations of human movement and behavior in the wild. As Rizzo and colleagues discuss, these trackers can advance the goal of examining performance, strategies, tactics, interactions, and errors in humans engaged in real-world tasks, drawing from established principles of ethology. Besides issues of device development and sensor choice and placement for various classes of devices (outside looking inside, inside looking inside, inside looking outside), taxonomies are needed for classifying likely behavior from sensor output. Different sensor array implementations will provide unique data on how the brain interacts with diverse environments and systems, at work and at play, and in health, disease, or fatigue states.

Such measurement techniques are likely to be important in the assessment of stress and fatigue at work. Fatigue on the job is common in our 24-hour-service society. Operational demands in round-the-clock industries such as health care and transportation and industries that require shift work inevitably cause fatigue from sleep loss and circadian displacement, which contributes to increased cognitive errors and risk of adverse events (such as medical errors and airplane crashes). Predicting and mitigating the risks posed by physiologically based variations in sleepiness and alertness at work are key functions of neuroergonomics, as discussed in chapter 14. Near-term advances will depend on more unobtrusive tools for detecting fatigue and a better understanding of the effects of sleep-wake cycles and circadian biology on human performance.

Cross-Fertilization of Fields

Evidence from many of the above-mentioned neuroergonomic areas will continue to converge with evidence from other established fields such as animal physiology and human neuropsychology. De Pontbriand (2005) also envisaged cross-fertilization between neuroergonomics and rapidly growing fields such as biotechnology. Neuroergonomics will also continue to provide a unique source of evidence in its own right. For example, Maguire (chapter 9) explains how neuroergonomics can lead to a better understanding of plasticity and dynamics within the brain's navigation systems; she foresees an increasingly fruitful exchange whereby technological and environmental improvements will be driven by an informed understanding of how the brain finds its way in the real world.

In this spirit of interdisciplinary cross-fertilization, Grafman (chapter 11) outlines a representational framework for understanding executive functions that underpin work-related performance in the real world. This performance depends on key structures in human prefrontal cortex that mediate decision making and implementation, planning and judgment, social cognition, tactics, strategy, foresight, goal achievement, and risk evaluation. Further, Bechara recognizes that real-world decisions are critically affected by emotions, in a process that reconciles cold cognition in the cortex with processes in the more primitive limbic brain. Emotion systems are a key factor in interaction between environmental conditions and human cognitive processes at work and elsewhere in the real world. These systems provide implicit or explicit signals that recruit cognitive processes that are adaptive and advantageous for survival. Understanding the neural underpinnings and regulation of emotions and feelings is crucial for many aspects of human behavior and their disorders, including performance at work, and can help provide a model for the design of new generations of computers that interact with humans, as in chapter 18.

The Longer-Term Future

Breazeal and Picard (chapter 18) explain how findings in neuroscience, cognitive science, and human behavior inspire the development of robots with social-emotional skills. Currently in an early stage of development, these relational robots hold promise in future applications in work productivity and education, where the human user may perform better in cooperation with the robot. For example, embedding affective technologies in learning interactions with automated systems (robotic learning companions) should reveal what affective states are most conducive to learning and how they vary with teaching styles, and this information will hone the robot's ability to learn from a person. Thus, humans will learn from machines and vice versa, parallel to the vision of Grafman (chapter 11) in which cognitive neuroscience applications will inform and improve training and evaluation methods currently employed by human factors experts.

Before they augmented our minds, machines amplified our muscles. Machines began relieving stress on human muscles as soon as humans discovered wedges, sledges, wheels, fulcrums, and pulleys. Hancock and Szalma (chapter 13) point out that machines will continue to replace human muscle power to minimize stress and fatigue as much as possible. In this futuristic vision of neuroergonomics, human intentions, indexed by interpretable brain states, will connect directly to these machines to bring intention to action. The effector might be a body part-sized electromechanical prosthesis such as an artificial limb or hand, or a substantial machine such as an exoskeleton robot, easily capable of bone-crushing labor. Additional neuroergonomic tools will monitor operators, identify signs of impending performance failure, and adjust system behavior and workload to mitigate stress.

Neuroergonomic countermeasures to stress, fatigue, and sleepiness can include systems for adaptive automation in which the user and the system can initiate changes in the level of automation, triggered in part by psychophysiological measures. Operators may come to think of these adaptive systems more as coworkers (rather than tools, machines, or computer programs) and even expect them to behave like humans. To design these adaptive systems, developers will need better information about human-machine task sharing, methods for communicating goals and intentions, and assessment of operator states of mind, including trust of and annoyance with robotic systems. These adaptive automation systems will be advantageously applied in settings where safety concerns surround the operator, system, and recipient of system services. Other potential applications might include a personal assistant, butler, secretary, or receptionist; an adaptive house; and systems aimed at training and skill development, rehabilitative therapy, and entertainment.

As discussed by Scerbo, adaptive automation systems are challenged when the wishes of two or more of the operators have conflicting goals. Such conflicts may arise in health care and can be examined at several levels using a cognitive work analysis framework in research aimed at improving medical quality and reducing errors. Addressing such complexities will depend on a better understanding of how multiple operators and their brains behave in social situations in differently organized work environments.

Neural engineering is a dimension of neuroergonomics related to the establishment of direct interactions between the nervous system and artificial devices. In this arena, technology is applied to improve biological function while lessons from biology inform technology, with cross-fertilization between molecular biology, electrophysiology, mathematics, signal processing, physics, and psychology. Brain-machine and brain-computer interfaces (BCIs) can interact with telecommunication technologies for controlling remote devices or for transmitting and receiving information across large distances, enabling operations at a distance.

As Mussa-Ivaldi and colleagues (chapter 19) discuss, advances in neuroengineering will depend on better knowledge and models of brain functions, at all levels ranging from synapses to systems, and on learning how to represent these models in terms of discrete or continuous variables. In this vein, Poggel and colleagues (chapter 21) address how the brain encodes visual information to learn how to "talk" to the brain and restore vision in patients with visual impairment due to retinal lesions. Meeting this goal will depend on a better understanding of plasticity, perception, low-level signal processing, and top-down influences such as attention.

Creating a retinal implant, a visual cortical implant, or any other neural implant is a challenge for cognitive scientists, surgeons, electrical engineers, material scientists, biochemists, cell biologists, and computer scientists. For decades to come, scientists will be busy developing algorithms to emulate neural functions and control neuroprostheses;

semiconductor chips to implement the algorithms; microelectrode arrays to match the organic cytoarchitectonic scaffold of the nerves, spinal cord, or brain; and procedures to tune a device to a given patient. Ideally, human neural signals could be read remotely via transduction of electromechanical signals without surgically invading the body. These human-machine interfaces could be used to control a variety of external devices to improve human function and mobility.

Riener (chapter 22) reviews neurorehabilitation robotics and neuroprosthetics and shows how robots can be applied to support improved recovery in patients recovering from upper motor neuron and lower motor neuron lesions due to stroke, trauma, and other causes. Robots will make motion therapy more efficient by increasing training duration and number of training sessions, while reducing the number of therapists required per patient. Patient-cooperative control strategies have the potential to further increase the efficiency of robot-aided motion therapy. Neuroprosthetic function will be improved by applying closed-loop control components, computational models, and better BCIs.

Better BCIs will allow a user to interact with the environment without muscular activity (such as hand, foot, or mouth movement) and will require specific mental activity and strategy to modify brain signals in a predictive way. BCIs can be useful in augmented cognition applications (as described above) and may also help patients with paralysis due to amyotrophic lateral sclerosis, spinal cord injury, or other conditions. Pfurtscheller and colleagues (chapter 20) review BCIs that use EEG signals. The challenges are to record, analyze, and classify brain signals and transform them in real time into a control signal at the output of the BCI. This process is highly subject specific and requires rigorous training or learning sessions.

Guidelines, Standards, and Policy

Rapid development of devices, techniques, and applications in the field of neuroergonomics (neuroengineering, neural prostheses, augmented cognition, relational robots, and so on) have outpaced the development of guidelines, which are recommendations of good practice that rely on their authors' credibility and reputation, and of standards, which are formal documents published by standards-making bodies that are developed through extensive consultation, consensus, and formal voting. Guidelines and standards help establish best practices and provide a common frame of reference across new device designs over time to foster communication between different user groups, without hindering innovation. They allow reviewers to assess research quality, compare results across studies for quality and consistency, and identify inconsistencies and the need for further guidelines and standards.

Making fair guidelines and standards involves science, logic, policy, politics, and diplomacy to overcome entrenched interests and opinions, and to avoid favoritism toward or undue influence by certain persons or groups. It may be easier to propose standards than to apply them in practice, yet it is important to develop standards proactively before they become externally imposed. Relevant to neuroergonomics, the Food and Drug Administration is charged with administering standards for implantable brain devices, such as deep brain stimulators to mitigate Parkinson's disease, spinal cord stimulators for pain relief, cochlear implants for hearing, cardiac pacemakers to treat heart rhythm abnormalities, and vagal nerve stimulators to treat epilepsy or, lately, depression.

The Centers for Medicare and Medicaid Services will weigh in on efficacy and standards when it is asked to reimburse providers for rendering services that may as yet lack sufficient evidence to support clinical use (such as VR for phobias). Professional groups such as the American Academy of Neurology (AAN), American Academy of Neurosurgery, American Psychological Association, the Special Interest Group on Computer-Human Interaction of the Association for Computing Machinery, and the Human Factors and Ergonomics Society may also intervene to establish evidence-based quality standards when a practice, device, or treatment becomes professionally relevant and widespread.

For example, the Quality Standard Subcommittee of the AAN and similar groups in other medical subspecialties write clinical practice guidelines to assist their members in clinical decision making related to the prevention, diagnosis, prognosis, and treatment of medical disorders, which may come to include neuroergonomic applications or devices (such as neural implants to treat blindness or paralysis). The AAN guidelines make specific practice recommendations based upon a rigorous review of all available scientific data. Key goals are to improve health outcomes, determine if practice follows current best evidence, identify research priorities based on gaps in the literature, promote efficient use of resources, and influence related public policy. Standards and guidelines in specific application areas discussed in this book, such as VR systems, have been published (Stanney, 2002). For a more comprehensive examination of standards across all areas of human factors and ergonomics, see Karwowski (2006).

Ethical Issues

Issues of privacy have been at the forefront since the advent of neuroergonomics (Hancock & Szalma, 2003) and are likely to continue to be so in the future too. Workers may be helped by automated systems that detect when fatigue, stress, or anxiety increase to levels that threaten performance and safety. But is there a dark side to such methods? For example, could those who seem especially stress or anxiety prone based on highly variable physiological measures and inaccurate predictive models be unfairly excluded from new opportunities or promotions? How will workers behave and what are their rights to privacy when it becomes possible to record seemingly everything they do, all the time, from multiple channels of data on brain and body states? The proliferation of embedded monitoring devices will expose events and behaviors that were once hidden behind cultural or institutional barriers. What could be done with these data beyond their intended purpose of enhancing safety, reducing stress, improving performance and health, and preventing injury?

In a similar vein, surveillance countermeasures to perceived terrorist threats include software intended to predict the intent of individuals bent on mayhem based on body movements, fidgeting, facial expression, eye glances, and other physiological indices. In chapter 13, Hancock and Szalma emphasize that we must be wary when private thoughts, opinions, and ideas—regardless of whether we like them or not—are unfairly threatened by other individuals, corporations, politicians, or arms of the state.

Remarkable ethical issues may unfold over the next century concerning the cooperative relationship between humans and machines at physical, mental, and social levels. How much should a person trust an automated adaptive assistant, avatar, virtual human, or affective computer that is in the loop with the human operator and potentially serving as a counselor, physician, friend, coworker, or supervisor? Who is the boss? How shall we mitigate concerns of control over human minds, bodies, and institutions by implants, robots, and software?

Modern discourse on somatic cell nuclear transfer (cloning) has sparked enormous hope and controversy at the borders between science, faith, policy, and politics. Emerging applications in neuroergonomics that are capable of reading human brain activity, predicting human failures, creating emotive robots, virtual human coworkers, companions, and bosses—and of hybridizing human and machine—may face similar trials (see also Clark, 2003). Nevertheless, we must face these challenges and not bury our heads in the sand like ostriches and hope they go away. Unless we (scientists and engineers) *ourselves* consider the ethical and privacy questions, others outside of science will decide the issues for us. The developments in neuroergonomics may lead to extraordinary opportunities for improved human-machine and human-human interaction. Others have noted that such developments may well represent the next major step in human evolution (Clark, 2003; Hancock & Szalma, 2003). Neuroergonomic technologies should be developed to serve humans and to help them engage in enjoyable and purposeful activity to ensure a well-evolved future.

Conclusion

As an interdisciplinary endeavor, neuroergonomics will continue to benefit from and grow alongside aggressive developments in neuroscience, ergonomics, psychology, engineering, and other fields. This ongoing synthesis will significantly advance our understanding of brain function underlying human performance of complex, real-world tasks. This knowledge can be put to use to design technologies for safer and more efficient operation in various work and home environments and in diverse populations of users. The basic enterprise of human factors and ergonomics—how humans design, interact with, and use technology—can be considerably enriched if we also consider the human brain that makes such activities possible. As the chapters in this volume show, there have already been considerable achievements in basic research and application in neuroergonomics. The future is likely to yield more such advances.

References

Clark, A. (2003). *Natural born cyborgs: Minds, technologies, and the future of human intelligence*. New York: Oxford University Press.

De Pontbriand, R. (2005). Neuro-ergonomics support from bio- and nano-technologies. In *Proceedings of the Human Computer Interaction International Conference*. Las Vegas, NV: HCI International.

Gevins, A., Leong, H., Du, R., Smith, M., Le, J., DuRousseau, D., et al. (1995). Towards measurement of brain function in operational environments. *Biological Psychology, 40,* 169–186.

Hancock, P. A., & Szalma, J. L. (2003). The future of neuroergonomics. *Theoretical Issues in Ergonomics Science, 4,* 238–249.

Karwowski, W. (2006). *Handbook of standards and guidelines in ergonomics and human factors.* Mahwah, NJ: Erlbaum.

Parasuraman, R. (2003). Neuroergonomics: Research and practice. *Theoretical Issues in Ergonomics Science, 1–2,* 5–20.

Stanney, K. (Ed.). (2002). *Handbook of virtual environments.* Mahwah, NJ: Erlbaum.

St. John, M., Kobus, D. A., Morrison, J. G., & Schmorrow, D. (2004). Overview of the DARPA augmented cognition technical integration experiment. *International Journal of Human-Computer Interaction, 17,* 131–149.

Vertegaal, R. (2002). Designing attentive user interfaces. In *Proceedings of the Symposium on Eye Tracking Research and Applications* (pp. 23–30). New Orleans, Louisiana: SIGCHI.

Glossary

adaptable automation Systems in which changes in the state of automation are initiated by the user.

adaptive automation Systems in which changes in the state of automation can be initiated by either the user or the system.

adverse event Any undesirable outcome in the course of medical care. An adverse event need not imply an error or poor care.

appraisal An assessment of internal and external events made by an individual, including attributions of causality, personal relevance, and potential for coping with the event.

arousal A hypothetical construct representing a nonspecific (general) indicator of the level of stimulation and activity within an organism.

attention The act of restricting mental activity to consideration of only a small subset of the stimuli in the environment or a limited range of potential mental contents.

attentional narrowing Increased selective attention to specific cues in the environment. It can take the form of focusing on specific objects or events or scanning of the environment such that a wide spectrum of events is attended but not processed deeply.

augmented cognition Systems that aim to enhance user performance and cognitive capabilities through adaptive assistance. These systems can employ multiple psychophysiological sensors and measures to monitor a user's performance and regulate the information presented to the user to minimize stress, fatigue, and information overload (i.e., perceptual, attentional, and working memory bottlenecks). See also **adaptive automation.**

augmented reality Setups that superimpose or otherwise combine real and artificial stimuli, generally with the aim of improving human performance and creativity.

automation A machine agent capable of carrying out functions normally performed by a human.

avatar Digital representation of real humans in virtual worlds.

basal ganglia Central brain structures that are associated with motor learning, motor procedures, and reward.

biomathematical model of fatigue Application of mathematics to the circadian and sleep homeostatic processes underlying waking alertness and cognitive performance.

brain-based adaptive automation Systems that follow the neuroergonomics approach and use psychophysiological indices to trigger changes in the automation.

cerebral laterality Differences in left and right hemisphere specialization for processing diverse forms of information.

cochlear implant Electronic device implanted in the primary auditory organ (the cochlea) that stimulates the auditory terminals in the inner ear so as to generate a sense of sound and partially restore hearing in people that are severely deaf. The electrical stimulus is arranged spatially so as to reproduce the natural distribution of frequency bands—or tonotopy—over the cochlear membrane.

cognitive work analysis (CWA) A multilevel systems analysis framework to classify and analyze errors, and identify several potential levels for interventions. CWA comprises five layers, each analyzing a different aspect of an application domain (work domain, control tasks, strategies, social-organizational structure, and worker competencies). CWA has been successfully applied to aviation safety and provides a framework for improving health care safety.

C1 The first event-related potential (ERP) component representing the initial cortical response to visual stimuli, with a peak latency of about 60–100 milliseconds after stimulus onset. Whether C1 is modulated by attention is a controversial topic in cognitive neuroscience studies of visual selective attention.

continuous wave instrument Instrument for optical imaging based on light sources that are constant in intensity rather than being modulated in intensity or pulsed.

cortical plasticity See **neuroplasticity**.

covert attention Attention that is shifted without a movement of the eyes, head, or body.

critical incidents Key events that have potentially harmful consequences. These can be either near misses or essential steps in the chain of events that lead to harm. They may provide clues to the root causes of repeated disasters.

cross-modal interaction Exchange of information between and mutual influence of sensory modalities, especially in association areas of the brain that respond to input from different modalities, such as the activation of visual areas in blind patients during tactile tasks like reading Braille.

Doppler effect Change in the frequency of a moving sound or light source relative to a stationary point.

dwell Also called *fixation* or *gaze*. A series of consecutive fixations (pauses) of the eyes between saccadic eye movements within a single area of interest. Visual information is extracted from the scene during fixations.

ecological interface design (EID) A systems design framework that complements the systems analysis framework of a cognitive work analysis (CWA). EID takes information requirements identified by CWA models and turns them into specific interface designs that make it easier for workers to understand what they need to get a job done. Providing feedback in the interface has the potential to minimize errors and facilitate error detection and, thus, error recovery.

EEG artifact A component of the recorded electroencephalogram (EEG) signal that arises from a source other than the electrical activity of the brain. Some EEG artifacts are of physiological origin, including electrical potentials generated by the heart, muscle tension, or movements of the eyes. Others are from instrumental sources such as ambient electrical noise from equipment or electrical potentials induced by movement of an electrode relative to the skin surface.

electroencephalogram, electroencephalography (EEG) A time series of measurements of electrical potentials associated with momentary changes in brain electrical activity in collections of neurons resulting from stimulation or specific tasks. EEGs are usually recorded as a difference in voltage between two electrodes placed on the scalp.

electromyogram (EMG) A record of electric currents associated with muscle contractions.

electrooculogram (EOG) Electrical potentials recorded by electrodes placed at the canthi (for monitoring horizontal eye movements) or at, above, or below the eyes (for monitoring vertical eye movements, such as blinks).

emotion A collection of changes in the body proper involving physiological modifications that range from changes that are hidden from an external observer (e.g., changes in heart rate) to changes that are perceptible to an external observer (e.g., facial expression).

episodic memory A type of memory store for holding memories that have a specific spatial and temporal context.

error An act of commission or omission (doing something wrong or failing to act) that increases the risk of harm. Errors can be subclassified (e.g., as *slip* or *mistake* or other) under different classification schemes (taxonomies) of error.

error-related negativity (ERN) An event-related potential (ERP) component that is observed at about 100–150 milliseconds after the onset of erroneous responses relative to correct responses. The ERN is not a stimulus-locked ERP but is a response-locked component. The amplitude of the ERN is related to perceived accuracy, or the extent to which participants realize their errors.

ethology The study of animal behavior in natural settings, involving direct observations of behavior, including descriptive and quantitative methods for coding and recording behavioral events.

event rate The rate of presentation of neutral nonsignal stimuli in a vigilance task in which critical targets for detection are embedded.

event-related optical signal (EROS) A transient and localized change in the optical properties of activated cortical tissue compared to baseline values. EROS is thought to depend on scattering phenomena that accompany neural activity. Activation corresponds to increases in the phase delay light parameter (i.e., photons' time of flight) compared to baseline.

event-related potential (ERP) The summated neural response to a stimulus, motor, or cognitive event as measured at the scalp by signal averaging the EEG over many such events. Consists of a series of components with different onset and peak latencies and scalp distribution.

executive functions A set of cognitive processes that help manage intentional human behavior including dividing and controlling attention, maintaining information in mind, developing and executing plans, social and moral judgment, and reasoning.

eye field The region of the visual field within which an eye movement is required to process two spatially separated stimuli. The stationary field is the region within which no movement is needed to process the two stimuli. The head field is the region within which a movement of the head or body is needed to process both items.

fast Fourier transform Consists of a decomposition of a complex waveform into its component elementary parts (e.g., sine waves). The fast Fourier transform algorithm reduces the number of computations needed for N points in a complex waveform from $2N^2$ to $2N \lg N$, where lg is the base-2 logarithm.

feeling Physiological modifications in the body proper during an emotion send signals toward the brain itself, which produce changes that are mostly perceptible to the individual in whom they were enacted, thus providing the essential ingredients for what is ultimately perceived as a feeling. Feelings are what the individual senses or subjectively experiences during an emotional reaction.

Fitts' law A model of human psychomotor behavior developed by Paul Fitts. The law defines an index of difficulty in reaching a target, which is related to the logarithm of the movement distance from starting point to center of target and width of that target.

fixation A dwell between saccadic eye movements. This is the time during which visual information is extracted from the scene.

functional electrical stimulation (FES) A technology based on the direct delivery of electrical stimuli to the muscles of a paralyzed patient, so as to restore the ability to generate and control movements. A major challenge in this technology is the reproduction of the muscle activation patterns that occur naturally during motor activities such as gait or manipulation.

functional field of view (FFOV) The region surrounding fixation from which information is gathered during the course of a single fixation. Sometimes referred to as the *useful field of view* (UFOV), *perceptual span*, *visual span*, or *visual lobe*.

functional magnetic resonance imaging (fMRI) A technique for collecting images of the brain based on blood oxygenation levels of brain tissue so that activation of brain regions in response to specific tasks can be mapped.

functional neuroimaging Brain imaging procedures that allow an investigator to visualize the human brain at work while it performs a task and includes such techniques as functional magnetic resonance imaging (fMRI) and positron emission tomography (PET).

general adaptation syndrome A set of physiological responses (e.g., heart rate, blood pressure, etc.) to noxious stimulation that serve as physiological defense against the adverse effects of such stimuli.

heads-up display (HUD) An information display located or projected on a transparent surface along a pilot or driver's line of sight to the external world.

hedonomics The branch of science that facilitates the pleasant or enjoyable aspects of human-technology interaction.

Heinrich's triangle Relationships between different performance factors and safety errors can be represented by an imaginary triangle. This simple model can be applied to errors that lead to injuries in diverse settings, including transportation, nuclear power, and health care. Visible safety errors (at the apex of the triangle, or "above the waterline") include errors resulting in fatality or serious injury. Submerged below the waterline (toward the base of the triangle) are events that are theoretically related to injury outcome and occur more frequently but do not lead to harm.

hemovelocity The speed at which blood flows through a blood vessel.

hippocampus A brain structure located in the medial temporal lobes of the brain, which is very important for the formation of memories of daily episodes. The name hippocampus means *seahorse* and refers to the shape of this brain structure.

homeostatic sleep drive A physiological process that ensures one obtains the amount of sleep needed to provide for a stable level of daytime alertness. It increases with wakefulness and is reduced with sleep.

human prefrontal cortex (HPFC) The region of the cerebral cortex that is anterior to the motor cortex and most evolved in humans. The prefrontal cortex is involved in many higher-level executive functions, including planning, decision making, and coordination of multiple task performance.

immersion The degree of a participant's engagement in a virtual reality (VR) experience or task.

independent component analysis (ICA) A data-driven analytical technique for decomposing a complex waveform or time series of data. The technique assumes the complex waveform to be a linear mixture of independent signals, which it sorts into maximally independent components.

jet lag Psychophysiological disturbance induced by a rapid shift across time zones resulting in a phase difference between the brain's circadian pacemaker (i.e., suprachiasmatic nucleus) and environmental time.

lateralized readiness potential (LRP) An event-related potential (ERP) that occurs several hundreds of milliseconds before a hand or other limb movement. The LRP is an average of two difference waves obtained by subtracting the readiness potential (RP) of the ipsilateral site from that of the contralateral site, for left- and right-hand responses, respectively. LRP provides an estimate of the covert motor preparation process, for example, whether and when a motor response is selected.

light absorption A type of light-matter interaction that results in the transfer of the light energy to the matter. It typically depends on the wavelength of the light and the type of substance. Substances in bodily tissue (such as water, melanin, and hemoglobin) can be distinguished because they absorb light of different wavelengths.

light scattering A type of light-matter interaction that results in the random deviation of the direction of motion of the light (photons) through matter. Within the near-infrared (NIR) wavelength range (600–1000 nm), the dominant form of interaction between light and tissue is scattering (diffusion) rather than absorption. Within this range, light can penetrate several centimeters into tissue, approximating a diffusion process.

magnetoencephalography (MEG) Technique for measuring magnetic signals produced by electrical activity in the brain in response to stimulation or specific tasks.

mental resources An economic or thermodynamic metaphor for the energetic and structural capacities required for perceptual and cognitive activity.

mental workload The demands that a task imposes on an individual's limited capacity to actively process information and make responses in a timely fashion. Optimal performance typically occurs in tasks that neither underload nor overload an individual's mental resources.

microsleep A period of sleep lasting a few seconds. Microsleeps become extremely dangerous when they occur during situations when continual alertness is demanded, such as driving a motor vehicle.

mismatch negativity (MMN) An event-related potential (ERP) that is a difference wave obtained by subtracting ERPs of standard stimuli (usually auditory) from those of stimuli that differ physically from the standard (e.g., in pitch or loudness). The MMN has a peak latency at about 150 milliseconds. The MMN is considered to be an index of automatic processing in the auditory modality.

motion correction Estimating and correcting for the effect of subject motion on an fMRI data set.

motor activity-related cortical potential (MRCP) The electroencephalogram-derived brain potential associated with voluntary movements.

motor cortex The cortical area of the human brain (Brodmann's area 4) that regulates motor movements.

naturalistic True to life, as in a real-life task performed in a real-world setting. The setting may be somewhat constrained by an experimenter completely unstructured with the observer hidden and the subject totally unaware of being observed (the most "natural" setting).

near-infrared spectroscopy (NIRS) Measurements of the concentration of specific substances in living tissue, capitalizing on the differential absorption spectra of different substances. For example, it is possible to estimate both absolute and relative (percentage change) concentrations of oxy- and deoxyhemoglobin in the tissue with this approach.

nerve growth factor (NGF) Complex molecule, with three polypeptide chains, which stimulates and guides the growth of nerve cells. NGF has been used to guide the growth of nerve cells inside glass microelectrodes, thus establishing a stable physical connection between brain tissue and electronic elements of brain-machine interfaces.

neuroergonomics The study of brain and behavior at work together in naturalistic settings. This interdisciplinary area of research and practice merges the disciplines of neuroscience and ergonomics (or human factors) in order to maximize the benefits of each.

neuroplasticity The process of reorganization of the cortex, which indicates the ability of the brain to learn, adapt to new experience, and recover from brain injury. Also refers to a set of phenomena that characterize the variability in neuronal connectivity and neuronal response properties as a consequence of previously experienced inputs and activities.

neurovascular coupling The relationship between neuronal activity and the related hemodynamic activity as revealed by neuroimaging (PET and fMRI) measures. Analyses of fMRI and O^{15} PET data assume this relationship to be linear.

N1 An event-related potential (ERP) negative component, whose peak latency, scalp distribution, and brain localization changes according to the location of recording site. N1 is sensitive to attentional modulation, with attended stimuli eliciting a larger N1 than unattended stimuli.

operant conditioning The modification of behavior resulting from the behavior's own consequences (e.g., positive reinforcement of a behavior generally will increase that behavior).

optical imaging methods A large class of imaging methods that exploit the properties of reflectance and diffusion of light through biological tissue such as that found in the brain.

overt attention Attention that is shifted via a movement of the eyes, head, or body.

people tracker A device using a combination of sensors (such as accelerometers, global positioning systems, videos, and others) for making synchronous observations of human movement, physiology, and behavior in real-world settings.

perclos Percent eye closure. An objective, real time, alertness monitoring metric based on percentage of time in which slow eyelid closures occur.

person-environment transactions Mutual interactions between individuals and the environment such that environmental events impact individuals via appraisal mechanisms and, based on these processes, individuals act on or modify the environment.

phase shift Occurs when the peak or trough of the circadian rhythm has been advanced or delayed in time. This may result in an individual experiencing wakefulness when he or she would normally be sleeping.

phosphene Circumscribed light perception induced, for example, by external stimulation of the retina or visual cortex.

photons' time of flight The time it takes for photons emitted into tissue by a source to reach a detector. This parameter can only be obtained with time-resolved instrumentation.

P1 The first positive event-related potential (ERP) component with a peak latency for visual stimuli of 70–140 milliseconds. P1 is sensitive to both voluntary and involuntary allocation of attention.

positron emission tomography (PET) A computerized radiographic technique used to examine the metabolic activity in various tissues (especially in the brain).

presence The degree to which a person feels a part of, or engaged in, a virtual reality (VR) environment.

primary inducers Environmental stimuli that evoke an innately driven or learned response, whether pleasant or aversive. Once they are present in the immediate environment, primary inducers automatically, quickly, and obligatorily elicit an emotional response.

primary motor cortex Region of the cerebral cortex immediately anterior to the central sulcus. Also known as Brodmann's area 4, or M1. It contains the largest projection from the brain to the motor neurons in the spinal cord via the pyramidal tract. Neurons in the primary motor cortex (and in other motor areas) associated with movements of the arm tend to be mostly active when the hand moves in a particular direction, a characteristic known as a *tuning property*.

principal component analysis (PCA) A data-driven analytic technique for decomposing a complex waveform or time series of data into independent, orthogonal, elementary components. See also **independent component analysis (ICA)**.

proprioception The sense of self in sensory motor behavior. Proprioception refers to the perception of one's position and configuration in space, as derived from a variety of (nonvisual) sensory sources, such as the muscle spindles that inform the nervous system about the state of length of the muscles, the Golgi tendon organs, and "copies" of the motor commands that drive the muscles. Proprioception is the biological basis for feedback control of movements.

psychomotor vigilance test (PVT) A test of behavioral alertness that measures reaction times in a high-signal-load sustained-attention task that is largely independent of cognitive ability or learning.

P300 A slow, positive brain potential with a peak latency of about 300–700 milliseconds. P300 amplitude is sensitive to the probability of a task-defined category and to the amount of attentional resources allocated to the task, and the latency of P300 reflects the time needed for perceptual processing and categorization, independent of response selection and execution.

receptive field Area of representation of external space by a neuron. The stimulation of that area activates the neuron representing this area, for example, a particular region within the visual field.

retinal prosthesis Device for electrical stimulation of cells in the retina to create visual perception, implanted either under the lesioned retina (subretinal approach) or attached to the retinal surface (epiretinal approach).

retinotopy Principle of organization of the visual system architecture according to the topography of stimulation and neural connections on the retina.

root cause analysis A process for identifying key factors underlying adverse events of critical incidents. Harmful events often involve a concatenation of factors (personnel, training, equipment, work load, procedures, communication) related to the system and individual.

saccade A rapid, ballistic eye movement that shifts the observer's gaze from one point of interest to another. Saccades may be reflexive, voluntary, or memory driven.

saccadic suppression A sharp decrease in visual sensitivity that occurs during a saccade.

search asymmetry More rapid detection of the presence of a distinguishing feature while searching in an array of stimuli than the absence.

secondary inducers Entities generated by the recall of a personal or hypothetical emotional event (i.e., thoughts and memories about the primary inducer), which, when they are brought to working memory, slowly and gradually begin to elicit an emotional response.

simulator adaptation syndrome Transient discomfort, often comprising visceral symptoms (such as nausea, sweating, sighing) triggered by exposure to a virtual reality (VR) environment. In extreme cases there can be "cybersickness" or simulator sickness, with vomiting.

simultaneous vigilance task A comparative judgment type of vigilance task in which all the information needed to detect critical signals is present in the stimuli themselves and there is no need to appeal to working memory in target detection.

sleep debt The cumulative effect on performance or physiological state of not getting enough sleep over a number of days. Unlike sleep debt, a sleep surplus cannot be accumulated.

sleepiness (somnolence, drowsiness) Difficulty in maintaining the wakeful state so that the individual falls asleep involuntarily if not actively kept alert.

social cognition That aspect of cognition involving social behavior and including attitudes, stereotypes, reflective social judgment, and social inference.

somatic state The collection of body-related responses that hallmark an emotion. From the Greek word *soma* meaning *body*.

source localization Identification of the neural sources of scalp-recorded EEG or ERP potentials. Based on the surface-recorded ERP data and certain parameters of the *volume conductor*—the head and intervening tissues between brain and scalp—the brain areas involved in specific cognitive processes can be obtained by using a dipole fit or other mathematical methods.

spatial normalization Estimating and applying a transformation that takes a brain image as collected and maps it into a standardized "atlas" space.

spatial resolution The resolution a particular brain imaging technique provides regarding the spatial localization of neural activity. Some imaging techniques such as fMRI have better spatial resolution (<1 cm) than others such as PET and ERPs.

stress A dynamic state within an individual arising from interaction between the person and the environment that is taxing to the individual and is appraised as a psychological or physical threat.

structured event complex Knowledge structures that contain two or more events and are extended in time from seconds to a lifetime and are the underlying representations for plans, procedures, scripts, stories, and similar stored knowledge.

successive vigilance task An absolute judgment type of vigilance task in which current stimuli must be compared against information stored in working memory in order to detect critical signals.

suprachiasmatic nucleus A discrete brain region lying within the hypothalamus and responsible for the generation of circadian rhythms in physiology and behavior.

temporal resolution The resolution a particular brain imaging technique provides regarding the temporal accuracy with which neural activity can be measured. Some techniques, such as ERPs, have better temporal resolution (<1 ms) than others, such as fMRI.

time-resolved (TR) instruments Methods for optical imaging based on light sources varying in intensity over time. They allow for the estimation of the photons' time of flight, which cannot be obtained with continuous wave (CW) instruments.

top-down regulation Influence of higher cognitive processes (e.g., attention) on early sensory processing (e.g., perception of light stimuli).

transcranial Doppler sonography (TCD) The use of ultrasound to provide continuous noninvasive measurement of blood flow velocities in the cerebral arteries under specified stimulus conditions.

transcranial magnetic stimulation (TMS) Technique for inducing electrical currents and thus neural activation in the brain by strong magnetic pulses delivered through the scalp and skull to the brain surface. Useful for temporarily inhibiting brain function in a circumscribed region.

ultrasound Sound with a frequency over 20000 Hz.

ultrasound window The temple area of the skull where ultrasound energy can easily penetrate.

vigilance Sustained attention, the ability to focus attention and detect critical signals over prolonged periods of time.

vigilance decrement Decline in signal detection over time during a vigilance task.

virtual environment An environment in which virtual reality (VR) scenarios, tasks, and experiments are implemented.

virtual reality (VR) The use of computer-generated stimuli and interactive devices to situate participants in simulated surroundings that resemble real or fantasy worlds.

working memory The mental capacity to hold and manipulate information for several seconds in a short-term memory area in the context of cognitive activity.

workload The level of perceptual, cognitive, or physical demand placed on an individual; the energetic and cognitive capacity consumed by a task.

Author Index

Aaslid, R., 82, 83, 85, 86, 88, 147, 148, 149
Abbas, J., 355
Abbas, J. J., 356
Abernethy, B., 95
Accreditation Council for Graduate Medical Education, 215
Achermann, P., 209, 216
Ackerman, D., 260
Adali, T., 52, 53, 55, 61
Adamczyk, M. M., 354, 355, 356
Adamson, L., 281
Adolphs, R., 182, 183
Agid, Y., 170, 171
Agran, J., 78
Aguirre, G. K., 131, 132
Ahlfors, S. P., 318
Aiken, L. H., 215
Aisen, M. L., 350
Aist, G., 279
Akerstedt, T., 214, 256
Akshoomoff, N. A., 61
Albright, T. D., 6
Alessi, S. M., 262
Alexander, G. E., 160
Alford, S., 306
Allen, G., 61
Allen, G. L., 135
Als, H., 281
Amassian, V. E., 337

Amedi, A., 330, 341
Amelang, M., 201
Amenedo, E., 45
Amess, P., 148
Amit, D. J., 305
Ammons, D., 254
Ancoli, S., 19, 20
Ancoli-Israel, S., 215
Andersen, G. J., 263
Andersen, R. A., 229, 301
Anderson, C. M., 116
Anderson, C. W., 300
Anderson, E., 98
Anderson, L., 266
Anderson, S., 263, 266
Anderson, S. W., 132, 162, 186
Andrews, B., 296
Anllo-Vento, L., 42
Annett, J., 150
Annunziato, M., 305
Anthony, K., 119
Apuzzo, M. L., 373
Arcos, I., 297
Arendt, J. T., 16
Arens, Y., 242
Argall, B. D., 341
Argyle, M., 278
Armington, J. C., 18
Armstrong, E., 160

Arnavaz, A., 149
Arnegard, R. J., 21, 90, 244
Arnett, P. A., 161
Arnolds, B. J., 87
Arroyo, S., 25
Arthur, E. J., 132
Ashe, J., 302
Asterita, M. F., 196
Astur, R. S., 137
Atchley, P., 101
Attwood, D. A., 150

Babikian, V. L., 82
Babiloni, C., 300
Babkoff, H., 209
Bacher, L. F., 116
Backer, M., 87, 89
Baddeley, A., 17, 161, 162
Badoni, D., 305
Bahri, T., 241
Bailey, N. R., 246, 247
Bajd, T., 353
Bak, M., 303
Bakay, R. A., 301
Bakeman, R., 117
Baker, S. N., 228
Balasubramanian, K. N., 125
Baldwin, C. L., 40
Balkany, T., 295
Ball, K., 99
Ballard, D. H., 52, 104
Banks, S., 213
Baranowski, T., 119
Barbas, H., 159
Barbeau, H., 349
Bar Code Label for Human Drug Products and Blood; Proposed Rule, 372
Bardeleben, A., 350
Bareket, T., 261
Barger, L. K., 215
Barlow, J. S., 19
Barnes, M., 241
Barr, F. M. D., 356
Barrash, J., 131, 132
Barrett, C., 255
Bartlett, F. C., 132
Bartolome, D., 244
Barton, J. J., 264, 335
Bartz, D., 265
Basheer, R., 208
Bashore, T. R., 39
Basso, G., 161, 168
Bateman, K., 261
Bateson, G., 224
Batista, A. P., 229

Baum, H. M., 214
Baumgartner, R. W., 87, 90
Bavelier, D., 270
Bay-Hansen, J., 87
Baynard, M., 208, 210, 211
Beatty, J., 39
Beauchamp, M. S., 341
Bechara, A., 162, 180, 182, 183, 184, 186, 187, 188, 189, 190, 255, 265, 268, 276
Becker, A. B., 147
Becker, T., 284
Beeman, M., 167
Behrens-Baumann, W., 338
Beilock, R., 214
Bekkering, H., 45
Belenky, G., 208, 211, 212, 213
Bell, A. J., 53, 55
Bellenkes, A. H., 97
Belopolsky, A., 33, 38
Benbadis, S. R., 119
Benko, H., 353
Bennett, C. L., 360
Bennett, K. B., 6
Benson, D. F., 59
Bent, L. R., 372
Benton, A. L., 114
Berbaum, K. S., 262
Berch, D. B., 152
Bereitschaftpotential, 44
Berg, P., 19
Berger, H., 18
Bergstrom, M., 16
Berman, K. F., 86
Berns, G. S., 189
Bernstein, P. S., 43
Bertini, M., 209
Bickmore, T., 279
Biering-Sorensen, F., 303
Billings, C. E., 241
Birbaumer, N., 5, 41, 300, 315, 316, 317, 321
Birch, G. E., 300
Bishop, C., 293
Biswal, B. B., 52
Blakemore, C., 97
Blankertz, B., 300, 301
Blaser, E., 98
Bleckley, M. K., 103
Bliss, T. V. P., 306
Blom, J. A., 19
Blumberg, B., 285, 288
Blumer, D., 59
Boas, D. A., 65, 68, 71
Bobrow, D. G., 146
Boden, C., 256

Bodner, M., 160
Bogart, E. H., 244
Bogner, M. S., 361
Bohbot, V. D., 136
Bohning, A., 88
Boland, M., 87, 89
Bolozky, S., 99
Bonnet, M. H., 215
Boom, H. B. K., 356
Boot, W. R., 95
Borbely, A. A., 209
Borisoff, J. F., 300
Botella-Arbona, C., 269
Botvinick, M., 164
Bower, J. M., 61
Boyle, L., 4
Boynton, G. M., 42, 61, 73
Bragden, H. R., 18
Brandt, S. A., 330, 338, 339
Brandt, T., 261
Brannan, J. R., 96
Braren, M., 39
Brashers-Krug, T., 231
Braun, C. H., 44
Braune, R., 39
Braver, E. R., 214
Braver, T. S., 17
Breazeal, C., 277, 279, 280, 281, 283, 285, 288
Brewster, R. M., 217
Brickner, M., 25
Brindley, G. S., 264, 330, 353, 354
Brochier, T., 228
Brody, C. D., 303
Brooks, B. M., 132
Brooks, F., 253, 254, 263, 264
Brouwer, W. H., 263
Brown, H., 18
Brown, J., 255
Brown, M. M., 372
Brown, V., 97, 99
Brownlow, S., 87
Brugge, J. F., 294
Brunia, C. H., 19
Brunner, C., 317
Bruno, J. P., 6
Bubb-Lewis, C., 248
Buchel, C., 61, 141, 340
Buchtel, H. A., 103
Buckner, R. L., 61, 168
Buettner, H. M., 293
Bulla-Hellwig, M., 86, 88, 89
Bunch, W. H., 353
Bunge, S. A., 18
Buonocore, M. H., 54
Burgar, C. G., 351

Burgess, N., 131, 132, 133, 134, 135, 139, 140, 141
Burgess, P. W., 162
Burgess, R. C., 318
Burgkart, R., 352
Burke, R., 285
Burns, C. M., 372
Burr, D. C., 97
Burton, A., 254
Burton, H., 340
Busch, C., 195
Buschbaum, D., 288
Buxton, R. B., 61
Byrne, E. A., 33, 244

Cabeza, R., 6, 34, 51, 52
Cacciopo, J. T., 5
Cadaveira, F., 45
Caggiano, D., 35, 39, 42, 146, 148, 201
Cajochen, C., 208
Calhoun, G., 316
Calhoun, V. D., 7, 25, 52, 53, 54, 55, 59, 61
Callaway, E., 19
Caltagirone, C., 88
Calvo, M., 200
Caminiti, R., 302
Campa, J. H., 353
Campbell, F. W., 96
Campos, J., 288
Cannarsa, C., 87
Cannon, W., 196
Cao, C. G. L., 373
Caplan, A. H., 301
Caplan, L. R., 150
Caramanos, Z., 18
Carbonell, J. R., 103
Carciofini, J., 255, 260
Carenfelt, C., 263
Carlin, C., 269
Carlin, D., 161
Carmena, J. M., 301, 302
Carmody, D. P., 95
Carney, P., 247
Carpenter, P. A., 17, 18, 97, 106
Carroll, R. J., 217
Carswell, C. M., 104
Castellucci, V., 306
Castet, E., 96
Catcheside, P., 213
Cates, C. U., 373
Chabris, C., 264
Chan, G. C., 356
Chan, H. S., 99
Chance, B., 68
Changeux, J. P., 163
Chapin, J. K., 301

Chapman, R. M., 18
Charnnarong, J., 350
Chatterjee, M., 296
Chelette, T. L., 87
Chen, L., 42, 45
Cheng, H., 338
Chesney, G. L., 39
Chiappalone, M., 306
Chiavaras, M. M., 160
Chignell, M. H., 6, 241, 242
Chin, D. N., 242
Chin, K., 196
Chino, Y. M., 338
Chisholm, C. D., 16
Chizeck, H. J., 354, 356
Cho, E., 65, 69
Cho, K., 139
Choi, J. H., 65, 71
Choi, Y. K., 139
Chou, T. C., 207, 208
Chow, A. Y., 332
Christal, R. E., 17
Christensen, L., 279
Chrysler, S. T., 132
Chun, M. M., 103
Churchland, P. S., 7, 293
Cincotti, F., 300
Cisek, P., 229
Clark, A., 6, 250, 294, 387
Clark, K., 161, 165
Clark, V. P., 36, 42
Clayton, N. S., 137
Clemence, M., 148
Cloweer, D. M., 229
Clynes, M. E., 294
Coda, B., 269
Cohen, J. D., 17, 162
Cohen, L. B., 68
Cohen, L. G., 340
Cohen, M. J., 300
Cohen, M. S., 18
Cohen, R., 135
Cohon, D. A. D., 302
Colcombe, A. M., 102
Coles, M. G., 34, 39, 40, 43, 44, 45
Coles, M. G. H., 43, 200
Colletti, L. M., 217
Collingridge, G. L., 306
Collison, E. K., 16
Colman, D. R., 222
Colombo, G., 349, 351, 352
Committeri, G., 141
Comstock, J. R., 21, 90, 244
Conrad, B., 88
Conway, A. R. A., 103

Conway, T. L., 119
Cook, R. I., 361, 372
Cooper, C. E., 148
Cooper, E. B., 353
Cooper, R., 18, 163
Copeland, B. J., 329
Coppola, R., 17
Corballis, P. M., 65, 66, 69, 77
Corbetta, M., 60, 105
Cordell, W. H., 16
Corneil, B. D., 301
Corrado, G., 161
Coull, J. T., 60, 147
Courchesne, E., 61
Courtney, A. J., 99
Courtney, S. M., 162
Cowell, L. L., 119
Cowey, A., 338
Coyne, J. T., 40
Cozens, J. A., 350
Crabbe, J., 9, 201
Crago, P. E., 353, 354, 355, 356
Craig, A. D., 179
Craik, F. I., 149
Crane, D., 280
Creaser, J., 255, 260
Cremer, J., 254, 265, 266
Crozier, S., 168
Crutcher, M. D., 160
Cruz-Neira, C., 137
Csibra, G., 45
Csikszentmihalyi, M., 201
Cummings, J. L., 160
Cupini, L. M., 88
Curran, E. A., 325
Curry, R. E., 242
Cutillo, B., 17
Cutting, J. E., 265
Czeisler, C. A., 207, 208
Czigler, I., 45

Daas, A., 338
Dagher, A., 136
Dainoff, M. J., 372
Dale, A. M., 42, 334
Dal Forno, G., 333
Damasio, A., 180, 286
Damasio, A. R., 114, 162, 163, 179, 182, 183, 184,
    185, 186, 187, 188, 189, 190, 255, 265, 276
Damasio, H., 114, 162, 180, 182, 183, 185, 186, 187,
    188, 265, 276
Damon, W., 279
Dana, M., 372
Dark, V. J., 72
Daroff, R., 261

Darrell, T., 283
Dascola, I., 98
Dautenhahn, K., 277
Davies, D. R., 148, 149
Davis, B., 280
Davis, H., 20
Davis, P. A., 20
Dawson, J., 261, 262, 263
Dawson, J. D., 132
Deary, I. J., 201
Deaton, J. E., 241
de Charms, R. C., 316
Dechent, P., 316
Dedon, M., 19
Deecke, L., 44, 225
Degani, A., 239, 240
Degenetais, E., 170
De Gennaro, L., 209
deGroen, P. C., 119
Dehaene, S., 8, 44, 163
Deibert, E., 341
de Lacy Costello, A., 162
Deliagina, T. G., 306
DeLong, M. R., 160
del Zoppo, G. J., 84
Dember, W. N., 146, 147, 148, 152
Dement, W. C., 213, 215
DeMichele, G., 297
Dence, C., 82
Dennerlein, J. T., 284
Dennett, D., 5
de Oliveira Souza, R., 8
De Pontbriand, R., 9, 384
Deppe, M., 82, 83, 84, 87, 89
Desmond, P. A., 195
DeSoto, M. C., 70, 72, 73
D'Esposito, M., 131, 132
Detast, O., 285
Deubel, H., 96, 98
Devalois, R., 96
Devinsky, O., 60
DeVoogd, T. J., 137
de Waard, D., 263
Diager, S. P., 332
Diamond, A., 160
Diderichsen, F., 263
Dien, J., 43
Dietz, V., 349
DiGirolamo, G. J., 60
Dijk, D. J., 207, 209
Dimitrov, M., 161, 162, 168, 170
Dinadis, N., 372
Dinges, D., 256
Dinges, D. F., 27, 208, 209, 210, 211, 213, 215, 216, 217

Dingus, T. A., 120, 121, 125, 263, 267
Dingwell, J. B., 303
Dirkin, G. R., 196
Ditton, T. B., 280
Dobelle, W. H., 330
Dobkins, K. R., 340
Dobmeyer, S., 60
Doerfling, P., 42
Dolan, R. J., 183
Donaldson, D. I., 61
Donaldson, N. de N., 356
Donchin, E., 4, 5, 32, 34, 39, 40, 41, 43, 44, 300, 316
Donnett, J. G., 133, 135, 139, 140
Donoghue, J. P., 301, 325
Doran, S. M., 209, 210
Dorneich, M., 255, 259, 270
Dorrian, J., 209, 210, 211, 212, 217
Dorris, M. C., 106
Dosher, B., 98
Dostrovsky, J., 132, 134
Dotson, B., 254
Douglas, R. M., 103
Dow, M., 353
Downie, M., 285
Downs, R. M., 132
Doyle, D. J., 372
Doyle, J. C., 19
Drachman, D. A., 364
Dreher, J. C., 170
Drevets, W. C., 318
Drews, F. A., 99
Driver, J., 98
Droste, D. W., 87, 89
Du, R., 18
Du, W., 19
Dubois, B., 170
Dumas, J. D., 261
Dunbar, F., 196
Dunlap, W. P., 96
Dureman, E., 256
Durfee, W. K., 355
Durmer, J. S., 208, 210, 211
Duschek, S., 82, 83, 84, 87, 88, 148
d'Ydewalle, G., 97

Easterbrook, J. A., 196
Ecuyer-Dab, I., 137
Edell, D. J., 296
Edwards, E., 365
Edwards, J., 99
Egelund, N., 256
Eggemeier, F. T., 200
Ehrosson, H. H., 223
Ehrsson, H. H., 228, 232
Eichenbaum, H., 131

Author Index

Eisdorfer, C., 195
Ekeberg, O., 228
Ekman, P., 284
Ekstrom, A., 134
El-Bialy, A., 354
Electronic Arts, 54
Elias, B., 115
Eliez, S., 61
Elliot, G. R., 195
Ellis, S. R., 253, 265
Ellsworth, L. A., 217
Elston, G. N., 160
Emde, R., 288
Endsley, M. R., 246, 247
Engel, A. K., 169
Engel, G. R., 99
Engel, J. J., 18
Engle, R. W., 17, 103
Epstein, R., 132, 139
Eriksen, C. W., 44
Erkelens, C. J., 97
Eskola, H., 334
Eslinger, P. J., 8, 162
Essa, I., 284
Estepp, J., 90
Eubank, S., 255
Evans, A., 160
Evans, A. C., 89
Evans, G. W., 132, 139
Evarts, E. V., 301, 302
Everling, S., 103
Eysel, U. T., 338
Eysenck, H. J., 201
Eysenck, M. W., 200, 201

Fabiani, M., 34, 65, 66, 69, 70, 71, 72, 77, 78
Fagergren, A., 228
Fagergren, E., 228
Fairbanks, R. J., 217
Fan, J., 9, 45
Fan, S., 36, 42
Fancher, P., 120
Fang, Y., 225, 232
Farah, M. J., 163
Feinman, S., 288
Feldman, E., 82
Fellows, M. R., 301, 325
Fernandez-Duque, D., 60
Ferrara, M., 209
Ferrarin, M., 356
Ferraro, M., 351
Ferwerda, J., 265
Fetz, E. E., 301, 306
Feyer, A. M., 16

Figueroa, J., 115
Findlay, J. M., 96, 97, 98, 99, 103, 107
Fine, I., 340
Finger, S., 335
Finke, K., 88
Finkelstein, A., 265
Finney, E. M., 340
Finomore, V., Jr., 90
Fischer, B., 103
Fischer, J., 265
Fisher, D. L., 101
Fisher, F., 19
Fitts, P. M., 4, 95, 96, 97, 230
Fitzpatrick, R. C., 372
Flach, J., 370
Fleming, K. M., 306
Fletcher, E., 37
Flickner, M., 284
Flitman, S., 167
Flotzinger, D., 321
Fogassi, L., 325
Folk, C. L., 101
Folkman, S., 195, 197
Fong, T., 277
Forde, E. M. E., 163
Foreman, N., 132
Forneris, C. A., 320
Forssberg, H., 223, 228
Forsythe, C., 246
Fossella, J., 6, 201
Fossella, J. A., 9
Foster, P. J., 254
Fournier, L. R., 21
Fowler, B., 42
Fox, P. T., 32, 82, 89
Foxe, J. J., 318
Frackowiack, R. J., 147
Frackowiack, R. S., 132, 133, 135, 140, 340
Frackowiack, R. S. J., 52
Frahm, J., 316
Franceschini, M. A., 65, 68, 71
Frank, J. F., 263
Frank, R., 114, 185
Frankenhaeuser, M., 196
Franklin, B., 353
Frauenfelder, B. A., 90
Frederiksen, E., 350
Freehafer, A. A., 355
Freeman, F. G., 40, 242, 244, 246, 247
Freeman, W. T., 282
Freund, H. J., 338
Frey, M., 352
Fried, M. P., 373
Friedman, A., 97, 106
Frigo, C., 356

Friston, K., 340
Friston, K. J., 52, 55, 56, 61, 77
Frith, C., 133, 139
Frith, C. D., 60, 132, 147
Fritsch, C., 232
Friz, J. L., 9
Fromherz, P., 305
Frost, S. B., 305
Frostig, R. D., 68
Fu, S., 35, 42, 45
Fu, W., 104
Fuhr, T., 352, 353, 356
Fulton, J. F., 88
Furukawa, K., 231
Fusi, S., 305
Fuster, J. M., 160, 162, 306

Gabrieli, J. D., 18
Gabrieli, J. D. E., 141
Gaddie, P., 233
Gaffan, D., 182
Gaillard, A. W. K., 200
Galaburda, A. M., 114, 185
Galinsky, T. L., 147
Gallagher, A. G., 373
Gallese, V., 325
Gander, P. H., 27
Gandhi, S. P., 42, 61, 73
Ganey, H. C. N., 196
Garau, M., 261
Garcia-Monco, J. C., 163
Garcia-Palacios, A., 269
Gardner, A. W., 119
Garness, S. A., 120
Garrett, D., 300
Garrett, E. S., 54
Gatenby, D., 288
Gatenby, J. C., 182
Gaulin, S. J., 137
Gaymard, B., 105
Gazzaniga, M. S., 6, 32, 87, 147
Geary, D. L., 70
Geddes, N. D., 242
Gehring, W. J., 43
Gerner, H. J., 315
Gevins, A., 16, 18, 19, 21, 25, 26, 225, 384
Gevins, A. S., 17, 18, 19, 20, 23
Ghaem, O., 132
Gibson, J. J., 221, 224, 263
Gielo-Perczak, K., 221, 224, 233
Gilbert, C. D., 68, 338
Gilchrist, I. D., 96, 97, 99
Gil-Egul, G., 280
Gioanni, Y., 170

Girelli, M., 33
Gitelman, D. R., 60
Givens, B., 6
Glick, T., 363, 364
Glover, G. H., 52, 61
Glowinski, J., 170
Gluckman, J. P., 242
Godjin, R., 106
Goel, V., 161
Gold, P. E., 151
Goldberg, M. E., 105
Goldman, R. I., 18
Goldman-Rakic, P., 17
Goldman-Rakic, P. S., 160, 162
Gomer, F., 4, 41
Gomez, C. R., 88
Gomez, S. M., 88
Gomez-Beldarrain, M., 163
Gooch, A., 265, 266
Good, C. D., 137
Goodkin, H. P., 61
Goodman, K. W., 295
Goodman, M. J., 121
Goodman-Wood, M. R., 70
Goodstein, L. P., 361
Gopher, D., 25, 40, 146, 200, 261
Gordon, G., 283
Gore, J. C., 182
Gorman, P. H., 354
Gormican, S., 151, 152
Goss, B., 43
Gothe, J., 330
Gottlieb, J. P., 105
Gottman, J. M., 117
Gotzen, A., 86
Gould, E., 137
Grabowski, T., 114, 185
Grace, R., 217
Grady, C. L., 333
Grafman, J., 161, 162, 163, 165, 167, 168, 169, 170, 171, 172
Grafton, S. T., 230
Graimann, B., 300, 323, 325
Granetz, J., 161
Grattarola, M., 305
Gratton, E., 65, 66, 70
Gratton, G., 34, 44, 65, 66, 69, 70, 71, 72, 73, 74, 77, 78
Gray, J., 288
Gray, W. D., 104
Graybiel, A., 262
Graziano, M. S., 137
Greaves, K., 119
Green, C., 270
Greenberg, D., 265

Greenberg, J., 263
Greenberg, L., 280
Greenwood, P. M., 6, 9, 35, 42, 153, 201
Greger, B., 301
Griffin, R., 75
Grillner, S., 306
Grill-Spector, K., 334
Grinvald, A., 68
Grishin, A., 230
Groeger, J., 52
Groenewegen, H. J., 160
Grön, G., 137
Gross, C. G., 137
Grossman, E., 229
Grubb, P. L., 152
Grunstein, R. R., 213
Grüsser, O. J., 337
Guazzelli, M., 168
Guerrier, J. H., 263
Guger, C., 300, 321
Guitton, D., 103, 105
Gunnar, M., 288
Gur, R. C., 87
Gur, R. E., 87
Gutbrod, K., 88
Guzman, A., 217

Haase, J., 303
Hackley, S. A., 45, 72
Hackos, J., 268
Hager, L. D., 103
Hahn, S., 101, 102, 263
Hailman, J. P., 117
Hajdukiewicz, J. R., 372
Hakkanen, H., 256
Hakkinen, V., 20
Halgren, E., 301, 334
Hall, I. S., 88
Hallet, P. E., 102, 103
Hallett, M., 161, 168, 169, 171
Hallt, M., 229
Halpern, A. R., 89
Hamalainen, M., 334
Hamann, G. F., 84, 87
Hambrecht, F. T., 354
Hamilton, P., 195, 196, 199
Hamilton, R., 341
Hammer, J. M., 242, 249
Hampson, E., 137
Hancock, P., 90
Hancock, P. A., 6, 132, 147, 152, 195, 196, 197, 198, 199, 200, 201, 202, 203, 241, 242, 387
Handy, T., 38
Hanes, D. P., 106
Hannen, M. D., 242, 243, 247, 249

Hanowski, R. J., 120
Hansen, J. C., 53
Hansen, L. K., 53
Hanson, C., 167
Hanson, S. E., 167
Haraldsson, P.-O., 263
Hardee, H. L., 263
Harders, A., 82
Harders, A. G., 87, 89
Harer, C., 90
Hari, R., 228, 334, 340
Harkema, S. J., 349
Haro, A., 284
Harper, R. M., 68
Harrison, Y., 16, 210, 256
Hart, S. G., 147, 152
Hartje, W., 86, 88
Hartley, T., 131, 134, 135, 136, 137, 141
Hartt, J., 341
Hartup, W., 279
Harvey, E. N., 20
Harwin, W., 350
Hasan, J., 20
Hashikawa, K., 228
HASTE (Human-Machine Interface and the Safety of Traffic in Europe), 267
Hatsopoulos, N. G., 301, 325
Hatwell, M. S., 356
Hatziparitelis, M., 137
Haugland, M., 303
Hausdorff, J. M., 355
Haviland-Jones, J., 284
Hawkins, F. H., 365
Haxby, J. V., 162, 333, 335
Hayhoe, M., 103
Hayhoe, M. M., 52, 104
Heasman, J. M., 324
Hebb, D. O., 8, 196
Heeger, D. J., 42, 61, 73
Heim, M., 253
Heinrich, H. W., 362
Heinze, H. J., 37
Hell, D., 90
Hellige, J. B., 88
Helmicki, A. J., 372
Helmreich, R., 373
Helton, W. S., 154, 155
Henderson, J. M., 98
Hendler, T., 341
Henik, A., 105
Henningsen, H., 89
Hentz, V. R., 296
Heriaud, L., 119
Hermer, L., 132
Hernandez, A., 303

Heron, C., 103
Hesse, S., 349, 350
Hicks, R. E., 106
Hilburn, B., 90, 240
Hill, D. K., 66
Hille, B., 68
Hillyard, S. A., 32, 33, 36, 41, 42, 45
Hims, M. M., 332
Hink, R. F., 32, 41
Hirkoven, K., 20
Hitchcock, E. M., 8, 86, 88, 89, 150, 151, 154
Ho, J., 70
Hobart, G., 20
Hochman, D., 68
Hockey, G. R. J., 195, 196, 197, 200
Hockey, R., 195, 196, 199
Hockey, R. J., 98
Hodges, A. V., 295
Hodges, L., 269
Hoffer, J. A., 298, 354
Hoffman, D. S., 302
Hoffman, H., 269, 270
Hoffman, J. E., 98
Hoffman, J. M., 360
Hogan, N., 350, 351, 352
Holcomb, H. H., 231
Hole, G., 99
Holland, F. G., 196
Holland, J. H., 324
Hollander, T. D., 152, 153, 154
Hollands, J. G., 6, 90, 96, 104, 146, 152, 195
Holley, D. C., 16
Hollnagel, C., 162, 165
Holmes, A. P., 56
Holmquest, M. E., 353
Holroyd, C. B., 43
Hono, T., 18
Hood, D., 65
Hood, D. C., 69
Hooge, I. T. C., 97
Hopfinger, J. B., 37, 54
Horch, K., 296, 354, 356
Horenstein, S., 88
Horne, J. A., 16, 210, 256
Hornik, R., 288
Horton, J. C., 339
Horvath, A., 280
Horvath, K., 280
Horwitz, B., 333
Hoshi, Y., 68
Houle, S., 149
Hovanitz, C. A., 196
Hoyt, M., 119
Hsiao, H., 254

Hsieh, K.-F., 288
Huang, Y., 42, 139
Hubel, D. H., 338
Huettel, S. A., 137
Huey, D., 99
Huf, O., 87
Huggins, J. E., 300, 325
Humayan, M., 337
Humayan, M. S., 332
Humphrey, D., 41
Humphrey, D. R., 301
Humphreys, G. W., 163
Hunt, K. J., 356
Hurdle, J. F., 360
Hutchins, E., 6
Hwang, W. T., 215
Hyde, M. L., 302
Hyman, B. T., 182
Hyönä, J., 96

Iaria, G., 136
Ifung, L., 282
Ikeda, A., 318, 319
Illi, M., 96
Inagaki, T., 241
Inanaga, K., 18
Inglehearn, C. F., 332
Inglis, J. T., 372
Inhoff, A. W., 97
Inmann, A., 303
Inouye, T., 18
Insel, T. R., 160
Insko, B., 254, 264
Institute of Medicine, 360, 362
International Ergonomics Association, 221
Interrante, V., 266
Ioffe, M., 230
Iogaki, H., 18
Irlbacher, K., 330
Irwin, D. E., 98, 100, 101, 102
Ishii, R., 18
Isla, D., 285
Isokoski, P., 96
Isreal, J. B., 39
Ito, M., 306
Itoh, J., 372
Itti, L., 103, 106, 107
Ivanov, Y., 285
Ivry, R., 147
Ivry, R. B., 87
Izard, C., 284

Jackson, C., 78
Jacob, J. K., 95, 97
Jacobs, A. M., 99, 100

Jacobs, L. F., 137
Jacobsen, R. B., 18
Jaeger, R. J., 356
Jaffe, K., 168
Jagacinski, R., 370
Jamaldin, B., 233
James, B., 296
James, T. W., 341
James, W., 88, 179, 256
Jameson, A., 40
Jamieson, G. A., 372
Janca, A., 54
Jancke, L., 139
Jang, R., 233
Jansen, A., 87
Jansen, C., 245
Jansma, J. M., 17
Janz, K., 114
Janzen, G., 139, 140
Jarvis, R., 114
Jasper, H. H., 25
Jenmalm, P., 223
Jensen, L., 349
Jensen, S. C., 332
Jerison, H. J., 149, 152
Jermeland, J., 120
Jessell, T. M., 6, 82
Jezernik, S., 349, 352
Jiang, Q., 154
Jiang, Y., 103
Joffe, K., 103
Johannes, S., 88, 149
Johannesen, L. J., 361
Johanson, R. S., 223, 228
John, E. R., 39
Johns, M. W., 256
Johnson, M. D., 215
Johnson, O., 281
Johnson, P. B., 302
Johnson, R., Jr., 39
Johnson, T., 284
Johnston, J. C., 101
Johnston, W. A., 72, 99
Joint Commission on Accreditation of Healthcare Organizations, 370
Jones, D., 195
Jones, D. M., 132
Jones, K. S., 316
Jones, M., 284
Jones, R. E., 4, 95
Jonides, J., 101, 102, 162
Jörg, M., 349
Joseph, R. D., 20
Jousmaki, V., 340
Jovanov, E., 119

Juang, B., 322
Jung, T. P., 19
Junque, C., 161
Jurado, M. A., 161
Just, M., 97, 106
Just, M. A., 17, 18

Kaas, J. H., 338
Kaber, D. B., 246
Kahn, R. S., 17
Kahneman, D., 146, 178
Kakei, S., 302
Kalaska, J. F., 229, 231, 302
Kalbfleisch, L. D., 16
Kalitzin, S., 317
Kandel, E. R., 6, 82, 306
Kane, M., 17
Kane, M. J., 103
Kanki, B., 373
Kanowitz, S. J., 296
Kantor, C., 356
Kantrowitz, A., 353
Kanwisher, N., 132, 139
Kapoor, A., 284
Kapur, S., 149
Karn, K. S., 95, 97
Karniel, A., 306
Karnik, A., 18
Karwowski, W., 221, 222, 224, 232, 233, 387
Kasten, E., 335, 338, 339
Kataria, P., 356
Katz, R. T., 263
Kaufman, J., 137
Kawasaki, H., 162
Kawato, M., 231
Kazennikov, O., 230
Kearney, J., 254, 266
Keating, J. G., 61
Kecklund, G., 256
Keith, M. W., 325, 355
Kelley, R. E., 88, 89, 90
Kellison, I. L., 4
Kellogg, R. S., 262
Kelly, J., 372
Kelly, S., 332
Kelly, T., 209
Kelso, J. A. S., 116
Kennard, C., 337, 339
Kennedy, P. R., 301
Kennedy, R. S., 96, 262
Kerkhoff, G., 338
Kertzman, C., 229
Kessler, C., 88
Kessler, G., 269
Keynes, R. D., 66, 68

Khalsa, S. B. S., 208
Kidd, C., 280
Kiehl, K. A., 53
Kieras, D. E., 163
Kilner, J. M., 228, 232
Kim, Y. H., 60
Kimberg, D. Y., 163
Kimmig, H., 105
King, J. A., 141
Kingstone, A., 6, 34
Kinoshita, H., 228, 232
Kinsbourne, M., 106
Kirn, C. L., 217
Klauer, S. G., 121
Klee, H. I., 261
Klein, B. E., 332
Klein, R., 332
Klein, R. M., 98, 106
Kleitman, N., 210
Klimesch, W., 18, 317
Kline, D. W., 99
Kline, N. S., 294
Klingberg, T., 18
Klingelhofer, J., 88, 148
Klinnert, M., 288
Knake, S., 87
Knapp, J., 266
Knecht, S., 82, 84, 87, 89
Knoblauch, V., 208
Knudsen, G. M., 87
Kobayashi, M., 333, 336
Kobetic, R., 353, 355
Kobus, D. A., 78, 245, 383
Koch, C., 103, 106, 107, 293
Koechlin, E., 161, 169, 170, 171
Koelman, T. W., 349
Koetsier, J. C., 349
Kokaji, S., 305
Kolb, B., 131
Kompf, D., 88
Konishi, S., 61
Konrad, M., 350
Koonce, J. M., 263
Koppel, R., 360
Korisek, G., 320
Kornhuber, H. H., 44, 225
Korol, D. L., 151
Kort, B., 279
Kositsky, M., 306
Koski, L., 18
Kosslyn, S. M., 337
Kovacs, G. T., 296
Kowler, E., 98
Krajnik, J., 353
Kralj, A., 353

Kramer, A. F., 3, 33, 38, 39, 40, 41, 95, 96, 97, 98, 101, 102, 103, 146
Krauchi, K., 208
Krausz, G., 320, 321
Kraut, M., 55, 341
Krebs, H. I., 350, 351
Krebs, J. R., 137
Kremen, S., 341
Kribbs, N. B., 210
Kristensen, M. P., 68
Kristeva-Feige, R., 232
Kroger, J. K., 160
Krueger, G. P., 217
Krull, K. R., 16
Kubler, A., 300
Kübler, A., 321
Kuiken, T., 297, 298
Kuiken, T. A., 305
Kumar, R., 6, 201
Kundel, H. L., 95, 96
Kuo, J., 372
Kupfermann, I., 306
Kurokawa, H., 305
Kurtzer, I., 230
Kussmaul, C. L., 37
Kusunoki, M., 105
Kutas, M., 32, 39, 44
Kwakkel, G., 349
Kyllonen, P. C., 17

LaBar, K. S., 182
Laborde, G., 89
Lack, L., 213
Lacourse, M. G., 300, 301
LaFollette, P. S., 95, 96
La France, M., 277, 279
Lamme, V. A., 42
Lammertse, P., 350
Lan, N., 356
Land, M. F., 103
Landis, T., 337
Landowne, D., 68
Landrigan, C. P., 215
Lane, N. E., 262
Langham, M., 99
Langmoen, I. A., 83
Lankhorst, G. J., 349
Lanzetta, T. M., 152
Larsen, J., 53
Lavenex, P., 137
Lawrence, K. E., 300
Lazarus, R. S., 195, 197
Leaver, E., 78
LeDoux, J., 184, 276
LeDoux, J. E., 182

Lee, D. S., 340
Lee, D. W., 137
Lee, G. P., 162, 182
Lee, J. D., 261, 370
Lee, K. E., 341
Lee, W. G., 233
LeGoualher, G., 160
Lehner, P. N., 115
Lemon, R. N., 228, 229
Lemus, L., 303
Leong, H. M., 19
Levanen, S., 340
Leveson, N. G., 372
Levi, D., 96
Levine, B., 172
Levine, J. A., 119
Levine, S. P., 300, 315, 325
Levy, R., 160, 162
Lewin, W. S., 330
Lewis, M., 284
Lezak, M. D., 8
Liberson, W. T., 353
Lieke, E., 68
Lieke, E. E., 68
Lilienthal, M. G., 262
Linde, L., 16
Lindegaard, K. F., 83
Lintern, G., 263
Litvan, I., 162, 168, 172
Liu, J., 227
Liu, J. Z., 225, 227
Liu, K., 281
Lockerd, A., 284, 288
Loeb, G. E., 294, 296, 298, 340
Loewenstein, J., 332
Loewenstein, J. I., 330, 331, 339
Loftus, G. R., 97
Lohmann, H., 84
Lok, B., 264
Lombard, M., 280
Loomis, A. L., 20
Loomis, J., 266
Lopes da Silva, F., 300
Lopes da Silva, F. H., 317, 318, 319
Lorenz, C., 285
Loschky, L. C., 96
Lotze, M., 316
Loula, P., 20
Lounasmaa, O. V., 334
Low, K., 78
Low, K. A., 70, 78
Luck, S., 38, 40, 42
Luck, S. J., 33
Lucking, C. H., 232
Lüders, H. O., 318

Lufkin, T., 222
Lum, P. S., 351
Luo, Y., 42
Luppens, E., 89
Lyman, B. J., 99
Lynch, K., 132, 139

Machado, L., 105
Macko, R. F., 119
Mackworth, N. H., 97
MacLean, A. W., 16
Maclin, E., 66
Maclin, E. L., 71, 72, 77
Macmillan, M., 162
MacVicar, B. A., 68
Maddock, R. J., 183
Maeda, H., 87
Maffei, L., 96
Magliano, J. P., 135
Magliero, A., 39
Maguire, E. A., 131, 132, 133, 134, 135, 137, 138, 139, 140
Mai, N., 335
Maier, J. S., 66
Maislin, G., 208, 213, 217
Majmundar, M., 351
Makeig, S., 34
Malach, R., 341
Malezic, M., 353
Malin, J. T., 241
Malkova, L., 182
Mallis, M., 217, 256
Mallis, M. M., 214, 216, 217
Malmivuo, J., 334
Malmo, H. P., 222
Malmo, R. B., 222
Malonek, D., 68
Mangold, R., 87
Mangun, G. R., 33, 37, 42, 54, 87, 147
Manivannan, P., 263
Mantulin, W., 66
Marescaux, J., 308
Marg, E., 330
Margalit, E., 330
Marini, C., 87
Mark, L. S., 372
Markosian, L., 265
Markowitz, R. S., 301
Markowitz, S., 115
Markus, H. S., 87, 89
Markwalder, T. M., 82
Marr, D., 293
Mars, R. B., 45
Marshall, S. J., 119
Marsolais, E. B., 353, 355

Martin, A., 341
Martinez, A., 42, 73
Martinoia, S., 305
Mason, S. G., 300
Masson, G. S., 96
Masterman, D. L., 160
Mathis, J., 87
Matin, E., 96
Matsuoka, S., 18
Matteis, M., 88
Matthews, G., 87, 195, 197, 198, 199, 200, 201
Mattle, H. P., 87, 88
Matzander, B., 88
Mavor, A., 8
Mavor, A. S., 90
May, J. G., 96
Maycock, G., 362
Mayleben, D. W., 148, 149, 150, 154
Maynard, E. M., 301, 303, 326, 330
McCallum, W. C., 18
McCane, L. M., 320
McCarley, J. S., 95, 96, 98, 103
McCarley, R. W., 208
McCarthy, G., 39
McConkie, G., 95, 99
McConkie, G. W., 96
McCormick, E. F., 9
McCreery, D. B., 296
McCroskey, J., 279
McDonald, R. J., 136
Mcenvoy, L., 225
McEvoy, L., 18, 21
McEvoy, L. K., 16, 19, 21
McEvoy, R. D., 213
McEvoy, S., 263
McFadyen, B. J., 372
McFarland, D. J., 41, 300, 301, 320
McGaugh, J. L., 136
McGee, J. P., 90
McGehee, D., 263
McGehee, D. V., 261
McGinty, V. B., 54
McGovern, K. T., 373
McGovern, L. T., 373
McGown, A., 214
McGrath, J. J., 196
McKenzie, T. L., 119
McKeown, M. J., 53
McKinney, W. M., 89
McMillan, G., 316
McMillen, D. L., 263
McNeal, D. R., 353, 355
Meadows, P., 355
Mechelli, A., 139

Meehan, M., 254, 256
Mehta, A. D., 42
Mejdal, S., 216
Mellet, E., 141
Meltzoff, M., 288
Menon, R. S., 68
Menon, V., 61
Menzel, K., 279
Merabet, L. B., 330, 333, 336, 341, 342
Merboldt, K. D., 316
Meredith, M. A., 339
Merton, P. A., 97
Merzenich, M. M., 294
Mesulam, M. M., 60
Meuer, S. M., 332
Meyer, B. U., 330
Meyer, D. E., 43, 163
Meyer, E., 89
Middendorf, M., 316
Miezin, F. M., 60
Mignot, E., 207, 210
Mikulka, P. J., 242, 244, 246
Milea, D., 96
Milenkovic, A., 119
Milgram, P., 372
Millan, J. R., 300
Miller, C. A., 242, 243, 247, 248, 250
Miller, D. J., 284
Miller, D. L., 27
Miller, E. K., 162
Miller, J. C., 27
Miller, L. E., 302
Milliken, G. W., 305
Milner, P., 201
Miltner, W. H. R., 44
Milton, J. L., 4, 95
Mintun, M. A., 82
Miyake, A., 106
Miyasato, L. E., 137
Miyata, Y., 18
Mizuki, Y., 18
Moffatt, S. D., 137
Mohler, C. W., 338
Moll, J., 8
Mollenhauer, M., 262
Molloy, G. J., 365
Molloy, R., 21, 240, 247
Monta, K., 372
Montague, P. R., 189
Montezuma, S. R., 330, 331
Montgomery, P. S., 119
Moon, Y., 247
Moore, C. M., 45
Moore, J. K., 296
Moore, M. K., 288

Moore, R. Y., 207
Moosmann, M., 18
Morari, M., 349, 352
Moray, N., 39, 99, 146, 241
Moray, N. P., 43
Morgan, N. H., 19, 20
Morrell, M. J., 60
Morren, G., 71
Morris, D. S., 301
Morrison, J. G., 78, 241, 242, 245, 383
Morrone, M. C., 97
Mortimer, J. T., 353, 354
Morton, H. B., 97
Moscovitch, M., 149
Mostow, J., 279
Mota, S., 282, 283
Mouloua, M., 21, 45, 90, 240
Mourant, R. R., 95, 256
Mourino, J., 300
Moxon, K. A., 301
Mozer, M. C., 248, 249, 250
Mulder, A. J., 356
Mulholland, T., 18, 25, 320
Müller, G., 321
Müller, G. R., 315, 321, 323
Müller-Oehring, E. M., 338, 339
Müller-Putz, G. R., 321
Mullington, J. M., 208
Munakata, Y., 8
Munih, M., 356
Munoz, D. P., 106
Munt, P. W., 16
Munte, T. F., 42, 88, 149
Murata, S., 305
Müri, R., 96
Muri, R. M., 105
Murphy, L. L., 201
Murray, E. A., 182
Murri, L., 87
Musallam, S., 301, 305
Mussa-Ivaldi, F. A., 231, 302, 303, 306

Naatanen, R., 37, 41, 42, 45
Nadel, L., 131, 132, 141
Nadler, E., 195
Nadolne, M. J., 132
Nagel, D. C., 4
Naik, S., 264
Naitoh, P., 209
Nakai, R. J., 355
Naqvi, N., 255
Nass, C., 247, 277
Nathanielsz, P., 115
Nathanielsz, P. W., 115
National Institutes of Health, 269

Navalpakkam, V., 107
Navon, D., 25, 40, 146, 200
Neale, V. L., 121
Neat, G. W., 320
Nebeker, J. R., 360
Nef, T., 351
Nelson, D. R., 16
Neri, D. F., 217
Neuman, M. R., 354
Neuper, C., 5, 41, 316, 317, 318, 320, 321, 323
Newell, D. W., 88
Nezafat, R., 231
Nguyen, T. T., 216
Ni, R., 263
Nichelli, P., 161, 165, 167
Nicolelis, M. A., 5, 301, 315
Niedermeyer, E., 300
Nilsson, L., 261
Nishijima, H., 18
Nishimura, T., 228
Nishino, S., 207
Njemanze, P. C., 88, 89
Nobre, A. C., 60
Nobumasa, K., 84
Nodine, C. F., 95
Nordhausen, C. T., 330
Norman, D. A., 43, 146, 162, 163, 367
Normann, R. A., 301, 303, 330
Nornes, H., 82, 83
Noser, H., 373
Nourbakshsh, I., 277
Nudo, R. J., 305
Nuechterlein, K., 154
Nunes, L. M., 96
Nyberg, L., 51, 52
Nygren, A., 263
Nystrom, L. E., 162

Obermaier, B., 300, 321, 322
O'Boyle, C. A., 365
O'Donnell, R. D., 200
Offenhausser, A., 305
Ogilvie, R., 256
O'Keefe, J., 131, 132, 133, 134, 139, 141
Oku, N., 228
Olds, J., 201
O'Leary, D. S., 89
O'Neil, C., 99
Opdyke, D., 269
Ordidge, R. J., 148
O'Reilly, R. C., 8
Orlandi, G., 87
Orlovsky, G. N., 306
Oron-Gilad, T., 200, 201

Ortiz, M. L., 137
Oshercon, D., 182
Osman, A., 45
Otto, C., 119
Owsley, C., 101
Oyung, R. L., 217

Pacheco, A., 263
Packard, M. G., 136
Palmer, S. E., 261, 265
Pambakian, L., 337, 339
Pandya, D. N., 159, 160
Paninski, L., 301, 325
Panksepp, J., 276
Panzer, S., 161
Parasuraman, R., 3, 6, 9, 21, 33, 35, 39, 41, 42, 45, 46, 90, 95, 105, 106, 146, 147, 148, 149, 150, 152, 153, 154, 155, 172, 195, 196, 199, 200, 201, 221, 222, 224, 239, 241, 242, 244, 247, 248, 329, 381
Pardo, J. V., 89
Pardue, M. T., 332
Parrish, T. B., 60
Parsons, L., 182
Parsons, O. A., 16
Partiot, A., 167, 168
Pascual-Leone, A., 163, 169, 171, 330, 333, 336, 340, 341
Pascual-Marqui, R. D., 37, 38
Pashler, H., 40, 153
Patel, S. N., 137
Patterson, E. S., 372
Patterson, P. E., 137
Paul, A., 4
Paunescu, L. A., 65, 71
Paus, T., 18, 147
Pavan, E., 356
Payne, S. J., 132
Pazo-Alvarez, P., 45
Pearlson, G. D., 52, 54, 55, 59, 61
Peckham, P. H., 353, 354, 355
Pejtersen, A. M., 361
Pekar, J. J., 52, 53, 55, 59, 61
Peled, S., 341
Pellouchoud, E., 21, 25
Pelz, J. B., 104
Penfield, W., 25
Penney, T. B., 65
Pentland, A., 282
Pepe, A., 201
Peponis, J., 139
Peres, M., 7, 60
Perreault, E. J., 302
Perry, J., 353
Pertaub, D., 261

Peters, B. A., 217
Peters, T. M., 373
Petersen, D., 362
Petersen, S. E., 32, 60
Peterson, D. A., 300
Peterson, M. S., 98, 103
Petit, L., 162
Petrides, M., 136, 159, 160
Petrofsky, J. S., 353
Pew, R., 8
Pfurtscheller, G., 5, 18, 41, 300, 315, 316, 317, 318, 319, 320, 321, 323, 324, 325
Pfurtscheller, J., 315
Phelps, E., 279
Phelps, E. A., 182
Phillips, C. A., 353
Phipps, M., 162, 168, 170
Piaget, J., 6
Piantadosi, S., 269
Picard, R., 279, 284
Picard, R. W., 224, 276, 277, 279, 281, 282, 283, 284, 288
Picton, T. W., 32
Pierard, C., 7
Pierrot-Deseilligny, C., 96, 105, 106
Pietrini, P., 161, 168, 341
Pike, B., 136
Pilcher, J. J., 195
Pillai, S. B., 119
Pillon, B., 170, 171
Pillsbury, H. C., 3rd, 329
Pimm-Smith, M., 60
Pinsker, H., 306
Plant, G. T., 339
Platz, T., 349
Plaut, D. C., 164
Plautz, E. J., 305
Plomin, R., 9, 201
Ploner, C. J., 105
Plumert, J., 254, 266
Plutchik, R., 288
Pocock, P. V., 18
Poe, G. R., 68
Poggel, D. A., 338, 339
Pohl, P. S., 230
Pohlmeyer, E. A., 302
Poldrack, R. A., 136
Polich, J., 39
Polkey, C. E., 353
Pollatsek, A., 98, 99
Polson, M., 106
Pomplun, M., 100
Ponds, R. W., 263
Pope, A. T., 244, 245
Porro, C. A., 316

Posner, M. I., 6, 8, 9, 18, 32, 39, 44, 60, 98, 146, 154
Potter, S., 306
Powell, J. W., 217
Pregenzer, M., 300, 319, 321, 325
Prencipe, M., 87
Prete, F. R., 115
Preusser, C. W., 214
Preusser, D. F., 214
Price, C., 340
Price, J. L., 169
Price, K., 265
Price, W. J., 16
Pringle, H. L., 101, 103
Prinzel, L. J., 244
Pröll, T., 352
Prud'homme, M., 302
Puhl, J., 119
Punwani, S., 148
Purves, D., 87

Qi, Y., 284
Quinlan, P. T., 153
Quintern, J., 352, 354, 356

Rabiner, L., 322
Radach, R., 96, 97
Rafal, R. D., 105
Raichle, M. E., 32, 82, 84, 88, 89, 147
Rainville, P., 255
Ramsey, N. F., 17
Ranganathan, V., 227, 232
Ranganathan, V. K., 225
Ranney, T. A., 52
Rasmussen, J., 361, 365, 367, 368, 369, 371
Rasmussen, T., 87
Rastogi, E., 87, 89
Rauch, S. L., 59
Rauschecker, J. P., 294, 295
Raven, T., 87
Raymond, J. E., 99
Rayner, K., 95, 96, 97, 98, 99, 101
Razzaque, S., 261
Reason, J., 365, 367
Reason, J. T., 43, 367
Recarte, M. A., 96
Rector, D. M., 68
Redish, J., 268
Reduta, D. D., 217
Rees, G., 59
Reeves, A. J., 137
Reger, B. D., 306
Regian, J. W., 132
Reichle, E. D., 18
Reilly, R., 279
Reilly, R. E., 297

Reingold, E. M., 96, 100, 103
Reinkensmeyer, D. J., 349
Reiss, A. L., 61
Remington, R. W., 101
Render, M. L., 372
Renz, C., 208
Reyner, L. A., 256
Reynolds, C., 284
Rheingold, H., 253, 260
Rich, C., 282
Richards, A., 269
Richards, T., 269
Richmond, V., 279
Riddoch, M. J., 163
Riener, R., 351, 352, 353, 355, 356
Riepe, M. W., 137
Ries, B., 266
Riggio, L., 98
Riggs, L. A., 97
Rihs, F., 87, 88, 89
Riley, J. M., 246
Riley, V., 3, 239
Rilling, J. K., 160
Ringelstein, E. B., 82, 84, 89
Rinne, T., 71, 72
Risberg, J., 82, 83, 88, 147
Riskind, J. H., 279
Risser, M., 256
Rivaud, S., 105
Rizzo, J. F., 330, 331
Rizzo, J. F., 3rd., 330, 332, 337, 339
Rizzo, M., 4, 120, 132, 261, 262, 263, 264
Rizzolatti, G., 98, 325
Ro, T., 105
Robbins, T. W., 160
Robert, M., 137
Roberts, A. E., 87, 89
Roberts, D., 288
Roberts, R. J., 103
Robertson, S. S., 115, 116
Robinson, S. R., 115, 116, 117
Rockwell, T. H., 95, 256
Rodrigue, J. R., 135
Rogers, A. E., 215
Rogers, N. L., 208, 210, 211, 217
Rogers, W., 259
Rojas, E., 68
Roland, P. E., 89
Rollnik, J. D., 44
Rolls, E. T., 114
Romero, D. H., 300
Romo, R., 303
Rooijers, T., 263
Roos, N., 362
Rorden, C., 98

Roricht, S., 330
Rosa, R. R., 147, 215
Roscoe, S. N., 263
Rose, F. D., 132
Rosekind, M. R., 27
Rosen, J. M., 296
Ross, J., 97
Rossignol, S., 349
Rossini, P. M., 333
Rotenberg, I., 99
Roth, E. M., 6
Rothbart, M. K., 18
Rothbaum, B., 269
Rothwell, A., 16
Rousche, P. J., 303
Rouse, S. H., 242
Rouse, W. B., 241, 242
Rovamo, J., 96
Roy, C. S., 88, 146
Rubinstein, J. T., 294
Ruchkin, D. S., 161
Ruddle, R. A., 132
Rudiak, D., 330
Rueckert, L., 165
Ruiter, B., 350
Rupp, R., 315
Rushton, D. N., 353
Russell, C. A., 245

Sabel, B. A., 330, 337, 338, 339
Sabes, P. N., 229, 231
Sable, J. J., 71
Sadato, N., 161, 167, 168, 340
Sadowski, W., 266
Sahgal, V., 225, 227, 232
Saidpour, A., 263
Sakthivel, M., 137
Sakuma, A., 372
Salamon, A., 305
Saleh, M., 301
Salenius, S., 228
Salgian, G., 52
Salimi, I., 228
Sallis, J. F., 119
Salmelin, R., 334
Sanchez-Vives, M., 254, 258
Sander, D., 88, 148
Sanders, A. F., 99, 100
Sanders, M. S., 9
Sandstrom, N. J., 137
Sanguineti, V., 306
Santalucia, P., 82
Santoro, L., 97
Saper, C. B., 207, 210
Sarnacki, W. A., 41

Sarno, A. J., 66
Sarter, M., 6
Sarter, N., 6, 240
Sarter, N. B., 361
Sartorius, N., 54
Sasaki, Y., 338
Sasson, A. D., 137
Satava, R. M., 373
Sato, S., 103
Saucier, D. M., 137
Sawyer, D., 288
Scammell, T. E., 207, 208
Scassellati, B., 285
Scerbo, M. W., 90, 152, 241, 242, 244, 245, 246, 248, 249
Schabes, Y., 282
Schacter, D. L., 168
Schaffer, R. E., 18, 19
Schandry, R., 82, 83, 84, 87, 88, 148
Schärer, R., 352
Scheffers, M. K., 43, 44
Scheidt, R. A., 303
Scherberger, H., 301
Scherer, K. R., 199
Scherer, R., 320, 321, 323
Scherg, M., 19, 37
Schier, M. A., 61
Schlaug, G., 139
Schleicher, A., 160
Schlögl, A., 320, 321
Schmid, C., 335
Schmidt, E. A., 87
Schmidt, E. M., 301
Schmidt, G., 356
Schmidt, P., 87
Schmitz, C., 223, 224, 228, 229, 231
Schmorrow, D., 78, 245, 383
Schneider, W., 60
Schneider, W. X., 98
Schnittger, C., 88, 149
Schoenfeld, V. S., 152
Schreckenghost, D. L., 241
Schreier, R., 349
Schroeder, C. E., 42
Schroth, G., 88
Schuepback, D., 90
Schulte-Tigges, G., 350
Schultheis, H., 40
Schwab, K., 172
Schwartz, A., 269
Schwartz, A. B., 301
Schwartz, J. H., 82
Schwartz, M. F., 163
Schwartz, U., 229
Schwent, V. L., 32

Scialfa, C. T., 99, 103
Scot, D., 353
Scott, A. J., 16
Scott, H., 281
Scott, L. A., 246
Scott, L. D., 215
Scott, S. H., 229, 302
Scott, W. B., 241
Scott Osberg, J., 213
See, J. W., 147
Segal, L. D., 263
Seiffert, A. E., 42
Sejnowski, T. J., 7, 53, 55, 293
Sekuler, R., 99
Seligman, M. E. P., 201
Selye, H., 196
Semendeferi, K., 160
Senders, J., 103
Senders, J. W., 43, 103
Sergio, L. E., 229
Serrati, C., 90
Serruya, M. D., 301, 305, 325
Severson, J., 120, 265
Seymour, N. E., 373
Shadmehr, R., 231, 303
Shallice, T., 43, 162, 163
Shannon, R. V., 294, 295, 296
Shapiro, D., 99
Sharar, S. R., 269
Sharit, J., 215
Sharon, A., 350
Sharp, T. D., 372
Shaw, C., 254
Sheehan, J., 270
Sheer, D. E., 18
Sheffield, R., 262
Shell, P., 17
Shelton, A. L., 141
Shen, J., 100
Shenoy, K. V., 301, 305
Shepherd, M., 98
Sheridan, T. B., 241
Sherrington, C. S., 88, 146
Sherry, D. F., 137
Shi, Q., 132
Shibasaki, H., 318
Shih, R. A., 54
Shinoda, H., 52
Shire, D., 332
Shoham, S., 301
Shor, P. C., 351
Shors, T. J., 137
Shulman, G. L., 60
Shumway-Cook, A., 347
Siegel, A. W., 132

Siegrist, K., 119
Sieminski, D. J., 119
Siemionow, V., 225, 227, 232
Siemionow, W., 221, 225, 227, 232
Sifuentes, F., 305
Silvestrini, M., 88
Simeonov, P., 254, 258
Simons, D., 264
Simpson, G. V., 318
Singer, M. J., 132
Singer, W., 169
Singh, I. L., 21, 247
Singhal, A., 42
Sinkjaer, T., 303
Sirevaag, E. J., 39, 40, 44
Sirigu, A., 166, 168, 170, 171
Skolnick, B. E., 87
Skreczek, W., 86
Slater, M., 254, 258, 261
Sluming, V., 139
Small, R. L., 242, 249
Smith, A. M., 228
Smith, C., 281
Smith, E. E., 162
Smith, E. G., 338
Smith, G., 255
Smith, L. T., 16
Smith, M. E., 16, 17, 18, 19, 20, 21, 22, 23, 25, 26, 225
Smotherman, W. P., 115, 117
Smulders, T. V., 137
Snyder, L. H., 229
Soares, A. H., 161
Solla, S. A., 302
Solopova, I., 230
Somers, D. C., 42, 330
Sommer, T., 9
Sorteberg, W., 83
Source, J., 288
Spekreijse, H., 42
Spelke, E. S., 132
Spelsberg, B., 88
Spencer, D. D., 182
Spencer, K. M., 5, 41, 300, 316
Sperry, R. W., 222
Spicer, M. A., 373
Spiers, H., 131, 135, 138, 140
Spiers, H. J., 132, 133, 134, 140
Spitzer, M., 137
Spohrer, J., 259
St. John, M., 78, 245, 383
St. Julien, T., 254
Stablum, F., 172
Stafford, S. C., 200
Stampe, D. M., 96

Stanney, K., 387
Stasheff, S. F., 335
Staszewski, J., 217
Staveland, L. E., 152
Stea, D., 132
Steele, M. A., 137
Stein, B. E., 339
Steinbrink, J., 65, 68, 71
Steinmetz, H., 139
Stenberg, C., 288
Stepnoski, R. A., 68
Sterman, M. B., 320
Stern, C., 373
Stern, J. M., 18
Stevens, M., 53
Stierman, L., 262
Stinard, A., 70
Stokes, M. J., 325
Stoll, M., 87, 88
Storment, C. W., 296
Strasser, W., 265
Strayer, D. L., 99
Strecker, R. E., 208
Strick, P. L., 302
Stringer, A. Y., 132
Strohecker, C., 282
Stroobant, N., 83, 86, 87, 88, 89, 90, 148
Stucki, P., 373
Sturzenegger, M., 87, 88
Stuss, D. T., 162
Stutts, J. C., 213, 214
Subramaniam, B., 98
Sudweeks, J., 121
Suffczynski, P., 317
Suihko, V., 334
Sunderland, T., 9
Super, H., 42
Surakka, V., 96
Sustare, B. D., 117
Sutherland, R. J., 137
Sutton, S., 39
Suzuki, R., 231
Svejda, M., 288
Swain, C. R., 21
Swanson, K., 288
Swanson, L. W., 222, 223
Sweeney, J. A., 105
Syre, F., 71
Szalma, J. L., 196, 199, 200, 202, 387

Tadafumi, K., 84
Taheri, S., 207
Takae, Y., 241
"Taking Neuroscience beyond the Bench," 5
Talairach, J., 55

Talis, V., 230
Talwar, S. K., 303
Tamura, M., 68
Tan, H. Z., 282
Tanaka, M., 18
Tanaka, Y., 18
Tatum, W. O., 119
Taylor, D. M., 300, 301
Taylor, J. L., 372
Teasdale, J. D., 197, 200
Temple, J. G., 154, 155
Tepas, D. I., 16
Terao, Y., 105
Thach, W. T., 61, 222
Thacker, P., 255
Thakkar, M. M., 208
Thaut, M. H., 300
Theeuwes, J., 101, 102, 103, 106
Theoret, H., 333
Theunissen, E. R., 266
Thierry, A. M., 170
Thilo, K. V., 97
Thomas, C., 89
Thomas, F., 281
Thompson, P., 131
Thompson, W., 266
Thompson, W. D., 301
Thoroughman, K. A., 303
Thrope, G. B., 353, 354
Thropp, J. E., 200
Tien, K.-R., 65
Tillery, S. I., 301
Timmer, J., 232
Tinbergen, N., 285
Tingvall, C., 263
Tippin, J., 4
Tlauka, M., 132
Tomczak, R., 137
Toole, J. F., 83, 148
Tootell, R. B., 42
Toroisi, E., 88
Toronov, V., 65, 71, 72, 148
Torrance, M. C., 282
Torres, F., 340
Totaro, R., 87
Tournoux, P., 55
Tourville, J., 372
Towell, M. E., 115
Townsend, J., 61
Townsend, J. T., 73
Tranel, D., 161, 162, 182, 183, 187, 188, 265, 276
Trappenberg, T. P., 106, 107
Treisman, A., 103, 152
Treisman, A. M., 151

Treserras, P., 161
Trevarthen, C., 281
Tripp, L. D., 87
Trohimovich, B., 372
Troisi, E., 87
Tronick, E., 281
Troyk, P. R., 297
Tse, C.-Y., 65, 71
Tsirliganis, N., 253
Ts'o, D. Y., 68
Tu, W., 355
Tucker, D. M., 44
Tudela, P., 146
Tuholski, S., 17
Tulving, E., 149
Turing, A. M., 261
Turk, R., 353
Turkle, S., 285
Tversky, A., 178
Tversky, B., 167

Uc, E. Y., 132
Uhlenbrock, D., 349
Ulbert, I., 42
Ullman, S., 103
Ulmer, J. L., 52
Ulmer, R., 214
Umiltà, C., 98
UNEC & IFR (United Nations Economic Commission and the International Federation of Robotics), 276
Ungerleider, L. G., 162, 333, 335
Urbano, A., 302
Ustun, T. B., 54
Uylings, H. B., 160

Vais, M. J., 103
Valentine, E. R., 139
Van De Moortele, P. F., 7, 55
Van den Berg-Lensssen, M. M., 19
Van der Linde, R. Q., 350
van der Loos, M., 351
van Diepen, P. M. J., 97
Van Dongen, H., 217
Van Dongen, H. P., 208, 211, 212, 213
Van Dongen, H. P. A., 208, 209, 216
Vanesse, L., 40
Van Essen, D., 335
Van Hoesen, G. W., 160
Van Nooten, G., 87
Van Schie, H. T., 45
van Turennout, M., 139, 140
Van Voorhis, 41
Van Wolffelaar, P. C., 263
VanZomeren, A. H., 263

Vargha-Khadem, F., 131, 134, 141
Varnadore, A. E., 89
Varri, A., 20
Vaughn, B. V., 213
Veitch, E., 162
Veltink, P. H., 354, 356
Veltman, H. J. A., 245
Vendrell, P., 161
Vermersch, A. I., 105
Verplank, W. L., 241
Vertegaal, R., 383
Ververs, P., 255, 260
Vetter, T., 305
Vicente, K. J., 367, 368, 369, 370, 371, 372
Vidoni, E. D., 95
Viglione, S. S., 20
Villringer, A., 68
Vingerhoets, G., 83, 86, 87, 88, 89, 90, 148
Viola, P., 284
Voermans, N. C., 136
Vogt, B. A., 60
Vollmer, J., 86
Vollmer-Haase, J., 88
Volpe, B. T., 350, 351
von Cramon, D., 335, 338
Von Neumann, J., 293
Von Reutern, G. M., 87

Wachs, J., 168
Wada, W., 87
Wagenaar, R. C., 349
Waldrop, M. M., 116
Walker, R., 107
Walsh, J. K., 213, 215
Walsh, V., 97, 336
Walter, H., 25, 52, 54, 56, 61
Walter, K. D., 87
Wandell, B. A., 96, 334
Wang, R. F., 98
Wanger, L., 265
Ward, J. L., 103
Wardman, D. L., 372
Warm, J. S., 87, 146, 147, 148, 149, 152, 195, 196, 197, 198, 199, 200
Warren, D. J., 303
Washburn, D., 87
Wassermann, E. M., 169, 336
Watanabe, A., 84
Waters, R. C, 282
Waters, R. L., 353
Watson, B., 269
Weaver, J. L., 196
Weber, T., 3, 33, 39, 146
Wechsler, L. R., 82
Wee, E., 70

Weeks, R., 340
Weil, M., 261
Weiller, C., 141
Weinberger, D. R., 86
Weinger, M. B., 215
Weir, C. R., 360
Weir, R. P. F., 296, 297
Weiskopf, N., 315
Welk, G., 119
Well, A. D., 99
Wells-Parker, E., 263
Werner, C., 350
Werth, E., 209
Wessberg, J., 306, 325
Westbury, C., 18
Westling, G., 228
Whalen, C., 77
Whalen, P. J., 182
Wharton, C. M., 161
Whishaw, I. Q., 131
White, C. D., 61
White, N. M., 136
White, S. H., 132
Whiteman, M. C., 201
Whitlow, S., 255, 260
Whitton, M., 254, 264
Wickens, C. D., 6, 25, 33, 39, 40, 90, 95, 96, 97, 104, 106, 146, 152, 195, 200, 240
Wiederhold, B. K., 269
Wiederhold, M. D., 269
Wiener, E., 373
Wiener, E. L., 4, 150, 240, 241
Wierwille, W., 256
Wierwille, W. W., 120, 122, 217, 263
Wiesel, T. N., 68, 338
Wijesinghe, R., 5, 41, 300, 316
Wild, K., 167
Wilde, G. J., 16
Wilensky, R., 242
Wilkie, F. L., 263
Wilkins, J. W., 213
Willemsen, P., 265, 266
Williams, D. E., 103
Williams, L. G., 103
Williams, M. C., 96
Williamson, A. M., 16
Wilson, G., 90
Wilson, G. F., 19, 21, 245
Wilson, H. R., 96
Wilson, P. N., 132
Winkler, I., 45
Winstein, C. J., 230
Winterhoff-Spurk, P., 87
Wirz-Justice, A., 208
Wise, B. M., 305

Witmer, B., 266
Witmer, B. G., 132
Witney, A. G., 231
Wittich, I., 88, 148
Wolbers, T., 141
Woldorff, M. G., 45
Wolf, M., 65, 71
Wolf, U., 65, 71, 72, 74
Wolfe, J. M., 103
Wolpaw, J. R., 41, 300, 301, 316, 320, 323
Wong, C. H., 115
Wong, E. C., 61
Woodfill, J., 283
Woods, D., 361
Woods, D. D., 6, 240, 241, 361
Woods, R. P., 53
Woollacott, M. H., 347
Worden, M., 60
Worsley, K. J., 52, 55
Wreggit, S. S., 217
Wright, N., 214
Wu, Y., 9
Wunderlich, A. P., 137
Wurtz, R. H., 96, 106, 338
Wüst, S., 338
Wyatt, J., 332
Wynne, J. H., 349

Xiong, F., 225
Xu, W., 372

Yadrick, R. M., 132
Yamamoto, S., 18
Yan, L., 284
Yang, M., 372
Yantis, S., 101, 102
Yarbus, A. L., 103
Yates, F. A., 139
Yeager, C. L., 19, 20
Yeh, Y. Y., 200
Yingling, C. D., 19
Yoshida, K., 354, 356
Yu, C.-H., 356
Yu, D., 18, 225
Yue, G. H., 225, 227, 232
Yue, H. G., 225

Zacks, J. M., 167
Zahn, T. P., 161, 162
Zainos, A., 303
Zalla, T., 168, 169, 170, 171, 182
Zatorre, R. J., 89
Zeck, G., 305
Zeffiro, T. A., 229
Zeitlin, G. M., 19, 20

Zelenin, P. V., 306
Zhuo, Y., 42
Zigmond, M. J., 222
Zihl, J., 335, 338
Zilles, K., 160
Zimring, C., 139
Zinni, M., 90

Zohary, E., 341
Zrenner, E., 303, 330, 332
Zubin, J., 39
Zucker, R. S., 306
Zunker, P., 89
Zwahlen, H. T., 125
Zyda, M., 270

Subject Index

Page numbers followed by *f* indicate figures. Page numbers followed by *t* indicate tables.

AAN. *See* American Academy of Neurology
ACA. *See* anterior cerebral artery
ACC. *See* anterior cingulate cortex
accelerometry, 119–120
ACHE. *See* adaptive control of home environment
activities of daily living (ADL), 349
adaptable automation
   definition, 389
   distinction between adaptive automation and, 241
adaptive automation, 239–252
   adaptive strategies, 242
   definition, 389
   distinction between adaptable automation and, 241
   examples of adaptive automation systems, 242–246
      associate systems, 242–243, 243*f*
      brain-based systems, 243–246, 244*f*
      human-computer etiquette, 247–248
      living with, 248–249, 249*f*
      overview, 250
      workload and situation awareness, 246–247
         situation awareness, 246–247
         workload, 246
adaptive control of home environment (ACHE), 248–249, 249*f*

ADL. *See* activities of daily living
adverse event, definition, 389
aircraft
   adaptive automation and, 259–260
   vigilance decrement with signal cueing, 150
ALS. *See* amyotrophic lateral sclerosis
American Academy of Neurology (AAN), 386–387
American Academy of Neurosurgery, 386
American Psychological Association, 386
amygdala
   disturbances after brain damage, 181–184, 181*f*
   primary inducers, 182
amyotrophic lateral sclerosis (ALS), EEG-based brain-computer interface and, 315, 321, 322*f*
anosognosia, 183
anterior cerebral artery (ACA), examination with TCD, 83, 83*f*
anterior cingulate cortex (ACC), event-related potentials and, 44
AOI. *See* area of interest
appraisal, definition, 389
AR. *See* augmented reality
area of interest (AOI), 97
ARMin, 351, 352*f*
arousal
   definition, 389
   stress and, 196–197

Subject Index

artificial vision, 329–345
  brain plasticity, 337–342
    cross-modal plasticity, 339–342, 341f
    visual system plasticity and influence of cognitive factors, 337–339
  overview, 342
  prosthetic devices, 331f
  systems perspective and role of new technologies, 333
  technologies, 333–337
    electroencephalography, 334
    functional magnetic resonance imaging, 333–334, 334f
    lesion studies, 334–336
    magnetoencephalography, 334
    systems approach, 337
    transcranial magnetic stimulation, 336–337
  visual prostheses, history, 330–333, 330f, 331f
ASRS. *See* Aviation Safety Reporting System
attention, definition, 389
attentional narrowing, definition, 389
augmented cognition, 373–374, 374f, 375f
  definition, 389
augmented reality (AR)
  definition, 389
  description, 258
automation
  definition, 389
  electroencephalogram and, 22, 22f
avatar, definition, 389
aviation, neuroergonomics research, 4
Aviation Safety Reporting System (ASRS), 364–365

bar codes, patient tracking and, 372
basal ganglia
  definition, 390
  human prefrontal cortex and, 169
BCI. *See* brain-computer interface
behavior
  human. *See* human behavior
  human prefrontal cortex and, 173
  role of emotions and feelings in behavioral decisions, 178–192
  saccadic. *See* saccadic behavior
  sleep and circadian control of neurobehavioral functions, 207–220
  stress and, 195–206
biomathematical model of fatigue
  definition, 390
  to predict performance capability, 215–216
biotechnology. *See also* robots
  neuroergonomics and, 9
blood flow
  baseline measurement, 86–87
  linguistics and, 89–90
  velocity and vigilance tasks, 149, 149f

blood oxygenated level-dependent (BOLD) signal, 51–52
  optical imaging of brain function and, 68
blood oxygenation, functional magnetic resonance imaging and, 51–52
blue eyes camera, 284, 285f
BOLD. *See* blood oxygenated level-dependent signal
brain-based adaptive automation
  criticisms, 245–246
  definition, 390
brain-computer interface (BCI), 4
  application for paralyzed patients, 315, 321, 322f
  -based control of functional electrical stimulation in tetraplegic patients, 324–325, 324f
  -based control of spelling systems, 321–324, 323f
  components, 315–316, 316f
  description, 315
  EEG-based, 315–328
  event-related potentials and, 41
  event-related synchronization, 317
  future perspectives, 325–326, 386
  mental strategy and, 316
  motor imagery used as control strategy, 316–320
    desynchronization and synchronization of sensorimotor rhythms, 317–318, 318f, 319f
    synchronization of central beta oscillations, 319–320, 319t
  neural engineering and, 299–303, 300f
  overview, 326
  training, 320–321
    with a basket paradigm, 320–321
    with feedback, 320
brain damage
  amygdala damage, 181–182, 181f
  damage to the insular or somatosensory cortex, 182–183
  disturbances after, 181–184, 181f
  lesions of the orbitofrontal and anterior cingulate cortex, 183–184
brain function
  activity in motor control tasks, 225–231, 226f
  brain-based adaptive automation systems, 243–246, 244f
  cerebral hemodynamics, 146–158
  chronic fatigue syndrome and, 227
  in control of muscular performance, 222–223, 223f, 225f
  future prospects, 384
  imaging technologies, 132
  in natural/naturalistic settings, 113–128
  optical imaging of, 65–81
  slow-wave sleep, 208
  training and, 139
  work environment and, 224–225

C1, definition, 390
CDAS. *See* Cognitive Decision Aiding System
Centers for Medicare and Medicaid Services, 386
central nervous system (CNS). *See also* neurorehabilitation robotics and neuroprosthetics
    functions, 222–223, 223*f*, 224*f*
    human brain in control of muscular performance and, 222
    as source of control signals, 299–303, 300*f*
cerebral hemodynamics, 146–158
    abbreviated vigilance, 154–155, 154*f*, 155*f*
    brain systems, 147
    overview, 155–156
    transcranial cerebral oximetry and, 148
    transcranial Doppler sonography and, 147–148
    vigilance decrement with signal cueing, 150–151, 151*f*
    visual search, 151–154, 153*f*
    working memory, 148–150, 149*f*
cerebral laterality, definition, 390
CFS. *See* chronic fatigue syndrome
chronic fatigue syndrome (CFS), 227
CIM. *See* Cockpit Information Manager
circadian rhythm. *See also* sleep
    control of sleep and neurobehavioral functions, 207–220
cloning, 387
CNS. *See* central nervous system
cochlear implants, 329
    definition, 390
    neural engineering, 294–296, 295*f*, 303–304
Cockpit Information Manager (CIM), 242–243, 243*f*
cognition, 6
    augmented, 258, 259*f*
    cognitive-affective control system in robots, 284–288, 286*f*, 287*f*
    cognitive performance effects of sleep deprivation, 210–211, 210*t*
    EROS application of optical signals, 72–73
    human prefrontal cortex and, 165
    individual differences between individuals, 201
    medical safety and, 362–363, 363*f*
    monitoring of EEG and, 21–24, 22*f*, 23*f*
Cognitive Decision Aiding System (CDAS), 242–243
cognitive processing, 164
cognitive work analysis (CWA)
    definition, 390
    in medical safety, 367, 368*f*
    system design interventions and, 369*f*
color stereo vision, 283–284, 283*f*
Computer-Adaptive MGAT test, 23
computerized physician order entry (CPOE), 360
computers
    human-computer etiquette in adaptive automation, 247–248

simulation technology for spatial research, 132–133
conspicuity area, 99
contextual cueing, 103
continuous wave (CW) procedure, 84–85
    definition, 390
cortical plasticity. *See* neuroplasticity
covert attention, definition, 390
CPOE. *See* computerized physician order entry
critical incidents, definition, 390
cross-modal interaction, definition, 390
CW. *See* continuous wave procedure
CWA. *See* cognitive work analysis
cyborg, 294

DARPA. *See* Defense Advanced Research Projects Agency
decision making, human prefrontal cortex and, 173
Defense Advanced Research Projects Agency (DARPA), 242
degree of freedom (DOF) device, 350–351
Department of Transportation, 125
diseases, tracking human activity and, 118
DLPFC. *See* dorsolateral prefrontal cortex
DOF. *See* degree of freedom device
Doppler effect. *See also* transcranial Doppler sonography
    definition, 390
dorsolateral prefrontal cortex (DLPFC), 105
    event-related potentials and, 44
driving
    drowsy, 214
    functional magnetic resonance imaging (fMRI) and, 51–64
    GLM results, 56, 56*f*
    human prefrontal cortex and, 172–173
    ICA results, 56–59, 57*f*, 58*t*, 59*f*, 60*f*
    interpretation of imaging results, 60*f*
    neuroergonomics research, 4
    100-Car naturalistic Driving Study, 122–123
    paradigm, 53–54, 53*f*
    simulated, 52
    simulator adaptation and discomfort, 261, 261*f*
    sleepiness-related accidents, 210
    spatial navigation and, 137–139, 138*f*
    speeds, 57–58, 59*t*
    tracking human behavior and, 120–125, 121*t*, 122*f*, 123*f*, 124*f*, 125*f*, 126*t*
    virtual reality simulation, 267–269, 268*t*
drowsiness
    definition, 395
    detection, 217
dwell, 102, 102*f*
    definition, 390
    frequencies, 97

422   Subject Index

echo planar imaging (EPI), for spatial research, 132
ECoG. *See* electrocorticogram
ecological interface design (EID), definition, 390
EEG. *See* electroencephalogram; electroencephalography
EEG artifact, definition, 390
EID. *See* ecological interface design
electrocorticogram (ECoG), 315
electroculogram (EOG)
    definition, 391
    event-related potentials and, 34
electroencephalogram (EEG), 15–31
    artifact detection, 19
    artificial vision and, 334
    brain-computer interface and, 315–328
    cognitive state monitoring, 21–24, 22f, 23f
    definition, 390
    future prospects, 382
    in human factors research, 16
    measures of workload, 24–27, 26f
    in neuroergonomics, 15–31
    overview, 27–28
    performance, 16
    progress of development, 16–17
    sensitivity, 20–21
    signals, 18
    variation sensitivity, 17–18, 17f
    in virtual environments, 256t
electrogastrogram, in virtual environments, 256t
electromyogram (EMG)
    definition, 391
    of motor activity, 117f
    motor activity-related cortical potential and, 225, 226f
    in virtual environments, 256t
EMG. *See* electromyogram
emotions
    definition, 391
    distinction between feelings and, 179–181
    disturbances after focal brain damage, 181–184, 181f
        amygdala damage, 181–182, 181f
        damage to the insular or somatosensory cortex, 182–183
        lesions of the orbitofrontal and anterior cingulate cortex, 183–184
    evidence of guided decisions, 186–189
        emotional signals, 187–188
        Iowa gambling task, 186–187
        unconscious signals, 188–189
    interplay between feelings, decision making, and, 185–186
        Phineas Gage and, 185–186
        somatic marker hypothesis, 186
    neural systems development subserving feelings and, 184–185

neurology of, 179–181, 180f
overview, 191
robots and, 285–286
role in behavioral decisions, 178–192
role of emotion-inspired abilities in relational robots, 275–292
work environment and, 224–225
environment. *See also* virtual environment
    home, 248–249, 249f
    human brain and, 224–225
    prediction and detection of effects of sleep loss in operational, 215–217
    spatial navigation and, 139–140, 141f
EOG. *See* electroculogram
EPI. *See* echo planar imaging
episodic memory, definition, 391
ERD. *See* event-related desynchronization
ERN. *See* error-related negativity
EROS. *See* event-related optical signal
ERPs. *See* event-related potentials
error, definition, 391
error-related negativity (ERN), 33
    definition, 391
    in event-related potentials, 43–44
ethics
    cloning, 387
    future prospects, 387
    in neuroergonomics and stress, 202–203
    privacy and, 387
ethology, 115–116
    definition, 115, 391
event rate, definition, 391
event-related desynchronization (ERD)
    in EEG-based brain-computer interface, 317–318, 318f, 319f
event-related optical signal (EROS)
    comparison of fMRI responses in visual stimulation, 70f
    definition, 391
    in imaging of brain function, 71–72
event-related potentials (ERPs)
    amplitude, 34
    applications, 33
    attentional resources and the P1 and N1 components, 41–43
    automatic processing assessment, 45
    brain-computer inferface, 41
    cognitive processes and, 38
    definition, 391
    error detection and performance monitoring, 43–44
    fundamentals of, 34–38, 35f, 36f, 38f
    future prospects, 382–383
    LORETA algorithm, 37, 38f
    measurement, 36, 36f
    mental workload assessment, 39–41

naming components, 36–37
in neuroergonomics, 32–50
overview, 46
in relation to other neuroimaging techniques, 33–34
research, 33
response readiness, 44–45
signal averaging technique, 34
signal-to-noise ratio, 34–36, 35f
source localization, 37, 38f
temporal information, 37–38
time course of mental processes, 37
executive functions, definition, 391
exercise, tracking energy expenditure and, 119
expected utility theory, 178
eye field, 100
definition, 391
eye-mind assumption, 98
eye movements, 95–112
attentional breadth, 99–101, 100f
computational models of saccadic behavior, 106–107
control, 101–104, 101f, 102f
effort, 104
future prospects, 383
neurophysiology of saccadic behavior, 105–106
overt and covert attention shifts, 97–99
overview, 108
saccades and fixations, 96–97
saccadic behavior, 97
in virtual environments, 256t
eye tracking. See eye movements

fast Fourier transform, definition, 391
fatigue. See also sleep
detection, 217
driving and, 214
feelings
definition, 391
distinction between emotions and, 179–181
disturbances after focal brain damage, 181–184, 181f
amygdala damage, 181–182, 181f
damage to the insular or somatosensory cortex, 182–183
lesions of the orbitofrontal and anterior cingulate cortex, 183–184
emotional signals, 187–188
evidence of guided decisions, 186–189
interplay between emotions, decision making, and, 185–186
Iowa gambling task, 186–187
neural systems development subserving emotions and, 184–185
neurology of, 179–181, 180f
overview, 191

Phineas Gage and, 185–186
role in behavioral decisions, 178–192
somatic marker hypothesis, 186
unconscious signals, 188–189
work environment and, 224–225
FES. See functional electrical stimulation
FFOV. See functional field of view
fidelity
description, 260
in virtual environments, 260–261
Fitts' law, definition, 391
fixation.
definition, 391
duration, 97
fMRI. See functional magnetic resonance imaging
fovea, 96
foveola, 96
functional electrical stimulation (FES)
artificial vision and, 333–334, 334f
definition, 392
neuroprosthetics and, 352–353, 353f
prosthetics and, 296–297
functional field of view (FFOV), 99
definition, 392
functional magnetic resonance imaging (fMRI), 51–64
analysis challenges, 52
comparison of EROS responses in visual stimulation, 70f
data-driven approach, 52–53, 53f
definition, 392
driving paradigm, 53–54
essentials, 51–52
experiments and methods, 54–55
data analysis, 55
image acquisition, 54–55
participants, 54, 55f
results, 56–59
GLM results, 56, 56f, 57f
ICA results, 56–59, 57f, 58t, 59f, 60f
future prospects, 382
in neuroergonomic research and practice, 6
optical imaging of brain function and, 68
overview, 62
in simulated driving, 52
for spatial research, 132, 136
functional neuroimaging, definition, 392

Gait Trainer, 349
galvanic skin response (GSR), 196
gaze. See dwell
gaze-contingent paradigms, 99
general adaptation syndrome
definition, 392
stress and, 196

general linear model (GLM)
    in driving, 56, 56f
    in functional magnetic resonance imaging analysis, 52–54, 53f
genetics, neuroergonomics and, 9
GLM. *See* general linear model
grip, 227–229
GSR. *See* galvanic skin response

Haptic Master, 350
HASTE. *See* Human Machine Interface and the Safety of Traffic in Europe
HD. *See* Huntington's disease
headband devices, 85–86
head field, 100
head-mounted displays (HMDs)
    technological advances, 270
    in virtual environments, 259
heads-up display (HUD), definition, 392
health care
    legislation, 365
    medical safety and, 361–362, 361f, 362f
    reporting, 364–365
heart rate, in virtual environments, 256t
hedonomics
    definition, 392
    stress and, 201–202
Heinrich's triangle, definition, 392
helmet
    for optical recording, 75–76, 75f
    for transcranial Doppler sonography, 86
hemovelocity
    changes in transcranial Doppler sonography, 89
    definition, 392
hippocampus, definition, 392
HMDs. *See* head-mounted displays
homeostatic sleep drive, definition, 392
HPFC. *See* human prefrontal cortex
HUD. *See* heads-up display
human behavior
    clinical tests, self-report, and real-life behavior, 114–115
    data analysis strategies, 116–117, 117f
    environment and, 114–115
    ethology and remote tracking, 115–116
    measurement limitations, 115
    in natural/naturalistic settings, 113–128
    tracking
        applications, 118–119
        human movement and energy expenditure, 119–120
        over long distances, 120–125, 121t, 122f, 123f, 124f, 125f, 126t
        system taxonomies, 117–118
Human Factors and Ergonomics Society, 386

human factors research
    electroencephalogram and, 16
    P300 studies, 40–41
Human Machine Interface and the Safety of Traffic in Europe (HASTE), 255
human motor system, 222
    hierarchical model, 222–223, 223f, 224f
human prefrontal cortex (HPFC)
    anatomical organization, 159–161
    cognitive abilities, 165
    computational frameworks, 163–164
    definition, 392
    description, 159–161
    functional studies, 161
    functions, 159–177
        commonalities and weaknesses, 164
    memory and, 164–165
    neuroergonomic applications, 172–173
    neuropsychological framework, 161–163
        action models, 163
        attentional/control processes, 162
        social cognition and somatic marking, 162–163
        working memory, 161–162
    overview, 174
    process versus representation, 164–165
    relationship to basal ganglia functions, 169
    relationship to hippocampus and amygdala functions, 169–170
    relationship to temporal-parietal cortex functions, 170
    structured event complex, 165–166
        archetype, 165–166
        associative properties within functional region, 167
        binding, 169
        category specificity, 168
        event sequence, 166
        evidence for and against, 170–171
        frequency of use and exposure, 167
        future directions for the SEC model, 171–172
        goal orientation, 166
        hierarchical representation, 169
        memory characteristics, 166–167
        neuroplasticity, 168
        order of events, 167–168
        priming, 168–169
        representational format, 166–169
        representation versus process, 172
Huntington's disease (HD), 136–137

ICA. *See* independent component analysis
immersion, definition, 392
implantable myoelectric sensors (IMES), prosthetics and, 297–299, 298f, 299f

independent component analysis (ICA)
    definition, 392
    in driving, 56–59, 57f, 58t, 59f, 60f
    in functional magnetic resonance imaging, 53, 53f
information transfer rate (ITS), 325
instrumented vehicles (IVs), for tracking human behavior, 120–125, 121t, 122f, 123f, 124f, 125f, 126t
intelligence. *See also* learning
    of robots, 276
ITS. *See* information transfer rate

jet lag
    definition, 392
    sleep deprivation and, 214–215
joint cognitive systems, 6
judgments, human prefrontal cortex and, 173

knowledge-based mistakes, 367

lapses, 367
lateralized readiness potential (LRP), 33
    definition, 392
    event-related potentials and, 44–45
learning. *See also* intelligence
    motor, 231
    social, 185
        with robots, 288, 289f
    spatial navigation and, 141
lesion studies, artificial vision and, 334–336
light absorption, definition, 392
light scattering, definition, 393
linguistics, increased blood flow and, 89–90
Lokomat, 349, 350f
LORETA algorithm, 37, 38f
LRP. *See* lateralized readiness potential

magnetic resonance imaging (MRI), for spatial research, 132
magnetoencephalography (MEG)
    artificial vision and, 334
    definition, 393
    in neuroergonomic research and practice, 6
MATB. *See* Multi-Attribute Task Battery
MCA. *See* middle cerebral artery
medical safety, 360–378
    cognitive factors, 362–363, 363f
    development of a reporting tool, 365–369, 366f
        systemic factors, 367–369, 368f, 369t
        taxonomy of error, 367
    health care delivery and errors, 362, 362f
    health care system perspective, 361–362, 361f
    information-processing model, 362–363, 363f

neuroergonomic interventions, 370–374
    augmented reality, 373–374, 374f, 375f
    ecological interface design, 371–372
    expertise, 374
    patient tracking through bar codes, 372
    virtual reality, 373–374, 374f, 375f
    overview, 375–376
    safety management, 369–370, 371f
    tracking health care errors, 363–365
        closed claims analyses, 362f, 363–364
        mandatory reporting, 364
        voluntary reporting, 364–365
MEG. *See* magnetoencephalography
memory
    human prefrontal cortex and, 164–165
    spatial navigation and, 131–133
    working, 148–150, 149f, 161–162
mental chronometry, 32
mental resources, definition, 393
mental workload, definition, 393
MEPs. *See* motor evoked potentials
metamers, 264
microsaccade, 97
microsleep, definition, 393
middle cerebral artery (MCA), examination with transcranial Doppler sonography, 83, 83f
MIME. *See* Mirror Image Movement Enhancer
Mirror Image Movement Enhancer (MIME), 351
mismatch negativity (MMN), 33
    auditory modality and, 45
    characteristics, 45
    definition, 393
    event-related potentials and, 45
MIT-Manus, 350–351, 351f
MMN. *See* mismatch negativity (MMN)
monoparesis, 346
monoplegia, 346
motion correction, definition, 393
motion impairment, 348. *See also* movement restoration
motor activity, human brain activity and, 225–231, 226f
motor activity-related cortical potential (MRCP)
    definition, 393
    human brain activity in, 225–231
    muscle electromyograph and, 225, 226f
motor cortex, definition, 393
motor evoked potentials (MEPS), 230
motor learning, 231
movement restoration, 348
    human control in, 227–231
    reaching, 229
moving window, 99, 100f
MRCP. *See* motor activity-related cortical potential
MRI. *See* magnetic resonance imaging
Multi-Attribute Task Battery (MATB), 21, 22f, 23f
muscle fatigue, 227

426 Subject Index

N1
  definition, 393
  event-related potentials and, 41–43
  studies, 41–43
nanotechnology, neuroergonomics and, 9
National Weather Service, 125
naturalistic, definition, 393
near-infrared spectroscopy (NIRS)
  cerebral hemodynamics and, 148
  definition, 393
  future prospects, 383
  optical imaging of brain function and, 65–66
  in virtual environments, 256t
nerve growth factor (NGF), definition, 393
neural engineering, 293–312
  biomimetic and hybrid technologies, 305–306
  brain-computer interface, 299–303, 300f
  brain-machine interactions for understanding neural computation, 306, 307f
  central nervous system as source of control signals, 299–303, 300f
    EEG recordings, 300–301
    intracortical recordings, 301–303, 302f
  clinical impact, 303–305
    cochlear implants, 303–304
    EEG-based BCIs, 304
    implanted EEG recording systems, 304–305
    implanted nerve cuffs, 305
    implanted stimulation systems, 305
    implanted unitary recording systems, 305
  cochlear implants, 294–296, 295f, 303–304
  description, 293
  future prospects, 385
  learning and control, 303
  motor prosthetics, 296–299, 297f, 298f, 299f
  neural plasticity as a programming language, 306
  overview, 308–309
neuroergonomics
  adaptive automation, 239–252
  applications, 172–173
  artificial vision, 329–345
  cerebral hemodynamics, 146–158
  challenges, 381
  conceptual, theoretical, and philosophical issues, 5–6
  definition, 3, 195, 393
  EEG-based brain-computer interface, 315–328
  electroencephalography in, 15–31
  emotions and feelings, role in behavioral decisions, 178–192
  ethical issues, 202–203
  event-related potentials in, 32–50
  eye movements and, 95–112
  functional magnetic resonance imaging in, 51–64
  future prospects, 381–388
    the brain, 384
    cross-fertilization of fields, 384
    ethical issues, 387
    guidelines, standards, and policy, 386–387
    longer-term, 384–386
    physiological monitoring, 382–384
    simulation and virtual reality, 382
  genetics and, 9
  goal, 9
  human behavior in natural/naturalistic settings, 113–128
  human prefrontal cortex and, 159–177
  medical safety and, 360–378
    intervention examples, 370–374
  methods, 6–8, 7f
  neural engineering, 293–312
  neurorehabilitation robotics and neuroprosthetics, 346–359
  optical imaging of brain function and, 65–81
  overview, 3–12, 10–11
  physical, 221–235
  research, 4–5
    guidelines, 329
  role of emotion-inspired abilities in relational robots, 275–292
  sleep and circadian control of neurobehavioral functions, 207–220
  spatial navigation, 131–145
  stress and, 195–206
  techniques for ergonomic applications, 6–8, 7f
  transcranial Doppler sonography in, 82–94
  virtual reality and, 5, 253–274
neuroimaging techniques. *See individual techniques*
  motor learning and, 231
neurology
  brain-based model of robot decisions, 189–190
    primacy of emotion during development, 189–190
    willpower to endure sacrifices and resist temptations, 190
  role of emotions and feelings in behavioral decisions, 178–192
neuroplasticity
  in the blind, 339–342, 341f
  definition, 393
neuropsychology
  definition, 8
  human prefrontal cortex and, 161–163
    action models, 163
    attentional/control processes, 162
    social cognition and somatic marking, 162–163
    working memory, 161–162
  neuroergonomics and, 8

Subject Index 427

neurorehabilitation robotics and neuroprosthetics, 346–359
  automated training devices
    cooperative control strategies, 351–352
    gait-training, 349, 350f
    for the upper extremities, 350–352, 351f, 352f
  background, 352–353, 353f
  future prospects, 386
  human motor system
    anatomical structure and function, 346, 347f, 347t
    pathologies, 346, 348
  movement therapy rationale, 348–349
  natural and artificial mechanisms of movement restoration, 348
  neuroprostheses
    control, 355–356, 355f
    development challenges, 354
  overview, 356–357
  robot-aided training rationale, 349
  technical principles, 353–354
neurovascular coupling, definition, 393
NGF. *See* nerve growth factor
night shift work, 213–214
NIRS. *See* near-infrared spectroscopy
nonphotorealistic rendering (NPR) techniques, 264–266, 264f
NPR. *See* nonphotorealistic rendering techniques

object-related actions (ORAs), 104
oculomotor behavior, 107
oculomotor capture, 101–102, 101f
100-Car Naturalistic Driving Study, 122–123
operant conditioning, definition, 394
optical imaging methods
  of brain function, 65–81
  definition, 394
  future prospects, 383
  mathematical model of propagation of light, 66, 67f
  methodology, 75–77
    analysis, 77
    digitization and coregistration, 77
    recording, 75–77, 75f
  neuroergonomics considerations, 77–78
  optical signals, 66, 68–74, 69f, 70f, 71f, 74f
  overview, 78
  principles of noninvasive optical imaging, 66, 67f
optical signals, 66, 68–74, 69f, 70f, 71f, 74f
  fast, 69–73
    EROS applications in cognitive neuroscience, 72–73
    event-related, 69–72, 69f, 70f, 71f
  slow, 73–74, 74f
ORAs. *See* object-related actions
overt attention, definition, 394

P1
  definition, 394
  event-related potentials and, 41–43
  studies, 41–43
P300
  definition, 394
  event-related potentials and, 39–40
  latench, 40
  studies, 40–41
passive coping mode, 197
Patient Safety and Quality Improvement Act of 2005, 365, 371
PCA. *See* posterior cerebral artery; principal component analysis
people tracker, definition, 394
perceptual span, 99
perceptual systems, 282–284
  blue eyes camera, 284, 285f
  color stereo vision, 283–284, 283f
  mouse pressure sensor, 284
  sensor chair, 282–283, 282f
perclos, definition, 394
peripheral nervous system (PNS), 346
personal service robots, 276
person-environment transactions, definition, 394
PET. *See* positron emission tomography
phase shift, definition, 394
phosphene, definition, 394
photons' time of flight, definition, 394
physical neuroergonomics, 221–235
  human brain activity in motor control tasks, 225–231, 226f
    eccentric and concentric muscle activities, 225–226
    internal models and motor learning, 231
    load expectation, 233
    mechanism of muscle fatigue, 227
    motor control in human movements, 227–231
      control of extension and flexion movements, 227
      postural adjustments and control, 229–230
      power and precision grip, 227–229
      reaching movements, 229
      task difficulty, 230–231
    studies of muscle activation, 225
  human brain and the work environment, 224–225
  human brain in control of muscular performance, 222–223, 223f, 225f
  hierarchical model of the human motor system, 222–223, 223f, 224f
  human motor system, 222
  overview, 233
physiological monitoring, future prospects, 382–384
physiology, in virtual reality, 257, 257f

PNS. *See also* neurorehabilitation robotics and neuroprosthetics; peripheral nervous system
positron emission tomography (PET)
  definition, 394
  future prospects, 382
  in neuroergonomic research and practice, 6
  for spatial research, 133
posterior cerebral artery (PCA), examination with TCD, 83, 83*f*
posture, adjustments and control, 229–230
presence, definition, 394
primary inducers, definition, 394
primary motor cortex, definition, 394
principal component analysis (PCA), definition, 394
proprioception, definition, 394
prosthetics
  artificial vision and, 331*f*
  history, 330–333, 330*f*, 331*f*
  limitations, 296
  motor, 296–299, 297*f*, 303–304
  neuroelectric control, 296
psychological research
  hedonomics and, 201–202
  transcranial Doppler sonography and, 88–90
    basic perceptual effects, 88–89
    history, 88
    information processing, 89–90
psychomotor vigilance test (PVT), definition, 394
pulsed wave (PW) procedure, 84–85
pursuit movements, 96
PVT. *See* psychomotor vigilance test
PW. *See* pulsed wave procedure

questionnaires, for tracking human behavior, 113–114

receptive field, definition, 395
repetitive transcranial magnetic stimulation (rTMS), event-related potentials and, 44
respiration, in virtual environments, 256*t*
retinal prosthesis
  definition, 395
  future prospects, 385–386
retinotopy, definition, 395
robots
  brain-based model of decisions, 189–190
    primacy of emotion during development, 189–190
    willpower to endure sacrifices and resist temptations, 190
  design of relational robot for education, 278–280, 278*f*
    RoCo as a robotic learning companion, 279
  emotion system, 285–286
  intelligence, 276

  neurorehabilitation robotics and neuroprosthetics, 346–359
  overvview, 290
  personal service, 276
  role of emotion-inspired abilities in relational robots, 275–292
    as a scientific research tool, 277–278
    sensing and responding to human affect, 280–288
      cognitive-affective control system, 284–288, 286*f*, 287*f*
        affective appraisal of the event, 286
        characteristic display, 286
        modulation of cognitive and motor systems to motivate behavioral response, 288
        precipitating event, 286
      expressive behavior, 280–282
      perceptual systems, 282–284
        blue eyes camera, 284, 285*f*
        color stereo vision, 283–284, 283*f*
        mouse pressure sensor, 284
        sensor chair, 282–283, 282*f*
    social interaction with humans, 276–277
    social learning, 288, 289*f*
    social presence, 280
    social rapport, 279–280
root cause analysis, definition, 395
rTMS. *See* repetitive transcranial magnetic stimulation
rule-based mistakes, 367

SA. *See* situation awareness
saccades, 96
  definition, 395
  microsaccade, 97
  selectivity, 103
saccadic behavior
  computational models, 106–107
  fixations and, 96–97
  neurophysiology, 105–106
saccadic suppression, 96–97
  definition, 395
SCI. *See* spinal cord injuries
SCN. *See* suprachiasmatic nucleus
search asymmetry, definition, 395
secondary inducers, definition, 395
sensor chair, 282–283, 282*f*
SHEL model, 365
simulator adaptation syndrome, definition, 395
simultaneous vigilance task, definition, 395
situation awareness (SA), adaptive automation and, 246–247
skin, in virtual environments, 256*t*
sleep
  circadian control of neurobehavioral functions and, 207–220

inadequate, neurobehavioral and neurocognitive
consequences of, 209–213, 210t, 212f
individual differences in response to sleep deprivation, 213
rest time and effects of chronic partial sleep deprivation, 211–213, 212f
sleepiness versus performance during sleep deprivation, 213
intrusions, 210
operational causes of sleep loss, 213–215
fatigue and drowsy driving, 214
night shift work, 213–214
prolonged work hours and errors, 215
transmeridian travel, 214–215
overview, 207–208, 218
prediction and detection of effects of sleep loss in operational environments, 215–217
biomathematical models to predict performance capability, 215–216
fatigue and drowsiness detection, 217
technologies for detecting operator hypovigilance, 216–217
quality of, 207–208
slow-wave, 208
types of sleep deprivation, 208–209, 209f
partial, 208
sleep inertia, 208–209
total, 208, 209f
sleep debt, definition, 395
sleepiness, definition, 395
slips, 367
social cognition
definition, 395
learning and, 185
somatic marking and, 162–163
somatic state, definition, 395
somatosensory system
damage to the insular or somatosensory cortex, 182–183
neural engineering and, 303
somnolence, definition, 395
source localization, definition, 395
spatial navigation, 131–145
accuracy, 133–137, 133f, 134f, 135f, 136f
driving and, 137–139, 138f
environment and, 139–140, 141f
learning and, 141
memory and, 131–133
overview, 142
research, 131–133
spatial normalization, definition, 395
spatial resolution, definition, 395
Special Interest Group on Computer-Human Interaction of the Association for Computing Machinery, 386

spinal cord injuries (SCI), neurorehabilitation robotics and, 348
state instability, sleep and, 20–210
stationary field, 99
stock investments, human prefrontal cortex and, 173
strain mode, 197
stress, 195–206
appraisal and regulatory theories, 197–198, 198f, 199f
arousal theory, 196–197
commonalities and differences between individuals, 201
concepts of, 195
definition, 395
ethical issues in neuroergonomics and, 202–203
hedonomics and positive psychology, 201–202
measurement, 200–201
neuroergonomics research, 199–200
overview, 203–204
resoure theory, 200
stress-adaptation model, 197–198, 198f, 199f
validation of neuroergonomic stress measures, 200–201
structured event complex
definition, 395
human prefrontal cortex and, 159–177
successive vigilance task, definition, 395
suprachiasmatic nucleus (SCN)
circadian thythms and, 207–208
definition, 396

Task Load Index (TLX) scale, 152–153
taxonomies
of error, 367
of human behavior, 117–118
TCD. See transcranial Doppler sonography
temporal discounting, 189
temporal resolution, definition, 396
time-resolved (TR) instruments, definition, 396
TLX. See Task Load Index scale
TMS. See transcranial magnetic stimulation
tonotopy, 294–295
top-down regulation, definition, 396
TR. See time-resolved instruments
training, effects on the brain, 139
transcranial cerebral oximetry, cerebral hemodynamics and, 148
transcranial Doppler sonography (TCD), 82–94
cerebral hemodynamics and, 147–148
criteria for artery identification, 85t
definition, 396
exclusion criteria, 87
future prospects, 383
neuroergonomic implications, 90
in neuroergonomic research and practice, 6

transcranial Doppler sonography (TCD) (continued)
  overview, 90–91
  principles, 83–88
    Doppler fundamentals, 83–84, 83f
    instrumentation, 84–87, 84f, 85f, 85t, 86f
    methodological concerns, 87
    reliability and validity, 87–88
  psychological research and, 88–90
    basic perceptual effects, 88–89
    history, 88
    information processing, 89–90
transcranial magnetic stimulation (TMS)
  artificial vision and, 336–337
  definition, 396
  in the study of neuronal underpinnings of saccades, 105
transmeridian travel, sleep deprivation and, 214–215
triggers, tracking human behavior and, 121t

ultrasound, definition, 396
ultrasound window, definition, 396

vergence shifts, 96
video monitoring, for tracking energy expenditure, 119
vigilance
  abbreviated, 154–155, 154f, 155f
  brain systems and, 147
  definition, 396
vigilance decrement, definition, 396
virtual environment, definition, 396
virtual reality (VR)
  accuracy, 262
  augmented reality, 258–259, 258f
  definition, 396
  description, 253, 254f
  fidelity, 260–261
  future directions, 269–270
  future prospects, 382
  games, 132–133
  guidelines, standards, and clinical trials, 267–269, 268t
  history of, 253
  learning and, 141
  medical safety and, 373–374, 374f, 375f
  museum, 139–140, 140f
  neuroergonomics and, 5, 253–274
  nonrealistic environments, 264–266, 265f
  overview, 270–271
  physiology of the experience, 255–258, 256t, 257f
  representation, 262–264, 264f
  simulator adaptation and discomfort, 261–263, 262f
  for spatial research, 133, 133f, 134f, 135f
  validation and cross-platform comparisons, 265–267
vision. *See also* eye field; eye movements
  color stereo, 283–284, 283f
visual lobe, 99
visual span, 99
visual stimulation, comparison of fMRI and EROS responses, 70f
VR. *See* virtual reality

willpower
  decision output and, 190
  definition, 190
  emotional evaluation, 190
  input of information, 190
Wisconsin Card Sorting Test, 163
WM. *See* working memory
working memory (WM)
  definition, 396
  electroencephalogram and, 17–18
workload
  definition, 396
  measures of, 24–27, 26f

Yekes-Dodson law, 196

zeitgebers, 207